HANDBOOK
of
ELECTRICAL
DESIGN
DETAILS

OTHER ELECTRICAL CONSTRUCTION BOOKS OF INTEREST

Cadick – Electrical Safety Handbook

Croft & Summers – American Electrician's Handbook

Denno – Electric Protective Devices

Johnson – Electrical Contracting Business Handbook

Johnson – Successful Business Operations for Electrical Contractors

Johnson & Whitson – Electrical Contracting Forms and Procedures Manual

Kolstad – Rapid Electrical Estimating and Pricing

Kusko – Emergency/Standby Power Systems

Linden – Handbook of Batteries and Fuel Cells

Maybin – Low Voltage Wiring Handbook

McPartland – McGraw-Hill's National Electrical Code Handbook

McPartland – McGraw-Hill's Handbook of Electrical Construction Calculations

McPartland – Handbook of Practical Electrical Design

Smeaton – Switchgear and Control Handbook

Traister – Electrician's Pocket Reference

Traister – Electrician's Troubleshooting Pocket Guide

Traister – Security/Fire Alarm Systems: Design, Installation, Maintenance

Traister – McGraw-Hill's Illustrated Pocket Guide to the 1996 NEC Tables

Traister – McGraw-Hill's Illustrated Index to the 1996 National Electrical Code

Whitson – Handbook of Electrical Construction Tools and Materials

HANDBOOK
of
ELECTRICAL
DESIGN
DETAILS

John E. Traister

McGraw-Hill

New York San Francisco Washington, D.C. Auckland Bogotá
Caracas Lisbon London Madrid Mexico City Milan
Montreal New Delhi San Juan Singapore
Sydney Tokyo Toronto

McGraw-Hill

A Division of The **McGraw·Hill** Companies

1 2 3 4 5 6 7 8 9 0 DOC/DOC 9 0 2 1 0 9 8 7

ISBN-0-07-065330-5

National Electrical Code® and NEC® are registered trademarks of the National Fire Protection Association, Inc., Quincy, MA 02269

Printed and bound by R. R. Donnelley & Sons

This book is printed on acid-free paper.

Contents

Preface

The ability of those involved in the design and construction of electrical systems to accurately transmit technical information through the preparation and interpretation of working drawings and written specifications is vital to the development, construction, and use of such systems.

Ideally, electrical contractors and their employees should be able to bid and construct any electrical system from the information furnished by the engineer with no further questions concerning dimensions, materials, or the slightest detail in construction. To help make this possible, certain contract documents are necessary.

- **Specifications:** Specifications describe the work to be done, the method of construction, the standard of workmanship, the manner of conducting the work, and the quality of materials and equipment to be used.

- **Floor Plans:** A floor plan represents a cut horizontally through a building at approximately eye level showing the view from above. A separate drawing is made for each floor level, including the basement level or levels. Such plans show the arrangements and locations of walls, partitions, doors, windows, stairways, and all electrical equipment drawn to scale (where possible) using appropriate graphic symbols such as the ones covered in Chapter 1 of this book.

- **Elevations and Sections:** Elevations are drawings of head-on vertical views of a structure or equipment in a single plane. Sections are sliced open views of an area to reveal construction arrangements which cannot be shown by conventional elevation or plan views.

- **Detail Drawings:** Although all parts of an electrical system are usually described in floor plans, site plans, elevations, sections, and written specifications, certain construction conditions cannot be adequately shown in the scale in which these drawings are usually made. Therefore, larger-scaled drawings must be made of such items or areas to ensure the necessary information for proper construction. These detail drawings are found in nearly every set of electrical working drawings.

Electrical Design Details provides practical electrical detail drawings taken from actual working drawings that have been used on a variety of projects for the past decade or so.

All of the details in this book, therefore, have been tried and proven to be useful to electrical engineers, designers, drafters, contractors, and workers.

Electrical engineers and designers will find daily use for this book in preparing working drawings that are easily interpreted by the workers on the job; electrical drafters should find it extremely helpful as a desk reference; electrical contractors, estimators, and workers will want to refer to the various details for solving installation problems which are constantly developing in their work; while various trade apprentices, student engineers, and technicians will find the details invaluable when used in conjunction with theory text books to better visualize actual installations.

I would like to express my appreciation to many well-qualified persons who have been most helpful to me during the preparation of this book.

John E. Traister
Bentonville, Virginia
1997

Introduction

Electrical engineers, designers, technicians, and drafters constantly work on drawings and construction documents for electrical systems of every imaginable type. Much of the design for such systems is accomplished by the use of mathematics and good judgment. The designs are then conveyed to workers via drawings with accompanying written specifications.

Mathematical equations are used in calculating the amount of illumination for a given area, circuit and feeder wire sizes, the service-entrance size, and the like. Good judgment is gained from experience and is used to determine such items as the type of lighting fixtures to be used in a given area, types and quality of components, etc.

Electrical design professionals must also have a good knowledge of the *National Electrical Code®* (*NEC*) which has become the "bible" of the electrical construction industry. Consequently, an up-to-date copy of the *NEC* should be handy at all times for frequent reference.

Electrical drawings use lines, symbols, dimensions, and notations to convey the engineer's calculations and design to workers on the job. Therefore, any construction drawing is really a specialized language — a means of communication — and anyone involved in electrical construction in any capacity must master this language; that is the drawings must show a complete description of the project so that workers on the job can understand what is required them. Furthermore, the drawings must be detailed enough so that electrical estimators are able to make a complete material list and apply the appropriate labor units for bidding purposes.

Figure I-1: The NEC has become the "bible" of the electrical construction industry and a current copy should be handy at all times for frequent reference.

TYPES OF ELECTRICAL DRAWINGS

Many types of drawings are likely to be encountered on any electrical construction project. The most common types include:

- Pictorial
- Orthographic projections
- Single-line diagrams
- Schematic wiring diagrams

Pictorial Drawings

Pictorial drawings are objects shown in one view only; that is, three-dimensional effects are simulated on the flat plane of drawing paper by showing several faces of an object in a single view. This type of drawing is very useful to describe objects and convey information to those who are not trained in reading construction documents or to supplement conventional orthographic drawings in the more complex systems.

Examples of using pictorial drawings are power-riser diagrams, illustrations in training manuals, catalog illustrations, illustrations to accompany proposals to clients on certain projects, and to supplement troubleshooting information for new types of equipment.

The types of pictorial drawings most often found in the electrical field include:

- Isometric drawings
- Oblique drawings
- Perspective drawings

All of these types are more difficult to draw than conventional multiview drawings. Further disadvantages of pictorial drawings are that intricate parts cannot be pictured clearly and that they are difficult to dimension. Still, with the use of CAD systems with which much of the work is done automatically by the computer, pictorial drawings are being used more and more in electrical construction documents.

Isometric Drawings: A view projected onto a vertical plane in which all of the edges are foreshortened equally is called an *isometric projection.* Figure I-2 shows an isometric drawing of a cube. In this view, the edges are 120 degrees apart and are called the isometric axes, while the three surfaces shown are called the isometric planes. The lines parallel to the isometric axes are called the isometric lines.

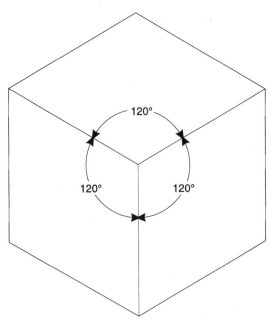

Figure I-2: An isometric drawing of a cube showing the isometric axis 120 degrees apart.

This type of pictorial drawing is usually preferred over the other two types mentioned for certain electrical details and wiring diagrams on working drawings, because it is possible to draw isometric lines to scale in the same manner as floor plans or multiview plans are drawn in orthographic views. When drawn manually, isometric drawings may be made with the 30-60-degree triangle. The better CAD programs are capable of producing isometric views from practically any type of drawing or even dimensions for that matter.

Oblique Drawings: The oblique drawing is similar to the isometric drawing in that one face of the object is drawn in its true shape and the other visible faces are shown by parallel lines drawn at the same angle (usually 45 – 30 degrees) with the horizontal. However, unlike an isometric drawing, the lines drawn at a 30-degree angle are shortened to preserve the appearance of the object and are therefore not drawn to scale. The drawing in Figure I-3 is an oblique drawing of a cube.

The two pictorial drawing methods described so far produce only approximate representations of objects as they appear to the eye, because each type produces some degree of distortion of any object. However, because of certain advantages, they are the types of pictorial drawings most often used for electrical drawings.

Perspective Drawings: Sometimes it is desired to draw an exact pictorial representation of an object as it actually appears to the eye. This type of drawing is called a *per-*

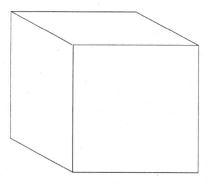

Figure I-3: Oblique drawings are another form of pictorial drawings used in working drawings.

spective drawing and is frequently used for catalog illustrations and also for project presentations to clients. The principles of perspective drawing with a CAD system appears in Figure I-4 and I-5, while a practical application is shown in Figure I-6 on the next page.

Orthographic Projection: This type of drawing represents the physical arrangement and views of specific objects. These drawings give all plan views, elevation views, dimensions, and other details necessary to construct the project or object. Although the pictorial drawing in Figure I-7 suggests the form of a room or building, it does not show the actual shape of the surfaces, nor does it show the dimensions of the object so that it may be constructed.

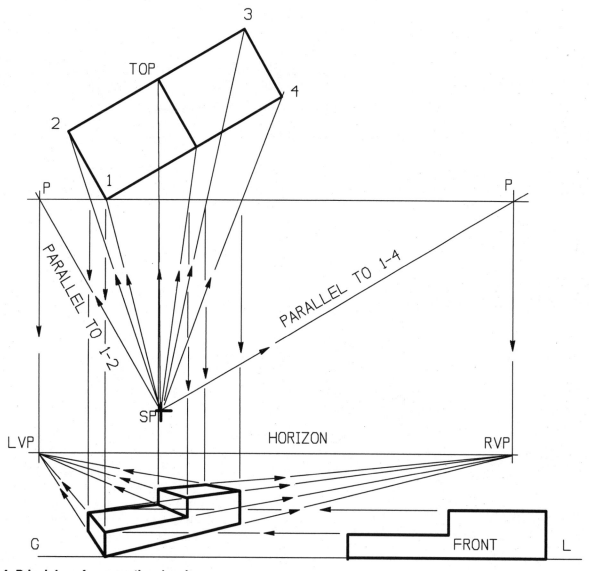

Figure I-4: Principles of perspective drawing.

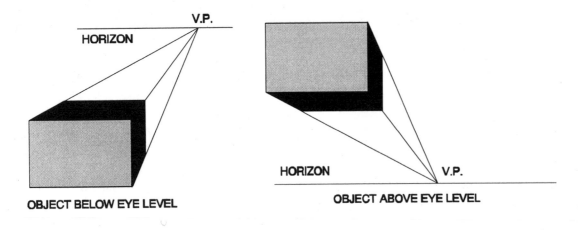

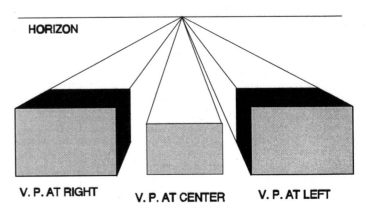

Figure I-5: Perspective views of an object from different angles.

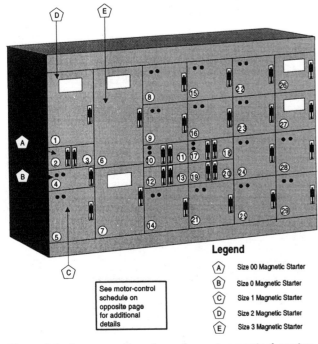

Figure I-6: A perspective view of a motor-control center.

Legend

- Ⓐ Size 00 Magnetic Starter
- Ⓑ Size 0 Magnetic Starter
- Ⓒ Size 1 Magnetic Starter
- Ⓓ Size 2 Magnetic Starter
- Ⓔ Size 3 Magnetic Starter

See motor-control schedule on opposite page for additional details

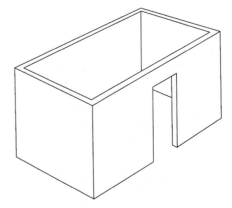

Figure I-7: Pictorial drawing of a transformer vault.

An orthographic projection of the vault in Figure I-7 is shown in Figure I-8. One of the drawings in this figure shows the vault as though the observer were looking straight at the left side; one, as though the observer were looking straight at the right side; one as though the observer were looking at the front, and one as though the observer were looking at the rear of the vault. The remaining

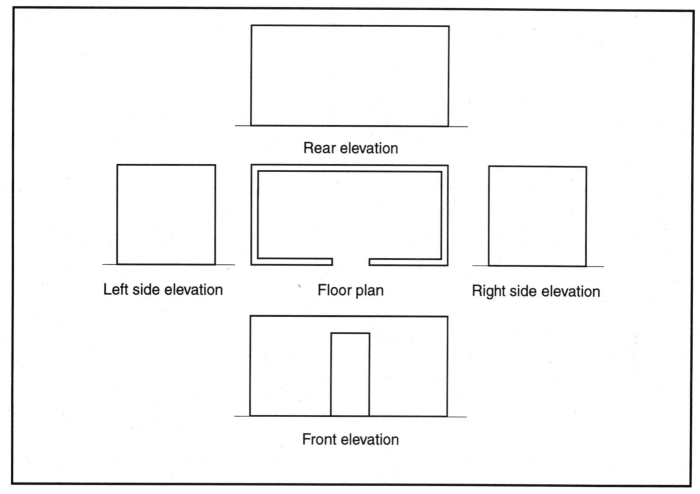

Figure I-8: Orthographic projection of the transformer vault in Figure I-7.

view is as if the observer were looking straight down on top of the vault. These views, when combined with dimensions, will allow the object to be constructed properly from materials called for in the written specifications.

Electrical Diagrams: These drawings are intended to show, in diagrammatic form, electrical components and their related connections. Such drawings are seldom drawn to scale, and show only the electrical association of the different components. In diagram drawings, symbols are used extensively to represent the various components and lines are used to connect them; these lines usually indicate the size, type and number of conductors in each circuit.

In general, the types of diagrams that will be used the most on electrical working drawings will include single-line diagrams (Figure 1-9), schematic wiring diagrams (Figure 1-10), and power-riser diagrams. Many other samples of each type are found throughout this book.

DIMENSIONING

Since it is the electrical engineer's responsibility to provide neat, detailed, and accurate drawings for contractors and workers, a means of expressing accurate dimensions must be used. Without definite dimensions, it would be impossible, on most drawings, to indicate clearly any design intent or to achieve successful fabrication of the building or system.

The ideal electrical drawing should indicate, beyond any question of a doubt, exactly what is required to construct the desired electrical system. It should never be necessary for the workers to assume any dimension to perform their work from drawings made by consulting engineers or their drafters.

The dimensioning practices described in this section are those recommended by the American National Stan-

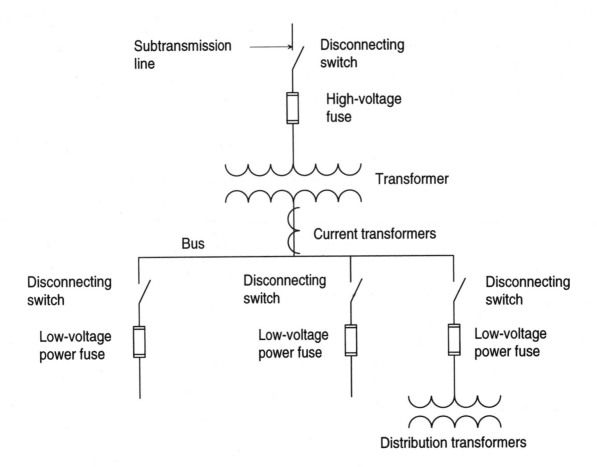

Figure I-9: Single-line diagram of a substation.

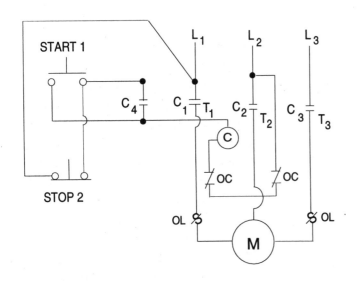

Figure I-10: Wiring diagram of an ac magnetic motor starter.

dards Institute (ANSI) and the U.S. Department of Defense — they are standard in almost all industries.

Fundamental Procedures

Dimensions on most electrical drawings usually appear in feet, inches, or fractional parts of an inch unless the metric system is used. If the metric system is used, a general note should be included on the drawings to indicate that dimensions are shown in meters, millimeters, etc.

Dimension lines are used with numerals and arrowheads to show the sizes or placement of an object or component on a drawing. These lines should be placed off the main views so as not to be confusing. Furthermore, in most cases, dimensions should not be repeated in other views of the same object. The light full line is used for dimension lines, with a space — usually in the middle — for the numerals that represent the size or dimension. These lines should be placed from $\frac{1}{4}$ in to $\frac{3}{4}$ in from the views, depending on the type and size. *See* Figure I-11. Some engineers and drafters, however, prefer to place the dimension above

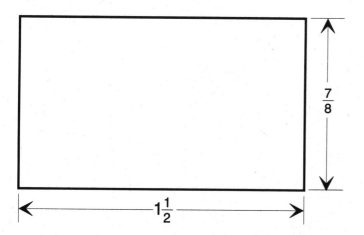

Figure I-11: Dimension lines should be placed between ¼ to ¾ inch from the views.

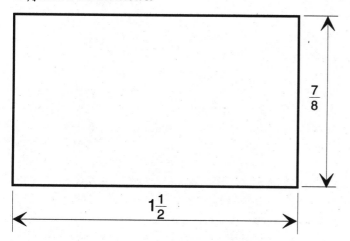

Figure I-12: Alternate method of placing dimensions on a drawing.

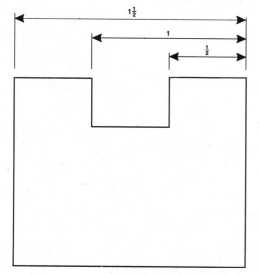

Figure I-13: Where several parallel dimension lines are drawn, they should be the same distance apart, with the dimensions staggered so that they will not be crowded.

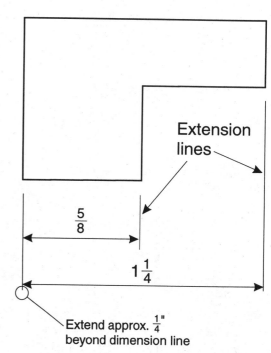

Figure I-14: Dimension lines should be placed between ¼ to ¾ inch from the views.

or beside the line rather than within the line as shown in Figure I-12.

When several parallel dimension lines are drawn, they should be the same distance apart, with the dimensions staggered so that they will not be crowded. *See* Figure I-13.

Extension lines are used to connect a part of a view with its dimension line. Again, use the light full line and begin about $\frac{1}{16}$ inch from the corners of the views and extend the extension line about $\frac{1}{8}$ in beyond the dimension line, as shown in Figure I-14.

Arrowheads or other marks are placed on the ends of dimension lines, and all such marks on a particular drawing should be uniform. Arrowheads, for example, are made with two freehand strokes from the barb to the point (Figure I-15). Sometimes a third stroke is made to join the two barbs and then the arrowhead is filled in as shown in Figure I-16.

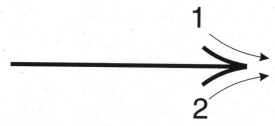

Figure I-15: Arrowheads are made with two freehand strokes from the barb to the point.

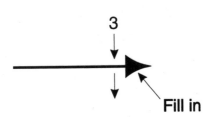

Figure I-16: Sometimes a third stroke is used to connect the two barbs with the resulting triangle being filled.

Computer generated arrowheads are quick and uniform with a large array to choose from, depending upon the CAD or DRAW program. When using a CAD system to prepare drawings, the arrowhead is the most popular style of end for dimension lines. *See* Figure I-17.

Another common style of end for dimension lines is the "tic mark" as shown in Figure I-18. A good tic-mark will be a full heavy line placed at a 45 degree angle, about $\frac{3}{16}$ in long, with the tic-mark centered on the intersection of the extension and dimension lines. Tic-marks commonly are drawn down from right to left on horizontal dimensions and from left to right on vertical dimensions. Because of its simplicity, the tic-mark style can be the fastest hand-drawn style and therefore, very popular for hand-drawn dimension lines. Computer generated tic-marks, however,

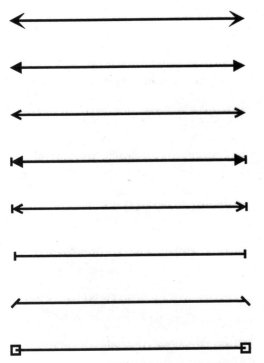

Figure I-17: Several styles of computer generated arrowheads.

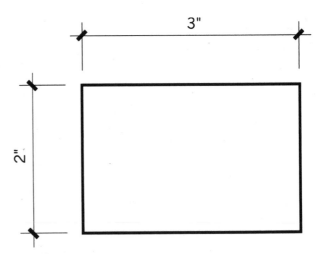

Figure I-18: Tic marks are very popular for manually generated dimension lines.

are sometimes too long and look awkward. They might also print the same line width as the dimension line.

In some cases, there may not be enough room between the extension lines to insert the dimension lines and numerals. If so, the dimension lines may then be drawn outside of the extension lines with the arrowheads reversed to point inward (Figure I-19). This arrangement permits the fractions to be placed either inside or outside of the extension lines so as to prevent crowding and also to allow the length of the arrowheads to be the same as others on the drawing.

If at all possible, dimension lines should not cross extension lines, as shown in Figure I-20. Rather, the shorter dimension should be placed nearest the view, with the larger dimensions placed outside the shorter dimensions as shown in Figure I-21. Also note that numerals in hori-

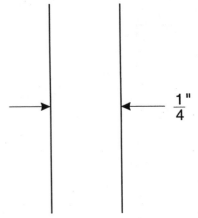

Figure I-19: In tight places, the dimension lines may be drawn outside the extension lines with the arrowheads reversed to point inward.

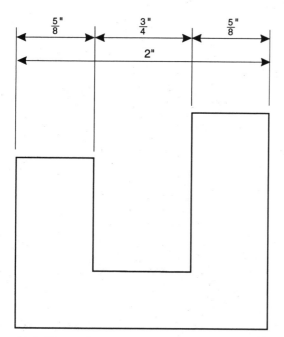

Figure I-20: Dimension lines should not cross extension lines as shown here.

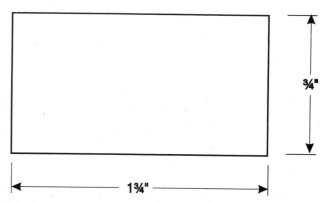

Figure I-22: Horizontal dimensions should be placed so they are read from the bottom edge of the drawing; vertical dimensions should appear on the right-hand edge of the drawing.

zontal dimension lines should be placed so that they may be read from the bottom edge of the drawing (Figure I-22), while vertical dimensions should be placed so that they may be read from the right-hand edge of the drawing.

Sloping dimension lines should also be drawn so that the numerals may be read from either the bottom or right-hand side of the drawing as shown in Figure I-23.

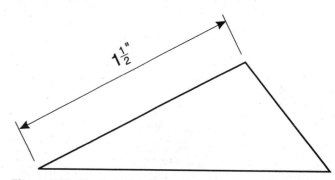

Figure I-23: Sloping dimension lines should be drawn so that the numerals may be read from either the bottom or right-hand side of the drawing, depending on which dimension is needed.

When arcs must be dimensioned on drawings, the center of the arc should be indicated by a small cross and the radius of the arc by a line drawn from the cross to the arc. The numerals indicating the dimension should be arranged as shown in Figure I-24.

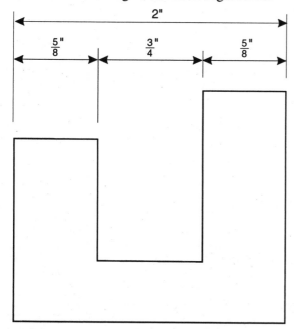

Figure I-21: Proper method of placing dimensions on a drawing.

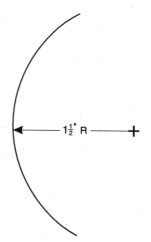

Figure I-24: Method of arranging the dimensions for the radius of an arc.

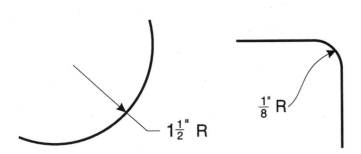

Figure I-25: Method of arranging the dimension for the corner of an instrument chassis.

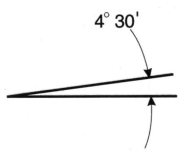

Figure I-27: For small angles, the arrowhead may be reversed to point inward.

Another way of dimensioning arcs is shown in Figure I-25 which shows the radius of a corner of an instrument chassis. Note that the radius dimension is always followed by the letter R placed above the dimension line or next to the numerator in the case of a fraction. Only one arrowhead, however is necessary; the point should touch the arc and is placed on the dimension line.

The dimension line for an angle is an arc as shown in Figure I-26. When arcs are drawn manually, they are usually drawn with a circle template. However, CAD and DRAW computer programs are capable of performing this dimension automatically. Except for large angles, the dimensions are placed to read from the bottom of the drawing. For small angles (Figure I-27), the arrowheads may be reversed to point inward.

Drawings indicating the centerline of knockouts (holes) in such items as outlet boxes, panelboard housings, instrument chassis, etc., should appear in the view in which the holes appear as circles (Figure I-28) and not in views where these holes appear as other than circles, as shown in Figure I-29. The centerlines are extended away from the object to serve as extension lines for dimensioning.

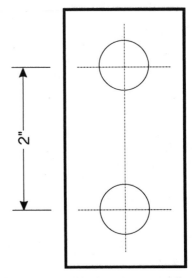

Figure I-28: Drawings indicating the centerline of holes should appear in the view in which the hole appears as a circle.

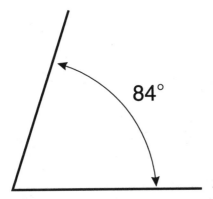

Figure I-26: The dimension line for an angle is an arc.

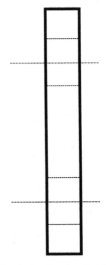

Figure I-29: Example of circle views where the holes appear other than circles. If at all possible, dimensions indicating the centerlines of holes should not appear on drawings of this type.

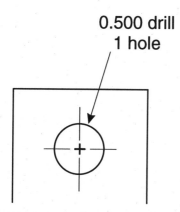

0.500 drill
1 hole

Figure I-30: In most cases, the diameter of a hole is expressed in decimal fractions for control and electronic equipment.

Drilled holes are dimensioned with a note stating the diameter of the drill and perhaps the number of holes of one particular size to be drilled in the object. In most cases of this type, the diameter of the hole is expressed in decimal fractions for control and electronic equipment (Figure I-30) and in common fractions for most electrical equipment (Figure I-31).

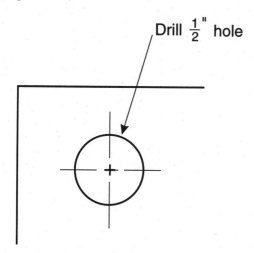

Drill $\frac{1}{2}$" hole

Figure I-31: Common fractions are used for dimensioning holes on drawings of electrical equipment.

Pictorial drawings are dimensioned with extension lines in much the same manner as orthographic drawings, but the dimensioning of pictorial drawings (like isometric drawings) is somewhat more difficult because less room is available on the single view. In dimensioning isometric drawings, the following rules should be followed:

- Dimension lines and numerals should be kept off the view itself wherever possible.

- The radius dimensions of an arc should be drawn from the center of the enclosing isometric square to the arc.

- Visible surfaces should be dimensioned in preference to hidden surfaces.

- Both extension lines for a dimension should be extended from isometric visible edges.

- Vertical numerals should be used in dimensioning; numerals and arrowheads should be made in isomeric.

- Horizontal and vertical dimension lines on the orthographic views should be parallel to isometric axes in the isometric view.

- Notes need not be lettered in isometric; rather, standard vertical single-stroke lettering may be used.

ARCHITECTURAL DIMENSIONING

The common rules that apply to other types of drawings also apply to architectural drawings and since most electrical engineers, designers, and drafters will be working with architectural drawings to some extent, a brief discussion of architectural dimensioning follows. Furthermore, electrical drafters will be required to use architectural dimensioning in showing the placement of certain lighting fixtures and other built-in electrical equipment. Here are some basic rules to follow:

- Keep all outside dimension lines well away from the building lines. They should be located a minimum of $\frac{3}{4}$ in away from the building lines and should be approximately $\frac{5}{16}$ to $\frac{3}{8}$ in apart.

- Masonry openings should be the centerlines of partitions or else are made to the outside walls. In any case, the wall thicknesses should also be shown.

- Dimensions should be provided to the centerlines of columns in both directions.

- Dimensions for openings are normally made to the centerline of the opening or else to the sides of the opening, as required.

Figures I-32 and I-33 show various methods of dimensioning architectural drawings in both plan and sectional views. Figure I-34 shows an example of using architectural dimensioning on an electrical drawing. In this case, the distance between outlets is indicated.

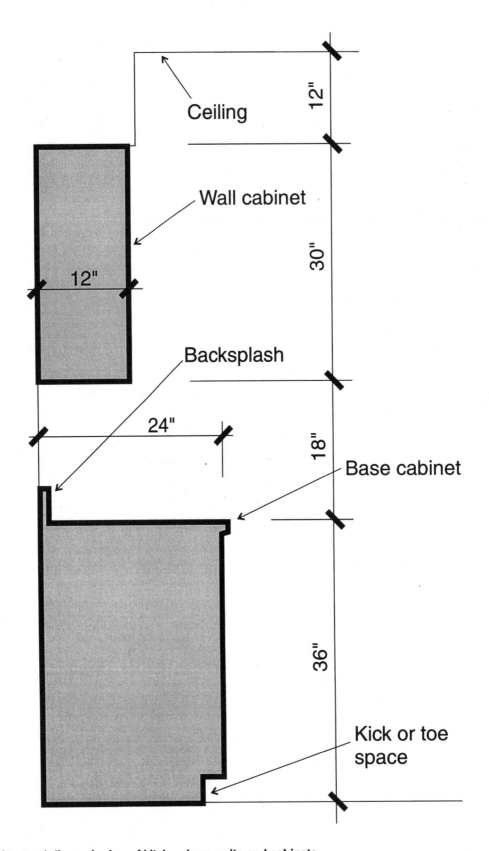

Figure I-32: Architectural dimensioning of kitchen base units and cabinets.

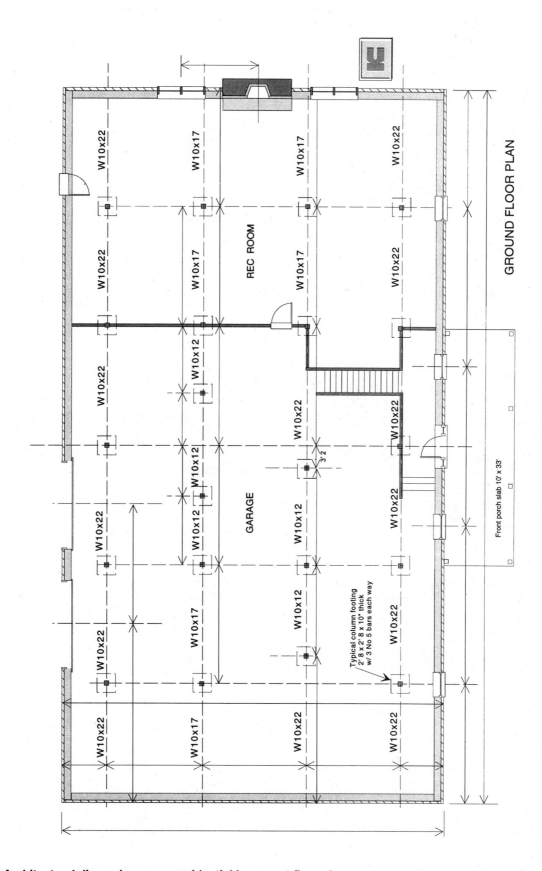

Figure I-33: Architectural dimensions on a residential basement floor plan.

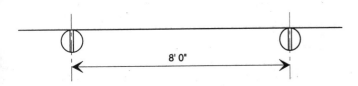

Figure I-34: Architectural dimensioning of receptacle outlets.

Notice that most of the dimensions on these drawings are from center to center of various objects. The reason is that lumber and other construction materials often vary slightly in size, making it impossible to indicate actual values of edges of structural members. Using the center to center method always gives a true location.

CAD COMMANDS

Dimensioning with computer software takes a lot of the headaches out of the dimensioning process. At the same time, because of its versatility, it can be tedious. Each type of software has its own peculiararities but there are some features common to many of them. Computed-Aided Drafting (CAD) or Computed-Aided Drafting/Design (CADD) programs have been developed specifically for engineering and architectural use. Most draw or graphics software (as opposed to CAD) has been developed to appeal more to the creative designer such as the advertising artist, illustrators, and photographic refinishers.

However, despite their artistic emphasis some Draw programs, such as MicrografxDesigner and CorelDraw, lend themselves nicely to electrical drafting. There is enough dimensional precision to these programs to produce an excellent electrical drawing. In fact, most of the drawings in this book were done using either Micrografx Designer or CorelDraw.

The variations in the style of dimensioning in graphics software is limited only by the imagination. Some of the common line ends simulate the same ones that have been the choice of table drafters for years.

Line weight is also flexible and generally easy to set. Most graphics software calibrate line widths in point sizes. Point size is an old measure of type size used for many years by the publishing industry.

Since most electrical drawings are drawn to a certain scale, the finished drawing is not resized, but printed as originally drawn on the screen. However, most of the better programs allow either proportional scaling (resizing) of line width or static line width. Proportional sizing increases or decreases the line width as the drawing is resized. Static line width keeps the line widths the same as when originally drawn regardless of the finished size or scale of the drawing.

For dimensioning distances that are not horizontal or vertical, most software has alignment tools. This feature allows accurate dimensions to be drawn at any angle.

Placement of dimension text and text size is also flexible. A common problem with most default dimension settings is the placement of dimension text within a distance that is too short to include both text and arrows. This problem is generally handled by offsetting the text and either placing it above the dimension line or to the side of the line.

Computer generated dimensions can be extremely accurate and very neat-looking. However, it takes a bit of experience to achieve this, since working with computer programs can be tedious. The great advantage of computer generated work is that the finished look, the technique of a drawing, will be uniform regardless of which drafter performed the work. The accuracy will not necessarily be any greater. A CADD program is only as good as the operator. Mistakes can still be as prevalent as in hand-drawn work.

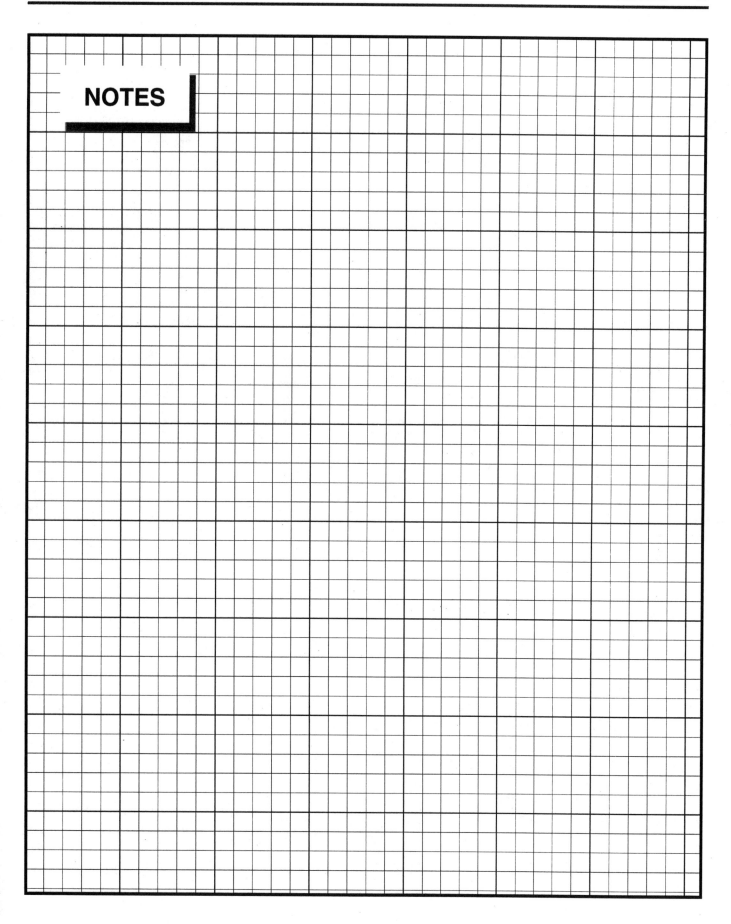

NOTES

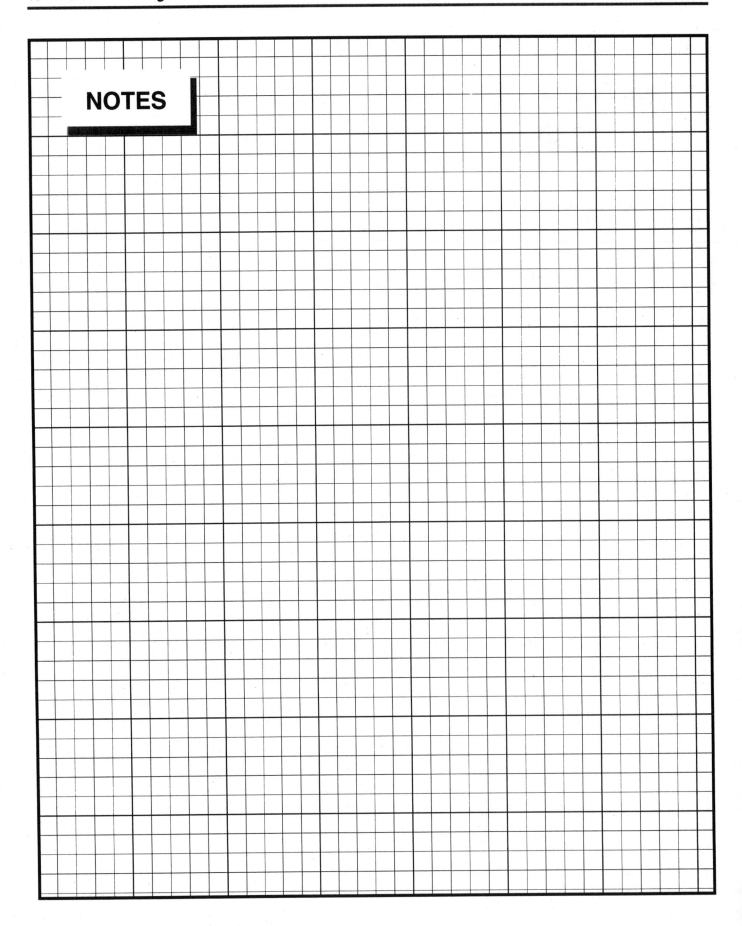

NOTES

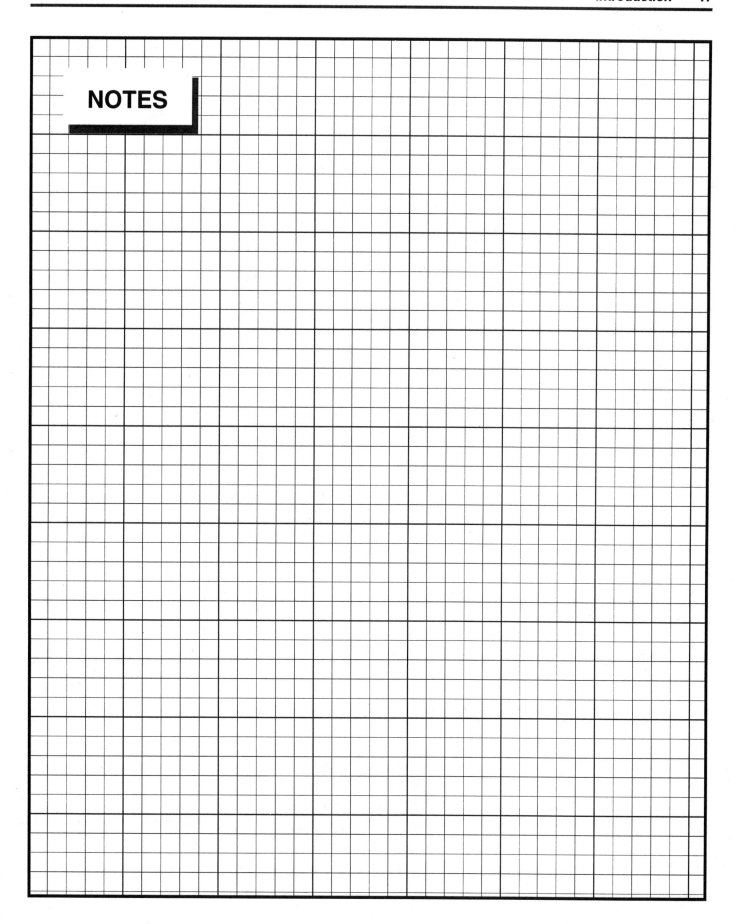

NOTES

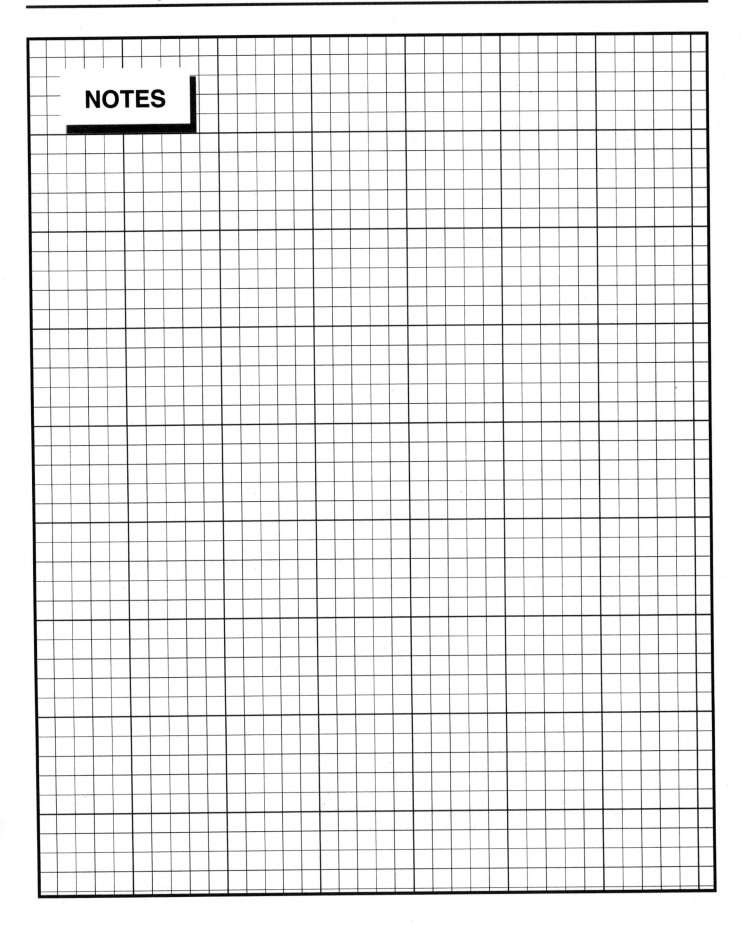

NOTES

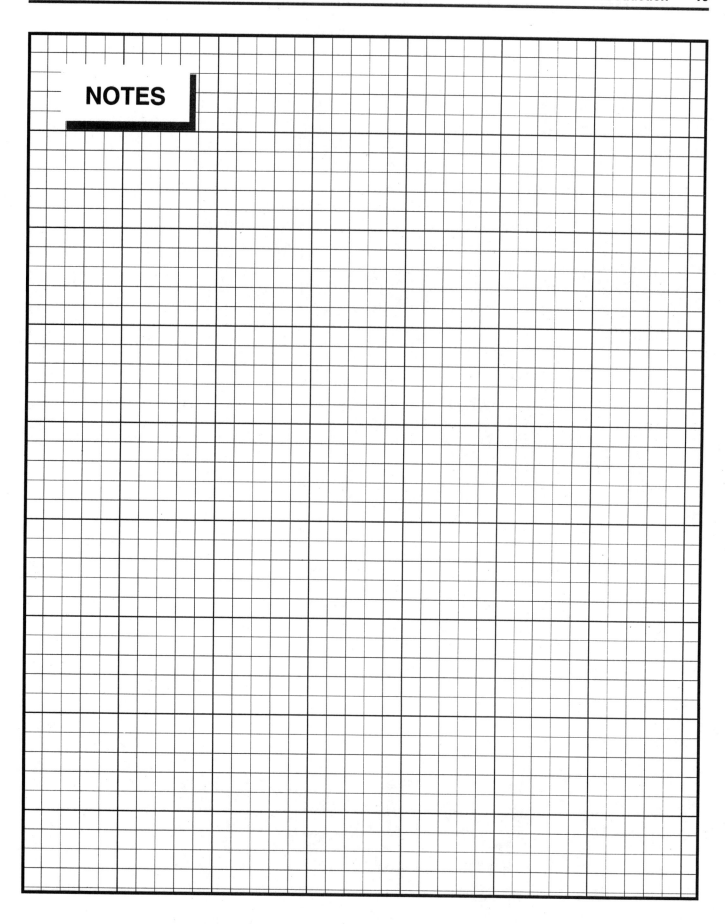

NOTES

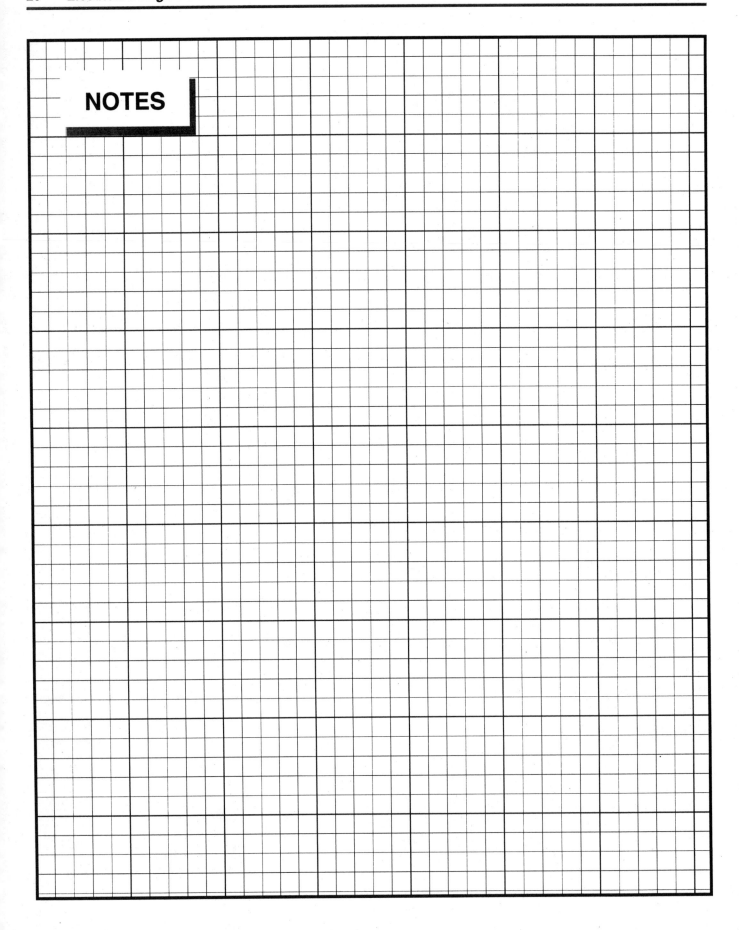

NOTES

## Chapter 1

General Provisions

In all large construction projects and in many of the smaller ones, an architect is commissioned to prepare complete working drawings and specifications for the project. These drawings usually include:

- A plot plan indicating the location of the building on the property.
- Floor plans showing the walls and partitions for each floor or level.
- Elevations of all exterior faces of the building.
- Several vertical cross sections to indicate clearly the various floor levels and details of the footings, foundation, walls, floors, ceilings, and roof construction.
- Large-scale detail drawings showing such construction details as may be required.

For projects of any consequence, the architect usually hires consulting engineers to prepare structural, electrical, and mechanical drawings.

Plot Plans

This type of plan of the building site is as if the site is viewed from an airplane and shows the property boundaries, the existing contour lines, the new contour lines (after grading), the location of the building on the property, new and existing roadways, all utility lines, and other pertinent details. Descriptive notes may also be found on the plot plan listing names of adjacent property owners, the land surveyor, and the date of the survey. A legend or symbol list is also included so that anyone who must work with site plans can readily read the information directly from the drawing without further instructions. *See* Figure 1-1.

Floor Plans

The plan view of any object is a drawing showing the outline and all details as seen when looking directly down on the object. It shows only two dimensions — length and width. The floor plan of a building is drawn as if a slice was taken through the building — about window height — and then the top portion removed to reveal the bottom part where the slice was taken. *See* Figure 1-2.

If it is desired to draw, say, a plan view of a commercial laundry building, the part of the building about the middle of the first-floor windows is imagined to be cut away. By looking down on the uncovered portion, every detail, partition, window and door opening, and the like can be seen. This would be called the first- floor plan. A cut through the second-floor windows (if applicable) would be the second-floor plan, etc. A single-floor building, as shown in Figure 1-3, will only have one basic floor plan, while a high-rise office building may contain a dozen or more floor plans. However, this floor plan will normally be duplicated for each separate trade; that is, electrical floor plan, plumbing floor plan, heating, ventilating, and air conditioning (HVAC) and so on.

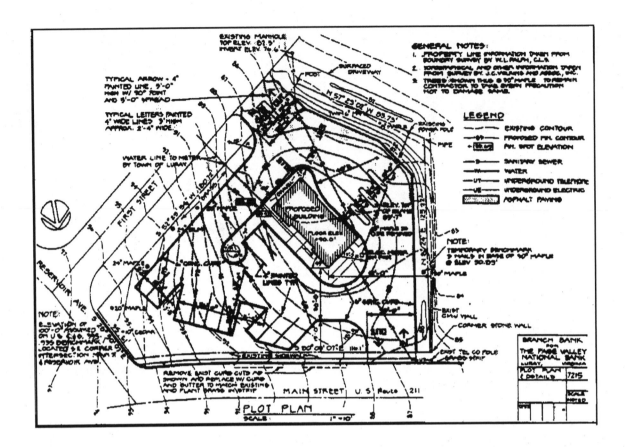

Figure 1-1: Typical plot plan.

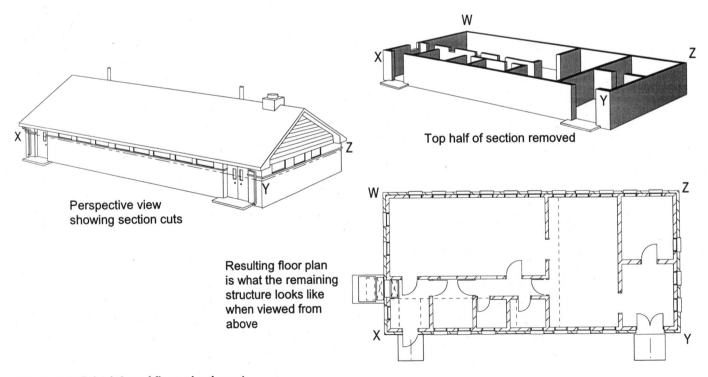

Perspective view
showing section cuts

Top half of section removed

Resulting floor plan
is what the remaining
structure looks like
when viewed from
above

Figure 1-2: Principles of floor-plan layout.

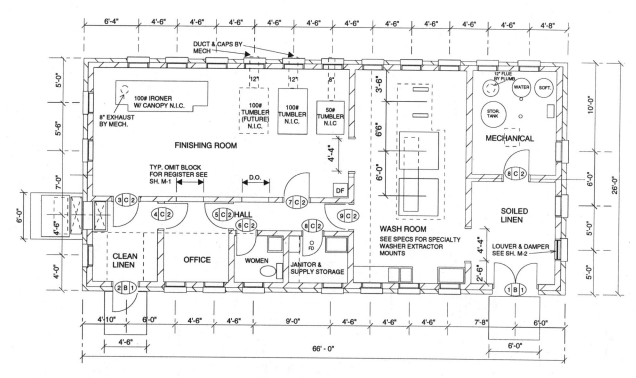

Figure 1-3: Floor plan of a commercial laundry building.

Elevations

A plan view may represent a flat surface, a curved surface, or a slanting one, but for clarification, it is usually necessary to refer to elevations and sections of the building. The *elevation* is an outline of an object that shows heights and may show the length or width of a particular side, but not depth. Figure 1-4 on the next page shows elevation drawings for a building. Note that these elevation drawings show the heights of windows, doors, porches, the pitch of roofs, etc. — all of which cannot be shown conveniently on floor plans.

Sections

A *section* or *sectional view* of an object is a view facing a point where a part of an object is supposed to be cut away, allowing the viewer to see the object's inside. The point on the plan or elevation showing where the imaginary cut has been made is indicated by the section line, which is usually a very heavy double dot-and-dash line. The section line shows the location of the section on the plan or elevation. It is, therefore, necessary to know which of the cutaway parts is represented in the sectional drawing when an object is represented as if it were cut in two. Arrow points are thus placed at the ends of the sectional lines.

In architectural drawings it is often necessary to show more than one section on the same drawing. The different section lines must be distinguished by letters, numbers, or other designations placed at the ends of the lines as shown in Figure 1-5 on page 25, in which one section is lettered A-A; detail section B, etc. These section letters are generally heavy and large so as to stand out on the drawings. To further avoid confusion, the same letter is usually placed at each end of the section line. The section is named according to these letters — that is, Section A-A, Detail Section B, and so forth.

A longitudinal section is taken lengthwise while a cross section is usually taken straight across the width of an object. Sometimes, however, a section is not taken along one straight line. It is often taken along a zigzag line to show important parts of the object.

A sectional view, as applied to architectural drawings, is a drawing showing the building, or portion of a building, as though cut through, as if by a saw, on some imaginary line. This line may be either vertical (straight up and down) or horizontal. Wall sections are nearly always made vertically so that the cut edge is exposed from top to bottom. In some ways the wall section is one of the most important of all the drawings to construction workers, because it answers the questions on how a structure is built. The floor plans of a

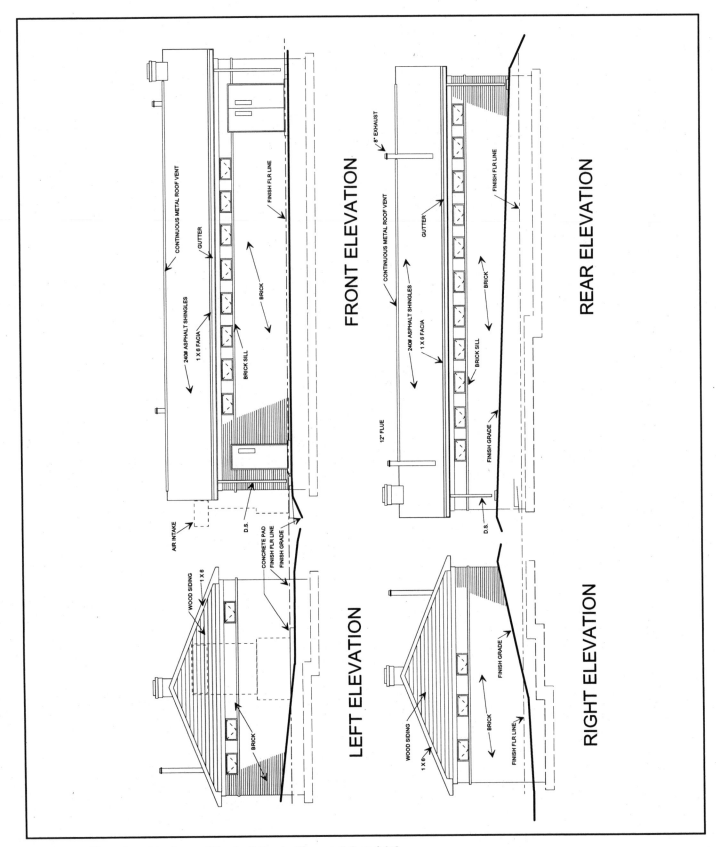

Figure 1-4: Elevation drawings of the building in Figures 1-2 and 1-3.

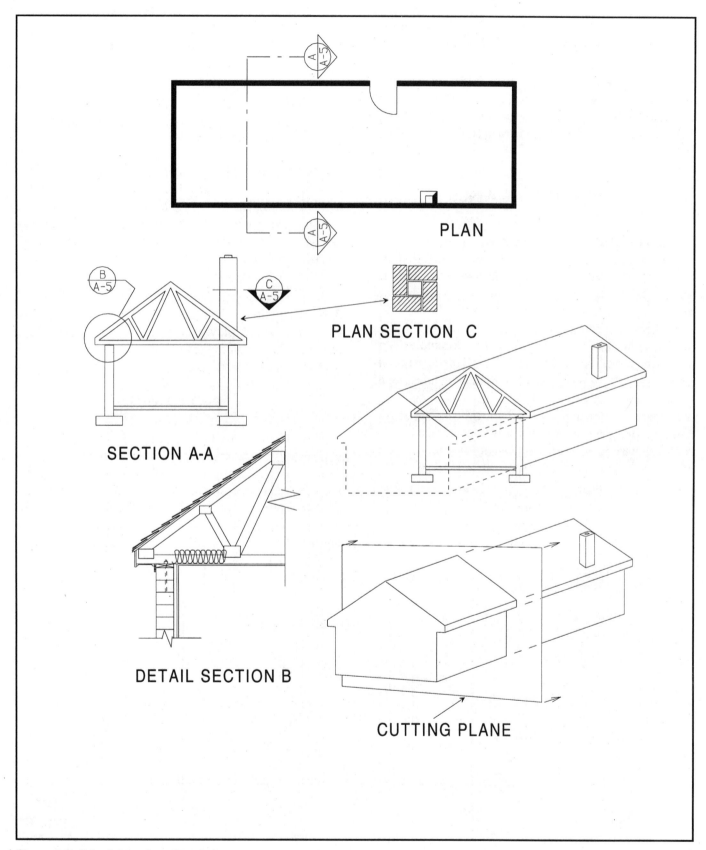

Figure 1-5: Principles of sectional views.

building show how each floor is arranged, but the wall sections tell how each part is constructed and usually indicate the material to be used. The electrician needs to know this information when determining wiring methods that comply with the *National Electrical Code*® *(NEC)*. Electrical engineers and designers also need to know this information to make sure that certain electrical components will fit in any given area.

ELECTRICAL DRAWINGS

The ideal electrical drawing should show in a clear, concise manner exactly what is required of the workers. The amount of data shown on such drawings should be sufficient, but not overdone. This means that a complete set of electrical drawings could consist of only one 8½ × 11 in sheet, or it could consist of several dozen 24 × 36 in (or larger) sheets, depending on the size and complexity of the given project. A shop drawing, for example, may contain details of only one piece of equipment, while a set of working drawings for an industrial installation may contain dozens of drawing sheets detailing the electrical system for lighting and power, along with equipment, motor controls, wiring diagrams, schematic diagrams, equipment schedules and a host of other pertinent data.

In general, electrical working drawings for a given project serve three distinct functions:

- To give electrical contractors an exact description of the project so that materials and labor may be estimated in order to form a total cost of the project for bidding purposes.
- To give workers on the project instructions as to how the electrical system is to be installed.
- To provide a "map" of the electrical system once the job is completed to aid in maintenance and troubleshooting for years to come.

Electrical drawings from consulting engineering firms will vary in quality from sketchy, incomplete drawings to neat, very complete drawings that are easy to understand. Few, however, will cover every exact detail of the electrical system. Therefore, a good knowledge of installation practices must go hand-in-hand with interpreting electrical working drawings.

Sometimes electrical contractors will have electrical drafters prepare special supplemental drawings for use by the contractors' employees. On certain projects, these supplemental drawings can save supervision time in the field once the project has begun.

PRINT LAYOUT

Most drawings used for building construction projects will be drawn on drawing paper in sizes from 11 × 17 in to 24 × 36 in. Each drawing sheet will have border lines framing the overall drawing and a title block as shown in Figure 1-6. Note that the type and size of title blocks vary with each firm preparing the drawings. In addition, some drawing sheets will also contain a revision block near the title

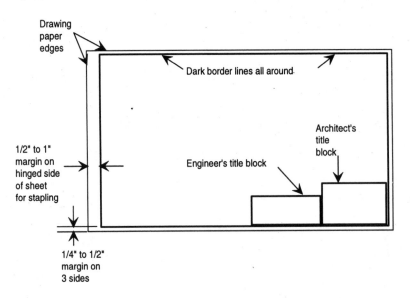

Figure 1-6: Typical drawing layout.

block, and perhaps an approval block. This information is normally found on each drawing sheet, regardless of the type of project or the information contained on the sheet.

Although a strong effort has been made to standardize drawing practices in the building construction industry, seldom will prints — prepared by different architectural or engineering firms — be identical. Similarities, however, will exist between most sets of prints, and with a little experience, you should have little trouble interpreting any set of drawings that might be encountered.

Title Block

The title block for a blueprint is usually boxed in the lower right-hand corner of the drawing sheet; the size of the block varies with the size of the drawing and also with the information required.

In general, the title block of an electrical drawing should contain the following:

- Name of the project
- Address of the project
- Name of the owner or client
- Name of the architectural and/or engineering firm
- Date of completion
- Scale(s)
- Initials of the drafter, checker, and designer, with dates under each if different from the drawing date

Figure 1-8: Stick-on decal being applied to a standard sheet of tracing paper.

- Job number
- Sheet number
- General description of the drawing

Every architectural/engineering firm has its own standard for drawing titles (*see* Figure 1-7), and they are often preprinted directly on the tracing paper or else printed on "stick-on" decals which are placed on the drawing. *See* Figure 1-8.

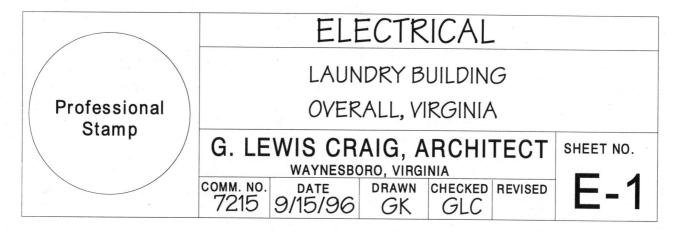

Figure 1-7: Architect's title block.

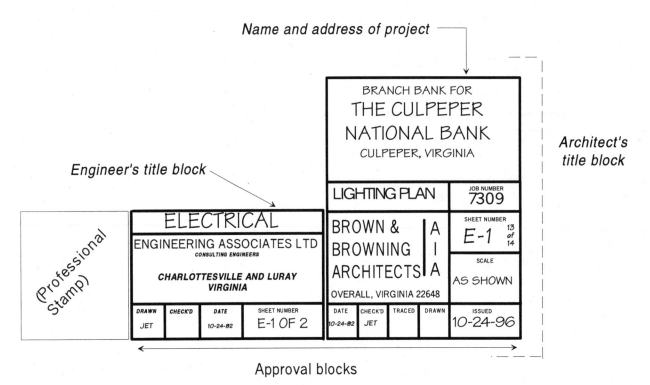

Name and address of project ⌐

Architect's title block

Engineer's title block

Figure 1-9: Combined engineer's and architect's title block.

Often the consulting engineering firm will also be listed, which means that an additional title block will be applied to the drawing — usually next to the architect's title block. Figure 1-9 shows completed architectural and engineering title blocks as they appear on an actual drawing for a commercial bank building. Although not shown, the architect's professional stamp appears above the title block.

Drawing title blocks range from very simple blocks with only the very essential information to ultra-fancy blocks that take up nearly a third of the drawing sheet. Most consulting firms, however, strive for a neat-looking title block that provides sufficient room for all essential information. Many consulting firms can be recognized by the drawing title block alone, without actually reading the name.

The sample title blocks appearing in Figures 1-10 through 1-19 are typical styles that are currently used by some consulting engineering firms. These samples should prove useful to consulting engineers who are designing a new title block or modifiying an existing one.

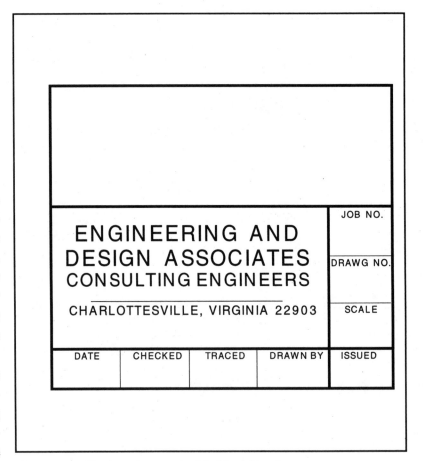

Figure 1-10: Engineer's sample title block No. 1.

ENGINEERING AND DESIGN ASSOCIATES				SHEET NO.
CONSULTING ENGINEERS				
CHARLOTTESVILLE, VIRGINIA 22903				
JOB NO.	DATE	DRAWN.	CHECKED	REVISED

Figure 1-11: Engineer's sample title block No. 2.

ENGINEERING AND DESIGN ASSOCIATES	DATE	SHEET NO.
CONSULTING ENGINEERS	DRAWN	
	CHECKED	
CHARLOTTESVILLE, VIRGINIA 22903	REVISED	JOB NO.

Figure 1-12: Engineer's sample title block No. 3.

		EDA ENGINEERING AND DESIGN ASSOCIATES CONSULTING ENGINEERS CHARLOTTESVILLE, VIRGINIA 22903	DATE	SHEET NO.
			DRAWN	
			CHECKED	
SEAL			REVISED	JOB NO.

Figure 1-13: Engineer's sample title block No. 4.

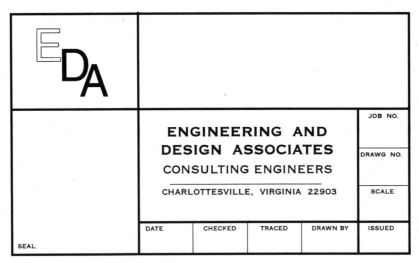

Figure 1-14: Engineer's sample title block No. 5.

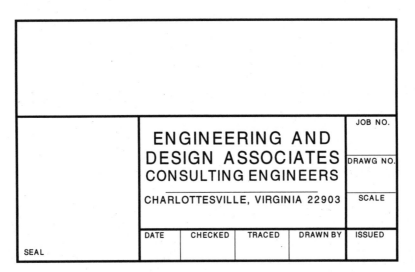

Figure 1-15: Engineer's sample title block No. 6.

| SHEET NO. | ENGINEERING AND DESIGN ASSOCIATES | CONSULTING ENGINEERS CHARLOTTESVILLE, VIRGINIA 22903 JOB: |

SHEET TITLE	JOB NO.	DATE	DRAWN	ISSUED	SCALE

Figure 1-16: Engineer's sample title block No. 7.

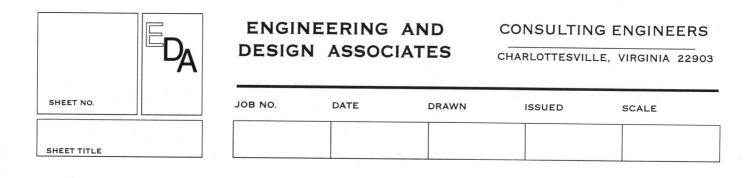

Figure 1-17: Engineer's sample title block No. 8.

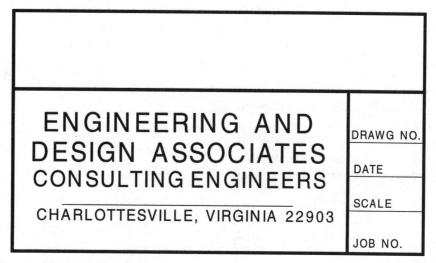

Figure 1-18: Engineer's sample title block No. 9.

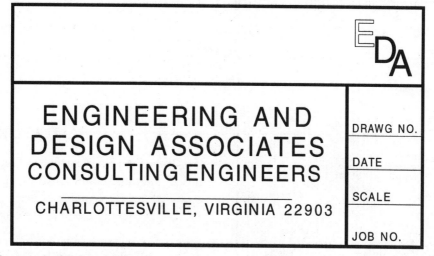

Figure 1-19: Engineer's sample title block No. 10.

Approval Block

The "approval block," in most cases, will appear on the drawing sheet as shown in Figure 1-20. The various types of approval blocks — *drawn, checked, etc.* — will be initialed by the appropriate personnel. This type of approval block is usually part of the title block and appears on each drawing sheet.

COMM. NO.	DATE	DRAWN	CHECKED	REVISED
7215	9/16/96	GK	GLC	

Figure 1-20: One type of approval block used on electrical drawings.

On some projects, authorized signatures are required before certain systems may be installed, or even before the project begins. An approval block such as the one shown in Figure 1-21 indicates that all required personnel has checked the drawings for accuracy, and that the set meets with everyone's approval. If a signature is missing, the project should not be started. Such an approval block usually appears on the front sheet of the blueprint set and may include:

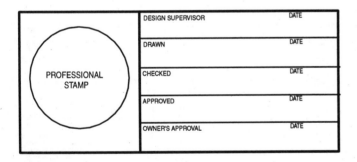

Figure 1-21: Alternate approval box.

- Professional stamp — registered seal of approval by the architect or consulting engineer.
- Design supervisor's signature — the person who is overseeing the design.
- Drawn (by) — signature or initials of the person who drafted the drawing and the date it was completed.

- Checked (by) — signature or initials of the person(s) who reviewed the drawing and the date of approval.
- Approved — signature of initials of the architect/engineer and the date of the approval.
- Owner's approval — signature of the project owner or the owner's representative along with the date signed.
- Any other pertinent data required to facilitate the reading and accuracy of the drawing.

Revision Block

Sometimes electrical drawings will have to be partially redrawn or modified during construction of a project. It is extremely important that such modifications are noted and dated on the drawings to ensure that the workers have an up-to-date set of drawings to work from. In some situations, sufficient space is left near the title block for dates and description of revisions as shown in Figure 1-22. In other cases, a revision block is provided (again, near the title block) as shown in Figure 1-23. But these two samples are by no means the only types or styles of revision blocks that will be seen on electrical working drawings — not by any means. Each architect/engineer/designer/drafter has his or her own method of showing revisions, so expect to find deviations from those shown.

Caution: *When a set of electrical working drawings has been revised, always make certain that the most up-to-date set is used for all future layout work. Either destroy the old, obsolete set of drawings or else clearly mark on the affected sheets, "Obsolete Drawing — Do Not Use." Also, when working with a set of working drawings and written specifications for the first time, thoroughly check each page to see if any revisions or modifications have been made to the originals. Doing so can save much time and expense to all concerned with the project.*

DRAFTING LINES

All drafting lines have one thing in common — they are all the same color. However, good easy-to-read contrasting lines can be made by varying the width of the lines or else "breaking" the lines in some uniform way.

Figure 1-24 on page 34 shows common lines used on architectural drawings. However, these lines can vary. Architects and engineers have striven for a common "stan-

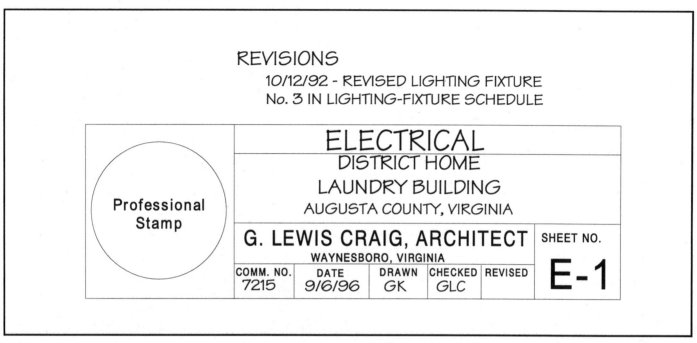

Figure 1-22: One method of showing revisions on electrical working drawings.

dard" for the past century, but unfortunately, their goal has yet to be reached. Therefore, you will find variations in lines and symbols from drawing to drawing, so always consult the legend or symbol list when referring to an architectural or electrical drawing. Also carefully inspect each drawing to ensure that line types are used consistently.

A brief description of the drafting lines shown in Figure 1-24 follows:

Light Full Line - This line is used for section lines, building background (outlines), and similar uses where the object to be drawn is secondary to the electrical system. Care should be taken in choosing the width of such lines, as they sometimes fade out on some printers.

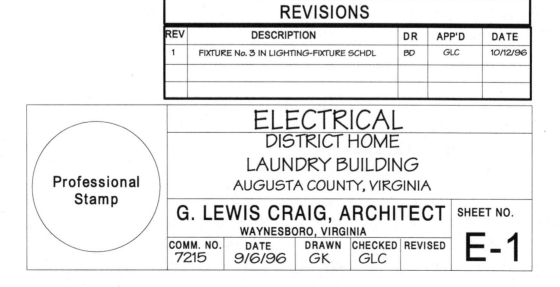

Figure 1-23: One method of showing revisions on electrical working drawings.

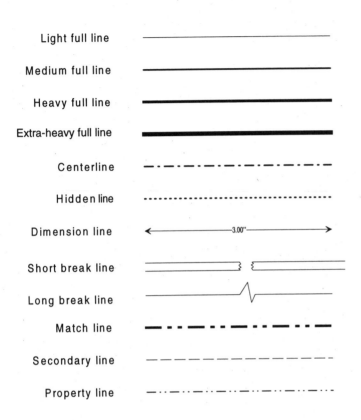

Light full line

Medium full line

Heavy full line

Extra-heavy full line

Centerline

Hidden line

Dimension line

Short break line

Long break line

Match line

Secondary line

Property line

Figure 1-24: Typical drafting lines.

Medium Full Line - This type of line is frequently used for hand lettering on drawings. It is further used for some drawing symbols, circuit lines, and the like.

Heavy Full Line - This line is used for borders around title blocks, schedules and for hand lettering drawing titles. Some types of symbols are frequently drawn with the heavy full line.

Extra-heavy Full Line - This line is used for border lines on architectural/engineering drawings.

Centerline - A centerline is a broken line made up of long and short dashes alternately spaced. It indicates the centers of objects such as holes, pillars, or fixtures. Sometimes, the centerline indicates the dimensions of a finished floor.

Hidden Line - A hidden line consists of a series of short dashes closely and evenly spaced. It shows the edges of objects that are not visible in a particular view. The object outlined by hidden lines in one drawing is often fully pictured in another drawing.

Dimension Lines - These are thinly drawn lines used to show the extent and direction of dimensions. The dimension is usually placed in a break inside of the dimension lines. Normal practice is to place the dimension lines outside the object's outline. However, sometimes it may be necessary to draw the dimensions inside the outline.

Short Break Line - This line is usually drawn freehand and is used for short breaks.

Long Break Line - This line which is drawn partly with a straightedge and partly with freehand zigzags, is used for long breaks.

Match Line - This line is used to show the position of the cutting plane. Therefore, it is also called cutting plane line. A match or cutting plane line is an extra-heavy line with long dashes alternating with two short dashes. It is used on drawings of large structures to show where one drawing stops and the next drawing starts.

Secondary Line - This line is frequently used to outline pieces of equipment or to indicate reference points of a drawing that is secondary to the drawing's purpose.

Property Line - This is a line made up of one long and two short dashes alternately spaced. It indicates land boundaries on the site plan.

Other uses of the lines just mentioned include the following:

Extension Lines - Extension lines are lightweight lines that start about $\frac{1}{16}$ inch away from an object's edge and extend out. A common use of extension lines is to create a boundary for dimension lines. Dimension lines meet extension lines with arrowheads, slashes, or dots. Extension lines that point from a note or other reference to a particular feature on a drawing are called leaders. They usually end in either an arrowhead or a dot and may include an explanatory note at the end.

Section Lines - These are often referred to as crosshatch lines. Drawn at a 45-degree angle, these lines show where an object has been cut away to reveal the inside.

Phantom Lines - Phantom lines are solid, light lines that show where an object will be installed. A future door opening or a future piece of equipment can be shown with phantom lines.

A summary of lines used on architectural drawings appear in Figure 1-25. These lines are frequently modified by architectural drafters to better convey design details to estimators and workers on the job.

Architectural Lines

Dimension lines

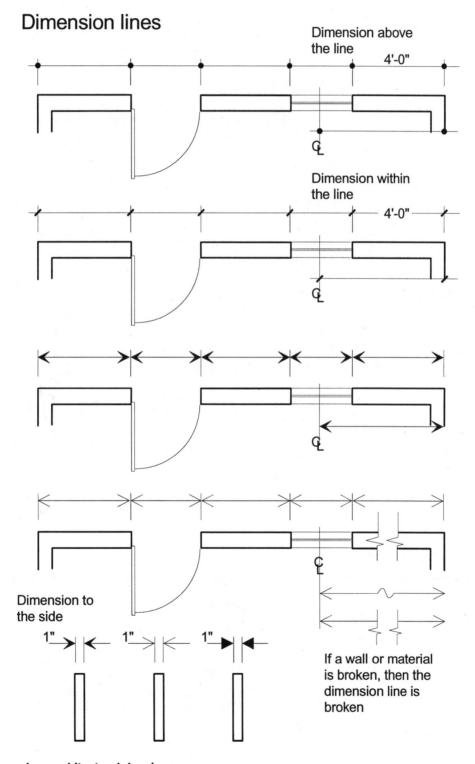

Figure 1-25: Lines used on architectural drawings.

Architectural Lines

Line widths

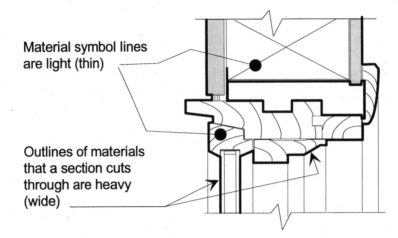

Material symbol lines are light (thin)

Outlines of materials that a section cuts through are heavy (wide)

Leaders

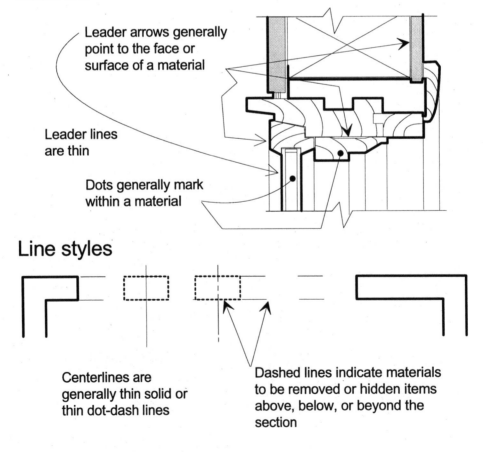

Leader arrows generally point to the face or surface of a material

Leader lines are thin

Dots generally mark within a material

Line styles

Centerlines are generally thin solid or thin dot-dash lines

Dashed lines indicate materials to be removed or hidden items above, below, or beyond the section

Figure 1-25: Lines used on architectural drawings. *(Cont.)*

Architectural Lines

Line styles

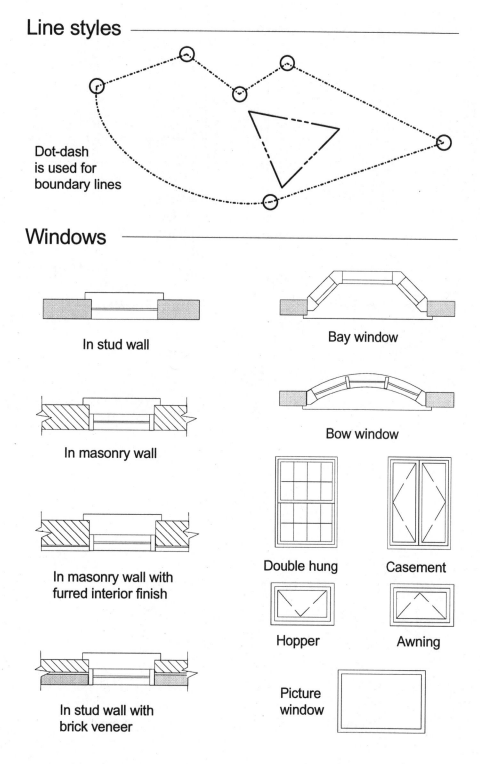

Dot-dash
is used for
boundary lines

Windows

In stud wall

In masonry wall

In masonry wall with
furred interior finish

In stud wall with
brick veneer

Bay window

Bow window

Double hung

Casement

Hopper

Awning

Picture
window

Figure 1-25: Lines used on architectural drawings. *(Cont.)*

Architectural Lines

Doors

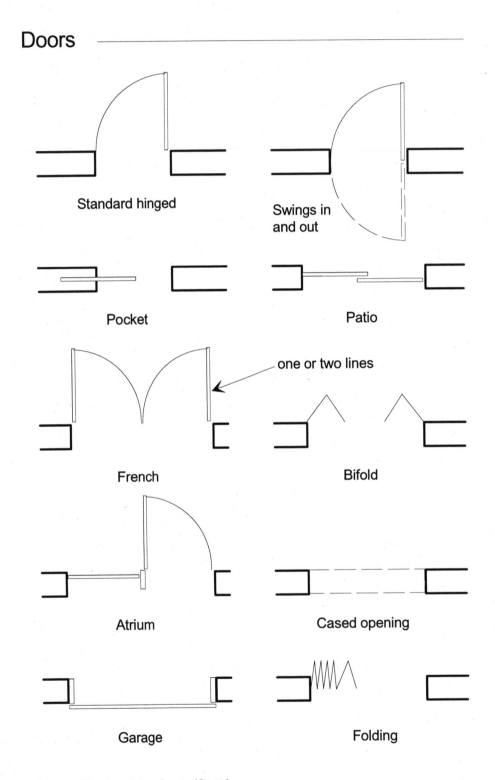

Figure 1-25: Lines used on architectural drawings. *(Cont.)*

Architectural Lines

Popular drawing symbols

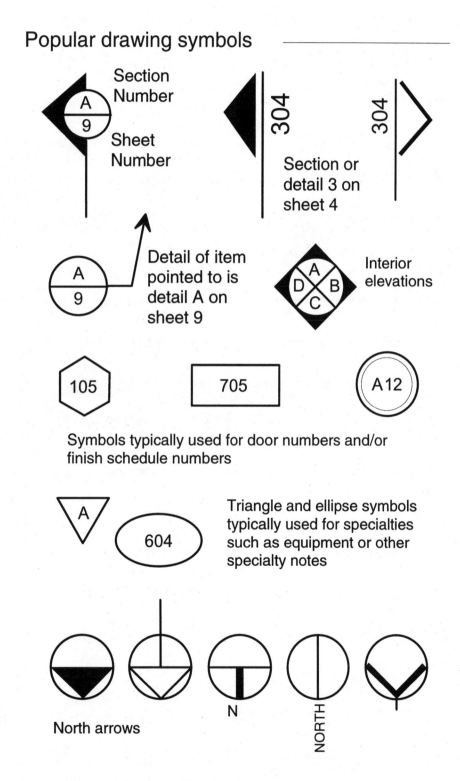

Section Number

Sheet Number

Section or detail 3 on sheet 4

Detail of item pointed to is detail A on sheet 9

Interior elevations

Symbols typically used for door numbers and/or finish schedule numbers

Triangle and ellipse symbols typically used for specialties such as equipment or other specialty notes

North arrows

N

NORTH

Figure 1-25: Lines used on architectural drawings. *(Cont.)*

Electrical Drafting Lines

Besides the architectural lines shown in Figures 1-24 and 1-25, consulting electrical engineers, designers, and drafters use additional lines to represent circuits and their related components. Again, these lines may vary from drawing to drawing, so check the symbol list or legend for the exact meaning of lines on the drawing with which you are working. Figure 1-26 shows lines used on some electrical drawings. Again, these lines may vary from drawing to drawing, so always verify the meaning of lines in the symbol list or legend.

ELECTRICAL SYMBOLS

Electrical workers must be able to correctly read and understand electrical working drawings which requires a thorough knowledge of electrical symbols and their appli-cation. Electrical engineers, designers and drafters, therefore, have the responsibility of providing meaningful symbols that are easily interpreted by workers on the job.

An electrical symbol is a figure, picture, or mark that stands for a component used in the electrical system. For example, Figure 1-27 shows a list of electrical symbols that are currently recommended by the American National Standards Institute (ANSI). It is evident from this list of symbols that many have the same basic form, but because of some slight difference, their meaning changes. For example, the outlet symbols in Figure 1-28 on page 42 each have the same basic form (a circle), but the addition of a line or an abbreviation gives each an individual meaning. A good procedure to follow in learning or designing new symbols is to first learn the basic form and then apply the variations for obtaining different meanings.

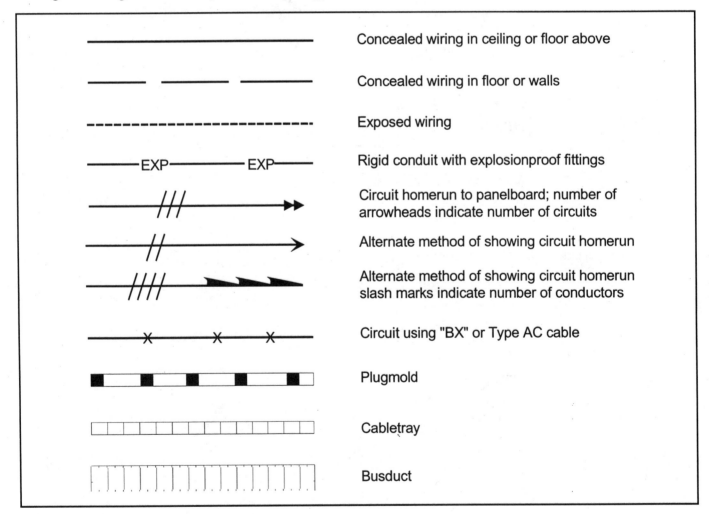

Figure 1-26: Lines used on electrical drawings.

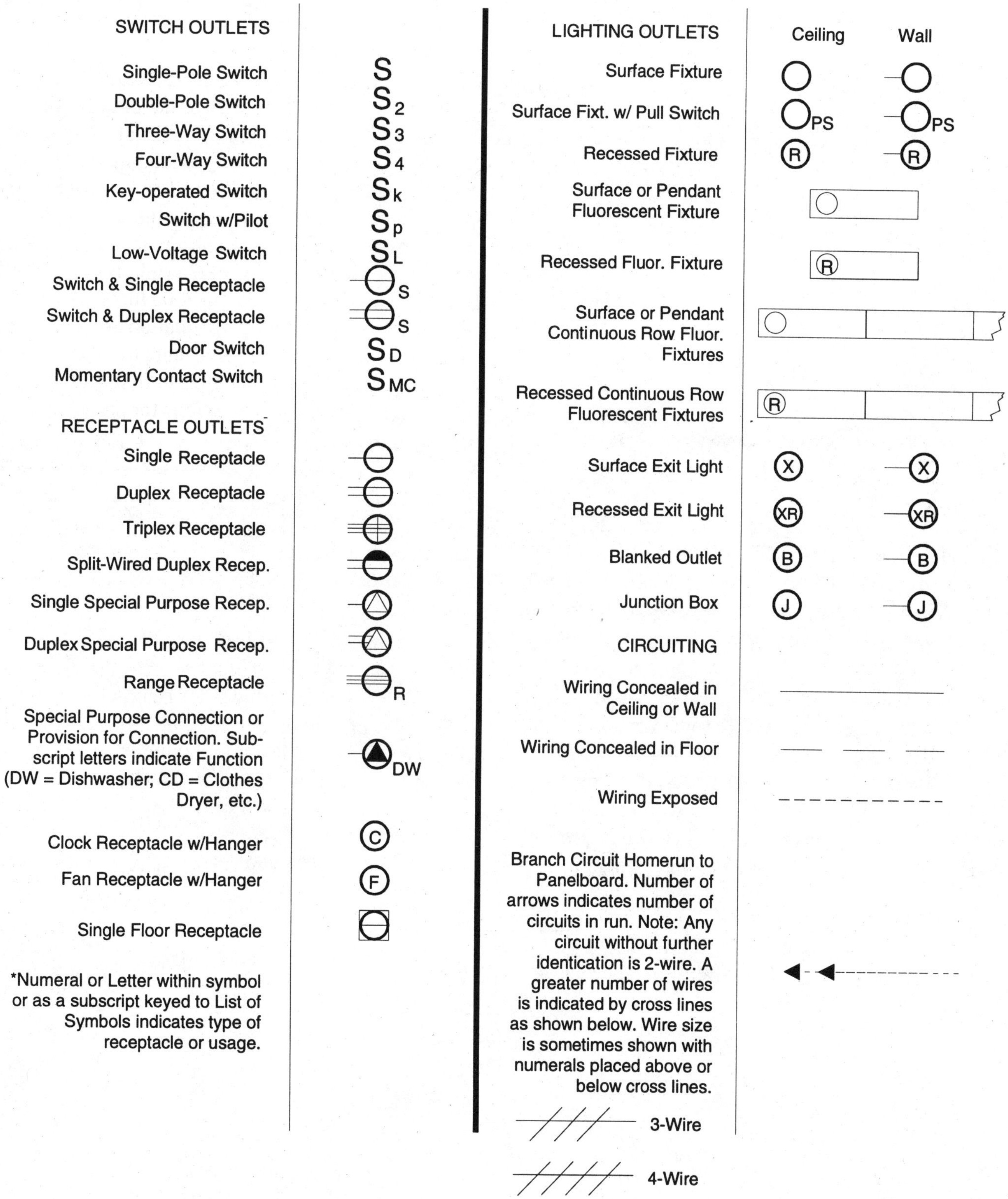

SWITCH OUTLETS

Single-Pole Switch	S
Double-Pole Switch	S_2
Three-Way Switch	S_3
Four-Way Switch	S_4
Key-operated Switch	S_k
Switch w/Pilot	S_p
Low-Voltage Switch	S_L
Switch & Single Receptacle	⊖$_S$
Switch & Duplex Receptacle	⊖$_S$
Door Switch	S_D
Momentary Contact Switch	S_{MC}

RECEPTACLE OUTLETS

Single Receptacle	⊖
Duplex Receptacle	⊖
Triplex Receptacle	⊕
Split-Wired Duplex Recep.	⊖
Single Special Purpose Recep.	⊖
Duplex Special Purpose Recep.	⊖
Range Receptacle	⊖$_R$
Special Purpose Connection or Provision for Connection. Subscript letters indicate Function (DW = Dishwasher; CD = Clothes Dryer, etc.)	⊖$_{DW}$
Clock Receptacle w/Hanger	Ⓒ
Fan Receptacle w/Hanger	Ⓕ
Single Floor Receptacle	⊖

*Numeral or Letter within symbol or as a subscript keyed to List of Symbols indicates type of receptacle or usage.

LIGHTING OUTLETS

	Ceiling	Wall
Surface Fixture	○	○
Surface Fixt. w/ Pull Switch	○PS	○PS
Recessed Fixture	Ⓡ	Ⓡ
Surface or Pendant Fluorescent Fixture	▭	
Recessed Fluor. Fixture	▭Ⓡ	
Surface or Pendant Continuous Row Fluor. Fixtures	▭	
Recessed Continuous Row Fluorescent Fixtures	▭Ⓡ	
Surface Exit Light	Ⓧ	Ⓧ
Recessed Exit Light	⊗R	⊗R
Blanked Outlet	Ⓑ	Ⓑ
Junction Box	Ⓙ	Ⓙ

CIRCUITING

Wiring Concealed in Ceiling or Wall ————————

Wiring Concealed in Floor — — — —

Wiring Exposed – – – – – – – –

Branch Circuit Homerun to Panelboard. Number of arrows indicates number of circuits in run. Note: Any circuit without further identication is 2-wire. A greater number of wires is indicated by cross lines as shown below. Wire size is sometimes shown with numerals placed above or below cross lines.

◄—◄ – – – – – – –

⫫⫫⫫ 3-Wire

⫫⫫⫫⫫ 4-Wire

Figure 1-27: Electrical symbols recommended by ANSI.

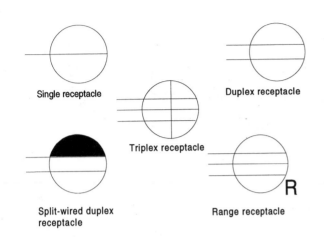

Figure 1-28: Various types of receptacle symbols used on electrical drawings.

It would be much simpler if all architects, engineers, electrical designers, and drafters used the same symbols. However, this is not the case. Although standardization is getting closer to a reality, existing symbols are still modified and new symbols are created for almost every new project.

The electrical symbols described in the following paragraphs represent those found on actual electrical working drawings throughout the United States and Canada. Many are similar to those recommended by ANSI and the Consulting Engineers Council/US; others are not. Understanding how these symbols were devised will help you to interpret unknown electrical symbols in the future.

Some of the symbols used on electrical drawings are abbreviations, such as WP for weatherproof and AFF for above finished floor. Others are simplified pictographs, such as "A" in Figure 1-29 for a double floodlight fixture or like "B" in Figure 1-29 for an infrared electric heater with two quartz lamps.

In some cases, the symbols are combinations of abbreviation and pictograph, such as "C" in Figure 1-29 for fusible safety switch, "D" for a double-throw safety switch, and "E" for a nonfusible safety switch. In each example, a pictograph of a switch enclosure has been combined with an abbreviation, F (fusible), DT (double throw), and NF (nonfusible), respectively. The numerals indicate the bus-bar capacity in amperes.

Lighting-outlet symbols have been devised that represent incandescent, fluorescent, and high-intensity discharge lighting; a circle usually represents an incandescent fixture; and a rectangle is used to represent a fluorescent fixture. All these symbols are designed to indicate the physical shape of a particular fixture and while the circles representing incandescent lamps are frequently enlarged somewhat, symbols for fluorescent fixtures are usually drawn as close to scale as possible.

The type of mounting used for all lighting fixtures is usually indicated in a lighting-fixture schedule, which is shown on the drawings or in the written specifications.

The type of lighting fixture is identified by a numeral placed inside a triangle or other symbol, and placed near the fixture to be identified. A complete description of the fixture identified by the symbols must be given in the lighting-fixture schedule and should include the manufacturer, catalog number, number and type of lamps, voltage,

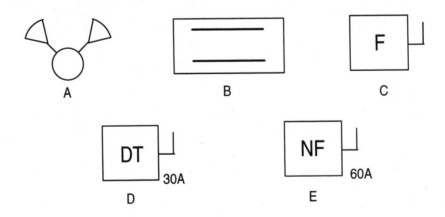

Figure 1-29: Pictographs and notations used to form electrical symbols.

finish, mounting, and any other information needed for a proper installation of the fixtures.

Switches used to control lighting fixtures are also indicated by symbols — usually the letter "S" followed by numerals or letters to define the exact type of switch. For example, S_3 indicates a three-way switch; S_4 identifies a four-way switch; S_P indicates a single-pole switch with a pilot light, etc.

Main distribution centers, panelboards, transformers, safety switches, and other similar electrical components are indicated by electrical symbols on floor plans and by a combination of symbols and semi-pictorial drawings in riser diagrams.

A detailed description of the service equipment is usually given in the panelboard schedule or in the written specifications. However, on small projects the service equipment is sometimes indicated only by notes on the drawings.

Circuit and feeder wiring symbols are getting closer to being standardized. Most circuits concealed in the ceiling or wall are indicated by a solid line; a broken line is used for circuits concealed in the floor or ceiling below; and exposed raceways are indicated by short dashes or else the letter "E" placed in the same plane with the circuit line at various intervals.

The number of conductors in a conduit or raceway system may be indicated in the panelboard schedule under the appropriate column, or the information may be shown on the floor plan.

Symbols for communication and signal systems, as well as symbols for light and power, are drawn to an appropriate scale and accurately located with respect to the building; this reduces the number of references made to the architectural drawings. Where extreme accuracy is required in locating outlets and equipment, exact dimensions are given on larger-scale drawings and shown on the plans.

Each different category in an electrical system is usually represented by a distinguishing basic symbol. To further identify items of equipment or outlets in the category, a numeral or other identifying mark is placed within the open basic symbol. In addition, all such individual symbols used on the drawings should be included in the symbol list or legend. The symbols shown in Figure 1-30 are those recommended by the Consulting Engineers Council/US.

RESIDENTIAL OCCUPANCIES

Signaling system symbols for use in identifying standardized residential-type signal system items on residential drawings where a descriptive symbol list is not included on the drawing. When other signal system items are to be identified, use the above basic symbols for such items together with a descriptive symbol list.

Pushbutton	
Buzzer	
Bell	
Combination Bell - Buzzer	
Chime	
Annunciator	
Electric Door Opener	
Maid's Signal Plug	
Interconnection Box	
Bell-Ringing Transformer	
Outside Telephone	
Interconnecting Telephone	
Television Outlet	

Figure 1-30: Recommended electrical symbols.

SWITCH OUTLETS

Single Pole Switch	S
Double Pole Switch	S_2
Three-Way Switch	S_3
Four-Way Switch	S_4
Key-Operated Switch	S_K
Switch and Fusestat Holder	$S_F H$
Switch and Pilot Lamp	S_P
Fan Switch	S_F
Switch for Low-Voltage Switching System	S_L
Master Switch for Low-Voltage Switching System	S_{LM}
Switch and Single Receptacle	S
Switch and Duplex Receptacle	S
Door Switch	S_D
Time Switch	S_T
Momentary Contact Switch	S_{MC}
Ceiling Pull Switch	S
"Hand-Off-Auto" Control Switch	HOA
Multi-Speed Control Switch	M
Pushbutton	•

RECEPTACLE OUTLETS

Where weatherproof, explosionproof, or other specific types of devices are to be required, use the uppercase subscript letters to specify. For example, weatherproof single or duplex receptacles would have the uppercase WP subscript letters noted alongside of the symbols. All outlets must be grounded.

Single Receptacle Outlet	
Duplex Receptacle Outlet	
Triplex Receptacle Outlet	
Quadruplex Receptacle Outlet	
Duplex Receptacle Outlet Split Wired	
Triplex Receptacle Outlet Split Wired	
250 Volt Receptacle Single Phase Use Subscript Letter to Indicate Function (DW - Dishwasher, RA - Range) or Numerals (with explanation in symbols schedule)	
250 Volt Receptacle Three Phase	
Clock Receptacle	C
Fan Receptacle	F
Floor Single Receptacle Outlet	
Floor Duplex Receptacle Outlet	
Floor Special-Purpose Outlet	*
Floor Telephone Outlet - Public	
Floor Telephone Outlet - Private	

* Use numeral keyed explanation of symbol usage

Figure 1-30: Recommended electrical symbols. *(Cont.)*

Example of the use of several floor outlet symbols to identify a 2, 3, or more gang outlet:

Underfloor Duct and Junction Box for Triple, Double or Single Duct System as indicated by the number of parallel lines

Example of use of various symbols to identify location of different types of outlets or connections for underfloor duct or cellular floor systems:

Cellular Floor Header Duct

CIRCUITING

Wiring Exposed (not in conduit) ——E——

Wiring Concealed in Ceiling or Wall

Wiring Concealed in Floor

Wiring Existing*

Wiring Turned Up

Wiring Turned Down

Branch Circuit Homerun to Panelboard

2 1

Number of arrows indicates number of circuits. (A number at each arrow may be used to identify circuit number.)**

NOTE: *See* top of next page for explanation of asterisks.

BUS DUCTS AND WIREWAYS

Trolley Duct*** — | T | | T |

Busway (Service, Feeder or Plug-in)*** — | B | | B |

Cable Trough Ladder or Channel*** — | C | | C |

Wireway*** — | W | | W |

PANELBOARDS, SWITCHBOARDS AND RELATED EQUIPMENT

Flush Mounted Panelboard and Cabinet***

Surface Mounted Panelboard and Cabinet***

Switchboard, Power Control Center, Unit Substation (Should be drawn to scale)***

Flush Mounted Terminal Cabinet (In small scale drawings the TC may be indicated alongside the symbol)*** — TC

Surface Mounted Terminal Cabinet (In small scale drawings the TC may be indicated alongside the symbol)*** — TC

Pull Box (Identify in relation to Wiring System Section and Size)

Motor or Other Power Controller (May be a starter or contactor)***

Externally Operated Disconnection Switch***

Combination Controller and Disconnection Means***

Figure 1-30: Recommended electrical symbols. *(Cont.)*

*Note: Use heavy-weight line to identify service and feeders. Indicate empty conduit by notation CO.
**Note: Any circuit without further identification indicates two-wire circuit. For a greater number of wires, indicate with cross lines, e.g.:
3 wires   4 wires, etc.
 Neutral wire may be shown longer. Unless indicated otherwise, the wire size of the circuit is the minimum size required by the specification. Identify different functions of wiring system, e.g., signalling system by notation or other means.
***Identify by Notation or Schedule

POWER EQUIPMENT

Electric Motor (HP as Indicated)	1/4
Power Transformer	
Pothead (Cable Termination)	
Circuit Element E.g., Circuit Breaker	CB
Circuit Breaker	
Fusible Element	
Single-Throw Knife Switch	
Double-Throw Knife Switch	
Ground	
Battery	
Contactor	C
Photoelectric Cell	PE
Voltage Cycles, Phase	EX: 480/60/3
Relay	R
Equipment Connection (as noted)	

REMOTE CONTROL STATIONS FOR MOTORS OR OTHER EQUIPMENT

Pushbutton Station	PB
Float Switch - Mechanical	F
Limit Switch - Mechanical	L
Pneumatic Switch - Mechanical	P
Electric Eye - Beam Source	
Electric Eye - Relay	
Temperature Control Relay Connection (3 Denotes Quantity)	R₃
Solenoid Control Valve Connection	S
Pressure Switch Connection	P
Aquastat Connection	A
Vacuum Switch Connection	V
Gas Solenoid Valve Connection	G
Flow Switch Connection	F
Timer Connection	T
Limit Switch Connection	L

LIGHTING OUTLETS

	Ceiling	Wall
Incandescent Fixture (Surface or Pendant)		
Incandescent Fixture with Pull Chain (Surface or Pendant)	PC	PC

Figure 1-30: Recommended electrical symbols. *(Cont.)*

Exit Light
(Surface or Pendant)

Ceiling	Wall

Blanked Outlet

Junction Box

Recessed Incandescent Fixture

Individual Fluorescent Fixture
(Surface or Pendant)

Continuous Row Fluorescent Fixture
(Surface or Pendant)

Letter indicating controlling switch → A

Fixture No.
Wattage

Symbol not needed at each fixture

Bare-Lamp Fluorescent Strip*

ELECTRIC DISTRIBUTION OR
LIGHTING SYSTEM, AERIAL

Pole**

Street or Parking Lot Light
and Bracket

Transformer**

Primary Circuit**

Secondary Circuit**

Down Guy

Head Guy

Sidewalk Guy

Service Weatherhead**

ELECTRIC DISTRIBUTION OR
LIGHTING SYSTEM, UNDERGROUND

Manhole

Handhole

Transformer Manhole
or Vault

Transformer Pad

Underground Direct Burial Cable
(Indicate type, size and number
of conductors by notation
or schedule)

Underground Duct Line
(Indicate type, size and
number of ducts by cross-
section identification of each
run by notation or schedule.
Indicate type, size and
number of conductors by
notation or schedule.)

Street Light Standard Fed
From Underground Circuit**

*In the case of continuous-row bare-lamp
fluorescent strip above an area-wide diffusing
means, show each fixture run using the standard symbol; indicate area of diffusing means
and type by light shading and/or drawing
notation.
**Identify by Notation or Schedule

Figure 1-30: Recommended electrical symbols. *(Cont.)*

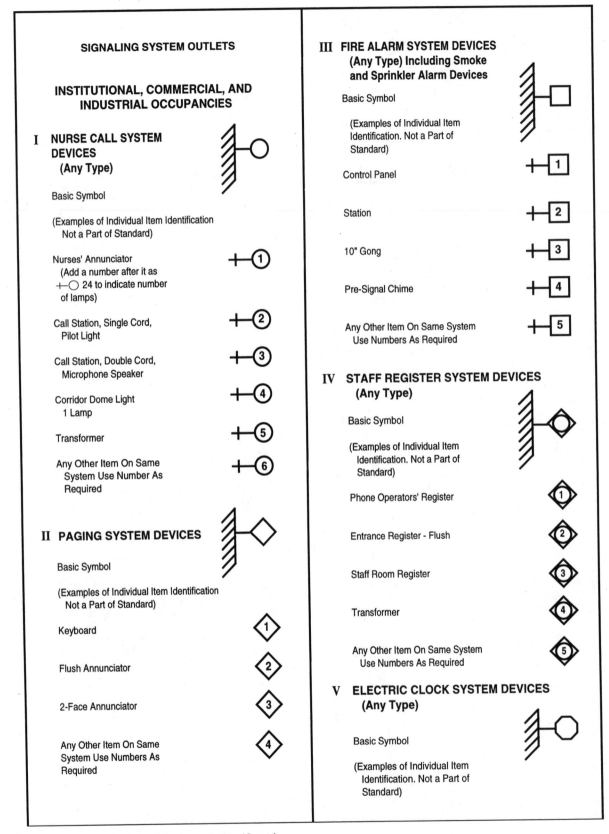

Figure 1-30: Recommended electrical symbols. *(Cont.)*

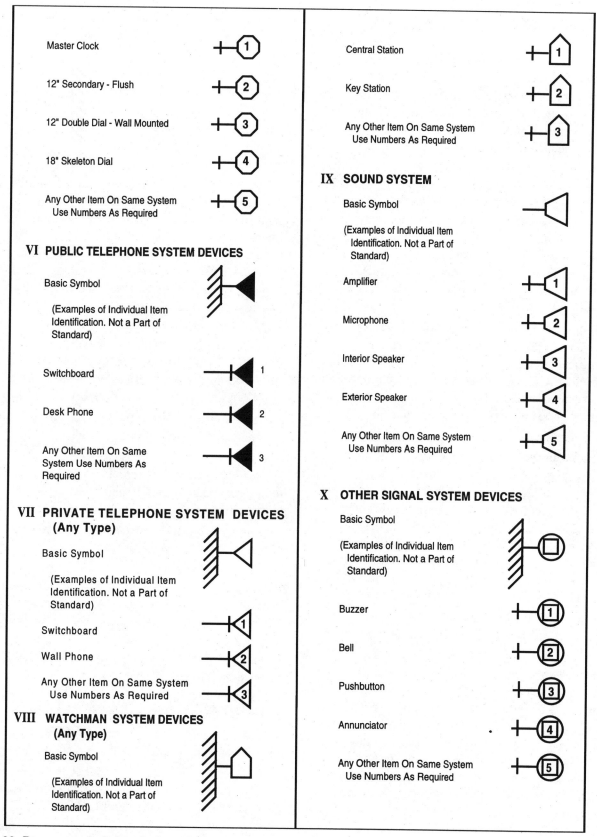

Figure 1-30: Recommended electrical symbols. *(Cont.)*

ONE-LINE DRAWING SYMBOLS

The symbols used for single-line electrical drawings are somewhat different from those used on floor plans and in power-riser diagrams. Single-line drawing symbols are used in conjunction with lines to show the component and equipment connections to an electrical system and not necessarily the physical location of such equipment.

A list of the most commonly used symbols are shown under the heading "POWER EQUIPMENT" in Figure 1-30. Note that some of the symbols are abbreviated idioms, like "CB" for circuit breaker or "R" for relay. Other symbols are simplified pictographs, like a stator and brushes for motor or a triangle for a pothead. In some cases there are combinations of idioms and pictographs, as in electric motors where the horsepower is indicated by a numeral.

Single-line electrical drawing symbols have evolved over the years to their present state after much discussion with electrical engineers, electrical drafters, electrical estimators, electricians and others who are required to interpret electrical drawings. It is felt that the current list represents a good set of symbols in that they are:

- Easy to draw
- Easily interpreted by workers
- Sufficient for most applications

The use of "standard" symbols for single-line diagrams has been attempted for almost 100 years, but the standard symbols are frequently modified to suit a particular need. Consequently, if a deviation is made from the standard, a legend or symbol list normally appears on the working drawings or in the written specifications.

Refer again to Figure 1-30 to see how some of them might be modified on drawings that will be used for specific projects.

Electric motor: The symbol shown for an electric motor is a circle which represents the motor's stator. Two short diagonal lines represent motor brushes. The rated horsepower is indicated by numerals inside the circle. On some drawings, the diagonal lines are omitted and the letter "M" is inserted inside the circle while the horsepower rating is in numerical form (outside the circle) or else coded and the type and horsepower of the motor indicated in a motor schedule on the drawings or in the written specifications.

Power transformer: The symbol shown differs slightly from conventional transformer symbols in that the cooling tubes or radiator are depicted in a power-transformer symbol; they are omitted for other types of transformers. In many cases, only two coils will be shown, representing the primary and secondary of the transformer. Other symbols will also show core lines between the primary and secondary coils.

Pothead: The symbol for a pothead is a triangle which is the general shape of a pothead. The symbol is rarely modified from the one shown.

Circuit element: When a box is used in a single-line electrical drawing, some identifying letters or numerals are normally used inside the box. For example, "CB" stands for circuit breaker. This type of circuit breaker, however, represents the huge outdoor oil-immersed circuit breaker used on high-voltage systems rather than the plug-in type of circuit breaker used in panelboards and load centers.

Circuit breaker: This symbol is normally used for lower-voltage thermal circuit breakers of 600 V and below. Other symbols found on electrical drawings representing circuit breakers are also described in this chapter.

Fusible element: The symbol shown is one of many that is used to represent overcurrent protection in an electrical system. Modification of this symbol are also covered in this chapter.

Single-throw knife switch: This is the standard symbol for a disconnecting switch, regardless of the voltage, and few modifications will be found on electrical drawings.

Double-throw knife switch: There are many variations of this symbol and the more common modifications are shown in this chapter.

Ground: This is the standard symbol for ground and is used in all types of drawings from wiring diagrams, ladder diagrams, schematic diagrams, as well as single-line electrical diagrams.

Battery: This is also the standard symbol for a battery. Sometimes the symbol is modified to indicate the number of battery cells; that is, the long and short lines are repeated for the number of cells in the battery.

Contactor: The symbol shown is frequently used as a circuit element in a high-voltage system, although two separated short lines are (—·—) are also common.

Photoelectric cell: There are many modifications for this symbol and most are described later in this chapter.

Voltage, cycle, phase: These electrical characteristics are most often represented by the numerals separated by slash marks. Sometimes letters are used in conjunction with the numerals to further clarify the intent; that is, 480V/60HZ/3Phase.

Relay: The symbol shown (a square box with the letter R inside) is frequently used on one-line power diagrams. A circle with an R inside is sometimes used on schematic or ladder diagrams.

Equipment connections (as noted): This symbol is used to describe a wide variety of electrical connections. Notes usually accompany the symbol or else a legend or symbol list is used to denote certain connections. For example, this symbol with the letter "W" next to it could specify an outlet for an electric welding machine or it could indicate a special outlet for a washing machine (clothes washer). Always check the symbol list or legend for an explanation of this type of symbol.

ONE-LINE DIAGRAMS

In general, a one-line diagram is never drawn to scale. Such drawings show the major components in an electrical system and then utilize only one drawing line to indicate the connections between these components. Even though only one line is used between components, this single line may indicate a raceway of two, three, four, or more conductors. Notes, symbols, tables, and detailed drawings are used to supplement and clarify a one-line diagram.

The electrical symbols shown in Figure 1-31 are typical of those used on one-line drawings to convey operational connections of electrical systems. If there is any question about these symbols being properly interpreted by workers, a symbol list or legend should be provided.

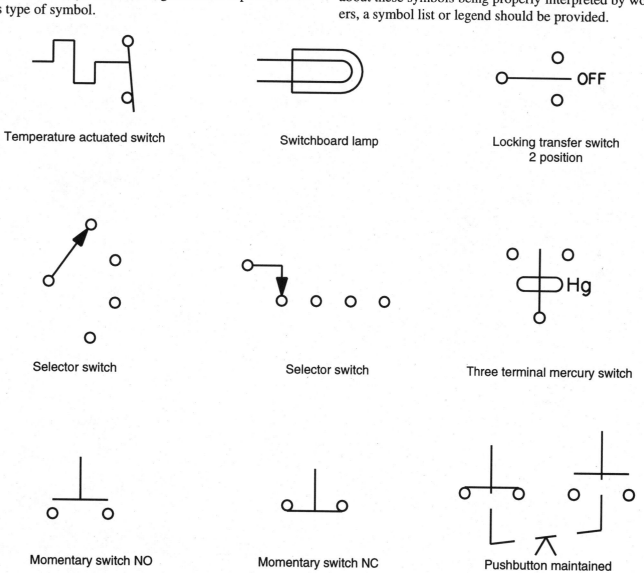

Temperature actuated switch	Switchboard lamp	Locking transfer switch 2 position
Selector switch	Selector switch	Three terminal mercury switch
Momentary switch NO	Momentary switch NC	Pushbutton maintained

Figure 1-31: Electrical symbols for one-line diagrams.

A
ASSEMBLY, SUBASSEMBLY
It is an assembly of items that is mounted and prewired as a unit, which can not be identified in a specific group or which may contain items made up of other parts.

AMPLIFER
AR

B
MOTOR
1. General (fam, blower)

B

2. Series Field

3. Application: Engine Starting Motor

SER FLD   **B**

BATTERY
BT

The long line always positive, polarity must be identified in addition (shown as multicell).

BT

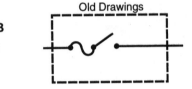

C
CAPACITOR
If it is necessary to identify the capacitor electrodes, the curved element shall represent the outside electrode in a fixed paper-dielectric and ceramic dielectric capacitors, and the low potential element in feed through capacitor.

1. General

2. Polarized Electrolytic Capascitor

---

3. Feed-through Capacitor (with terminals shown on feed-through element for clarity.)
     Commonly used for bypassing high frequency current to chassis.

CB
CIRCUIT BREAKER
1. General   Old Drawings

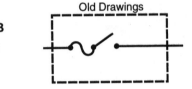

2. Circuit Breaker with thermal overload device.

3. Circuit Breaker with magnetic overload device.

4. Circuit Breaker with thermal magnetic overload device.

5. Application: 3-pole circuit breaker w/thermal magnetic overload device in each pole and trip coil (shown with boundary lines).

CB

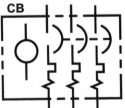

Figure 1-31: Electrical symbols for one-line diagrams. *(Cont.)*

CR

RECTIFIER, DIODE

Triangle points in direction in which rectifier conducts current easily.

1. Diode, Metallic Rectifier, Electrolytic Rectifier, Asymmetrical Varisitor

2. Application: Full-Wave Bridge Type Recitfier

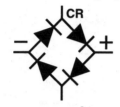

3. Controlled Rectifier (SCR)

4. Bidirectional Diode (Suppressor)

5. Zener Diode

6. Tunnel Diode

CT

CURRENT TRANSFORMER

1. General

2. Current Transformer with polarity marking. Instantaneous direction of current into one polarity mark corresponds to current out of the other polarity mark.

DS

SIGNALING DEVICE

Except Meter or Thermometer
1. Audible Signaling Device.
1.1 Bell

1.2 Buzzer

1.3 Howler

2. Visual Signaling Device (indicating, pilot, signal or illuminating lamp)
2.1 Incandescent Lamp

2.2 Neon Lamp
2.3 Alternating-Current Type

2.3.1 Direct-Current Type
NOTE: Polarity mark is not part of the symbol.

E

ELECTRICAL SHIELDING, PERMANENT MAGNET, SPARK PLUG, MISC. ELECTRICAL PARTS

1. Electrical shield (short dashes) normally used for electric or magnetic shielding. When used for other shielding, a note should so indicate.

2. Permanent Magnet

E
PM

3. Spark Plug

Figure 1-31: Electrical symbols for one-line diagrams. *(Cont.)*

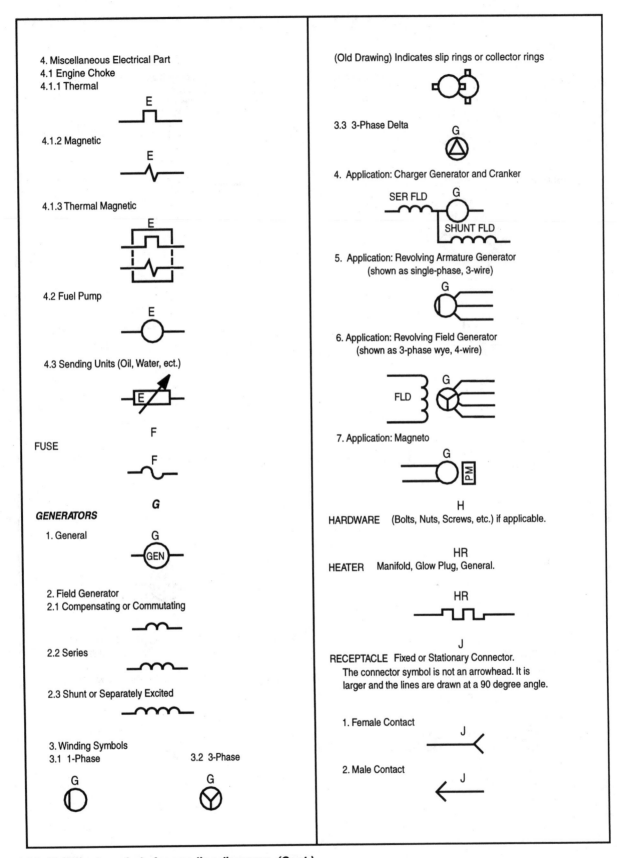

4. Miscellaneous Electrical Part
4.1 Engine Choke
4.1.1 Thermal

4.1.2 Magnetic

4.1.3 Thermal Magnetic

4.2 Fuel Pump

4.3 Sending Units (Oil, Water, ect.)

FUSE

GENERATORS

1. General

2. Field Generator
2.1 Compensating or Commutating

2.2 Series

2.3 Shunt or Separately Excited

3. Winding Symbols
3.1 1-Phase   3.2 3-Phase

(Old Drawing) Indicates slip rings or collector rings

3.3 3-Phase Delta

4. Application: Charger Generator and Cranker
SER FLD   SHUNT FLD

5. Application: Revolving Armature Generator
(shown as single-phase, 3-wire)

6. Application: Revolving Field Generator
(shown as 3-phase wye, 4-wire)
FLD

7. Application: Magneto

HARDWARE   (Bolts, Nuts, Screws, etc.) if applicable.

HEATER   Manifold, Glow Plug, General.

RECEPTACLE Fixed or Stationary Connector.
The connector symbol is not an arrowhead. It is
larger and the lines are drawn at a 90 degree angle.

1. Female Contact

2. Male Contact

Figure 1-31: Electrical symbols for one-line diagrams. *(Cont.)*

3. Application: Charger Generator and Cranker

or

if no confusion results from its use by disregarding the type of contacts in the receptacle, it may be shown as

4. Receptacles of the type commonly used for power supply purposes (convenience outlets)
4.1 Female Contact

4.2 Male Contact

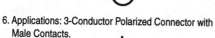

5. Applications: 3-Conductor Polarized Connector with Female Contacts.

6. Applications: 3-Conductor Polarized Connector with Male Contacts.

RELAYS, CONTACTORS, SOLENOID   (Electrically or Thermally operated)

1. Coil
1.1 Basic Operating

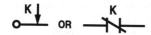

1.2 Time Delay

2. Contacts
2.1 Basic Contact Assemblies
2.1.1 Closed Contact (break)

2.1.2 Open Contact (Make)

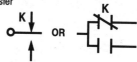

2.1.3 Transfer

2.2 Contacts with Time Delay Feature
2.2.1 Closed Contact, Time Delay Opening

2.2.2 Open Contact, Time Delay Closing

2.2.3 Closed Contact, Time Delay Closing

Note: Contacts at left are for wiring diagrams. Contacts at right for schematic diagrams & wiring diagrams of contractors.

2.2.4 Open Contact, Time Delay Opening

3. Application: Relay with Transfer Contacts

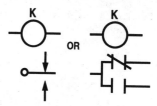

INDUCTOR, REACTOR

1. Air Core

2. Iron Core (If desired to distinguish magnetic-core inductors)

Figure 1-31: Electrical symbols for one-line diagrams. *(Cont.)*

3. Saturating Core

4. Saturable-Core Inductor (Reactor)
Note: Explanatory words & arrow are not part of the
    symbol shown.   **DC WINDING**

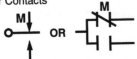

M

METERS, GAUGES, CLOCKS with calibrated dials

1. Clock, Electric Timer
1.1 Motor

1.2 Transfer Contacts

 OR

2. Indicating Meters, Gauges, etc.

M

(*)

* Replace the asterisk by one of the following letter
combinations, depending on the function of the meter.

A   **Ammeter**
AH   **Ampere-hour**
F   **Frequency Meter**
MA   **Milliammeter**
OP   **Oil Pressure**
PF   **Power Factor**
T   **Temperature**
TT   **Total Time (Running Time)**
V   **Voltmeter**
W   **Wattmeter**
WH   **Watthour Meter**

MP

MECHANICAL Including Nameplates- if applicable
PARTS
            P

PLUG-Affixed to a Cable, Cord or Wire
    The connector symbol is not an arrowhead. It is larger
    and the lines are drawn at a 90 degree angle.

1. Female Contact

                P

2. Male Contact

                P

3. Application: 4-Conductor connector with 3-male
    contacts and 1-female contact with individual
    contact designations.

OR

If no confusion results from its use by disregarding
the type of contacts in the plug, it may be shown as

4. Plugs of the type commonly used for power-supply
    purposes (mating connectors)
4.1 Female Contact

4.2 Male Contact

5. Application: 3-Conductor polarized connector with
    Female contacts   **P**

6. Application: 3-Conductor polarized connector with
    Male contacts   **P**

                Q

TRANSISTOR

1. General
1.1 NPN

1.2 PNP

2. Unijunction
2.1 N-Type Base

Figure 1-31: Electrical symbols for one-line diagrams. *(Cont.)*

2.2 P-Type Base

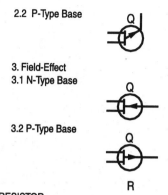

3. Field-Effect
3.1 N-Type Base

3.2 P-Type Base

RESISTOR

Note: Do not use both styles of symbols on the same diagram.
1. General (Fixed)

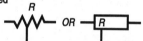

2. Tapped

3. Adjustable Contact

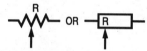

4. Rotary Type Adjustable
The preferred method of terminal indentification is to designate with the letters "CW" the terminal adjacent to the movable contact when it is in an extreme clockwise position as viewed from the knob end.

Rheostat

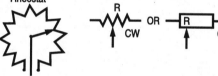

5. Nonlinear

RT

THERMISTOR, THERMAL RESISTOR
"T" indicates that the primary characteristic of the element within the circle is a function of temperature

RT

VARISTOR, Symmetrical
Resistor, Voltage Sensitive (Silicon Carbide, etc.)

S

SWITCH

1. Thermal cutout, thermal flasher

S

2. Switch
2.1 Momentary-Fixed Contact on Momentary Switch

2.1.1 Open Contact (Make) (Ignition Points)

S

2.1.2 Closed Contact (Break)

S

2.1.3 2-Open Contacts (Make)

S   OFF

2.1.4 Pushbutton, Open Contact (Make)

S

2.1.5 Pushbutton, Closed Contact (Break)

S

2.2 Locking or Maintained-Fixed Contact for Maintained Switch

2.2.1 Open Contact (Make)

S

2.2.2 Closed Contact (Break)

S

2.2.3 2-Open Contact (Make)

S   OFF

Figure 1-31: Electrical symbols for one-line diagrams. *(Cont.)*

2.3 Application: 3- Position, 1-Pole; Circuit Closing (Make), off, Momentary Circuit Closing (Make).

2.4 Application: 2- Position, 1-Pole; Momentary Circuit Closing (Make), Circuit Closing.

2.5 Selector Switch
2.5.1 4-Position with Nonshorting Contacts ;

 OR

2.5.2 4-Position with Shorting Contacts

 OR

2.6 Master or Control Switch
A table of contact operation must be shown on the diagram. A typical table is shown below.

DETACHED CONTACTS SHOWN ELSEWHERE
ON DIAGRAM

CONTACT	POSITION		
	A	B	C
1-2			X
3-4	X		
5-6			X
7-8	X		

X Indicates contacts closed

FOR WIRING DIAGRAM

2.7 Flow Actuated Switch
2.7.1 Closing On Increase Flow

2.7.2 Opening On Increase Flow

2.8 Liquid Level Actuated Switch
2.8.1 Closing On Rising Level

2.8.2 Open On Rising Level

2.9 Temperature Actuated Switch (Thermostat)
2.9.1 Closing On Rising Temperature

2.9.2 Open On Rising Temperature

2.10 Pressure Or Vacuum Actuated Switch
2.10.1 Closing On Rising Pressure

2.10.2 Open On Rising Pressure

2.11 Centrifugal Actuated Switch (Overspeed)
2.11.1 Closing On Speed

2.11.2 Open On Speed

Figure 1-31: Electrical symbols for one-line diagrams. (Cont.)

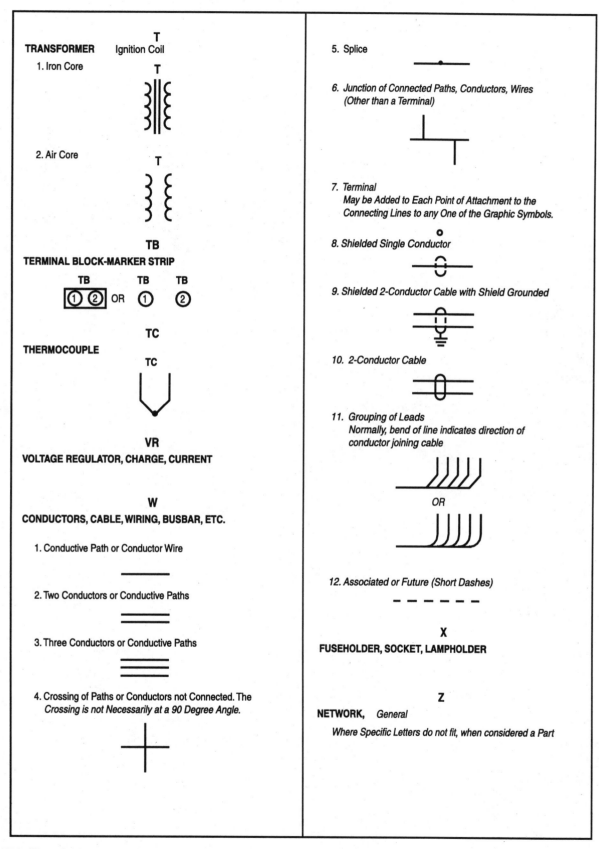

TRANSFORMER   Ignition Coil

1. Iron Core

2. Air Core

TB

TERMINAL BLOCK-MARKER STRIP

TC

THERMOCOUPLE

VR

VOLTAGE REGULATOR, CHARGE, CURRENT

W

CONDUCTORS, CABLE, WIRING, BUSBAR, ETC.

1. Conductive Path or Conductor Wire

2. Two Conductors or Conductive Paths

3. Three Conductors or Conductive Paths

4. Crossing of Paths or Conductors not Connected. The *Crossing is not Necessarily at a 90 Degree Angle.*

5. Splice

6. *Junction of Connected Paths, Conductors, Wires (Other than a Terminal)*

7. *Terminal*
   May be Added to Each Point of Attachment to the Connecting Lines to any One of the Graphic Symbols.

8. *Shielded Single Conductor*

9. *Shielded 2-Conductor Cable with Shield Grounded*

10. *2-Conductor Cable*

11. *Grouping of Leads*
    Normally, bend of line indicates direction of conductor joining cable

    OR

12. *Associated or Future (Short Dashes)*

X

FUSEHOLDER, SOCKET, LAMPHOLDER

Z

NETWORK,   *General*

     Where Specific Letters do not fit, when considered a Part

Figure 1-31: Electrical symbols for one-line diagrams. *(Cont.)*

ELECTRONIC SYMBOLS

Solid-state electronic control devices are rapidly replacing electro-mechanical controls in all areas of the electrical industry. Some of the most prominent include:

- Motor controls
- HVAC controls
- Lighting controls
- Security/fire-alarm systems
- Measuring and warning instruments

Therefore, anyone involved in the electrical industry — in any capacity — should have a basic knowledge of electronics in order to design, install, maintain, and troubleshoot electronic devices.

Electronic design begins with the purpose for the device; that is, what the finished device is supposed to do,

followed by a detailed circuit schematic diagram which includes all devices, components, connections, and hardware — including manufacturers' data sheets indicating the pin arrangement for all devices used.

When designing or manufacturing circuit boards, the layout should produce a board that has a uniform and symmetrical distribution of parts that can be readily wired onto the boards and interconnected with the chassis-mounted components. Consequently, one major objective is the overall appearance of the component placement, as well as their accessibility to the off-board components. Another important objective is to position the components so that they may be properly wired with the shortest paths possible.

Using the electronic symbols shown in Figure 1-32, layout drawings can be made to any appropriate scale — either manually or electronically with a computer using a CAD or other graphics system.

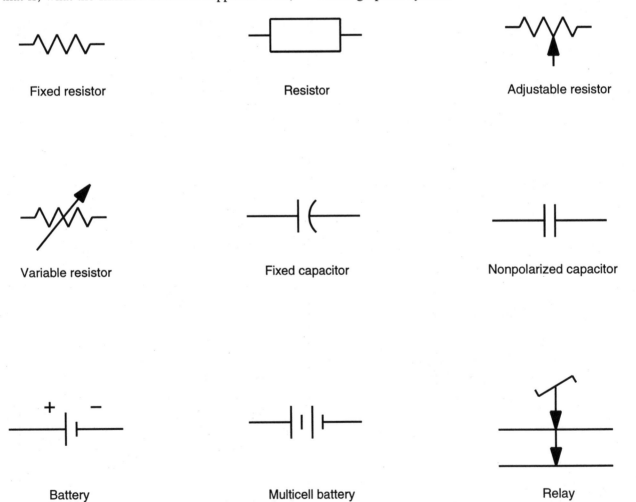

Figure 1-32: Electronic symbols.

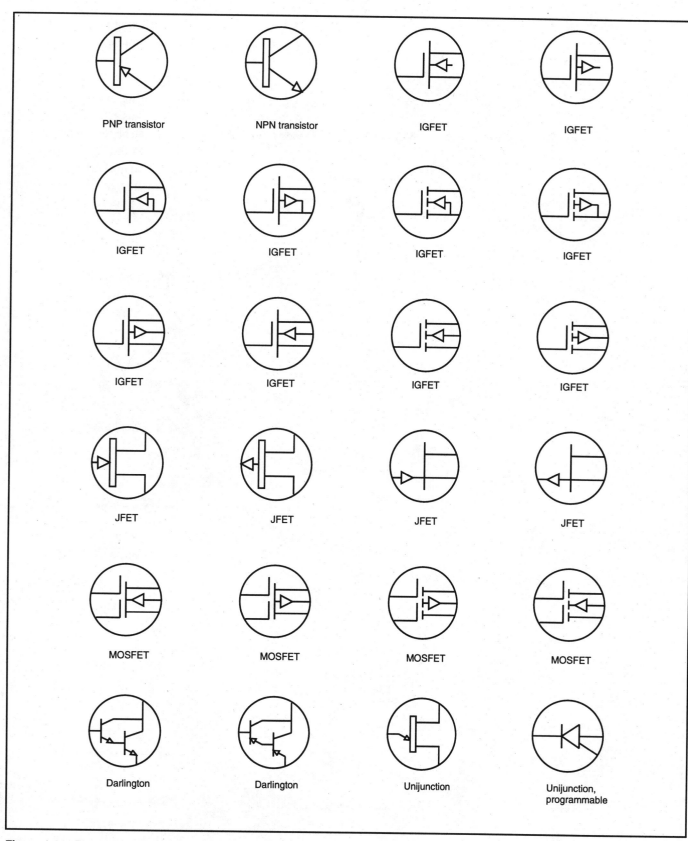

Figure 1-32: Electronic symbols. *(Cont.)*

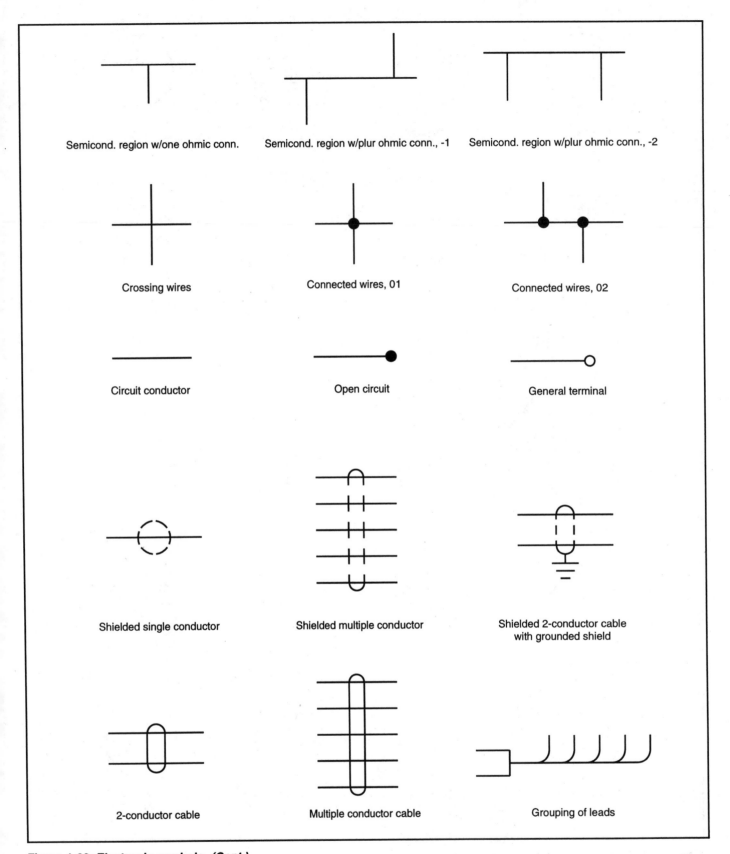

Figure 1-32: Electronic symbols. *(Cont.)*

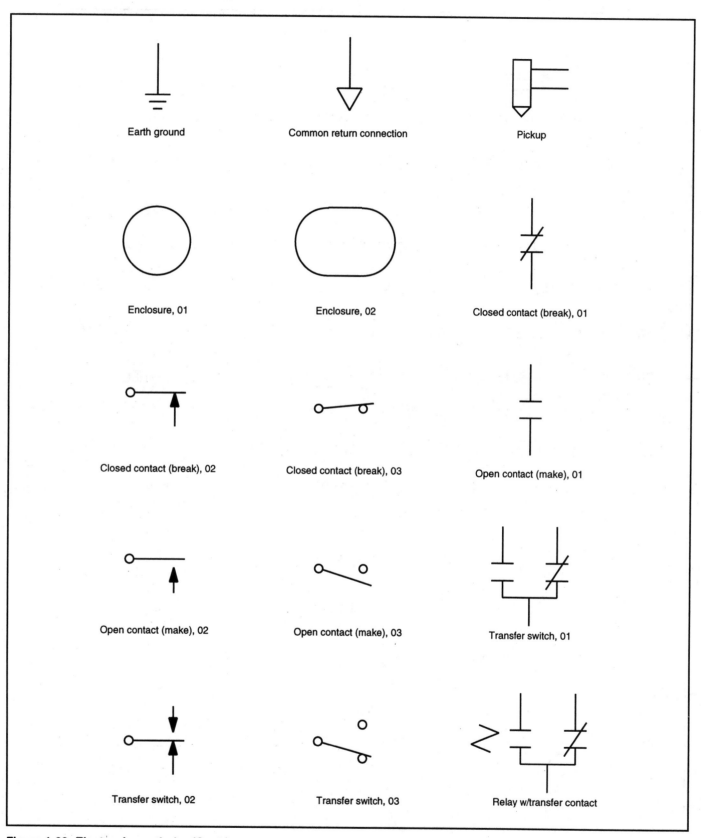

Figure 1-32: Electronic symbols. *(Cont.)*

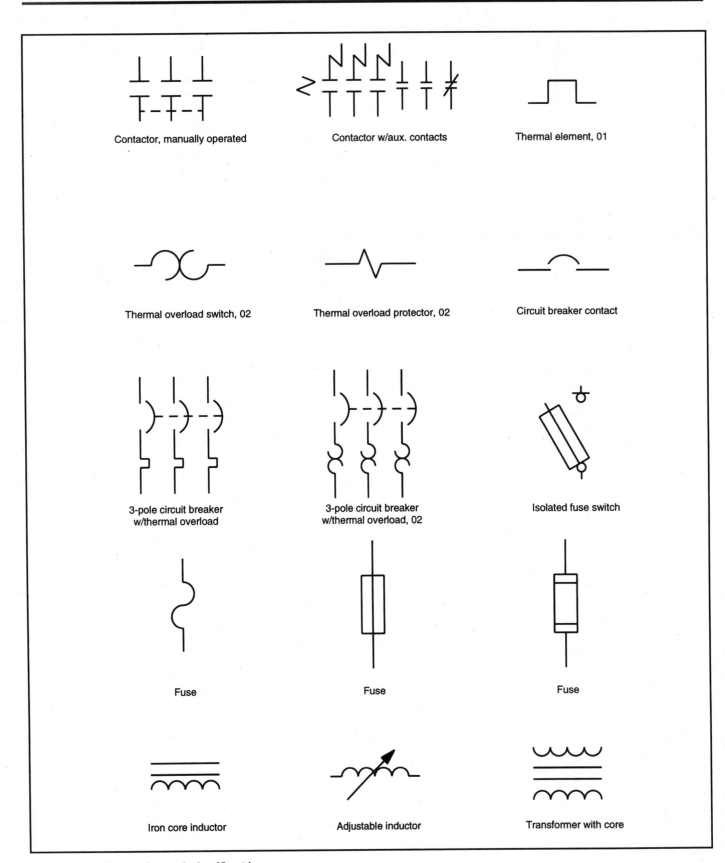

Figure 1-32: Electronic symbols. *(Cont.)*

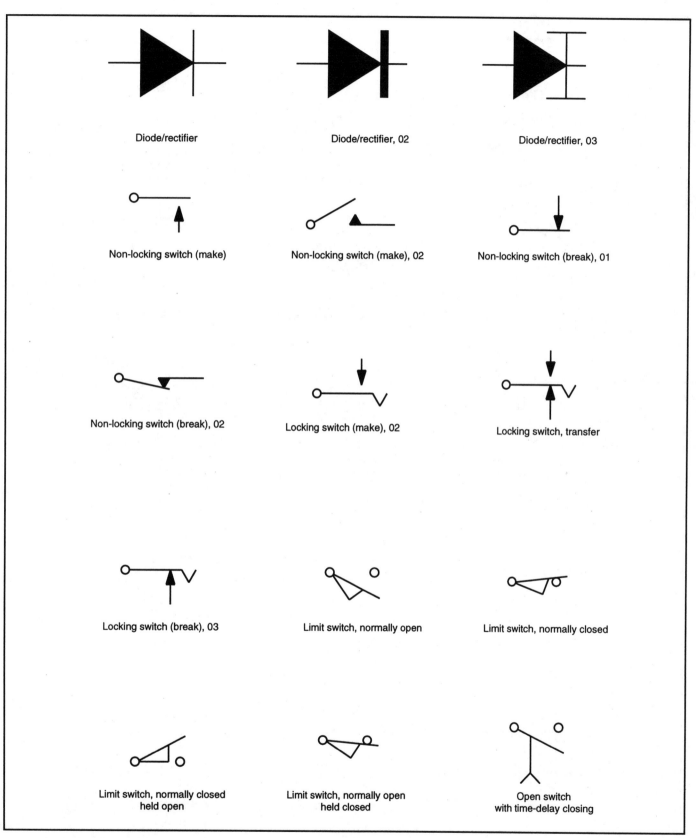

Figure 1-32: Electronic symbols. *(Cont.)*

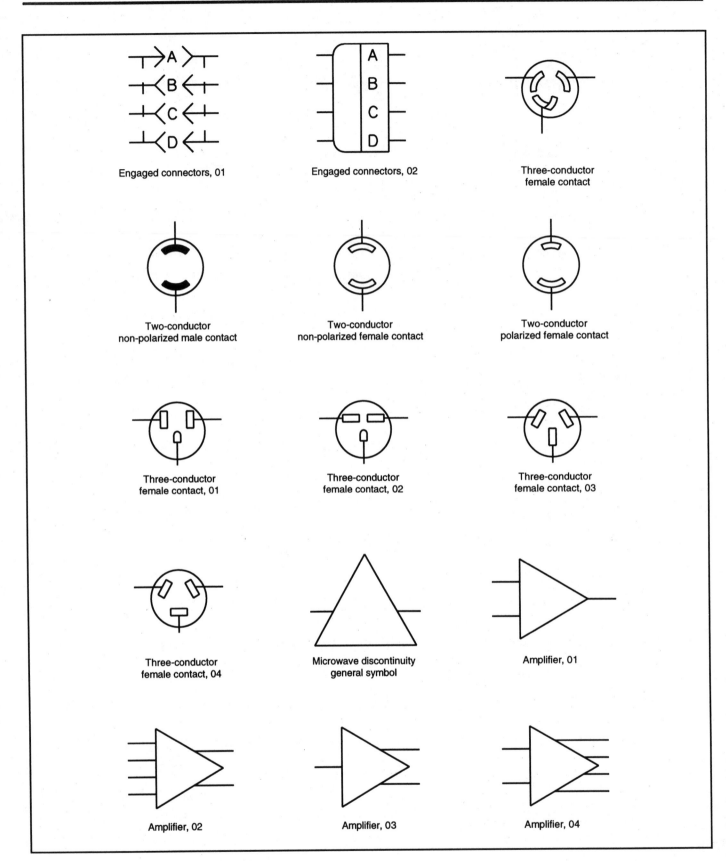

Engaged connectors, 01

Engaged connectors, 02

Three-conductor
female contact

Two-conductor
non-polarized male contact

Two-conductor
non-polarized female contact

Two-conductor
polarized female contact

Three-conductor
female contact, 01

Three-conductor
female contact, 02

Three-conductor
female contact, 03

Three-conductor
female contact, 04

Microwave discontinuity
general symbol

Amplifier, 01

Amplifier, 02

Amplifier, 03

Amplifier, 04

Figure 1-32: Electronic symbols. *(Cont.)*

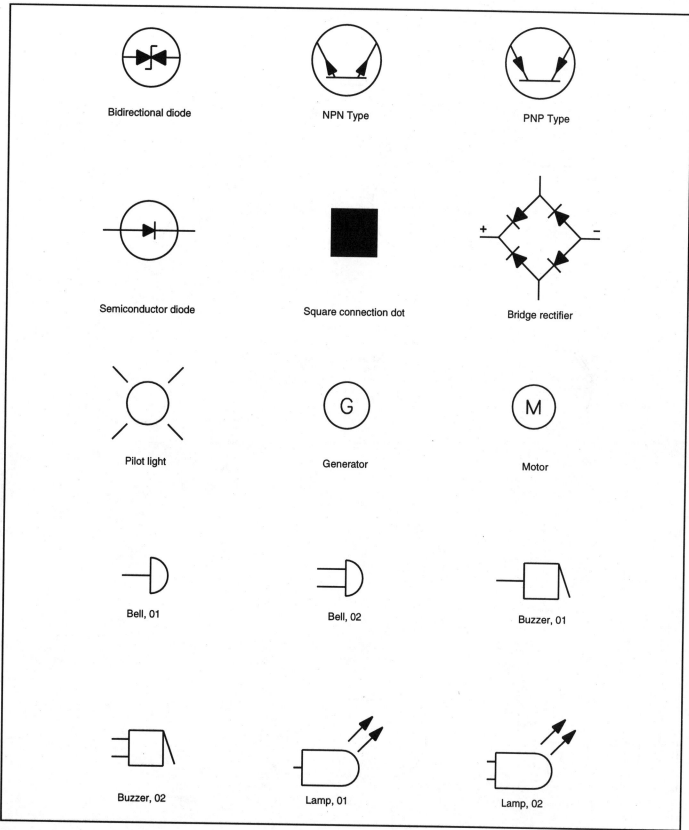

Figure 1-32: Electronic symbols. *(Cont.)*

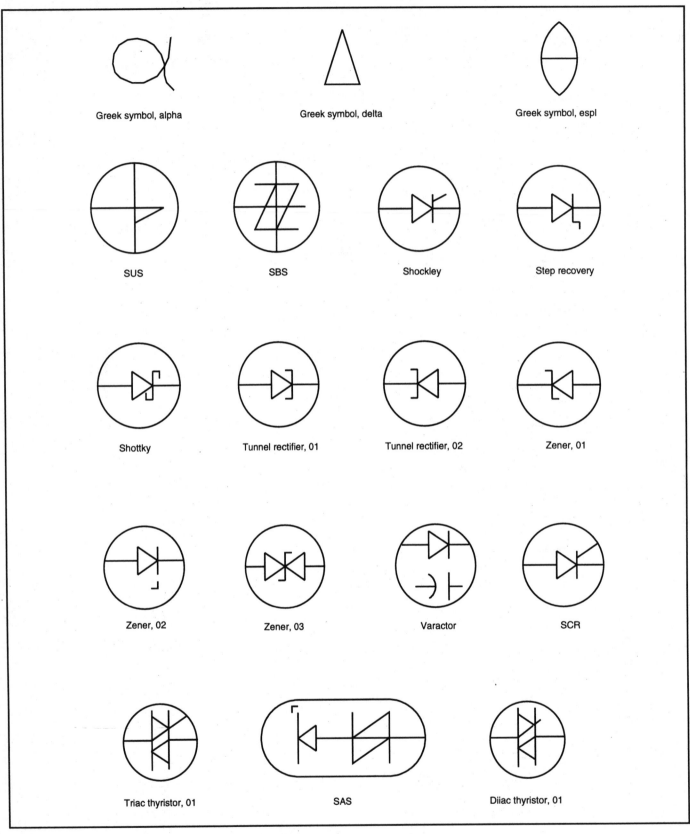

Figure 1-32: Electronic symbols. *(Cont.)*

Figure 1-32: Electronic symbols. *(Cont.)*

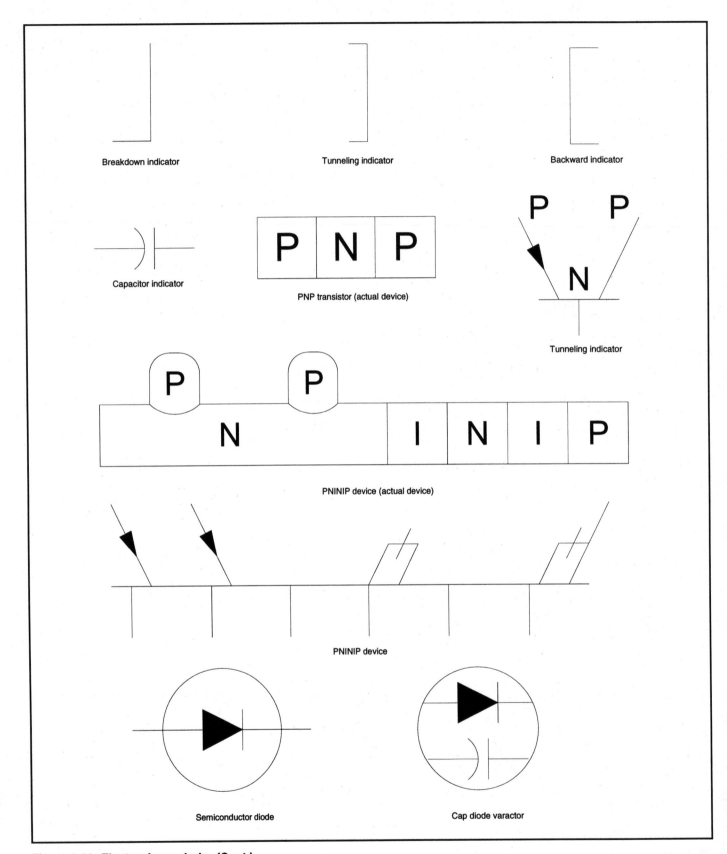

Breakdown indicator

Tunneling indicator

Backward indicator

Capacitor indicator

PNP transistor (actual device)

Tunneling indicator

PNINIP device (actual device)

PNINIP device

Semiconductor diode

Cap diode varactor

Figure 1-32: Electronic symbols. *(Cont.)*

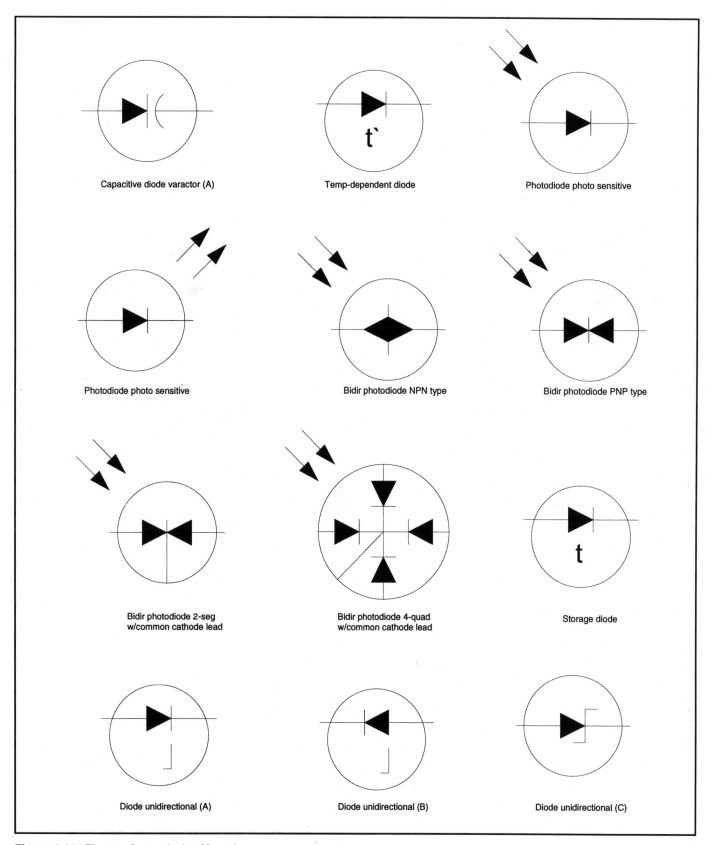

Figure 1-32: Electronic symbols. *(Cont.)*

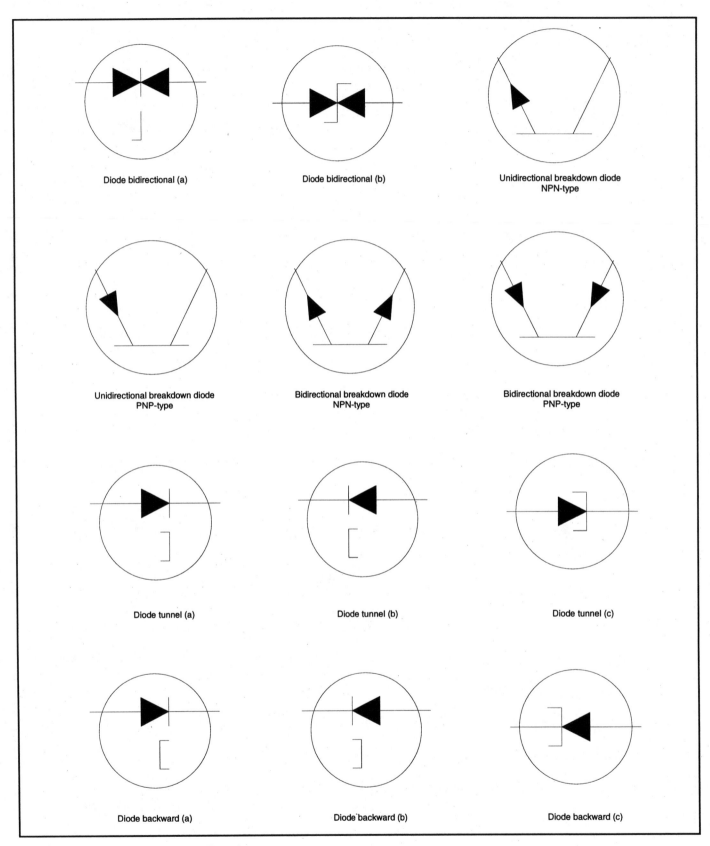

Figure 1-32: Electronic symbols. *(Cont.)*

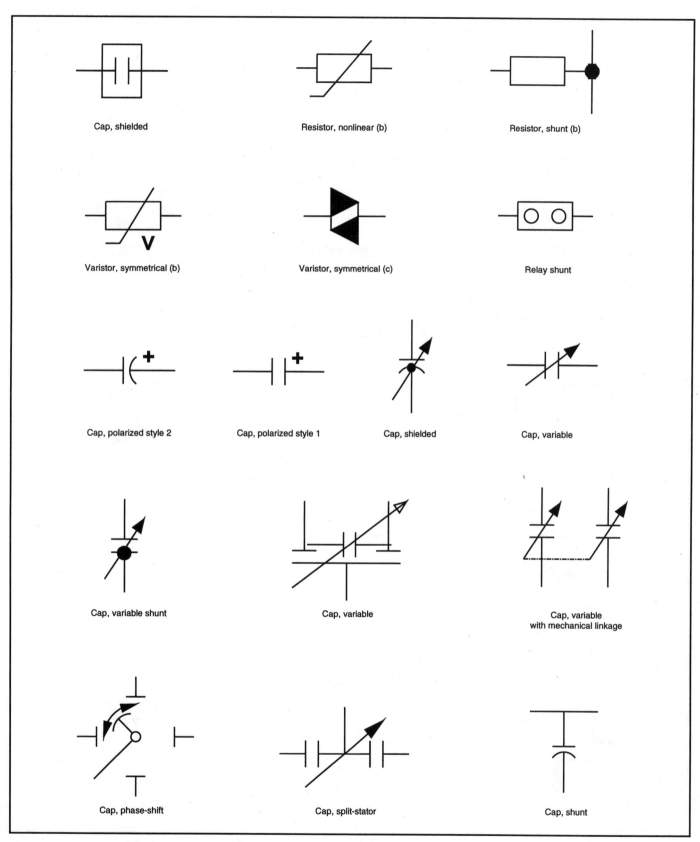

Figure 1-32: Electronic symbols. *(Cont.)*

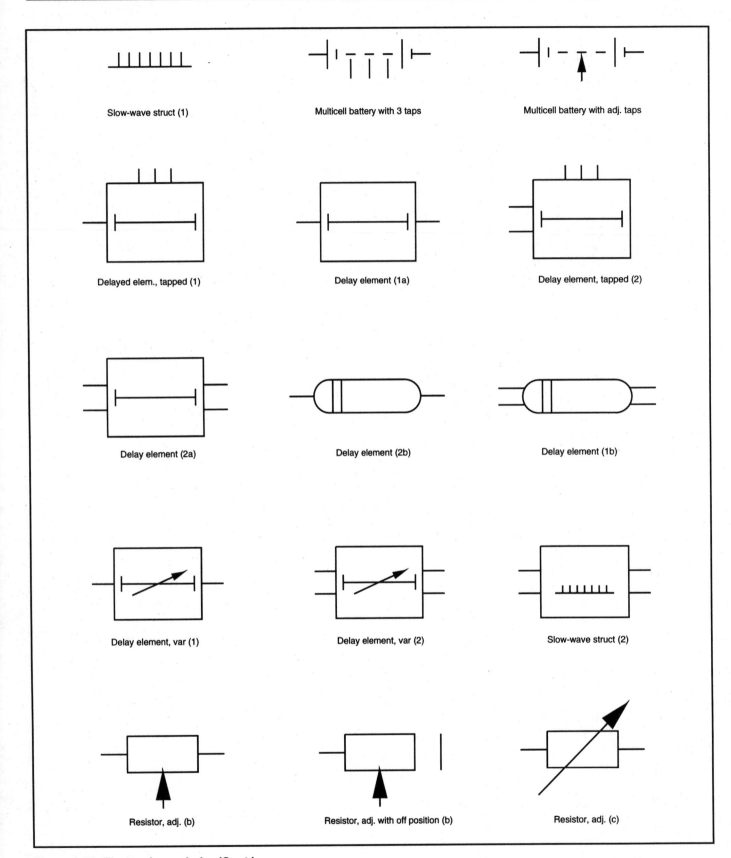

Figure 1-32: Electronic symbols. *(Cont.)*

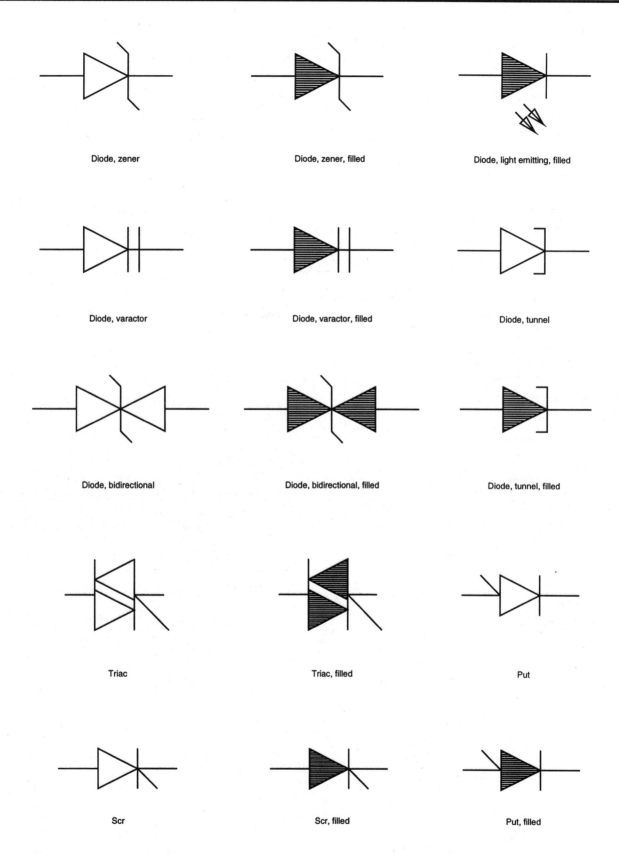

Figure 1-32: Electronic symbols. *(Cont.)*

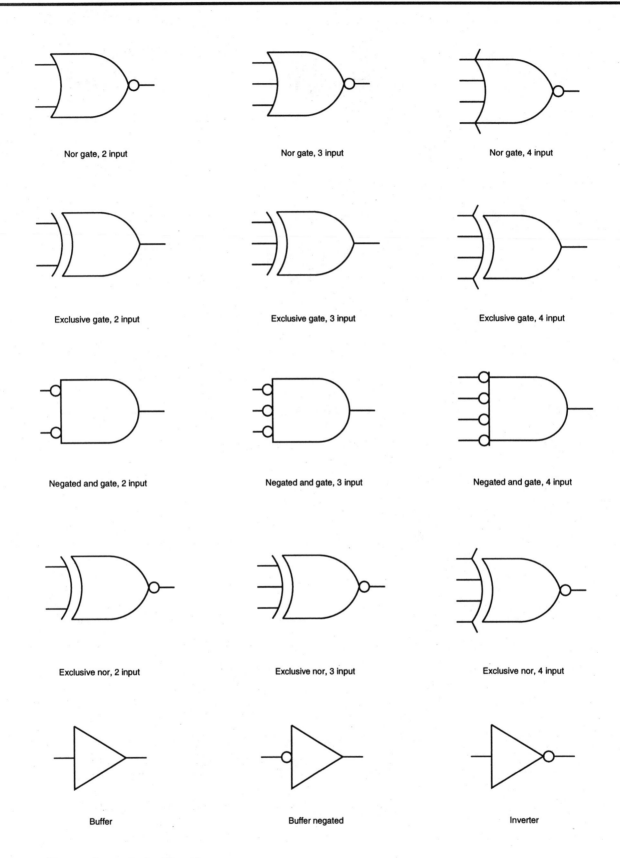

Figure 1-32: Electronic symbols. *(Cont.)*

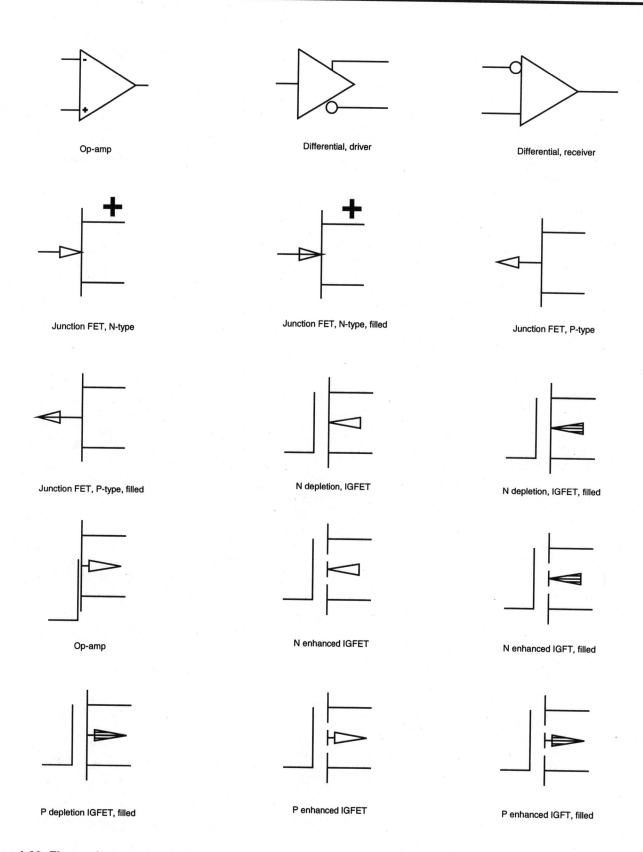

Figure 1-32: Electronic symbols. *(Cont.)*

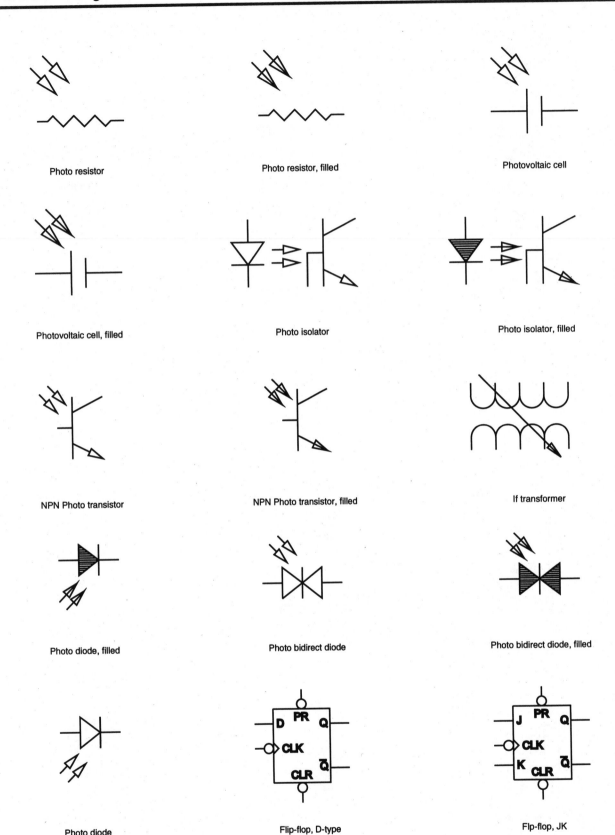

Figure 1-32: Electronic symbols. *(Cont.)*

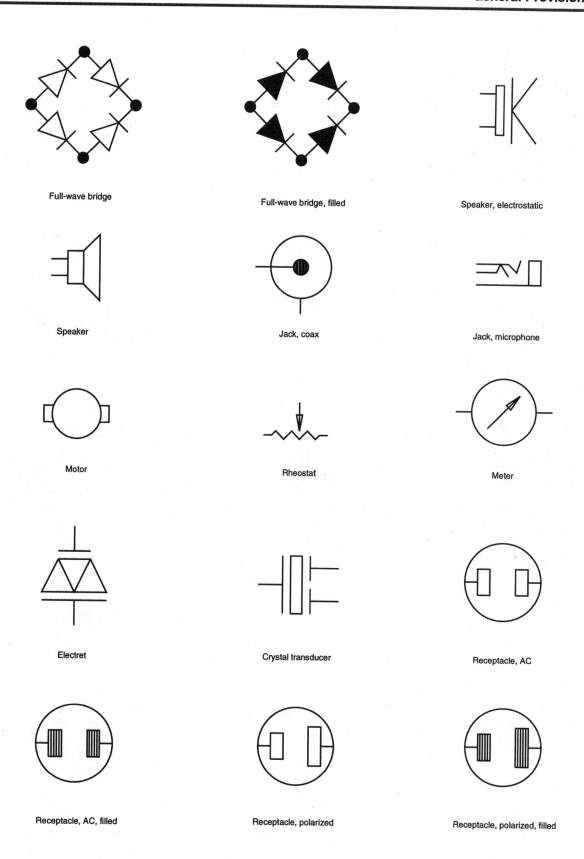

Figure 1-32: Electronic symbols. *(Cont.)*

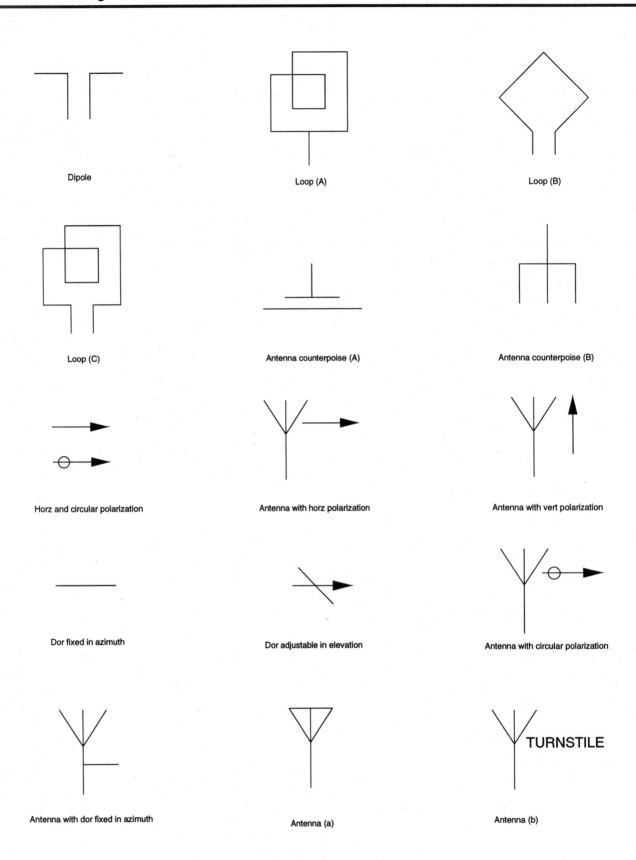

Dipole

Loop (A)

Loop (B)

Loop (C)

Antenna counterpoise (A)

Antenna counterpoise (B)

Horz and circular polarization

Antenna with horz polarization

Antenna with vert polarization

Dor fixed in azimuth

Dor adjustable in elevation

Antenna with circular polarization

Antenna with dor fixed in azimuth

Antenna (a)

TURNSTILE

Antenna (b)

Figure 1-32: Electronic symbols. *(Cont.)*

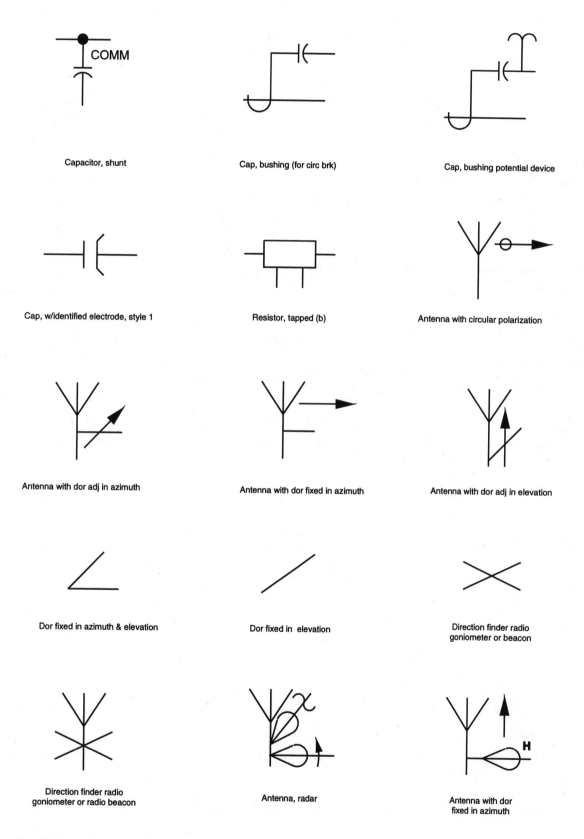

Figure 1-32: Electronic symbols. *(Cont.)*

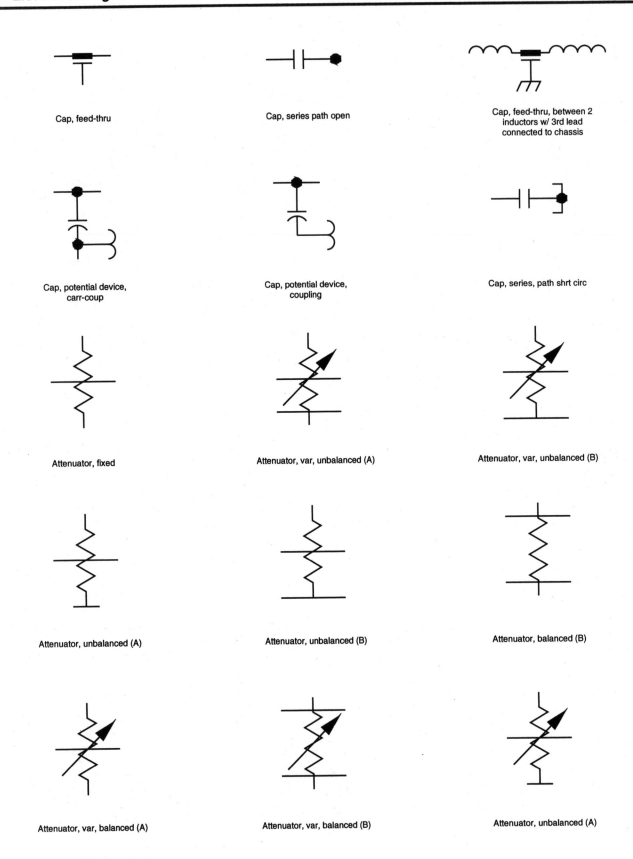

Figure 1-32: Electronic symbols. *(Cont.)*

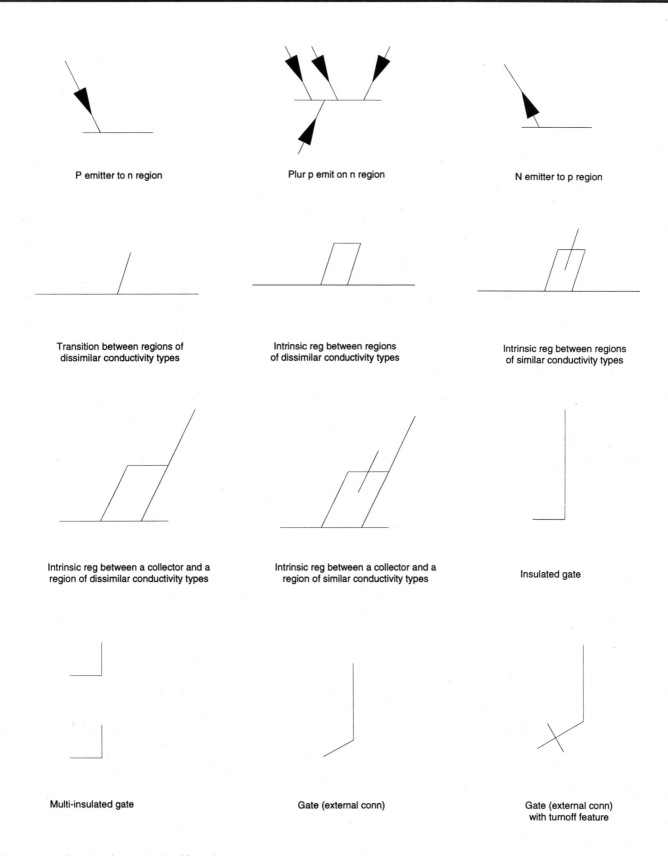

P emitter to n region

Plur p emit on n region

N emitter to p region

Transition between regions of
dissimilar conductivity types

Intrinsic reg between regions
of dissimilar conductivity types

Intrinsic reg between regions
of similar conductivity types

Intrinsic reg between a collector and a
region of dissimilar conductivity types

Intrinsic reg between a collector and a
region of similar conductivity types

Insulated gate

Multi-insulated gate

Gate (external conn)

Gate (external conn)
with turnoff feature

Figure 1-32: Electronic symbols. *(Cont.)*

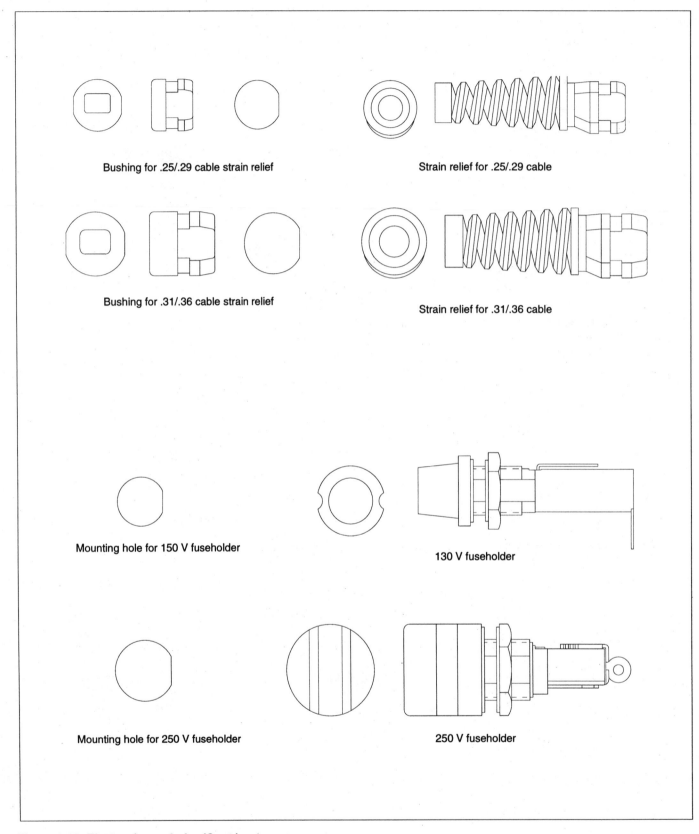

Bushing for .25/.29 cable strain relief

Strain relief for .25/.29 cable

Bushing for .31/.36 cable strain relief

Strain relief for .31/.36 cable

Mounting hole for 150 V fuseholder

130 V fuseholder

Mounting hole for 250 V fuseholder

250 V fuseholder

Figure 1-32: Electronic symbols. *(Cont.)*

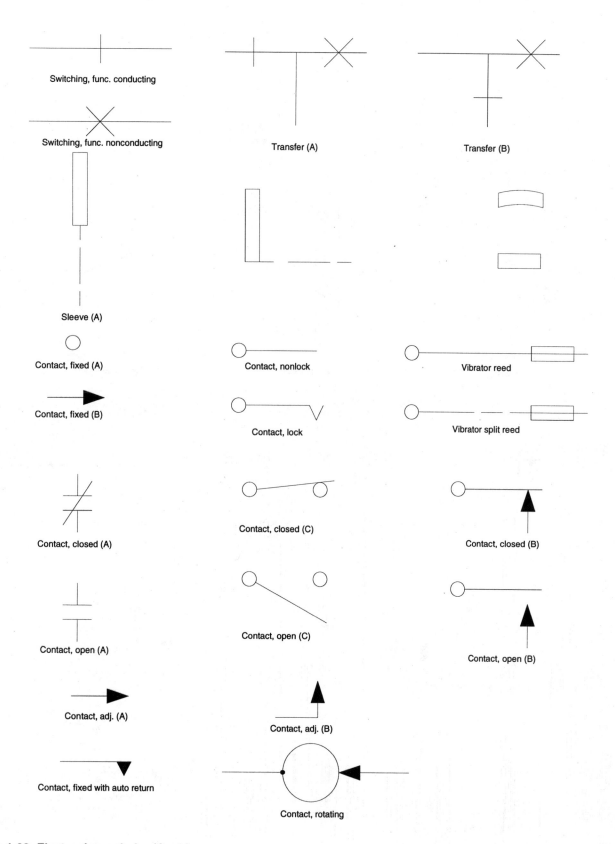

Figure 1-32: Electronic symbols. *(Cont.)*

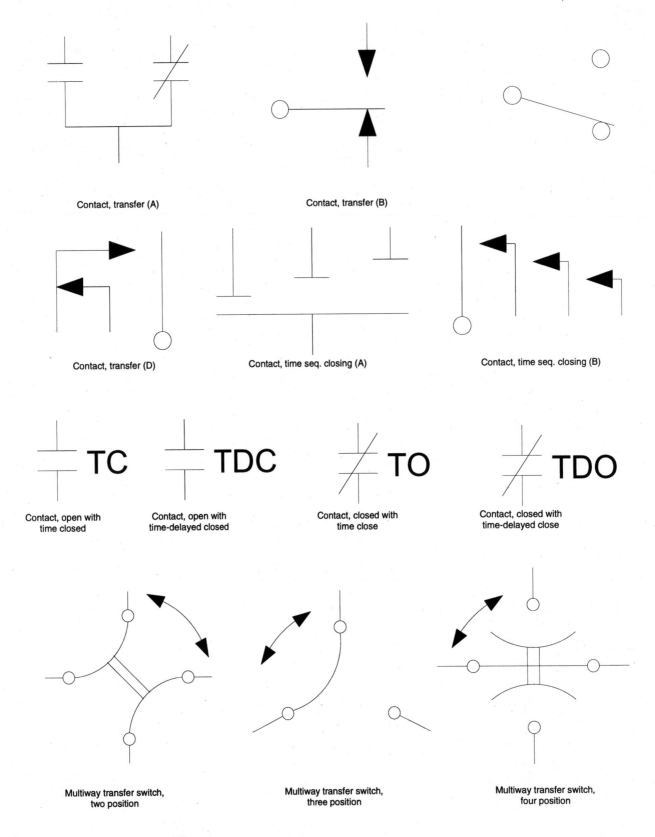

Contact, transfer (A)

Contact, transfer (B)

Contact, transfer (D)

Contact, time seq. closing (A)

Contact, time seq. closing (B)

TC — Contact, open with time closed

TDC — Contact, open with time-delayed closed

TO — Contact, closed with time close

TDO — Contact, closed with time-delayed close

Multiway transfer switch, two position

Multiway transfer switch, three position

Multiway transfer switch, four position

Figure 1-32: Electronic symbols. *(Cont.)*

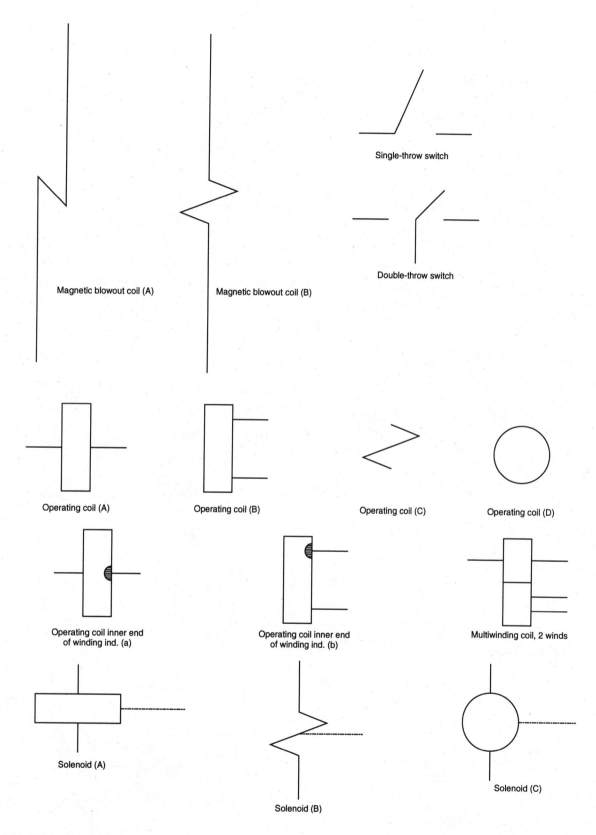

Figure 1-32: Electronic symbols. *(Cont.)*

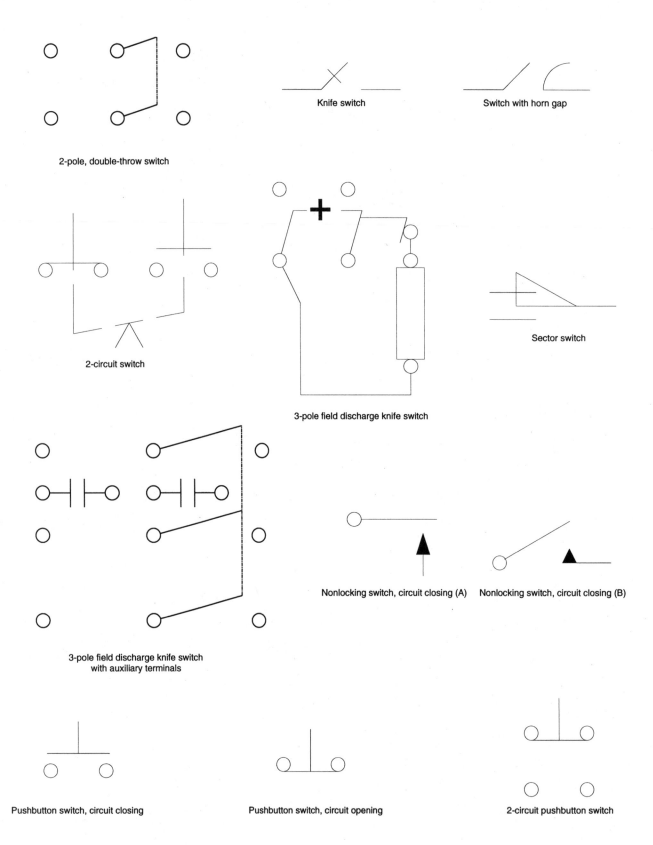

2-pole, double-throw switch

Knife switch

Switch with horn gap

2-circuit switch

3-pole field discharge knife switch

Sector switch

3-pole field discharge knife switch
with auxiliary terminals

Nonlocking switch, circuit closing (A)   Nonlocking switch, circuit closing (B)

Pushbutton switch, circuit closing

Pushbutton switch, circuit opening

2-circuit pushbutton switch

Figure 1-32: Electronic symbols. *(Cont.)*

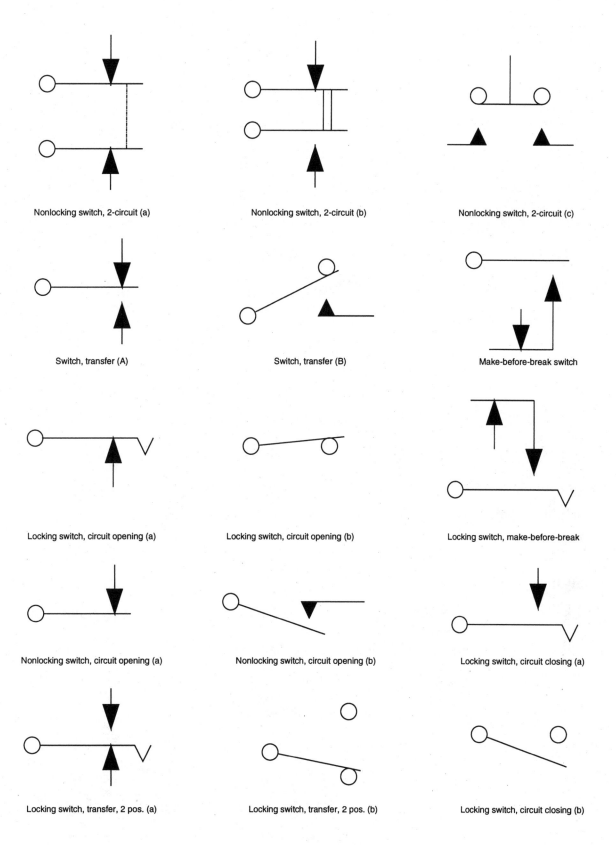

Figure 1-32: Electronic symbols. *(Cont.)*

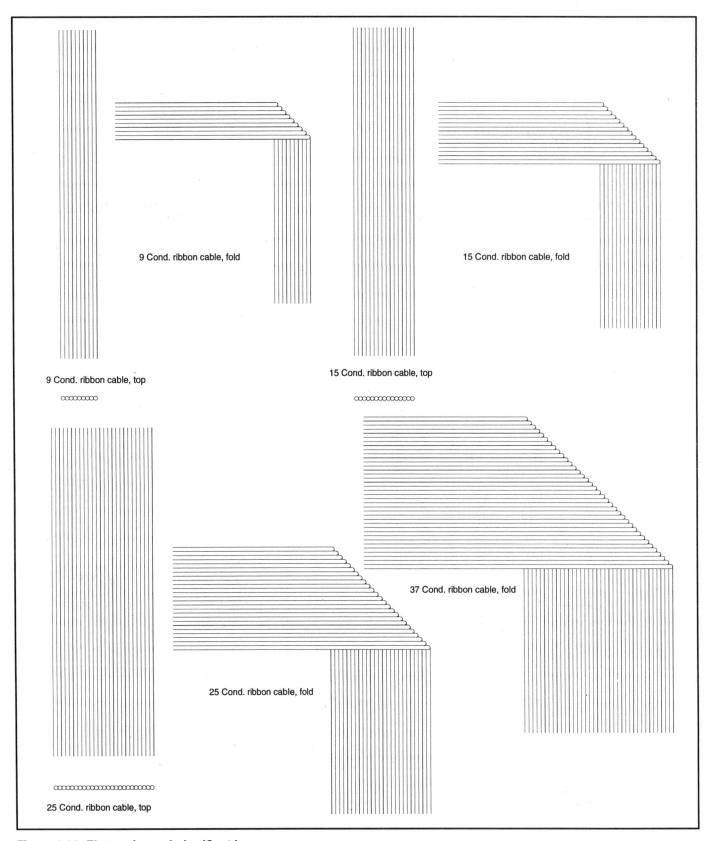

9 Cond. ribbon cable, fold

15 Cond. ribbon cable, fold

9 Cond. ribbon cable, top

15 Cond. ribbon cable, top

37 Cond. ribbon cable, fold

25 Cond. ribbon cable, fold

25 Cond. ribbon cable, top

Figure 1-32: Electronic symbols. *(Cont.)*

DRAWING SCHEDULES

A schedule is a systematic method of presenting notes or lists of equipment on a drawing in tabular form. When properly organized and thoroughly understood, schedules are not only powerful timesaving devices for those preparing the drawings, but they can also save the workers on the job much valuable time.

For example, the lighting-fixture schedule shown in Figure 1-33 lists the fixture type and identifies each fixture type on the drawing by number. The manufacturer and catalog number of each type are given along with the number, size, and type of lamp for each. A blank form is shown in Figure 1-34 for immediate use on any electrical drawings. Merely fill in the missing information for the project at hand — either manually or with a computer graphics program.

Sometimes all of the same information found in schedules will be duplicated in the written specifications, but combing through page after page of written specifications can be time consuming and workers do not always have access to the specifications at the job site, whereas they usually do have access to the working drawings. There-

fore, the schedule is an excellent means of providing essential information in a clear and accurate manner, allowing the workers to carry out their assignments in the least amount of time.

Other schedules that are frequently found on electrical working drawings include:

- Connected load schedule
- Panelboard schedule
- Electric-heat schedule
- Kitchen-equipment schedule
- Schedule of receptacle types

Some of the above schedules are shown in Figures 1-35 through 1-40.

There are also other schedules found on electrical drawings, depending upon the type of project. Most, however, will deal with lists of equipment; that is, such items as motors, motor controllers, service equipment, industrial machines, and the like. Many other sample schedules appear under the appropriate category throughout this book.

LIGHTING-FIXTURE SCHEDULE

SYMBOL	TYPE	MANUFACTURER	CATALOG NUMBER	MOUNTING	LAMPS
	A	Lightolier	10234	Wall	2-40W T-12WWX
	B	Lightolier	10420	Surface	2-40W T-12WWX
	C	Alkco	RPC-210-6E	Surface	2-8W T-5
	D	BriteLite	P7S-2936	Wall	1-100W A
	E	BriteLite	P7S-110	Surface	1-100W A

Figure 1-33: Typical lighting-fixture schedule.

FIXTURE TYPE	MANUFACTURER'S DESCRIPTION	LAMPS		VOLTS	MOUNTING	REMARKS
		No.	TYPE			

LIGHTING-FIXTURE SCHEDULE

Figure 1-34: Typical lighting-fixture schedule that may be used directly on any electrical drawing.

PANELBOARD SCHEDULE

PANEL No.	TYPE CABINET	PANEL MAINS			BRANCHES					
		AMPS	VOLTS	PHASE	1P	2P	3P	PROT	FRAME	Item Fed or Remarks

Figure 1-35: Sample panelboard schedule No. 1.

PANEL SCHEDULE

Panel No.		Designer			Checked		Date	
MAINS		Ø		W	Volts		Ampere	

Circuit No.	Switch or Breaker			SERVES		Connected KVA	Demand Factor	Demand KVA
	Pole	Frame	Trip or Fuse					
1								
2								
3								
4								
5								
6								
7								
8								
9								
10								
11								
12								
13								
14								
15								
16								
17								
18								
19								
20								
21								
22								
23								
24								
25								
26								
27								
28								
29								
30								
31								
32								
33								
34								
35								
36								
37								
38								
39								
40								
41								
42								
DEMAND I				TOTALS				

DESIGN I   PERMISSIBLE Ed.   FEEDER SIZE   Ed/1000 A.F.

FEEDER LENGTH   AMP. FT.   M.   ACTUAL Ed.

Figure 1-36: Sample panelboard schedule No. 2.

PANELBOARD SCHEDULE

Panel _____ 3 φ 4 Wire _____ Mounted _____ Ampere Main

Location _____ _____ Ampere Bus

CT No.	φA	φB	φC	DESCRIPTION	CB Pole	AIC	A	B	C	CB Pole	AIC	DESCRIPTION	φA	φB	φC	CT No.
1							●									2
3								●								4
5									●							6
7							●									8
9								●								10
11									●							12
13							●									14
15								●								16
17									●							18
19							●									20
21								●								22
23									●							24
25							●									26
27								●								28
29									●							30
31							●									32
33								●								34
35									●							36
37							●									38
39								●								40
41									●							42

	Total VA/ φ A
	Total VA/ φ B
	Total VA/ φ C
	Total VA
	Line Amperes

Figure 1-37: Sample panelboard schedule No. 3. Three-phase, 4-wire.

PANELBOARD SCHEDULE

Panel _____ 1 φ3 Wire _____ Mounted _____ Ampere Main

Location _____ _____ Ampere Bus

CT No.	φA	φB	DESCRIPTION	CB Pole	CB AIC	Phase A	Phase C	CB Pole	CB AIC	DESCRIPTION	φA	φB	CT No.
1						●							2
3							●						4
5						●							6
7							●						8
9						●							10
11							●						12
13						●							14
15							●						16
17						●							18
19							●						20
21						●							22
23							●						24
25						●							26
27							●						28
29						●							30
31							●						32
33						●							34
35							●						36
37						●							38
39							●						40

	Total VA/ φ A
	Total VA/ φ B
	Total VA
	Line Amperes

Figure 1-38: Sample panelboard schedule No. 4, Single-phase, 3-wire.

ELECTRIC-HEAT SCHEDULE

HEATER TYPE	MANUFACTURER'S DESCRIPTION	DIMENSIONS	VOLTS	MOUNTING	WATTAGE/REMARKS
⊠ 500 va	ElecTro-Heat Catalog No. 08531	3' x 5" x 2½"	240	Baseboard	500
⊠ 1000 va	ElecTro-Heat Catalog No. 08531	6' x 5" x 2½"	240	Baseboard	1000
⊠	ElecTro-Heat Catalog No. 08531	8' x 5" x 2½"	240	Baseboard	1500
1500 va	ElecTro-Heat Catalog No. 08531	16" x 12" x 3½"	240	Baseboard	1500 w/thermostat

Figure 1-39: Electric-heat schedule.

KITCHEN - EQUIPMENT SCHEDULE

EQUIP No.	DESCRIPTION	HP or Kw	VOLTS	CONNECTION			FURNISHED BY	REMARKS
				Wire	Conduit	Prot.		

Figure 1-40: Kitchen-equipment schedule.

ELECTRICAL DETAILS AND DIAGRAMS

Electrical diagrams are drawings that are intended to show, in diagrammatic form, electrical components and their related connections. They are seldom, if ever, drawn to scale, and show only the electrical association of the different components; that is, the various devices, and their connection to each other. Electrical diagrams include:

- Schematic diagrams
- Single-line diagrams
- Power-riser diagrams

Power-Riser Diagrams

Single-line block diagrams are used extensively to show the arrangement of electric service equipment. The power-riser diagram in Figure 1-41, for example, was used on an office/warehouse building (Figure 1-42) and is typical of such drawings. The drawing shows all pieces of electrical equipment as well as the connecting lines used to indicate service-entrance conductors and feeders. Notes are used to identify the equipment, indicate the size of conduit necessary for each feeder, and the number, size, and type of conductors in each conduit.

A panelboard schedule (Figure 1-43) on page 101 is included with the power-riser diagram to indicate the exact components contained in each panelboard. This panelboard schedule is for the main distribution panel. Schedules are also shown on actual drawings (not here) for the other two panels (PNL A and PNL B).

In general, panelboard schedules usually indicate the panel number, the type of cabinet (either flush- or surface-mounted), the panel mains (ampere and voltage rating), the phase (single- or 3-phase), and the number of wires. A 4-wire panel, for example, indicates that a solid neutral exists in the panel. Branches indicate the type of overcurrent protection; that is, the number of "poles," the trip rating,

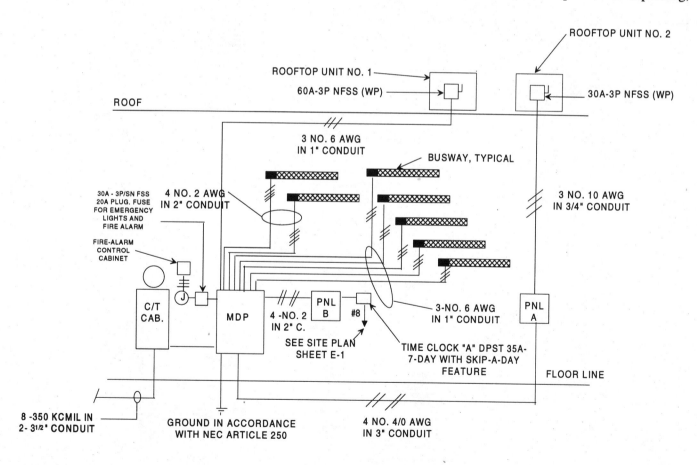

Figure 1-41: Typical power-riser diagram.

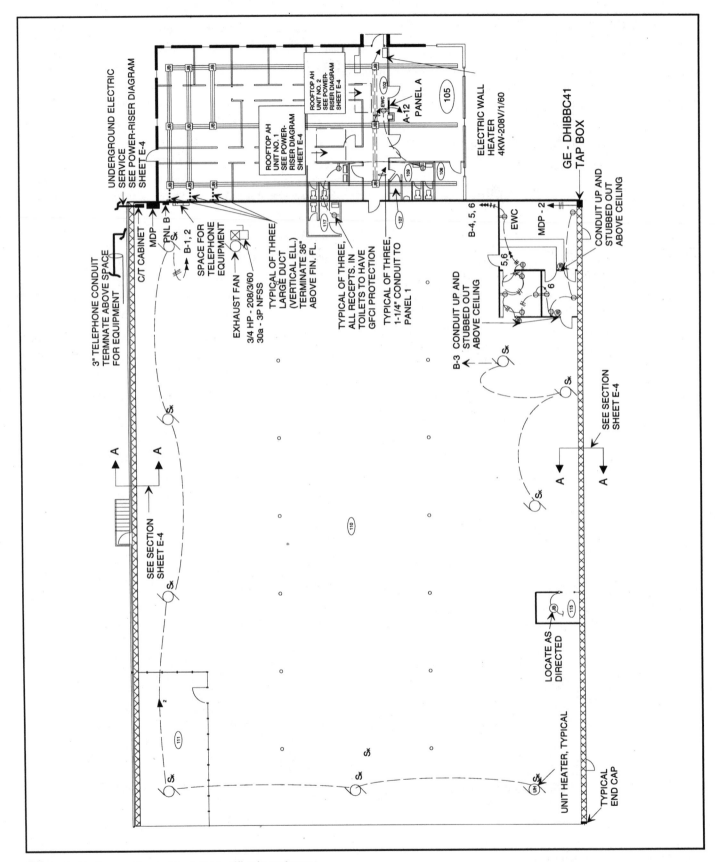

Figure 1-42: Power floor plan for an office/warehouse.

PANELBOARD SCHEDULE

PANEL No.	CABINET TYPE	PANEL MAINS			BRANCHES					
		AMPS	VOLTS	PHASE	1P	2P	3P	PROT.	FRAME	ITEMS FED OR REMARKS
MDP	SURFACE	600A	120/208	3ø, 4-W	—	—	1	225A	25,000	PANEL "A"
					—	—	1	100A	18,000	PANEL "B"
					—	—	1	100A		POWER BUSWAY
					—	—	1	60A		LIGHTING BUSWAY
					—	—	1	70A		ROOFTOP UNIT #1
					—	—	1	70A	18,000	SPARE
					—	—	1	600A	42,000	MAIN CIRCUIT BREAKER

Figure 1-43: Panelboard schedule used in conjunction with the power-riser diagram in Figure 1-41.

and the frame size. The items that each overcurrent device feeds is also indicated in one of the columns.

Schematic Diagrams

Complete schematic wiring diagrams are normally used only in highly unique and complicated electrical systems, such as control circuits. Components are represented by symbols, and every wire is either shown by itself or included in an assembly of several wires which appear as one line on the drawing. Each wire should be numbered when it enters an assembly and should keep the same number when it comes out again to be connected to some electrical component in the system. Figure 1-44 shows a complete schematic wiring diagram for a three-phase, ac magnetic nonreversing motor starter.

Note that this diagram shows the various devices in symbol form and indicates the actual connections of all wires between the devices. The three-wire supply lines are indicated by L_1, L_2, and L_3; the motor terminals of motor M are indicated by T_1, T_2, and T_3. Each line has a thermal overload-protection device (OL) connected in series with normally open line contactors C_1, C_2, and C_3, which are controlled by the magnetic starter coil, C. Each contactor

has a pair of contacts that close or open during operation. The control station, consisting of start pushbutton 1 and stop pushbutton 2, is connected across lines L_1 and L_2. An auxiliary contactor (C4) is connected in series with the stop

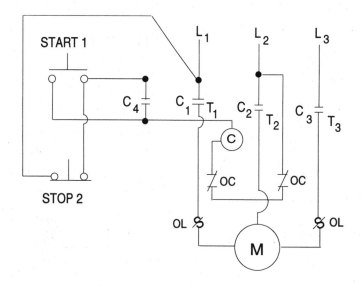

Figure 1-44: Typical schematic diagram of a motor-starting circuit.

pushbutton and in parallel with the start pushbutton. The control circuit also has normally closed overload contactors (OC) connected in series with the magnetic starter coil (C).

Any number of additional pushbutton stations may be added to this control circuit similarly to the way three- and four-way switches are added to control a lighting circuit. In adding pushbutton stations, the stop buttons are always connected in series and the start buttons are always connected in parallel.

Drawing Details

A detail drawing is a drawing of a separate item or portion of an electrical system, giving a complete and exact description of its use and all the details needed to show the worker exactly what is required for its installation. The

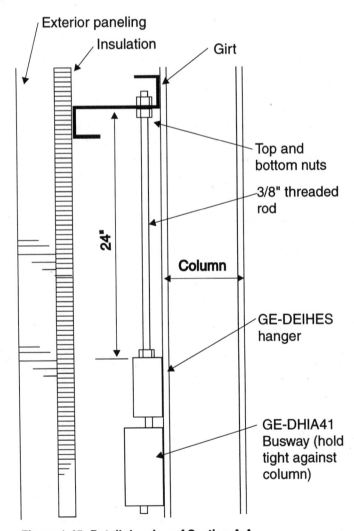

Figure 1-45: Detail drawing of Section A-A.

power floor plan for an office/warehouse (Figure 1-42) has a sectional cut (Section A-A) through the busduct. This is a good example of where an extra, detailed drawing is desirable. This section is shown in Figure 1-45, and provides additional details for installing the busduct.

A set of electrical drawings will sometimes require large-scale drawings of certain areas that are not indicated with sufficient clarity on the small-scale drawings. For example, a site plan may show exterior pole-mounted lighting fixtures that are to be installed by the contractor. The site plan itself will probably show only small circles to indicate the approximate location of these fixtures. In most cases, more detail is required for a correct installation.

TEMPORARY ELECTRICAL SERVICE

Traditionally, temporary electric power to job sites has been handled in a haphazard way that showed little signs of organization. In recent years, however, both OSHA and the *NEC* have established specific regulations governing the installation of temporary electric services to job sites. In general, all job-site power must be supplied using ground-fault circuit-interrupters and specified types of receptacles or outlets fitting only approved cord-cap types.

Most electrical contractors have developed their own style of furnishing temporary power to job sites. In many cases, the temporary-service arrangement is removed — nearly intact — at the completion of one project, and moved to another site. For example, the 100-A temporary-service detail in Figure 1-46 is typical of those used on residential and small commercial projects. At the completion of one project, the overhead service conductors are disconnected, the grounding conductor disconnect from the ground clamp attached to the ground rod, and then the pole, with all of the service equipment still attached, is pulled and moved to another site. Once the pole is firmly mounted at the new site, all that is necessary for the local power company to make connection is to drive a new ground rod and attach the grounding conductor to an attached ground clamp. In some areas, two grounding electrodes are required for even temporary power. If a metallic cold-water pipe is close by, the grounding conductor must be attached to the cold-water pipe and also to at least one additional grounding electrode, such as a ground rod, ground plate, etc.

The temporary-service details shown in Figures 1-46 through 1-48 should cover the majority of all electrical construction from small residential to high-rise commercial

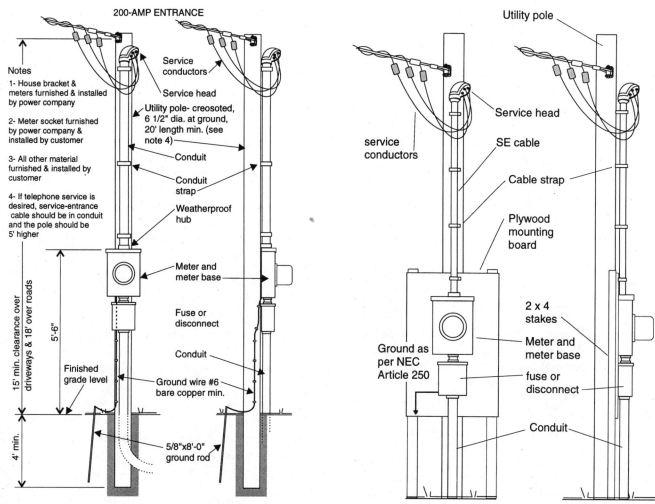

200-AMP ENTRANCE

Notes

1- House bracket & meters furnished & installed by power company

2- Meter socket furnished by power company & installed by customer

3- All other material furnished & installed by customer

4- If telephone service is desired, service-entrance cable should be in conduit and the pole should be 5' higher

Service conductors

Service head

Utility pole- creosoted, 6 1/2" dia. at ground, 20' length min. (see note 4)

Conduit

Conduit strap

Weatherproof hub

Meter and meter base

Fuse or disconnect

Conduit

Ground wire #6 bare copper min.

15' min. clearance over driveways & 18' over roads

5'-6"

4' min.

Finished grade level

5/8"x8'-0" ground rod

Figure 1-46: Detail of 200-A temporary service, utilizing rigid or PVC conduit.

Utility pole

service conductors

Service head

SE cable

Cable strap

Plywood mounting board

2 x 4 stakes

Meter and meter base

Ground as per NEC Article 250

fuse or disconnect

Conduit

Figure 1-48: Detail of 100-A temporary service, utilizing Type SE cable.

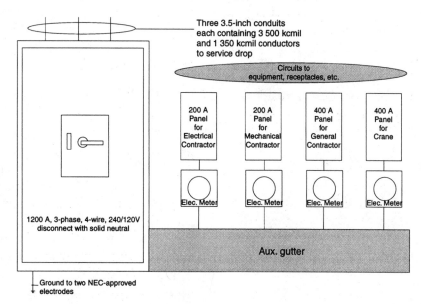

Three 3.5-inch conduits each containing 3 500 kcmil and 1 350 kcmil conductors to service drop

Circuits to equipment, receptacles, etc.

1200 A, 3-phase, 4-wire, 240/120V disconnect with solid neutral

| 200 A Panel for Electrical Contractor | 200 A Panel for Mechanical Contractor | 400 A Panel for General Contractor | 400 A Panel for Crane |

Elec. Meter   Elec. Meter   Elec. Meter   Elec. Meter

Aux. gutter

Ground to two NEC-approved electrodes

Figure 1-47: Power-riser diagram of a 1200-A temporary service for a large commercial project.

projects of all types. These details range in size of 120/240-V, 100-A, single-phase services to 120/240-V, 1200-A, three-phase services.

ELECTRICAL TESTS

Many consulting engineering firms require that a final testing of the electrical system be made in the presence of an electrical engineer or other member from the consulting firm. The extent of this test can vary from job to job and from engineer to engineer, but the following is typical of most electrical specifications:

Testing:

A. The electrical contractor shall take certain voltage and current readings, record all values, and submit in triplicate to the engineer. Two complete sets of readings are required, one under no load and one under maximum available load. The current and voltage shall be recorded on each phase (plus voltage between phases) at main panelboard and at each branch circuit panelboard. Additional spot readings shall be made if required. Resistance of grounding conductors shall be tested and recorded. Forms for submitting this report may be obtained from the engineer's office. *See* Figure 1-49.

B. The electrical contractor shall also take voltage and amperage readings on each phase of each motor circuit and each resistance heater circuit installed under the contract, and the same must be recorded as described previously. Also, record motor nameplate data, actual motor heater protective device, and all other data necessary for selection of heater device. A typical form is shown in Figure 1-50.

The requirements in the previous specification may seem rigid to those contractors familiar only with residential and small commercial projects; for those involved in industrial installation perhaps it seems lenient. The requirement of an electrical specification for a typical motor read is as follows:

A. It shall be the responsibility of the electrical contractor to connect the electrical loads to provide minimum phase unbalance throughout the building. The electrical contractor shall operate the building under full heating and other load conditions, with full lighting and provide a record of the amperage per phase for each feeder installed to the main distribution panel.

B. As soon as electric power is available and connected to serve the equipment in the building, and everything is ready for final testing and placing in service, a complete operational test shall be made. The electrical contractor shall furnish all necessary instruments and testing equipment to make all tests, adjustments, and trial operations required to place the system in a balanced and satisfactory operational condition. The contractor shall further furnish all necessary assistance and instruction to properly instruct the owner's authorized personnel in the operation and care of the system.

C. Prior to testing the system, the feeders and branch circuits shall be continuous from the main feeders to the main panels, to subpanels to outlets, with all circuit breakers and fuses in place. The system shall be tested free from shorts and ground faults. Such tests shall be made in the presence of the engineer or his or her representative.

D. No circuits shall be energized without the owner's approval.

Hi-Pot Tests

Suitable dielectric tests are commonly applied to determining the condition of insulation in cables, transformers, rotating machinery, etc. Properly conducted tests will indicate such faults as cracks, discontinuity, thin spots or voids in the insulation, excessive moisture or dirt, faulty splices, faulty potheads, etc. An experienced operator cannot only frequently predict the expected breakdown voltage of the item under test, but often can make a good estimate of the future operating life.

The use of direct current (dc) has several important advantages over alternating current (ac). The test equipment itself may be much smaller, lighter in weight and lower in price. Properly used and interpreted, dc tests will give

ELECTRICAL TEST DATA REPORT

Engineering and Design Associates 4756 Overall Drive Bentonville, Virginia 22610	Project:_____ _____ — _____ _____ —

Electrical Contractor:	

Date Tests were made:		Date Submitted:

Electrical Characteristics:	volts	Phase	Wires

Type voltmeter used:		When Calibrated:

Type Ammeter used:		When Calibrated:

Service Ground — Resistance in Ohms:	Note: Resistance test must be made with a hand crank, magneto type, megger.

Panel	Voltage – No Load		Voltage – Max. Load		Maximum Load Amperage			
	Phase to Phase	Phase to Ground	Phase to Phase	Phase to Ground	Phase A	Phase B	Phase C	Neutral

Figure 1-49: Electrical test data report form.

Motor Number	Machine Driven	HP	Voltage	Phase	Serial Number	Frame	Amperes	RPM	CODE	Make	NEMA Rating	Coil Number	Size

MOTOR DATA FOR:

Motor Starter

Figure 1-50: Motor data report and test form.

much more information than is obtainable with ac testing. There is far less chance of damage to equipment and less uncertainty in interpreting results. The high capacitance current frequently associated with ac testing is not present to mask the true leakage current, nor is it necessary to actually break down the material being tested to obtain information regarding its condition. Though dc testing may not simulate the operating conditions as closely as ac testing, the many other advantages of using dc make it well worthwhile.

The great majority of high-voltage cable testing is done with direct current — primarily due to the smaller size of machines and the ability to easily provide quantifiable values of charging, leakage, and absorption currents. While both ac and dc testers may damage a cable during tests, the damage in a dc test only occurs if increasing current is not interrupted soon enough and the capacitance stored is discharged across the insulation.

Hi-pot testers for high-voltage cable will also have a microammeter and a range dial to allow measurement of up to 5 milliamps, at which level the circuit breaker will trip or else the reactor will collapse.

Hi-pot testing may be divided into the following broad categories:

- *Design Tests* — These are the tests usually made in the laboratory to determine proper insulation levels prior to manufacturing.

- *Factory Tests* — These are the tests made by the manufacturer to determine compliance with the design or production requirements.

- *Acceptance Tests* — These are the tests made immediately after installation, but prior to putting the equipment or cables into service.

- *Proof Tests* — These are the high-voltage (HV) tests made soon after the equipment has been put into service and during the guarantee period.

- *Maintenance Tests* — Maintenance tests are those performed during normal maintenance operations or after servicing or repair of equipment or cables.

- *Fault Locating* — Fault locating tests are made to determine the location of a specific fault in a cable installation. *See* Figures 1-51 through 1-55.

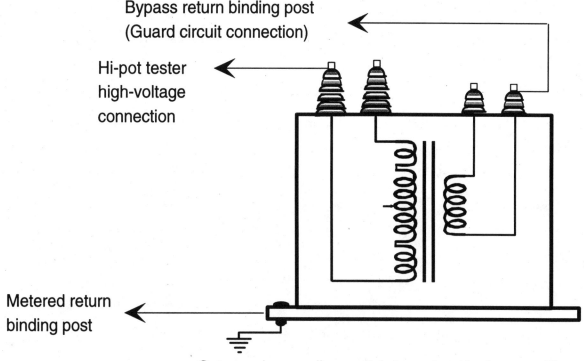

Bypass return binding post
(Guard circuit connection)

Hi-pot tester
high-voltage
connection

Metered return
binding post

Set panel grounding switch to metered return position

Figure 1-51: High-voltage wiring to grounded core or case.

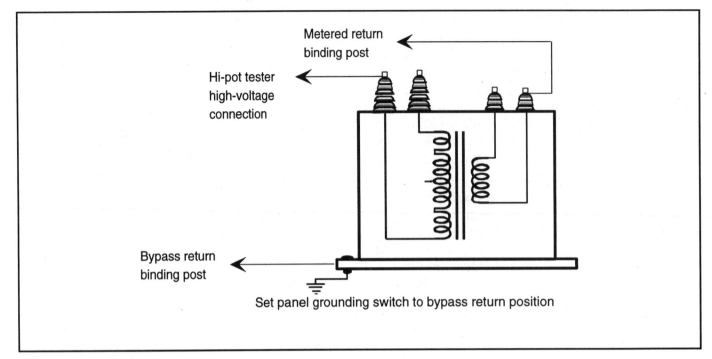

Figure 1-52: High-voltage winding to low-voltage winding.

The maximum test voltage, the testing techniques, and the interpretation of the test results will vary somewhat depending upon the particular type of test. Unfortunately, in many cases the specifications do not spell out the test voltage, nor do they outline the test procedure to be followed.

Therefore, it is necessary to apply a considerable amount of common sense and draw strongly upon past experience in making these tests. There are certain generally accepted procedures, but the specific requirements of the organization requiring the tests are the ones that should govern. It is

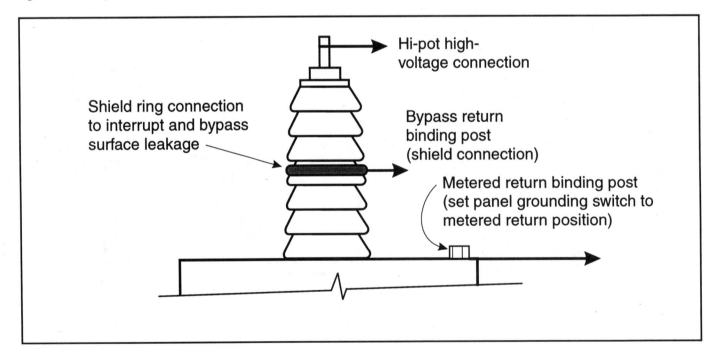

Figure 1-53: Measuring busing leakage.

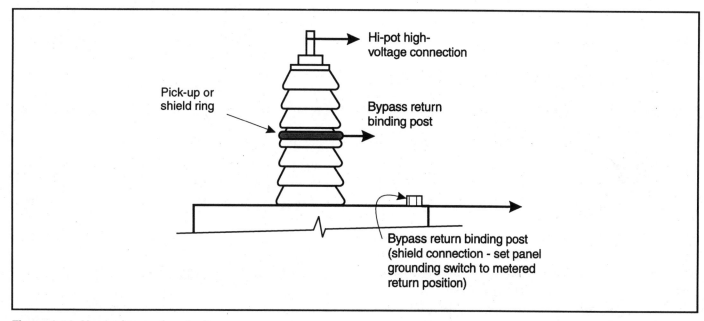

Figure 1-54: Measuring surface leakage.

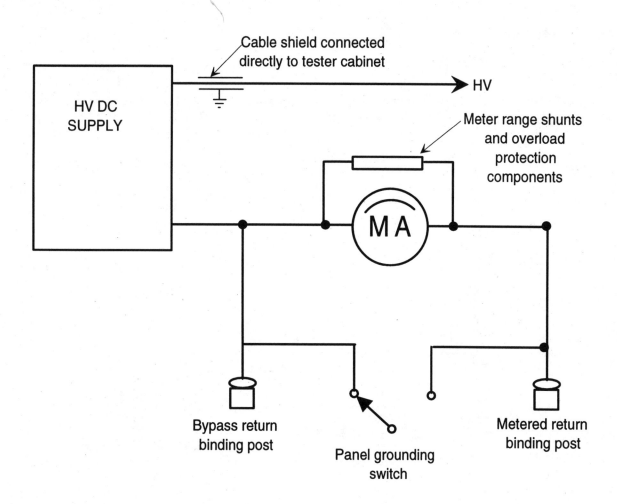

Figure 1-55: Simplified hi-pot tester output circuit diagram.

obvious that an acceptance test must be much more severe than a maintenance test, while a test made on equipment that is already faulted would be conducted in a manner different from a test being conducted on a piece of equipment in active service.

As a rough rule of thumb, acceptance tests are made at about 80% of the original factory voltage. Proof tests are usually made at about 60% of the factory test voltage. The maximum voltage used in maintenance testing depends on the age, previous history and condition of the equipment, but an acceptable value would be approximately 50 to 60% of the factory test voltage.

Special testing instruments are available for all of the tests listed above. However, testing instruments will be used mostly by electricians to perform routine production tests and acceptance tests for newly installed systems.

The general criteria for acceptance of a cable system is a consistent leakage for each voltage step (linear) and a decrease in current over time at the final voltage. The level of charging current should also be consistent for each step.

Ground Resistance

Figures 1-56 and 1-57 show two basic methods for testing the resistance of earth. The direct or two-terminal test (Figure 1-56) consists of connecting terminals $P1$ and $C1$ of the megger to the ground under test, and terminals $P2$ and $C2$ to an all-metallic water-pipe system. If the water-pipe system covers a large area, its resistance should only be a fraction of an ohm and, therefore, the meter reading will be that of the ground or electrode under test.

The direct method is the simplest way to make an earth-resistance test. With this method, resistance of two electrodes in series is measured — the driven ground rod and the water system. But there are some important limitations:

- The water-pipe system must cover enough area to have a negligible resistance.
- The water-pipe system must be metallic throughout, without any insulating couplings or flanges.
- The driven ground rod or electrode under test must be far enough away from the water-pipe system — about 10 times the radius of the electrode — to be outside its sphere of influence.

The fall-of-potential method (Figure 1-57 on the next page) consists of using terminals $C1$, $C2$, and $P2$ of the instrument. The $P1$ and $C1$ terminals on the instrument are

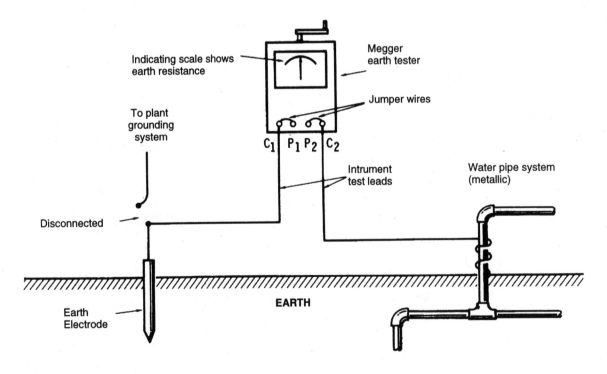

Figure 1-56: Direct method of earth-resistance testing.

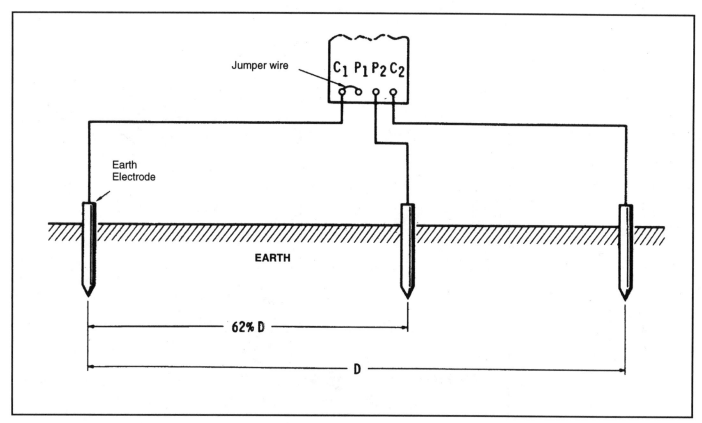

Figure 1-57: Fall-of-potential or three-terminal earth-resistance test.

jumped (connected) together and connected to the electrode or ground under test. The driven reference rod connected to terminal *C2* should be placed as far from the ground electrode under test as practical.

The potential-reference rod connected to terminal *P2* is then driven into the ground at a number of points roughly on a straight line between the ground under test and the rod connected to terminal *C2*. Resistance readings are logged for each of the points. A curve of resistance versus distance (Figure 1-58 on the next page) is then drawn. Correct earth resistance is read from the curve for the distance that is about 62 percent of the total distance from the earth electrode *C2*; that is, if the total distance is *D*, then the 62 percent distance is 0.62 *D*. For example, if *D* is 120 ft, the distance value for earth resistance is 0.62 × 120 or 74 ft.

Certain problems may arise when measuring with the fall-of-potential methods:

- Natural currents in the soil caused by electrolytic action. These cause the voltmeter to read either high or low, depending on polarity.

- Induced currents in the soil, instrument, or electrical leads. These may cause vibration of the meter pointer, interfering with readability.

- Resistance in the auxiliary electrode and in the electrical leads. These introduce error into the voltmeter reading.

Most meggers use a null balancing metering system. Unlike the separate voltmeter and ammeter method, this instrument provides a readout directly in ohms, thus eliminating calculation. Although the integrated systems of the megger are sophisticated, they still perform the basic functions for fall-of-potential testing.

The megger also has a current supply circuit. *See* Figure 1-59. This current may be traced from grounding electrode X through terminal X, potentiometer R1, the secondary of power transformers T1, terminal 2, and auxiliary current electrode 2. This function produces a current in the earth between electrodes X and 2.

When switch S1 is closed, battery B energizes the coil of vibrator V. Vibrator reed V1 begins oscillating, thereby producing an alternating current in the primary and secon-

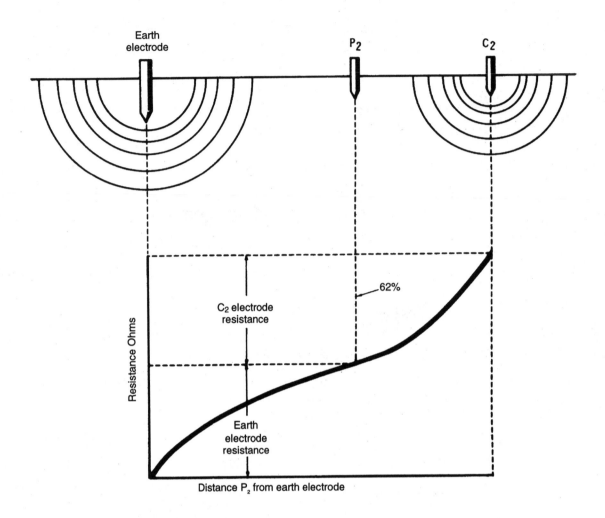

Figure 1-58: Resistance versus distance curve.

dary windings of T1. The negative battery terminal is connected alternately across first one and then the other half of the primary winding.

The voltage circuit in Figure 1-59 can be traced from grounding electrode X through terminal X, the T2 secondary, switch S2, resistors R2 and R3 (paralleled by the meter and V2 contacts), capacitor C, terminal 1, and auxiliary potential electrode 1.

Current in the earth between grounding electrode X and auxiliary current electrode 2 creates a voltage drop due to the earth's resistance. With auxiliary potential electrode 1 placed at any distance between grounding electrode X and auxiliary current electrode 2, the voltage drop causes a current in the voltmeter circuit through balanced resistors R2 and R3. The voltage drop across these resistors causes galvanometer M to deflect from zero center scale.

Vibrator reed V2 operates at the same frequency as V1, thereby functioning as a mechanical rectifier for galvanometer M. The vibrator is tuned to operate at 97.5 Hz, a frequency unrelated to commercial power line frequencies and their harmonics. Thus, currents induced in the earth by power lines are rejected by most meggers and have virtually no effect on accuracy. Stray direct current in the earth is blocked out of the voltmeter circuit by capacitor C which is also shown in the circuit drawing.

The current in the primary of T2 can be adjusted with potentiometer R1. Primary current in T2 induces a voltage in the secondary of T2 which is opposite in polarity to the voltage drop caused by current in the voltmeter circuit.

With R1 adjusted so the primary and opposing secondary voltages of T2 are equal, current in the voltmeter circuit is zero and the galvanometer reads zero. The resistance

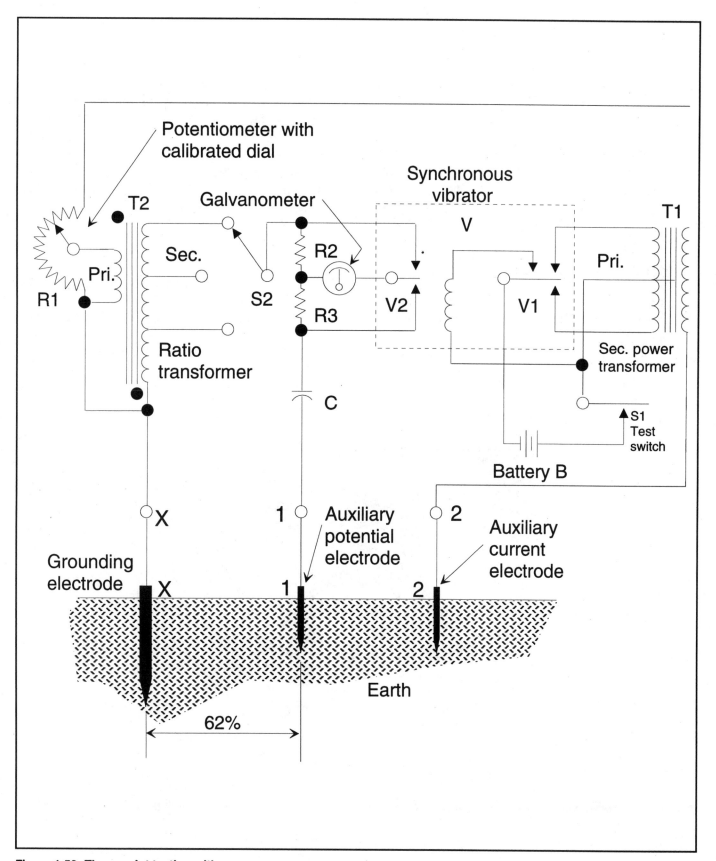

Figure 1-59: Three-point testing with megger.

of ground electrode X can then be read on the calibrated dial of the potentiometer.

With no current in the voltmeter circuit, the lead resistance of the auxiliary potential electrode 1 has no bearing on accuracy; that is, with no current, there is no voltage drop in the leads. Resistance to earth of the current electrode results only in a reduction of current and consequently a loss of sensitivity. Therefore, the auxiliary electrodes need only be inserted into the earth six to eight in to make sufficient contact. In some locations where the soil is very dry, it may be necessary to pour water around the current electrode to lower the resistance to a practical usable value.

IDENTIFICATION OF ELECTRICAL SYSTEMS

During the course of any electrical project, some changes in the original design are commonplace. It is usually the electrical contractor's responsibility to provide two (2) complete sets of marked-up drawings to the architect or engineer that show any deviations in work as actually installed from work indicated on the original drawings.

Furthermore, laminated plastic or rigid phenolic plastic nameplates with engraved letters are required to be mounted on each motor starter, disconnecting switch, pushbutton station, power and lighting panels, and at each ceiling or wall access panel to electrical work.

The engraving tool shown in Figure 1-60 is used extensively in the electrical industry to engrave plastic, metal tags, signs, panel markers and the like.

Abbreviations

One of the problems that frequently exists is identifying the meanings of abbreviations. The following list of abbreviations are those that have appeared on construction drawings over the past 50 years or so. However, the exact abbreviation for a word still varies considerably.

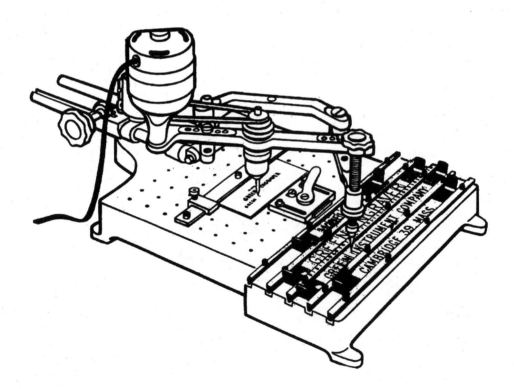

Figure 1-60: Typical engraver used to engrave metal tags, signs, and panel markers for identifying electrical equipment.

— A —

Abbreviation	Term
A	Air
A	Ampere
A	Area
A	Astragal
AB	Anchor Bolt
AB SW	Air-break Switch
ABBR	Abbreviate
ABCB	Air-break Circuit Breaker
ABRSV	Abrasive
ABRSV RES	Abrasive-resistant
ABS	Absolute
ABSORB	Absorption
ABSTR	Abstract
ABV	Above
AC	Alternating Current
AC	Armored Cable
AC UNIT	Package Air Conditioning Unit
ACCEL	Accelerate
ACCESS	Accessory
ACIRC	Air Circulating
ACLD	Air-cooled
ACR	Across
ACS	Access
ACST	Acoustic
ACT	Activity
ACTE	Actuate
ACTL	Actual
ACTR	Actuator
ACTVT	Activate
ACTVTR	Activator
ACU	Air Conditioning Unit
AD	Air Drawn
AD	Area Drain
ADD	Addition
ADDT	Additive
ADH	Adhesive
ADJ	Adjacent
ADJT	Adjustable

Abbreviation	Term
ADPTR	Adapter
ADSORB	Adsorbent
ADV	Advance
AE	Air Extractor
A/E	Architect/Engineer
AGT	Agent
AHR	Anchor
AHU	Air Handling Unit (AC or H&V)
AIL	Aileron
AINDTN	Air Induction
Air BB	Air Baseboard
AIR COND	Air-Condition(er)
AIRCLNR	Air Cleaner
AL	Air Lock
AL	Aluminum
ALC	Alcohol
ALIGN	Alignment
ALK	Alkaline
ALLOW	Allowance
ALTN	Alternate
ALTRN	Alteration
ALY	Alloy
ALY STL	Alloy Steel
AMB	Ambient
AMNA	Ammonia
AMP	Ampere
AMPTD	Amplitude
AMR	Advance Material Request
AMT	Amount
ANCFIL	Anchored Filament
ANDA	Anodize
ANLR	Angular
ANLR	Annular
ANN	Annunciator
ANS	American National Standard
ANT	Antenna
AO	Access Opening
AP	Acidproof

Abbreviation	Term
AP	Air Passage
APA	Axial Pressure Angle
APC	Acoustical Plastic Ceiling
APERT	Aperture
APP	Appearance
APP	Approximate
APPL	Application
APPROX	Approximate
APPV	Approve
APPVL	Approval
APPX	Appendix
APRCH	Approach
APT	Apartment
APU	Auxiliary Power Unit
APVD	Approved
APVD EQL	Approved Equal
AQL	Acceptable Quality Level
AR	As Required
ARCH	Architect
ARCW	Arc Weld
ARM	Armature
ARM	Armored
ARST	Arrester
ARTF	Artificial
AS	Automatic Sprinkler
ASB	Asbestos
ASPH	Asphalt
ASPHRS	Asphalt Roof Shingles
ASPRTR	Aspirator
ASTR	Astronomical
ASV	Angle Stop Valve
ASW	Auxiliary Switch
ASWG	American Steel Wire Gauge
AT	Acoustical Tile
AT	Ashphalt Tile
ATB	Asphalt-Tile Base
ATCH	Attachment
ATF	Asphalt-tile Floor
ATTEN	Attenuation

Abbreviation	Term
ATTEN	Attenuator
AUTO	Automatic
AUTO OVLD	Automatic Overload
AUTOSTR	Automatic Starter
AUX	Auxiliary
AUXR	Auxiliary Register
AVDP	Avoirdupois
AVE	Avenue
AVG DIA	Average Diameter
AW	Above Water
AWG	American Wire Gauge
AWN	Awning
AX FL	Axial Flow
AXP	Axial Pitch
AZ	Azimuth

— B —

B	Base
B	Bath(room)
B	Bit
B TO B	Back-to-Back
B&S	Bell and Spigot
BARR	Barrier
BAT	Batter
BATT	Batten
BB	Baseboard
BB	Brass Bolt
BC	Beginning of Curve
BC	Between Centers
BC	Bookcase
BC	Bottom Chord
BCL	Basic Contour Line
BCL	Broom Closet
BD	Board
BDO	Bow Door
BDRY	Boundary
BDY	Boundary
BDZR	Bulldozer

Abbreviation	Term
BE	Back End
BEV	Bevel
BF	Backface
BF	Board Foot
BF	Both Faces
BF	Bottom Face
BH	Brinell Harness
BHD	Bulkhead
BHN	Brinell Harness Number
BI	Black Iron
BITUM	Bituminous
BK	Book
BKGD	Background
BL	Base line
BL	Bend Line
BL	Billet
BL	Blade
BL	Bottom Layer
BL	Boundary Line
BL	Building Line
BL STL	Billet Steel
BLDG	Building
BLDR	Bleeder
BLK	Blank
BLK	Block
BLKG	Blocking
BLKT	Blanket
BLO	Blower
BLR	Boiler
BLT	Bolt
BLTIN	Built-in
BLVD	Boulevard
BLW	Below
BLWT	Blowout
BM	Beam
BM	Bench Mark
BMVP	Barrier, Moisture Vapor-proof

Abbreviation	Term
BNCH	Bench
BND	Bonded
BNSH	Burhish
BOD	Blackout Door
BOT	Bottom
BP	Back Pressure
BP	Base Plate
BP	Between Perpendiculars
BP	Bolted Plate
BPL	Bearing Plate
BR	Bedroom
BR	Bend Radius
BR	Braided
BR	Branch
BR	Brush
BRDG	Bridge
BRG	Bearing
BRK	Break
BRK	Brick
BRKR	Breaker
BRKT	Bracket
BRT	Bright
BS	Backsight
BS	Both Sides
BSC	Basic
BSHG	Bushing
BSMT	Basement
BSP	Ball Stop
BT	Bath Tub
BTU	British Thermal Unit
BTWLD	Butt Weld
BU	Blower Unit
BU	Bushel
BUR	Built-up Roofing
BVC	Beginning Vertical Curve
BWP	Barrier, Waterproof
BWS	Beveled Wood Siding

Abbreviation	Term
BYP	Bypass

— C —

Abbreviation	Term
°C	Degrees Celsius
C	Carbon
C	Centigrade
C	Courses
C	Cycle
C TO C	Center-to-Center
CA	Cable
CA	Cold Air
CAB	Cabinet
CAL	Caliber
CAL	Calibrate
CAL	Calorie
CAN	Canopy
CANTIL	Cantilever
CAP	Capacitor
CAP	Capacity
CAP	Capital
CARP	Carpentry
CARP	Carpet
CAT	Catalogue
CAV	Canvas
CAV	Cavity
CB	Catch Basin
CB	Circuit Breaker
CBAL	Counterbalance
CBD	Carbide
CBORE	Counterbore
CC	City Council
CC	Color Code
CC	County Council
CCA	Circuit Court of Appeals
CCB	Concrete Block
CCF	Concrete Floor
CD	Cable Duct
CD	Ceiling Diffuser

Abbreviation	Term
CD	Cold-Drawn
CD	Condensate Drain
CE	Civil Engineer
CEL	Celsius
CEM	Cement
CENT	Centigrade
CER	Ceramic
CF	Cement Floor
CF	Cooling Fan
CFLG	Counterflashing
CFM	Cubic Feet Per Minute
CFS	Cubic Feet Per Second
CH	Cabinet Heater
CH	Case Harden
CH	Chain
CH	Church
CHAM	Chamfer
CHAN	Channel
CHAP	Chapter
CHD	Chord
CHG	Charge
CHM	Chimney
CI	Cast Iron
CI	Circuit Interrupter
CIP	Cast-Iron Pipe
CIR	Circle
CIRC	Circular
CIV	Civil
CJ	Construction Joint
CKT	Circuit
CL	Center line
CL	Clearance
CL	Close
CL	Crane Load
CL GL	Clear Glass
CLD	Cleared
CLE	Closed End
CLG	Ceiling
CLJ	Control Joint

Abbreviation	Term
CLKJ	Caulked Joint
CLO	Closet
CLOS	Closure
CLP	Clamp
CLPR	Caliper
CLR	Clear
CLR	Collar
CLR	Cooler
CLT	Cleat
CM	Common Meter
CMBSTR	Combustor
CMET	Coated Metal
CMPS	Compass
CNCL	Concealed
CNCV	Concaved
CND	Conduit
CNDCT	Conductor
CNDS	Condensate
CNTR	Counter
CNVR	Conveyor
CO	Cased Opening
CO	Change Order
CO	Cleanout
CO	Colbalt
CO	Company
CO	County
CO	Cutoff
COAX	Coaxial
COD	Cash (or collect) on Delivery
COEF	Coefficient
COL	Colonial
COL	Column
COLL	Collector
COM	Common
COMB	Combination
COMB	Combustion
COMBL	Combustible
COML	Commercial
COMP	Compacted

Abbreviation	Term
COMP	Compound
COMPA	Compressed Air
COMPL	Complete
COMPOS	Composition
CONC	Concrete
COND	Condenser
CONST	Construction
CONT	Continuous
CONT	Contact
CONTR	Contractor
CONV	Convector
COOL	Coolant
COR	Corner
COR	Cornice
CORR	Corrugate
COT	Cotter
COT	Cotter Pin
COV	Cutoff Valve
CP	Clay Pipe
CP	Conrete Pipe
CPA	Certified Public Accountant
CPC	Cement Plaster Ceiling
CPL	Cement Plaster
CPLG	Coupling
CPM	Critical Path Method
CPNTR	Carpenter
CPPD	Capped
CPT	Critical Path Technique
CR	Chromium
CR	Cold-Rolled
CR	Credit
CRCLT	Circulate
CRCMF	Circumference
CRE	Corrosion-Resistant
CRES	Corrosion-Resistant Steel
CRG	Carriage
CRK	Crank
CRN	Crane
CRN	Crown

Abbreviation	Term
CRS	Coarse
CRS	Cold-Rolled Steel
CRSHD STN	Crushed Stone
CRSN	Corrosion
CRSV	Corrosive
CRTG	Centering
CRV	Curve
CS	Carbon Steel
CS	Case
CS	Cast Steel
CS	Control Switch
CSB	Concrete Splash Block
CSG	Casing
CSK	Countersink
CT	Ceramic Tile
CTB	Ceramic Tile Base
CTD	Coated
CTF	Ceramic Tile Floor
CTG	Coating
CTL	Central
CTLST	Catalyst
CTR	Center
CTR	Contour
CTWALK	Catwalk
CU	Copper
CU	Cubic
CU IN	Cubic Inches
CU YD	Cubic Yards
CULV	Culvert
CUR	Current
CUTS	Cut Stone
CV	Check Valve
CVNTL	Conventional
CVX	Convex
CW	Cold Water
CW	Continuous Wave
CWP	Circulating Water Pump
CYL	Cylindrical
CSMT	Casement

Abbreviation	Term

— D —

Abbreviation	Term
°DEG	Degree
D	Deep
D	Density
D	Dryer
D&M	Dressed and Matched
DAD	Double-Acting Door
DAT	Datum
DB	Decibel
DB	Distribution Box
DB	Dry Bulb
DBL	Double
DBL ACT	Double-Acting
DBLF	Double Face
DBLW	Double-Wall
DC	Direct Current
DCL	Door Closer
DD	Deep Drawn
DD	Disconnecting Device
DD	Dutch Door
DDR	Direct Drive
DE	Double End
DEC	Decimal
DECAL	Decalcomania
DECR	Decrease
DED	Dedendum
DEG	Degree
DENS	Density
DEPR	Depression
DEPT	Department
DET	Detail
DET	Detent
DEV	Development
DEV LG	Developed Length
DF	Douglas Fir
DF	Drinking Fountain
DF	Drive Fit
DF	Drop Forge

Abbreviation	Term
DFT	Draft
DFT	Drift
DFTG	Drafting
DFTSMN	Draftsman
DHMY	Dehumidify
DHW	Double-Hung Windows
DIA	Diameter
DIAG	Diagonal
DIAG	Diagram
DIAL PHTH	Diallyl Phthalate
DIAPH	Diaphragm
DIFF	Differential
DIFFUS	Diffusing
DIL	Dilute
DIM	Dimension
DIR	Direct
DIR	Direction
DISC	Disconnect
DISCH	Discharge
DIST	Distance
DIV	Divide
DIV	Division
DK	Deck
DL	Dead Load
DL	Drawing List
DLO	Daylight Opening
DLY	Delay
DMGZ	Demagnetize
DML	Demolition
DMPR	Damper
DMR	Dimmer
DN	Down
DNDFT	Downdraft
DNG	Dining
DNG R or DR	Dining Room
DNTRD	Denatured
DO	Ditto
DOZ	Dozen
DP	Dampproofing

Abbreviation	Term
DP	Depth
DP	Dew Point
DP	Diameter Pitch
DP	Dripproof
DP	Dual-Purpose
DP SW	Double-Pole Switch
DPDT	Double-Pole Double-Throw
DPDT SW	Double-Pole Double-Throw Switch
DPNL	Distribution Panel
DPST	Double-Pole Single-Throw
DPST SW	Double-Pole Single-Throw Switch
DPT	Dew Point Temperature
DPV	Dry Pipe Valve
DR	Door
DR	Drain
DR	Drill
DR	Drill Rod
DR	Drive
DRK	Derrick
DRM	Dormer
DRS	Dressed (lumber)
DS	Disconnect Switch
DS	Downspout
DS	Draft Stop
DSG	Double Strength Glass
DSGN	Design
DSPEC	Design Specification
DST	Door Stop
DSW	Door Switch
DT	Drain Tile
DT	Dust-Tight
DTCH	Detached
DTY CY	Duty Cycle
DUP	Duplicate
DVL	Develop
DVTL	Dovetail
DW	Dishwasher
DW	Double Weight
DW	Drywall

Abbreviation	Term
DW	Dumbwaiter
DWEL	Dwelling
DWG	Drawing
DWL	Design Water Line
DWL	Dowel
DWR	Drawer
DX	Duplex
DYNMT	Dynamite

— E —

Abbreviation	Term
E	East
E	Enamel
E TO E	End-To-End
E&SP	Equipment and Spare Parts
EA	Each
EA	Exhaust Air
EAT	Entering Air Temperature
ECC	Eccentric
ECD	Estimated Completion Date
ECO	Electronic Checkout
ECO	Engineering Change Order
EDGW	Edgewise
EF	Each Face
EF	Exhaust Fan
EF	Extra Fine (threads)
EFF	Effective
EFF	Efficiency
EHP	Effective Horsepower
EJCTR	Ejector
EJN	Ejection
EL	Electroluminescent
EL	Elevation
ELAS	Elastic
ELAS	Elasticity
ELB	Elbow
ELCTC	Electric Contact
ELCTD	Electrode
ELCTLT	Electrolytic

Abbreviation	Term
ELEC	Electric
ELEK	Electronic
ELEM	Elementary
ELEV	Elevate
ELEV	Elevator
ELEX	Electronics
ELL	Elbow
ELMCH	Electromechanical
ELONG	Elongation
ELP	Elliptical
ELPNEU	Electropneumatic
EM	Electromagnetic
EM	Expanded Metal
EMB	Emboss
EMER	Emergency
EMF	Electromotive Force
EMT	Electrical Metallic Tubing
EMUL	Emulsion
ENAM	Enamel
ENCL	Enclose
ENCL	Enclosure
ENCSD	Encased
ENG	Engine
ENGR	Engineer
ENGRG	Engineering
ENLG	Enlarge
ENLGD	Enlarged
ENRGZ	Energize
ENT	Entrance
ENTR	Entrance
ENV	Envelope
ENVIR	Environment
EP	Electric Panel
EP	Explosion-Proof
EPD	Electric Power Distribution
EPT	External Pipe Thread
EPWR	Emergency Power
EQ	Equal
EQL	Equally Spaced

Abbreviation	Term
EQPT	Equipment
EQUIP	Equipment
EQUIV	Equivalent
ERCG	Erecting
ERCR	Erector
ERECT	Erection
ES	Electrostatic
ES	Equal Section
ESC	Escutcheon
EST	Estimate
ET	Edge Thickness
EVAP	Evaporator
EVLTN	Evaluation
EWC	Electric Water Cooler
EX	Example
EX	Extra
EXC	Excavate
EXCH	Exchange
EXEC	Executive
EXH	Exhaust
EXHV	Exhaust Vent
EXIST	Existing
EXP	Expand
EXP	Expansion
EXP	Exposed
EXP JT	Expansion Joint
EXPLD	Explode
EXPSR	Exposure
EXST	Existing
EXT	Extension
EXT	Exterior
EXT	Extinguish
EXT	Extinguisher
EXT GR	Exterior Grade
EXTD	Extrude
EXTR	Extractor
EXTR	Extrude
EYLT	Eyelet
EYPC	Eyepiece

Abbreviation	Term
EL	Elastic Limit

— F —

Abbreviation	Term
°F	Fahrenheit
4P	Four-Pole
4PDT SW	Four-Pole Double-Throw Switch
4PST SW	Four-Pole Single-Throw Switch
4PSW	Four-Pole Switch
4W	Four-Wire
4WAY	Four-Way
F	Fahrenheit
F	Farads
F	Flat
F/D	Face or Field of Drawing
F DR	Fire Door
F TO F	Face-To-Face
F1S	Finish One Side
F2S	Finish Two Sides
FAB	Fabricate
FABL	Fire Alarm Bell
FABX	Fire Alarm Box
FAO	Finish All Over
FB	Face Brick
FB	Flat Bar
FB	Fuse Block
FBCK	Firebrick
FBDC	Fiberboard, Corrugated
FBDS	Fiberboard, Solid
FBM	Feet Board Measure
FBR	Fiber
FBRBD	Fiberboard
FBRS	Fibrous
FC	Fireclay
FC	Flat Cable
FC	Footcandle
FC	Furred Ceiling
FCC	Flat Conductor Cable
FCG	Facing
FD	Fire Damper

Abbreviation	Term	Abbreviation	Term
FD	Floor Drain	FLD	Field
FD	Forced Draft	FLDG	Folding
FDB	Forced-Draft Blower	FLDT	Floodlight
FDFL	Fluid Flow	FLG	Flange
FDN	Foundation	FLG	Flashing
FDPL	Fluid Pressure Line	FLG	Flooring
FDR	Feeder	FLGSTN	Flagstone
FDR	Fire Door	FLH	Flat Head
FDWL	Fiberboard, Double Wall	FLM	Flame
FED	Federal	FLMB	Flammable
FEM	Female	FLMPRF	Flameproof
FEX	Flexible	FLMT	Flush Mount
FEXT	Fire Extinguisher	FLN	Fuel Line
FFILH	Flat Fillister Head	FLR	Filler
FFL	Female Flared	FLRD	Flared
FG	Finish Grade	FLRG	Flaring
FGD	Forged	FLRT	Flow Rate
FH	Fire Hose	FLSW	Floor Switch
FH	Flat Head	FLT	Filter
FHC	Fire Hose Cabinet	FLTG	Floating
FHP	Fractional Horsepower	FLTR	Filter
FHR	Fire Hose Rack	FLUOR	Fluorescent
FHY	Fire Hydrant	FLW	Flat Washer
FIG	Figure	FM	Flow Meter
FIL	Filament	FM	Frequency Modulation
FIL	Fillet	FMAN	Foreman
FIL	Fillister	FNSH	Finish
FILH	Fillister Head	FNSH FL	Finished Floor
FIN	Finish	FORG	Forging
FIR	Fired	FP	Faceplate
FIX	Fixture	FP	Fireplace
FL	Flashing	FP	Flat Point
FL	Flat	FP	Freezing Point
FL	Floor	FPL	Fire Plug
FL	Floor Line	FPM	Feet Per Minute
FL	Flow	FPS	Feet Per Second
FL	Fluid	FPRF	Firepoof
FL	Flush	FR	Frame
FLA	Full Load Amps	FR	From

Abbreviation	Term
FR	Front
FRAC	Fractional
FRAG	Fragmentation
FREQ	Frequency
FRES	Fire-Resistant
FRWK	Framework
FRZR	Freezer
FS	Far Side
FS	Field Specification
FSBL	Fusible
FSC	Full Scale
FSN	Federal Stock Number
FST	Forged Steel
FSTNR	Fastner
FT	Feet
FT	Foot
FT LB	Foot Pound Force
FTG	Fitting
FTG	Footing
FTHRD	Female Thread
FTK	Fuel Tank
FTR	Finned Tube Radiation
FTR	Fixed Transom
FTR	Flat Tile Roof
FU	Fuse
FUBX	Fuse Box
FUHLR	Fuse Holder
FUNL	Funnel
FUR	Furring
FURN	Furnish
FUS	Fuselage
FUT	Future
FV	Front View
FV	Full Voltage
FW	Fire Wall
FW	Fresh Water
FWD	Forward
FWP	Fresh-Water Pump
FX WDW	Fixed Window

Abbreviation	Term
FXD	Fixed
FXTR	Fixture

— G —

Abbreviation	Term
G	Gas
G	Girder
G	Grid
G	Grounded (outlet)
GA	Gauge
GABD	Gauge Board
GAL	Gallon
GAll	Gallery
GALV	Galvanize
GALVI	Galvanized Iron
GALVS	Galvanized Steel
GAR	Garage
GARPH	Graphic
GBG	Garbage
GDR	Guard Rail
GEN	Generator
GEN CONT	General Contractor
GENL	General
GFCI	Ground-Fault Circuit Interrupter
GFU	Glazed Facing Unit
GGL	Ground Glass
GL	Glass
GL	Grade Line
GLB	Glass Block
GLV	Globe Valve
GLZ	Glaze
GND	Ground
GRD	Ground
GNLTD	Granulated
GOVT	Government
GP	General-Purpose
GPC	Gypsum-Plaster Ceiling
GPH	Gallon Per Hour
GPH	Graphite
GPM	Gallon Per Minute

Abbreviation	Term
GPW	Gypsum-Plaster Wall
GR	Gear
GR	Grab Rod
GR	Grade
GR	Grain
GR	Gross
GRAD	Gradient
GRAN	Granite
GRD	Grind
GRDTN	Graduation
GRK	Gear Rack
GRL	Grill
GROM	Grommet
GRTG	Grating
GRV	Groove
GRVD	Grooved
GRVG	Grooving
GRVR	Groover
GRWT	Gross Weight
GSB	Gypsum Sheathing Board
GSFU	Glazed Structural Facing Unit
GSKT	Gasket
GSV	Globe Stop Valve
GSWR	Galvanized Steel Wire Rope
GT	Grease Trap
GTV	Gate Valve
GUT	Gutter
GVL	Gravel
GVW	Gross Vehicle Weight
GWB	Gypsum Wallboard
GWT	Glazed Wall Tile
GYM	Gymnasium
GYP	Gypsum

— H —

Abbreviation	Term
½H	Half-Hard
½RD	Half-Round
H	Hard
H	Henries

Abbreviation	Term
H	High
H GALV	Hot-Galvanize
H&V	Heating and Ventilating
HA	Hot Air
HAZ	Hazardous
HB	Hose Bibb
HBD	Hardboard
HC	High Carbon
HC	Hollow Core
HC	Hose Clamp
HCL	Horizontal Center Line
HCS	High Carbon Steel
HCSHT	High Carbon Steel, Heat-Treated
HD	Hand (comb form)
HD	Hard
HD	Hard-Drawn
HD	Head
HD	Heavy-Duty
HDDRN	Hard-Drawn
HDL	Handle
HDLS	Headless
HDN	Harden
HDR	Header
HDW	Hardware
HDWD	Hardwood
HE	Heat Exchange
HEX	Hexagon
HEX HD	Hexagonal Head
HF	High Frequency
HGBN	Herringbone
HGR	Hanger
HGT	Height
HI HUM	High Humidity
HIMP	High Impact
HINT	High Intensity
HLCL	Helical
HM	Hollow Metal
HMD	Humidity
HNDRL	Hand Rail

Abbreviation	Term
HNDWL	Handwheel
HNG	Hanging
HNG	Hinge
HNYCMB	Honeycomb
HOA	Hand-Off-Auto
HOL	Hollow
HOR	Horizontal
HP	High Pressure
HP	Horsepower
HPS	High-Pressure Steam
HPT	High Point
HR	Hook Rail
HR	Hose Rack
HR	Hot Rolled
HR	Hour
HRS	Hot Rolled Steel
HS	High Speed
HSG	Housing
HSS	High-Speed Steel
HST	Hoist
HSTH	Hose Thread
HT	Heat
HT	High Tension
HT	Hollow Tile
HT RES	Heat-Resisting
HT SHLD	Heat Shield
HT TR	Heat Treat
HTCI	High-Tensile Cast Iron
HTD	Heated
HTG	Heating
HTNSL	High Tensile
HTR	Heater
HTS	High Tensile Strength
HTS	High-Tensile Steel
HV	High Voltage
HVAC	Heating/Ventilating/Air Conditioning
HVR	High-Voltage Regulator
HVRNG	Hovering
HVY	Heavy

Abbreviation	Term
HW	Hot Water
HWL	High-Water Line
HWT	Hot Water Tank
HWY	Highway
HYD	Hydraulic
HYDR	Hydraulic
HYDRELC	Hydroelectric
HYDRO	Hydrostatic
HYDTD	Hydrated
HZ	Hertz

— I —

I	Invert
I	Iron
I&O	Inlet and Outlet
IACS	International Copper Standard
IC	Intergrated Circuit
ICS	International Control Station
ICTCD	Insecticide
ID	Induced Draft
ID	Inside Diameter
ID	Internal Diameter
IDCTR	Inductor
IDL	Idler
IF	Intermediate Frequency
IGN	Ignition
ILLUM	Illumination
IMP	Impedance
IMPG	Impregnate
IN	Inch(es)
IN LB	Inch-Pound
INBD	Inboard
INC	Incorporated
INCAND	Incandescent
INCL	Include; inclusive
INCOMP	Incomplete
INCR	Increase
IND	Indicate
IND	Industry

Abbreviation	Term
INL	Inlet
INRT	Inert
INRTG	Inert Gas
INS	Inside
INSP	Inspect
INSR	Insert
INSTL	Install; Installation
INSUL	Insulate; Insulation
INT	Integral
INT	Intensifier
INTCHG	Interchangeable
INTERCOM	Intercommunication
INTL	Internal
INTMD	Intermediate
INTMT	Intermittent
INTR	Interior
INTRPT	Interrupt
IP	Iron Pipe
IPS	Inch(es) Per Second
IPS	International Pipe Standard
IPS	Iron Pipe Size
IPT	Internal Pipe Thread
IPT	Iron Pipe Thread
IR	Infrared
IR	Inside Radius
IRREG	Irregular
ISO	Isometric
ISS	Issue
IW	Indirect Waste

— J —

Abbreviation	Term
J	Joiner or Joist
J&P	Joists and Planks
JAP	Japanned
JB	Jamb
JC	Janitor's Closet
JCT	Junction
JCTBX	Junction Box
JFET	Junction Field-effect Transistor

Abbreviation	Term
JK	Jack
JKSCR	Jackscrew
JMB	Jamb
JNL	Journal
JO	Job Order
JOUR	Journeyman
JR	Junior
JT	Joint

— K —

Abbreviation	Term
K	Keel
K	Key
K	Kilohm
K	Kip (1000 lb)
KAL	Kalamein
KALD	Kalamein Door
KB	Knee Brace
KC	Kilocycle
KD	Kiln-Dried
KD	Knocked Down
KINE	Kinescope
KIT	Kitchen
KN	Knot
KN SW	Knife Switch
KO	Knockout
KPL	Kickplate
KST	Keyseat
KV	Kilovolt
KVAH	Kilovolt-Ampere Hour
KVAHM	Kilovolt-Ampere Hour Meter
KVAM	Kilovolt-Ampere Meter
KWH	Kilowatt Hour
KWY	Keyway

— L —

Abbreviation	Term
L	Lavatory
L	Left
L	Left-Hand Position

Abbreviation	Term	Abbreviation	Term
L	Line	LK WASH	Lock Washer
L	Long	LKG	Locking
LA	Lightning Arrester	LKNT	Locknut
LAB	Laboratory	LL	Live Load
LAD	Ladder	LM	List of Materials
LAG	Lagging	LNG	Lining
LAM	Laminate	LNTL	Lintel
LAQ	Lacquer	LOA	Length Over All
LAT	Lateral	LOC	Locate
LAT	Leaving Air Temperature	LOS	Line-of-Sight
LAU	Laundry	LP	Low Pressure
LAV	Lavatory	LPG	Lapping
LBL	Label	LPW	Lumen Per Watt
LBR	Lumber	LR	Ladder Rung
LBRY	Library	LR	Living Room
LC	Laundry Chute	LRA	Locked Rotor Amps
LC	Lead Covered	LS	Left Side
LC	Low Carbon	LS	Limestone
LCH	Latch	LS	Loudspeaker
LCL	Linen Closet	LS	Low Speed
LCL	Local	LT	Laundry Tray
LDG	Landing	LT	Light
LDR	Leader	LT	Low Tension
LE	Leading Edge	LTC	Lattice
LF	Linoleum Floor	LTEMP	Low Temperature
LF	Low Frequency	LTH	Lath
LG	Length	LTQ	Low Torque
LG	Long	LTR	Letter
LGE	Large	LTSW	Light Switch
LH	Laten Heat	LUB	Lubricate; Lubricator
LH	Left Hand	LV	Low Voltage
LHR	Latent Heat Ratio	LVL	Level
LIM	Limit	LWC	Lightweight Concrete
LIM SW	Limit Switch	LWIC	Lightweight Insulating Concrete
LIN	Linear	LWR	Lower
LINOL	Linoleum	LWST	Lowest
LIQ	Liquid	LYR	Layer
LITHO	Lithograph		
LK	Link		

Abbreviation	Term
	— M —
M	Magnaflux
M	Meter (Measure of Length)
M	Meter (Instrument)
M&F	Male & Female
MA	Master
MA	Metal Anchor
MA	Mixed Air
MACH	Machine
MAG	Magnesium
MAG	Magnetic
MAH	Mahogany
MAINT	Maintenance
MAL	Malleable
MALL	Malleable
MAN	Manual
MAN OVLD	Manual Overload
MANF	Manifold
MAR	Marine
MAS	Metal Anchor Slots
MASU	Machined Surface
MAT	Matrix
MATL	Material
MATW	Metal Awning-Type Window
MAX	Maximum
MB	Mailbox
MB	Model Block
MBH	Thousand British Thermal Units
MBL	Mobile
MBR	Member
MC	Manhole Cover
MC	Medicine Cabinet
MC	Megacycle
MC	Metal-clad Cable
MCD	Metal Covered Door
MCHRY	Machinery
MDC	Motor Direct-Connected

Abbreviation	Term
MDL	Middle
MDM	Maximum Design Meter
MDM	Medium
MDN	Median
MDRL	Mandrel
MEAS	Measure
MECH	Mechanical
MED	Median
MED	Medical
MEG	Megohm
MEL	Melamine
MEMB	Membrane
MEMO	Memorandum
MER	Meridian
MET	Metal
METB	Metal Base
METD	Metal Door
METF	Metal Flashing
METG	Metal Grill
METJ	Metal Jalousie
METP	Metal Partition
METR	Metal Roof
METS	Metal Strip
MEZZ	Mezzanine
MF	Mastic Floor
MF	Metered Flow
MFD	Manufactured
MFG	Manufacturing
MFR	Manufacture
MFSFU	Matt-Finish Structural Facing Units
MG	Magnetic Armature
MGL	Mogul
MH	Manhole
MH	Millihenries
MHD	Masthead
MI	Malleable Iron
MI	Mile
MI	Mineral-insulated Cable
MIC	Micrometer

Abbreviation	Term
MIL	Military
MIN (')	Minute
MIN	Minimum
MIN	Minor
MIR	Mirror
MISC	Miscellaneous
MIT	Miter
MIX	Mixture
MJ	Mastic Joint
MK	Mark
MKR	Marker
MKUP	Makeup
ML	Material List
ML	Mold Line
ML	Monolithic
MLD	Molded
MLDG	Molding
MLP	Metal Lath and Plaster
MM	Millimeter
MN	Main
MNL OPR	Manually Operated
MNRL	Mineral
MO	Masonry Opening
MO	Month
MO	Motor-operated
MOD	Model
MOD	Modification
MOD	Modify
MON	Monument
MOR	Mortar
MOR T	Morse Taper
MOS	Mosaic
MOT	Motor
MP	Melting Point
MPE	Maximum Permissible Exposure
MPG	Miles per Gallon
MPH	Miles per Hour
MPT	Male Pipe Thread
MR	Marble

Abbreviation	Term
MRD	Metal Rolling Door
MRF	Marble Floor
MRT	Marble Threshold
MRT	Mildew-Resistant Thread
MS	Machine Steel
MSCR	Machine Screw
MSL	Mean Sea Level
MSNRY	Masonry
MSR	Mineral-Surface Roof
MST	Machine Steel
MSTC	Mastic
MSW	Master Switch
MT	Maximum Torque
MT	Metal Threshold
MT	Mount
MTCHD	Matched
MTD	Mounted
MTG	Mounting
MTHRD	Male Threaded
MTLC	Metallic
MTRDN	Motor Driven
MTRS	Mattress
MTWF	Metal Through-Wall Flashing
MTZ	Motorized
MULT	Multiple
MVBL	Movable
MWG	Music Wire Gage
MWO	Modification Work Order
MWP	Maximum Working Pressure
MWP	Membrane Waterproofing
MWV	Maximum Working Voltage
MXT	Mixture

— N —

Abbreviation	Term
N	Noon
N	North
NA	Naval Architect
NA	Not Applicable

Abbreviation	Term
NAR	Narrow
NAS	National Aircraft Standards
NAT	Natural
NATL	National
NBS	National Bureau of Standards
NBS	New British Standard (Imperial Wire Gage)
NC	National Coarse (thread)
NC	Normally Closed
NCH	Notched
NCM	Noncorrosive Metal
NCOMBL	Noncombustible
NEC	National Electrical Code
NEF	National Extra Fine (thread)
NEG	Negative
NESC	National Electrical Safety Code
NEUT	Neutral
NF	National Fine (thread)
NF	Near Face
NIC	Not In Contract
NIP	Nipple
NM	Nonmetallic
NMAG	Nonmagnetic
NO	Normally Open
NO	Number
NOM	Nominal
NONFLAMB	Nonflammable
NORM	Normal
NOS	Nosing
NOZ	Nozzle
NP	Name Plate
NP	National Pipe Thread
NPL	Nameplate
NPRN	Neoprene
NPT	National Taper Pipe (thread)
NRCP	Nonreinforced-Concrete Pipe
NS	National Special (thread)
NS	Near Side
NST	Nonslip Tread
NTS	Not to Scale

Abbreviation	Term
NTWT	Net Weight
NYL	Nylon

— O —

OA	Outside Air
OA	Overall
OB	Obscure
OBJV	Objective
OBS	Obsolete
OBW	Observation Window
OC	On Center
OC	Outside Circumference
OC	Overcurrent
OCB	Oil Circuit Breaker
OCC	Occupy
OCLD	Oil-Cooled
OCR	Overcurrent Relay
OCT	Octagon
OD	Outside Diameter
OE	Open End
OF	Outside Face
OFCE	Office
OFF	Office
OGL	Obscure Glass
OHM	Ohmmeter
OI	Oil-Insulated
OP	Operator
OPA	Opaque
OPN	Operation
OPNG	Opening
OPNL	Operational
OPP	Opposite
OPR	Operate
OPRT	Operator Table
OPT	Optical
OPT	Optimum
OR	Outside Radius
ORD	Ordinance

Abbreviation	Term
ORIG	Origin
ORIG	Original
ORLY	Overload Relay
OSC	Oscillator
OSP	Operating Steam Pressure
OUT	Outlet
OUT	Output
OUT	Outside
OUTBD	Outboard
OV	Outlet Velocity
OV	Over
OVFL	Overflow
OVH	Oval Head
OVHD	Overhead
OVHG	Overhanging
OVHL	Overhaul
OVLD	Overload
OVRD	Override
OWJ	Open Web Joist
OWU	Open Window Unit
OXD	Oxidized
OXY	Oxygen
OZ	Ounce

— P —

Abbreviation	Term
P	Page
P	Pilaster
P	Pitch
P	Plate (electron tube)
P	Pole
P	Porch
P	Port
P	Pump
P-P	Peak-To-Peak
P&O	Paints and Oils
P&T	Posts and Timbers
PA	Pressure Angle
PAN	Pantry

Abbreviation	Term
PANB	Panic Bolt
PAR	Parallel
PARA	Paragraph
PART	Partial
PASS	Passage
PAT	Patent
PATT	Pattern
PB	Painted Base
PB	Pull Box
PB	Pushbutton
PB SW	Pull-Button Switch
PBD	Paperboard
PBD	Pressboard
PC	Personal Carrier
PC	Piece
PC	Pitch Circle
PC	Pressure Controlled
PC MK	Piece Mark
PCC	Point Of Compound Curve
PCH	Punch
PD	Pitch Diameter
PD	Pivoted Door
PD	Potential Difference
PD	Pressure Drop
PED	Pedestal
PEIM	Perimeter
PEMB	Permeability
PEN	Penetration
PERF	Perfect
PERF	Perforate
PERM	Permanent
PERP	Perpendicular
PF	Power Factor
PF	Profile
PF	Pump, Fixed Displacement
PFD	Preferred
PFN	Prefinished
PG	Pressure Gauge
PGMT	Pigment

Abbreviation	Term	Abbreviation	Term
PH	Phase	PMU	Plaster Mockup
PH BRZ	Phosphor Bronze	PNEU	Pneumatic
PHEN	Phenolic	PNH	Pan Head
PHH	Phillips Head	PNL	Panel
PHR	Preheater	PNT	Paint
PI	Point Of Intersection	PNTGN	Pentagon
PIN	Pinion	PO	Production Order
PK	Pack	POB	Point Of Beginning
PK	Peck	POL	Polish
PKG	Packing	POLYEST	Polyester
PKT	Pocket	POP	Popping
PKWY	Parkway	PORC	Porcelain
PL	Padlock	PORT	Portable
PL	Place	POS	Position
PL	Plain	POSN SW	Position Switch
PL	Plate	POT	Potential
PL	Plug	POT	Potentiometer
PL	Property Line	POTW	Potable Water
PLAS	Plaster	PP	Panel Point
PLATF	Platform	PP	Piping
PLB	Pull Button	PP	Push-Pull
PLBLK	Pillow Block	PPLN	Pipeline
PLD	Plated	PR	Pair
PLG	Piling	PR	Pipe Rail
PLGL	Plate Glass	PRCST	Precast
PLGR	Plunger	PREFAB	Prefabricated
PLK	Plank	PREP	Prepare
PLMB	Plumbing	PRESS	Pressure
PLR	Pillar	PRI	Primary
PLR	Pliers	PRIM	Primary
PLRT	Polarity	PRL	Parallel
PLSTC	Plastic	PRM	Priming
PLT	Pilot	PRMR	Primer
PLTG	Plating	PROC	Process
PLVRZD	Pulverized	PROD	Production
PLYWD	Plywood	PROJ	Project
PM	Phase Modulation	PROP	Propelling
PM	Pulse Modulation	PROP	Property
PMP	Pump	PROP	Proposed

Abbreviation	Term
PROT	Protective
PRS	Press
PRSD MET	Pressed Metal
PRV	Pressure-Reducing Valve
PS	Polystyrene
PS	Pressure Switch
PS	Pull Switch
PSI	Pounds per Square Inch
PSL	Pipe Sleeve
PSU	Power Supply Unit
PSVTV	Preservative
PSW	Potential Switch
PT	Part
PT	Pint
PT	Pipe Tap
PT	Point
PTD	Painted
PTN	Partition
PU	Pick Up
PUB	Publication
PUBN	Publication
PUL	Pulley
PUR	Purchase
PV	Plan View
PV	Pump, Variable Displacement
PVC	Polyvinyl Chloride
PVT	Pivot
PW	Plain Washer
PW	Projected Window
PWM	Pulse-Width Modulation
PWR	Power
PWR SPLY	Power Supply

— Q —

Abbreviation	Term
QA	Quick-Acting
QC	Quality Control
QDRNT	Quadrant
QRY	Quarry

Abbreviation	Term
QT	Quarry Tile
QT	Quart
QTB	Quarry-Tile Base
QTF	Quarry-Tile Floor
QTR	Quarry Tile Roof
QTR	Quarter
QTY	Quantity
QUAD	Quadrant
QUAL	Quality
QUOT	Quotation

— R —

Abbreviation	Term
R	Radius
R	Resistance
R	Right
R	Riser
RA	Return Air
RAACT	Radioactive
RAB	Rabbet
RACT	Reverse-Acting
RAD	Radial
RAD	Radius
RB	Roller Bearing
RB	Rubber Base
RBN	Ribbon
RBR	Rubber
RC	Rate of Change
RC	Reinforced Concrete
RCD	Reverse-Current Device
RCHT	Ratchet
RCP	Reinforced-Concrete Culvert Pipe
RCP	Reinforced-Concrete Pipe
RCPT	Receptacle
RCPTN	Reception
RCS	Remote Control System
RCVG	Receiving
RD	Road
RD	Roof Drain

Abbreviation	Term	Abbreviation	Term
RD	Root Diameter	REQD	Required
RD	Round	REQT	Requirement
RDC	Reduce	RES	Resistor
RDCR	Reducer	RESIL	Resilient
RDG	Ridge	RET	Retard
RDH	Round Head	RET	Return
RDL	Radial	REV	Reverse
RDTR	Radiator	REV	Revise
RE	Reel	REV	Revolution
REAC	Reactive	RF	Radio Frequency
REASM	Reassemble	RF	Raised Face
REC	Recess	RF	Roof
REC	Record	RFG	Roofing
RECD	Received	RFGT	Refrigerant
RECHRG	Recharger	RFRC	Refractory
RECIRC	Recirculate	RGD	Rigid
RECP	Receptacle	RGH	Rough
RECT	Rectangle	RGLTR	Regulator
RECT	Rectifier	RGNG	Rigging
RED	Reduce	RGTR	Register
RED	Reducer	RH	Relative Humidity
REDWN	Redrawn	RH	Right Hand
REF	Reference	RH	Rockwell Hardness
REF L	Reference Line	RHC	Reheat Coil
REFL	Reference Line	RHEO	Rheostat
REFLD	Reflected	RI	Reflective Insulation
REFR	Refrigerator	RIB	Ribbed
REG	Register	RINSUL	Rubber Insulation
REG	Regulator	RIV	Rivet
REINF	Reinforce	RL	Roof Leader
REL	Relay	RLD	Rolled
REL	Relief	RLF	Relief
REM	Remove	RLG	Railing
REM COV	Removable Cover	RLR	Roller
REN	Renewable	RM	Ream
REPL	Replace	RM	Room
REPRO	Reproduce	RMR	Reamer
REQ	Require	RMS	Root Mean Square
REQ	Requisition	RND	Round

Abbreviation	Term
RNDM	Random
RNG	Range
RNWBL	Renewable
RO	Rough Opening
RPM	Revolution Per Minute
RPQ	Request for Price Quotation
RPS	Revolution Per Second
RPVNTV	Rust Preventative
RR	Railroad
RR	Roll Roofing
RS	Rough Sawn
RS	Rubble Stone
RSD	Raised
RST	Reinforcing Steel
RSTPF	Rustproof
RT	Raintight
RTANG	Right Angle
RTD	Retard
RTF	Rubber-Tile Floor
RTL	Reinforced Tile Lintel
RTN	Return
RTNG	Retaining
RTR	Rotor
RTTL	Rattail
RV	Rear View
RV	Relief Valve
RVA	Reactive Volt-Ampere Meter
RVLG	Revolving
RVLV	Revolve
RVM	Reactive Voltmeter
RVS	Reverse
RVSBL	Reversible
RVT	Rivet
RW	Right-Of-Way
RWC	Rainwater Conductor
RWD	Redwood
RY	Railway

— S —

Abbreviation	Term
S	Scuttle
S	Side
S	Sink
S	Soft
S	South
S CHG	Supercharge
S1S	Surfaced or Dressed One Side
S1S1E	Surfaced or Dressed One Side And one Edge
S2S	Surfaced or Dressed Two Sides
S4S	Surfaced or Dressed Four Sides
SA	Stress Anneal
SA	Supply Air
SAC	Sprayed Acoustical Ceiling
SAF	Safety
SAN	Sanitary
SAPC	Suspended Acoustical-Plaster Ceiling
SAT	Saturate
SATC	Suspended Acoustical-Tile Ceiling
SB	Sleeve Bearing
SB	Splash Block
SB	Stove Bolt
SBW	Steel Basement Window
SC	Scale
SC	Sill Cock
SC	Smooth Contour
SC	Solid Core
SCC	Single Conductor Cable
SCD	Screen Door
SCDR	Screwdriver
SCH	Schedule
SCH	Socket Head
SCHED	Schedule
SCHEM	Schematic
SCR	Screw
SCR GT	Screen Gate
SCRN	Screen

Abbreviation	Term
SCSH	Structural Carbon Steel, Hard
SCSM	Structural Carbon Steel, Medium
SCSS	Structural Carbon Steel, Soft
SCT	Structural Clay Tile
SCUP	Scupper
SD	Shower Drain
SDG	Siding
SDL	Saddle
SE	Service-Entrance Cable
SE	Single-End
SE	Special Equipment
SEC	Second
SEC	Secondary
SECT	Section
SEJ	Sliding Expansion Joint
SEL	Select
SELF CL	Self-Closing
SEP	Separate
SEP	Separator
SER	Serial
SERR	Serrate
SERV	Servo
SEW	Sewage
SEW	Sewer
SEW	Sewer
SF	Semi-Finished
SF	Soffit
SF	Spot Faced
SFB	Solid Fiberboard
SFT	Shaft
SFXD	Semi-Fixed
SG	Structural Glass
SGL	Single
SGSFU	Salt-Glazed Structural Facing Units
SGW	Security Guard Window
SH	Shackle
SH	Sheet
SH	Sheeting
SH	Sensible Heat

Abbreviation	Term
SH	Shower
SH & T	Shower and Toilet
SH&RD	Shelf and Rod
SHELV	Shelving
SHGL	Shingle
SHK	Shake
SHK	Shank
SHL	Shell
SHL	Shellac
SHLD	Shielding
SHLD	Shoulder
SHLDR	Shoulder
SHORT	Short Circuit
SHRD	Shroud
SHTHG	Sheathing
SHTR	Shutter
SHV	Sheave
SI LT	Side Light
SIF	Single-Face
SIG	Signal
SILS	Silver Solder
SIM	Similar
SJ	Slip Joint
SK	Sink
SK	Sketch
SKT	Skirt
SKT	Socket
SL	Sea Level
SL	Sliding
SLD	Sealed
SLD	Sliding Door
SLDR	Solder
SLDR	Soldering
SLFCLN	Self-Cleaning
SLFLKG	Self-Locking
SLFSE	Self-Sealing
SLFTPG	Self-Tapping
SLOT	Slotted
SLP	Slope

Abbreviation	Term
SLT	Skylight
SLT	Spotlight
SLT or S	Slate
SLV	Sleeve
SLVT	Solvent
SM	Sheet Metal
SM	Small
SMLS	Seamless
SNDPRF	Soundproof
SNM	Shielded Nonmetallic Cable
SO	Shop Order
SOC	Socket
SOL	Solenoid
SOL PLT	Solenoid Controlled, Pilot Operated
SOV	Shut Off Valve
SOV	Shutoff Valve
SP	Static Pressure
SP	Split Phase
SP	Shear Plate
SP	Single Pole
SP	Soil Pipe
SP	Space
SP	Spare
SP	Special Purpose
SP	Specific
SP	Speed
SP	Splashproof
SP	Standpipe
SP GR	Specific Gravity
SP SW	Single-Pole Switch
SPC	Suspended Plaster Ceiling
SPCL	Special
SPCR	Spacer
SPDT	Single-Pole Double-Throw
SPDT SW	Single-Pole Double-Throw Switch
SPEC	Specification
SPG	Spring
SPH	Space Heater

Abbreviation	Term
SPHER	Spherical
SPHN	Siphon
SPK	Spike
SPL	Special
SPL	Spiral
SPLC	Splice
SPLN	Spline
SPLY	Supply
SPNSN	Suspension
SPR	Sprinkler
SPRDR	Spreader
SPST	Single-Pole Single-Throw
SPST SW	Single-Pole Single-Throw Switch
SQ	Square
SQ CG	Squirrel Cage
SQ FT	Square Foot
SQ IN	Square Inch
SQ YD	Square Yard
SQH	Square Head
SR	Slip Ring
SR	Split Ring
SRPR	Scraper
SS	Semi-Steel
SS	Service Sink
SS	Set Screw
SS	Slop Sink
SSAC	Suspended Sprayed Acoustical Ceiling
SSBR	Smooth-Surface Built-Up Roof
SSCR	Set Screw
SSD	Subsoil Drain
SSG	Single Strength Glass
SSK	Soil Stack
SST	Stainless Steel
SSTU	Seamless Steel Tubing
ST	Single-Throw
ST	Stairs
ST	Street
ST PR	Static Pressure
STA	Station

Abbreviation	Term	Abbreviation	Term
START	Starter	SUPVR	Supervisor
STBD	Starboard	SURF	Surface
STD	Standard	SURV	Survey
STDF	Standoff	SUSP	Suspend
STGR	Stringer	SV	Safety Valve
STIF	Stiffener	SVCE	Service
STIR	Stirrup	SW	Short Wave
STK	Stack	SW	Spot-Weld
STK	Stock	SW	Switch
STL	Steel	SW	Switched
STN	Stained	SWBD	Switchboard
STN	Stainless	SWG	Swage
STN	Stone	SWGD	Swinging Door
STNLS	Stainless	SWGR	Switchgear
STOR	Storage	SWP	Safe Working Pressure
STP	Strip	SWSG	Security Window Screen and Guard
STR	Straight	SWT	Sweat
STR	Strainer	SYM	Symbol
STR	Strength	SYM	Symmetrical
STR	Strip	SYN	Synchronous
STR	Structural	SYN	Synthetic
STRBK	Strongback	SYS	System
STRL	Structural		
STRUCT	Structure		
STS	Special Treatment Steel		**— T —**
STW	Storm Water		
STWY	Stairway	T	T-Bar
SUB	Substitute	T	Tee
SUBMG	Submergence	T	Teeth
SUBST	Substitute	T	Time
SUBSTR	Substructure	T	Toilet
SUM	Summary	T	Tooth
SUP	Supply	T	Top
SUPERSTR	Superstructure	T	Truss
SUPP	Supplement	T&B	Top and Bottom
SUPRSTR	Superstructure	T&BB	Top and Bottom Bolt
SUPSD	Supersede	T&G	Tongue and Groove
SUPT	Superintendent	TAB	Tabulate
SUPV	Supervise	TACH	Tachometer
		TAN	Tangent
		TARP	Tarpaulin

Abbreviation	Term
TB	Tile Base
TBE	Thread Both Ends
TBG	Tubing
TBLR	Tumbler
TC	Terra Cotta
TC	Thermocouple
TC	Thread Cutting
TC	Top Chord
TC	Tray Cable
TCH	Temporary Construction Hole
TD	Tile Drain
TD	Time Delay
TDM	Tandem
TE	Trailing Edge
TECH	Technical
TEL	Telephone
TELB	Telephone Booth
TEM	Temper
TEMP	Temperature
TEMP	Template
TEMP	Temporary
TEMPL	Template
TENS	Tension
TER	Terazzo
TERB	Terazzo Base
TERM	Terminal
TEV	Thermal Expansion Valve
TF	Tile Floor
TFR	Top of Frame
TGL	Toggle
TH	Threshold
TH	Toilet-Paper Holder
TH	Total Heat
THD	Thread
THERMO	Thermostat
THK	Thick
THKNS	Thickness
THRM	Thermal
THRT	Throat

Abbreviation	Term
THRU	Through
TIR	Total Indicator Reading
TKL	Tackle
TLLD	Total Load
TLPC	Tailpiece
TM	Technical Manual
TMBR	Timber
TMPD	Tempered
TNG	Tongue
TNG	Training
TNL	Tunnel
TNSL	Tensile
TNSN	Tension
TOB BRZ	Tobin Bronze
TOL	Thermal Overload
TOL	Tolerance
TOPG	Topping
TOT	Total
TPG	Tapping
TPI	Teeth per Inch
TPI	Threads per Inch
TPL	Triple
TPLW	Triple Wall
TPR	Taper
TR	Technical Report
TR	Towel Rack or Rod
TR	Transom
TR	Truss
TRANS	Transfer
TRANS	Transformer
TRANS	Transparent
TRANS	Transporation
TRANSV	Transverse
TRAV	Traversing
TRH	Truss Head
TRK	Track
TRLR	Trailer
TRMR	Trimmer
TRNBKL	Turnbuckle

Abbreviation	Term
TRND	Turned
TRNGL	Triangle
TRQ	Torque
TRSN	Torsion
TRTD	Treated
TRX	Triplex
TS	Tensile Strength
TS	Tip Speed
TS	Tool Steel
TSR	Tile-Shingle Roof
TT	Tile Threshold
TUB	Tubing
TV	Television
TW	Tile Wainscot
TW	Twisted
TYP	Typical

— U —

Abbreviation	Term
U	Urinal
U&L	Upper and Lower
UC	Undercut
UCMT	Unglazed Ceramic Mosaic Tile
UGND	Underground
UH	Unit Heater
UHF	Ultra-High Frequency
UHPFB	Untreated Hard-Pressed Fiberboard
ULT	Ultimate
UN	Union
UNC	Unified Coarse Thread
UND	Under
UNEF	Unified Extra Fine Thread
UNEX	Unexacavated
UNEXC	Unexcavated
UNF	Unified Fine Thread
UNFIN	Unfinished
UNIV	Universal
UNS	Unified Special Thread
UOS	Unless Otherwise Specified

Abbreviation	Term
UF	Underground Feeder Cable
UPDFT	Updraft
UPR	Upper
UR	Urinal
UR	Utility Room
URWC	Urinal Water Closet
US	Undersize
US	Utility Set
USASI	United States of America Standards Institute
USE	Underground Service-entrance Cable
USFU	Unglazed Structural Facing Units
USG	United States Gage
USS	United States Standard
USUB	Unglazed Structural Unit Base
UTIL	Utility
UTRTD	Untreated
UV	Unit Ventilator
UV	Ultraviolet
UWTR	Underwater

— V —

Abbreviation	Term
V	Valve
V	Velocity
V	Vent
V	Volt
V	Voltage
VAC	Vacuum
VAL	Valley
VAPPRF	Vapor Proof
VAR	Variance
VAR	Variation
VARN	Varnish
VBR	Vibration
VC	Vitrified Clay
VCP	Vitreous Clay Pipe
VCT	Vitrified Clay Tile
VD	Vandyke

Abbreviation	Term
VD	Void
VD	Voltage Drop
VE	Ventilating Equipment
VENT	Ventilate
VENT	Ventilator
VERN	Vernier
VERT	Vertical
VEST	Vestibule
VF	Video Frequency
VH	Vent Hole
VHF	Very-High Frequency
VIB	Vibrate
VIT	Vitreous
VLT	Volute
VM	Volt per Meter
VM	Voltmeter
VOL	Volume
VP	Velocity Pressure
VP	Vent Pipe
VR	Voltage Regulator
VRLY	Voltage Relay
VS	Vent Stack
VS	Voltmeter Switch
VSBL	Visible
VSTM	Valve Stem
VT	Vacuum Tube
VT	Vaportight
VT or VTILE	Vinyl Tile
VTR	Vent Through Roof
VTVM	Vacuum Tube Voltmeter
VULC	Vulcanize

— W —

Abbreviation	Term
W	Wall
W	Waste
W	Watt
W	West
W	Wide

Abbreviation	Term
W	Width
W	Wire
W/	With
WA	Wainscot
WA	Warm Air
WASH	Washer
WATW	Wood Awning-Type Window
WB	Wet Bulb Temperature
WB	Wheel Base
WB	Wood Base
WBF	Wood Block Floor
WBL	Wood Blocking
WC	Water Closet
WCW	Wood Casement Window
WD	Width
WD	Wood
WD	Wood Door
WDF	Wood Door and Frame
WDF	Woodruff
WDO	Window
WDP	Wood Panel
WF	Wide Flange
WFS	Wood Furring Strips
WG	Window Guard
WG	Wire Gauge
WGL	Wire Glass
WH	Wall Hydrant
WH	Water Heater
WH	Watthour
WH	Weep Hole
WHSE	Warehouse
WI	Wrought Iron
WIC	Walk-In Closet
WJ	Wood Jalousie
WK	Week
WL	Water Line
WL	Wide Load
WLB	Wallboard
WLD	Welded

Abbreviation	Term
WLDR	Welder
WLDS	Weldless
WM	Washing Machine
WM	Water Meter
WM	Wire Mesh
WMGR	Warm Gear
WN	Winch
WNDR	Winder
WO	Without
WP	Waste Pipe
WP	Weatherproof (insul)
WP	White Pine
WPFC	Waterproof Fan-Cooled
WPG	Waterproofing
WPR	Working Pressure
WR	Wall Receptacle
WR	Washroom
WR	Water Resistant
WR	Weather-Resistant
WR	Wrench
WRT	Wrought
WS	Waste Stack
WS	Weather Stripping
WSHR	Washer
WSL	Weather Seal
WSP	Working Steam Pressure
WSR	Wood-Shingle Roof
WT	Weight
WT	Wood Threshold
WTHPRF	Weatherproof
WTR	Water
WTRPRF	Waterproof

Abbreviation	Term
WTRTT	Watertight
WU	Window Unit
WV	Wall Vent
WV	Working Voltage
WW	Wireway
WWF	Welded Wire Fabric

— X —

X HVY	Extra Heavy
X STR	Extra Strong
XDCR	Transducer
XFMR	Transformer
XSECT	Cross Section
XSTR	Transistor
XX STR	Double Extra Strong

—Y—

YD	Yard
YR	Year
YP	Yellow Pine

Z

Z	Zone

Chapter 2

Basic Materials and Methods

A wide range of electrical equipment and components are covered in this chapter. The basic items include conduits, conductors, outlet boxes, wiring devices, fittings, connectors, apparatus, and the like.

Electrical symbols (*see* Chapter 1) are normally used to indicate these items on construction working drawings. Electrical-detail drawings are required only when a special condition exists or where the architect, engineer, or owner requires an electrical component to be installed in a specific manner or in an unusual location.

AC MOTORS

Specifying a motor for new applications or identifying a motor for replacement purposes can be done easily if the correct information is known. This includes:

- Nameplate data
- Mechanical characteristics
- Motor types
- Electrical characteristics and connections

Much of this information consists of standards defined by the National Electrical Manufacturers Association (NEMA). These standards are widely used throughout North America and many of these standards — in tabular form — are included in this chapter for reference. In other parts of the world, the standards of the International Electromechanical Commission (IEC) are most often used.

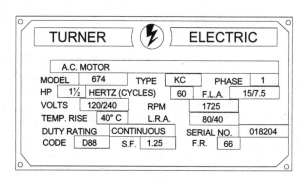

Figure 2-1: Typical motor nameplate.

Nameplate

Nameplate data is the critical first step in determining motor replacement. Much of the information needed can generally be obtained from the nameplate. Record all nameplate information; it can save time and confusion.

A typical motor nameplate is shown in Figure 2-1. Note that the manufacturer's name and logo is at the top of the plate; the placement of each item, however, will change with each manufacturer. The line directly below the manufacture's name identifies the motor for use on ac systems as opposed to dc or ac-dc systems. The model number identifies the particular motor from any other. The type or class specifies the insulation used to ensure that the motor will perform at the rated horsepower and service-factor load. The phase indicates whether the motor has been designed for single- or three-phase service.

Horsepower (hp) on the nameplate defines the rated output capacity of the motor; hertz (cycles) indicates the ac frequency at which the motor is designed to operate. The F.L.A. section gives the amperes of current the motor draws at full load. When two values are shown on the nameplate, the motor usually has a dual voltage rating. Volts and amps are inversely proportional; the higher the voltage, the lower the amperes, and vice versa. The higher amp value corresponds to the lower voltage rating on the nameplate. Two-speed motors will also show two ampere ratings.

The term "Volts" on the nameplate is the electrical potential "pressure" for which the motor is designed to operate. Sometimes two voltages are listed on the nameplate, such as 120/240. In this case the motor is intended for use on either a 120- or 240-V circuit. Special instructions are furnished for connecting the motor for each of the different voltages.

The rpm inscription represents revolutions per minute; that is, the motor speed. The rpm reading on motors is the approximate full-load speed. Temp. rise designates the maximum air temperature immediately surrounding the motor. Forty degrees centigrade is the NEMA maximum ambient temperature.

L.R.A. stands for locked rotor amps. It relates to starting current and selection of fuse or circuit breaker size. When two values are shown, the motor usually has a dual-voltage rating. The duty rating designates the duty cycle of a motor. "Continuous" means that the motor is designed for around-the-clock operation.

Each motor is usually given a different serial number for identification and tracking purposes. The code designation is a serial data code used by the manufacturer. In Figure 2-1, the first letter identifies the month and the last two numbers identify the year of manufacture (D88 is April 88).

A service factor (S.F.) is a multiplier which, when applied to the rated horsepower, indicates a permissible horsepower loading which may be carried continuously when the voltage and frequency are maintained at the value specified on the nameplate, although the motor will operate at an increased temperature rise.

The frame (F.R.) designation specifies the shaft height and motor mounting dimensions and provides recommendations for standard shaft diameters and usable shaft extension lengths.

Besides the information just mentioned, many larger motors also contain plates with wiring diagrams to facilitate connections, maintenance, and repairs.

MAJOR MOTOR TYPES

Alternating current (ac) induction motors are divided into two electrical categories based on their power source:

- Single phase
- Polyphase (three phase)

AC Single-Phase Motors

Single-phase ac motors are usually limited in size to about two or three horsepower. For residential and small commercial applications, these motors will be found in both central and individual room air conditioning units, fans, ventilating units, and refrigeration units such as household refrigerators and the larger units used to cool produce and other foods in market places.

Since there are so many applications of electric motors, there are many types of single-phase motors in use. Some of the more common types are repulsion, universal, and single-phase induction motors. This latter type includes split-phase, capacitor, shaded-pole, and repulsion-induction motors.

Split-Phase Motors

Split-phase motors are fractional-horsepower units that use an auxiliary winding on the stator to aid in starting the motor until it reaches its proper rotation speed (*see* Figure 2-2). This type of motor finds use in small pumps, oil burners, and other applications.

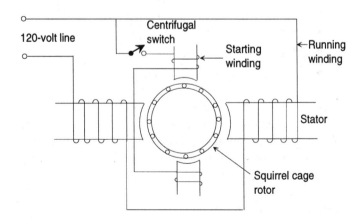

Figure 2-2: Diagram of a split-phase ac motor.

In general, the split-phase motor consists of a housing, a laminated iron-core stator with embedded windings forming the inside of the cylindrical housing, a rotor made up of copper bars set in slots in an iron core and connected to each other by copper rings around both ends of the core, plates that are bolted to the housing and contain the bearings that support the rotor shaft, and a centrifugal switch inside the housing. This type of rotor is often called a squirrel cage rotor since the configuration of the copper bars resembles an actual cage. These motors have no windings as such, and a centrifugal switch is provided to open the circuit to the starting winding when the motor reaches running speed.

To understand the operation of a split-phase motor, look at the wiring diagram in Figure 2-2 on the preceding page. Current is applied to the stator windings, both the main winding and the starting winding, which is in parallel with it through the centrifugal switch. The two windings set up a rotating magnetic field, and this field sets up a voltage in the copper bars of the squirrel-cage rotor. Because these bars are shortened at the ends of the rotor, current flows through the rotor bars. The current-carrying rotor bars then react with the magnetic field to produce motor action. When the rotor is turning at the proper speed, the centrifugal switch cuts out the starting winding since it is no longer needed.

Capacitor Motors

Capacitor motors are single-phase ac motors ranging in size from fractional horsepower (hp) to perhaps as high as 15 hp. This type of motor is widely used in all types of single-phase applications such as powering air compressors, refrigerator compressors, and the like. This type of motor is similar in construction to the split-phase motor, except a capacitor is wired in series with the starting winding, as shown in Figure 2-3.

The capacitor provides higher starting torque, with lower starting current, than does the split-phase motor, and although the capacitor is sometimes mounted inside the motor housing, it is more often mounted on top of the motor, encased in a metal compartment.

In general, two types of capacitor motors are in use:

- Capacitor-start motor
- Capacitor start-and-run motor

As the name implies, the former utilizes the capacitor only for starting; it is disconnected from the circuit once the motor reaches running speed, or at about 75 percent of the motor's full speed. Then the centrifugal switch opens to cut the capacitor out of the circuit.

The capacitor start-and-run motor keeps the capacitor and starting winding in parallel with the running winding, providing a quiet and smooth operation at all times.

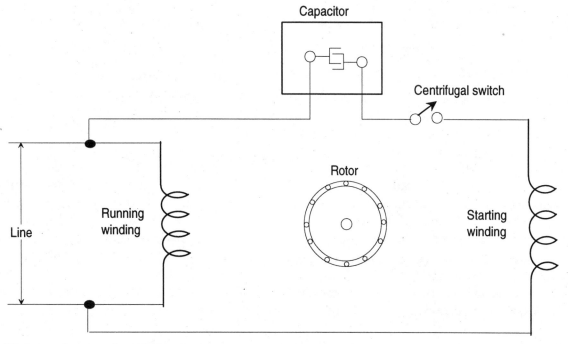

Figure 2-3: Diagram of a capacitor-start motor.

Capacitor split-phase motors require the least maintenance of all single-phase motors, but they have a very low starting torque, making them unsuitable for many applications. Its high maximum torque, however, makes it especially useful in HVAC systems to power slow-speed direct-connected fans.

Repulsion-Type Motors

Repulsion-type motors are divided into several groups, including the following:

- Repulsion-start, induction-run motors
- Repulsion motors
- Repulsion-induction motors

The repulsion-start, induction-run motor is of the single-phase type, ranging in size from about 0.10 hp to as high as 20 hp. It has high starting torque and a constant-speed characteristic, which makes it suitable for such applications as commercial refrigerators, compressors, pumps, and similar applications requiring high starting torque.

The repulsion motor is distinguished from the repulsion-start, induction-run motor by the fact that it is made exclusively as a brush-riding type and does not have any centrifugal mechanism. Therefore, this motor both starts and runs on the repulsion principle. This type of motor has high starting torque and a variable-speed characteristic. It is reversed by shifting the brush holder to either side of the neutral position. Its speed can be decreased by moving the brush holder farther away from the neutral position.

The repulsion-induction motor combines the high starting torque of the repulsion-type and the good speed regulation of the induction motor. The stator of this motor is provided with a regular single-phase winding, while the rotor winding is similar to that used on a dc motor. When starting, the changing single-phase stator flux cuts across the rotor windings and induces currents in them; thus, when flowing through the commutator, a continuous repulsive action on the stator poles is present.

This motor starts as a straight repulsion-type and accelerates to about 75 percent of normal full speed when a centrifugally operated device connects all the commutator bars together and converts the winding to an equivalent squirrel-cage type. The same mechanism usually raises the brushes to reduce noise and wear. Note that, when the machine is operating as a repulsion-type, the rotor and stator poles reverse at the same instant, and that the current in the commutator and brushes is ac.

This type of motor will develop four to five times normal full-load torque and will draw about three times normal full-load current when starting with full-line voltage applied. The speed variation from no load to full load will not exceed 5 percent of normal full-load speed.

The repulsion-induction motor is used to power air compressors, refrigeration (compressor and fans), pumps, stokers, and the like. In general, this type of motor is suitable for any load that requires a high starting torque and constant-speed operation. Most motors of this type are less than 5 hp.

Universal Motors

This type of motor is a special adaptation of the series-connected dc motor, and it gets its name "universal" from the fact that it can be connected on either ac or dc and operates the same. All are single-phase motors for use on 120 or 240 volts.

In general, the universal motor contains field windings on the stator within the frame, an armature with the ends of its windings brought out to a commutator at one end, and carbon brushes that are held in place by the motor's end plate, allowing them to have a proper contact with the commutator.

When current is applied to a universal motor, either ac or dc, the current flows through the field coils and the armature windings in series. The magnetic field set up by the field coils in the stator react with the current-carrying wires on the armature to produce rotation.

Universal motors are frequently used on small fans.

Shaded-Pole Motors

A shaded-pole motor is a single-phase induction motor provided with an uninsulated and permanently short-circuited auxiliary winding displaced in magnetic position from the main winding. The auxiliary winding is known as the shading coil and usually surrounds from one-third to one-half of the pole (*see* Figure 2-4). The main winding surrounds the entire pole and may consist of one or more coils per pole.

Applications for this motor include small fans, timing devices, relays, instrument dials, or any constant-speed load not requiring high starting torque.

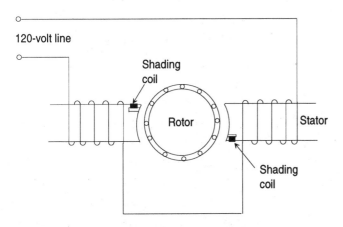

120-volt line

Shading coil

Rotor

Stator

Shading coil

Figure 2-4: Wiring diagram of a shaded-pole motor.

POLYPHASE MOTORS

Polyphase (three-phase) induction motors have a high starting torque, power factor, high efficiency, and low current. They do not use a switch, capacitor, relays, etc., and are suitable for use on large commercial and industrial applications.

Polyphase induction motors are specified by their electrical design type; that is, B, C, or D, as defined by NEMA. These designs are suited to particular classes of applications based upon the load requirements typical of each class. The table in Figure 2-5 can be used to help determine which design type is best suited for a given application.

NEMA Frame/Shaft Sizes

Frame numbers are not intended to indicate electrical characteristics such as horsepower. However, as a frame number becomes higher, so in general does the physical size of the motor and the hp. There are many motors of the same hp built in different frames. NEMA frame size refers to mounting only and has no direct bearing on the motor's body diameter.

In any standard frame number designation, there are either two or three numbers. Typical examples are frame numbers 48, 56, 145, and 215. The number relates to the "D" dimension (distance from center of shaft to center bottom of mount). For example, in the two-digit 56 frame, the "D" dimension is 3.5 in, 56 divided by 16 = 3.5 in. For the "D" dimension of a three-digit frame number, consider only the first two digits and use the divisor 4. In frame number 145, for example, the first two digits divided by the constant 4 is equal to the "D" dimension. Consequently, 14 divided by 4 = 3.5 in. Similarly, the "D" dimension of a 213 frame motor is 5.25 in, 21 divided by 4 = 5.25 in. *See* Figure 2-6 on the next page for a summary of NEMA standard dimensions.

NEMA ELECTRICAL DESIGN STANDARDS					
Classification	**Starting Torque (% Rated Load Torque)**	**Breakdown Torque (%Rated Load Torque)**	**Starting Current**	**Slip**	**Typical Application**
Design B: normal starting torque and normal starting current	100-200%	200-250%	Normal	<5%	Fans, blowers, centrifugal pumps and compressors, etc., where starting torque requirements are relatively low.
Design C: high starting torque and normal starting current	200-250%	200-250%	Normal	<5%	Conveyors, stirring machines, crushers, agitators, reciprocating pumps and compressors, etc., where starting under load is required.
Design D: high starting torque and high slip	275%	275%	Low	>5%	High peak loads, loads with flywheels such as punch press, shears, elevators, extractors, winches, hoists, oil well pumping and wire-drawing machines.

Figure 2-5: NEMA Electrical Design Standards for electric motors.

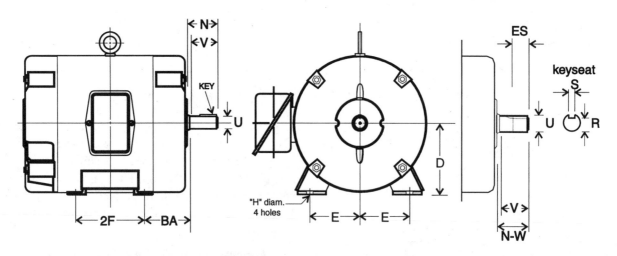

Frame	D	E	2F	H	U	BA	N-W	V min.	R	ES min.	S
48	3	2.12	2.75	.34	.5	2.5	1.5	—	.453	—	Flat
56	3.5	2.44	3	.34	.625	2.75	1.88	—	.517	1.41	.188
143	3.5	2.75	4	.34	.75	2.25	2	1.75	.643	1.41	.188
143T	3.5	2.75	4	.34	.875	2.25	2.25	2	.771	1.41	.188
145	3.5	2.75	5	.34	.75	2.25	2	1.75	.643	1.41	.188
145T	3.5	2.75	5	.34	.875	2.25	2.25	2	.771	1.41	.188
182	4.5	3.75	4.5	.41	.875	2.75	2.25	2	.771	1.41	.188
182T	4.5	3.75	4.5	.41	1.125	2.75	2.75	2.5	.986	1.78	.25
184	4.5	3.75	5.5	.41	.875	2.75	2.25	2	.771	1.41	.188
184T	4.5	3.75	5.5	.41	1.125	2.75	2.75	2.5	.986	1.78	.25
203	5	4	5.5	.41	.75	3.12	2.25	2	.643	1.53	.188
204	5	4	6.5	41	.75	3.12	2.25	2	.643	1.53	.188
213	5.25	4.25	5.5	.41	1.125	3.5	3	2.75	.986	2.03	.25
213T	5.25	4.25	5.5	.41	1.375	3.5	3.38	3.12	1.201	2.41	.312
215	5.25	4.25	7	.41	1.125	3.5	3	2.75	.986	2.03	.25
215T	5.25	4.25	7	.41	1.375	3.5	3.38	3.12	1.201	2.41	.312
224	5.5	4.5	6.75	.41	1	3.5	3	2.75	.857	2.03	.25
225	5.5	4.5	7.5	.41	1	3.5	3	2.75	.857	2.03	.25
254	6.25	5	8.25	.53	1.125	4.25	3.37	3.12	.986	2.03	.25
254U	6.25	5	8.25	.53	1.375	4.25	3.75	3.5	1.201	2.78	.312
254T	6.25	5	8.25	.53	1.625	4.25	4	3.75	1.416	2.91	.375

Figure 2-6: NEMA frame dimensions for ac motors and generators.

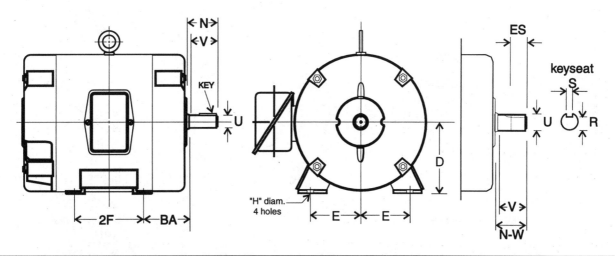

Frame	D	E	2F	H	U	BA	N-W	V min.	R	ES min.	S
256U	6.25	5	10	.53	1.375	4.25	3.75	3.5	1.201	2.78	.312
256T	6.25	5	10	.53	1.625	4.25	4	3.75	1.416	2.91	.375
284	7	5.5	9.5	.53	1.25	4.75	3.75	3.5	.986	2.03	.25
284U	7	5.5	9.5	.53	1.625	4.75	4.88	4.62	1.416	3.78	.375
284T	7	5.5	9.5	.53	1.875	4.75	4.62	4.38	1.591	3.28	.5
284TS	7	5.5	9.5	.53	1.625	4.75	3.25	3	1.416	1.91	.375
286U	7	5.5	11	.53	1.625	4.75	4.88	4.62	1.416	3.78	.375
286T	7	5.5	11	.53	1.875	4.75	4.62	4.38	1.591	3.28	.5
286TS	7	5.5	11	.53	1.625	4.75	3.25	3	1.416	1.91	.375
324	8	6.25	10.5	.66	1.625	5.25	4.87	4.62	1.416	3.78	.375
324U	8	6.25	10.5	.66	1.875	5.25	5.62	5.38	1.591	4.28	.5
324S	8	6.25	10.5	.66	1.625	5.25	3.25	3	1.416	1.91	.375
324T	8	6.25	10.5	.66	2.125	5.25	5.25	5	1.845	3.91	.5
324TS	8	6.25	10.5	.66	1.875	5.25	3.75	3.5	1.591	2.03	.5
326	8	6.25	12	.66	1.625	5.25	4.87	4.62	1.416	3.78	.375
326U	8	6.25	12	.66	1.875	5.25	5.62	5.38	1.591	4.28	.5
326S	8	6.25	12	.66	1.625	5.25	3.25	3	1.416	1.91	.375
326T	8	6.25	12	.66	2.125	5.25	5.25	5	1.845	3.91	.5
326TS	8	6.25	12	.66	1.875	5.25	3.75	3.5	1.591	2.03	.5
364	9	7	11.25	.66	1.875	5.88	5.62	5.38	1.591	4.28	.5
364S	9	7	11.25	.66	1.625	5.88	3.25	3	1.416	1.91	.375
364U	9	7	11.25	.66	2.125	5.88	6.37	6.12	1.845	5.03	.5

Figure 2-6: NEMA frame dimensions for ac motors and generators. *(Cont.)*

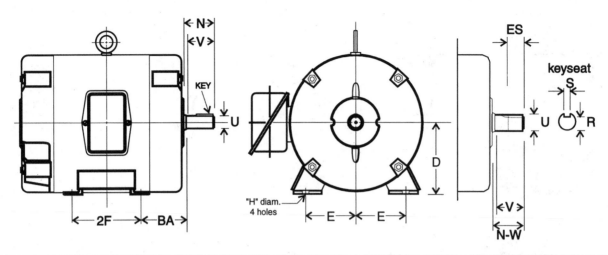

Frame	D	E	2F	H	U	BA	N-W	V min.	R	ES min.	S
364US	9	7	11.25	.66	1.875	5.88	3.75	3.5	1.591	2.03	.5
364T	9	7	11.25	.66	2.375	5.88	5.88	5.62	2.01	4.28	.625
364TS	9	7	11.25	.66	1.875	5.88	3.75	3.5	1.591	2.03	.5
365	9	7	12.25	.66	1.875	5.88	5.62	5.38	1.591	4.28	.5
365S	9	7	12.25	.66	1.625	5.88	3.25	3	1.416	1.91	.375
365U	9	7	12.25	.66	2.125	5.88	6.37	6.12	1.845	5.03	.5
365US	9	7	12.25	.66	1.875	5.88	3.75	3.5	1.591	2.03	.5
365T	9	7	12.25	.66	2.375	5.88	5.88	5.62	2.021	4.28	.625
365TS	9	7	12.25	.66	1.875	5.88	3.75	3.5	1.591	2.03	.5
404	10	8	12.25	.81	2.125	6.62	6.37	6.12	1.845	5.03	.5
404S	10	8	12.25	.81	1.875	6.62	3.75	3.5	1.591	2.03	.5
404U	10	8	12.25	.81	2.375	6.62	7.12	6.88	2.021	5.53	.625
404US	10	8	12.25	.81	2.125	6.62	4.25	4	1.845	2.78	.5
404T	10	8	12.25	.81	2.875	6.62	7.25	7	2.45	5.6	.75
404TS	10	8	12.25	.81	2.125	6.62	4.25	4	1.845	2.78	.5
405	10	8	13.75	.81	2.125	6.62	6.37	6.12	1.845	5.03	.5
405S	10	8	13.75	.81	1.875	6.62	3.75	3.5	1.591	2.03	.5
405U	10	8	13.75	.81	2.375	6.62	7.12	6.88	2.021	5.53	.625
405US	10	8	13.75	.81	2.125	6.62	4.25	4	1.845	2.78	.5
405T	10	8	13.75	.81	2.875	6.62	7.25	7	2.45	5.65	.75
405TS	10	8	13.75	.81	2.125	6.62	4.25	4	1.845	2.78	.5
444	11	9	14.5	.81	2.375	7.5	7.12	6.88	2.021	5.53	.625

Figure 2-6: NEMA frame dimensions for ac motors and generators. *(Cont.)*

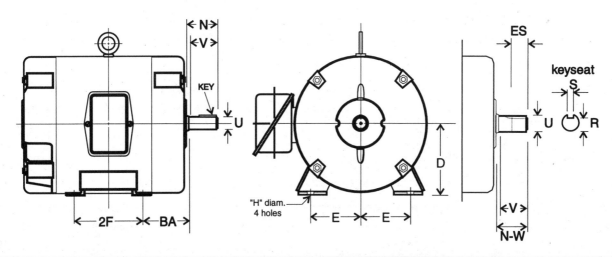

Frame	D	E	2F	H	U	BA	N-W	V min.	R	ES min.	S
444S	11	9	14.5	.81	2.125	7.5	4.25	4	1.845	2.78	.5
444U	11	9	14.5	.81	2.875	7.5	8.62	8.38	2.45	7.03	.75
444US	11	9	14.5	.81	2.125	7.5	4.25	4	1.845	2.78	.5
444T	11	9	14.5	.81	3.375	7.5	8.5	8.25	2.88	6.91	.875
444TS	11	9	14.5	.81	2.375	7.5	4.75	4.5	2.021	3.03	.625
445	11	9	16.5	.81	2.375	7.5	7.12	6.88	2.021	5.53	.625
445S	11	9	16.5	.81	2.125	7.5	4.25	4	1.845	2.78	.5
445U	11	9	16.5	.81	2.875	7.5	8.62	8.38	2.45	7.03	.75
445US	11	9	16.5	.81	2.125	7.5	4.25	4	1.845	2.78	.5
445T	11	9	16.5	.81	3.375	7.5	8.5	8.25	2.88	6.91	.875
445TS	11	9	16.5	.81	2.375	7.5	4.75	4.5	2.021	3.03	.625
447TS	11	9	20	DIMENSIONS VARY WITH MANUFACTURER							
449TS	11	9	25	DIMENSIONS VARY WITH MANUFACTURER							
504U	12.5	10	16	.94	2.875	8.5	8.62	8.38	2.45	7.28	.75
505	12.5	10	18	.94	2.875	8.5	8.62	8.38	2.45	7.28	.75
505S	12.5	10	18	.94	2.125	8.5	4.25	4	1.845	2.78	.5

NOTE

Letters or numbers appearing in front of the NEMA frame numbers are those of the manufacturer. They have no NEMA frame significance. The significance from one manufacturer to another will vary. For example, the letter in front of Leeson's (Leeson Electric Motors), L56, indicates the overall length of the motor.

Figure 2-6: NEMA frame dimensions for ac motors and generators. *(Cont.)*

NEMA FRAME SUFFIXES		
C	=	NEMA C face mounting (specify with or without rigid base).
D	=	NEMA D flange mounting (specify with or without rigid base).
H	=	Indicates a frame with a rigid base having an F dimension larger than that of the same frame without the suffix H. For example, combination 56H base motors have mounting holes for NEMA 56 and NEMA 143-5T and a standard NEMA 56 shaft.
J	=	NEMA C face, threaded shaft pump motor.
JM	=	Close-coupled pump motor with specific dimensions and bearings.
JP	=	Close-coupled pump motor with specific dimensions and bearings.
M	=	6.75 in flange (oil burner).
N	=	7.25 in flange (oil burner).
T, TS	=	Integral hp NEMA standard shaft dimensions if no additional letters follow the "T" or "TS."
TS	=	Motor with NEMA standard "short shaft" for belt-driven loads.
Y	=	Non-NEMA standard mount; a detailed drawing is required to be sure of dimensions. Can indicate a special base, face, or flange.
Z	=	Non-NEMA standard shaft; a detailed drawing is required to be sure of dimensions.

Figure 2-6: NEMA frame dimensions for ac motors and generators. *(Cont.)*

Motor Mounting

Unless specified otherwise, motors can be mounted in any position or any angle. However, unless a drip cover is used for shaft-up or shaft-down applications, dripproof motors must be mounted in the horizontal or sidewall position to meet the enclosure definition. Mount motor securely to the mounting base of equipment or to a rigid, flat surface, preferably metallic.

Types of Mounts

Rigid Base: This type of base is bolted, welded, or cast on the main frame and allows the motor to be rigidly mounted on equipment.

Resilient Base: A resilient base has isolation or resilient rings between the motor mounting hubs and the base to absorb vibrations and noise.

NEMA C Face Mount: This is a machined face with a pilot on the shaft end which allows direct mounting with the pump or other direct coupled equipment. Bolts pass through the mounted part to the threaded hole in the motor face.

NEMA D Flange Mount: This mount is a machined flange with a rabbet for mountings. Bolts pass through the motor flange to a threaded hole in the mounted part. NEMA C face motors are by far the most popular and most readily available. NEMA D flange kits are stocked by some motor manufacturers.

Type M or N Mount: This mount has a special flange for direct attachment to fuel atomizing pump or an oil burner. In recent years, this type of mounting has become widely used on auger drives in poultry feeders.

Extended Through-Bolt Motors: These motors have bolts protruding from the front or rear of the motor by which it is mounted. This is usually used on small direct-drive fans or blowers.

IEC Frame Sizes

A summary of International Electromechanical Commission (IEC) standard dimensions for polyphase motors are shown in Figure 2-7. These dimensions are especially useful when specifying or installing motors of foreign manufacture.

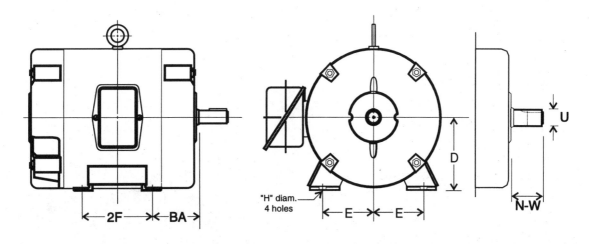

IEC FRAME DIMENSIONS FOR THREE-PHASE MOTORS														
Frame	**D**		**2E**		**2F**		**BA**		**H**		**U**		**N-W**	
	MM	IN	MM	IN	MM	IN	MM	IN	MM	IN	MM	IN	MM	IN
56	56	2.2	90	3.54	71	2.8	36	1.42	5.8	.228	9	.354	20	.787
63	63	2.48	100	3.94	80	3.15	40	1.57	7	.276	11	.433	23	.905
71	71	2.8	112	4.41	90	3.54	45	1.77	7	.276	14	.551	30	1.18
80	80	3.15	125	4.92	100	3.94	50	1.97	10	.394	19	.748	40	1.57
90S	90	3.54	140	5.51	100	3.94	56	2.2	10	.394	24	.945	50	1.97
90L	90	3.54	140	5.51	125	4.92	56	2.2	10	.394	24	.945	50	1.97
100L	100	3.94	160	6.3	140	5.5	63	2.48	12	.472	28	1.1	60	2.36
112M	112	4.41	190	7.48	140	5.5	70	2.75	12	.472	28	1.1	60	2.36
132S	132	5.2	216	8.5	140	5.5	89	3.5	12	.472	38	1.5	80	3.15
132M	132	5.2	216	8.5	178	7	89	3.5	12	.472	38	1.5	80	3.15
160M	160	6.3	254	10	210	8.27	108	4.25	15	.591	MANUFACTURERS DO NOT AGREE BEYOND THE 132 FRAME			
160L	160	6.3	254	10	254	10	108	4.25	15	.591				
250S	250	9.84	406	16	311	12.25	168	6.62	24	.945				
250M	250	9.84	406	16	349	13.75	168	6.62	24	.945				
280S	280	11	457	18	368	14.5	190	7.48	24	.945				
280M	280	11	457	18	419	16.38	190	7.48	24	.945				
315S	315	12.4	508	20	406	16	216	8.5	28	1.102				
315M	315	12.4	508	20	457	18	216	8.5	28	1.102				
315L	315	12.4	508	20	508	20	216	8.5	28	1.102				

Figure 2-7: Frame dimensions of IEC three-phase motors.

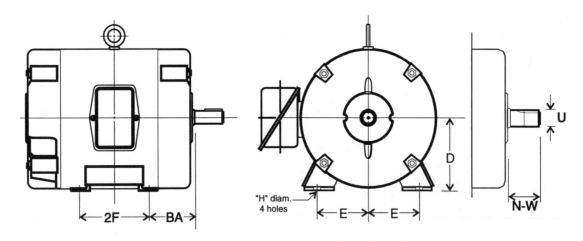

Figure 2-7: Frame dimensions of IEC three-phase motors. *(Cont.)*

Frame	D		2E		2F		BA		H		U		N-W	
	MM	IN	MM	IN	MM	IN	MM	IN	MM	IN	MM	IN	MM	IN
355S	355	14	610	24	500	19.69	254	10	28	1.102				
355M	355	14	610	24	560	22	254	10	28	1.102				
355L	355	14	610	24	630	24.8	254	10	28	1.102				
400M	400	15.75	686	27	630	24.8	280	11	35	1.378				
400L	400	15.75	686	27	710	27.95	280	11	35	1.378				

Table title: **IEC FRAME DIMENSIONS FOR THREE-PHASE MOTORS**

Application Mounting

For direct-coupled applications, the shaft and coupling of motors must be carefully aligned, using shims as required under the motor base. A flexible coupling should be used, if possible, but not as a substitute for good alignment practices.

Pulleys, sheaves, sprockets, and gears should be generally mounted as close as possible to the bearing on the motor shaft, thereby lessening the bearing load.

The center point of the belt, or system of V-belts, should not be beyond the end of the motor shaft.

The inner edge of the sheave or pulley rim should not be closer to the bearing than the shoulder on the shaft, but should be as close to this point as possible.

The outer edge of a chain sprocket or gear should not extend beyond the end of the motor shaft.

To obtain the minimum pitch diameters of flat-belt, timing-belt, chain, and gear drives, the multiplier given in

MULTIPLIER TABLE	
Drive	**Multiplier**
Flat belt	1.33
Timing belt	0.9
Chain sprocket	0.7
Spur gear	0.75
Helical gear	0.85

Figure 2-8: Multiplier table for flat-belt, timing-belt, chain, and gear drives.

Figure 2-8 should be applied to the narrow V-belt sheave pitch diameters in NEMA MG1-14.4441 for ac, general-purpose motors, or to the V-belt sheave pitch diameters as determined from NEMA MG 1-14.67 for industrial dc motors.

Frame Sizes by Horsepower

The table in Figure 2-6 gives the NEMA frame sizes by frame number. The table in Figure 2-9 acts as a cross-reference and gives frame numbers by hp rating. To use this table, first locate the hp rating and frame designation in the table in Figure 2-9. Then refer to the table in Figure 2-6 for the appropriate dimensions. For example, if it is desired to determine the frame designation for a 50 hp motor with a revolution of 1200 rpm and a 1964 rerate, scan down the left-hand column of the table in Figure 2-9 until 50 hp is reached; scan across this row until the column under 1800 rpm and 1964 rerate are found. Read the frame designation as 326T.

RPM	3600			1800			1200			900		
NEMA Program HP	Orig.	1952 Rerate	1964 Rerate	Orig.	1952 Rerate	1964 Rerate	Orig.	1952 Rerate	1964 Rerate	Orig.	1952 Rerate	1964 Rerate
1	—	—	—	203	182	143T	204	184	145T	225	213	182T
1.5	203	182	143T	204	184	145T	224	184	182T	254	213	184T
2	204	184	145T	224	184	145T	225	213	184T	254	215	213T
3	224	184	145T	225	213	182T	254	215	213T	284	254U	215T
5	225	213	182T	254	215	184T	284	254U	215T	324	256U	254T
7.5	254	215	184T	284	254U	213T	324	256U	254T	326	284U	256T
10	284	254U	213T	324	256U	215T	326	284U	256T	364	286U	284T
15	324	256U	215T	326	284U	254T	364	324U	284T	365	326U	286T
20	326	284U	254T	364	286U	256T	365	326U	286T	404	364U	324T
25	364S	286U	256T	364	324U	284T	404	364U	324T	405	365U	326T
30	364S	324S	284TS	365	326U	286T	405	365U	326T	444	404U	364T
40	365S	326S	286TS	404	364U	324T	444	404U	364T	445	405U	365T
50	404S	364US	324TS	405S	365US	326T	445	405U	365T	504U	444U	404T
60	405S	365US	326TS	444S	404US	364TS	504U	444U	404T	505	445U	405T
75	444S	404US	364TS	445S	405US	365TS	505	445U	405T	—	—	444T
100	445S	405US	365TS	504S	444US	404TS	—	—	444T	—	—	445T
125	504S	444US	404TS	505S	445US	405TS	—	—	445T	—	—	—
150	505S	445US	405TS	—	—	444TS	—	—	—	—	—	—
200	—	—	444TS	—	—	445TS	—	—	—	—	—	—
250	—	—	445TS	—	—	—	—	—	—	—	—	—

THREE-PHASE FRAME SIZES FOR OPEN, GENERAL-PURPOSE MOTORS

Figure 2-9: Three-phase frame sizes for open, general-purpose motors.

RPM	3600			1800			1200			900		
NEMA Program HP	Orig.	1952 Rerate	1964 Rerate	Orig.	1952 Rerate	1964 Rerate	Orig.	1952 Rerate	1964 Rerate	Orig.	1952 Rerate	1964 Rerate
1	—	—	—	203	182	143T	204	184	145T	225	213	182T
1.5	203	182	143T	204	184	145T	224	184	182T	254	213	184T
2	204	184	145T	224	184	145T	225	213	184T	254	215	213T
3	224	184	182T	225	213	182T	254	215	213T	284	254U	215T
5	225	213	184T	254	215	184T	284	254U	215T	324	256U	254T
7.5	254	215	213T	284	254U	213T	324	256U	254T	326	284U	256T
10	284	254U	215T	324	256U	215T	326	284U	256T	364	286U	284T
15	324	256U	254T	326	284U	254T	364	324U	284T	365	326U	286T
20	326	286U	256T	364	286U	256T	365	326U	286T	404	364U	324T
25	365S	324U	284TS	365	324U	284T	404	364U	324T	405	365U	326T
30	404S	326S	286TS	404	326U	286T	405	365U	326T	444	404U	364T
40	405S	364US	324TS	405	364U	324T	444	404U	364T	445	405U	365T
50	444S	365US	326TS	444S	365US	326T	445	405U	365T	504U	444U	404T
60	445S	405US	364TS	445S	405US	364TS	504U	444U	404T	505	445U	405T
75	504S	444US	365TS	504S	444US	365TS	505	445U	405T	—	—	444T
100	505S	445US	405TS	505S	445US	405TS	—	—	444T	—	—	445T
125	—	—	444TS	—	—	444TS	—	—	445T	—	—	—
150	—	—	445TS	—	—	445TS	—	—	—	—	—	—

THREE-PHASE FRAME SIZES FOR TEFC, GENERAL-PURPOSE MOTORS

Figure 2-10: Three-phase frame sizes for TEFC, general-purpose motors.

Belt Tensioning

Manufacturers of belts can provide recommended tensioning values and instruments for precisely determining belt tension. Particularly in very high-speed, very high-torque or very high-horsepower applications, critical belt tensioning can be important. For most industrial applications, however, these general belt tensioning procedures are usually adequate:

1. The best tension is typically the lowest at which the belt will not slip under peak load.

2. Over-tensioning will shorten belt/bearing life.

3. After installing a new belt, it is important to check the tension often during the first 24 to 48 operating hours, and to re-tension as necessary.

4. Periodically inspect and re-tension the belt over the course of operation.

As a general rule, the correct belt tension can be gauged by deflecting the belt at mid-span with your thumb while the motor is stopped. You should be able to deflect the belt approximately .5 in with light to moderate pressure on single-ribbed belts. Multiple ribs will require additional pressure.

Two methods of checking belt tension while the motor is operating include visually assessing whether there is any belt flutter, or listening for belt squeal. Either can occur as a result of inadequate belt tension.

ELECTRICAL CHARACTERISTICS AND CONNECTIONS OF AC MOTORS

Voltage, frequency and phase of power supply should be consistent with the motor nameplate rating. A motor will operate satisfactorily on voltage within 10% of nameplate value, or frequency within 5%, or combined voltage and frequency variation not to exceed 10%.

Voltage

Common 60 hz voltages for single-phase motors are 115 V, 230 V, and 115/230 V.

Common 60 hz voltage for three-phase motors are 230 V, 460 V and 230/460V. Sometimes, 200-V and 575-V motors are also encountered. In prior NEMA standards these voltages were listed as 208 or 220/440 or 550 V. Motors with these voltages on the nameplate can safely be replaced by motors having the current standard markings of 200 or 208, 230/460 or 575 V, respectively.

Motors rated 115/208-230 V and 208-230/460 V, in most cases, will operate satisfactorily at 208 V, but the torque will be 20%-25% lower. Operating below 208 V may require a 208 V (or 200 V) motor or the use of the next higher horsepower, standard voltage motor.

Phase

Single-phase motors account for up to 80% of the motors used in the United States but are used mostly in homes and in auxiliary low-horsepower industrial applications such as fans and on farms.

Three-phase motors are used on larger commercial and industrial equipment.

Current (Amperes)

In comparing motor types, the full-load A and/or service-factor A are key parameters for determining the proper loading on the motor. For example, never replace a PSC type motor with a shaded-pole type as th' shaded-pole amperage will normally be 50%-60% higher. Compare PSC with PSC, capacitor start with capacitor start, and so forth.

Hertz/Frequency

In North America 60 hz (cycles) is the common power source. However, most of the rest of the world is supplied with 50 hz power.

Horsepower

Exactly 746 watts of electrical power will produce 1 hp if a motor could operate at 100% efficiency, but of course no motor is 100% efficient. A 1 hp motor operating at 84% efficiency will have a total watt consumption of 888 W. This amounts to 746 W of usable power and 142 W loss due to heat, friction, etc. (888 x .84 = 746 = 1 hp).

Horsepower can also be calculated if torque is known, using one of the following equations:

$$HP = \frac{Torque\ (lb\ /\ ft)\ x\ RPM}{5{,}250}$$

$$HP = \frac{Torque\ (oz\ /\ ft)\ x\ RPM}{84{,}000}$$

$$HP = \frac{Torque\ (in\ /\ lb)\ x\ RPM}{63{,}000}$$

Speeds

The approximate rpm at rated load for small and medium motors operating at 60 hz and 50 hz at rated voltage are as follows:

	60 Hz	50 Hz.	Synch. Speed
2 Pole	3450	2850	3600
4 Pole	1725	1425	1800
6 Pole	1140	950	1200
8 Pole	850	700	900

Synchronous speed (no-load) can be determined by the following equation:

$$\text{Synchronous speed} = \frac{Frequency\ (Hz) \times 120}{Number\ of\ Poles}$$

Insulation Class

Insulation systems are rated by standard NEMA classifications according to maximum allowable operating temperatures. They are as follows:

Class	Maximum Allowed Temperature	
A	105°C	221°F
B	130°C	266°F
F	155°C	311°F
H	180°C	356°F

Generally, replace a motor with one having an equal or higher insulation class. Replacement with one of lower temperature rating could result in premature failure of the motor. Each 10°C rise above these ratings can reduce the motor's service life by one half.

Service Factor

The service factor (SF) is a measure of continuous overload capacity at which a motor can operate without overload or damage, provided the other design parameters such as rated voltage, frequency and ambient temperature are within norms. Example: a ¾ hp motor with a 1.15 SF can operate at .86 hp, (.75 hp × 1.15 = .862 hp) without overheating or otherwise damaging the motor if rated voltage and frequency are supplied at the motor's leads. Some motors have higher service factors than the NEMA standard.

It is not uncommon for the original equipment manufacturer (OEM) to load the motor to its maximum load capability (service factor). For this reason, do not replace a motor with one of the same nameplate horsepower but with a lower service factor. Always make certain that the replacement motor has a maximum hp rating (rated hp × SF) equal to or higher than that which it replaces. Multiply the horsepower by the service factor for determining maximum potential loading.

For easy reference, standard NEMA service factors for various horsepower motors and motor speeds are shown in the following table:

HP	For Dripproof Motors Service Factor Synchronous Speed (RPM)			
	3600	1800	1200	900
⅙, ¼, ⅓	1.35	1.35	1.35	1.35
½	1.25	1.25	1.25	1.25
¾	1.25	1.25	1.15	1.15
1	1.25	1.15	1.15	1.15
1½ up	1.15	1.15	1.15	1.15

Capacitors

Capacitors are used on all fractional hp induction motors except shaded-pole, split-phase and polyphase. Start capacitors are designed to stay in circuit a very short time (3-5 seconds), while run capacitors are permanently in circuit. Capacitors are rated by capacity and voltage. Never use a capacitor with a voltage less than that recommended with the replacement motor. A higher voltage is acceptable.

When power factor correction capacitors are used, the total corrective kVAR on the load side of the motor controller should not exceed the value required to raise the no-load power factor to unity. Corrective kVAR in excess of this value may cause over excitation that results in high transient voltage, current and torque that can increase safety hazards to personnel and possibly damage the motor or driven equipment.

> **WARNING!**
>
> Power factor correction capacitors must not be connected at motor terminals on elevator motors, multispeed motors, plugging or jogging applications or open transition, wye-delta, autotransformer starting and some part-winding start motors.

If possible, capacitors should be located at position No. 2 in Figure 2-11. Doing so does not change the current flowing through the motor overload protectors. Connection of the capacitors at position No. 3 requires a change of overload protectors. Capacitors should be located at position No. 1 for the following:

- Elevator motors
- Multispeed motors
- Plugging or jogging applications
- Open transition, wye-delta, autotransformer starters
- Part-winding start motors

Make certain that the bus power factor is not increased above 95 percent under all loading conditions to avoid over excitation.

The table in Figure 2-12 on the next page allows the determination of corrective kVAR required where capacitors are individually connected at motor loads. These values should be considered the maximum capacitor rating when the motor and capacitor are switched as a unit. The figures given are for 3-phase, 60 Hz, NEMA Class B motors to raise the full-load power factor to 95%.

Efficiency

A motor's efficiency is a measurement of useful work produced by the motor versus the energy it consumes (heat and friction). An 84% efficient motor with a total watt draw of 400 W produces 336 W of useful energy (400 x .84 = 336 W). The 64 W lost (400 - 336 = 64 W) becomes heat.

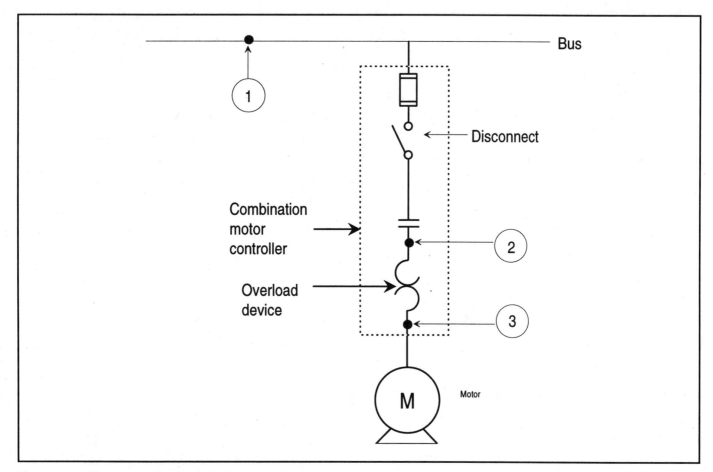

Figure 2-11: Placement of capacitors in motor circuits.

Induction Motor HP Rating	3600		1800		1200		900		720		600	
MOTOR POWER FACTOR CORRECTION TABLE												
Nominal Motor Speed in RPM												
	Capacitor Rating kVAR	Line Current Reduction %	Capacitor Rating kVAR	Line Current Reduction %	Capacitor Rating kVAR	Line Current Reduction %	Capacitor Rating kVAR	Line Current Reduction %	Capacitor Rating kVAR	Line Current Reduction %	Capacitor Rating kVAR	Line Current Reduction %
3	1.5	14	1.5	15	1.5	20	2	27	2.5	35	3.5	41
5	2	12	2	13	2	17	3	25	4	32	4.5	37
7½	2.5	11	2.5	12	3	15	4	22	5.5	30	6	34
10	3	10	3	11	3.5	14	5	21	6.5	27	7.5	31
15	4	9	4	10	5	13	6.5	18	8	23	9.5	27
20	5	9	5	10	6.5	12	7.5	16	9	21	12	25
25	6	9	6	10	7.5	11	9	15	11	20	14	23
30	7	8	7	9	9	11	10	14	12	18	16	22
40	9	8	9	9	11	10	12	13	15	16	20	20
50	12	8	11	9	13	10	15	12	19	15	24	19
60	14	8	14	8	15	10	18	11	22	15	27	19
75	17	8	16	8	18	10	21	10	26	14	32.5	18
100	22	8	21	8	25	9	27	10	32.5	13	40	17
125	27	8	26	8	30	9	32.5	10	40	13	47.5	16
150	32.5	8	30	8	35	9	37.5	10	47.5	12	52.5	15
200	40	8	37.5	8	42.5	9	47.5	10	60	12	65	14
250	50	8	45	7	52.5	8	57.5	9	70	11	77.5	13
300	57.5	8	52.5	7	60	8	65	9	80	11	87.5	12
350	65	8	60	7	67.5	8	75	9	87.5	10	95	11
400	70	8	65	6	75	8	85	9	95	10	105	11
450	75	8	67.5	6	80	8	92.5	9	100	9	110	11
500	77.5	8	72.5	6	82.5	8	97.5	9	107.5	9	115	10

Figure 2-12: Motor power factor correction table.

Thermal Protection (Overload)

A thermal protector, automatic or manual, mounted in the end frame or on a winding, is designed to prevent a motor from getting too hot, causing possible fire or damage to the motor. Protectors are generally current- and temperature-sensitive. Some motors have no inherent protector, but they should have protection provided in the overall system's design for safety.

Never bypass a protector because of nuisance tripping. This is generally an indication of some other problem, such as overloading or lack of proper ventilation.

Never replace nor choose an automatic-reset thermal overload protected motor for an application where the driven load could cause personal injury if the motor should restart unexpectedly. Only manual-reset thermal overloads should be used in such applications.

Basic types of overload protectors include:

Automatic Reset: After the motor cools, this line-interrupting protector automatically restores power. It should not be used where unexpected restarting would be hazardous.

Manual Reset: This line-interrupting protector has an external button that must be pushed to restore power to the motor. Use where unexpected restarting would be hazardous, as on saws, conveyors, compressors and other machinery.

Resistance Temperature Detectors: Precision-calibrated resistors are mounted in the motor and are used in conjunction with an instrument supplied by the customer to detect high temperatures.

Independent Overload Relays: Independent overload relays are intended for use with separately mounted magnetic contactors. They are applied where space considerations or design considerations necessitate the mounting of overload relays apart from the contactor itself. They are also utilized where one contactor is employed in operating two or more motors. In this case, independently mounted overload relays are utilized in series to protect the equipment when any motor is affected. Units are available in the open or the NEMA 1 enclosures.

Selecting Thermal Overload Protection

Thermal overload protection (heater units) are selected for various motors using the following method:

1. Obtain the full-load current and service factor from the motor nameplate or from

manufacturer's specifications. Do not estimate full-load current from horsepower tables.

2. Determine the number of overload relays needed; that is, 1, 2, or 3.

3. Select the proper heater unit from tables according to class, size, type of enclosure and number of overload relays. Full-load motor currents should be within the Min.–Max. ratings shown for the correct number of overload relays being used.

4. Do not use conventional table for hermetically sealed or submersible motors; refer to the motor manufacturer's specifications.

Individual Branch Circuit Wiring

All wiring and electrical connections should comply with the National Electrical Code (*NEC*) and with local codes and practices. Undersized wire between the motor and the power source will limit the starting and load carrying abilities of the motor.

The basic elements that must be accounted for in any motor circuit are shown in Figure 2-13. Although these elements are shown separately in this illustration, there are certain cases where the *NEC* permits a single device to serve more than one function. For example, in some cases, short-circuit protection and overload protection can be combined in a single circuit breaker or set of fuses.

While reviewing the drawing in Figure 2-13, note that the basic *NEC* rule for sizing conductors supplying a single-speed motor used for continuous duty must have a current-carrying capacity of not less than 125% of the motor full-load current rating.

Conductors on the line side of the controller supplying multispeed motors must be based on the highest of the full-load current ratings shown on the motor nameplate.

Conductors between the controller and the motor must have a current-carrying rating based on the current rating for the speed of the motor that each set of conductors is feeding. Also be aware that it is often necessary to increase the size of conductors; that is, beyond the sizes determined using calculations as required by the *NEC*. The increased size of conductors, when necessary, compensates for voltage drop and other power losses in the circuit.

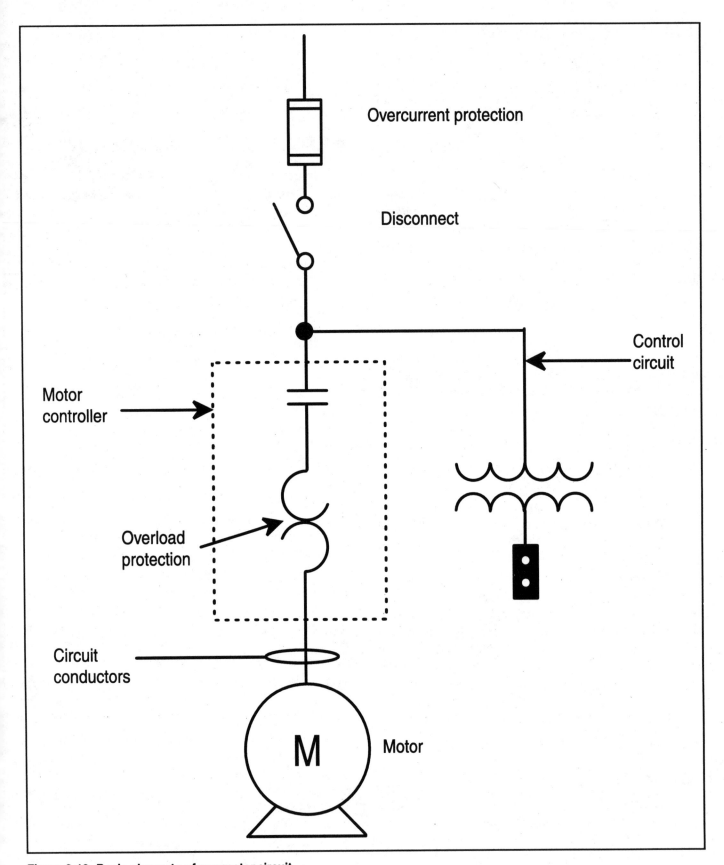

Figure 2-13: Basic elements of any motor circuit.

SINGLE-PHASE MOTORS — 230 V						
Transformer		Distance – Motor to Transformer (ft)				
HP	kVA	100	150	200	300	500
1.5	3	10	8	8	6	4
2	3	10	8	8	6	4
3	5	8	8	6	4	2
5	7.5	6	4	4	2	0
7.5	10	6	4	3	1	0

Figure 2-14: Copper wire and transformer size for single-phase, 230-V motors.

The recommended copper wire and transformer sizes for motor connections are shown in the charts in Figures 2-14 and 2-15.

Motor Connections

Stator windings of three-phase motors are connected in either wye or delta. Some stators are designed to operate both ways; that is, some motors are started as a wye-connected motor to help reduce starting current, and then it is changed to a delta connection for running. Many three-phase motors have dual-voltage stators. These stators are designed to be connected to either, say, 240 V or 480 V. The leads of a dual-voltage stator use a standard numbering system.

When a dual-voltage motor is operated at 240 V, the current draw of the motor is double the current of a 480-V connection. For example, if a motor draws 10 A when connected to 240 V, it will draw only 5 A when connected to 480 V. The reason for this is the difference of impedance in the windings between a 240-V connection and a 480-V connection. In dual-voltage motors, the low-voltage windings are always connected in parallel, while the high-voltage windings are connected in series.

THREE-PHASE MOTORS – 240 AND 480 VOLTS							
Transformer			Distance – Motor to Transformer (ft)				
HP	Volts	kVA	100	150	200	300	500
1.5	230	3	12	12	12	12	10
1.5	230	3	12	12	12	12	10
1.5	460	3	12	12	12	12	12
2	230	3	12	12	12	10	8
2	460	3	12	12	12	12	12
3	230	5	12	10	10	8	6
3	460	5	12	12	12	12	10
5	230	7.5	10	8	8	6	4
5	460	7.5	12	12	12	10	8
7.5	230	10	8	6	6	4	2
7.5	460	10	12	12	12	10	8
10	230	15	6	4	4	4	1
10	460	15	12	12	12	10	8
15	230	20	4	4	4	2	0
15	460	20	12	10	10	8	6

Figure 2-15: Copper wire and transformer size for three-phase, 240- and 480-V motors.

THREE-PHASE MOTORS – 240 AND 480 VOLTS							
Transformer			Distance – Motor to Transformer (ft)				
HP	Volts	kVA	100	150	200	300	500
20	230	Consult Local Power Company	4	2	2	1	0
20	460		10	8	8	6	4
25	230		2	2	2	0	0
25	460		8	.8	6	6	4
30	230		2	1	1	0	0
30	460		8	6	6	4	2
40	230		1	0	0	0	300 kcmil
40	460		6	6	4	2	6
50	230		1	0	0	0	300 kcmil
50	460		4	4	2	2	0
30	230		1	0	0	250 kcmil	500 kcmil
60	460		4	2	2	0	0
75	230		0	0	0	300 kcmil	500 kcmil
75	460		4	2	2	0	0

Figure 2-15: Copper wire and transformer size for three-phase, 240- and 480-V motors. *(Cont.)*

Basic motor connections for three-phase motors are shown in Figures 2-16 through 2-22. These connections will account for the majority of all electric motors encountered in the industry.

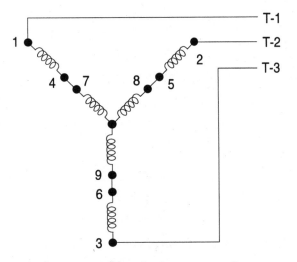

Wye-connected motor windings (series connected)

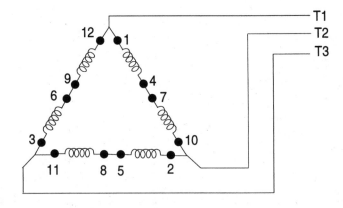

Delta-connected motor windings (series connected)

Figure 2-16: Two types of windings found in three-phase motors.

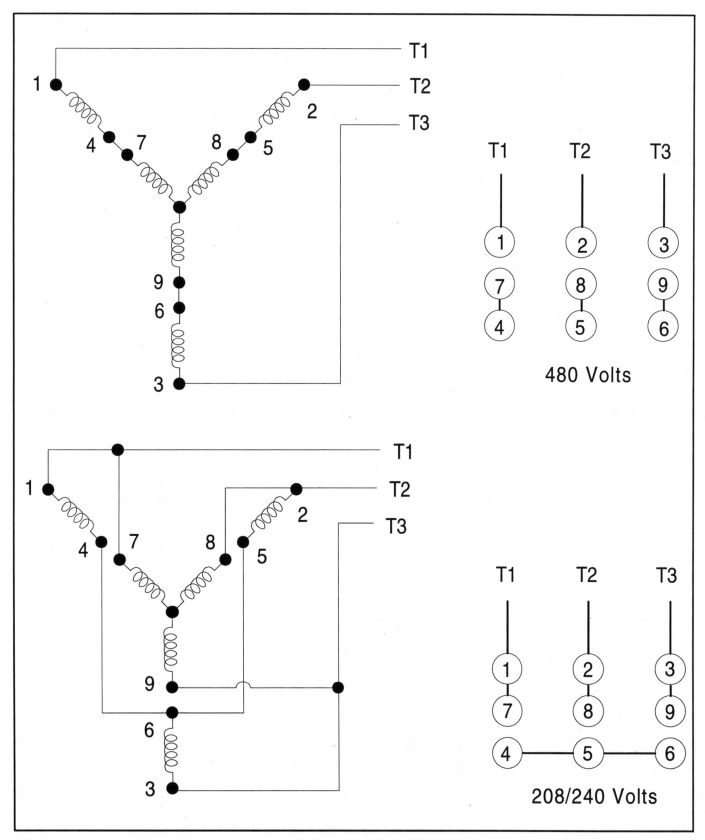

Figure 2-17: High- and low-voltage connections for wye-connected three-phase motors.

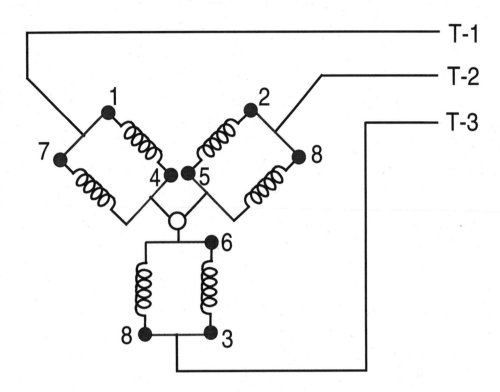

Figure 2-18: Lead connected in parallel for the lower voltage in a dual-voltage motor winding.

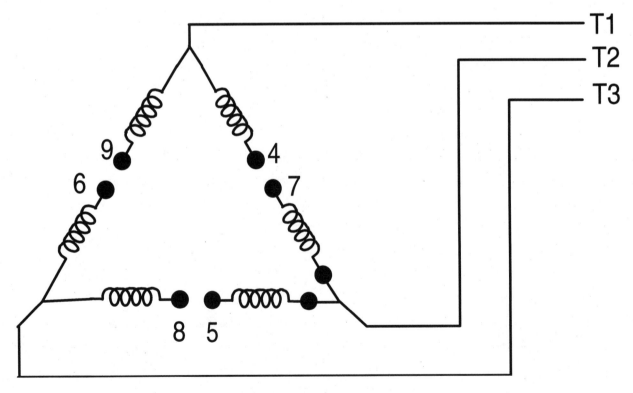

Figure 2-19: Arrangement of leads in a 9-lead, delta-wound, dual-voltage motor.

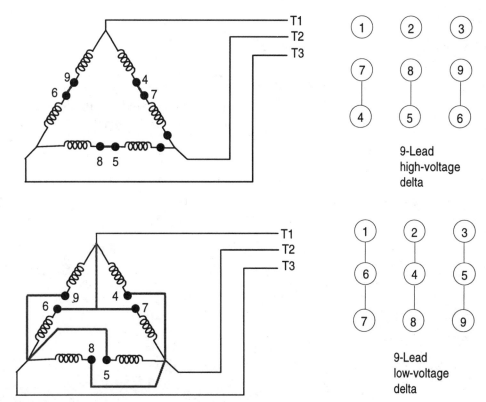

Figure 2-20: Lead connections for a three-phase, dual-voltage, delta-wound motor.

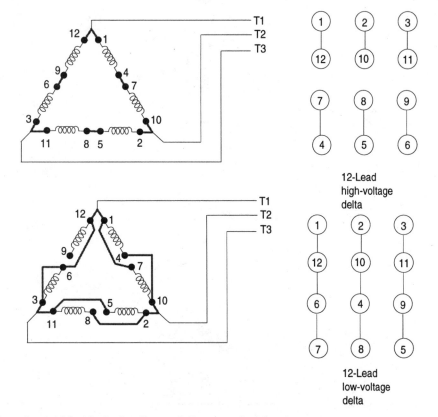

Figure 2-21: Connections for a 12-lead, dual-voltage, delta-wound motor.

AC Motor Starters

As their name implies, motor starters apply electric power to a motor to begin its operation. They also remove power to stop the motor. Beyond merely switching power on and off, starters include overload protection as required by the *NEC*. The *NEC* also usually requires a disconnect and short circuit protection on motor branch circuits. Fused disconnects and circuit breakers provide this and are often incorporated into a motor starter enclosure, resulting in a unit referred to as a combination starter. Figure 2-23 gives electrical ratings for ac magnetic contactors and motor starters, and the illustrations through Figure 2-31 give typical wiring diagrams.

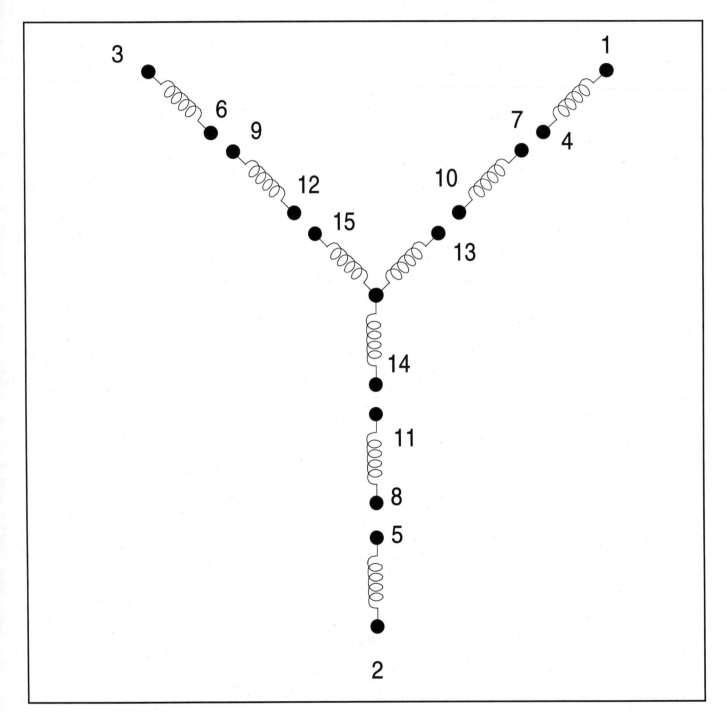

Figure 2-22: Fifteen-lead motor connection.

Electrical Ratings for AC Magnetic Contactors and Starters

NEMA SIZE	Volts	Maximum HP Nonplugging and Nonjogging Duty		Maximum HP Rating Plugging and Jogging Duty		Continuous Current Rating in amperes 600 V Max.	Service-Limit Current Rating Amperes	Tungsten and Infrared Lamp Load, Amperes 250 V Max.	Resistance Heating Loads in kW, other than Infrared Lamps Loads		kVA Rating for Switching Transformer Primaries at 50 or 60 Cycles		3-Phase Rating for Switching Capacitor
		Single Phase	Poly-Phase	Single Phase	Poly-Phase				Single Phase	Poly-Phase	Single Phase	Poly-Phase	kVAR
00	115	⅓	—	—	—	9	11	5	—	—	—	—	—
	200	—	1½	—	—	9	11	5	—	—	—	—	—
	230	1	1½	—	—	9	11	5	—	—	—	—	—
	380	—	1½	—	—	9	11	—	—	—	—	—	—
	460	—	2	—	—	9	11	—	—	—	—	—	—
	575	—	2	—	—	9	11	—	—	—	—	—	—
0	115	1	—	½	—	18	21	10	—	—	0.9	1.2	—
	200	—	3	—	1½	18	21	10	—	—	—	1.4	—
	230	2	3	1	1½	18	21	10	—	—	1.4	1.7	—
	380	—	5	—	1½	18	21	—	—	—	—	2.0	—
	460	—	5	—	2	18	21	—	—	—	1.9	2.5	—
	575	...	5	...	2	18	21	...	...	...	1.9	2.5	...
1	115	2	—	1	—	27	32	15	3	5	1.4	1.7	—
	200	—	7½	—	3	27	32	15	—	9.1	—	3.5	—
	230	3	7½	2	3	27	32	15	6	10	1.9	4.1	—
	380	—	10	—	5	27	32	—	—	16.5	—	4.3	—
	460	—	10	—	5	27	32	—	12	20	3	5.3	—
	575	—	10	—	5	27	32	—	15	25	3	5.3	—
1P	115	3	—	1½	—	36	42	24	—	—	—	—	—
	230	5	—	3	—	36	42	24	—	—	—	—	—
2	115	3	—	2	—	45	52	30	5	8.5	1.0	4.1	—
	200	—	10	—	7½	45	52	30	—	15.4	—	6.6	11.3
	230	7½	15	5	10	45	52	30	10	17	4.6	7.6	13
	380	—	25	—	15	45	52	—	—	28	—	9.9	21
	460	—	25	—	15	45	52	—	20	34	5.7	12	26
	575	—	25	—	15	45	52	—	25	43	5.7	12	33
3	115	7½	—	—	—	90	104	60	10	17	4.6	7.6	—
	200	—	25	—	15	90	104	60	—	31	—	13	23.4
	230	15	30	—	20	90	104	60	20	34	8.6	15	27
	380	—	50	—	30	90	104	—	—	56	—	19	43.7
	460	—	50	—	30	90	104	—	40	68	14	23	53
	575	—	50	—	30	90	104	—	50	86	14	23	67
4	200	—	40	—	25	135	156	120	—	45	—	20	34
	230	—	50	—	30	135	156	120	30	52	11	23	40
	380	—	75	—	50	135	156	—	—	86.7	—	38	66
	460	—	100	—	60	135	156	—	60	105	22	46	80
	575	—	100	—	60	135	156	—	75	130	22	46	100

Figure 2-23: Electrical ratings for ac magnetic contactors and starters.

NEMA SIZE	Volts	Maximum HP Nonplugging and Nonjogging Duty		Maximum HP Rating Plugging and Jogging Duty		Continuous Current Rating in amperes 600 V Max.	Service-Limit Current Rating Amperes	Tungsten and Infrared Lamp Load, Amperes 250 V Max.	Resistance Heating Loads in kW, other than Infrared Lamps Loads		kVA Rating for Switching Transformer Primaries at 50 or 60 Cycles		3-Phase Rating for Switching Capacitor
		Single Phase	Poly-Phase	Single Phase	Poly-Phase				Single Phase	Poly-Phase	Single Phase	Poly-Phase	kVAR
5	200	—	75	—	60	270	311	240	—	91	—	40	69
	230	—	100	—	75	270	311	240	60	105	28	46	80
	380	—	150	—	125	270	311	—	—	173	—	75	132
	460	—	200	—	150	270	311	—	120	210	40	91	160
	575	—	200	—	150	270	311	—	150	260	40	91	200
6	200	—	150	—	125	540	621	480	—	182	—	79	139
	230	—	200	—	150	540	621	480	120	210	57	91	160
	380	—	300	—	250	540	621	—	—	342	—	148	264
	460	—	400	—	300	540	621	—	240	415	86	180	320
	575	—	400	—	300	540	621	—	300	515	86	180	400
7	230	—	300	—	—	810	932	720	180	315	—	—	240
	460	—	600	—	—	810	932	—	360	625	—	—	480
	575	—	600	—	—	810	932	—	450	775	—	—	600
8	230	—	450	—	—	1215	1400	1080	—	—	—	—	360
	460	—	900	—	—	1215	1400	—	—	—	—	—	720
	575	—	900	—	—	1215	1400	—	—	—	—	—	900

Figure 2-23: Electrical ratings for ac magnetic contactors and starters. *(Cont.)*

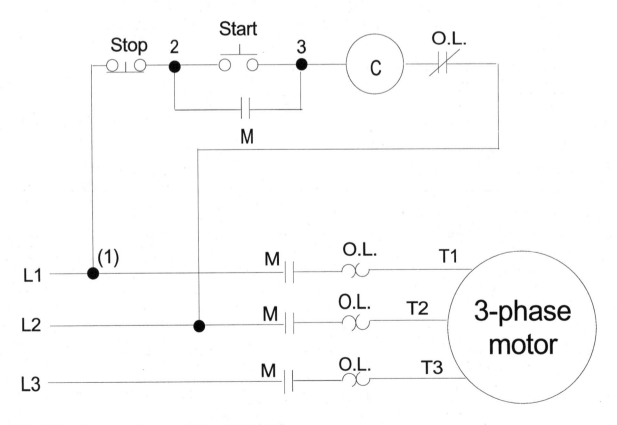

Figure 2-24: Wiring diagram of basic motor-control circuit.

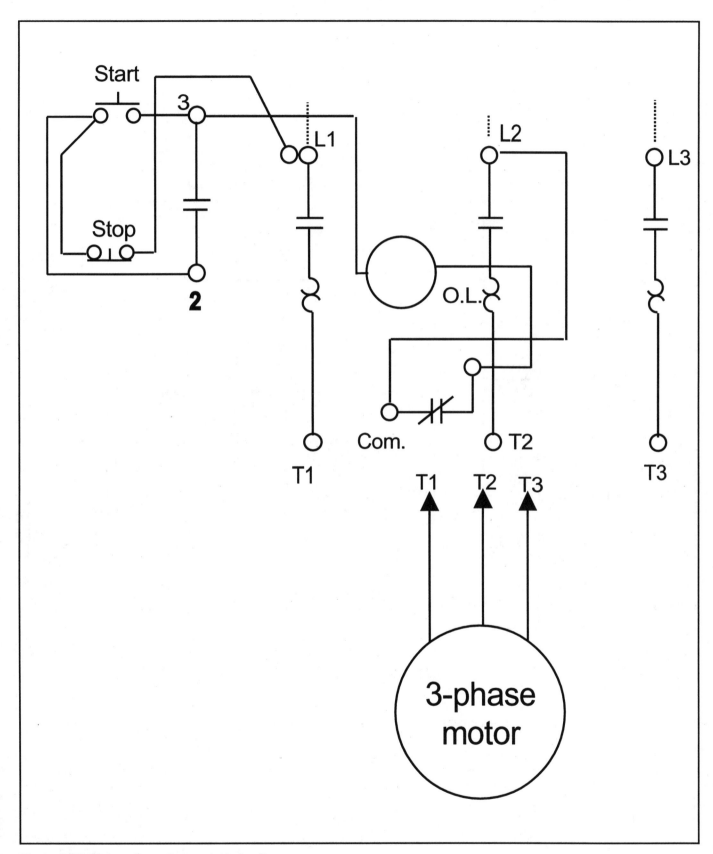

Figure 2-25: Three-wire motor control circuit.

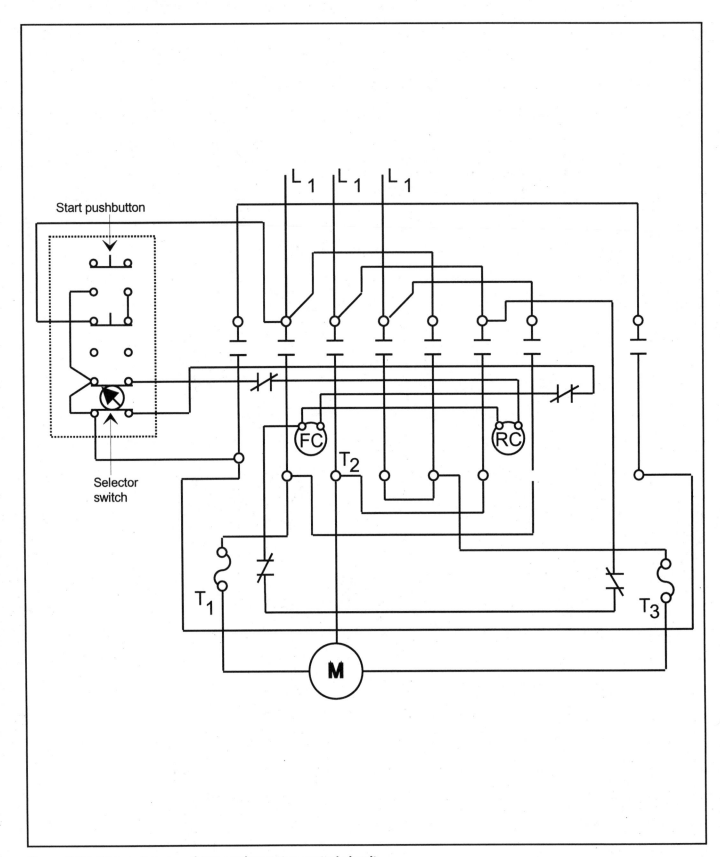

Figure 2-26: Wiring diagram of a reversing motor control circuit.

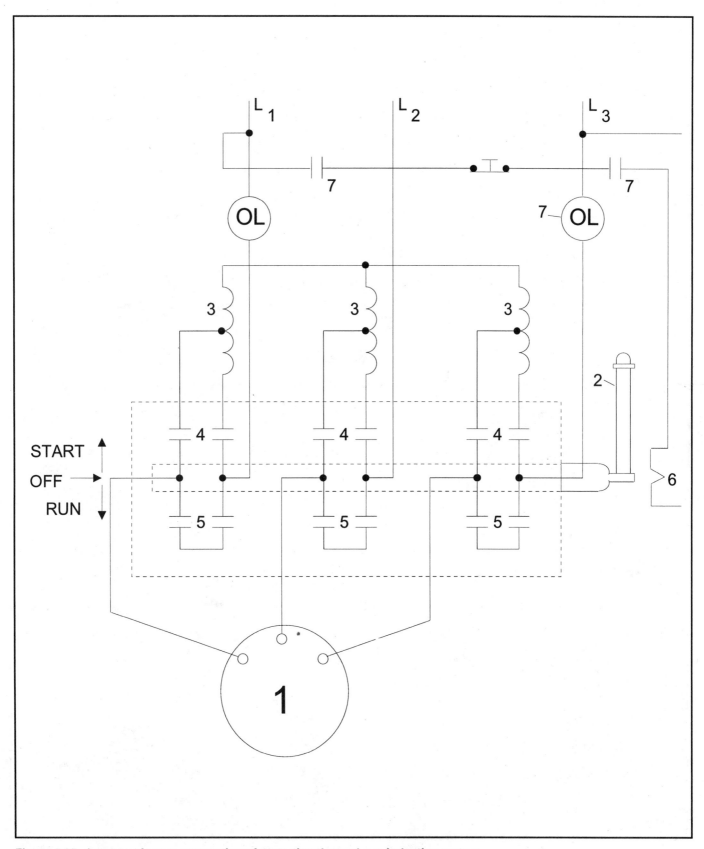

Figure 2-27: Autotransformer connections for starting three-phase induction motors.

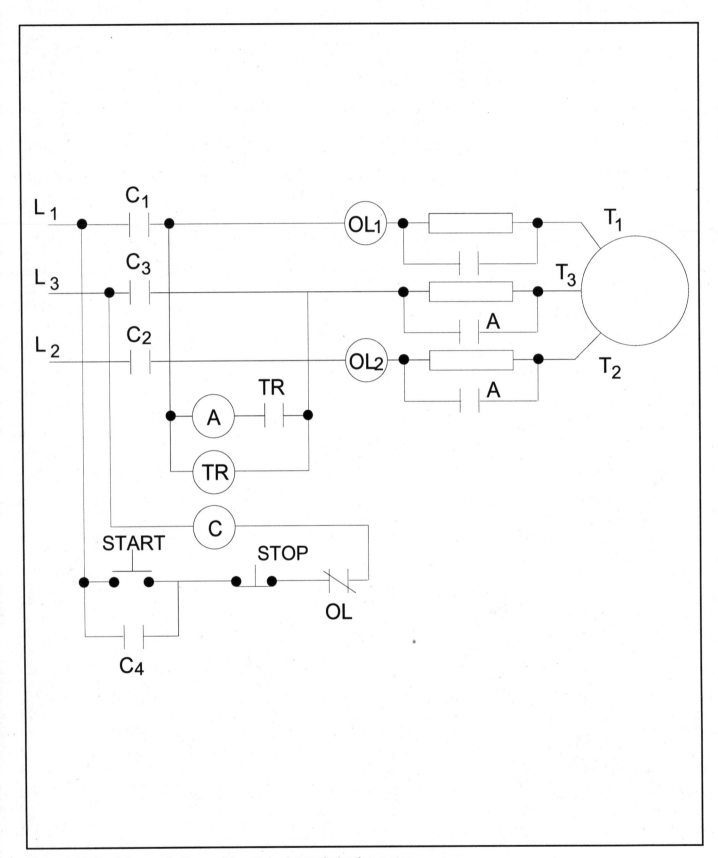

Figure 2-28: Resistance starter used for squirrel-cage induction motor.

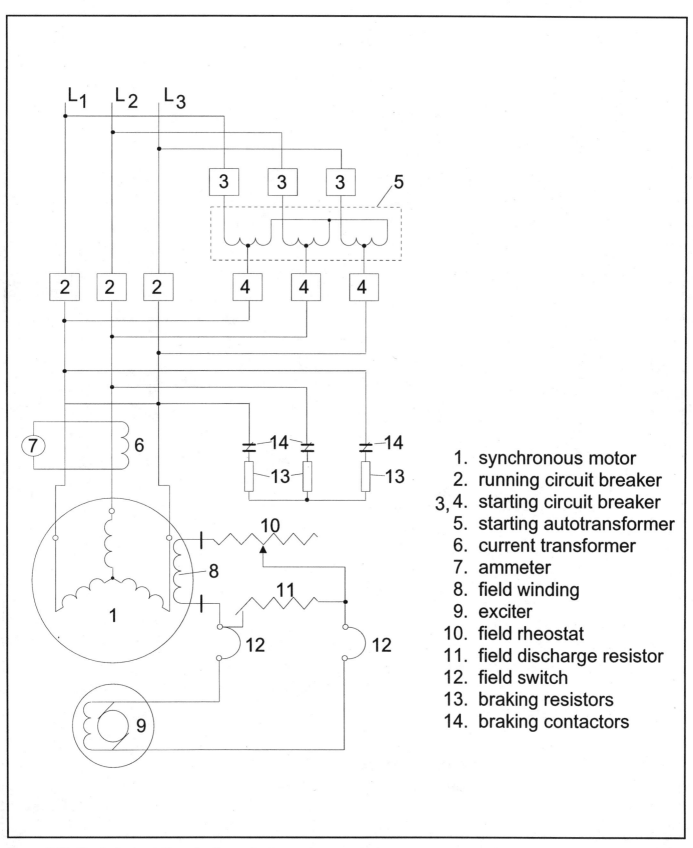

Figure 2-29: Control connections for three-phase synchronous motor.

1. synchronous motor
2. running circuit breaker
3, 4. starting circuit breaker
5. starting autotransformer
6. current transformer
7. ammeter
8. field winding
9. exciter
10. field rheostat
11. field discharge resistor
12. field switch
13. braking resistors
14. braking contactors

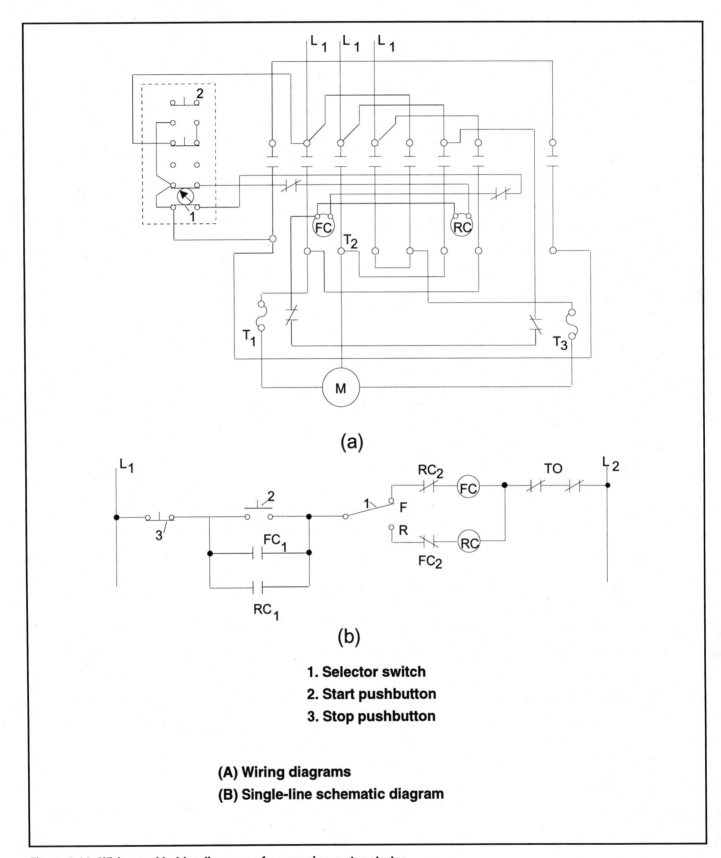

(a)

(b)

1. Selector switch
2. Start pushbutton
3. Stop pushbutton

(A) Wiring diagrams
(B) Single-line schematic diagram

Figure 2-30: Wiring and ladder diagrams of a reversing motor starter.

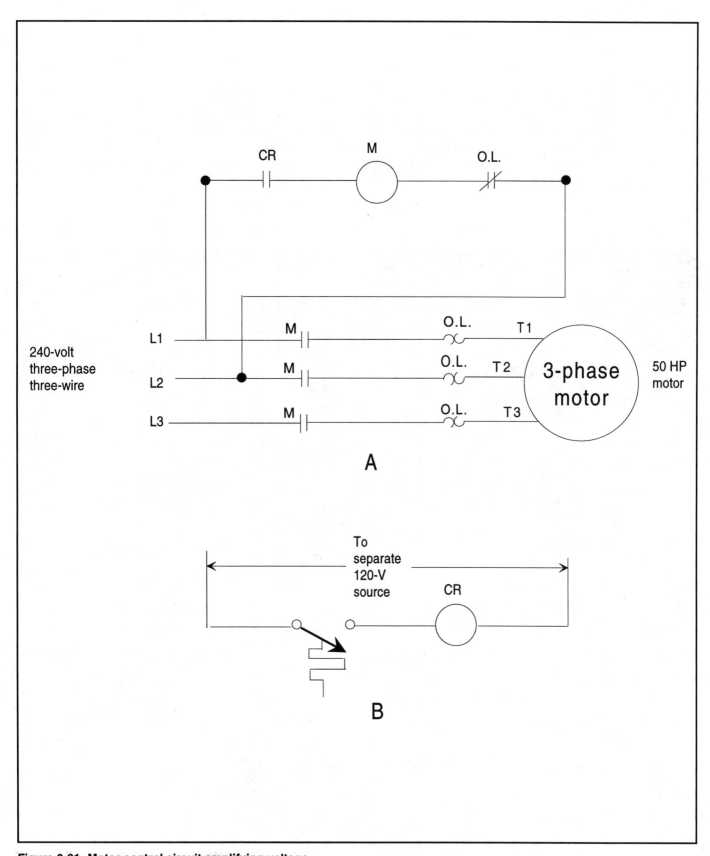

Figure 2-31: Motor control circuit amplifying voltage.

ALUMINUM CONDUCTORS

Aluminum and copper-clad aluminum conductors are being used to an ever-increasing extent. Although aluminum is lighter and easier to bend than copper, an AWG No. 12 aluminum wire has greater resistance than an equal length of AWG No. 12 copper wire. Therefore, a larger diameter aluminum conductor must be used for a given load or current demand. *See* Figures 2-32 through 2-37.

In general, aluminum conductors are used mainly to replace the larger sizes of copper; that is, No. 1/0 and larger, or where long runs are encountered.

The use of aluminum conductors with brass or copper terminals can lead to electrolytic action which causes corrosion, high resistance, loose connections, and overheating. However, much progress has been made in solving these problems, using AL-CU connectors and copper-clad aluminum conductors.

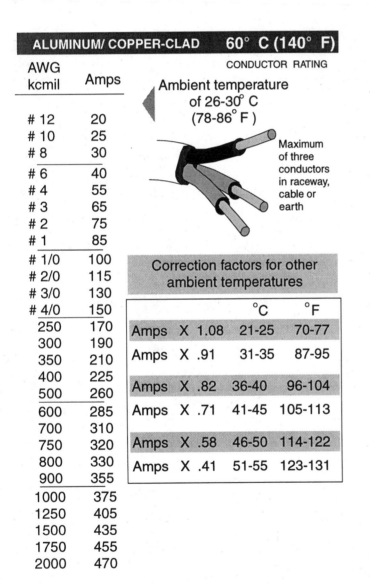

ALUMINUM/ COPPER-CLAD   60° C (140° F)

AWG kcmil	Amps
# 12	20
# 10	25
# 8	30
# 6	40
# 4	55
# 3	65
# 2	75
# 1	85
# 1/0	100
# 2/0	115
# 3/0	130
# 4/0	150
250	170
300	190
350	210
400	225
500	260
600	285
700	310
750	320
800	330
900	355
1000	375
1250	405
1500	435
1750	455
2000	470

CONDUCTOR RATING
Ambient temperature of 26-30° C (78-86° F)

Maximum of three conductors in raceway, cable or earth

Correction factors for other ambient temperatures

		°C	°F
Amps X	1.08	21-25	70-77
Amps X	.91	31-35	87-95
Amps X	.82	36-40	96-104
Amps X	.71	41-45	105-113
Amps X	.58	46-50	114-122
Amps X	.41	51-55	123-131

Figure 2-32: Current-carrying capacity of insulated aluminum or copper-clad conductors with insulation rated at 60°C — 0 to 2000 V — in cable or raceway systems.

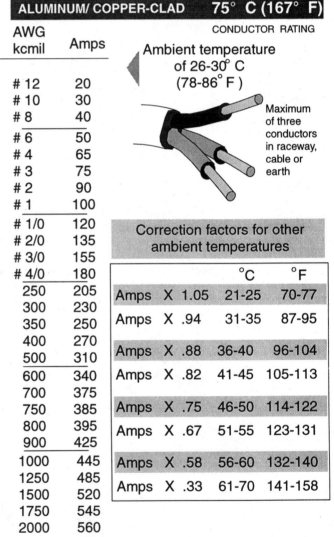

ALUMINUM/ COPPER-CLAD   75° C (167° F)

AWG kcmil	Amps
# 12	20
# 10	30
# 8	40
# 6	50
# 4	65
# 3	75
# 2	90
# 1	100
# 1/0	120
# 2/0	135
# 3/0	155
# 4/0	180
250	205
300	230
350	250
400	270
500	310
600	340
700	375
750	385
800	395
900	425
1000	445
1250	485
1500	520
1750	545
2000	560

CONDUCTOR RATING
Ambient temperature of 26-30° C (78-86° F)

Maximum of three conductors in raceway, cable or earth

Correction factors for other ambient temperatures

		°C	°F
Amps X	1.05	21-25	70-77
Amps X	.94	31-35	87-95
Amps X	.88	36-40	96-104
Amps X	.82	41-45	105-113
Amps X	.75	46-50	114-122
Amps X	.67	51-55	123-131
Amps X	.58	56-60	132-140
Amps X	.33	61-70	141-158

Figure 2-33: Current-carrying capacity of insulated aluminum or copper-clad conductors with insulation rated at 75°C — 0 to 2000 V — in cable or raceway systems.

As with other types of conductors, the rated current-carrying capacity of aluminum is based on the size of conductors, type of insulation, ambient temperature and the number of conductors in a raceway.

The metallic conductor itself is not harmed by heat. However, if the insulation is overheated, it is harmed in various ways, depending on the amount of overheating and the type of insulation. In all cases, insulation loses its usefulness if overheated, leading to breakdowns and fires.

The tables in Figures 2-32 through 2-37 give the current-carrying capacity of aluminum and copper-clad conductors with various types of insulation.

The conductors in Figures 2-32 through 2-37 are for voltages up to, and including, 2000 V; that is, 0 to 2000 V. For ratings of higher voltages, *see* Chapter 4.

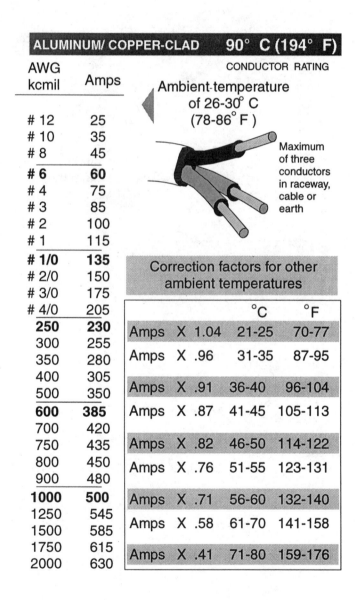

ALUMINUM/ COPPER-CLAD   90° C (194° F)

AWG kcmil	Amps
# 12	25
# 10	35
# 8	45
# 6	60
# 4	75
# 3	85
# 2	100
# 1	115
# 1/0	135
# 2/0	150
# 3/0	175
# 4/0	205
250	230
300	255
350	280
400	305
500	350
600	385
700	420
750	435
800	450
900	480
1000	500
1250	545
1500	585
1750	615
2000	630

CONDUCTOR RATING

Ambient-temperature of 26-30° C (78-86° F)

Maximum of three conductors in raceway, cable or earth

Correction factors for other ambient temperatures

		°C	°F
Amps X 1.04		21-25	70-77
Amps X .96		31-35	87-95
Amps X .91		36-40	96-104
Amps X .87		41-45	105-113
Amps X .82		46-50	114-122
Amps X .76		51-55	123-131
Amps X .71		56-60	132-140
Amps X .58		61-70	141-158
Amps X .41		71-80	159-176

ALUMINUM/ COPPER-CLAD   60° C (140° F)

AWG kcmil	Amps
# 12	25
# 10	35
# 8	45
# 6	60
# 4	80
# 3	95
# 2	110
# 1	130
# 1/0	150
# 2/0	175
# 3/0	200
# 4/0	235
250	265
300	290
350	330
400	355
500	405
600	455
700	500
750	515
800	535
900	580
1000	625
1250	710
1500	795
1750	875
2000	960

CONDUCTOR RATING

Ambient temperature of 26-30° C (78-86° F)

single insulated conductors

Correction factors for other ambient temperatures

		°C	°F
Amps X 1.08		21-25	70-77
Amps X .91		31-35	87-95
Amps X .82		36-40	96-104
Amps X .71		41-45	105-113
Amps X .58		46-50	114-122
Amps X .41		51-55	123-131

Figure 2-34: Current-carrying capacity of insulated aluminum or copper-clad conductors with insulation rated at 90°C — 0 to 2000 V — in cable or raceway systems.

Figure 2-35: Current-carrying capacity of insulated aluminum or copper-clad conductors with insulation rated at 60°C — 0 to 2000 V — in free air.

ALUMINUM CONDUIT

Rigid aluminum conduit is supplied with both ends threaded and in standard 10-ft lengths, and is lighter and easier to install than rigid steel conduit. One coupling is supplied with each length of conduit. This type of conduit is used in some electrical installations due to its resistance to corrosive atmospheres and chemicals. Aluminum conduit, however, is not suitable for direct burial in the earth or concrete encasement due to the electrolytic reaction on the conduit.

Aluminum rigid conduit is available in inside diameters sizes from ½-in through 6 in. The table in Figure 2-38 gives important data for rigid aluminum conduit.

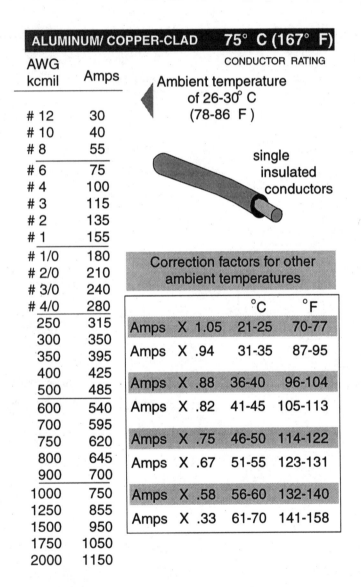

Figure 2-36: Current-carrying capacity of insulated aluminum or copper-clad conductors with insulation rated at 75°C — 0 to 2000 V — in free air.

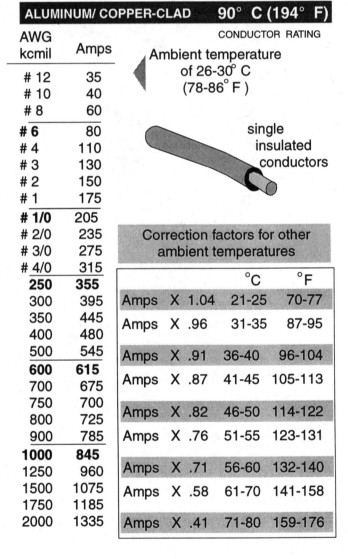

Figure 2-37: Current-carrying capacity of insulated aluminum or copper-clad conductors with insulation rated at 90°C — 0 to 2000 V — in free air.

ALUMINUM RIGID CONDUIT — NOMINAL 10-FT LENGTHS							
Trade Size	Lb per 100 ft	No. per Bundle	Lb per Bundle	Master Shipping Package			
				Bundles	Pieces	Feet	Pounds
½	20.8	10	29.8	20	200	2,000	596
¾	39.8	10	39.8	20	200	2,000	796
1	58.9	10	58.9	10	100	1,000	589
1¼	79.8	5	39.9	10	50	500	399
1½	95.6	5	47.8	10	50	500	478
2	128.8	5	64.4	9	45	450	580
2½	204.7	1	20.47	Loose	30	300	614
3	268.0	1	26.8	Loose	20	200	536
3½	321.3	1	32.13	Loose	20	200	642
4	382.1	1	38.21	Loose	20	200	764
5	521.5	1	52.15	Loose	8	80	417
6	677.5	1	67.75	Loose	6	60	406

Figure 2-38: Specifications of aluminum rigid conduit.

BOXES

Boxes used in electrical installations are classified as outlet, device, junction, or pull boxes, depending on their use. Boxes with supports are also required in vertical raceways where the weight of the cable would place an excessive strain on the conductor terminals. Such conduit-support boxes are common in high-rise buildings where heavy feeder conductors are involved.

When conduit-type conductor supports with split taper are used, hard fiber bushings are available that fit over the end of the conduit risers. The size of the support box is determined the same as any junction or pull box. Where the box or cabinet is to be used as a combined splice and support box, there should be ample space above the support units to make the splices and install the required insulation.

The majority of boxes used in electrical installations are made of steel with a galvanized finish, but boxes are also constructed of cast iron, sheet metal, nonmetallic PVC, and other materials.

Outlet, device and junction boxes are sized in accordance with *NEC* Section 370-16.

Junction and Pull Boxes

Junction and pull boxes provide access points for pulling and feeding conductors into a raceway system. Their use is mandatory in conduit runs where the number of bends between outlets exceed the maximum permitted by the *NEC*. The following detail drawings of junction and pull boxes, along with other types, are those enclosures currently in common use throughout the electrical industry.

The larger the box, the easier it is to pull, install, and position conductors, along with any supports. Furthermore, splices or taps are also easier made if the box is slightly oversized.

Boxes and conduit bodies used as pull or junction boxes must comply with paragraphs (a) through (d) of *NEC* Section 370-28. In general, when the box is used in straight pulls and where the conduit enters and leaves on opposite sides of the box, the length of the box must be at least 8 times the nominal diameter of the largest raceways. Good practice, however, suggests a minimum length of 12 times the diameter to allow some excess capacity and efficient installation.

Figures 2-39 through 2-43 cover the majority of pull and junction boxes in common use.

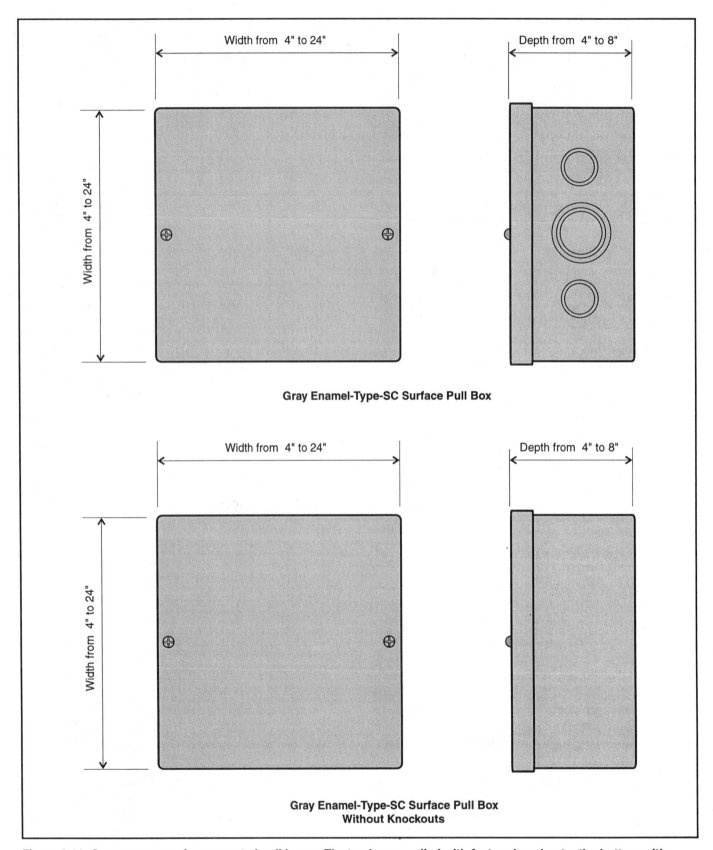

Width from 4" to 24"

Depth from 4" to 8"

Width from 4" to 24"

Gray Enamel-Type-SC Surface Pull Box

Width from 4" to 24"

Depth from 4" to 8"

Width from 4" to 24"

**Gray Enamel-Type-SC Surface Pull Box
Without Knockouts**

Figure 2-39: Screw-cover surface-mounted pull boxes. The top box supplied with factory knockouts; the bottom with none.

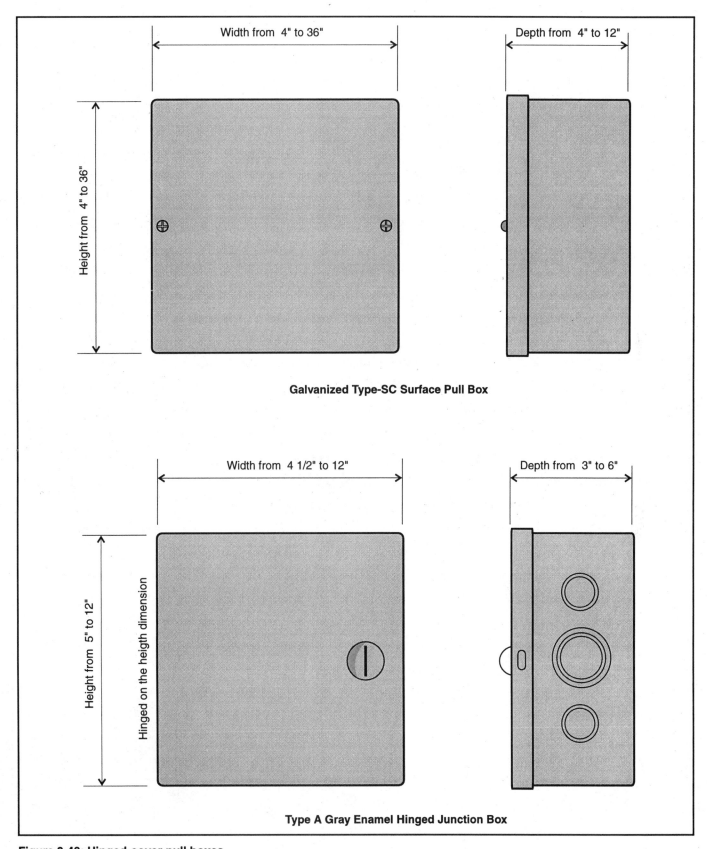

Width from 4" to 36"

Depth from 4" to 12"

Height from 4" to 36"

Galvanized Type-SC Surface Pull Box

Width from 4 1/2" to 12"

Depth from 3" to 6"

Height from 5" to 12"

Hinged on the heigth dimension

Type A Gray Enamel Hinged Junction Box

Figure 2-40: Hinged-cover pull boxes.

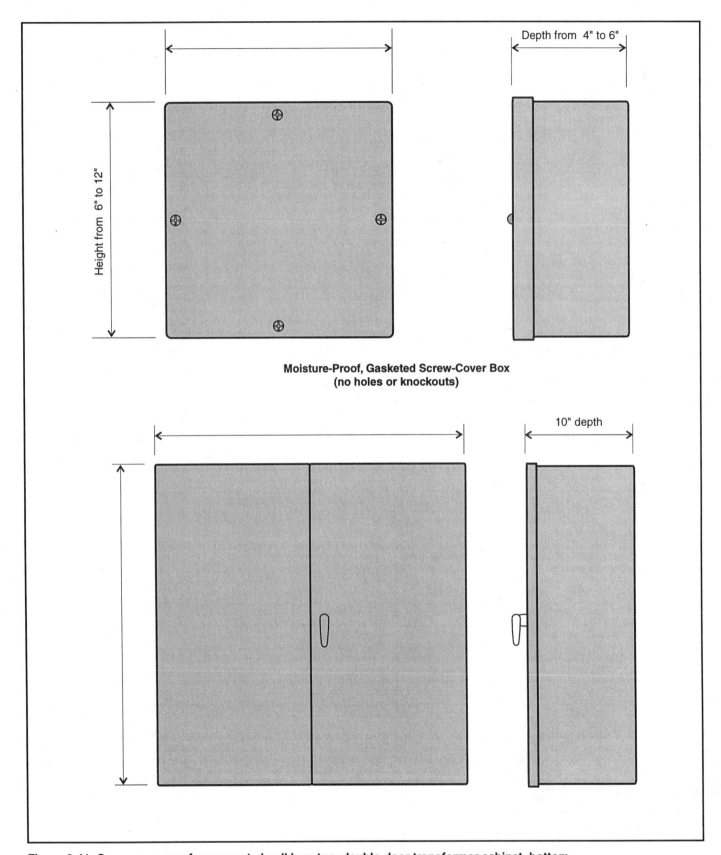

Height from 6" to 12"

Depth from 4" to 6"

Moisture-Proof, Gasketed Screw-Cover Box
(no holes or knockouts)

10" depth

Figure 2-41: Screw-cover surface-mounted pull box, top; double-door transformer cabinet, bottom.

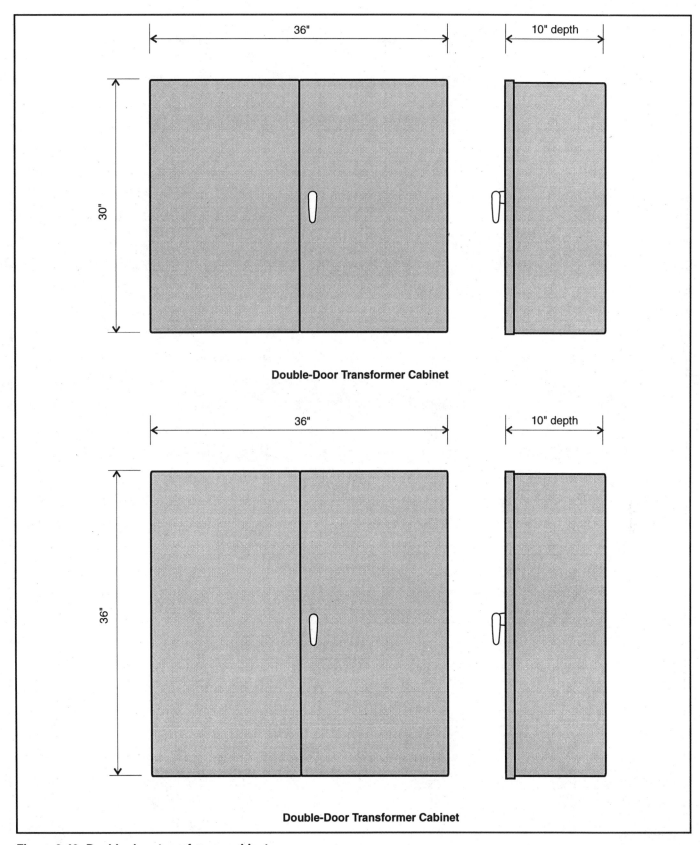

Double-Door Transformer Cabinet

Double-Door Transformer Cabinet

Figure 2-42: Double-door transformer cabinets.

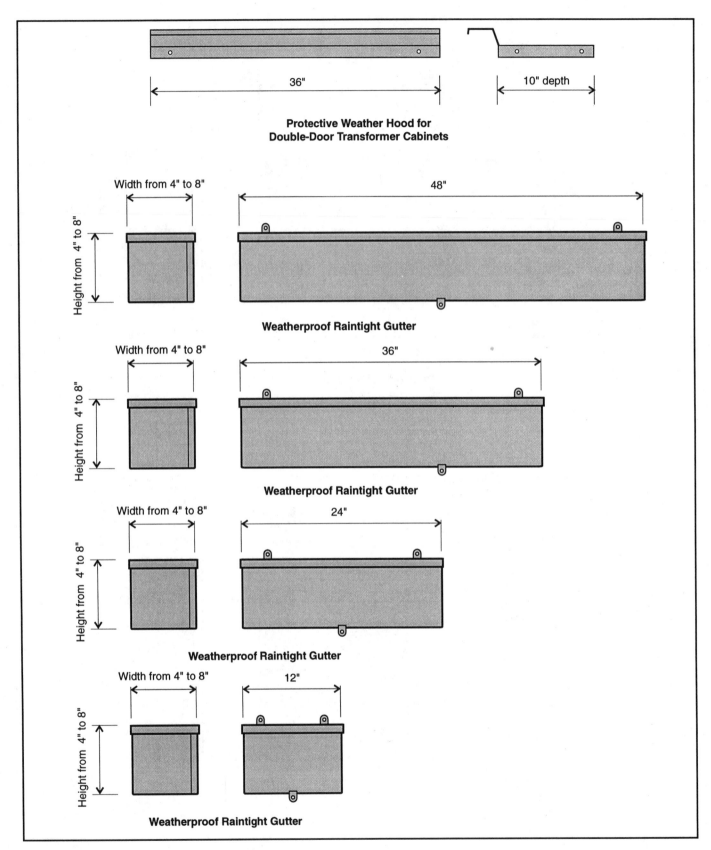

Protective Weather Hood for
Double-Door Transformer Cabinets

36"

10" depth

Width from 4" to 8"

48"

Height from 4" to 8"

Weatherproof Raintight Gutter

Width from 4" to 8"

36"

Height from 4" to 8"

Weatherproof Raintight Gutter

Width from 4" to 8"

24"

Height from 4" to 8"

Weatherproof Raintight Gutter

Width from 4" to 8"

12"

Height from 4" to 8"

Weatherproof Raintight Gutter

Figure 2-43: Dimensions of raintight auxiliary gutters.

Outlet Boxes

A box or fitting must be installed at:

- Each conductor splice point
- Each outlet, switch point, or junction point
- Each pull point for the connection of conduit and other raceways

Furthermore, boxes or other fittings are required when a change is made from conduit to open wiring. Electrical workers also install pull boxes in raceway systems to facilitate the pulling of conductors.

In each case — raceways, outlet boxes, pull and junction boxes — the *NEC* specifies specific maximum fill requirements; that is, the area of conductors in relation to the box, fitting, or raceway system.

Sizing Outlet Boxes

In general, the maximum number of conductors permitted in standard outlet boxes is listed in *NEC* Table 370-16(a). These figures apply where no fittings or devices such as fixture studs, cable clamps, switches, or receptacles are contained in the box and where no grounding conductors are part of the wiring within the box. Obviously, in all modern residential wiring systems there will be one or more of these items contained in the outlet box. Therefore, where one or more of the above mentioned items are present, the number of conductors must be one less than shown in the tables. For example, a deduction of two conductors must be made for each strap containing a device such as a switch or duplex receptacle; a further deduction of one conductor must be made for one or more grounded conductors entering the box. A 3- × 2- × 2¾-in box for example, is listed in the table as containing a maximum number of six No. 12 wires. If the box contains cable clamps and a duplex receptacle, three wires will have to be deducted from the total of six — providing for only three No. 12 wires. If a ground wire is used, only two No. 12 wires may be used.

Figure 2-44 illustrates one possible wiring configuration for outlet boxes and the maximum number of conductors permitted in them as governed by *NEC* Section 370-

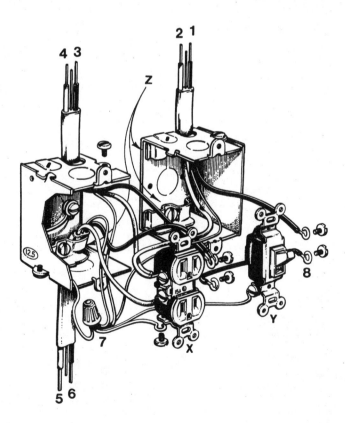

Figure 2-44: One possible outlet box configuration for a ganged switch and receptacle.

16. This example shows two single-gang switch boxes joined or "ganged" together to hold a single-pole toggle switch and a duplex receptacle. This type of arrangement is likely to be found above kitchen countertops where the duplex receptacle is provided for small appliances and the single-pole switch could be used to control a garbage disposal. This arrangement is also useful above a workbench — the receptacle for small power tools and the switch to control lighting over the bench.

Since Table 370-16(a) gives the capacity of one $3 \times 2 \times 2\frac{1}{4}$-in device box as 12.5 in^3, the total capacity of both boxes in Figure 2-44 is 25 in^3. These two boxes have a capacity to allow 10 No. 12 AWG conductors, or 12 No. 14 AWG conductors, less the following deductions:

- Two conductors must be deducted for each strap-mounted device. Since there is one duplex receptacle (X) and one single-pole toggle switch (Y), four conductors must be deducted from the total number stated in the above paragraph.

- Since the combined boxes contain one or more cable clamps (Z), another conductor must be deducted. Note that only one deduction is made for similar clamps, regardless of the number. However, any unused clamps may be removed to facilitate the electrical worker s job; that is, allowing for more work space.

- The equipment grounding conductors, regardless of the number, count as one conductor only.

Therefore, to comply with the *NEC*, and considering the combined deduction of six conductors, only four No. 12 AWG conductors (six No. 14 AWG conductors) may be installed in the outlet-box configuration in Figure 2-44.

Three nonmetallic-sheathed (NM) cables, designated 12/2 with ground, enters the ganged outlet boxes. This is a total of six current-carrying conductors and three ground wires, for a total of nine. Therefore, the arrangement shown is in violation of the *NEC*. However, if No. 14 AWG conductors were installed rather than No. 12, the configuration would comply with the 1996 *NEC*. Another alternative is to go to $3 \times 2 \times 3\frac{1}{2}$-in device boxes which would then have a total of 36 in^3 for the two boxes.

Also note the jumper wire in Figure 2-44; this is numbered "8" in the drawing. Conductors that both originate and end in the same outlet box are exempt from being counted against the allowable capacity of an outlet box. This jumper wire (8) taps off one terminal of the duplex receptacle to furnish a "hot wire" to the single-pole toggle switch. Therefore, this wire originates and terminates in the same set of ganged boxes and is not counted against the total number of conductors. By the same token, the three grounding conductors extending from the wire nut to the individual grounding screws on the devices originate and terminate in the same set of boxes. These conductors are also exempt from being counted with the total. Incidentally, the wire nut has a crimp connector beneath; wire nuts alone are not allowed to connect equipment grounding conductors.

If, say, two No. 12 AWG conductors were installed in $\frac{1}{2}$-inch EMT and terminating into an outlet box containing one duplex receptacle, what size outlet box will meet *NEC* requirements?

The first step is to count the total number of conductors and equivalents that will be used in the box — following the requirements specified in *NEC* Section 370-16.

Step 1. Calculate the total number of conductors and equivalents.

One receptacle	= 2 conductors
Two #12 conductors	= 2 conductors
Total #12 conductors	= 4

Step 2. Determine amount of space required for each conductor.

NEC Table 370-16(b) gives the box volume required for each conductor:

$$\text{No. 12 AWG} \quad = \quad 2.25 \text{ in}^3$$

Step 3. Calculate the outlet-box space required by multiplying the number of cubic inches required for each conductor by the number of conductors found in No. 1 above.

$$4 \times 2.25 = 9.00 \text{ in}^3$$

Once you have determined the required box capacity, again refer to *NEC* Table 370-16(a) and note that a $3 \times 2 \times 2\frac{1}{4}$-in box comes closest to our requirements. This box size is rated for 10.5 in^3. Of course, any box with a larger capacity is permitted.

The illustrations beginning with Figure 2-45 give dimensions and capacity of common outlet boxes.

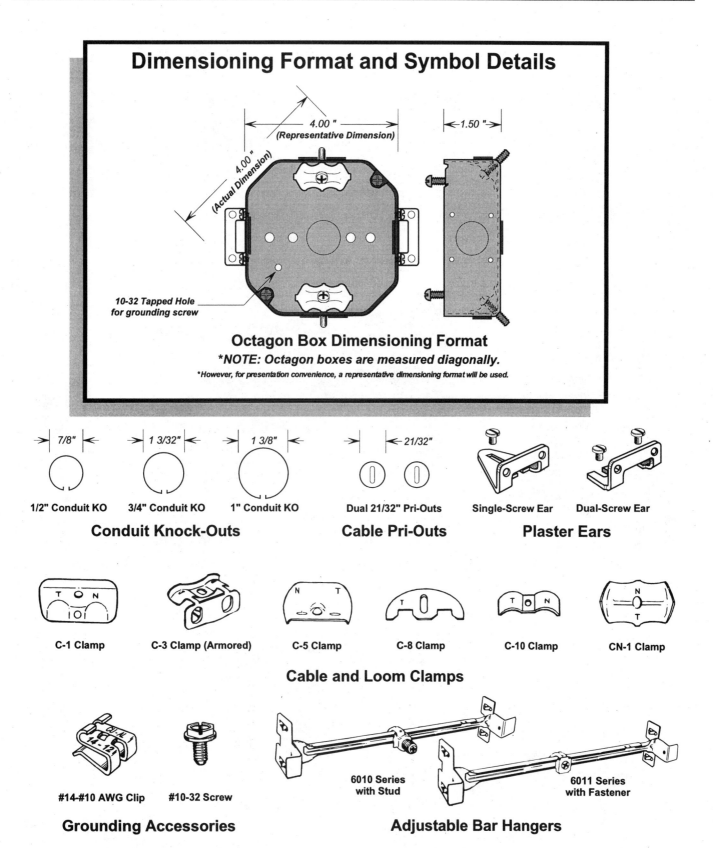

Figure 2-45: Dimensioning format and accessories for outlet boxes.

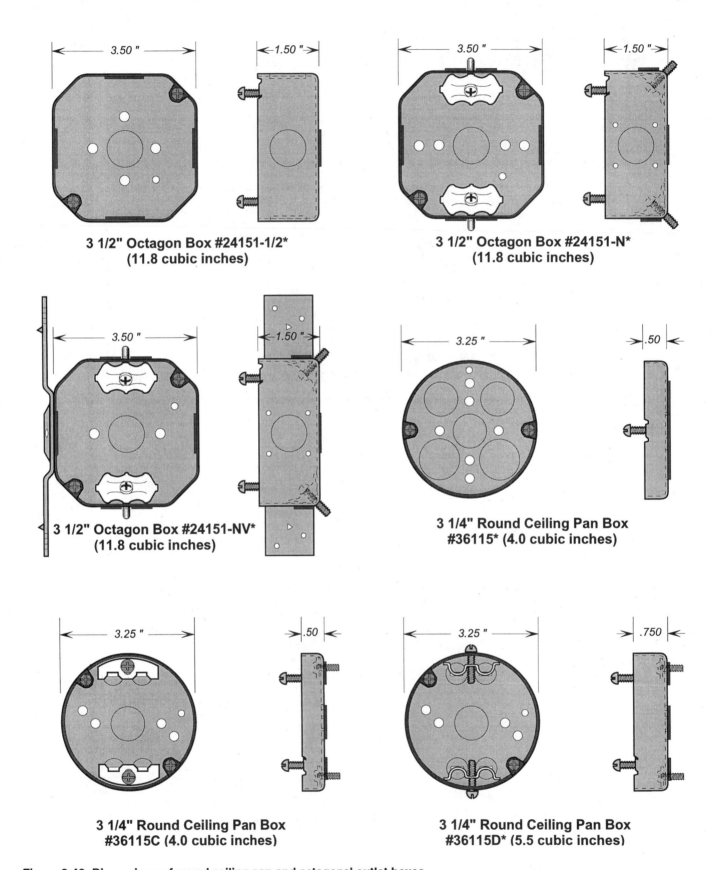

3 1/2" Octagon Box #24151-1/2*
(11.8 cubic inches)

3 1/2" Octagon Box #24151-N*
(11.8 cubic inches)

3 1/2" Octagon Box #24151-NV*
(11.8 cubic inches)

3 1/4" Round Ceiling Pan Box
#36115* (4.0 cubic inches)

3 1/4" Round Ceiling Pan Box
#36115C (4.0 cubic inches)

3 1/4" Round Ceiling Pan Box
#36115D* (5.5 cubic inches)

Figure 2-46: Dimensions of round ceiling pan and octagonal outlet boxes.

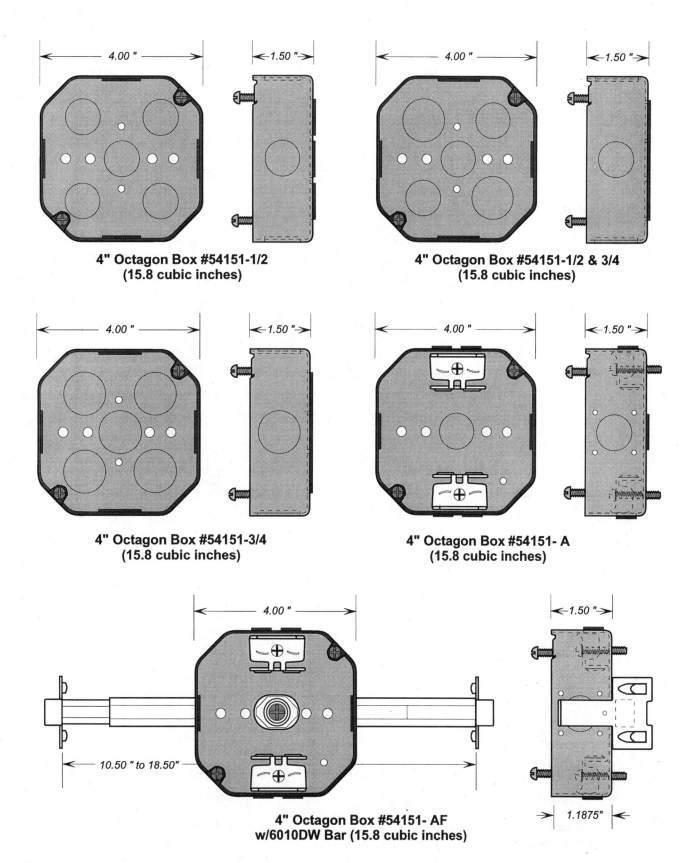

4" Octagon Box #54151-1/2
(15.8 cubic inches)

4" Octagon Box #54151-1/2 & 3/4
(15.8 cubic inches)

4" Octagon Box #54151-3/4
(15.8 cubic inches)

4" Octagon Box #54151- A
(15.8 cubic inches)

4" Octagon Box #54151- AF
w/6010DW Bar (15.8 cubic inches)

Figure 2-46: Dimensions of round ceiling pan and octagonal outlet boxes. *(Cont.)*

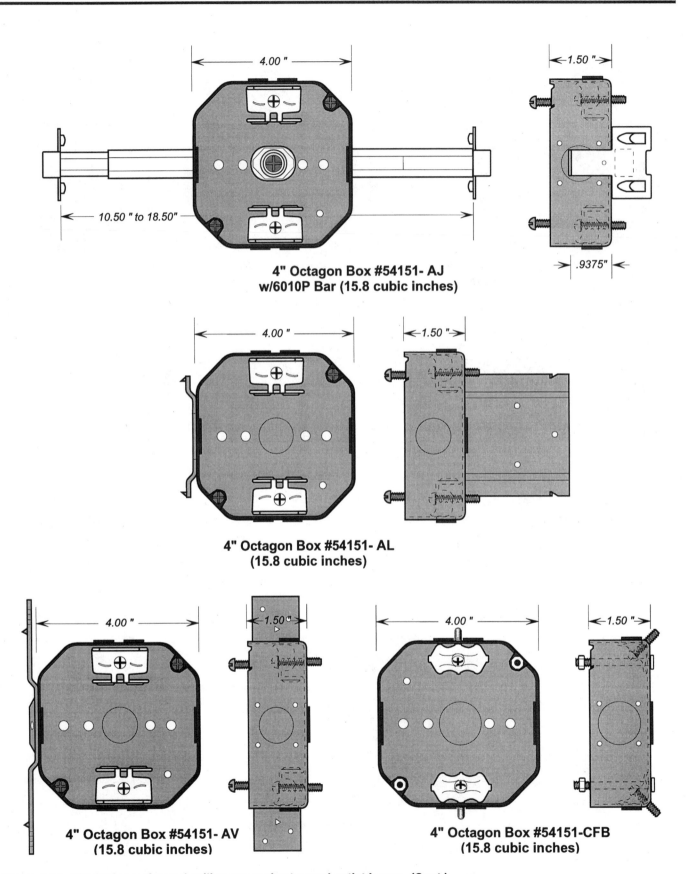

4" Octagon Box #54151- AJ
w/6010P Bar (15.8 cubic inches)

4" Octagon Box #54151- AL
(15.8 cubic inches)

4" Octagon Box #54151- AV
(15.8 cubic inches)

4" Octagon Box #54151-CFB
(15.8 cubic inches)

Figure 2-46: Dimensions of round ceiling pan and octagonal outlet boxes. *(Cont.)*

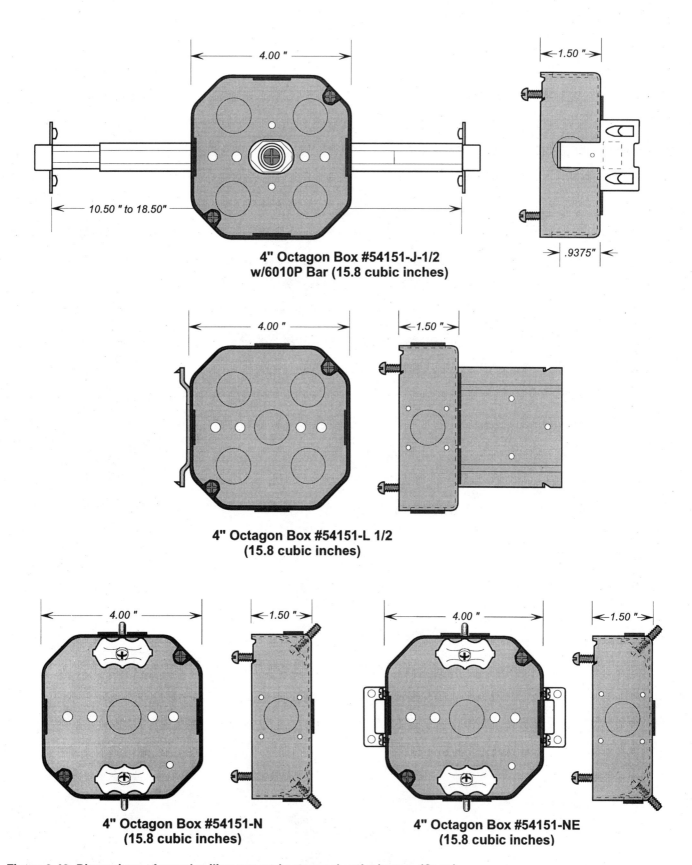

4" Octagon Box #54151-J-1/2
w/6010P Bar (15.8 cubic inches)

4" Octagon Box #54151-L 1/2
(15.8 cubic inches)

4" Octagon Box #54151-N
(15.8 cubic inches)

4" Octagon Box #54151-NE
(15.8 cubic inches)

Figure 2-46: Dimensions of round ceiling pan and octagonal outlet boxes. *(Cont.)*

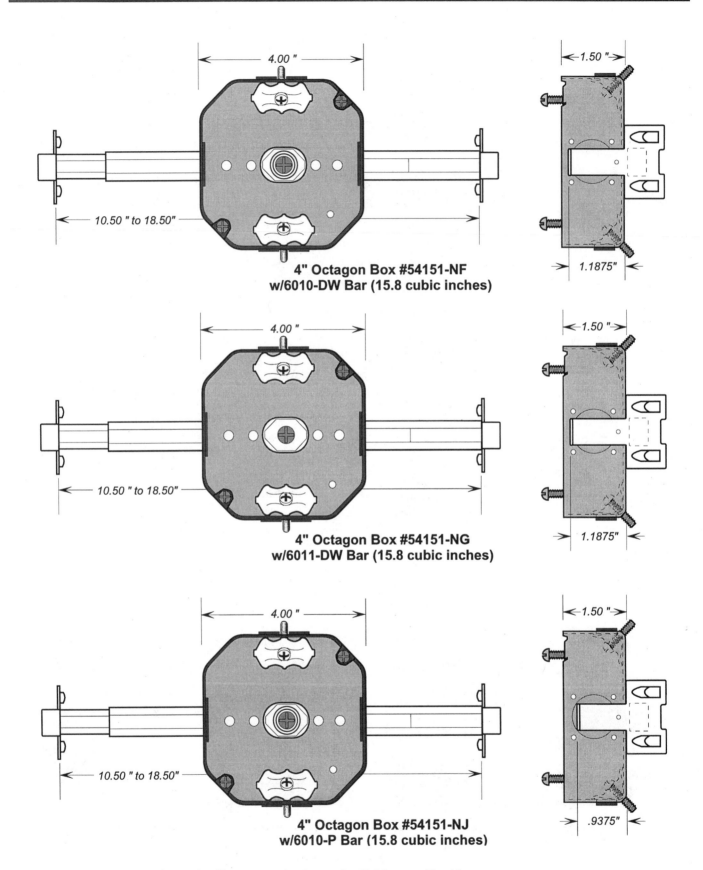

4.00 "

1.50 "

10.50 " to 18.50"

1.1875"

**4" Octagon Box #54151-NF
w/6010-DW Bar (15.8 cubic inches)**

4.00 "

1.50 "

10.50 " to 18.50"

1.1875"

**4" Octagon Box #54151-NG
w/6011-DW Bar (15.8 cubic inches)**

4.00 "

1.50 "

10.50 " to 18.50"

.9375"

**4" Octagon Box #54151-NJ
w/6010-P Bar (15.8 cubic inches)**

Figure 2-46: Dimensions of round ceiling pan and octagonal outlet boxes. *(Cont.)*

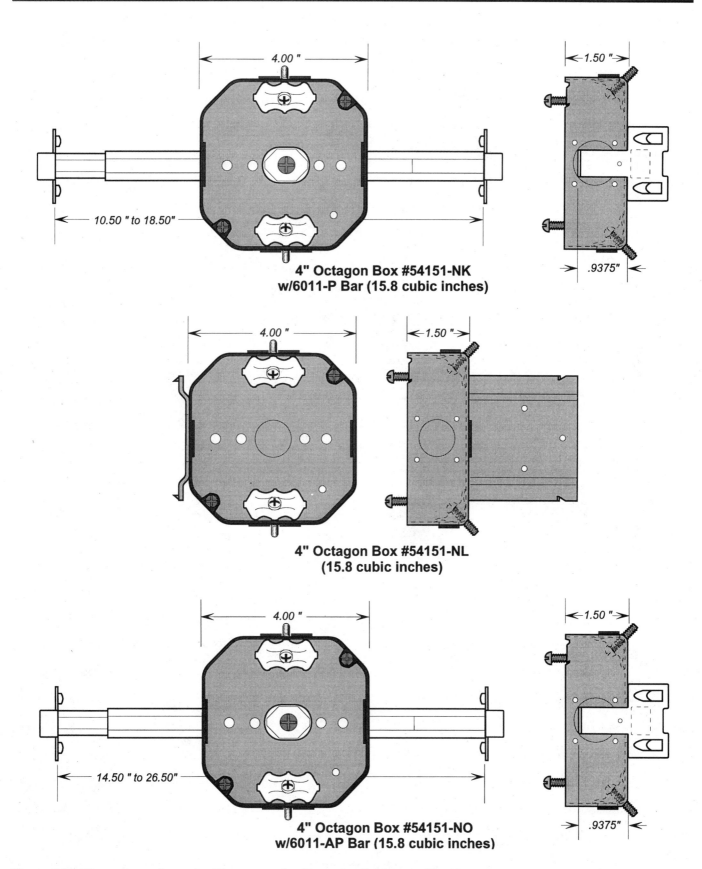

Figure 2-46: Dimensions of round ceiling pan and octagonal outlet boxes. *(Cont.)*

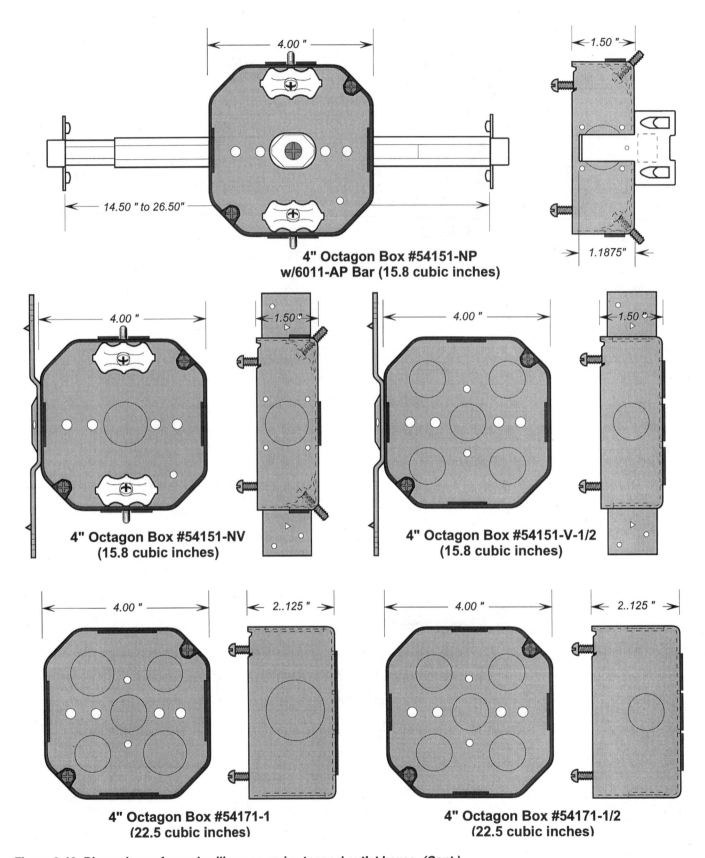

Figure 2-46: Dimensions of round ceiling pan and octagonal outlet boxes. *(Cont.)*

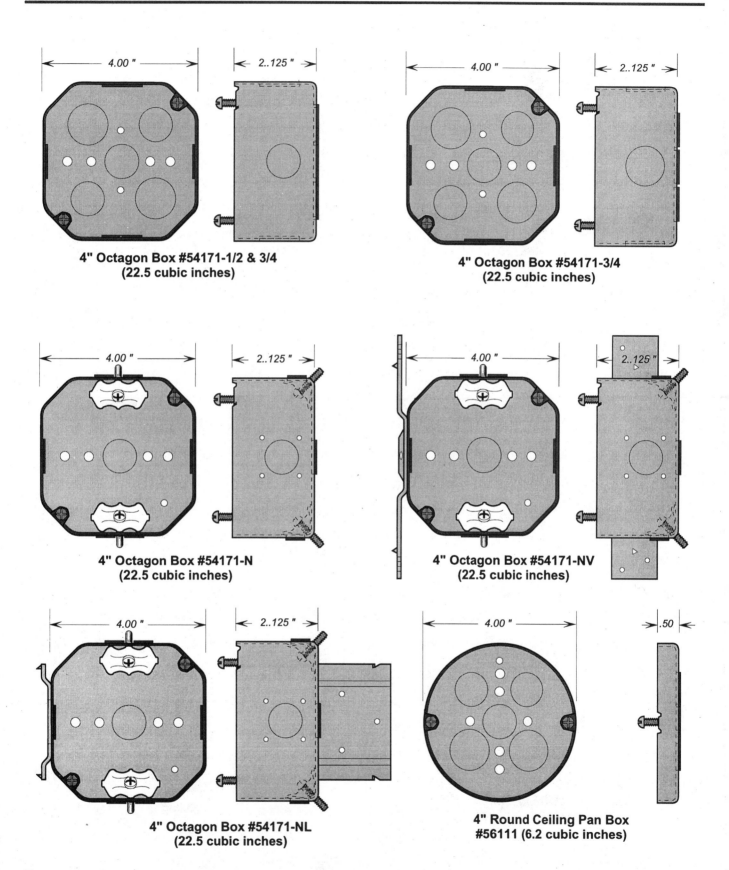

4" Octagon Box #54171-1/2 & 3/4
(22.5 cubic inches)

4" Octagon Box #54171-3/4
(22.5 cubic inches)

4" Octagon Box #54171-N
(22.5 cubic inches)

4" Octagon Box #54171-NV
(22.5 cubic inches)

4" Octagon Box #54171-NL
(22.5 cubic inches)

4" Round Ceiling Pan Box
#56111 (6.2 cubic inches)

Figure 2-46: Dimensions of round ceiling pan and octagonal outlet boxes. *(Cont.)*

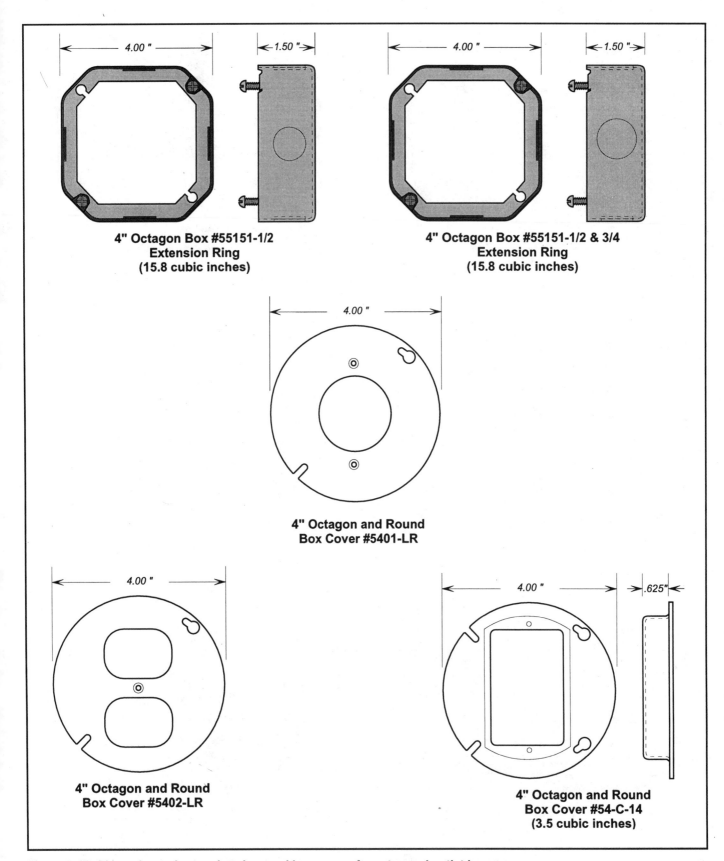

4" Octagon Box #55151-1/2
Extension Ring
(15.8 cubic inches)

4" Octagon Box #55151-1/2 & 3/4
Extension Ring
(15.8 cubic inches)

4" Octagon and Round
Box Cover #5401-LR

4" Octagon and Round
Box Cover #5402-LR

4" Octagon and Round
Box Cover #54-C-14
(3.5 cubic inches)

Figure 2-47: Dimensions of extension rings and box covers for octagonal outlet boxes.

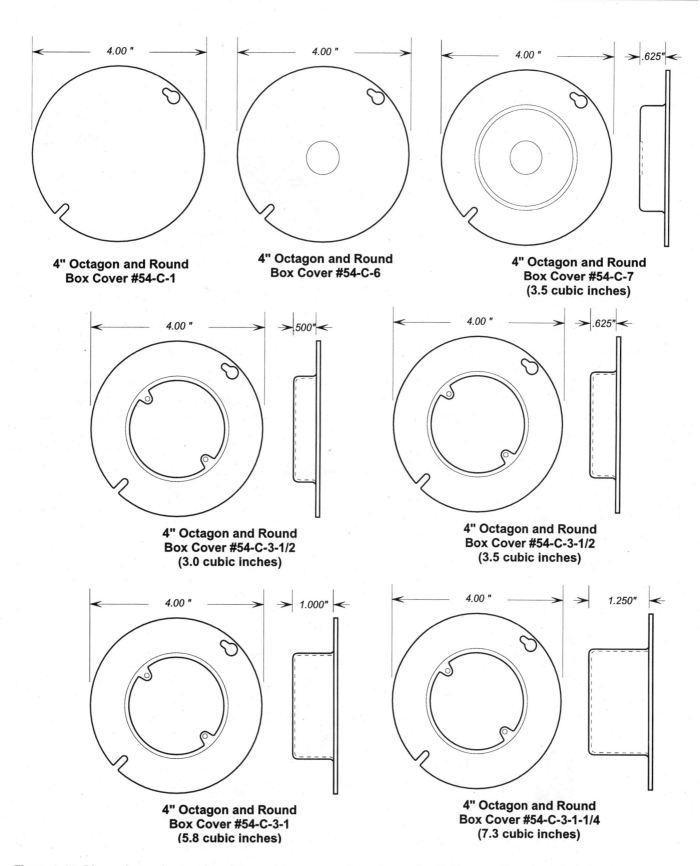

<- 4.00 " ->

**4" Octagon and Round
Box Cover #54-C-1**

<- 4.00 " ->

**4" Octagon and Round
Box Cover #54-C-6**

<- 4.00 " -> |.625"|<-

**4" Octagon and Round
Box Cover #54-C-7
(3.5 cubic inches)**

<- 4.00 " -> ->|500"|<-

**4" Octagon and Round
Box Cover #54-C-3-1/2
(3.0 cubic inches)**

<- 4.00 " -> ->|.625"|<-

**4" Octagon and Round
Box Cover #54-C-3-1/2
(3.5 cubic inches)**

<- 4.00 " -> ->| 1.000" |<-

**4" Octagon and Round
Box Cover #54-C-3-1
(5.8 cubic inches)**

<- 4.00 " -> | 1.250" |<-

**4" Octagon and Round
Box Cover #54-C-3-1-1/4
(7.3 cubic inches)**

Figure 2-47: Dimensions of extension rings and box covers for octagonal outlet boxes. *(Cont.)*

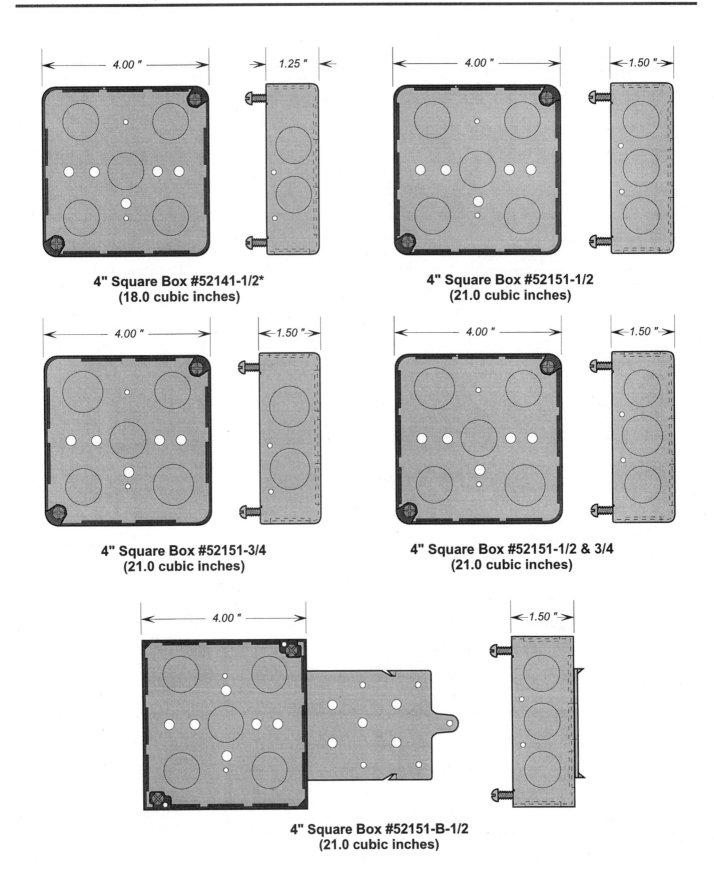

4" Square Box #52141-1/2*
(18.0 cubic inches)

4" Square Box #52151-1/2
(21.0 cubic inches)

4" Square Box #52151-3/4
(21.0 cubic inches)

4" Square Box #52151-1/2 & 3/4
(21.0 cubic inches)

4" Square Box #52151-B-1/2
(21.0 cubic inches)

Figure 2-48: Dimensions of 4-in square outlet boxes.

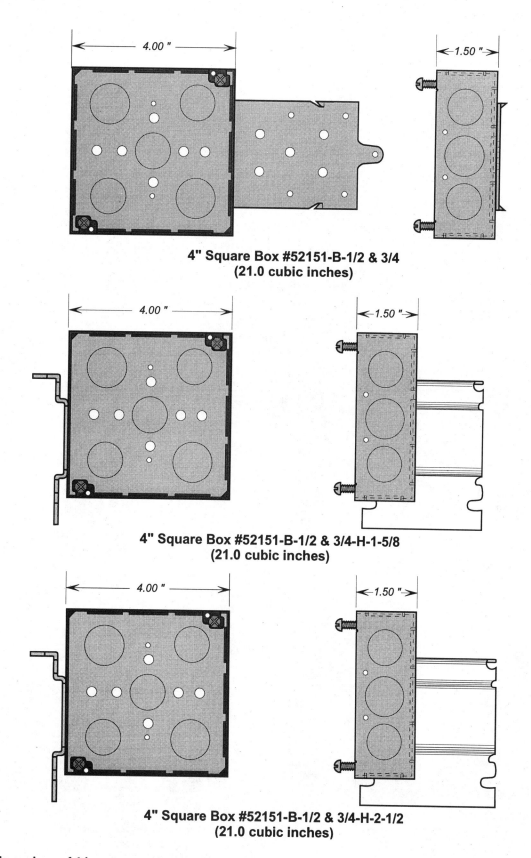

4" Square Box #52151-B-1/2 & 3/4
(21.0 cubic inches)

4" Square Box #52151-B-1/2 & 3/4-H-1-5/8
(21.0 cubic inches)

4" Square Box #52151-B-1/2 & 3/4-H-2-1/2
(21.0 cubic inches)

Figure 2-48: Dimensions of 4-in square outlet boxes. *(Cont.)*

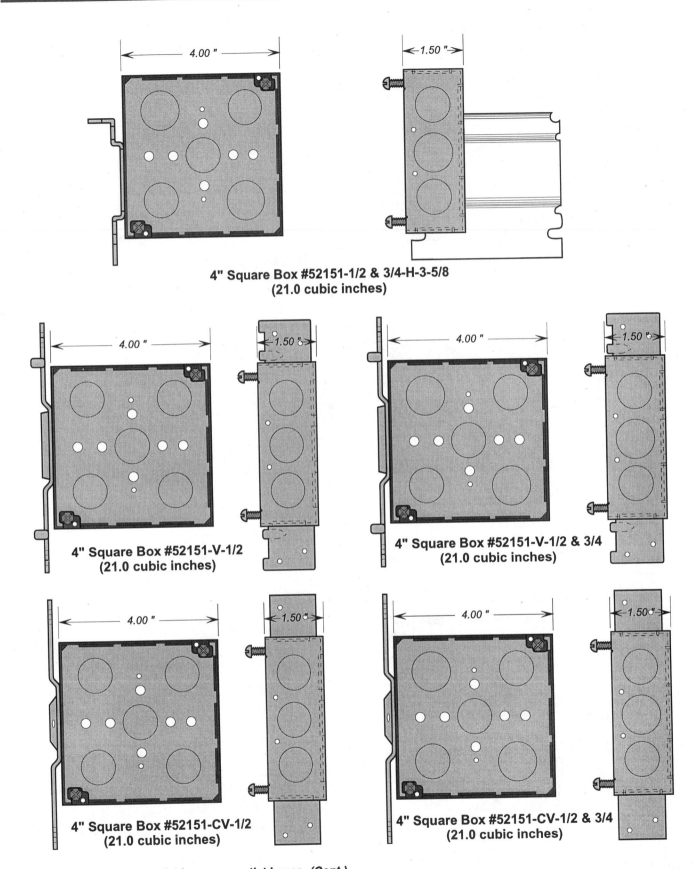

4" Square Box #52151-1/2 & 3/4-H-3-5/8
(21.0 cubic inches)

4" Square Box #52151-V-1/2
(21.0 cubic inches)

4" Square Box #52151-V-1/2 & 3/4
(21.0 cubic inches)

4" Square Box #52151-CV-1/2
(21.0 cubic inches)

4" Square Box #52151-CV-1/2 & 3/4
(21.0 cubic inches)

Figure 2-48: Dimensions of 4-in square outlet boxes. *(Cont.)*

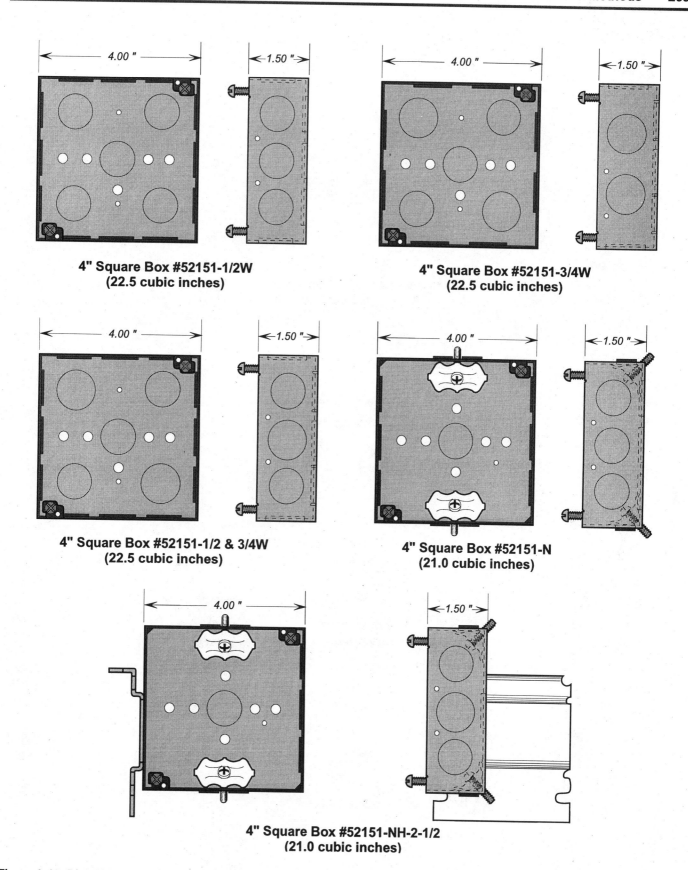

4" Square Box #52151-1/2W
(22.5 cubic inches)

4" Square Box #52151-3/4W
(22.5 cubic inches)

4" Square Box #52151-1/2 & 3/4W
(22.5 cubic inches)

4" Square Box #52151-N
(21.0 cubic inches)

4" Square Box #52151-NH-2-1/2
(21.0 cubic inches)

Figure 2-48: Dimensions of 4-in square outlet boxes. *(Cont.)*

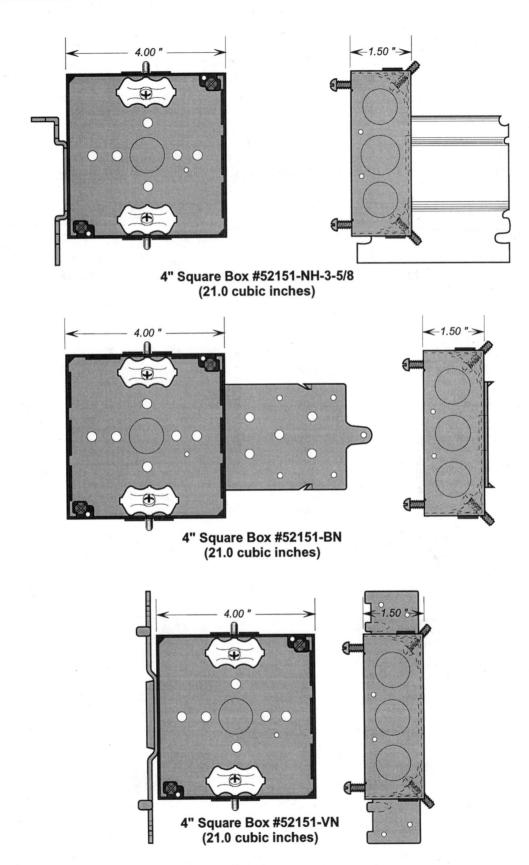

4" Square Box #52151-NH-3-5/8
(21.0 cubic inches)

4" Square Box #52151-BN
(21.0 cubic inches)

4" Square Box #52151-VN
(21.0 cubic inches)

Figure 2-48: Dimensions of 4-in square outlet boxes. *(Cont.)*

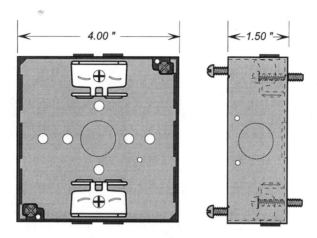

4" Square Box #52151-X
(21.0 cubic inches)

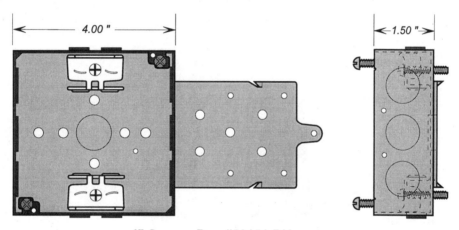

4" Square Box #52151-BX
(21.0 cubic inches)

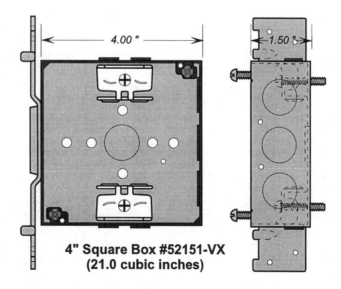

4" Square Box #52151-VX
(21.0 cubic inches)

Figure 2-48: Dimensions of 4-in square outlet boxes. *(Cont.)*

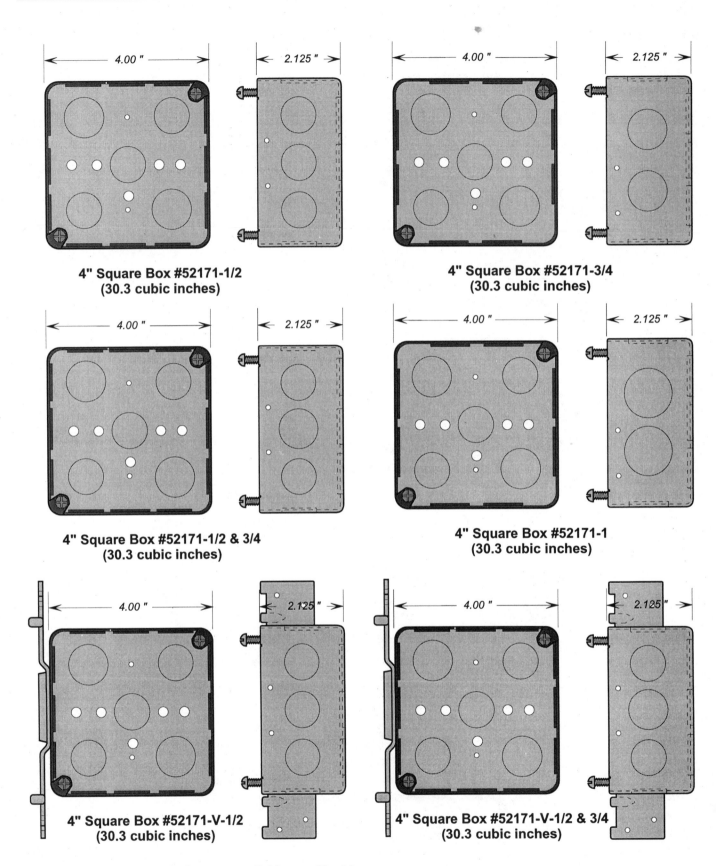

4" Square Box #52171-1/2
(30.3 cubic inches)

4" Square Box #52171-3/4
(30.3 cubic inches)

4" Square Box #52171-1/2 & 3/4
(30.3 cubic inches)

4" Square Box #52171-1
(30.3 cubic inches)

4" Square Box #52171-V-1/2
(30.3 cubic inches)

4" Square Box #52171-V-1/2 & 3/4
(30.3 cubic inches)

Figure 2-48: Dimensions of 4-in square outlet boxes. *(Cont.)*

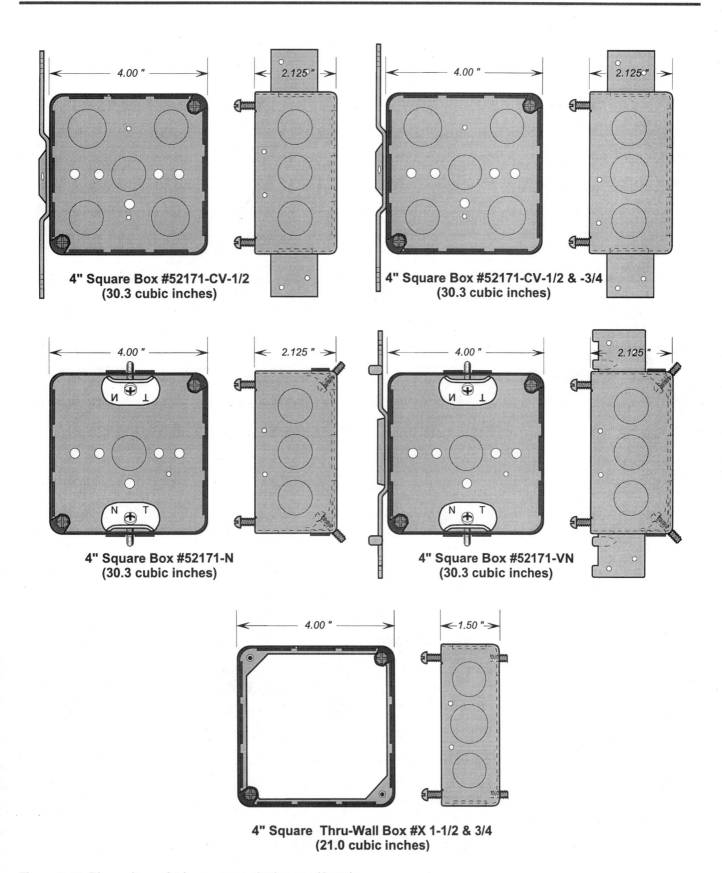

4" Square Box #52171-CV-1/2
(30.3 cubic inches)

4" Square Box #52171-CV-1/2 & -3/4
(30.3 cubic inches)

4" Square Box #52171-N
(30.3 cubic inches)

4" Square Box #52171-VN
(30.3 cubic inches)

4" Square Thru-Wall Box #X 1-1/2 & 3/4
(21.0 cubic inches)

Figure 2-48: Dimensions of 4-in square outlet boxes. *(Cont.)*

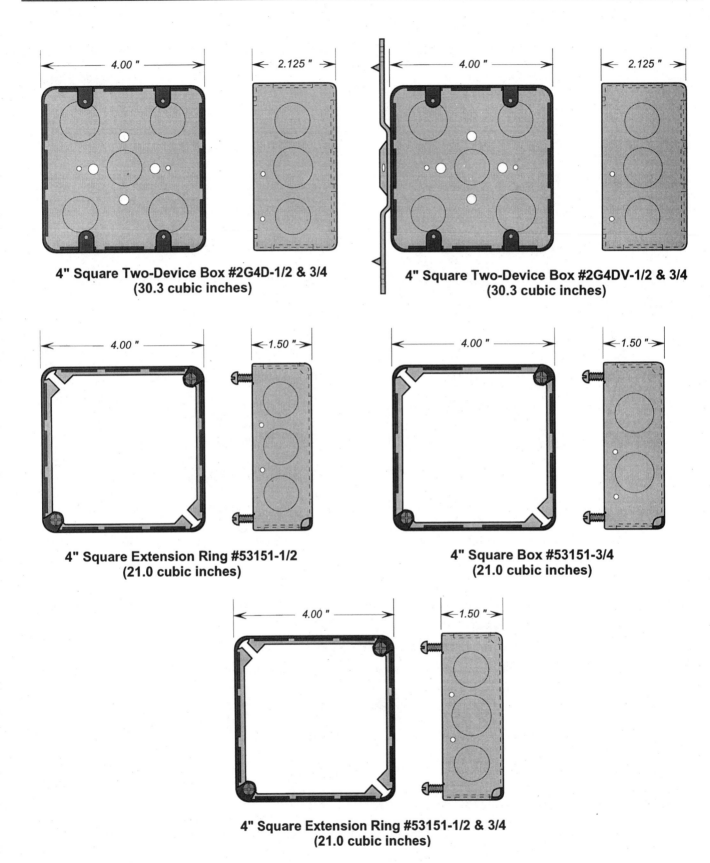

4" Square Two-Device Box #2G4D-1/2 & 3/4
(30.3 cubic inches)

4" Square Two-Device Box #2G4DV-1/2 & 3/4
(30.3 cubic inches)

4" Square Extension Ring #53151-1/2
(21.0 cubic inches)

4" Square Box #53151-3/4
(21.0 cubic inches)

4" Square Extension Ring #53151-1/2 & 3/4
(21.0 cubic inches)

Figure 2-49: Dimensions of 4-in square outlet boxes and related extension rings.

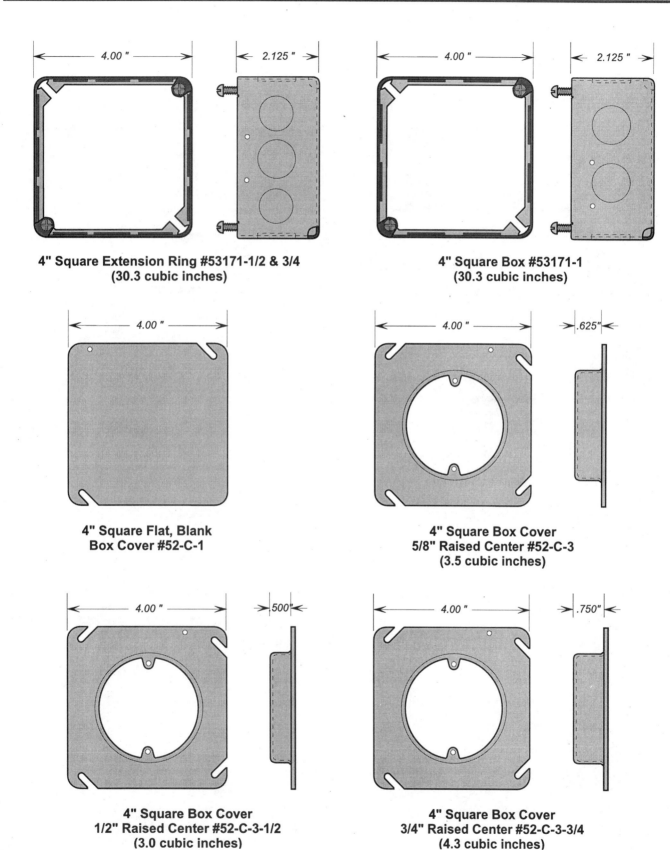

4" Square Extension Ring #53171-1/2 & 3/4
(30.3 cubic inches)

4" Square Box #53171-1
(30.3 cubic inches)

4" Square Flat, Blank
Box Cover #52-C-1

4" Square Box Cover
5/8" Raised Center #52-C-3
(3.5 cubic inches)

4" Square Box Cover
1/2" Raised Center #52-C-3-1/2
(3.0 cubic inches)

4" Square Box Cover
3/4" Raised Center #52-C-3-3/4
(4.3 cubic inches)

Figure 2-50: Dimensions of 4-in square extension rings and box covers.

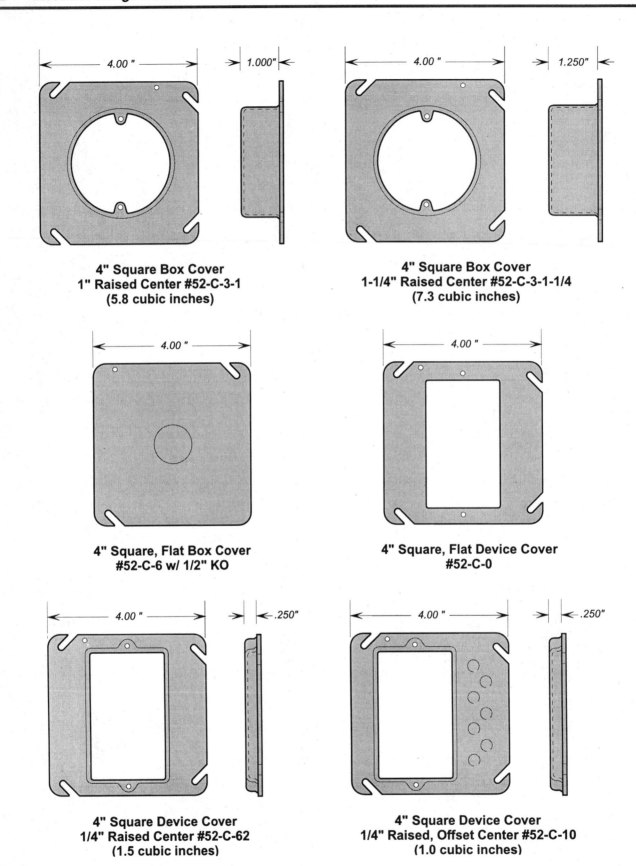

4" Square Box Cover
1" Raised Center #52-C-3-1
(5.8 cubic inches)

4" Square Box Cover
1-1/4" Raised Center #52-C-3-1-1/4
(7.3 cubic inches)

4" Square, Flat Box Cover
#52-C-6 w/ 1/2" KO

4" Square, Flat Device Cover
#52-C-0

4" Square Device Cover
1/4" Raised Center #52-C-62
(1.5 cubic inches)

4" Square Device Cover
1/4" Raised, Offset Center #52-C-10
(1.0 cubic inches)

Figure 2-50: Dimensions of 4-in square extension rings and box covers. *(Cont.)*

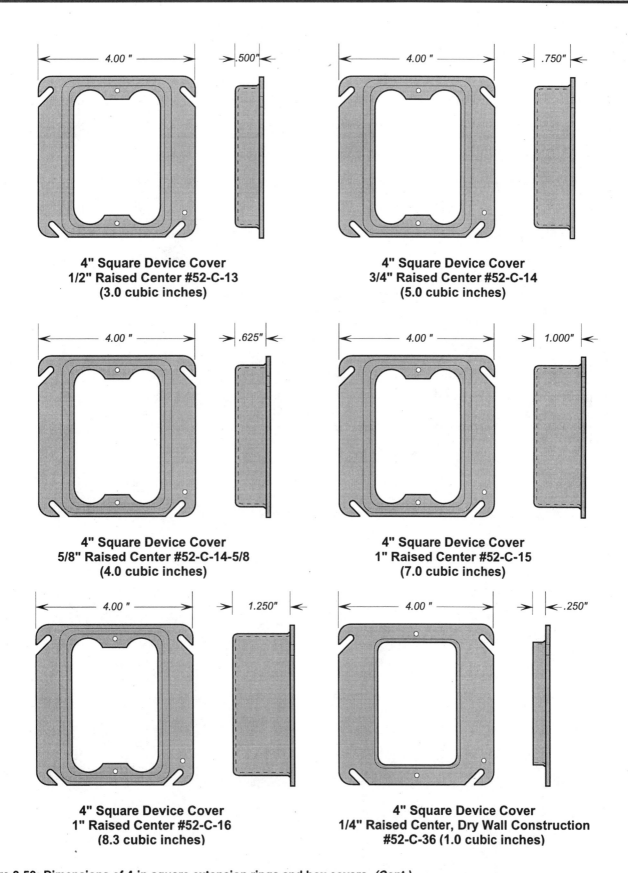

4" Square Device Cover
1/2" Raised Center #52-C-13
(3.0 cubic inches)

4" Square Device Cover
3/4" Raised Center #52-C-14
(5.0 cubic inches)

4" Square Device Cover
5/8" Raised Center #52-C-14-5/8
(4.0 cubic inches)

4" Square Device Cover
1" Raised Center #52-C-15
(7.0 cubic inches)

4" Square Device Cover
1" Raised Center #52-C-16
(8.3 cubic inches)

4" Square Device Cover
1/4" Raised Center, Dry Wall Construction
#52-C-36 (1.0 cubic inches)

Figure 2-50: Dimensions of 4-in square extension rings and box covers. *(Cont.)*

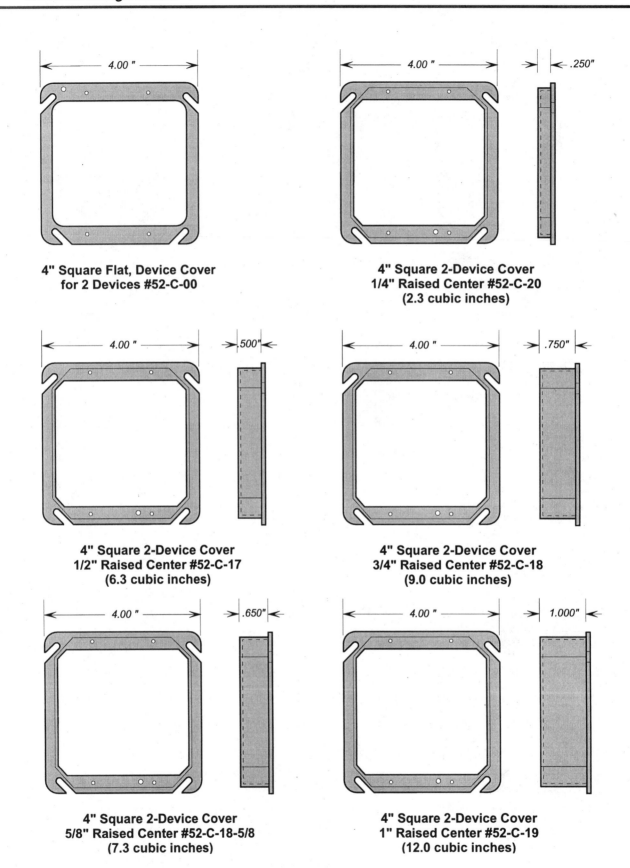

**4" Square Flat, Device Cover
for 2 Devices #52-C-00**

**4" Square 2-Device Cover
1/4" Raised Center #52-C-20
(2.3 cubic inches)**

**4" Square 2-Device Cover
1/2" Raised Center #52-C-17
(6.3 cubic inches)**

**4" Square 2-Device Cover
3/4" Raised Center #52-C-18
(9.0 cubic inches)**

**4" Square 2-Device Cover
5/8" Raised Center #52-C-18-5/8
(7.3 cubic inches)**

**4" Square 2-Device Cover
1" Raised Center #52-C-19
(12.0 cubic inches)**

Figure 2-50: Dimensions of 4-in square extension rings and box covers. *(Cont.)*

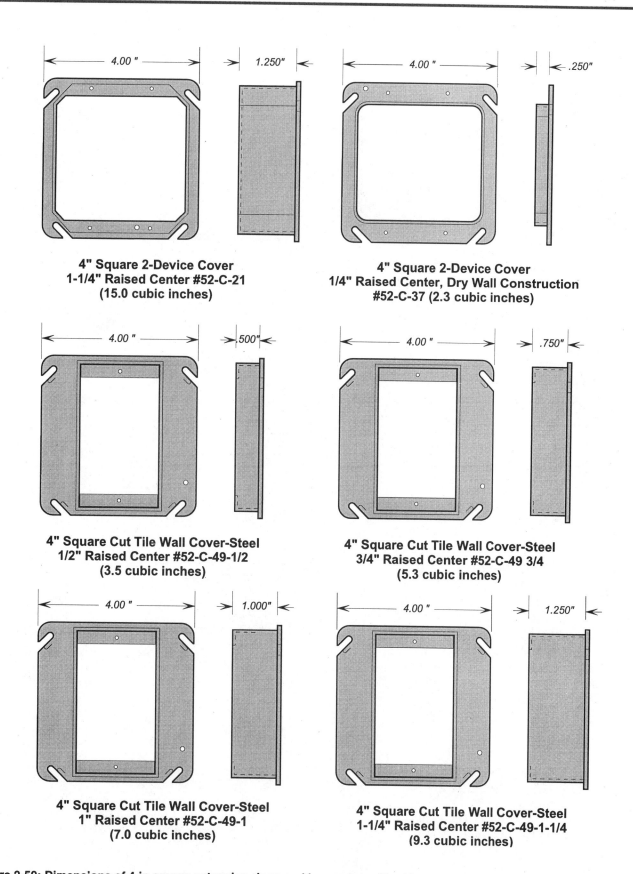

4" Square 2-Device Cover
1-1/4" Raised Center #52-C-21
(15.0 cubic inches)

4" Square 2-Device Cover
1/4" Raised Center, Dry Wall Construction
#52-C-37 (2.3 cubic inches)

4" Square Cut Tile Wall Cover-Steel
1/2" Raised Center #52-C-49-1/2
(3.5 cubic inches)

4" Square Cut Tile Wall Cover-Steel
3/4" Raised Center #52-C-49 3/4
(5.3 cubic inches)

4" Square Cut Tile Wall Cover-Steel
1" Raised Center #52-C-49-1
(7.0 cubic inches)

4" Square Cut Tile Wall Cover-Steel
1-1/4" Raised Center #52-C-49-1-1/4
(9.3 cubic inches)

Figure 2-50: Dimensions of 4-in square extension rings and box covers. *(Cont.)*

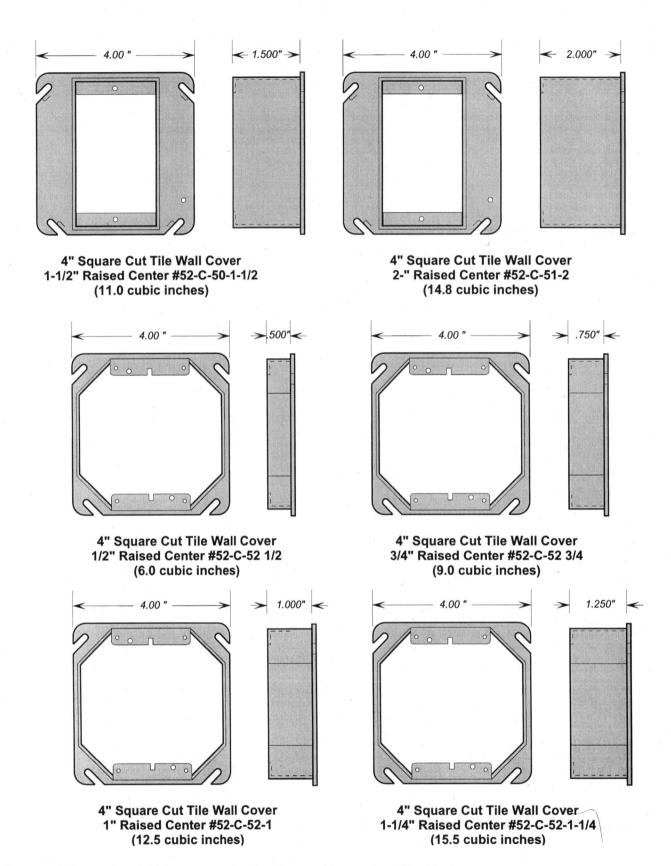

4" Square Cut Tile Wall Cover
1-1/2" Raised Center #52-C-50-1-1/2
(11.0 cubic inches)

4" Square Cut Tile Wall Cover
2-" Raised Center #52-C-51-2
(14.8 cubic inches)

4" Square Cut Tile Wall Cover
1/2" Raised Center #52-C-52 1/2
(6.0 cubic inches)

4" Square Cut Tile Wall Cover
3/4" Raised Center #52-C-52 3/4
(9.0 cubic inches)

4" Square Cut Tile Wall Cover
1" Raised Center #52-C-52-1
(12.5 cubic inches)

4" Square Cut Tile Wall Cover
1-1/4" Raised Center #52-C-52-1-1/4
(15.5 cubic inches)

Figure 2-50: Dimensions of 4-in square extension rings and box covers. *(Cont.)*

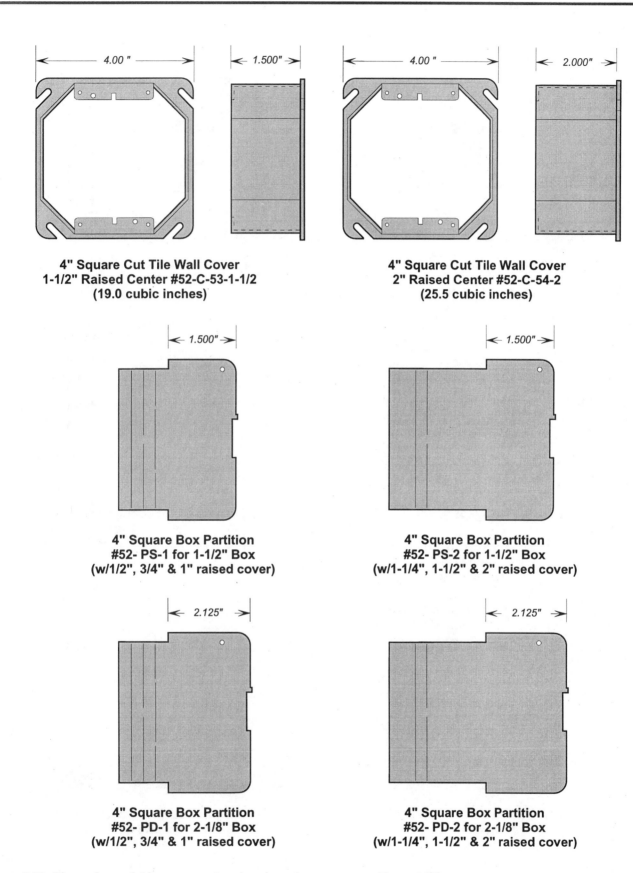

4" Square Cut Tile Wall Cover
1-1/2" Raised Center #52-C-53-1-1/2
(19.0 cubic inches)

4" Square Cut Tile Wall Cover
2" Raised Center #52-C-54-2
(25.5 cubic inches)

4" Square Box Partition
#52- PS-1 for 1-1/2" Box
(w/1/2", 3/4" & 1" raised cover)

4" Square Box Partition
#52- PS-2 for 1-1/2" Box
(w/1-1/4", 1-1/2" & 2" raised cover)

4" Square Box Partition
#52- PD-1 for 2-1/8" Box
(w/1/2", 3/4" & 1" raised cover)

4" Square Box Partition
#52- PD-2 for 2-1/8" Box
(w/1-1/4", 1-1/2" & 2" raised cover)

Figure 2-51: Dimensions of 4-in square extension rings, box covers and box partitions.

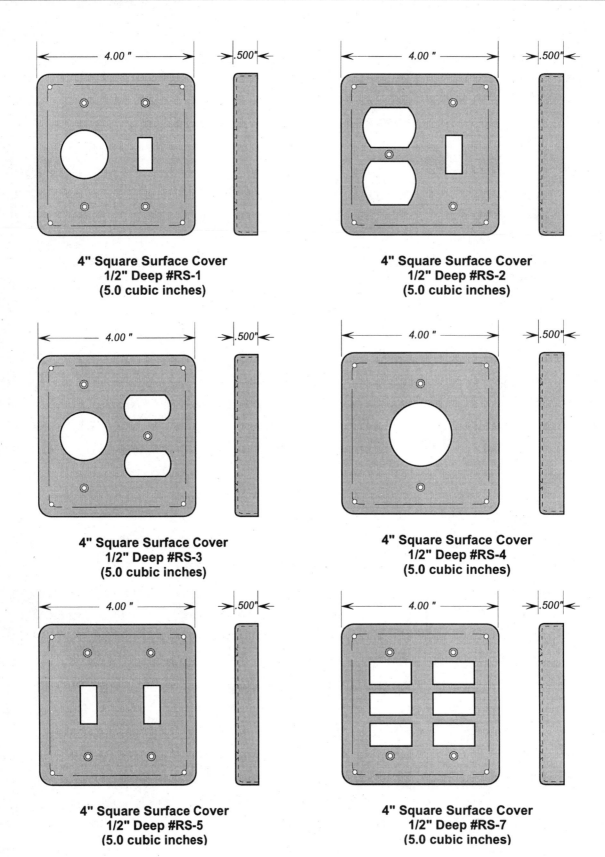

Figure 2-52: Dimensions of 4-in square box covers.

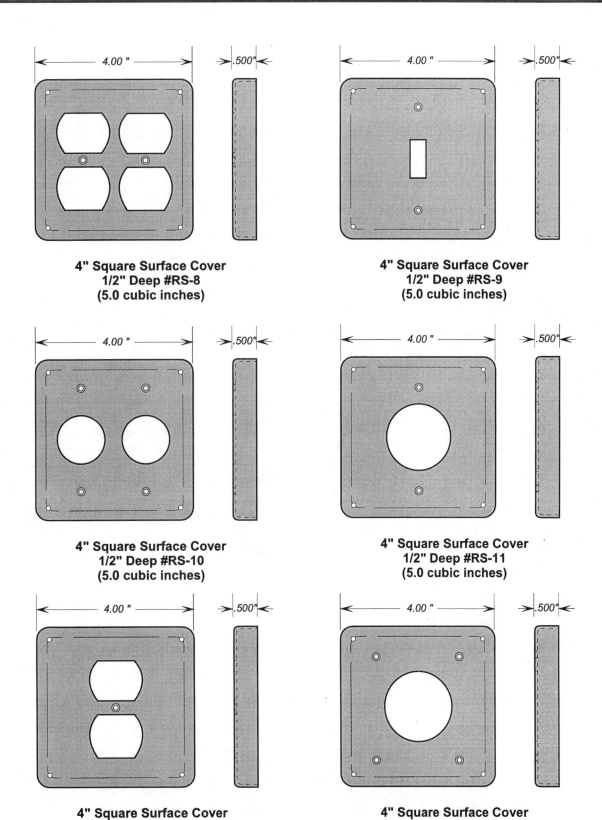

4" Square Surface Cover
1/2" Deep #RS-8
(5.0 cubic inches)

4" Square Surface Cover
1/2" Deep #RS-9
(5.0 cubic inches)

4" Square Surface Cover
1/2" Deep #RS-10
(5.0 cubic inches)

4" Square Surface Cover
1/2" Deep #RS-11
(5.0 cubic inches)

4" Square Surface Cover
1/2" Deep #RS-12
(5.0 cubic inches)

4" Square Surface Cover
1/2" Deep #RS-13
(5.0 cubic inches)

Figure 2-52: Dimensions of 4-in square box covers. *(Cont.)*

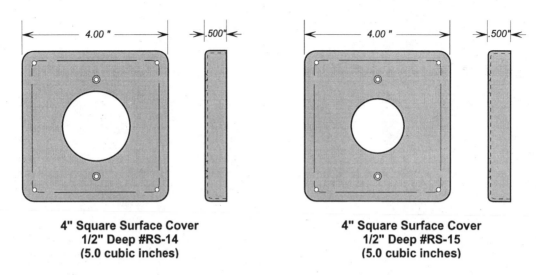

4" Square Surface Cover
1/2" Deep #RS-14
(5.0 cubic inches)

4" Square Surface Cover
1/2" Deep #RS-15
(5.0 cubic inches)

Figure 2-52: Dimensions of 4-in square box covers. *(Cont.)*

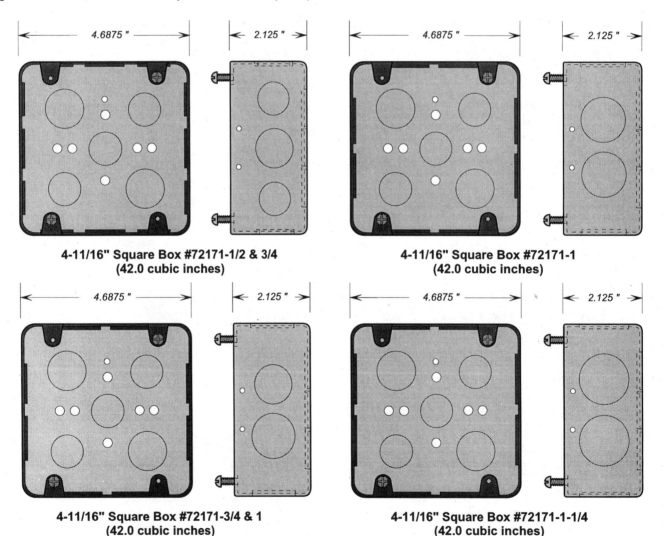

4-11/16" Square Box #72171-1/2 & 3/4
(42.0 cubic inches)

4-11/16" Square Box #72171-1
(42.0 cubic inches)

4-11/16" Square Box #72171-3/4 & 1
(42.0 cubic inches)

4-11/16" Square Box #72171-1-1/4
(42.0 cubic inches)

Figure 2-53: Dimensions of 4¹¹⁄₁₆-in square boxes.

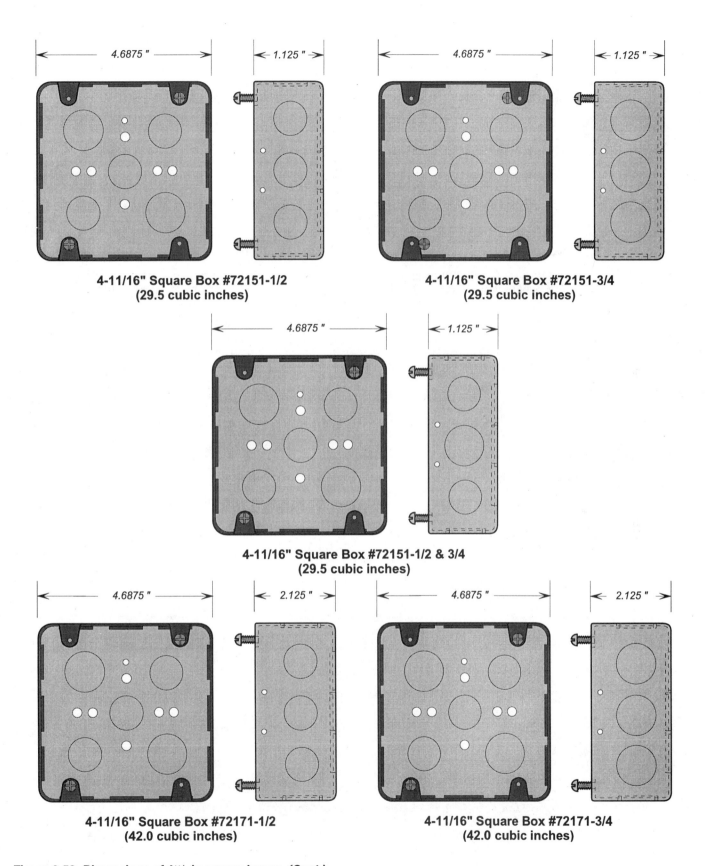

4-11/16" Square Box #72151-1/2
(29.5 cubic inches)

4-11/16" Square Box #72151-3/4
(29.5 cubic inches)

4-11/16" Square Box #72151-1/2 & 3/4
(29.5 cubic inches)

4-11/16" Square Box #72171-1/2
(42.0 cubic inches)

4-11/16" Square Box #72171-3/4
(42.0 cubic inches)

Figure 2-53: Dimensions of 4¹¹⁄₁₆-in square boxes. *(Cont.)*

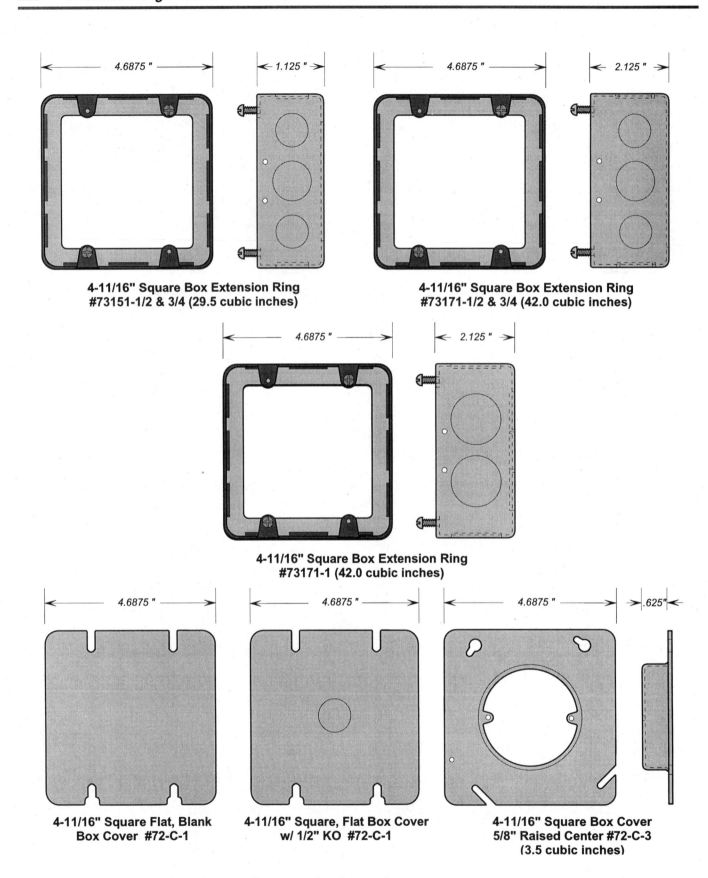

← 4.6875 " →   ← 1.125 " →   ← 4.6875 " →   ← 2.125 " →

**4-11/16" Square Box Extension Ring
#73151-1/2 & 3/4 (29.5 cubic inches)**

**4-11/16" Square Box Extension Ring
#73171-1/2 & 3/4 (42.0 cubic inches)**

← 4.6875 " →   ← 2.125 " →

**4-11/16" Square Box Extension Ring
#73171-1 (42.0 cubic inches)**

← 4.6875 " →   ← 4.6875 " →   ← 4.6875 " →   →.625"←

**4-11/16" Square Flat, Blank
Box Cover #72-C-1**

**4-11/16" Square, Flat Box Cover
w/ 1/2" KO #72-C-1**

**4-11/16" Square Box Cover
5/8" Raised Center #72-C-3
(3.5 cubic inches)**

Figure 2-54: Dimensions of 4¹¹⁄₁₆-in square-box extension rings and covers.

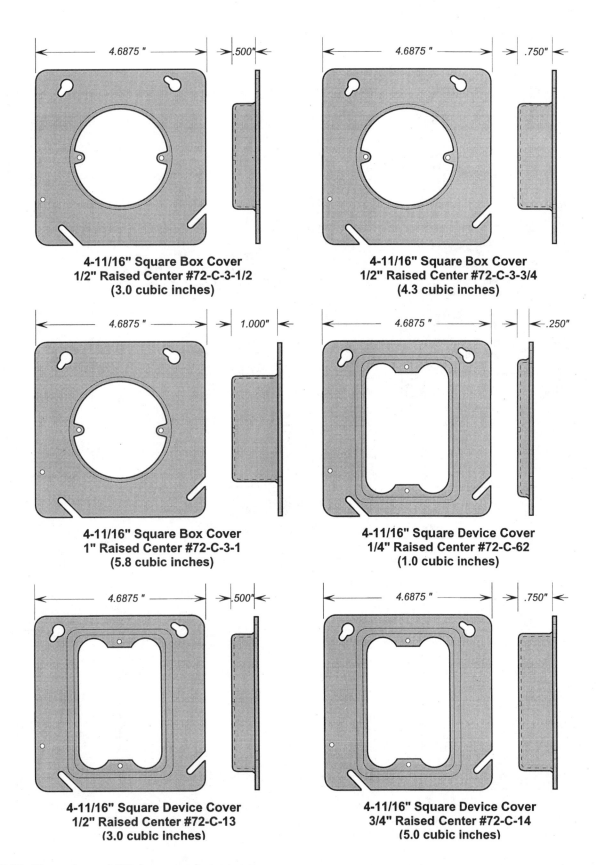

4-11/16" Square Box Cover
1/2" Raised Center #72-C-3-1/2
(3.0 cubic inches)

4-11/16" Square Box Cover
1/2" Raised Center #72-C-3-3/4
(4.3 cubic inches)

4-11/16" Square Box Cover
1" Raised Center #72-C-3-1
(5.8 cubic inches)

4-11/16" Square Device Cover
1/4" Raised Center #72-C-62
(1.0 cubic inches)

4-11/16" Square Device Cover
1/2" Raised Center #72-C-13
(3.0 cubic inches)

4-11/16" Square Device Cover
3/4" Raised Center #72-C-14
(5.0 cubic inches)

Figure 2-55: Dimensions of 4¹¹⁄₁₆-in square-box covers.

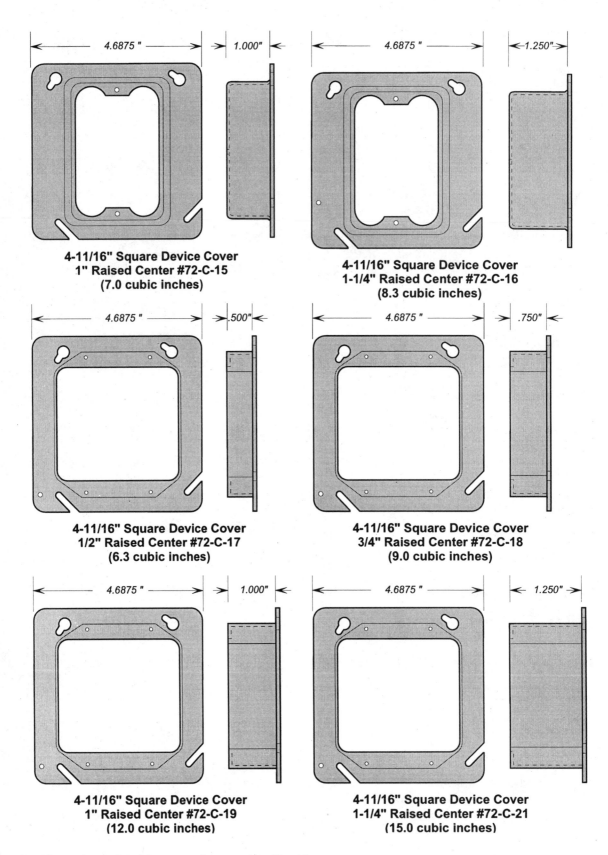

4-11/16" Square Device Cover
1" Raised Center #72-C-15
(7.0 cubic inches)

4-11/16" Square Device Cover
1-1/4" Raised Center #72-C-16
(8.3 cubic inches)

4-11/16" Square Device Cover
1/2" Raised Center #72-C-17
(6.3 cubic inches)

4-11/16" Square Device Cover
3/4" Raised Center #72-C-18
(9.0 cubic inches)

4-11/16" Square Device Cover
1" Raised Center #72-C-19
(12.0 cubic inches)

4-11/16" Square Device Cover
1-1/4" Raised Center #72-C-21
(15.0 cubic inches)

Figure 2-55: Dimensions of 4¹¹⁄₁₆-in square-box covers. *(Cont.)*

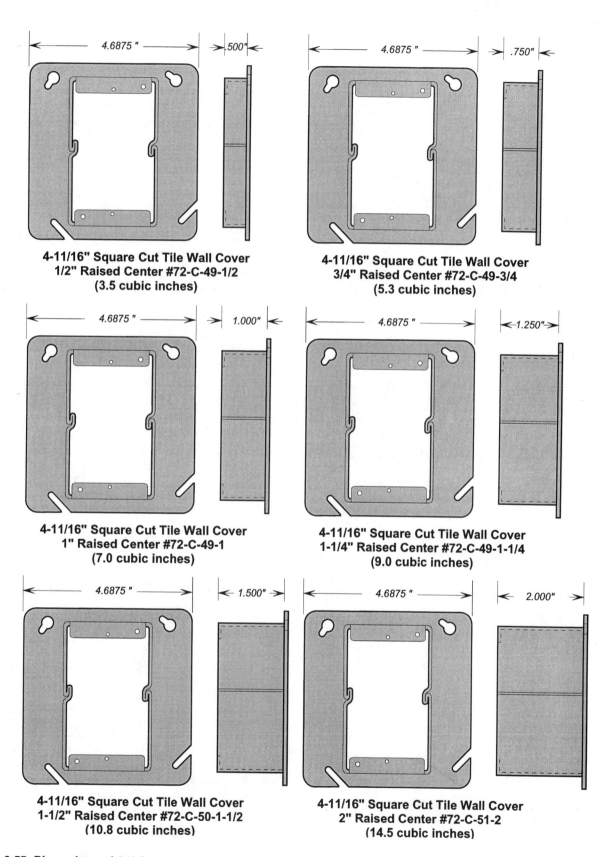

4-11/16" Square Cut Tile Wall Cover
1/2" Raised Center #72-C-49-1/2
(3.5 cubic inches)

4-11/16" Square Cut Tile Wall Cover
3/4" Raised Center #72-C-49-3/4
(5.3 cubic inches)

4-11/16" Square Cut Tile Wall Cover
1" Raised Center #72-C-49-1
(7.0 cubic inches)

4-11/16" Square Cut Tile Wall Cover
1-1/4" Raised Center #72-C-49-1-1/4
(9.0 cubic inches)

4-11/16" Square Cut Tile Wall Cover
1-1/2" Raised Center #72-C-50-1-1/2
(10.8 cubic inches)

4-11/16" Square Cut Tile Wall Cover
2" Raised Center #72-C-51-2
(14.5 cubic inches)

Figure 2-55: Dimensions of 4$\frac{11}{16}$-in square-box covers. *(Cont.)*

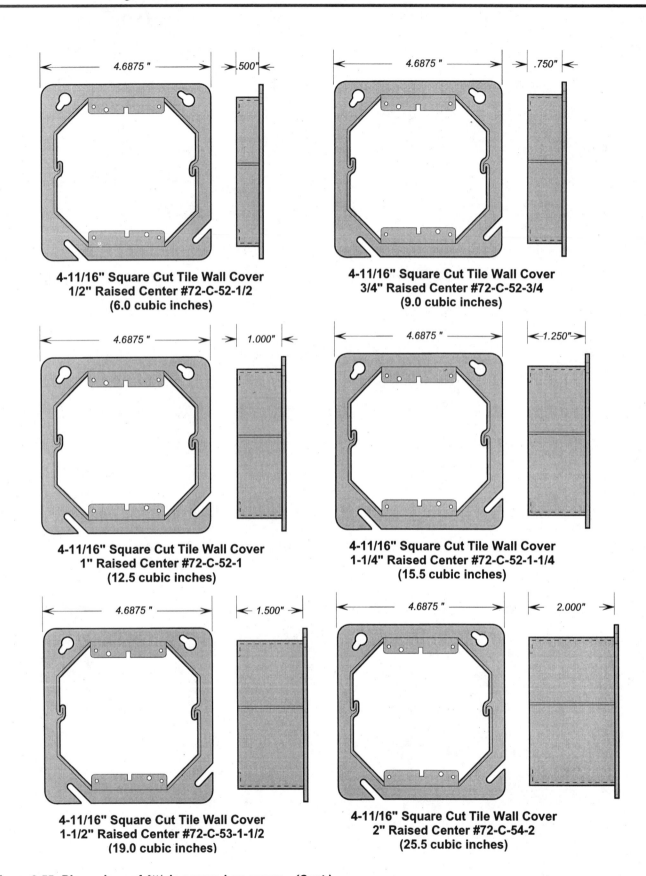

4-11/16" Square Cut Tile Wall Cover
1/2" Raised Center #72-C-52-1/2
(6.0 cubic inches)

4-11/16" Square Cut Tile Wall Cover
3/4" Raised Center #72-C-52-3/4
(9.0 cubic inches)

4-11/16" Square Cut Tile Wall Cover
1" Raised Center #72-C-52-1
(12.5 cubic inches)

4-11/16" Square Cut Tile Wall Cover
1-1/4" Raised Center #72-C-52-1-1/4
(15.5 cubic inches)

4-11/16" Square Cut Tile Wall Cover
1-1/2" Raised Center #72-C-53-1-1/2
(19.0 cubic inches)

4-11/16" Square Cut Tile Wall Cover
2" Raised Center #72-C-54-2
(25.5 cubic inches)

Figure 2-55: Dimensions of 4¹¹/₁₆-in square-box covers. *(Cont.)*

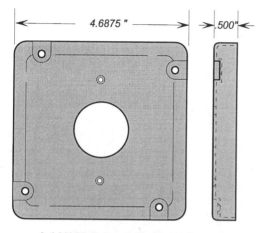

**4-11/16" Square Surface Cover
1/2" Deep #RSL-4
(7.5 cubic inches)**

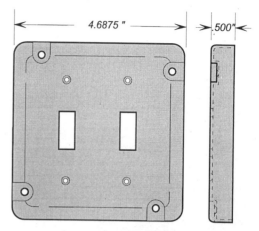

**4-11/16" Square Surface Cover
1/2" Deep #RSL-5
(7.5 cubic inches)**

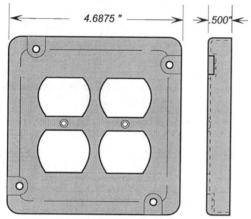

**4-11/16" Square Surface Cover
1/2" Deep #RSL-8
(7.5 cubic inches)**

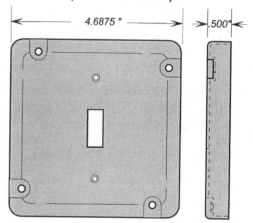

**4-11/16" Square Surface Cover
1/2" Deep #RSL-9
(7.5 cubic inches)**

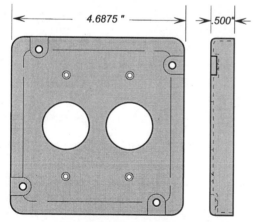

**4-11/16" Square Surface Cover
1/2" Deep #RSL-10
(7.5 cubic inches)**

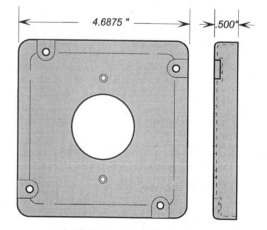

**4-11/16" Square Surface Cover
1/2" Deep #RSL-11
(7.5 cubic inches)**

Figure 2-55: Dimensions of 4¹¹⁄₁₆-in square-box covers. *(Cont.)*

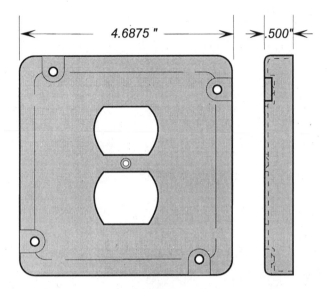

4-11/16" Square Surface Cover
1/2" Deep #RSL-12
(7.5 cubic inches)

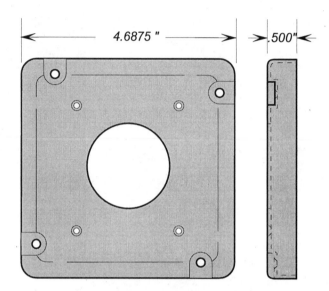

4-11/16" Square Surface Cover
1/2" Deep #RSL-13
(7.5 cubic inches)

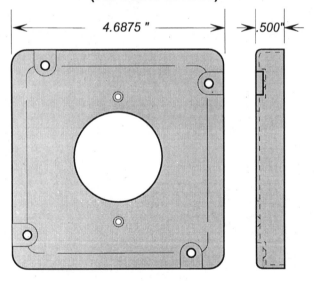

4-11/16" Square Surface Cover
1/2" Deep #RSL-14
(7.5 cubic inches)

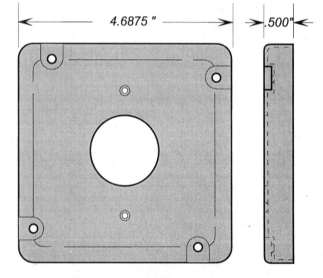

4-11/16" Square Surface Cover
1/2" Deep #RSL-15
(7.5 cubic inches)

Figure 2-55: Dimensions of 4¹¹⁄₁₆-in square-box covers. *(Cont.)*

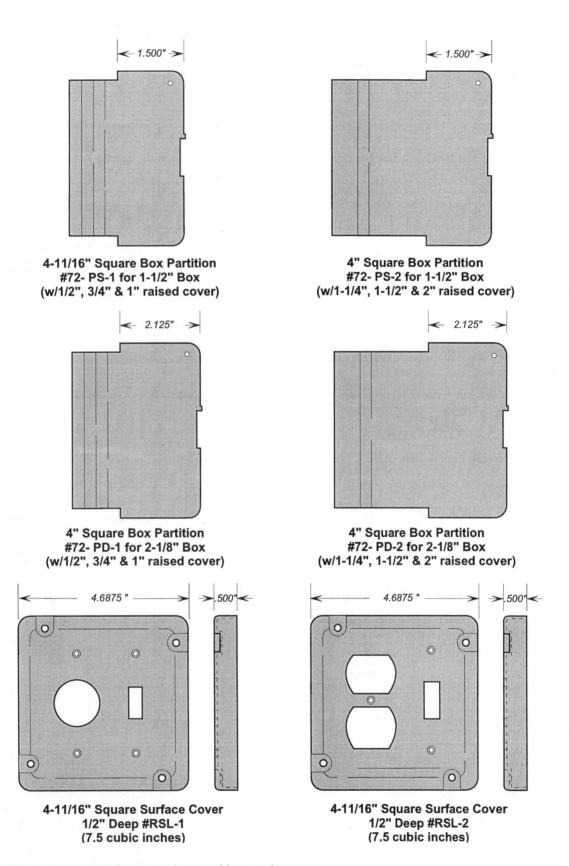

Figure 2-56: Dimensions of 4¹¹⁄₁₆-in square-box partitions and covers.

LXW-806 Switch Box Series Configurations

Box No.	Clamps	Ears	Brackets	Each Side Conduit	Each End Conduit	Cable	Bottom Conduit
LXW *(illustrated)*	C-5	Yes	----	1 - 1/2"	1 - 1/2"	2	1 - 1/2"
LXWOW	C-5	Yes	----	1 - 1/2"	1 - 1/2"	2	1 - 1/2"
LXWOW-12C	C-5	Yes	----	1 - 1/2"	1 - 1/2"	2	1 - 1/2"
LXWLE	C-5	----	----	1 - 1/2"	1 - 1/2"	2	1 - 1/2"
LXWOWE	C-5	Yes	----	1 - 1/2"	1 - 1/2"	2	1 - 1/2"
LXWSV	C-5	----	V	1 - 1/2"	1 - 1/2"	2	1 - 1/2"
LXWV	C-5	----	V	1 - 1/2"	1 - 1/2"	2	1 - 1/2"
LXWV-2G	C-5	----	V	1 - 1/2"	2 - 1/2"	4	2 - 1/2"
806-SW	C-5	----	S	1 - 1/2"	1 - 1/2"	2	1 - 1/2"
806-SW-1/4	C-5	----	S	1 - 1/2"	1 - 1/2"	2	1 - 1/2"

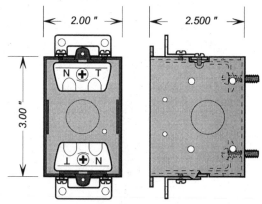

**3"x2" Switch Box #LXW-806 Series
2-1/2" deep gangable
w/conduit KO's & PO's
(12.5 cubic inches)**

LC-802 Switch Box Series Configurations

Box No.	Clamps	Ears	Brackets	Each Side Conduit	Each End Conduit	Cable	Bottom Conduit
LC *(illustrated)*	C-1	Yes	----	---------	1 - 1/2"	2	1 - 1/2"
LCOW	C-1	Yes	----	---------	1 - 1/2"	2	1 - 1/2"
LCOWE	C-1	Yes	----	---------	1 - 1/2"	2	1 - 1/2"
LCLE	C-1	----	----	---------	1 - 1/2"	2	1 - 1/2"
LCV	C-1	----	V	---------	1 - 1/2"	2	1 - 1/2"
LCNOW	C-1	Yes	----	---------	1 - 1/2"	2	1 - 1/2"
802-S	C-1	----	S	---------	1 - 1/2"	2	1 - 1/2"
802-S1/4	C-1	----	S	---------	1 - 1/2"	2	1 - 1/2"
802-S3/8	C-1	----	S	---------	1 - 1/2"	2	1 - 1/2"

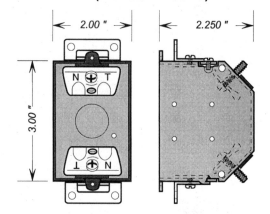

**3"x2" Switch Box #LC-802 Series
2-1/4" deep gangable
w/beveled corners & cable clamps
(10.5 cubic inches)**

LX-806 Switch Box Series Configurations

Box No.	Clamps	Ears	Brackets	Each Side Conduit	Each End Cable	Bottom Conduit
LX *(illustrated)*	C-3	Yes	----	1 - 1/2"	2	1 - 1/2"
LXOW	C-3	Yes	----	1 - 1/2"	2	1 - 1/2"
LXLE	C-3	----	----	1 - 1/2"	2	1 - 1/2"
LXV	C-3	----	S	1 - 1/2"	2	1 - 1/2"
806-S	C-3	----	V	1 - 1/2"	2	1 - 1/2"

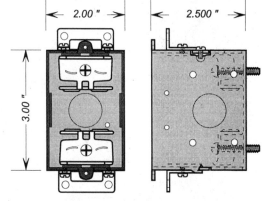

**3"x2" Switch Box #LX-806 Series
2-1/2" deep gangable
w/armored cable clamps
(12.5 cubic inches)**

Figure 2-57: Dimensions of switch (device) boxes.

CW Switch Box Series Configurations

Box No.	Ears	Brackets	Each Side Conduit	Each End Conduit	Bottom Conduit
CW-1/2 *(illustrated)*	Yes	----	1 - 1/2"	1 - 1/2"	1 - 1/2"
CW-3/4	Yes	----	1 - 3/4"	1 - 3/4"	1 - 3/4"
CWLE-1/2	----	----	1 - 1/2"	1 - 1/2"	1 - 1/2"
CWLE-3/4	----	----	1 - 3/4"	1 - 3/4"	1 - 3/4"
CWV-1/2	----	V	1 - 1/2"	1 - 1/2"	1 - 1/2"

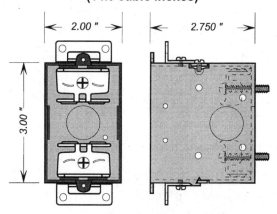

3"x2" Switch Box #CW Series
2-3/4" deep gangable
w/conduit KO's
(14.0 cubic inches)

CWX Switch Box Series Configurations

Box No.	Clamps	Ears	Brackets	Each Side Conduit	Each End Cable	Bottom Conduit
CWX *(illustrated)*	C-3	Yes	----	1 - 1/2"	2	1 - 1/2"
CWXLE	C-3	----	----	1 - 1/2"	2	1 - 1/2"
CWXV	C-3	----	V	1 - 1/2"	2	1 - 1/2"

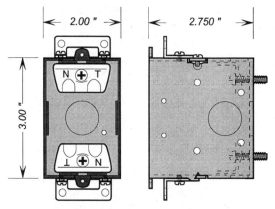

3"x2" Switch Box #CWX Series
2-3/4" deep gangable
w/conduit KO's & cable clamps
(14.0 cubic inches)

CWN Switch Box Series Configurations

Box No.	Clamps	Ears	Brackets	Each Side Conduit	Each End Cable	Bottom Conduit
CWN *(illustrated)*	C-5	Yes	----	1 - 1/2"	2	1 - 1/2"
CWNLE	C-5	----	----	1 - 1/2"	2	1 - 1/2"
CWNV	C-5	----	V	1 - 1/2"	2	1 - 1/2"

3"x2" Switch Box #CWN Series
2-3/4" deep gangable
w/conduit KO's & cable clamps
(14.0 cubic inches)

Figure 2-57: Dimensions of switch (device) boxes. *(Cont.)*

CY Switch Box Series Configurations

Box No.	Ears	Brackets	Each Side Conduit	Each End Conduit	Bottom Conduit
CY-1/2 *(illustrated)*	Yes	----	1 - 1/2"	1 - 1/2"	1 - 1/2"
CY-3/4	Yes	----	1 - 3/4"	1 - 3/4"	1 - 3/4"
CYLE-1/2	----	----	1 - 1/2"	1 - 1/2"	1 - 1/2"
CYLE-3/4	----	----	1 - 3/4"	1 - 3/4"	1 - 3/4"

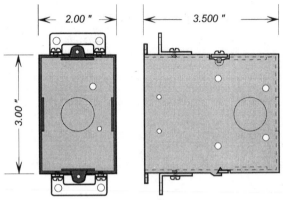

3"x 2" Switch Box #CY Series
3-1/2" deep gangable
w/conduit KO's
(18.0 cubic inches)

CX Switch Box Series Configurations

Box No.	Clamps	Ears	Brackets	Each Side Conduit	Each End Conduit	Cable	Bottom Conduit
CX *(illustrated)*	C-3	Yes	----	2 - 1/2"	1 - 1/2"	2	1 - 1/2"
CV	C-3	----	V	2 - 1/2"	1 - 1/2"	2	1 - 1/2"
				2 - 1/2"	1 - 1/2"	2	1 - 1/2"

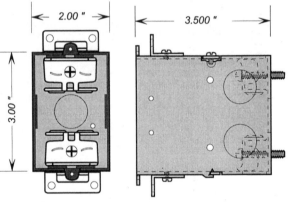

3"x 2" Switch Box #CX Series
3-1/2" deep gangable
w/conduit KO's & cable clamps
(18.0 cubic inches)

CXW Switch Box Series Configurations

Box No.	Clamps	Ears	Brackets	Each Side Conduit	Each End Conduit	Cable	Bottom Conduit
CXW *(illustrated)*	C-5	Yes	----	2 - 1/2"	1 - 1/2"	2	1 - 1/2"
CXWLE	C-5	----	----	2 - 1/2"	1 - 1/2"	2	1 - 1/2"
CXWOW	C-5	Yes	----	2 - 1/2"	1 - 1/2"	2	1 - 1/2"
CXWV	C-5	----	V	2 - 1/2"	1 - 1/2"	2	1 - 1/2"

3"x2" Switch Box #CXW Series
3-1/2" deep gangable
w/conduit KO's & cable clamps
(18.0 cubic inches)

Figure 2-57: Dimensions of switch (device) boxes. *(Cont.)*

A-1 Switch Box Series Configurations

Box No.	Clamps	Ears	Each End Conduit	Cable	Bottom Conduit
A-1-2 *(illustrated)*	C-1	----	1 - 1/2"	2	1 - 1/2"
A-1-2-E	C-1	Yes	1 - 1/2"	2	1 - 1/2"
A-16	C-1	----	1 - 1/2"	2	1 - 1/2"

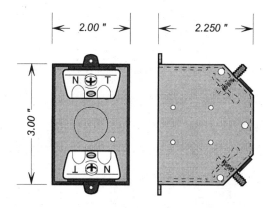

**3"x 2" Switch Box #A-1 Series
2-1/4" deep non-gangable
w/conduit KO's & PO's
(10.5 cubic inches)**

A-25 Switch Box Series Configurations

Box No.	Clamps	Each End Cable	Bottom Conduit
A-25-4 *(illustrated)*	C-5	2	1 - 1/2"
A-257	C-5	2	1 - 1/2"
A-258	C-5	2	1 - 1/2"

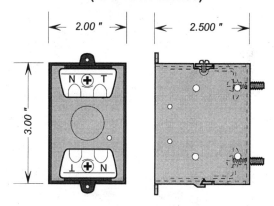

**3"x 2" Switch Box #A-25 Series
2-1/2" deep non-gangable
w/conduit KO's & PO's
(12.5 cubic inches)**

CD-804S Switch Box Series Configurations

Box No.	Ears	Brackets	Each Side Conduit	Each End Conduit	Bottom Conduit
CD *(illustrated)*	Yes	----	1 - 1/2"	1 - 1/2"	2 - 1/2"
CDOW	Yes	----	1 - 1/2"	1 - 1/2"	2 - 1/2"
CDLE	----	----	1 - 1/2"	1 - 1/2"	2 - 1/2"
CDV	----	V	1 - 1/2"	1 - 1/2"	2 - 1/2"
804-S	----	S	1 - 1/2"	1 - 1/2"	2 - 1/2"

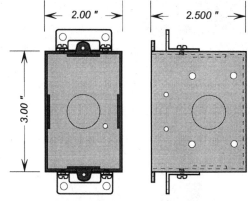

**3"x 2" Switch Box #CD Series
2-1/2" deep non-gangable
w/conduit KO's
(12.5 cubic inches)**

Figure 2-57: Dimensions of switch (device) boxes. *(Cont.)*

CD-16S Switch Box Series Configurations

Box No.	Ears	Brackets	Each Side Conduit	Each End Conduit	Bottom Conduit
16S (illustrated)	Yes	----	1 - 1/2"	1 - 1/2"	2 - 1/2"
16SCDOW	Yes	----	1 - 1/2"	1 - 1/2"	2 - 1/2"
16SCDLE	----	----	1 - 1/2"	1 - 1/2"	2 - 1/2"
16SCDV	----	V	1 - 1/2"	1 - 1/2"	2 - 1/2"
16S-S	----	S	1 - 1/2"	1 - 1/2"	2 - 1/2"

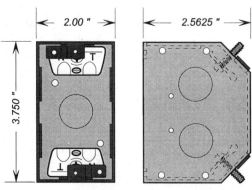

**3-3/4"x 2" Switch Box #16 Series
2-9/16" deep non-gangable
w/conduit KO's & PO's
(16.0 cubic inches)**

#16 Switch Box Series Configurations

Box No.	Clamps	Each Side Conduit	Each End Conduit	Cable	Bottom Conduit
16-WN (illustrated)	C-1	2 - 1/2"	1 - 1/2"	2	1 - 1/2"
16-DWN	C-1	2 - 1/2"	1 - 1/2"	2	1 - 1/2"

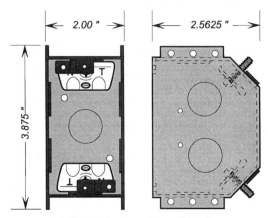

**3-7/8"x 2" Switch Box #16 Series
2-9/16" deep non-gangable
w/conduit KO's & PO's
(16.0 cubic inches)**

#18 Switch Box Series Configurations

Box No.	Clamps	Brackets	Each Side Conduit	Each End Conduit	Cable
18-GN (illustrated)	C-1	----	2 - 1/2"	1 - 1/2"	2
18-DGN	C-1	----	2 - 1/2"	1 - 1/2"	2
18-VGN	C-1	V	2 - 1/2"	1 - 1/2"	
18-S-1/4GN	C-1	S	2 - 1/2"	1 - 1/2"	

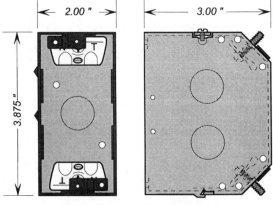

**3-7/8"x 2" Switch Box #18 Series
3" deep gangable
w/conduit KO's & PO's
(18.0 cubic inches)**

Figure 2-57: Dimensions of switch (device) boxes. *(Cont.)*

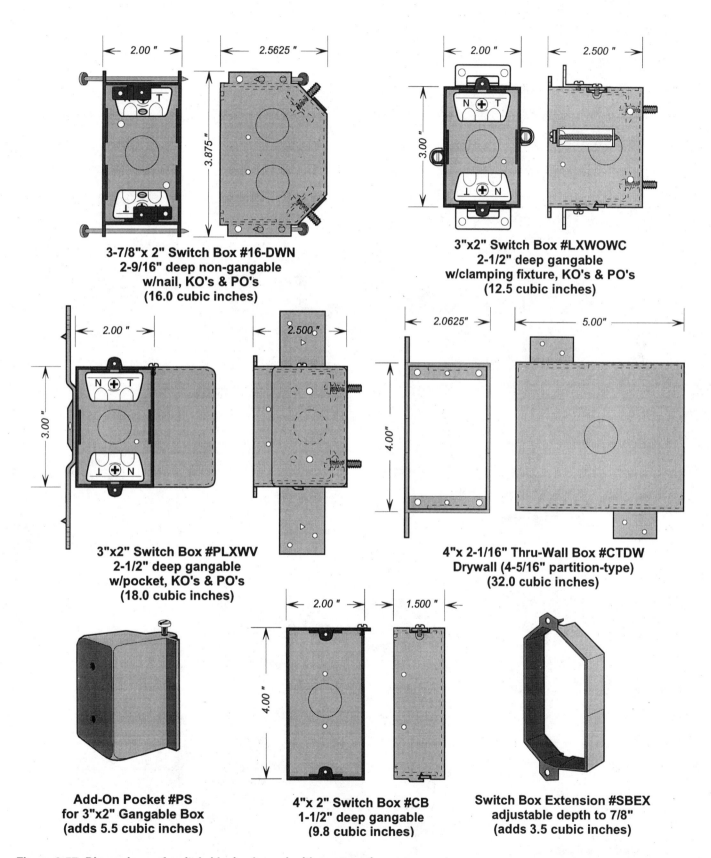

3-7/8"x 2" Switch Box #16-DWN
2-9/16" deep non-gangable
w/nail, KO's & PO's
(16.0 cubic inches)

3"x2" Switch Box #LXWOWC
2-1/2" deep gangable
w/clamping fixture, KO's & PO's
(12.5 cubic inches)

3"x2" Switch Box #PLXWV
2-1/2" deep gangable
w/pocket, KO's & PO's
(18.0 cubic inches)

4"x 2-1/16" Thru-Wall Box #CTDW
Drywall (4-5/16" partition-type)
(32.0 cubic inches)

Add-On Pocket #PS
for 3"x2" Gangable Box
(adds 5.5 cubic inches)

4"x 2" Switch Box #CB
1-1/2" deep gangable
(9.8 cubic inches)

Switch Box Extension #SBEX
adjustable depth to 7/8"
(adds 3.5 cubic inches)

Figure 2-57: Dimensions of switch (device boxes) with accessories.

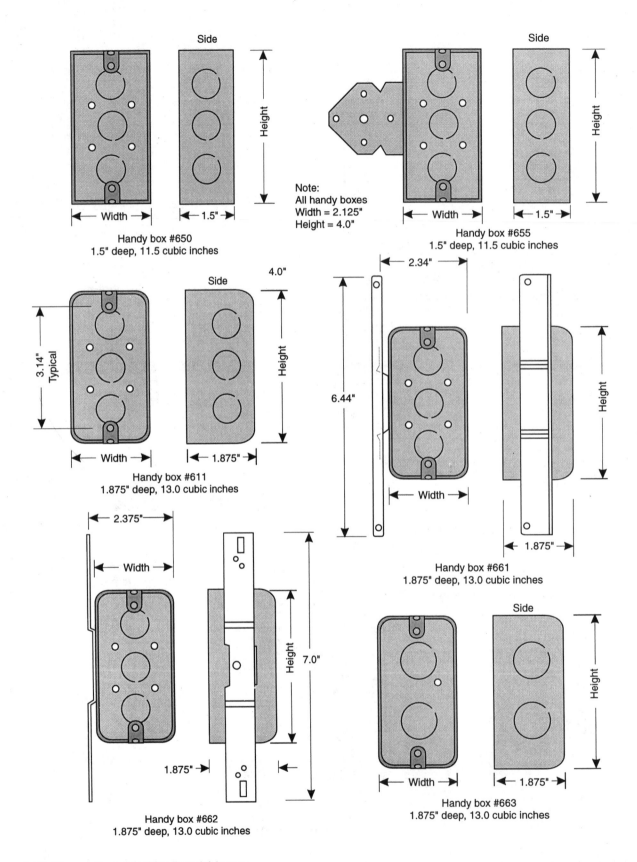

Side

Height

Width ←→ 1.5"

Handy box #650
1.5" deep, 11.5 cubic inches

Note:
All handy boxes
Width = 2.125"
Height = 4.0"

Side

Height

Width ←→ 1.5"

Handy box #655
1.5" deep, 11.5 cubic inches

Side

4.0"

Height

3.14"
Typical

Width ←→ 1.875"

Handy box #611
1.875" deep, 13.0 cubic inches

2.34"

6.44"

Width

1.875"

Height

Handy box #661
1.875" deep, 13.0 cubic inches

2.375"

Width

Height

7.0"

1.875"

Handy box #662
1.875" deep, 13.0 cubic inches

Side

Height

Width ←→ 1.875"

Handy box #663
1.875" deep, 13.0 cubic inches

Figure 2-58: Dimensions of utility (handy) boxes.

CONDUIT BODIES

Conduit bodies are constructed of malleable iron or copper-free aluminum. All are raintight when used with gaskets under their covers. Malleable iron bodies are available finished with zinc-plated aluminum enamel or hot dip and/or mechanically galvanized finish.

When specifying or installing conduit bodies, consult the requirements of *NEC* Sections 370-16 and 410-57. Section 370-16 details the *NEC* requirements for conductor fill; that is, the maximum conductor count or the minimum box size required for a particular situation. This *NEC* Section may also be used to figure how many additional conductors may be added without exceeding the Code-prescribed limit.

NEC Section 410-57 provides details on the *NEC* requirements for receptacles being used in wet locations.

In general, safe electrical practice demands that wires not be jammed into boxes because the possibility of nicks, abrasions, or other damage to the insulating material, creating the potential for ground faults or short circuits in the system.

LB Conduit Body

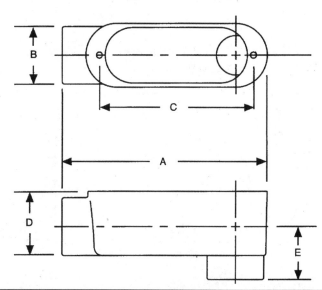

> **NOTE**
>
> Holding the conduit body like a pistol, the opening will be at the top of the body, the 90° hub takeoff will be the "handle," while the other hub opening (90° from the first) will be the "barrel."

DIMENSIONS (in inches) OF LB CONDUIT BODIES					
Conduit Size	**A**	**B**	**C**	**D**	**E**
½	4.5	1.31	3.16	1.38	1.5
¾	5.19	1.5	3.25	1.63	1.63
1	6.25	1.75	4.5	1.88	2.09
1¼	8.06	2.5	6.06	2.53	2.22
1½	8.06	2.5	6.06	2.75	2.38
2	10	3.13	8.13	3.44	2.63
2½	13	4.44	10.5	4.44	3.88
3	13	4.44	10.5	4.44	3.88
3½	15.75	5.5	13.25	5.5	4.56
4	15.75	5.5	13.25	5.5	4.59
5	20.25	7.38	16.19	7.25	5.25

Figure 2-59: Dimensions and characteristics of LB conduit bodies.

C Conduit Body

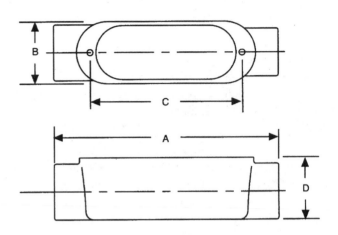

The Type C conduit body is a straight-through configuration used in line with the conduit run for splices and also to facilitate pulling conductors.

DIMENSIONS (in inches) OF C CONDUIT BODIES				
Conduit Size	A	B	C	D
½	5.19	1.31	3.13	1.38
¾	5.75	1.5	3.75	1.63
1	7.06	1.75	4.5	1.88
1¼	8.13	2.5	6	2.5
1½	8.13	2.5	6	2.75
2	10	3.13	8.06	3.44

Figure 2-60: Dimensions and characteristics of C conduit bodies.

E Conduit Body

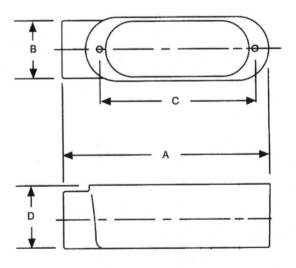

DIMENSIONS (in inches) OF E CONDUIT BODIES				
Conduit Size	A	B	C	D
½	4.5	1.31	3.16	1.38
¾	5.19	1.5	3.25	1.63
1	6.25	1.75	4.5	1.88

Figure 2-61: Dimensions and characteristics of E conduit bodies.

LL Conduit Body

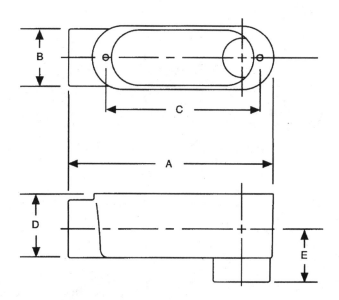

> **NOTE**
>
> Holding the conduit body like a pistol, the opening will be on the left of the body, the 90° hub takeoff will be the "handle," while the other hub opening (90° from the first) will be the "barrel."

DIMENSIONS (in inches) OF LL and LR CONDUIT BODIES					
Conduit Size	**A**	**B**	**C**	**D**	**E**
½	4.5	1.31	3.16	1.38	1.5
¾	5.19	1.5	3.25	1.63	1.63
1	6.25	1.75	4.5	1.88	2.09
1¼	8.06	2.5	6.06	2.53	2.22
1½	8.06	2.5	6.06	2.75	2.38
2	10	3.13	8.13	3.44	2.63

Figure 2-62: Dimensions and characteristics of LL conduit bodies.

LR Conduit Body

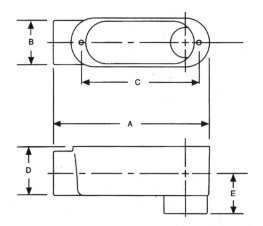

> **NOTE**
>
> Holding the conduit body like a pistol, the opening will be on the right of the body, the 90° hub takeoff will be the "handle," while the other hub opening (90° from the first) will be the "barrel."

Figure 2-63: Dimensions and characteristics of LR conduit bodies. Use dimensions shown in Figure 2-62.

T Conduit Body

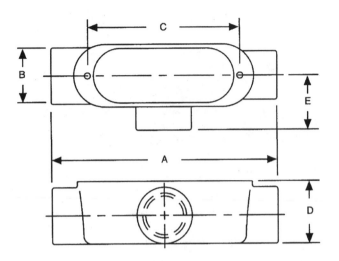

Type T conduit bodies have three hubs; one at each end in line with the conduit run, and one on the side at a right angle to allow a tap to be made in the wiring system.

DIMENSIONS (in inches) OF T CONDUIT BODIES					
Conduit Size	**A**	**B**	**C**	**D**	**E**
½	5.19	1.31	3.13	1.38	1.56
¾	5.75	1.5	3.75	1.63	1.63
1	7	1.75	4.5	1.88	2.13
1¼	8.69	2.5	6	2.38	2.25
1½	8.13	2.5	6	2.75	2.5
2	10	3.13	8.06	3.44	2.19

Figure 2-64: Dimensions and characteristics of T conduit bodies.

Flat Conduit-Body Cover

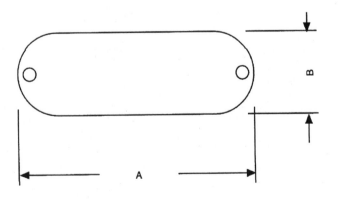

DIMENSIONS (in inches) OF FLAT BODY COVER		
Raco Catalog No.	**A**	**B**
3882, 3892	3.9	1.31
3883, 3893	4.63	1.5
3884, 3894	5.38	1.75
3886, 3896	7.25	2.5
3888, 3898	9.5	3.13
4082, 4092	12.25	4.44
4086, 4096	14.88	5.5
4090	18.25	7.31

Figure 2-65: Dimensions of flat conduit-body covers.

Domed-Top Conduit-Body Cover

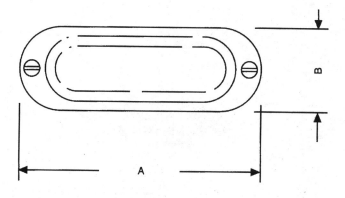

DIMENSIONS (in inches) OF DOMED BODY COVERS		
Raco Catalog No.	A	B
3872	3.88	1.31
3873	4.59	1.5
3874	5.38	1.75
3876	7.25	2.5
3878	9.56	3.13
4072	12.25	4.44
4076	14.88	5.5

Figure 2-66: Dimensions of domed-top conduit-body covers.

CONDUIT FITTINGS

When it is necessary to couple, connect, fasten, or ground conduit or cable systems, couplings and fittings are used. Several specification grades are available:

- Malleable iron
- Lightweight aluminum
- Die cast zinc

Most conduit fittings are designed in accordance with the standards established by Underwriters' Laboratories (UL) and are also certified by the Canadian Standards Association (CSA).

The following illustrations cover most conduit fittings in current use. Dimensional data listed is intended for general reference with broad tolerance limits.

Liquidtight connectors shown in this section are approved for use in the following hazardous locations:

- Class 1, Division 2 where volatile liquids are handled or stored (*NEC* Section 501-4b).
- Class 2, Division 1 where combustible dust is in the air (*NEC* Section 502-4b2).
- Class 2, Division 2 where dust deposits may accumulate (*NEC* Section 502-3a2).
- Class 3, Division 1 where easily ignitable fibers are manufactured or used (*NEC* Section 503-3a2).

- Class 3, Division 2 where easily ignitable fibers are handled or stored (*NEC* Section 503-3b).

The *NEC* does not permit liquidtight connectors for Class 1, Division 1 locations where volatile liquids and flammable gases are likely to be in the air. Explosionproof fittings for Class 1, Division 1 locations are covered in Chapter 7 — Special Systems.

EMT Connectors

In general, three types of EMT connectors are in current use:

- Steel and die cast set screw connectors
- Steel and die cast compression connectors
- Steel indenter connectors

Steel set screw connectors are used in dry locations to bond EMT conduit to a box or other enclosure. EMT couplings specified should be UL listed and comply with Standard 5148.

Steel compression connectors are for use in wet or dry locations and provide both raintight and concrete-tight connections up through 2 in. EMT compression connectors 2½ through 4 in are concrete-tight only. All EMT fitting should be zinc electroplated for corrosion protection. Furthermore, they should be UL listed and comply with Standard 5148.

EMT SET SCREW CONNECTORS

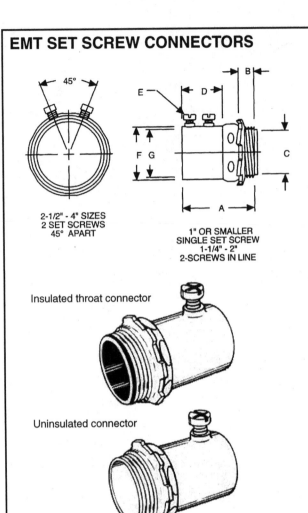

2-1/2" - 4" SIZES
2 SET SCREWS
45° APART

1" OR SMALLER
SINGLE SET SCREW
1-1/4" - 2"
2-SCREWS IN LINE

Insulated throat connector

Uninsulated connector

A = Overall length
B = Hub length, thread length and thread spec.
C = Throat inside diameter
D = Length to conduit stop
E = Screw spec. (type of head, length, thread spec.)
F = Body outside diameter
G = Body Inside diameter

EMT SET SCREW CONNECTORS DIMENSIONS A, B, AND C

Raco No.	A	B	Thd. Spec.	C
2122, 2002	1.57	.47	½-14 NPS	.63
2123, 2003	1.63	.34	¾-14 NPS	.81
2124, 2004	.88	.39	1-11½ NPS	1.03
2125, 2005	2.34	.63	1¼-11½ NPS	1.38
2126, 2006	2.59	.63	1½-11½ NPS	1..58
2128, 2008	2.47	.63	2-11½ NPS	2.06
2160, 2140	2.94	.94	2½-11½ NPS	2.56
2162, 2142	3.06	1	3-8 NPS	3.13
2164, 2144	3.38	1	3½-8NPS	3.75
2166, 2146	3.56	1	4-8NPS	4.13

EMT SET SCREW CONNECTORS DIMENSIONS D, E, F, AND G

Raco No.	D	E Set Screw	F	G
2122, 2002	.78	Tri-drive, 12-24 × .25 in	2.22	.72
2123, 2003	.91	Tri-drive, 12-24 × .25 in	1.06	.94
2124, 2004	1.16	Tri-drive, 12-24 × .25 in	1.28	1.19
2125, 2005	1.41	Hex Head/Slot .25-28 × .31	172	11.22
2126, 2006	1.5	Hex Head/Slot .25-28 × .31	1.94	1.75
2128, 2008	1.78	Hex Head/Slot .25-28 × .31	2.42	2.22
2160, 2140	1.88	Hex Head .31-24 × .34	3.22	2.92
2162, 2142	1.88	Hex Head .31-24 × .34	3..84	3.05
2164, 2144	2.13	Hex Head .31-24 × .34	4.36	4.05
2166, 2146	2.38	Hex Head .31-24 × .34	4.86	4.56

Figure 2-67: Dimensional data for EMT set screw connectors.

EMT COMPRESSION CONNECTORS

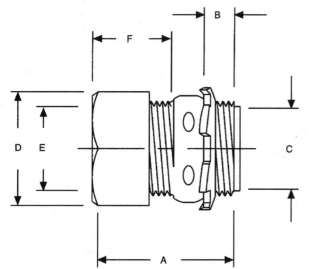

A = Overall length
B = Hub length, thread length and thread spec.
C = Throat inside diameter
D = Compression nut outside diameter
E = Compression nut inside diameter
F = Length of conduit stop

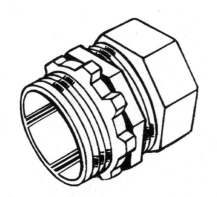

EMT COMPRESSION CONNECTORS DIMENSIONS A, B, AND C				
Raco No.	A	B	Thd. Spec.	C
2912	2.88	.44	½-14 NPS	.63
2902	2.75	.44	½-14 NPS	.63
2913, 2903	1.44	.34	¾-14 NPS	83
2914, 2904	2	.39	1-11.5 NPS	1.03
2915, 2905	2	.56	1¼-11.5 NPS	1.38
2916, 2906	2.31	.63	1½-11.5 NPS	1.61
2918, 2908	2.5	.63	2-11.5 NPS	2.06
2960, 2940	3.38	.94	2½-8 NPS	2.56
2962, 2942	3.5	1	3 - 8 NPS	3.07
2964, 2944	3.88	1	3½-8 NPS	3.75
2966, 2946	4	1	4 - 8 NPS	4.13

EMT COMPRESSION CONNECTORS DIMENSIONS D, E, AND F			
Raco No.	D	E	F
2912, 2902	1.13	.72	1.25
2913, 2903	1.38	.94	.94
2914, 2904	1.67	1.17	1.06
2915, 2905	2.19	1.53	.94
2916, 2906	2.44	1.75	1.06
2918, 2908	2.88	2.22	1.75
2960, 2940	3.94	2.94	1.69
2962, 2942	4.66	3.56	2
2964, 2944	5.19	4.06	2.25
2966, 2946	5.75	4.56	2.5

Figure 2-68: Dimensional data for EMT compression connectors.

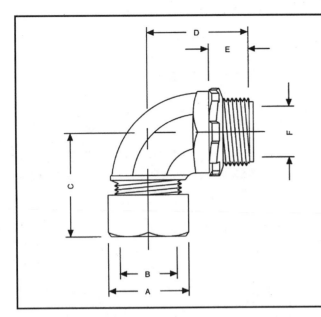

EMT SHORT 90° COMPRESSION CONNECTOR				
Raco No.	A	B	C	D
2072	1.13	.78	1.28	1.22
2073	1.38	.94	1.59	1.47
2074	1.67	1.69	1.97	1.56

EMT SHORT 90° COMPRESSION CONNECTOR			
Raco No.	E	Thread Spec.	F
2072	.5	½ - 14 NPT	.63
2073	.56	¾ - 14 NPT	.81
2074	.5	1- 11.5 NPT	1.03

Figure 2-69: Dimensional data for EMT short 90° compression connector.

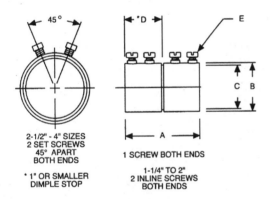

A = Overall length
B = Outside diameter of body
C = Inside diameter of body
D = Length to center stop
E = Screw spec.

EMT Set Screw Coupling

EMT SET SCREW COUPLING					
Raco No.	A	B	C	D	Thread Specs.
2022	1.75	.92	.78	.88	Tri-Drive 12 - 24 × ¼
2023	1.97	1.08	1.31	.97	Tri-Drive 12 - 24 × ¼
2024	1.88	1.28	1.19	.88	Tri-Drive 12 - 24 × ¼
2025	2.69	1.70	1.53	1.30	Hex Head/Slot: ¼ -28 × .31 in
2026	2.88	1.94	1.75	1.39	Hex Head/Slot: ¼ -28 × .31 in
2028	3.44	2.42	2.22	1.67	Hex Head/Slot: ¼ -28 × .31 in
2150	3.86	3.22	2.45	1.5	Hex Head/Slot: .31 in -24 × .34 in
2152	3.81	3.84	3.55	1.5	Hex Head/Slot: .31 in -24 × .34 in
2154	4.63	4.36	4.05	2.19	Hex Head/Slot: .31 in -24 × .34 in
2156	4.94	4.86	4.56	2.09	Hex Head/Slot: .31in -24 × .34 in

Figure 2-70: Dimensional data for EMT set screw couplings.

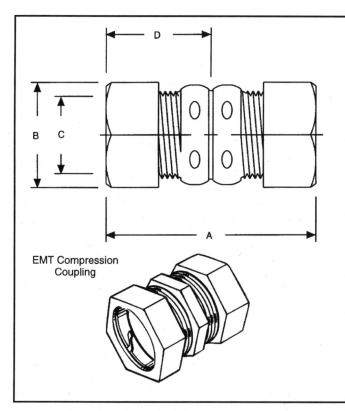

EMT COMPRESSION COUPLINGS				
Raco No.	A	B	C	D
2922	1.66	1.13	.72	.75
2923	1.72	1.38	.94	69
2924	1.97	1.67	1.17	.94
2925	2.13	2.19	1.53	1
2926	2.56	2.44	1.75	1.13
2928	2.88	2.88	2.22	1.28
2950	4.75	3.94	2.94	2
2952	4.72	4.66	3.56	2
2954	5.5	5.14	4..06	2.69
2956	5.81	5.75	4.56	2.59

A = Overall length
B = Compression nut outside diameter
C = Compression nut inside diameter
D = Length to center of stop

EMT Compression Coupling

Figure 2-71: Dimensional data for EMT compression couplings.

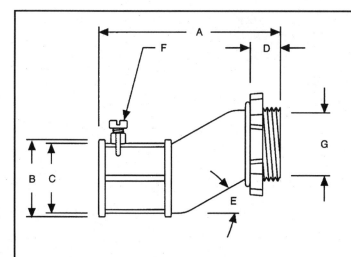

EMT SET SCREW OFFSET CONNECTOR					
Raco No.	A	B	C	D	E
1762	2.1	.95	.72	.38	30°
1763	3	1.16	.94	.44	30°

EMT SET SCREW OFFSET CONNECTOR			
Raco No.	F	G	Screw Type
1762	10-24 × .31 in	.63	½ - 14 NPS

A = Overall length
B = Outside diameter of body
C = Inside diameter of body
D = Thread length
E = Angle of offset
F = Screw spec.
G = Throat inside diameter

EMT Set Screw Offset Connector

Figure 2-72: Dimensional data for EMT set screw offset connectors.

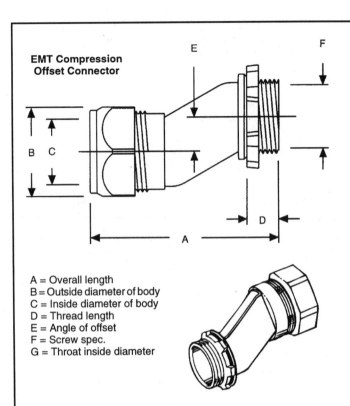

EMT Compression Offset Connector

A = Overall length
B = Outside diameter of body
C = Inside diameter of body
D = Thread length
E = Angle of offset
F = Screw spec.
G = Throat inside diameter

EMT SET SCREW OFFSET CONNECTOR				
Raco No.	A	B	C	D
1952	2.16	1.09	.72	.38
1953	2.89	1.31	.95	.44
1954	3.03	1.63	1.19	.56

EMT SET SCREW OFFSET CONNECTOR			
Raco No.	E	F	THD. Spec.
1952	.88	.63	½ - 14 NPS
1953	.75	.81	¾ - 14 NPS
1954	.75	1.05	1 - 11½ NPS

Figure 2-73: Dimensional data for EMT compression offset connectors.

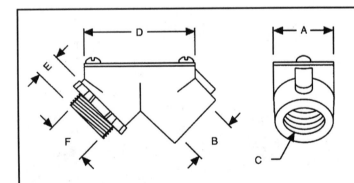

A = Width of body
B = Inside throat diameter
C = Screw spec.
D = Overall length
E = Thread length and spec.
F = Throat inside diameter

EMT HANDY ELL CONNECTORS		
Raco No.	F	Screw Type
2762	.63	¼-14NPS-Slotted
2763	.81	¼-14NPS-Slotted

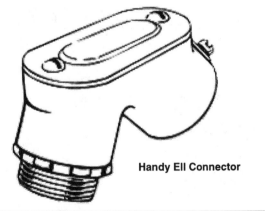

Handy Ell Connector

EMT HANDY ELL CONNECTORS					
Raco No.	A	B	C	D	E
2762	1.13	.72	10-24 × .31	1.88	.38
2763	1.34	.94	10-24 × .31	2.19	.44

Figure 2-74: Dimensional data for EMT 90° set screw "Handy-Ell" connectors.

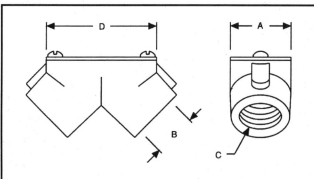

A = Width of body
B = Inside throat diameter
C = Screw spec.
D = Overall length

EMT HANDY-ELL COUPLINGS	
Raco No.	Screw Type
2752	Slotted, female to female
2753	Slotted, female to female
2764	Slotted, female to female

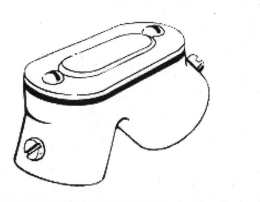

EMT HANDY-ELL COUPLINGS				
Raco No.	A	B	C	D
2752	1.13	.71	10-24 × .94	1.88
2753	1.34	.94	10-24 × .94	2.19
2754	1.63	1.19	10-24 × .94	3.13

Figure 2-75: Dimensional data for EMT 90° set screw "Handy-Ell" couplings.

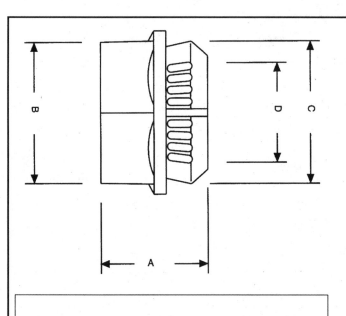

A = Overall length
B = Outside diameter of nut
C = Outside diameter of body
D = Inside diameter of throat

EMT TWO-PIECE CONNECTOR				
Raco No.	A	B	C	D
2702	.69	1.05	.94	.61
2703	.81	1.31	1.16	.81

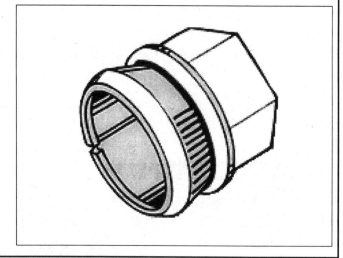

Figure 2-76: Dimensional data for EMT two-piece connectors.

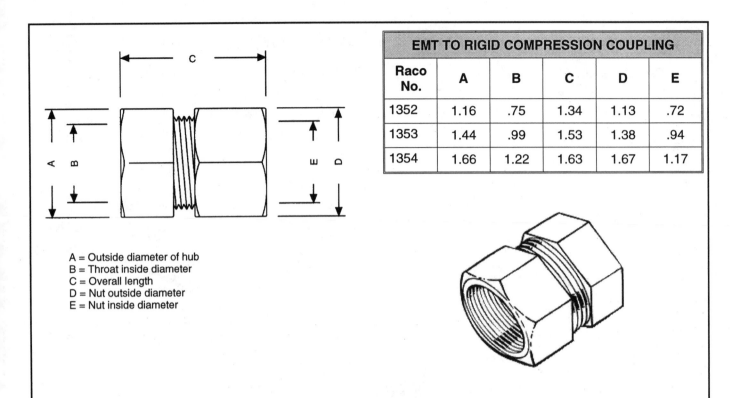

EMT TO RIGID COMPRESSION COUPLING					
Raco No.	A	B	C	D	E
1352	1.16	.75	1.34	1.13	.72
1353	1.44	.99	1.53	1.38	.94
1354	1.66	1.22	1.63	1.67	1.17

A = Outside diameter of hub
B = Throat inside diameter
C = Overall length
D = Nut outside diameter
E = Nut inside diameter

Figure 2-77: Dimensional data for EMT to rigid compression couplings.

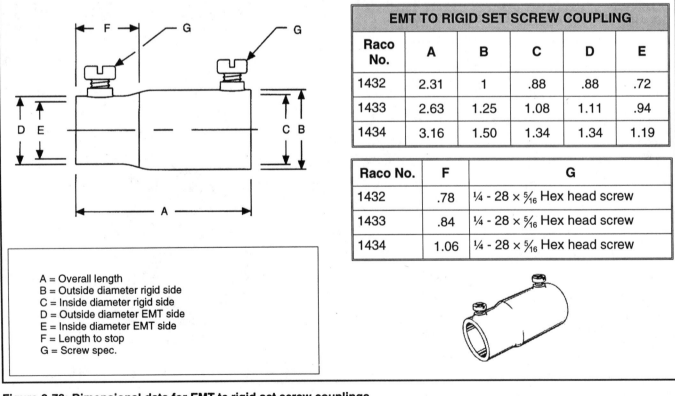

EMT TO RIGID SET SCREW COUPLING					
Raco No.	A	B	C	D	E
1432	2.31	1	.88	.88	.72
1433	2.63	1.25	1.08	1.11	.94
1434	3.16	1.50	1.34	1.34	1.19

Raco No.	F	G
1432	.78	¼ - 28 × ⁵⁄₁₆ Hex head screw
1433	.84	¼ - 28 × ⁵⁄₁₆ Hex head screw
1434	1.06	¼ - 28 × ⁵⁄₁₆ Hex head screw

A = Overall length
B = Outside diameter rigid side
C = Inside diameter rigid side
D = Outside diameter EMT side
E = Inside diameter EMT side
F = Length to stop
G = Screw spec.

Figure 2-78: Dimensional data for EMT to rigid set screw couplings.

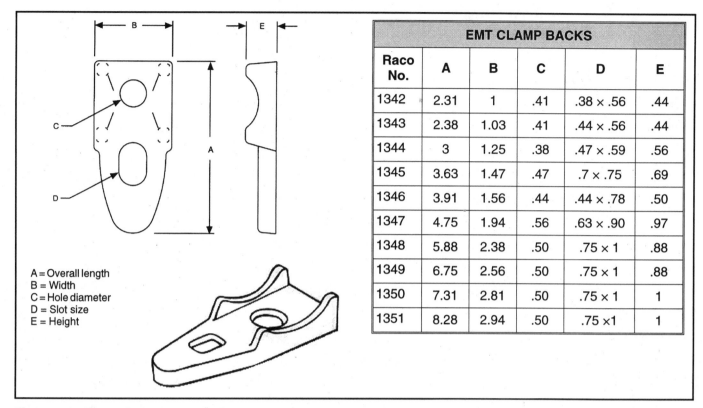

EMT CLAMP BACKS					
Raco No.	A	B	C	D	E
1342	2.31	1	.41	.38 × .56	.44
1343	2.38	1.03	.41	.44 × .56	.44
1344	3	1.25	.38	.47 × .59	.56
1345	3.63	1.47	.47	.7 × .75	.69
1346	3.91	1.56	.44	.44 × .78	.50
1347	4.75	1.94	.56	.63 × .90	.97
1348	5.88	2.38	.50	.75 × 1	.88
1349	6.75	2.56	.50	.75 × 1	.88
1350	7.31	2.81	.50	.75 × 1	1
1351	8.28	2.94	.50	.75 ×1	1

A = Overall length
B = Width
C = Hole diameter
D = Slot size
E = Height

Figure 2-79: Dimensional data for EMT clamp backs.

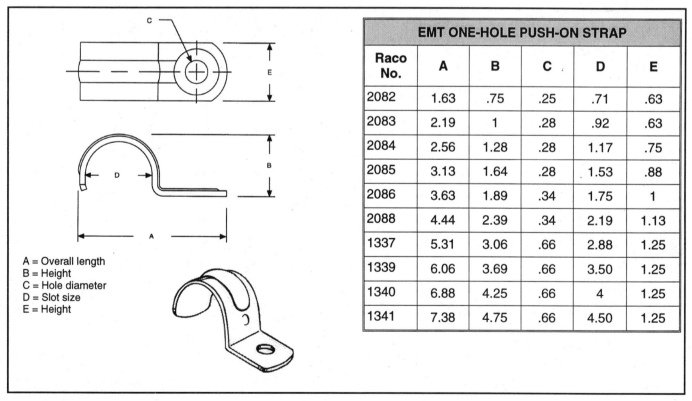

EMT ONE-HOLE PUSH-ON STRAP					
Raco No.	A	B	C	D	E
2082	1.63	.75	.25	.71	.63
2083	2.19	1	.28	.92	.63
2084	2.56	1.28	.28	1.17	.75
2085	3.13	1.64	.28	1.53	.88
2086	3.63	1.89	.34	1.75	1
2088	4.44	2.39	.34	2.19	1.13
1337	5.31	3.06	.66	2.88	1.25
1339	6.06	3.69	.66	3.50	1.25
1340	6.88	4.25	.66	4	1.25
1341	7.38	4.75	.66	4.50	1.25

A = Overall length
B = Height
C = Hole diameter
D = Slot size
E = Height

Figure 2-80: Dimensional data for EMT one-hole push-on straps.

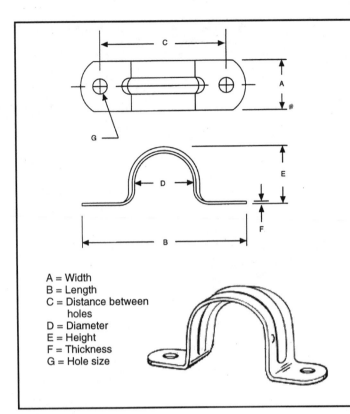

EMT TWO-HOLE STRAP				
Raco No.	A	B	C	D
2092	.625	2.140	1.625	.706
2093	.625	2.288	1.715	.922
2094	.750	2.750	2.140	1.163
2095	.750	3.530	2.750	1.510
2096	.750	3.770	3.141	1.740

Raco No.	E	F	G
2092	.750	.032 × .625	.187
2093	1.015	.032 × .625	.187
2094	1.235	.036 × .750	.250
2095	1.540	.036 × .750	.250
2096	1.785	.036 × .750	.250

A = Width
B = Length
C = Distance between
    holes
D = Diameter
E = Height
F = Thickness
G = Hole size

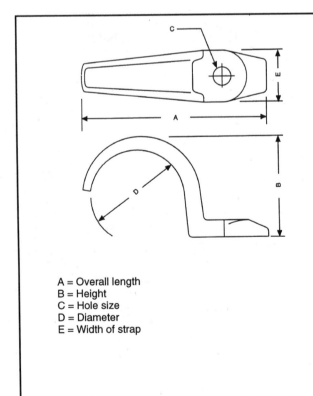

Figure 2-81: Dimensional data for EMT two-hole straps.

EMT MALLEABLE IRON STRAPS					
Raco No.	A	B	C	D	E
1310	6.44	3.28	.66	2.88	1.81
1312	7.50	4	.66	3.50	2
1314	8.13	4.5	.83	4	2.13
1316	8.88	5.56	.75	4.50	2.25

A = Overall length
B = Height
C = Hole size
D = Diameter
E = Width of strap

Figure 2-82: Dimensional data for EMT malleable iron straps.

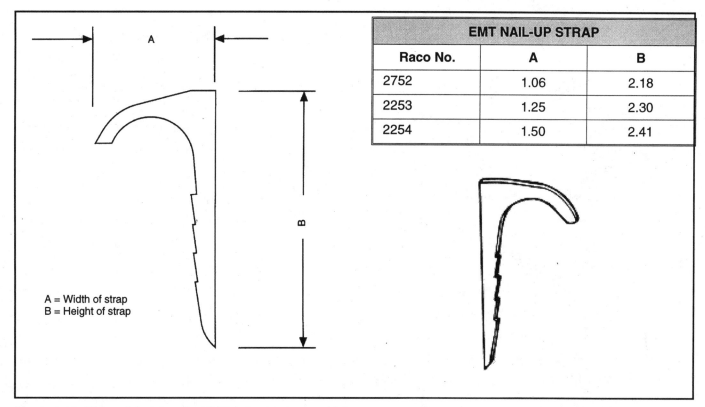

Figure 2-83: Dimensional data for EMT nail-up straps.

EMT NAIL-UP STRAP		
Raco No.	A	B
2752	1.06	2.18
2253	1.25	2.30
2254	1.50	2.41

LIQUIDTIGHT CONDUIT AND CORD CONNECTORS

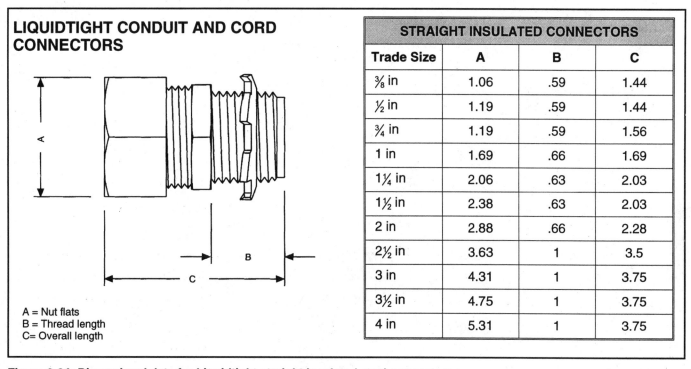

A = Nut flats
B = Thread length
C = Overall length

STRAIGHT INSULATED CONNECTORS			
Trade Size	A	B	C
⅜ in	1.06	.59	1.44
½ in	1.19	.59	1.44
¾ in	1.19	.59	1.56
1 in	1.69	.66	1.69
1¼ in	2.06	.63	2.03
1½ in	2.38	.63	2.03
2 in	2.88	.66	2.28
2½ in	3.63	1	3.5
3 in	4.31	1	3.75
3½ in	4.75	1	3.75
4 in	5.31	1	3.75

Figure 2-84: Dimensional data for Liquidtight straight insulated steel connectors.

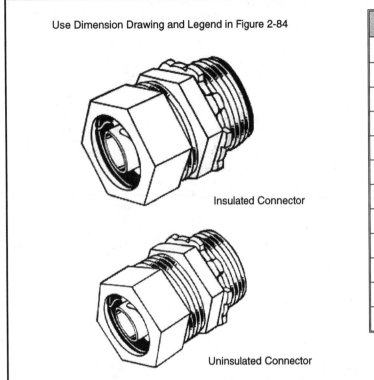

Use Dimension Drawing and Legend in Figure 2-84

Insulated Connector

Uninsulated Connector

STRAIGHT UNINSULATED CONNECTORS			
Trade Size	**A**	**B**	**C**
⅜ in	1.06	.59	1.44
½ in	1.19	.59	1.44
¾ in	1.19	.59	1.56
1 in	1.69	.66	1.69
1¼ in	2.06	.63	2.03
1½ in	2.38	.63	2.03
2 in	2.88	.66	2.28
2½ in	3.63	1	3.5
3 in	4.31	1	3.75
3½ in	4.75	1	3.75
4 in	5.31	1	3.75

Figure 2-85: Dimensional data for Liquidtight straight uninsulated steel connectors.

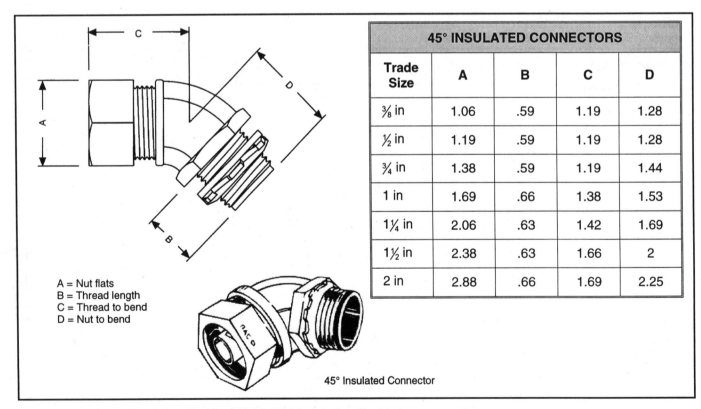

A = Nut flats
B = Thread length
C = Thread to bend
D = Nut to bend

45° Insulated Connector

45° INSULATED CONNECTORS				
Trade Size	**A**	**B**	**C**	**D**
⅜ in	1.06	.59	1.19	1.28
½ in	1.19	.59	1.19	1.28
¾ in	1.38	.59	1.19	1.44
1 in	1.69	.66	1.38	1.53
1¼ in	2.06	.63	1.42	1.69
1½ in	2.38	.63	1.66	2
2 in	2.88	.66	1.69	2.25

Figure 2-86: Dimensional data for Liquidtight 45° insulated malleable iron connectors.

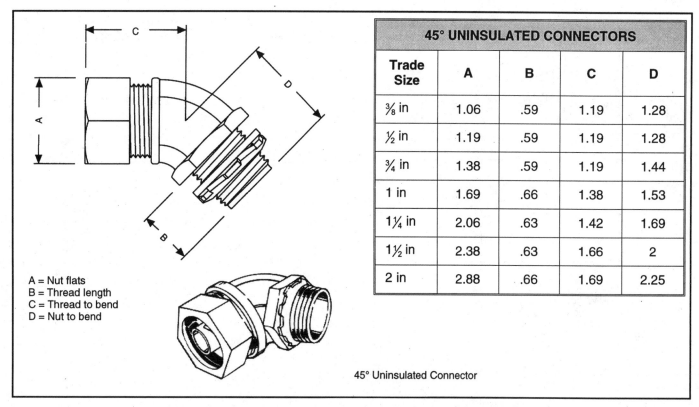

45° UNINSULATED CONNECTORS				
Trade Size	A	B	C	D
⅜ in	1.06	.59	1.19	1.28
½ in	1.19	.59	1.19	1.28
¾ in	1.38	.59	1.19	1.44
1 in	1.69	.66	1.38	1.53
1¼ in	2.06	.63	1.42	1.69
1½ in	2.38	.63	1.66	2
2 in	2.88	.66	1.69	2.25

A = Nut flats
B = Thread length
C = Thread to bend
D = Nut to bend

45° Uninsulated Connector

Figure 2-87: Dimensional data for Liquidtight 45° uninsulated malleable iron connectors.

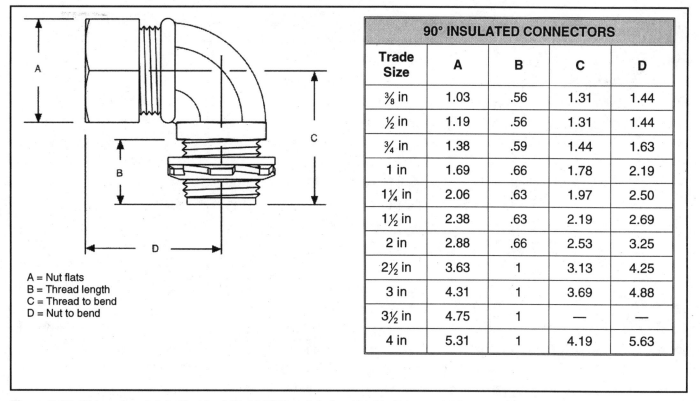

90° INSULATED CONNECTORS				
Trade Size	A	B	C	D
⅜ in	1.03	.56	1.31	1.44
½ in	1.19	.56	1.31	1.44
¾ in	1.38	.59	1.44	1.63
1 in	1.69	.66	1.78	2.19
1¼ in	2.06	.63	1.97	2.50
1½ in	2.38	.63	2.19	2.69
2 in	2.88	.66	2.53	3.25
2½ in	3.63	1	3.13	4.25
3 in	4.31	1	3.69	4.88
3½ in	4.75	1	—	—
4 in	5.31	1	4.19	5.63

A = Nut flats
B = Thread length
C = Thread to bend
D = Nut to bend

Figure 2-88: Dimensional data for Liquidtight 90° insulated malleable iron connectors.

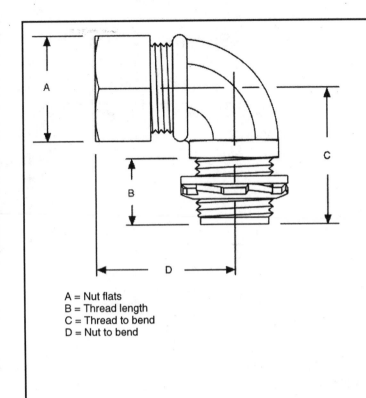

90° UNINSULATED CONNECTORS				
Trade Size	A	B	C	D
⅜ in	1.03	.56	1.31	1.44
½ in	1.19	.56	1.31	1.44
¾ in	1.38	.59	1.44	1.63
1 in	1.69	.66	1.78	2.19
1¼ in	2.06	.63	1.97	2.50
1½ in	2.38	.63	2.19	2.69
2 in	2.88	.66	2.53	3.25
2½ in	3.63	1	3.13	4.25
3 in	4.31	1	3.69	4.88
3½ in	4.75	1	—	—
4 in	5.31	1	4.19	5.63

A = Nut flats
B = Thread length
C = Thread to bend
D = Nut to bend

Figure 2-89: Dimensional data for Liquidtight 90° uninsulated malleable iron connectors.

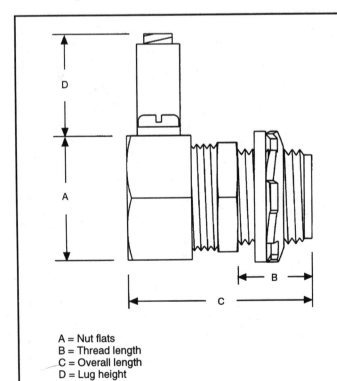

STRAIGHT INSULATED CONNECTORS WITH GROUND LUG				
Trade Size	A	B	C	D
⅜ in	1.06	.59	1.44	.78
½ in	1.19	.59	1.44	.78
¾ in	1.19	.59	1.56	.78
1 in	1.69	.66	1.69	.78
1¼ in	2.06	.63	2.03	.78
1½ in	2.38	.63	2.03	.78
2 in	2.88	.66	2.28	.78
2½ in	3.63	1	3.5	1.17
3 in	4.31	1	3.75	1.17
3½ in	4.75	1	3.75	1.89
4 in	5.31	1	3.75	1.89

A = Nut flats
B = Thread length
C = Overall length
D = Lug height

Figure 2-90: Dimensional data for Liquidtight straight insulated malleable iron connectors with ground lug.

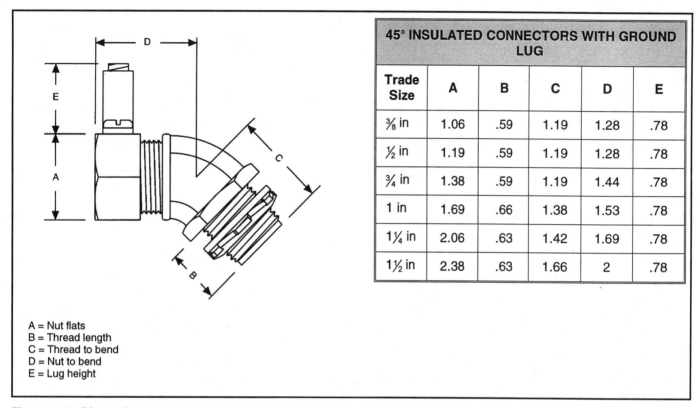

45° INSULATED CONNECTORS WITH GROUND LUG					
Trade Size	**A**	**B**	**C**	**D**	**E**
⅜ in	1.06	.59	1.19	1.28	.78
½ in	1.19	.59	1.19	1.28	.78
¾ in	1.38	.59	1.19	1.44	.78
1 in	1.69	.66	1.38	1.53	.78
1¼ in	2.06	.63	1.42	1.69	.78
1½ in	2.38	.63	1.66	2	.78

A = Nut flats
B = Thread length
C = Thread to bend
D = Nut to bend
E = Lug height

Figure 2-91: Dimensional data for Liquidtight 45° malleable iron connectors with ground lug.

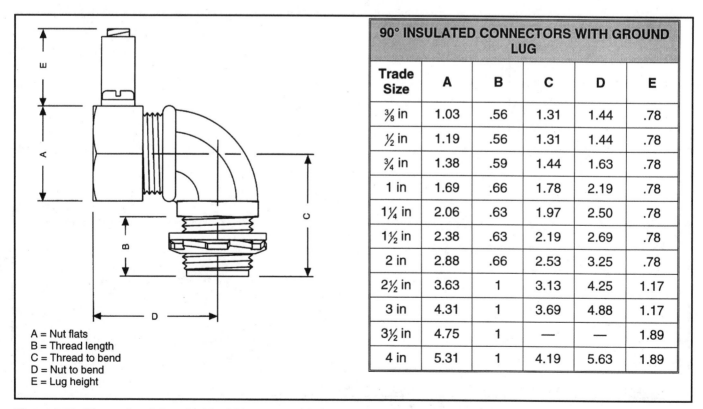

90° INSULATED CONNECTORS WITH GROUND LUG					
Trade Size	**A**	**B**	**C**	**D**	**E**
⅜ in	1.03	.56	1.31	1.44	.78
½ in	1.19	.56	1.31	1.44	.78
¾ in	1.38	.59	1.44	1.63	.78
1 in	1.69	.66	1.78	2.19	.78
1¼ in	2.06	.63	1.97	2.50	.78
1½ in	2.38	.63	2.19	2.69	.78
2 in	2.88	.66	2.53	3.25	.78
2½ in	3.63	1	3.13	4.25	1.17
3 in	4.31	1	3.69	4.88	1.17
3½ in	4.75	1	—	—	1.89
4 in	5.31	1	4.19	5.63	1.89

A = Nut flats
B = Thread length
C = Thread to bend
D = Nut to bend
E = Lug height

Figure 2-92: Dimensional data for Liquidtight 90° malleable iron connector with ground lug.

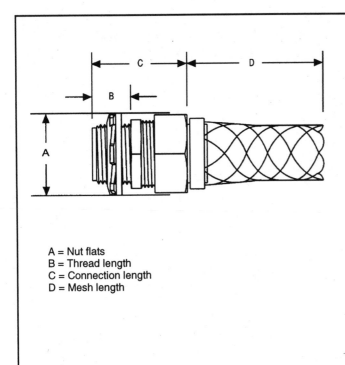

A = Nut flats
B = Thread length
C = Connection length
D = Mesh length

STRAIGHT MESH-INSULATED THROAT				
Hub Size	**A**	**B**	**C**	**D**
1/8 in	1.06	.59	1.44	3
1/2 in	1.19	.59	1.44	3.5
3/4 in	1.38	.59	1.56	4
1 in	1.69	.66	1.69	5
1 1/4 in	2.06	.63	2.03	6
1 1/2 in	2.38	.63	2.22	6.75
2 in	2.88	.66	2.28	8
2 1/2 in	3.63	1	3.5	9.75
3 in	4.31	1	3.75	11
4 in	5.31	1	3.75	14

Figure 2-93: Dimensional data for straight mesh-insulated throat.

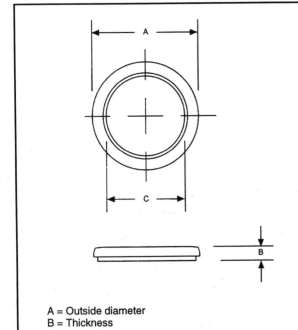

A = Outside diameter
B = Thickness
C = Inside diameter

SEALING WASHERS			
Conduit Size	**A**	**B**	**C**
2452	1.08	.08	.80
2453	1.34	.16	1
2454	1.62	.17	1.25
2455	2	.17	1.61
2456	2.36	.19	1.84
2458	2.83	.19	2.31

Sealing Washer

Figure 2-94: Dimensional data for sealing washers.

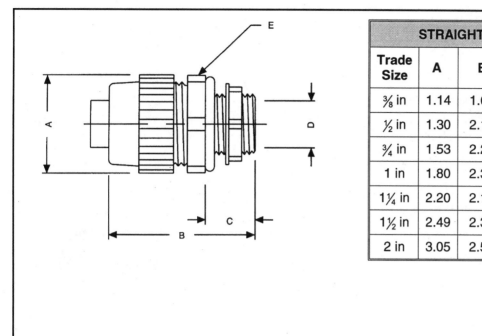

STRAIGHT NYLON CONNECTORS						
Trade Size	A	B	C	D	E A/C	E A/F
⅜ in	1.14	1.63	.57	.42	1.41	1.30
½ in	1.30	2.14	.57	.55	1.41	1.30
¾ in	1.53	2.22	.58	.74	1.85	1.53
1 in	1.80	2.32	.72	.96	1.94	1.80
1¼ in	2.20	2.15	.74	1.30	2.38	2.18
1½ in	2.49	2.35	.76	1.45	2.63	2.43
2 in	3.05	2.51	.79	1.90	3.13	2.93

Figure 2-95: Dimensional data for straight nylon connectors — used with Type B conduit.

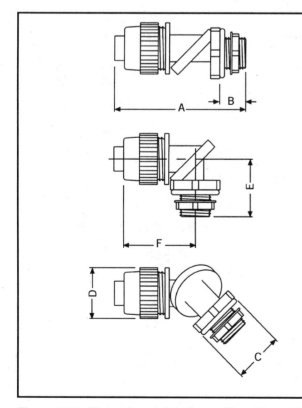

MULTIPOSITION NYLON CONNECTORS				
Trade Size	A	B	C	
			A/C	A/F
½ in	3.27	.57	1.41	1.30
¾ in	3.66	.58	1.65	1.53
1 in	4.00	.72	1.94	1.80

Trade Size	D	E	F
½ in	1.30	1.43	2.00
¾ in	1.53	1.59	2.23
1 in	1.80	1.84	2.30

Figure 2-96: Dimensional data for multiposition nylon connectors — used with Type B conduit.

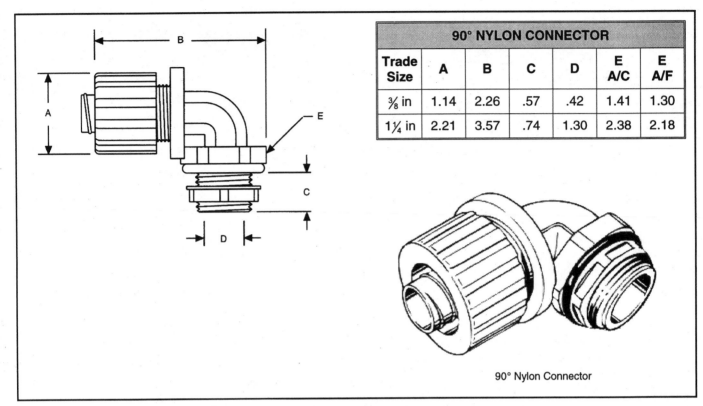

90° NYLON CONNECTOR						
Trade Size	A	B	C	D	E A/C	E A/F
⅜ in	1.14	2.26	.57	.42	1.41	1.30
1¼ in	2.21	3.57	.74	1.30	2.38	2.18

90° Nylon Connector

Figure 2-97: Dimensional data for 90° nylon connectors for use with Type B conduit.

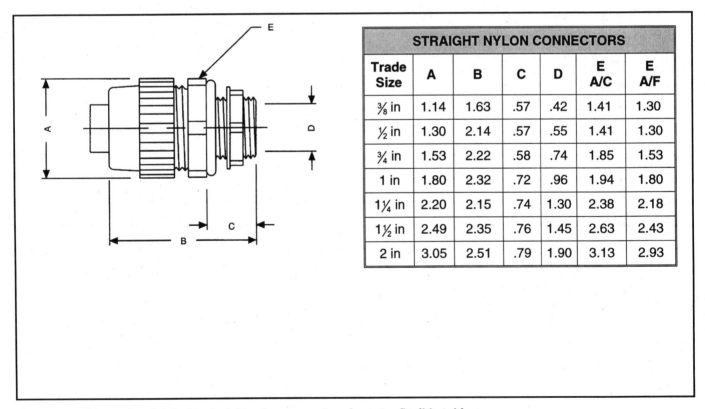

STRAIGHT NYLON CONNECTORS						
Trade Size	A	B	C	D	E A/C	E A/F
⅜ in	1.14	1.63	.57	.42	1.41	1.30
½ in	1.30	2.14	.57	.55	1.41	1.30
¾ in	1.53	2.22	.58	.74	1.85	1.53
1 in	1.80	2.32	.72	.96	1.94	1.80
1¼ in	2.20	2.15	.74	1.30	2.38	2.18
1½ in	2.49	2.35	.76	1.45	2.63	2.43
2 in	3.05	2.51	.79	1.90	3.13	2.93

Figure 2-98: Dimensional data for straight nylon connectors for extra-flexible tubing.

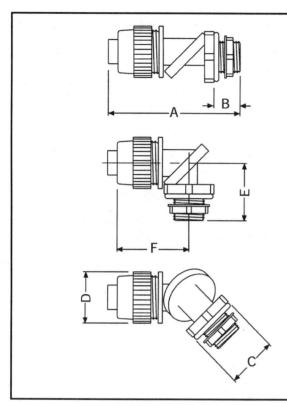

MULTIPOSITION NYLON CONNECTORS				
Trade Size	A	B	C	
			A/C	A/F
½ in	3.27	.57	1.41	1.30
¾ in	3.66	.58	1.65	1.53
1 in	4.00	.72	1.94	1.80

Trade Size	D	E	F
½ in	1.30	1.43	2.00
¾ in	1.53	1.59	2.23
1 in	1.80	1.84	2.30

Figure 2-99: Dimensional data for multiposition nylon connectors for extra-flexible tubing.

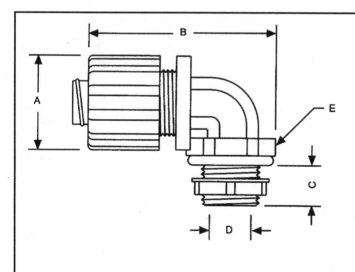

90° NYLON CONNECTOR						
Trade Size	A	B	C	D	E A/C	E A/F
⅜ in	1.14	2.26	.57	.42	1.41	1.30
1¼ in	2.21	3.57	.74	1.30	2.38	2.18

Figure 2-100: Dimensional data for 90° nylon connectors for extra-flexbile tubing.

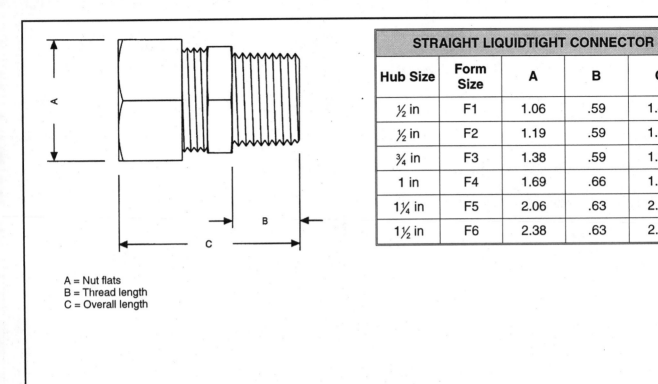

Figure 2-101: Dimensional data for straight Liquidtight uninsulated malleable iron connectors.

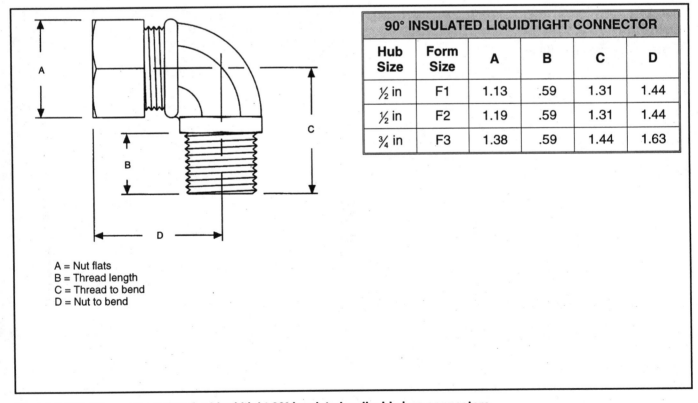

Figure 2-102: Dimensional data for Liquidtight 90° insulated malleable iron connectors.

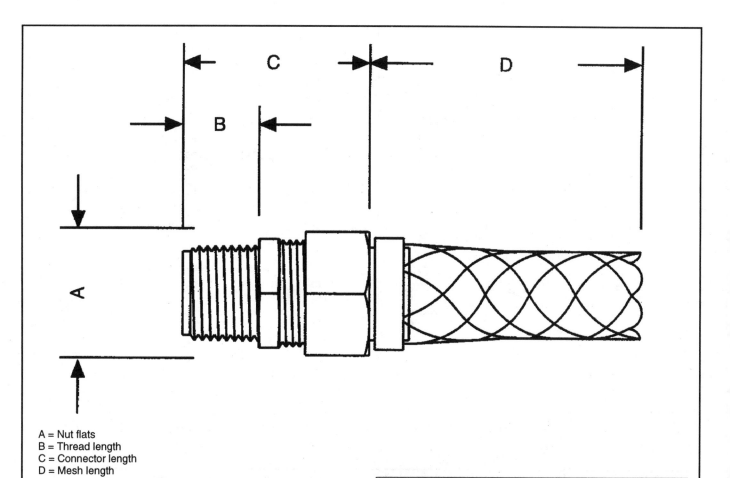

A = Nut flats
B = Thread length
C = Connector length
D = Mesh length

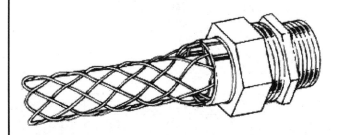

STRAIGHT LIQUIDTIGHT MESH CONNECTOR				
Hub Size	Form Size	A	B	C
½ in	F1	1.06	.59	1.44
½ in	F2	1.19	.59	1.44
¾ in	F3	1.38	.59	1.56
1 in	F4	1.69	.66	1.69

STRAIGHT LIQUIDTIGHT MESH CONNECTOR	
Cable Range	D Mesh Length
.125 - .250	2.12@ .24 dia.
.250 - .375	3.62@ .37 dia.
.375 - .425	4.12@ .50 dia.
.425 - .500	4.12@ .50 dia.
.375 - .500	4.12@ .50 dia.
,500 - .600	4.40@ .62 dia.
.600 - .650	5.00@ .68 dia.
.650 - .700	5.00@ .68 dia.
.700 - .850	6.04@ .85 dia.
.600 - .700	5.00@ .68 dia.
.700 - .850	6.04@ .85 dia.
.850 - 1.000	6.00@ 1.00 dia.

Figure 2-103: Dimensional data for Liquidtight straight-mesh steel/malleable iron cord connector.

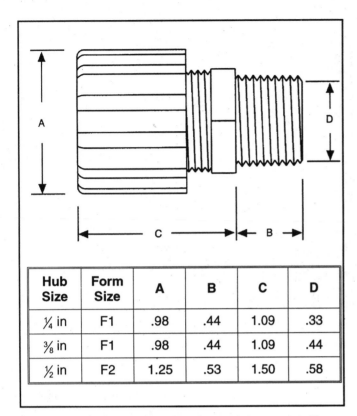

Hub Size	Form Size	A	B	C	D
¼ in	F1	.98	.44	1.09	.33
⅜ in	F1	.98	.44	1.09	.44
½ in	F2	1.25	.53	1.50	.58

Figure 2-104: Dimensional data for straight nonmetallic connectors.

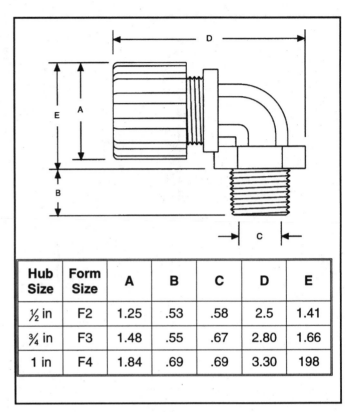

Hub Size	Form Size	A	B	C	D	E
½ in	F2	1.25	.53	.58	2.5	1.41
¾ in	F3	1.48	.55	.67	2.80	1.66
1 in	F4	1.84	.69	.69	3.30	198

Figure 2-105: Dimensional data for Liquidtight 90° nonmetallic connectors.

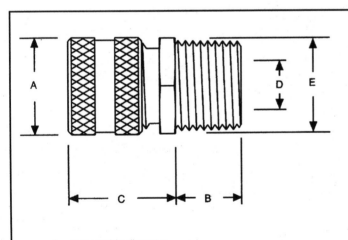

A = Nut outside diameter
B = Thread length
C = Length to shoulder
D = Throat inside diameter
E = Hex
A/F = Across corners
A/F = Across flats

LIQUIDTIGHT STRAIGHT NYLON CONNECTORS				
NPT	**Form Size**	**A Dia.**	**B Ref.**	**C**
½ - 14	F1	.88	.46	.90
½ - 14	F2	1.13	.55	1.10
½ - 14	F3	1.38	.55	1.50
¾-14	F2	1.13	.55	1.10
¾-14	F3	1.38	.55	1.50
1-11½	F4	1.75	.71	1.60

NPT	Form Size	D	E	
½ - 14	F1	.44	1.00 A/C	.88 A/F
½ - 14	F2	.64	1.11 A/C	1.00 A/F
½ - 14	F3	.64	1.40 A/C	1.25 A/F
¾-14	F2	.64	1.29 A/C	1.13 A/F
¾-14	F3	.82	1.40 A/C	1.25 A/F
1-11½	F4	1.02	1.81 A/C	1.62 A/F

Figure 2-106: Dimensional data for Liquidtight straight nylon connectors.

FLEXIBLE METAL CONDUIT FITTINGS

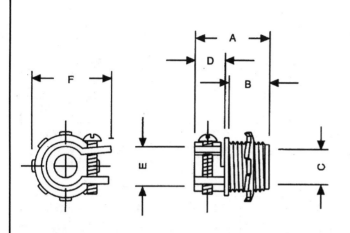

Raco No.	A	B	B Thread Spec.	C
2191	.94	.41	½-14	.34
2192	1.44	.41	½-14	.63
2193	1.56	.44	¾ - 14	.81
2194	1.75	.56	1 - 11½	1.03
2195	2	.63	1¼ - 11½	1.31
2196	2.06	.69	1½ - 11½	1.59
2198	2.31	.63	2 - 11½	2

STRAIGHT SQUEEZE CONNECTOR FOR FLEX

A = Overall length
B = Thread length and thread specifications
C = Throat inside diameter
D = Length to conduit stop
E = Clamping range
F = Maximum width of body

Dimensions continue in Figure 2-108

Figure 2-107: Dimensional data for straight squeeze insulated connectors for flexible metal conduit.

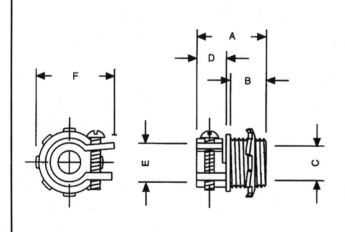

Raco No.	D	E	F
2191	.41	.53 - .66	1.06
2192	.50	.84 - .97	1.38
2193	.66	.97 - 1.16	1.59
2194	.56	1.19 - 1.44	1.84
2195	1.19	1.47 - 1.69	2.13
2196	1.31	1.75 - 2	2.38
2198	1.56	2.16 - 2.5	2.47

STRAIGHT SQUEEZE CONNECTOR FOR FLEX

A = Overall length
B = Thread length and thread specifications
C = Throat inside diameter
D = Length to conduit stop
E = Clamping range
F = Maximum width of body

See Figure 2-107 for additional dimensions

Figure 2-108: Dimensional data for straight squeeze uninsulated connectors for flexible metal conduit.

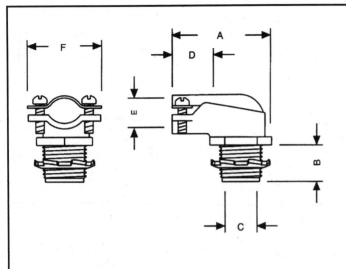

90° SQUEEZE CONNECTOR FOR FLEX			
Raco No.	A	B	Thread Specs.
3201, 2201	2	.56	½ - 14 NPT
3202, 2202	2	.56	½ - 14 NPT
3203, 2203	2.13	.50	¾ - 14 NPT
3204, 2204	2.88	.63	1 - 11½ NPT
3205, 2205	4.88	.63	1¼ - 11½ NPT
3206, 2206	5.06	.63	1½ - 11½ NPT
3208, 2208	6.5	.66	2 - 11½ NPT
2210	7.63	.94	2½ - 8 NPS
2212	8.81	1.03	3 - 8 NPS

A = Overall length
B = Thread length and thread specifications
C = Throat inside diameter
D = Length to conduit stop
E = Clamping range
F = Maximum width of body

See Figure 2-110 for additional dimensions

Figure 2-109: Dimensional data for 90° squeeze insulated connectors for flexible metal conduit.

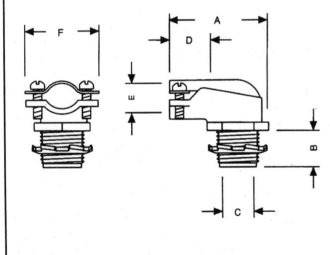

90° SQUEEZE CONNECTOR FOR FLEX *(Cont.)*			
Raco No.	C	D	E	F
3201, 2201	.622	.75		1.38
3202, 2202	.622	.88	.44	2
3203, 2203	.822	1		2.19
3204, 2204	1.05	1	to	3.38
3205, 2205	1.38	1.06		2.75
3206, 2206	1.61	1.63		3.13
3208, 2208	2.07	1.88	3.56	3.63
2210	2.50	2.25		4.38
2212	3.05	2.50		5

A = Overall length
B = Thread length and thread specifications
C = Throat inside diameter
D = Length to conduit stop
E = Clamping range
F = Maximum width of body

See Figure 2-109 for additional dimensions

Figure 2-110: Dimensional data for 90° squeeze uninsulated connectors for flexible metal conduit.

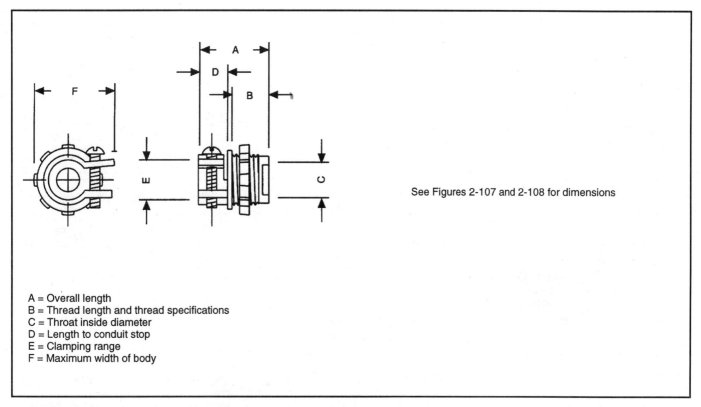

A = Overall length
B = Thread length and thread specifications
C = Throat inside diameter
D = Length to conduit stop
E = Clamping range
F = Maximum width of body

See Figures 2-107 and 2-108 for dimensions

Figure 2-111: Dimensional data for straight squeeze uninsulated die cast connectors for flexible metal conduit.

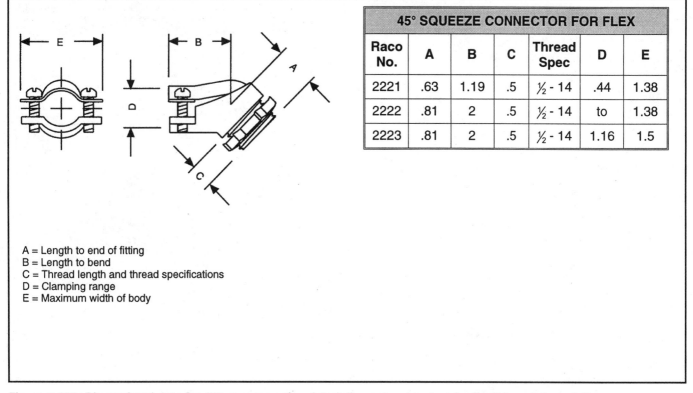

A = Length to end of fitting
B = Length to bend
C = Thread length and thread specifications
D = Clamping range
E = Maximum width of body

45° SQUEEZE CONNECTOR FOR FLEX						
Raco No.	A	B	C	Thread Spec	D	E
2221	.63	1.19	.5	½ - 14	.44	1.38
2222	.81	2	.5	½ - 14	to	1.38
2223	.81	2	.5	½ - 14	1.16	1.5

Figure 2-112: Dimensional data for 45° squeeze uninsulated die cast connectors for flexible metal conduit.

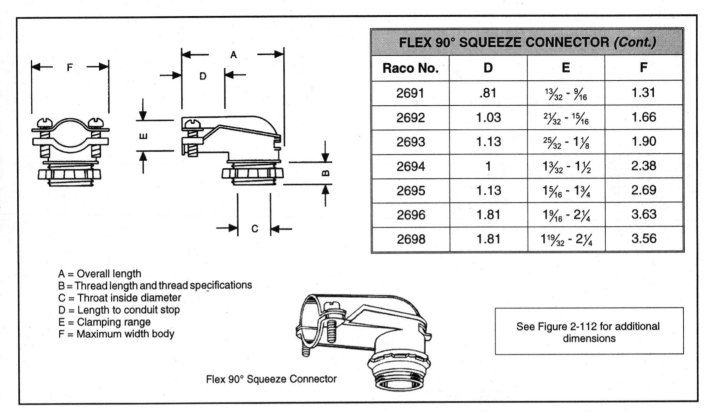

FLEX 90° SQUEEZE CONNECTOR *(Cont.)*			
Raco No.	**D**	**E**	**F**
2691	.81	$\frac{13}{32}$ - $\frac{9}{16}$	1.31
2692	1.03	$\frac{21}{32}$ - $\frac{15}{16}$	1.66
2693	1.13	$\frac{25}{32}$ - $1\frac{1}{8}$	1.90
2694	1	$1\frac{3}{32}$ - $1\frac{1}{2}$	2.38
2695	1.13	$1\frac{5}{16}$ - $1\frac{3}{4}$	2.69
2696	1.81	$1\frac{9}{16}$ - $2\frac{1}{4}$	3.63
2698	1.81	$1\frac{19}{32}$ - $2\frac{1}{4}$	3.56

A = Overall length
B = Thread length and thread specifications
C = Throat inside diameter
D = Length to conduit stop
E = Clamping range
F = Maximum width body

Flex 90° Squeeze Connector

See Figure 2-112 for additional dimensions

Figure 2-113: Dimensional data for 90° squeeze uninsulated die cast connectors for flexible metal conduit.

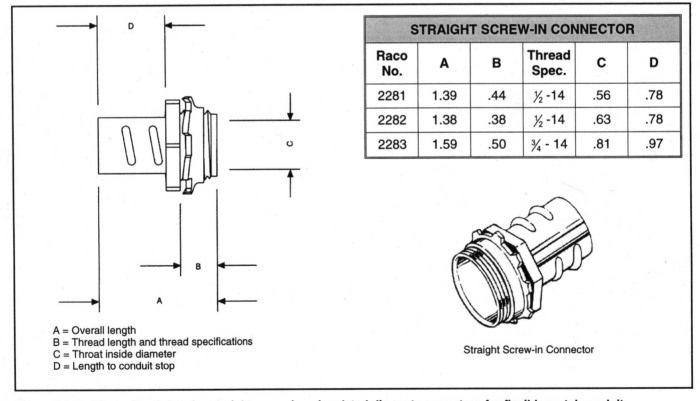

STRAIGHT SCREW-IN CONNECTOR					
Raco No.	**A**	**B**	**Thread Spec.**	**C**	**D**
2281	1.39	.44	$\frac{1}{2}$ -14	.56	.78
2282	1.38	.38	$\frac{1}{2}$ -14	.63	.78
2283	1.59	.50	$\frac{3}{4}$ - 14	.81	.97

A = Overall length
B = Thread length and thread specifications
C = Throat inside diameter
D = Length to conduit stop

Straight Screw-in Connector

Figure 2-114: Dimensional data for straight screw-in uninsulated die cast connectors for flexible metal conduit.

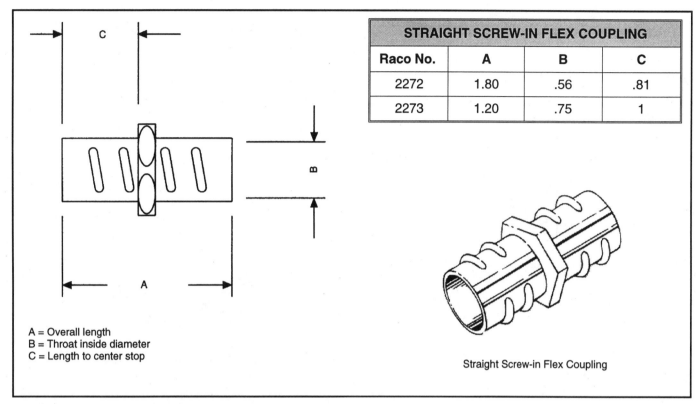

STRAIGHT SCREW-IN FLEX COUPLING			
Raco No.	A	B	C
2272	1.80	.56	.81
2273	1.20	.75	1

A = Overall length
B = Throat inside diameter
C = Length to center stop

Straight Screw-in Flex Coupling

Figure 2-115: Dimensional data for straight screw-in uninsulated die cast couplings for flexible metal conduit.

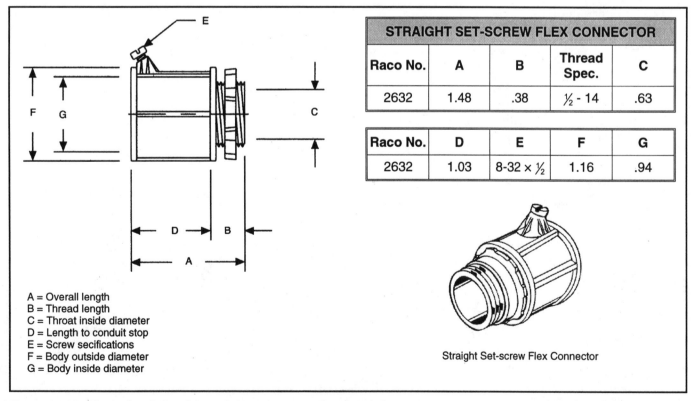

STRAIGHT SET-SCREW FLEX CONNECTOR				
Raco No.	A	B	Thread Spec.	C
2632	1.48	.38	½ - 14	.63

Raco No.	D	E	F	G
2632	1.03	8-32 × ½	1.16	.94

A = Overall length
B = Thread length
C = Throat inside diameter
D = Length to conduit stop
E = Screw secifications
F = Body outside diameter
G = Body inside diameter

Straight Set-screw Flex Connector

Figure 2-116: Dimensional data for straight set-screw uninsulated die cast connectors for flexible metal conduit.

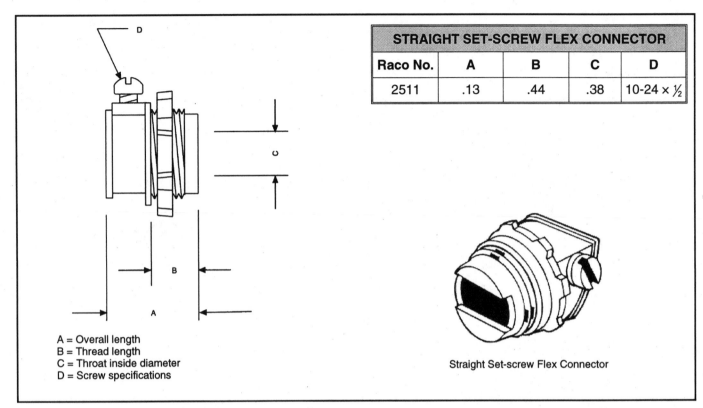

STRAIGHT SET-SCREW FLEX CONNECTOR				
Raco No.	**A**	**B**	**C**	**D**
2511	.13	.44	.38	10-24 × ½

A = Overall length
B = Thread length
C = Throat inside diameter
D = Screw specifications

Straight Set-screw Flex Connector

Figure 2-117: Dimensional data for straight set-screw uninsulated die cast connectors for flexible metal conduit.

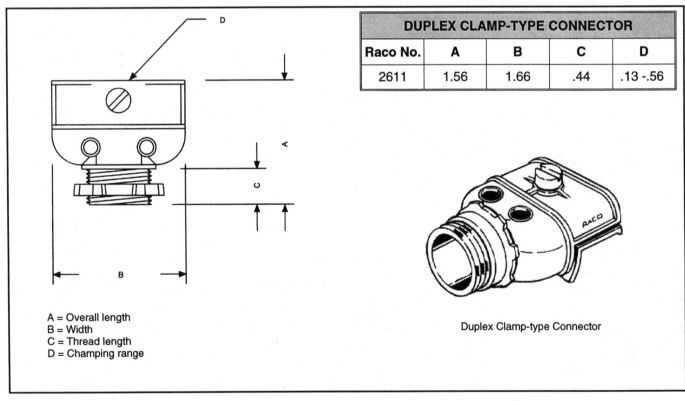

DUPLEX CLAMP-TYPE CONNECTOR				
Raco No.	**A**	**B**	**C**	**D**
2611	1.56	1.66	.44	.13 -.56

A = Overall length
B = Width
C = Thread length
D = Champing range

Duplex Clamp-type Connector

Figure 2-118: Dimensional data for duplex clamp-type uninsulated die cast connectors for flexible metal conduit.

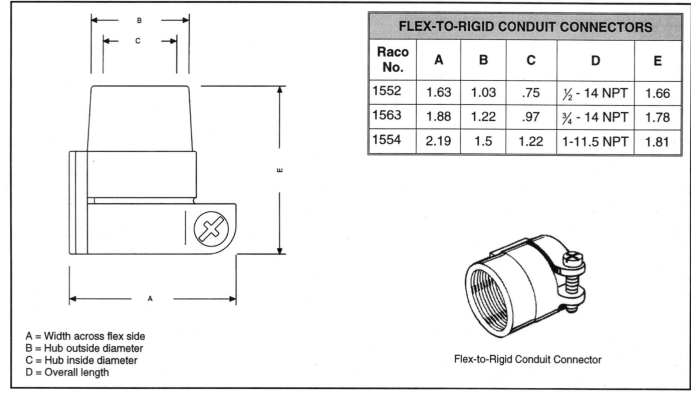

FLEX-TO-RIGID CONDUIT CONNECTORS					
Raco No.	A	B	C	D	E
1552	1.63	1.03	.75	½ - 14 NPT	1.66
1563	1.88	1.22	.97	¾ - 14 NPT	1.78
1554	2.19	1.5	1.22	1-11.5 NPT	1.81

A = Width across flex side
B = Hub outside diameter
C = Hub inside diameter
D = Overall length

Flex-to-Rigid Conduit Connector

Figure 2-119: Dimensional data for flex-to-rigid conduit connectors.

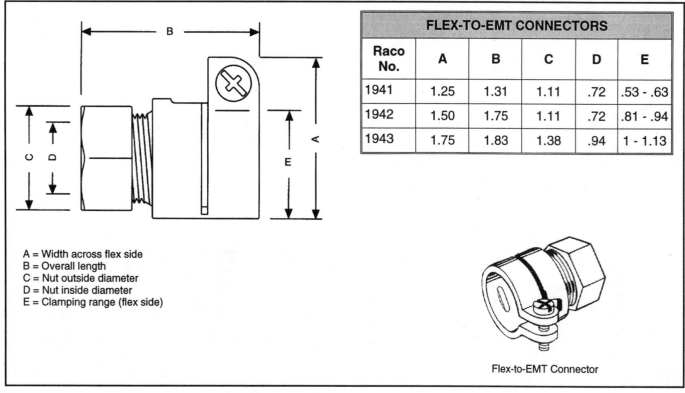

FLEX-TO-EMT CONNECTORS					
Raco No.	A	B	C	D	E
1941	1.25	1.31	1.11	.72	.53 - .63
1942	1.50	1.75	1.11	.72	.81 - .94
1943	1.75	1.83	1.38	.94	1 - 1.13

A = Width across flex side
B = Overall length
C = Nut outside diameter
D = Nut inside diameter
E = Clamping range (flex side)

Flex-to-EMT Connector

Figure 2-120: Dimensional data for flex-to-EMT connectors.

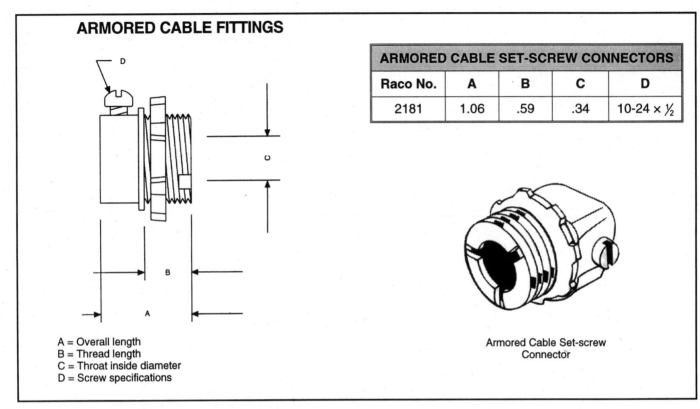

ARMORED CABLE FITTINGS

ARMORED CABLE SET-SCREW CONNECTORS				
Raco No.	A	B	C	D
2181	1.06	.59	.34	10-24 × ½

A = Overall length
B = Thread length
C = Throat inside diameter
D = Screw specifications

Armored Cable Set-screw
Connector

Figure 2-121: Dimensional data for set-screw armored cable connectors.

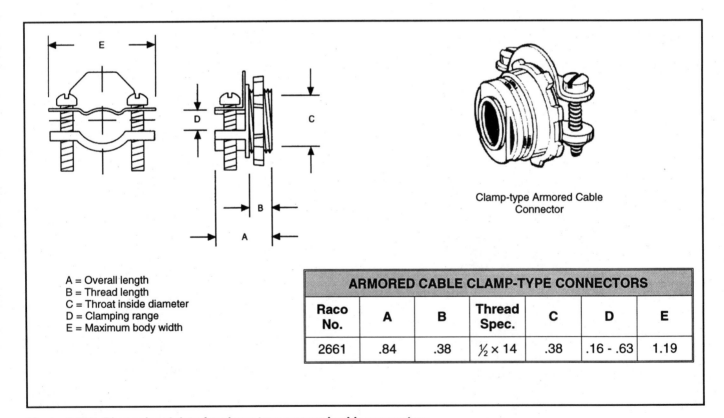

Clamp-type Armored Cable
Connector

A = Overall length
B = Thread length
C = Throat inside diameter
D = Clamping range
E = Maximum body width

ARMORED CABLE CLAMP-TYPE CONNECTORS						
Raco No.	A	B	Thread Spec.	C	D	E
2661	.84	.38	½ × 14	.38	.16 - .63	1.19

Figure 2-122: Dimensional data for clamp-type armored cable connectors.

DC MOTORS AND GENERATORS

A dc motor is a machine for converting dc electrical energy into rotating mechanical energy. The principle underlying the operation of a dc motor is called motor action and is based on the fact that, when a wire that is carrying current is placed in a magnetic field, a force is exerted on the wire — causing it to move through the magnetic field. There are three elements to motor action as it takes place in a dc motor:

- Many coils of wire are wound on a cylindrical rotor or armature on the shaft of the motor.

- A magnetic field necessary for motor action is created by placing fixed electromagnetic poles around the inside of the cylindrical motor housing. When current is passed through the fixed coils, a magnetic field is set up without the housing. Then, when the armature is placed inside the motor housing, the wires of the armature coils will be situated in the field of magnetic lines of force set up by the electromagnetic poles arranged around the stator. The stationary cylindrical part of the motor is called the stator.

- The shaft of the armature is free to rotate because it is supported at both ends of bearing brackets. Freedom of rotation is assured by providing clearance between the rotor and the faces of the magnetic poles.

Shunt-Wound DC Motors

In this type of motor, the strength of the field is not affected appreciably by a change in the load, so a relatively constant speed is obtainable. This type of motor may be used for the operation of machines that require an approximate constant speed and impose low starting torque and light overload on the motor. *See* Figure 2-123.

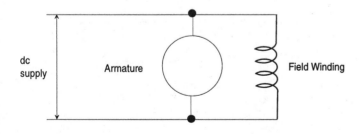

Figure 2-123: Shunt motor wiring diagram.

Series-Wound DC Motors

In motors of this type, any increase in load results in more current passing through the armature and the field windings. As the field is strengthened by this increased current, the motor speed decreases. Conversely, as the load is decreased, the field is weakened and the speed increases, and at very light loads speed may become excessive. For this reason, series-wound motors are usually directly connected or geared to the load to prevent runaway. The increase in armature current with an increasing load produces increased torque, so the series-wound motor is particularly suited to heavy starting duty and where severe overloads may be expected. Its speed may be adjusted by means of a variable resistance placed in series with the motor, but due to variation with load, the speed cannot be held at any constant value. This variation of speed with load becomes greater as the speed is reduced. Use of this motor is normally limited to traction and lifting service. *See* Figure 2-124.

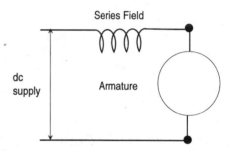

Figure 2-124: Series-wound dc motor.

Compound-Wound Motors

In this type of motor, the speed variation due to the load changes is much less in the series-wound motor, but greater than in the shunt-wound motor. It also has a greater starting torque than the shunt-wound motor and is able to withstand heavier overloads. However, it has a narrower adjustable-winding, the differential-compound winding being limited to special applications. They are used where the starting load is very heavy or where the load changes suddenly and violently, as with reciprocating pumps, printing presses, and punch presses. *See* Figure 2-125 on the next page.

Brushless DC Motors

The brushless dc motor was developed to eliminate commutator problems in missiles and spacecraft in operation above the earth's atmosphere. Two general types of

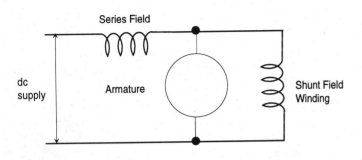

Figure 2-125: Compound-wound dc motor circuit.

brushless motors are in use: the inverter-induction motor and a dc motor with an electronic commutator.

The inverter-induction motor uses an inverter that uses the motor windings as the usual filter. The operation is square wave, and the combined efficiencies of the inverter and induction motor are at least as high as for a dc motor alone. In all cases, the motors must be designed for low starting current or else the inverter must be designed to saturate so that starting current is limited; otherwise, the transistors or silicon-controlled rectifiers in the inverter will be overloaded.

DC Motor and Generator Connections

The connections shown in Figures 2-126 and 2-127 are for counterclockwise rotation. For clockwise rotation, reverse connections to A1 and A2.

Some dc motors and generators have the interpole winding preconnected on the A2 side of the armature. When the shunt field is separately excited in dc motors, the same polarities must be observed for a given rotation.

When connecting dc generators, the shunt field may be either self-excited or separately excited. When self-excited, connections should be made as shown in Figure 2-127. When separately excited, the shunt field is isolated from the other windings, and the same polarities must be observed for a given rotation.

NEMA Frame Sizes

NEMA frame dimensions for dc machines are shown in Figure 2-128 while NEMA shaft dimensions are shown in Figure 2-129. Both charts are helpful for specifying new dc motors and generators, or for replacing existing ones in renovation projects.

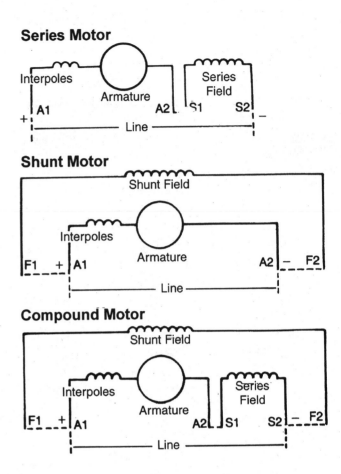

Figure 2-126: Terminal markings and connections for dc motors.

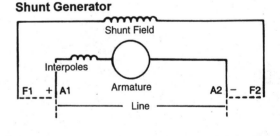

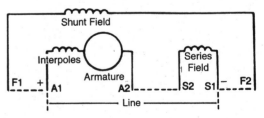

Figure 2-127: Terminal markings and connections for dc generators.

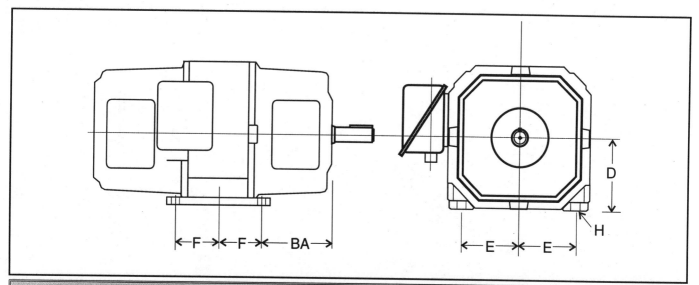

NEMA FRAME DIMENSIONS FOR DC MOTORS AND GENERATORS					
Frame Number	**D**	**E**	**2F**	**H**	**BA**
42	2.62	1.75	1.69	.28	2.06
48	3	2.12	2.75	.34	2.5
56	3.5	2.44	3	.34	2.75
56H	3.5	2.44	5	.34	2.75
142AT	3.5	2.75	3.5	.34	2.75
143AT	3.5	2.75	4	.34	2.75
144AT	3.5	2.75	4.5	.34	2.75
145AT	3.5	2.75	5	.34	2.75
146AT	3.5	2.75	5.5	.34	2.75
147AT	3.5	2.75	6.25	.34	2
148AT	3.5	2.75	7	.34	2
149AT	3.5	2.75	8	.34	2.75
141OAT	3.5	2.75	9	.34	2.75
1411AT	3.5	2.75	10	.34	2.75
1412AT	3.5	2.75	11	.34	2.75
162AT	4	3.12	4	.41	2.5
163AT	4	3.12	4.5	.41	2.5
164AT	4	3.12	5	.41	2.5
165AT	4	3.12	5.5	.41	2.5
166AT	4	3.12	6.25	.41	2.5

Figure 2-128: NEMA frame dimensions for dc motors and generators.

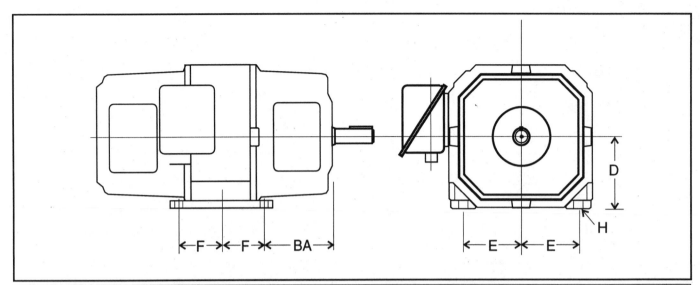

NEMA FRAME DIMENSIONS FOR DC MOTORS AND GENERATORS				
Frame Number	D	E	2F	H	BA
167AT	4	3.12	7	.41	2.5
168AT	4	3.12	8	.41	2.5
169AT	4	3.12	9	.41	2.5
1610AT	4	3.12	10	.41	2.5
182AT	4.5	3.75	4.5	.41	2.75
183AT	4.5	3.75	5	.41	2.75
184AT	4.5	3.75	5.5	.41	2
185AT	4.5	3.75	6.25	.41	2
186AT	4.5	3.75	7	.41	2.75
187AT	4.5	3.75	8	.41	2.75
188AT	4.5	3.75	9	.41	2.75
189AT	4.5	3.75	10	.41	2.75
181OAT	4.5	3.75	11	.41	2.75
213AT	5.25	4.25	5.5	.41	3.5
214AT	5.25	4.25	6.25	.41	3.5
215AT	5.25	4.25	7	.41	3.5
216AT	5.25	4.25	8	.41	3.5
217AT	5.25	4.25	9	.41	3
218AT	5.25	4.25	10	.41	3
219AT	5.25	4.25	11	.41	3.5

Figure 2-128: NEMA frame dimensions for dc motors and generators. *(Cont.)*

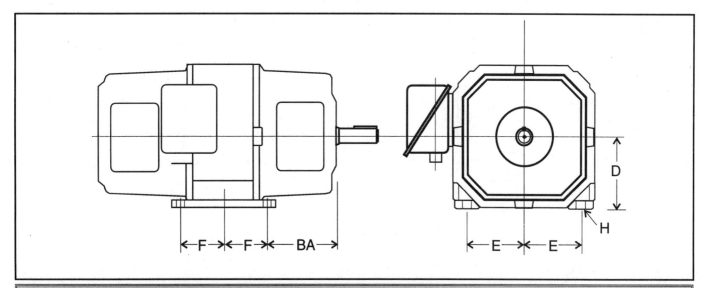

NEMA FRAME DIMENSIONS FOR DC MOTORS AND GENERATORS				
Frame Number	**D**	**E**	**2F**	**H**	**BA**
2110AT	5.25	4.25	12.5	.41	3.5
253AT	6.25	5	7	.53	4.25
254AT	6.25	5	8.25	.53	4.25
255AT	6.25	5	9	.53	4.25
256AT	6.25	5	10	.53	4
257AT	6.25	5	11	.53	4
258AT	6.25	5	12.5	.53	4.25
259AT	6.25	5	14	.53	4.25
283AT	7	5.5	8	.53	4.75
284AT	7	5.5	9.5	.53	4.75
285AT	7	5.5	10	.53	4.75
286AT	7	5.5	11	.53	4.75
287AT	7	5.5	12.5	.53	4.75
288AT	7	5.5	14	.53	4.75
289AT	7	5.5	16	.53	4.75
323AT	8	6.25	9	.66	5.25
324AT	8	6.25	10.5	.66	5.25
325AT	8	6.25	11	.66	5.25
326AT	8	6.25	12	.66	5.25
327AT	8	6.25	14	.66	5.25

Figure 2-128: NEMA frame dimensions for dc motors and generators. *(Cont.)*

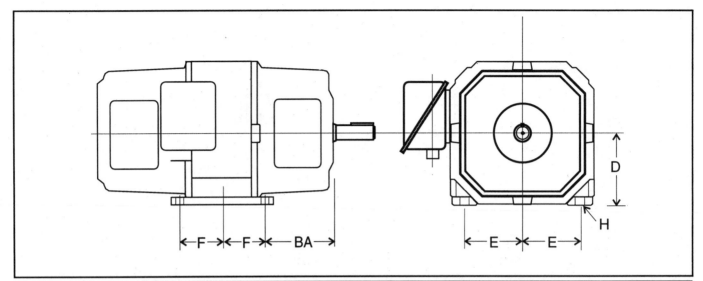

Frame Number	D	E	2F	H	BA
NEMA FRAME DIMENSIONS FOR DC MOTORS AND GENERATORS					
328AT	8	6.25	16	.66	5.25
329AT	8	6.25	18	.66	5.25
363AT	9	7	10	.81	5.88
364AT	9	7	11.25	.81	5.88
365AT	9	7	12.25	.81	5.88
366AT	9	7	14	.81	5.88
367AT	9	7	16	.81	5.88
368AT	9	7	18	.81	5.88
369AT	9	7	20	.81	5.88
403AT	10	8	11	.94	6.62
404AT	10	8	12.25	.94	6.62
405AT	10	8	13.75	.94	6.62
406AT	10	8	16	.94	6.62
407AT	10	8	18	.94	6.62
408AT	10	8	20	.94	6.62
409AT	10	8	22	.94	6.62
443AT	11	9	12.5	1.06	7.5
444AT	11	9	15	1.06	7.5
445AT	11	9	16.5	1.06	7.5
446AT	11	9	18	1.06	7.5

Figure 2-128: NEMA frame dimensions for dc motors and generators. *(Cont.)*

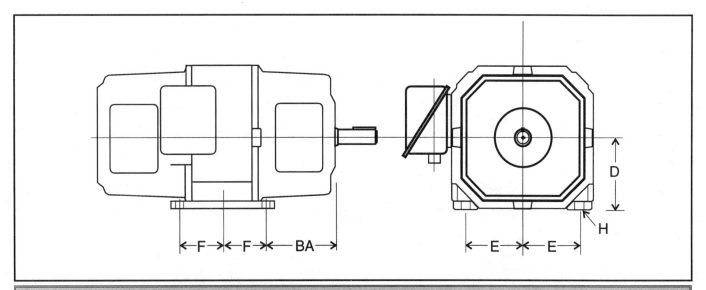

NEMA FRAME DIMENSIONS FOR DC MOTORS AND GENERATORS					
Frame Number	D	E	2F	H	BA
447AT	11	9	20	1.06	7.5
448AT	11	9	22	1.06	7.5
449AT	11	9	25	1.06	7.5
502AT	12.5	10	12.5	1.19	8.5
503AT	12.5	10	14	1.19	8.5
504AT	12.5	10	16	1.19	8.5
505AT	12.5	10	18	1.19	8.5
506AT	12.5	10	20	1.19	8.5
507AT	12.5	10	22	1.19	8.5
508AT	12.5	10	25	1.19	8.5
509AT	12.5	10	28	1.19	8.5
583A	14.5	11.5	16	1.19	10
584A	14.5	11.5	18	1.19	10
585A	14.5	11.5	20	1.19	10
586A	14.5	11.5	22	1.19	10
587A	14.5	11.5	25	1.19	10
588A	14.5	11.5	28	1.19	10
683A	17	13.5	20	1.19	11.5
684A	17	13.5	22	1.19	11.5
685A	17	13.5	25	1.19	11.5

Figure 2-128: NEMA frame dimensions for dc motors and generators. *(Cont.)*

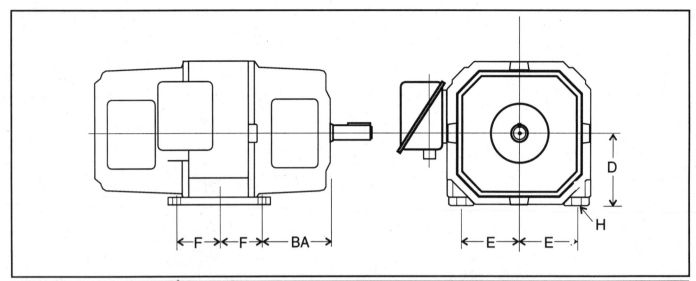

NEMA FRAME DIMENSIONS FOR DC MOTORS AND GENERATORS					
Frame Number	D	E	2F	H	BA
686A	17	13.5	28	1.19	11.5
687A	17	13.5	32	1.19	11.5
688A	17	13.5	36	1.19	11.5

Figure 2-128: NEMA frame dimensions for dc motors and generators. *(Cont.)*

NEMA SHAFT DIMENSIONS FOR DC MOTOR AND GENERATORS																	
Frame Range	Drive End — Belt						Drive End — Direct						Opposite Drive End — Straight					
	U	N-W	V Min.	R	ES Min.	S	U	N-W	V Min.	R	ES Min.	S	FU	FN-FW	FV Min.	FR	FES Min.	FS
42	.375	1.12	—	.328	—	Flat	.375	1-12	—	.328	—	Flat	—	—	—	—	—	—
48	.5	1.5	—	.453	—	Flat	.5	1.5	—	.453	—	Flat	—	—	—	—	—	—
56	.625	1.88	—	5.17	1.41	.188	6.25	1.88	—	.517	1.41	.188	—	—	—	—	—	—
56H	.625	1.88	.517	5.17	1.41	.188	6.25	1.88	—	.517	1.41	.188	—	—	—	—	—	—
142AT-1412AT	.875	1.75	1.5	.771	.91	.188	—	—	—	—	—	—	.625	1.25	1	.517	.66	.188
162AT-1610A	.875	1.75	1.5	.771	.91	.188	—	—	—	—	—	—	.625	1.25	1	.517	.66	.188
182AT-1810AT	1.125	2.25	2	.986	1.41	.25	—	—	—	—	—	—	.875	1.75	1.5	.771	.91	.188
213AT-2110AT	1.375	2.75	2.5	1.201	1.78	.312	—	—	—	—	—	—	1.125	2.25	2	.986	1.41	.25
253AT-259AT	1.625	3.25	3	1.416	2.28	.375	—	—	—	—	—	—	1.375	2.75	2.5	1.201	1.78	.312
283AT-289AT	1.875	3.75	3.5	1.591	2.53	.5	—	—	—	—	—	—	1.625	2,75	3	1.416	2.28	.375

Figure 2-129: NEMA shaft dimensions dc motors and generators; see illustration top of next page.

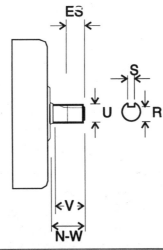

NEMA SHAFT DIMENSIONS FOR DC MOTOR AND GENERATORS																	
Frame Range	**Drive End — Belt**						**Drive End — Direct**						**Opposite Drive End — Straight**					
	U	N-W	V Min.	R	ES Min.	S	U	N-W	V Min.	R	ES Min.	S	FU	FN-FW	FV Min.	FR	FES Min.	FS
323AT - 329AT	2.125	4.25	4	1.845	3.03	.5	—	—	—	—	—	—	1.875	3.75	3.5	1.591	2.53	.5
363AT - 369AT	2.375	4.75	4.5	2.021	3.53	.625	—	—	—	—	—	—	2.125	4.25	4	1.845	3.03	.5
403AT - 409AT	2.625	5.25	5	2.275	4.03	.625	—	—	—	—	—	—	2.375	4.75	4.5	2.021	3.53	.625
443AT - 449AT	2.875	5.75	5.5	2.45	4.53	.75	—	—	—	—	—	—	2.625	5.25	5	2.275	4.03	.625
502AT - 509AT	3.25	6.5	6.25	2.831	5.28	.75	—	—	—	—	—	—	2.875	5.75	5.5	2.45	4.53	.75

Figure 2-129: NEMA shaft dimensions for dc motors and generators. *(Cont.)*

Drum Controllers

General-purpose drum controllers are designed for starting, speed regulating and reversing series, shunt- and compound-wound dc motors. Typical applications include machine tools, mill motors, and crane motors for bridge and trolley. Drum controllers can also be applied to control hoist motors when there is no overhauling load, either through the use of worm gearing or with an automatic mechanical brake. Forward and reverse armature points, off-position reset, and limit switch protection are standard on most controllers, while armature shunt and/or drift points are usually optional.

Dynamic-braking-lowering drum controllers are designed to control series or compound-wound crane hoist motors where power or dynamic lowering is required. Armature points in hoisting positions, dynamic lowering, off-position rest, and control limit switch protection are normally standard on this type of controller, while armature shunt points in the hoisting direction are optional. Typical wiring diagrams for both types are shown in Figure 2-130 on the next page.

Ac nonreversing drum controllers are used with wound rotor induction motors on pumps, blowers, crushers, kilns, and similar nonreversing applications. They control the

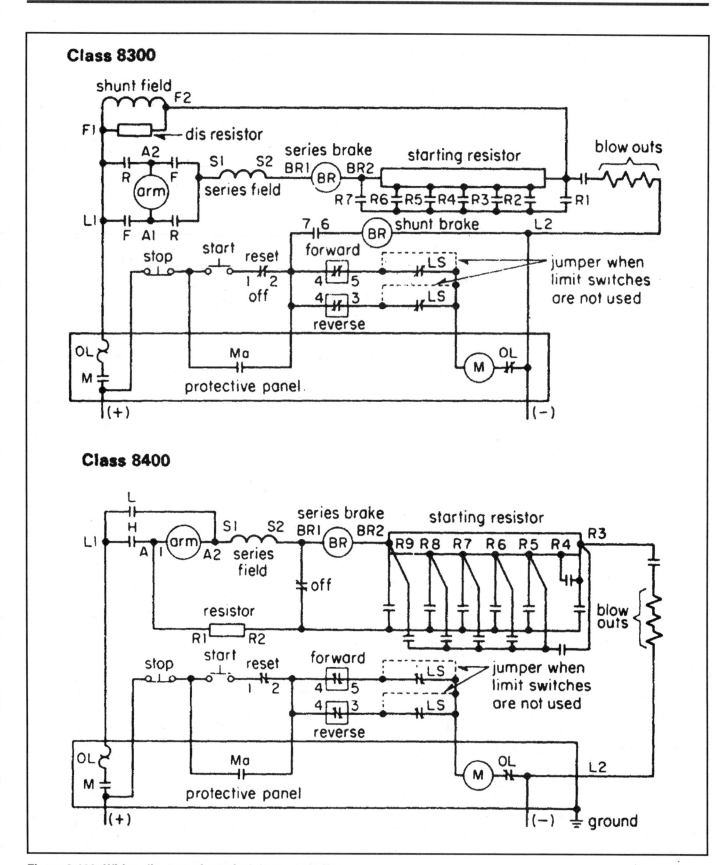

Figure 2-130: Wiring diagram of a typical drum controller.

Class 12-300

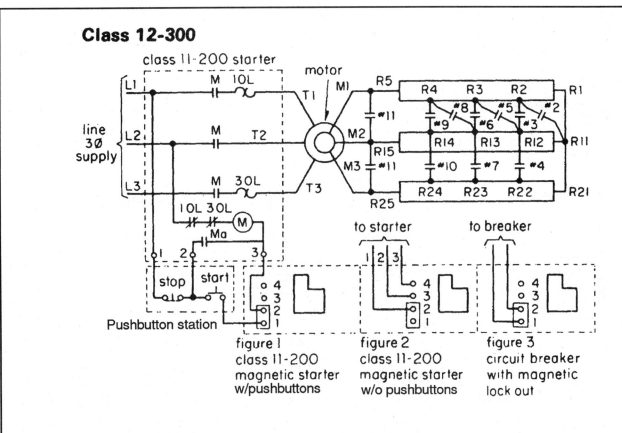

Class 12-700

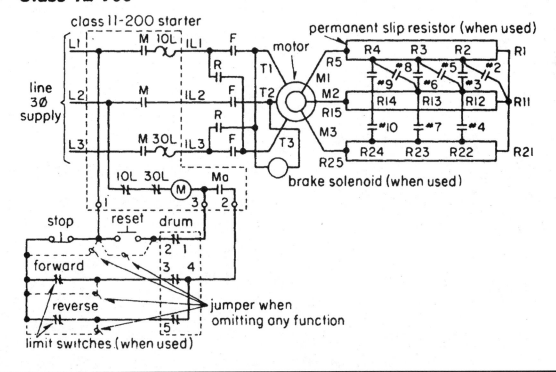

Figure 2-131: Wiring diagram of a nonreversing drum controller.

secondary circuit of the motor only and a suitable primary control must therefore be added.

For primary control, line-starters are normally used. Pilot circuits in the drum permit use of the starter with or without a pushbutton station as shown in Figure 2-131 on the preceding page. Provision is also made to control the magnetic lockout on a circuit-breaker primary, when used. In any case, it is impossible to start the motor unless the drum is in the "resistance-all-in" position.

DynAC® Controllers

This type of controller is manufactured by Westinghouse Electric Corporation and is a packaged electrical braking control for bringing ac motors to a smooth, quick stop. When the motor stop button is pressed, the control quickly opens the ac line and applies dc power to the ac motor winding. The amount of braking effort can be adjusted by varying the amount of dc power applied. These controllers are applicable where:

1. Motors are of the induction type and NEMA design B for three-phase, 60-Hz service.

2. The duty cycle is not greater than five starts and five stops per minute, and is not greater than 10 seconds of braking time per minute, or the equivalent.

3. The desired braking can be obtained with either:

    a. 100 percent average braking torque.
    b. 200 percent average braking torque.

Custom-design DynAC is available for special applications with NEMA D motors or other types of induction motors. DynAC can also be applied to wound rotor motors.

DynAC has been applied in such diverse industries as textile, dairy, materials handling, lumber, canning, automotive, and steel. Proper application of DynAC depends on the following:

- Motor
- Load
- Desired degree of braking
- Duty cycle

Most applications, especially in the lower horsepower ratings, can be made from charts supplied by the manufacturer. However, because of the wide range of possible motor characteristics and load conditions, a custom-designed controller may be required in many cases.

In operation, dc power for braking is furnished by a rectifier included as part of the unit. If a suitable dc bus is available, it will not be necessary to provide a dc supply in the control.

When only the braking unit is to be used with an existing control, a normally open and normally closed interlock must be provided on each line contactor to actuate the braking unit. A typical wiring diagram is shown in Figure 2-132.

ENCLOSURES

Many types of electrical components, apparatus, and equipment are housed in an enclosure. The enclosures will vary with the types of devices and the environment in which they are located.

Motor Enclosures

Electric motors differ in construction and appearance, depending on the type of service for which they are to be used. Open and closed frames are quite common. In the former enclosure, the motor's parts are covered for protection, but the air can freely enter the enclosure. Further designations for this type of enclosure include dripproof, weather-protected, and splash-proof.

Totally enclosed motors have an airtight enclosure. They may be fan cooled or self-ventilated. An enclosed motor equipped with a fan has the fan as an integral part of the machine, but external to the enclosed parts. In the self-ventilated enclosure, no external means of cooling is provided.

The type of enclosure to use will depend on the ambient temperature and surrounding conditions. In a drip-proof machine, for example, all ventilating openings are so constructed that drops of liquid or solid particles falling on the machine at an angle of not greater than 15 degrees from the vertical cannot enter the machine, even directly or by striking and running along a horizontal or inclined surface of the machine. The application of this machine would lend itself to areas where liquids are processed.

An open motor having all air openings that give direct access to live or rotating parts, other than the shaft, limited in size by the design of the parts or by screen to prevent accidental contact with such parts is classified as a dripproof, fully guarded machine. In such enclosures, openings shall not permit the passage of a cylindrical rod ½ in diameter, except where the distance from the guard to the live rotating parts is more than 4 in, in which case the

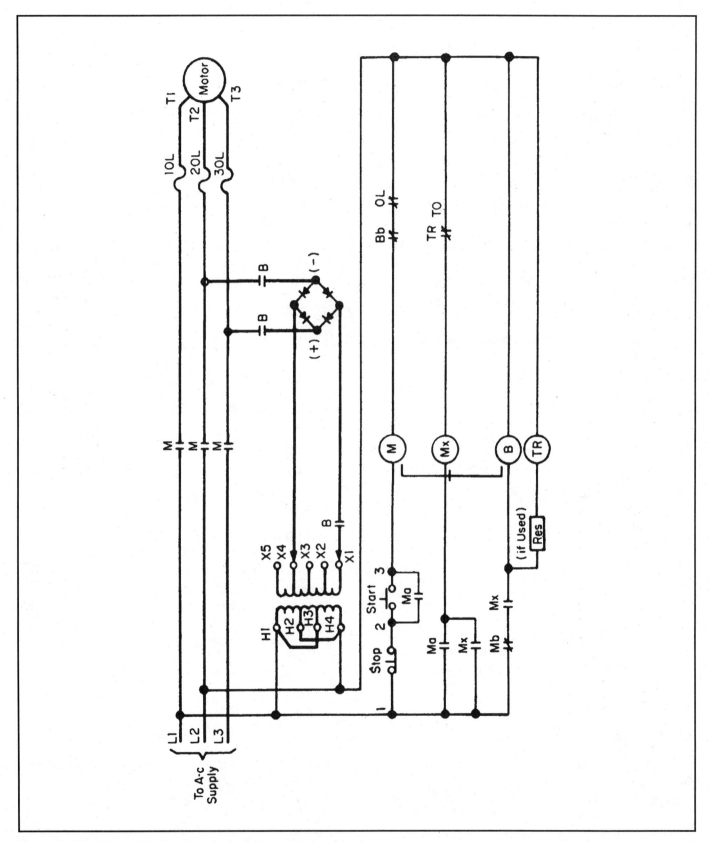

Figure 2-132: Schematic diagram of a braking control, using dc power.

openings must not permit the passage of a cylindrical rod ¾ in in diameter.

There are other types of dripproof machines for special applications such as externally ventilated and pipe ventilated, which as the names imply are either ventilated by a separate motor-driven blower or cooled by ventilating air from inlet ducts or pipes.

An enclosed motor whose enclosure is designed and constructed to withstand an explosion of a specified gas or vapor that may occur within the motor and to prevent the ignition of this gas or vapor surrounding the machine is designated "explosionproof" (XP) motors.

Hazardous atmospheres (requiring XP enclosures) of both a gaseous and dusty nature are classified by the *NEC* as follows:

- Class I, Group A: atmospheres containing acetylene.
- Class I, Group B: atmospheres containing hydrogen gases or vapors of equivalent hazards such as manufactured gas.
- Class I, Group C: atmospheres containing ethyl ether vapor.
- Class I, Group D: atmospheres containing gasoline, petroleum, naphtha, alcohols, acetone, lacquer-solvent vapors, and natural gas.
- Class II, Group E: atmospheres containing metal dust.
- Class II, Group F: atmospheres containing carbon-black, coal, or coke dust.
- Class II, Group G: atmospheres containing grain dust.

The proper motor enclosure must be selected to fit the particular atmosphere. However, explosionproof equipment is not generally available for Class I, Groups A and B, and it is therefore necessary to isolate motors from these classes of hazardous areas.

NEMA Enclosures

NEMA has established enclosure designations for safety switches, panelboards, motor starters, etc. because such items are used in so many different locations and in numerous environmental conditions.

A designation such as NEMA 12 indicates an enclosure type to fulfill requirements for a paticular application. NEMA enclosure designations were recently revised to

provide a clearer and more precise definition of the enclosure need to meet various standard requirements.

Some of the revisions covered in the NEMA designations are as follows:

- NEMA Type 1A (semi-dusttight) has been dropped.
- The NEMA 12 enclosure can now be substituted in many installations in place of the NEMA 5 enclosure. One advantage of this substitution is that the NEMA 12 enclosure is less expensive than the NEMA 5 enclosure.
- NEMA Type 3R enclosures are of lighter weight and are less expensive than many other types of weatherproof enclosures.

The table in Figure 2-133 gives brief descriptions of various NEMA enclosures.

FUSES

There are several different types of fuses, and although all operate in a similar fashion, all have slightly different characteristics. Each is described in the paragraphs to follow.

Non-Time Delay Fuses

The basic component of a fuse is the link. Depending upon the ampere rating of the fuse, the single-element fuse may have one or more links. They are electrically connected to the end blades (or ferrules) and enclosed in a tube or cartridge surrounded by an arc quenching filler material.

Under normal operation, when the fuse is operating at or near its ampere rating, it simply functions as a conductor. However, as illustrated in Figure 2-134 on page 286, if an overload current occurs and persists for more than a short interval of time, the temperature of the link eventually reaches a level which causes a restricted segment of the link to melt; as a result, a gap is formed and an electric arc established. However, as the arc causes the link metal to burn back, the gap becomes progressively larger. Electrical resistance of the arc eventually reaches such a high level that the arc cannot be sustained and is extinguished; the fuse will have then completely cut off all current flow in the circuit. Suppression or quenching of the arc is accelerated by the filler material.

Overload current normally falls within the region of between one and six times normal current — resulting in currents that are quite high. Consequently, a fuse may be

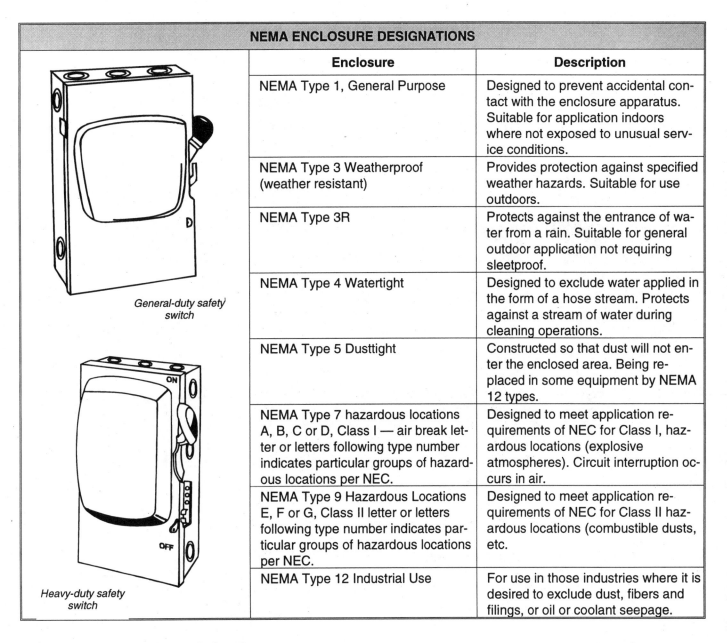

NEMA ENCLOSURE DESIGNATIONS		
	Enclosure	**Description**
General-duty safety switch	NEMA Type 1, General Purpose	Designed to prevent accidental contact with the enclosure apparatus. Suitable for application indoors where not exposed to unusual service conditions.
	NEMA Type 3 Weatherproof (weather resistant)	Provides protection against specified weather hazards. Suitable for use outdoors.
	NEMA Type 3R	Protects against the entrance of water from a rain. Suitable for general outdoor application not requiring sleetproof.
	NEMA Type 4 Watertight	Designed to exclude water applied in the form of a hose stream. Protects against a stream of water during cleaning operations.
	NEMA Type 5 Dusttight	Constructed so that dust will not enter the enclosed area. Being replaced in some equipment by NEMA 12 types.
	NEMA Type 7 hazardous locations A, B, C or D, Class I — air break letter or letters following type number indicates particular groups of hazardous locations per NEC.	Designed to meet application requirements of NEC for Class I, hazardous locations (explosive atmospheres). Circuit interruption occurs in air.
Heavy-duty safety switch	NEMA Type 9 Hazardous Locations E, F or G, Class II letter or letters following type number indicates particular groups of hazardous locations per NEC.	Designed to meet application requirements of NEC for Class II hazardous locations (combustible dusts, etc.
	NEMA Type 12 Industrial Use	For use in those industries where it is desired to exclude dust, fibers and filings, or oil or coolant seepage.

Figure 2-133: NEMA enclosure designations.

subjected to short-circuit currents of 30,000 or 40,000 amperes or higher. Response of current-limiting fuses to such currents is extremely fast. The restricted sections of the fuse link will simultaneously melt within a matter of two or three-thousandths of a second in the event of a high level fault current.

The high resistance of the multiple arcs, together with the quenching effects of the filler particles, results in rapid arc suppression and clearing of the circuit. Again, refer to Figure 2-134. Short-circuit current is cut off in less than a half-cycle — long before the short-circuit current can reach its full value.

Dual-Element Time-delay Fuses

Unlike single-element fuses, the dual-element time-delay fuse can be applied in circuits subject to temporary motor overloads and surge currents to provide both high performance short-circuit and overload protection. Oversizing to prevent nuisance openings is not necessary with this type of fuse. The dual-element time-delay fuse contains two distinctly separate types of elements. *See* Figure 2-135 on page 287. Electrically, the two elements are connected in series. The fuse links similar to those used in the non-time-delay fuse perform the short-circuit protection function; the overload element provides protection

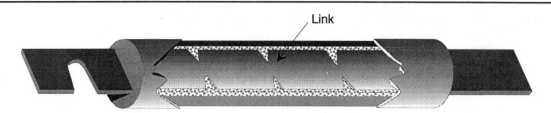

Cut-away view of single-element fuse.

Under sustained overload a section of the link
melts and an arc is established.

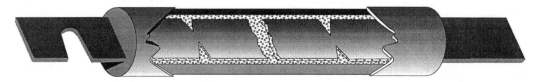

The "open" single-element fuse after opening a
circuit overload.

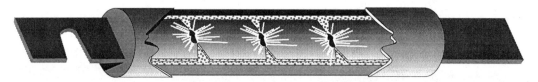

When subjected to a short-circuit, several sections
of the fuse link melt almost instantly.

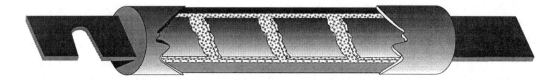

The appearance of an "open" single-element fuse
after opening a short-circuit.

Figure 2-134: Characteristics of single-element fuses.

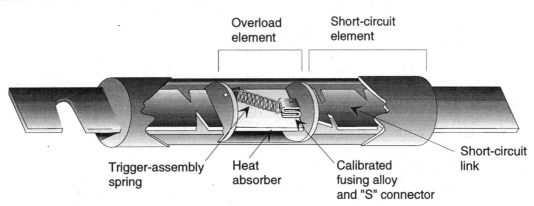

The true dual-element fuse has distinct and separate overload and short-circuit elements.

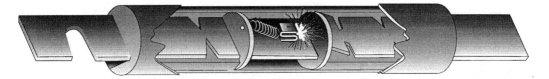

Under sustained overload conditions, the trigger spring fractures the calibrated fusing alloy and releases the "connector."

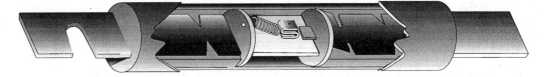

The "open" dual-element fuse after opening under an overload.

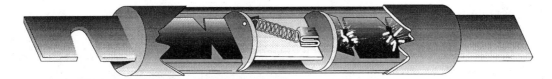

Like the single-element fuse, a short-circuit current causes the restricted portions of the short-circuit elements to melt and arcing to burn back the resulting gaps until the arcs are suppressed by the arc-quenching material and increased arc resistance.

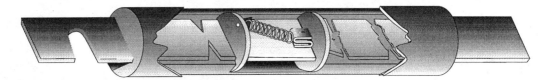

The "open" dual-element fuse after opening under a short-circuit condition.

Figure 2-135: Characteristics of dual-element time-delay fuses.

against low-level overcurrents or overloads and will hold an overload which is five times greater than the ampere rating of the fuse for a minimum time of 10 seconds.

As shown in Figure 2-135, the overload section consists of a copper heat absorber and a spring-operated trigger assembly. The heat absorber bar is permanently connected to the heat absorber extension and to the short-circuit link on the opposite end of the fuse by the S-shaped connector of the trigger assembly. The connector electrically joins the short-circuit link to the heat absorber in the overload section of the fuse. These elements are joined by a "calibrated" fusing alloy. An overload current causes heating of the short-circuit link connected to the trigger assembly. Transfer of heat from the short-circuit link to the heat absorbing bar in the mid-section of the fuse begins to raise the temperature of the heat absorber. If the overload is sustained, the temperature of the heat absorber eventually reaches a level which permits the trigger spring to "fracture" the calibrated fusing alloy and pull the connector free of the short-circuit link and the heat absorber. As a result, the short-circuit link is electrically disconnected from the heat absorber, the conducting path through the fuse is opened, and overload current is interrupted. A critical aspect of the fusing alloy is that it retains its original characteristic after repeated temporary overloads with degradation. The main purposes of dual-element fuses are as follows:

- Provide motor overload, ground-fault and short-circuit protection.

- Permits the use of smaller switches.

- Give a higher degree of short-circuit protection (greater current limitation) in circuits in which surge currents or temporary overloads occur.

- Simplify and improve blackout prevention (selective coordination).

U.L. Fuse Classes

Safety is the U.L. mandate. However, proper selection, overall functional performance and reliability of a product are factors that are not within the basic scope of U.L. activities. To develop its safety test procedures, U.L. does develop basic performance and physical specifications of standards of a product. In the case of fuses, these standards have culminated in the establishment of distinct classes of low-voltage (600 V or less) fuses, Classes FK 1, RK 5, G, L, T, J. H, and CC being the more important.

Class R fuses: U.L. Class R (rejection) fuses are high performance $\frac{1}{10}$ to 600 A units, 250 V and 600 V, having a high degree of current limitation and a short-circuit inter-

rupting rating of up to 200,000 A (rms symmetrical). This type of fuse is designed to be mounted in rejection type fuseclips to prevent older type Class H fuses from being installed. Since Class H fuses are not current limiting and are recognized by U.L. as having only a 10,000 A interrupting rating, serious damage could result if a Class H fuse were inserted in a system designed for Class R fuses. Consequently, *NEC* Section 240-60(b) requires fuseholders for current-limiting fuses to reject non-current limiting type fuses.

Figure 2-136 shows a standard Class H fuse (left) and a Class R fuse (right). A grooved ring in one ferrule of the Class R fuse provides the rejection feature of the Class R fuse in contrast to the lower interrupting capacity, non-rejection type. Figure 2-137 shows Class R type fuse rejection clips that accept only the Class R rejection type fuses.

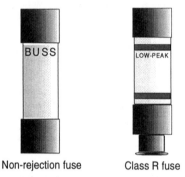

Non-rejection fuse   Class R fuse

Figure 2-136: Comparison of Class H and Class R fuses.

Class CC fuses: 600-V, 200,000-A interrupting rating, branch circuit fuses with overall dimensions of $\frac{15}{32}$ in $\times 1\frac{1}{2}$ in. Their design incorporates rejection features that allow them to be inserted into rejection fuse holders and fuse

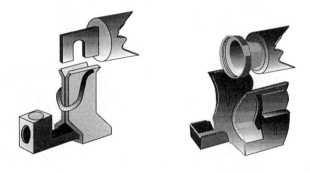

Figure 2-137: Class R fuse rejection clips that accept only Class R fuses.

blocks that reject all lower voltage, lower interrupting rating $^{15}\!/_{32}$ in × $1\!/_{2}$ in fuses. They are available from $^{1}\!/_{10}$ through 30 A.

Class G fuses: 300-V, 100,000 A interrupting rating branch circuit fuses that are size rejecting to eliminate overfusing. The fuse diameter is $^{13}\!/_{32}$ in while the length varies from $1^{5}\!/_{16}$ in to $2^{1}\!/_{4}$ in. These are available in ratings from 1 through 60 A.

Class H fuses: 250-V and 600-V, 10,000 A interrupting rating branch circuit fuses that may be renewable or non-renewable. These are available in ampere ratings of 1 through 600 A.

Class J fuses: These fuses are rated to interrupt 200,000 A ac. They are U.L. labeled as "current limiting", are rated for 600 V ac, and are not interchangeable with other classes.

Class K fuses: These are fuses listed by U.L. as K-1, K-5, or K-9 fuses. Each subclass has designated I^2t and Ip maximums. These are dimensionally the same as Class H fuses, (*NEC* dimensions) and they can have interrupting ratings of 50,000, 100,000, or 200,000 A. These fuses are current limiting, however, they are not marked "current limiting" on their label since they do not have a rejection feature.

Class L fuses: These fuses are rated for 601 through 6000 A, and are rated to interrupt 200,000 amperes ac. They are labeled "current limiting" and are rated for 600 V ac. They are intended to be bolted into their mountings and are not normally used in clips. Some Class L fuses have designed in time-delay features for all purpose use.

Class T fuses: A U.L. classification of fuses in 300- and 600-V ratings from 1 through 1200 A. They are physically very small and can be applied where space is at a premium. They are fast acting fuses, with an interrupting rating of 200,000 A RMS.

Branch-Circuit Listed Fuses

Branch circuit listed fuses are designed to prevent the installation of fuses that cannot provide a comparable level of protection to equipment. The characteristics of branch-circuit fuses are as follows:

- They just have a minimum interrupting rating of 10,000 A.
- They must have a minimum voltage rating of 125 V.

- They must be size rejecting such that a fuse of a lower voltage rating cannot be installed in the circuit.
- They must be size rejecting such that a fuse with a current rating higher than the fuseholder rating cannot be installed.

Medium-Voltage Fuses

Fuses above 600 V are classified under one of three classifications as defined in ANSI/IEEE 40-1981:

- General purpose current-limiting fuse
- Back-up current-limiting fuse
- Expulsion fuse

General purpose current-limiting fuse: A fuse capable of interrupting all currents from the rated interrupting current down to the current that causes melting of the fusible element in one hour.

Back-up current-limiting fuse: A fuse capable of interrupting all currents from the maximum rated interrupting current down to the rated minimum interrupting current.

Expulsion fuse: A vented fuse in which the expulsion effect of gasses produced by the arc and lining of the fuseholder, either alone or aided by a spring, extinguishes the arc.

One should note that in the definitions just given, the fuses are defined as either expulsion or current limiting. A current limiting fuse is a sealed, non-venting fuse that, when melted by a current within its interrupting rating, produces arc voltages exceeding the system voltage which in turn forces the current to zero. The arc voltages are produced by introducing a series of high resistance arcs within the fuse. The result is a fuse that typically interrupts high fault currents with the first $\frac{1}{2}$ cycle of the fault.

In contrast an expulsion fuse depends on one arc to initiate the interruption process. The arc acts as a catalyst causing the generation of de-ionizing gas from its housing. The arc is then elongated either by the force of the gasses created or a spring. At some point the arc elongates far enough to prevent a restrike after passing through a current zero. Therefore, it is not atypical for an expulsion fuse to take many cycles to clear.

Application of Medium-Voltage Fuses

Many of the rules for applying expulsion fuses and current-limiting fuses are the same, but because the current-limiting fuse operates much faster on high-fault, some additional rules must be applied.

Three basic factors must be considered when applying any fuse:

- Voltage
- Continuous current-carrying capacity
- Interrupting rating

Voltage: The fuse must have a voltage rating equal to or greater than the normal frequency recovery voltage which will be seen across the fuse under all conditions. On three-phase systems, it is a good rule-of-thumb that the voltage rating of the fuse be greater than or equal to the line-to-line voltage of the system.

Continuous current-carrying capacity: Continuous current values that are shown on the fuse represent the level of current the fuse can carry continuously without exceeding the temperature rises as specified in ANSI C37.46. An application that exposes the fuse to a current slightly above its continuous rating but below its minimum interrupting rating may damage the fuse due to excessive heat. This is the main reason overload relays are used in series with back-up current-limiting fuses for motor protection.

Interrupting rating: All fuses are given a maximum interrupting rating. This rating is the maximum level of fault current that the fuse can safely interrupt. Back-up current-limiting fuses are also given a minimum interrupting rating. When using back-up current-limiting fuses, it is important that other protective devices are used to interrupt currents below this level.

When choosing a fuse, it is important that the fuse be properly coordinated with other protective devices located upstream and downstream. To accomplish this, one must consider the melting and clearing characteristics of the devices. Two curves, the minimum melting curve and the total clearing curve, provide this information. To insure proper coordination, the following rules should be used:

- The total clearing curve of any downstream protective device must be below a curve representing 75% of the minimum melting curve of the fuse being applied.
- The total clearing curve, of the fuse being applied, must lie below a curve representing 75% of the minimum melting curve for any upstream protective device.

Current-Limiting Fuses

To insure proper application of a current-limiting fuse, it is important that the following additional rules be applied.

1. Current-limiting fuses produce arc voltages that exceed the system voltage. Care must be taken to make sure that the peak voltages do not exceed the insulation level of the system. If the fuse voltage rating is not permitted to exceed 140% of the system voltage, there should not be a problem. This does not mean that a higher rated fuse cannot be used, but points out that one must be assured that the system insulation level (BIL) will handle the peak arc voltage produced. BIL stands for basic impulse level which is the reference impulse insulation strength of an electrical system.

2. As with the expulsion fuse, current-limiting fuses must be properly coordinated with other protective devices on the system. For this to happen, the rules for applying an expulsion fuse must be used at all currents that cause the fuse to interrupt in 0.01 seconds or greater.

When other current-limiting protective devices are on the system, it becomes necessary to use I^2t values for coordination at currents causing the fuse to interrupt in less than 0.01 seconds. These values may be supplied as minimum and maximum values or minimum melting and total clearing I^2t curves. In either case, the following rules should be followed.

1. The minimum melting I^2t of the fuse should be greater than the total clearing I^2t of the downstream current-limiting device.

2. The total clearing I^2t of the fuse should be less than the minimum melting I^2t of the upstream current-limiting device.

The fuse-selection chart in Figure 2-138 should serve as a guide for selecting fuses on circuits of 600 V or less. The dimensions of various Buss fuses are shown beginning with Figure 2-139. All of these charts will prove invaluable on all types of projects — from residential to heavy industrial applications. Other valuable information may be found in catalogs furnished by manufacturers of overcurrent protective devices. These are usually obtainable from electrical supply houses or from manufacturers' reps. You may also write the various manufacturers for a complete list (and price, if any) for all reference materials offered by them.

Circuit	Load	Ampere Rating	Fuse Type	Symbol	Voltage Rating (ac)	UL Class	Interrupting Rating (K)	Remarks
Main, Feeder and Branch (Conventional dimensions)	All type load (optimum overcurrent protection)	0-600A	LOW-PEAK® (dual-element, time-delay)	LPN-RK	250V	RK1	200	All-purpose fuses. Unequaled for combined short-circuit and overload protection.
				LPS-RK	600V			
		601 to 6000A	LOW-PEAK® time delay	KRP-C	600V	L	200	
	Motors, welders, transformers, capacitor banks (circuits with heavy inrush currents)	0 to 600A	FUSETRON® (dual-element, time-delay)	FRN-R	250V	RK5	200	Moderate degree of current limitation. Time-delay passes surge currents.
				FRS-R	600V			
		601 to 4000A	LIMITRON® (time delay)	KLU	600V	L	200	All-purpose fuse. Time-delay passes surge-currents.
	Non-motor loads (circuits with no heavy inrush currents)	0 to 600A	LIMITRON® (fast-acting)	KTN-R	250V	RK1	200	Same short-circuit protection as LOW-PEAK fuses but must be sized larger for circuits with surge-currents; i.e., up to 300%.
				KTS-R	600V			
	LIMITRON fuses particular suited for circuit breaker protection	601 to 6000A	LIMITRON® (fast-acting)	KTU	600V	L	200	A fast-acting, high performance fuse.
	All type loads (optimum overcurrent protection)	0 to 600A	LOW-PEAK® (dual-element time-delay)	LPJ	600V	J	200	All-purpose fuses. Unequaled for combined short-circuit and overload protection.

Figure 2-138: Fuse selection chart (600 V or less).

Circuit	Load	Ampere Rating	Fuse Type	Symbol	Voltage Rating (ac)	UL Class	Interrupting Rating (K)	Remarks
Note: All shaded areas represent fuses with reduced dimensions for installation in restricted space	Non-motor loads (circuits with no heavy inrush currents)	0 to 600A	LIMITRON® (quick-acting)	JKS	600V	J	200	Very similar to KTS-R LIMITRON, but smaller.
		0 to 1200A	T-TRON™	JJN	300V	T	200	The space saver (⅓ the size of KTN-R/KTS-R).
				JJS	600V			
Branch	General purpose; i.e., lighting panelboards	0 to 60A	SC	SC	300V	G	100	Current limiting; ¹³⁄₃₂" dia. x varying lengths per amp rating.
	Misc.	0 to 600A	ONE-TIME	NON	250V	H or K5	10	Forerunners of the modern cartridge fuse.
				NOS	600V			
			SUPER-LAG® Renewable	REN	250V	H	10	
				RES	600V			
General Purpose (non-current limiting fuses)	Plug fuses can be used for branch circuits and small component protection)	0 to 30A	FUSTAT® (dual-element, time-delay)	S	125V	S	10	Base threads of Type S differ with amp ratings. T and W have Edison base. T & S fuses for motor circuits. W not for circuits with motor loads.
			Fusetron® (dual-element, time-delay)	T	125V	**	10	

Figure 2-138: Fuse selection chart (600 V or less). *(Cont.)*

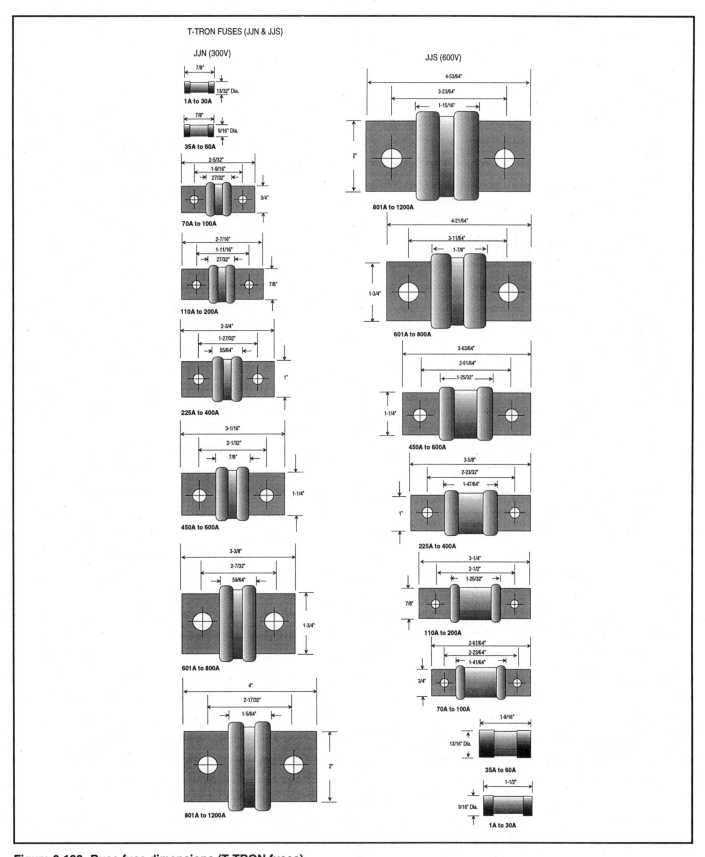

Figure 2-139: Buss fuse dimensions (T-TRON fuses).

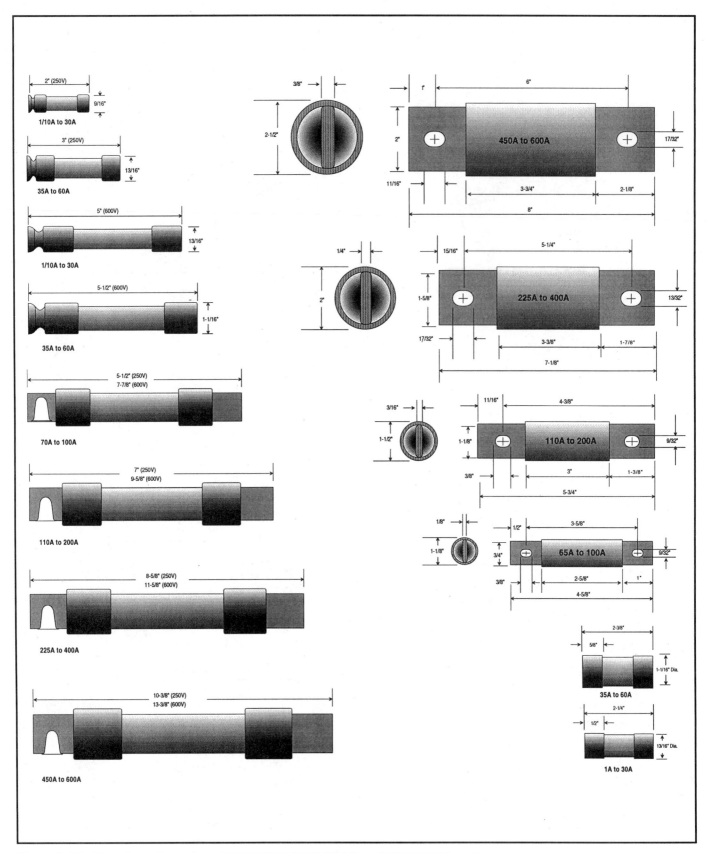

Figure 2-140: Buss fuse dimensions (FUSETRON, LOW-PEAK, and LIMITRON fuses).

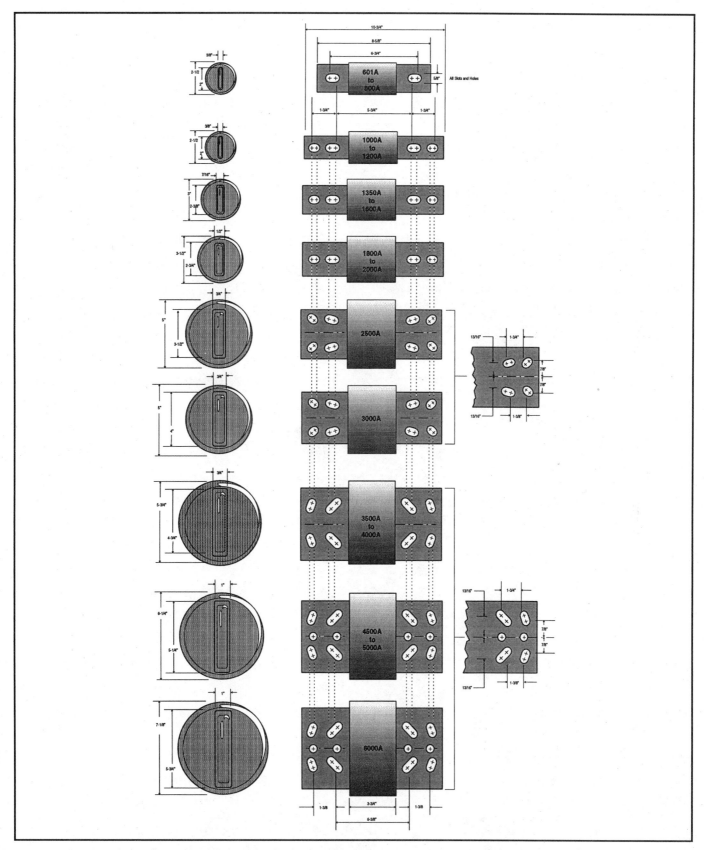

Figure 2-141: Buss fuse dimensions (LOW PEAK and LIMITRON KRP-C, KTU, and KLU).

Fuses for Selective Coordination

The larger the upstream fuse is relative to a downstream fuse (feeder to branch, etc.), the less possibility there is of an overcurrent in the downstream circuit causing both fuses to open. Fast action, non-time-delay fuses require at least a 3:1 ratio between the ampere rating of a large upstream, line-side time-delay fuse to that of the downstream, load-side fuse in order to be selectively coordinated. In contrast, the minimum selective coordination ratio necessary for dual-element fuses is only 2:1 when used with Low Peak loadside fuses.

The use of time-delay, dual-element fuses affords easy selective coordination — coordination hardly requires anything more than a routine check of a tabulation of required selectively ratios. Close sizing of dual-element fuses in the branch circuit for motor overload protection provides a large difference (ratio) in the ampere ratings between the feeder fuse and the branch fuse compared to the single-element, non-time-delay fuse.

Fuse Time-Current Curves

When a low-level overcurrent occurs, a long interval of time will be required for a fuse to open (melt) and clear the fault. On the other hand, if the overcurrent is large, the fuse will open very quickly. The opening time is a function of the magnitude of the level of overcurrent. Overcurrent levels and the corresponding intervals of opening times are logarithmically plotted in graph form as shown in Figure 2-142. Levels of overcurrent are scaled on the horizontal axis, time intervals on the vertical axis. The curve is therefore called a "time-current" curve.

The plot in Figure 2-142 reflects the characteristics of a 200 A, 600 V, dual-element fuse. Note that at the 1,000 A overload level, the time interval which is required for the fuse to open is 10 seconds. Yet, at approximately the 2200 A overcurrent level, the opening (melt) time of a fuse is only 0.01 seconds. It is apparent that the time intervals become shorter and shorter as the overcurrent levels become larger. This relationship is termed an inverse time-to-current characteristic. Time-current curves are published or are available on most commonly used fuses showing "minimum melt," "average melt" and/or "total clear" characteristics. Although upstream and downstream fuses are easily coordinated by adhering to simple ampere ratios, these time-current curves permit close or critical analysis of coordination.

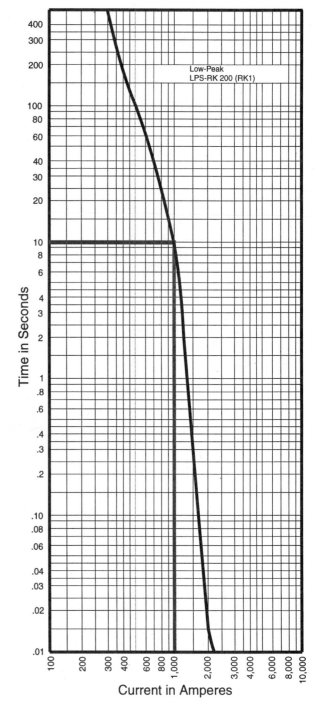

Figure 2-142: Typical time-current curve of a fuse.

Peak Let-Through Charts

Peak let-through charts enable you to determine both the peak let-through current and the apparent prospective rms symmetrical let-through current. Such charts are commonly referred to as *current limitation curves*. Figure 2-143 shows a simplified chart with explanations of the various functions. Figures 2-144 through 2-149 show current limi-

tation curves for several types of fuses commonly used in electrical systems.

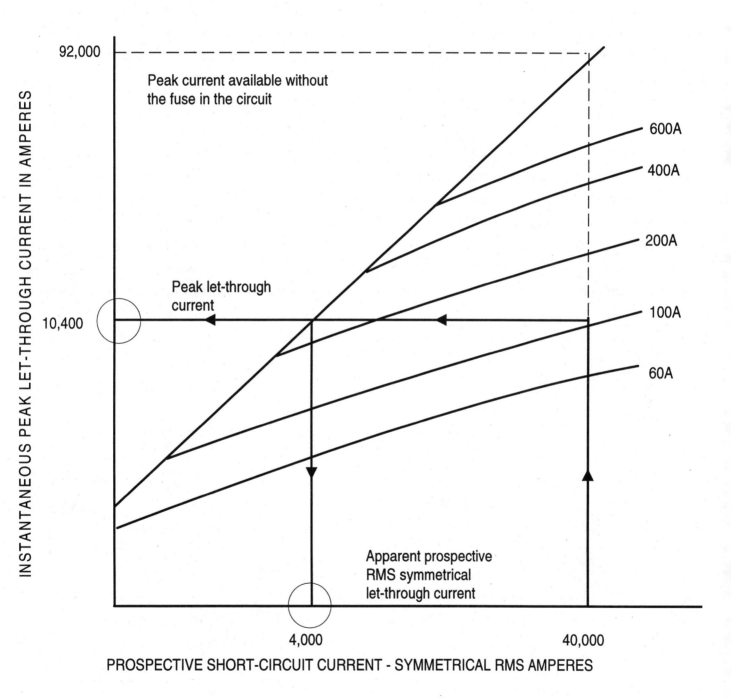

Figure 2-143: Principles of forming current limitation curves.

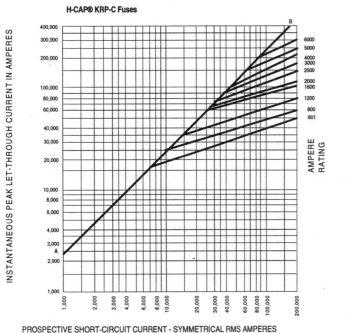

Figure 2-144: HI-CAP KRP-C fuses.

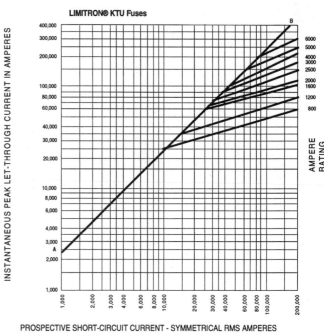

Figure 2-145: LIMITRON KTU fuses.

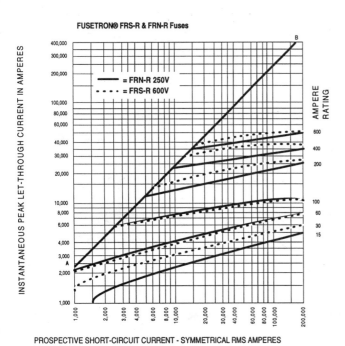

Figure 2-146: FUSETRON FRS-R & FRN-R fuses.

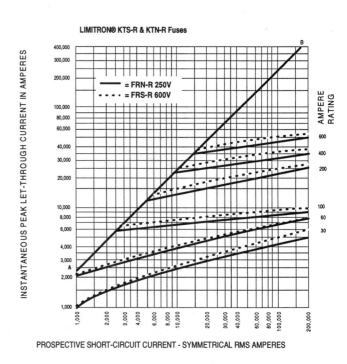

Figure 2-147: LIMITRON KTS-R & KTN-R fuses.

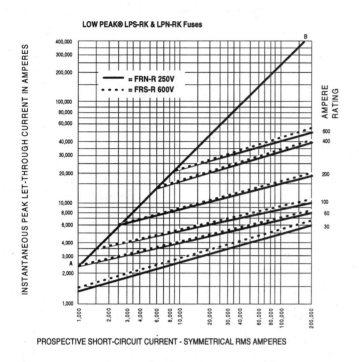

Figure 2-148: LOW PEAK LPS-RK & PLN-RK fuses.

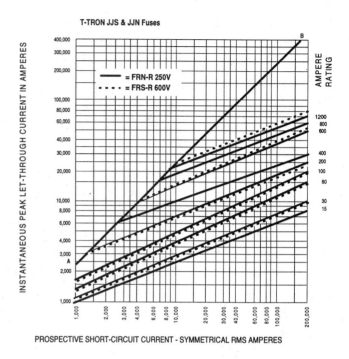

Figure 2-149: T-TRON JJS & JJN fuses.

Motor Overload and Short-Circuit Protection

When used in circuits with surge currents such as those caused by motors, transformers, and other inductive components, dual-element, time-delay fuses can be sized close to full-load amperes to give maximum overcurrent protection. For example, let's assume that a 10 hp, 208-V motor has a full-load current of 31 A. *NEC* Sections 430-32 and 430-52 require the following fuse types and sizes, as well as the switch size:

Fuse and Switch Sizing for 10 HP Motor (208V, 3Ø, 31 FLA)

Fuse Type	Maximum Fuse Size (Amperes)	Required Switch Size (Amperes)
Dual-element, time-delay	40A	60A
Single-element, non-time delay	90A	100A

The preceding table shows that a 40-A, dual-element fuse will protect the 31 A motor compared to the much larger, necessary 90-A, single-element fuse. It is apparent that if a sustained, harmful overload of 300% occurred in the motor circuit, the 90-A, single-element fuse would never open and the motor could be damaged. The non-time-delay fuse provides only ground fault and short-circuit protection — requiring separate overload protection as per the *NEC*.

In contrast, the 40-A dual-element fuse provides ground-fault, short-circuit protection plus a high degree of back-up protection against motor burnout from overload or single-phasing, should other overload protective devices fail. If thermal overloads, relays, or contacts should fail to operate, the dual-element fuses will act independently to protect the motor.

Motor overload relays and/or overcurrent protection should be sized according to the motor's current. The table in Figure 2-150 on the next page may be used to assist in sizing dual-element fuses. Also specify a labeling system on fuse clips to mark the type and ampere rating based on the actual current.

The charts in Figure 2-151 give a summary of fuse applications for all types of electrical systems.

Dual-Element Fuse Size	Motor Protection (Used without properly sized overload relays). Motor Full-Load Amps		Back-up Motor Protection (Used with properly sized overload relays). Motor Full-load Amps	
	Motor Service Factor of 1.15 or Greater or With Temp. Rise Not Over 40° C.	Motor Service Factor Less Than 1.15 or With Temp. Rise Not Over 40° C.	Motor Service Factor of 1.15 or Greater or With Temp. Rise Not Over 40° C.	Motor Service Factor of Less Than 1.15 or With Temp. Rise Not Over 40° C.
$\frac{1}{10}$	0.08 - 0.09	0.09 - 0.10	0 - 0.08	0 - 0.09
$\frac{1}{8}$	0.10 - 0.11	0.11 - 0.125	0.09 - 0.10	0.10 - 0.11
$\frac{15}{100}$	0.12 - 0.15	0.14 - 0.15	0.11 - 0.12	0.12 - 0.13
$\frac{2}{10}$	0.16 - 0.19	0.18 - 0.20	0.13 - 0.16	0.14 - 0.17
$\frac{1}{4}$	0.20 - 0.23	0.22 - 0.25	0.17 - 0.20	0.18 - 0.22
$\frac{3}{10}$	0.24 - 0.30	0.27 - 0.30	0.21 - 0.24	0.23 - 0.26
$\frac{4}{10}$	0.32 - 0.39	0.35 - 0.40	0.25 - 0.32	0.27 - 0.35
$\frac{1}{2}$	0.40 - 0.47	0.44 - 0.50	0.33 - 0.40	0.36 - 0.43
$\frac{6}{10}$	0.48 - 0.60	0.53 - 0.60	0.41 - 0.48	0.44 - 0.52
$\frac{8}{10}$	0.64 - 0.79	0.70 - 0.80	0.49 - 0.64	0.53 - 0.70
1	0.80 - 0.89	0.87 - 0.97	0.65 - 0.80	0.71 - 0.87
$1\frac{1}{8}$	0.90 - 0.99	0.98 - 1.08	0.81 - 0.90	0.88 - 0.98
$1\frac{1}{4}$	1.00 - 1.11	1.09 - 1.21	0.91 - 1.00	0.99 - 1.09
$1\frac{4}{10}$	1.12 - 1.19	1.22 - 1.30	1.01 - 1.12	1.10 - 1.22
$1\frac{1}{2}$	1.20 - 1.27	1.31 - 1.39	1.13 - 1.20	1.23 - 1.30
$1\frac{6}{10}$	1.28 - 1.43	1.40 - 1.56	1.21 - 1.28	1.31 - 1.39
$1\frac{8}{10}$	1.44 - 1.59	1.57 - 1.73	1.29 - 1.44	1.40 - 1.57
2	1.60 - 1.79	1.74 - 1.95	1.45 - 1.60	1.58 - 1.74
$2\frac{1}{4}$	1.80 - 1.99	1.96 - 2.17	1.61 - 1.80	1.75 - 1.96
$2\frac{1}{2}$	2.00 - 2.23	2.18 - 2.43	1.81 - 2.00	1.97 - 2.17

Figure 2-150: Chart for selecting dual-element fuses for motor protection.

Dual-Element Fuse Size	Motor Protection (Used without properly sized overload relays). Motor Full-Load Amps		Back-up Motor Protection (Used with properly sized overload relays). Motor Full-load Amps	
	Motor Service Factor of 1.15 or Greater or With Temp. Rise Not Over 40° C.	Motor Service Factor Less Than 1.15 or With Temp. Rise Not Over 40° C.	Motor Service Factor of 1.15 or Greater or With Temp. Rise Not Over 40° C.	Motor Service Factor of Less Than 1.15 or With Temp. Rise Not Over 40° C.
2⁶⁄₁₀	2.24 - 2.39	2.44 -2.60	2.01 - 2.24	2.18 -2.43
3	2.40 - 2.55	2.61 - 2.78	2.25 - 2.40	2.44 - 2.60
3²⁄₁₀	2.56 - 2.79	2.79 - 3.04	2.41 - 2.56	2.61 - 2.78
3½	2.80 - 3.19	3.05 - 3.47	2.57 - 2.80	2.79 - 3.04
4	3.20 - 3.59	3.48—3.91	2.81 - 3.20	3.05 - 3.48
4½	3.60 - 3.99	3.92 - 4.34	3.21 - 3.60	3.49 - 3.91
5	4.00 - 4.47	4.35 - 4.86	3.61 - 4.00	3.92 - 4.35
5⁶⁄₁₀	4.48 - 4.79	4.87 - 5.21	4.01 - 4.48	4.36 - 4.87
6	4.80 - 4.99	5.22 - 5.43	4.49 - 4.80	4.88 - 5.22
6¼	5.00 - 5.59	5.44 - 6.08	4.81 - 5.00	5.23 - 5.43
7	5.60 - 5.99	6.09 - 6.52	5.01 - 5.60	5.44 - 6.09
7½	6.00 - 6.39	6.53 - 6.95	5.61 - 6.00	6.10 - 6.52
8	6.40 - 7.19	6.96 - 7.82	6.01 - 6.40	6.53 - 6.96
9	7.20 - 7.99	7.83 - 8.69	6.41 - 7.20	6.97 - 7.83
10	8.00 - 9.59	8.70 - 10.00	7.21 - 8.00	7.84 - 8 70
12	9.60 - 11.99	10.44 - 12.00	8.01 - 9.60	8.71 - 10.43
15	12.00 - 13.99	13.05 - 15.00	9.61 - 12.00	10.44 - 13.04
17½	14.00 - 15.99	15.22 - 17.39	12.01 - 14.00	13.05 - 15.21
20	16.00 - 19.99	17.40 - 20.00	14.01 - 16.00	15.22 -17.39
25	20.00 - 23.99	21.74 - 25.00	16.01 - 20.00	17.40 - 21.74
30	24.00 - 27.99	26.09 - 30.00	20.01 - 24.00	21.75 - 26.09
35	28.00 - 31.99	30.44 - 34.78	24.01 - 28.00	26.10 - 30.43

Figure 2-150: Chart for selecting dual-element fuses for motor protection. *(Cont.)*

Dual-Element Fuse Size	Motor Protection (Used without properly sized overload relays). Motor Full-Load Amps		Back-up Motor Protection (Used with properly sized overload relays). Motor Full-load Amps	
	Motor Service Factor of 1.15 or Greater or With Temp. Rise Not Over 40° C.	Motor Service Factor Less Than 1.15 or With Temp. Rise Not Over 40° C.	Motor Service Factor of 1.15 or Greater or With Temp. Rise Not Over 40° C.	Motor Service Factor of Less Than 1.15 or With Temp. Rise Not Over 40° C.
40	32.00 - 35.99	34.79 - 39.12	28.01 - 32.00	30.44 - 37.78
45	36.00 - 39.99	39.13 - 43.47	32.01 - 36.00	37.79 - 39.13
50	40.00 - 47.99	43.48 - 50.00	36.01 - 40.00	39.14 - 43.48
60	48.00 - 55.99	52.17 - 60.00	40.01 - 48.00	43.49 - 52.17
70	56.00 - 59.99	60.87 - 65.21	48.01 - 56.00	52.18 - 60.87
75	60.00 - 63.99	65.22 - 69.56	56.01 - 60.00	60.88 - 65.22
80	64.00 - 71 .99	69.57 - 78.25	60.01 - 64.00	65.23 - 69.57
90	72.00 - 79.99	78.26 - 86.95	64.01 - 72.00	69.58 - 78.26
100	80.00 - 87.99	86.96 - 95.64	72.01 - 80.00	78.27 - 86.96
110	88.00 - 99.99	95.65 - 108.69	80.01 - 88.00	86.97 - 95.65
125	100.00 - 119.99	108.70 - 125.00	88.01 - 100.00	95.66 - 108.70
150	120.00 - 139.99	131.30 - 150.00	100.01 - 1 20.00	108.71 - 30.43
175	140.00 - 159.99	152.17 - 173.90	120.01 - 140.00	130.44 - 152.17
200	160.00 - 179.99	173.91 - 195.64	140.01 - 160.00	152.18 - 173.91
225	180.00 - 199.99	195.65 - 217.38	160.01 - 180.00	173.92 - 195.62
250	200.00 - 239.99	217.39 - 250.00	180.01 - 200.00	195.63 - 217.39
300	240.00 - 279.99	260.87 - 300.00	200.01 - 240.00	217.40 - 260.87
350	280.00 - 319.99	304.35 - 347.82	240.01 - 280.00	260.88 - 304.35
400	320.00 - 359.99	347.83 - 391.29	280.01 - 320.00	304.36 - 347.83
450	360.00 - 399.99	391.30 - 434.77	320.01 - 360.00	347.84 - 391.30
500	400.00 - 479.99	434.78 - 500.00	360.01 - 400.00	391.31 - 434.78
600	480.00 - 600.00	521.74 - 600.00	400.01 - 480.00	434.79 - 521.74

Figure 2-150: Chart for selecting dual-element fuses for motor protection.

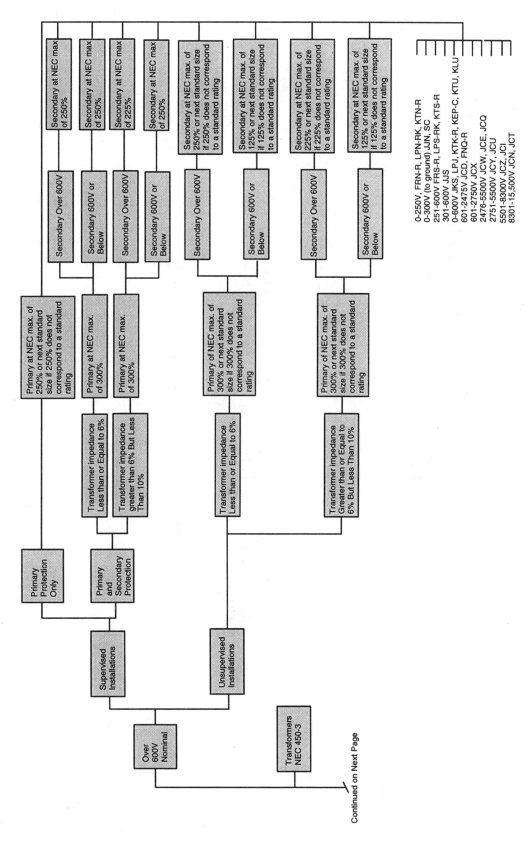

Figure 2-151: Summary of fuse applications for electrical systems.

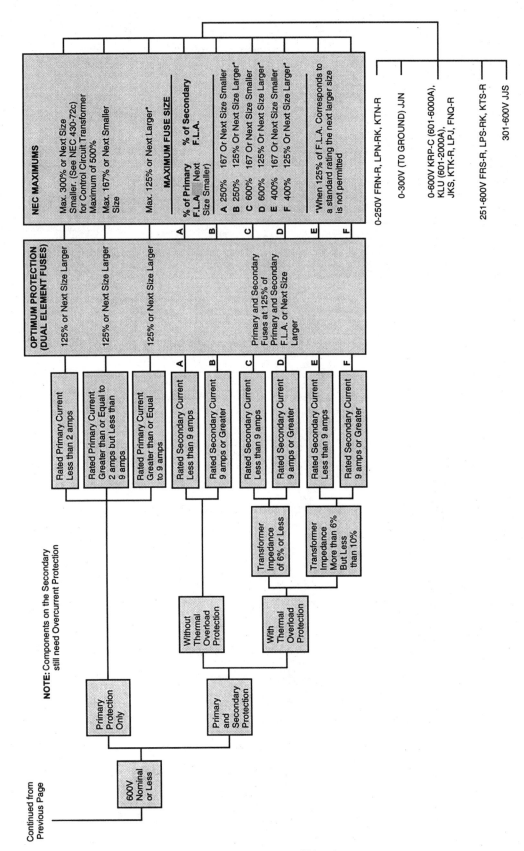

Figure 2-151: Summary of fuse applications for electrical systems. *(Cont.)*

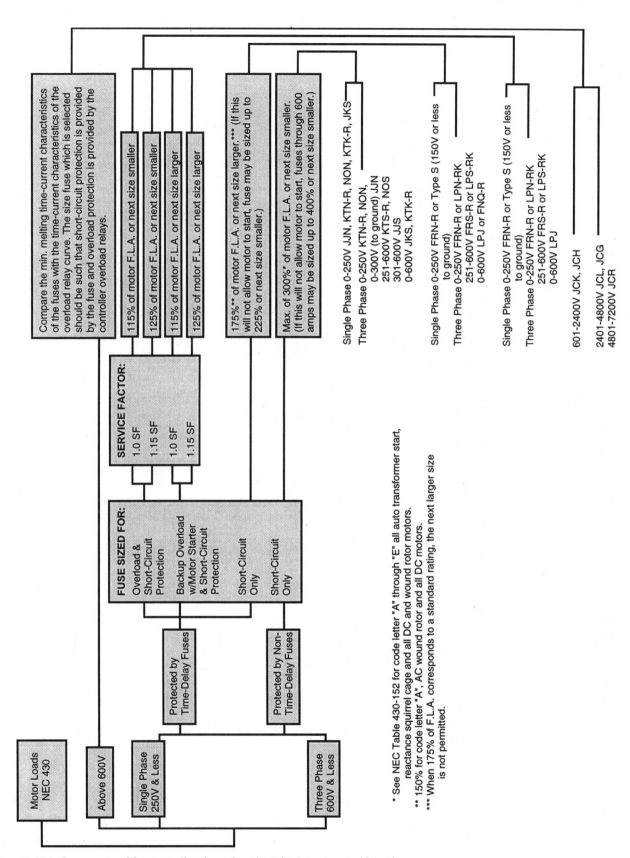

Figure 2-151: Summary of fuse applications for electrical systems. *(Cont.)*

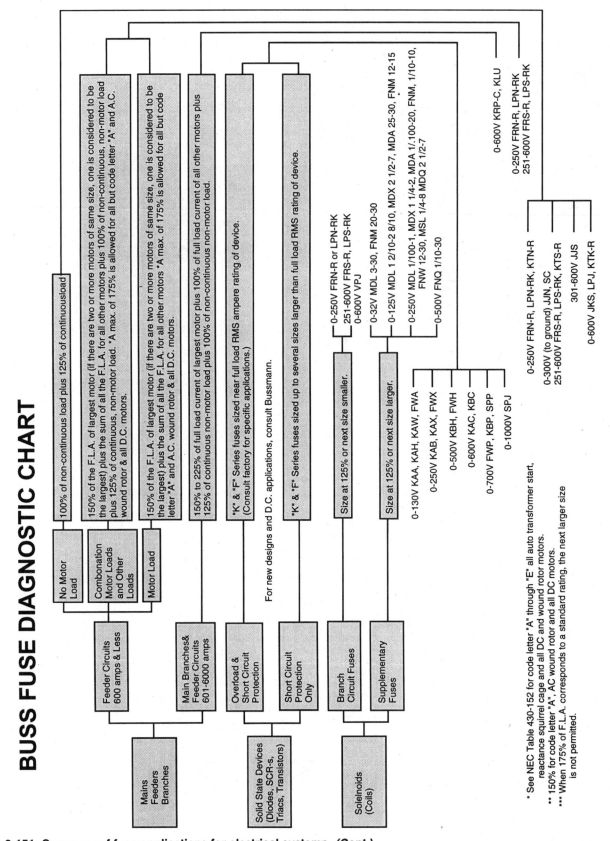

Figure 2-151: Summary of fuse applications for electrical systems. *(Cont.)*

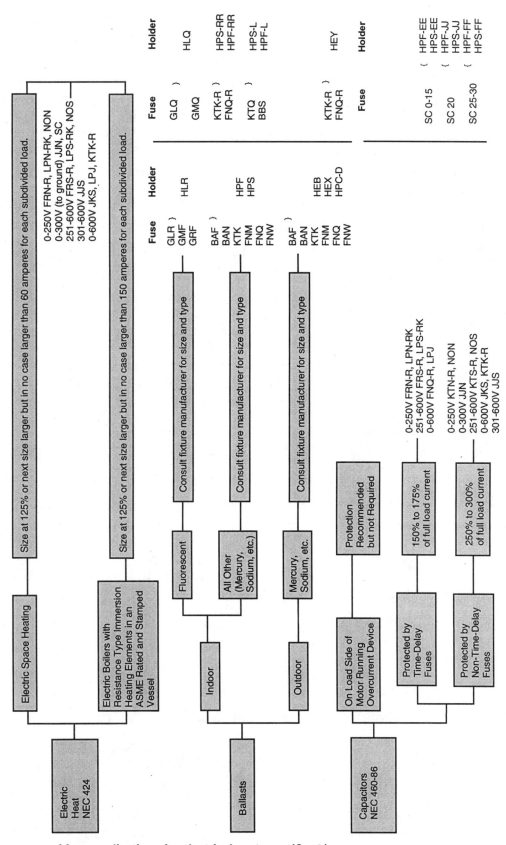

Figure 2-151: Summary of fuse applications for electrical systems. *(Cont.)*

GROUND FAULT PROTECTIONS

Ground-fault protection systems are designed to protect electrical equipment from destructive arcing ground faults and also to protect people from harmful (and deadly) electrical shocks. There are several different types to provide protection in a wide range of electrical power applications. One type is designed for outdoor use on construction sites, in mobile home parks, marinas, shipyards, industrial locations, and temporary power applications. Another is portable, to go where the work is. This type is used with power tools, appliances, and other equipment wired with either two- or three-wire plugs. Models are available to protect one branch circuit or a complete load center. Ground-fault protectors must be used to protect swimming-pool areas, kitchens, bathrooms, and other areas in residential, commercial, and industrial applications.

Ground-Fault Circuit Interrupters

Under certain conditions, the amount of current needed and the time it takes to open an overcurrent-protective device can be fatal. Because of this fact, the *NEC* requires ground-fault circuit-interrupters (GFCIs) to be installed on the following:

- Circuits feeding impedance heating units operating at voltages greater than 30 V. Such units are frequently used for deicing and snow-melting equipment.
- Circuits for electrically operated pool covers.
- Power or lighting circuits for swimming pools, fountains, and similar locations.
- Receptacles in both commercial and residential garages.
- Receptacles in bathrooms.
- Receptacles installed outdoors where there is direct grade level access.
- Receptacles installed in residential crawl spaces or unfinished basements.
- Kitchen or bar countertop receptacles.
- Receptacles installed in bathhouses.
- Receptacles installed in bathrooms of commercial, industrial, or any other building.
- Receptacles installed on roofs of any building except dwellings.
- Branch circuits derived from autotransformers.

GFCI circuit breakers require the same mounting space as standard single-pole circuit breakers and provide the same branch circuit wiring protection as standard circuit breakers. They also provide Class A ground-fault protection.

GFCI breakers that are UL listed are available in single and two-pole construction; 15, 20, 25 and 30-A ratings; and have a 10,000 A interrupting capacity. Single-pole units are rated 120 V ac; two-pole units 120/240 V ac.

GFCI circuit breakers not only can be used in load centers and panelboards, but they are also available factory installed in meter pedestals and power outlet panels for RV parks and construction sites.

The GFCI sensor continuously monitors the current balance in the ungrounded "hot" load conductor and the neutral load conductor. If the current in the neutral load wire becomes less than the current in the "hot" load wire, then a ground fault exists, since a portion of the current is returning to the source by some means other than the neutral load wire. When an imbalance in current occurs, the sensor, which is a differential current transformer, sends a signal to the solid-state circuitry which activates the ground trip solenoid mechanism and breaks the "hot" load connection. A current imbalance as low as 6 ma will cause the circuit breaker to interrupt the circuit. This will be indicated by the trip indicator as well as the position of the operating handle centered between "OFF" and "ON". *See* Figure 2-152.

The two-pole GFCI breaker (Figure 2-153) continuously monitors the current balance between the two "hot" conductors and the neutral conductor. As long as the sum of these three currents is zero, the device will not trip; that is, if there were 10 A current in the A load wire, 5 A in the neutral, and 5 A in the B load wire, then the sensor is balanced and will not produce a signal. A current imbalance from a ground fault condition as low as 6 ma will cause the sensor to produce a signal of sufficient magnitude to trip the device.

Single-Pole GFCI Circuit Breakers

The single-pole breaker has two load lugs and a white wire "pigtail" in addition to the line side plug-on or bolt-on connector. The line side "hot" connection is made by installing the GFCI breaker in the panel the same as you would install any other circuit breaker. The white wire "pigtail" is attached to the panel neutral (S/N) assembly. Both the neutral and "hot" wires of the branch circuit being protected are terminated in the GFCI breaker. These

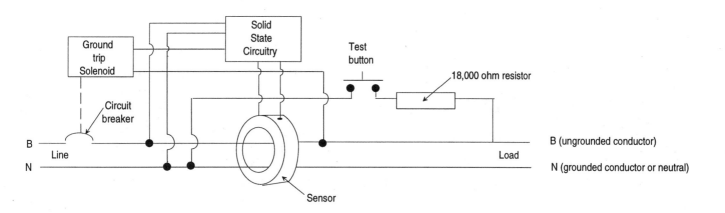

Figure 2-152: Operating circuitry of a typical GFCI.

two load lugs are clearly marked "LOAD POWER" and "LOAD NEUTRAL" by moldings in the breaker case. Also molded in the case is the identifying marking for the "pigtail", "PANEL NEUTRAL".

Single-pole GFCI circuit breakers must be installed on independent circuits. Circuits which employ a neutral common to more than one "hot" conductor cannot be protected against ground faults by a single-pole breaker because a common neutral cannot be split and retain the necessary "hot" wire — neutral wire balance under normal use to prevent the GFCI breaker from tripping.

Care should be exercised when installing GFCI breakers in existing panels to be sure the neutral wire for the branch circuit corresponds with the "hot" wire of the same circuit.

Always remember that unless the current in the neutral wire is equal to that in the "hot" wire (within 6 ma), the

GFCI breaker senses this as being a possible ground-fault and it will open the circuit to interrupt what it believes to be harmful current. *See* Figure 2-154.

Two-Pole GFCI Circuit Breakers

A two-pole GFCI circuit breaker can be installed on a 120/240-V ac 1φ-3W system, the 120/240-V ac portion of a 120/240-V ac 3φ-4W system, or two phases and neutral of a 120/208-V ac 3φ-4W system. Regardless of the application, the installation of the breaker is the same — connections made to two "hot" busses and the panel neutral assembly. When installed on these systems, protection is provided for two wire 240-V ac or 208-V ac circuits; three wire 120/240-V ac or 120/208-V ac circuits and 120-V ac multi-wire circuits.

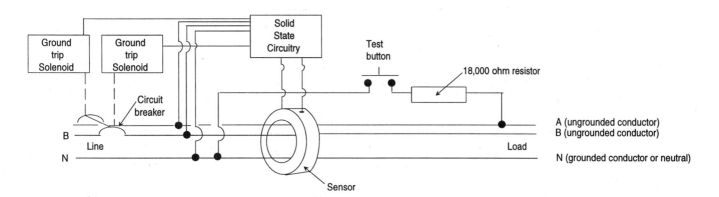

Figure 2-153: Operating characteristics of a two-pole GFCI.

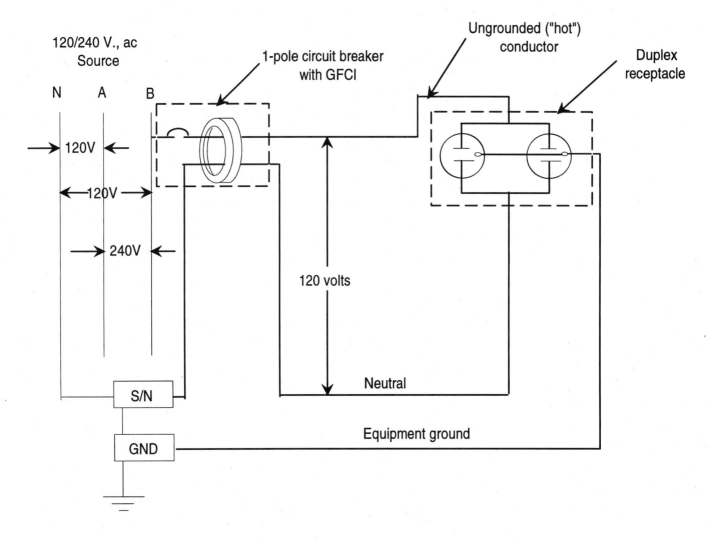

Figure 2-154: Operating characteristics of a single-pole circuit breaker with GFCI.

The circuit in Figure 2-155 can be used to illustrate the problems that would be encountered if a common load neutral were to be used for two single-pole GFCI breakers. Either or both breakers would trip when a load is applied at #2 duplex receptacle. The neutral current from #2 duplex receptacle would be flowing through breaker #1; this increase in neutral current through breaker #1 causes an imbalance in its sensor, thus causing it to produce a fault signal. At the same time, there is no neutral current flowing through breaker #2; therefore, it also senses a current imbalance. What happens if a load is applied at #1 duplex receptacle? As long as there is no load at #2 duplex receptacle, then neither breaker will trip because neither breaker will sense a current imbalance.

Wiring practices often used in junction boxes can also present problems when the junction box is used for taps for more than one branch circuit. Even though the circuits are not wired using a common neutral, sometimes all neutral conductors are connected together. Thus, parallel neutral paths would be established, producing an imbalance in each GFCI breaker sensor, causing them to trip.

Two-pole GFCI circuit breakers eliminate the problems encountered when trying to use two single-pole GFCI breakers with a common neutral. Because both "hot" currents and the neutral current pass through the same sensor, under normal load conditions, no imbalance in current occurs between the three currents and the breaker will not trip.

Direct-Wired GFCI Receptacles

Direct-wired GFCI receptacles provide Class A ground fault protection on 120-V ac circuits.

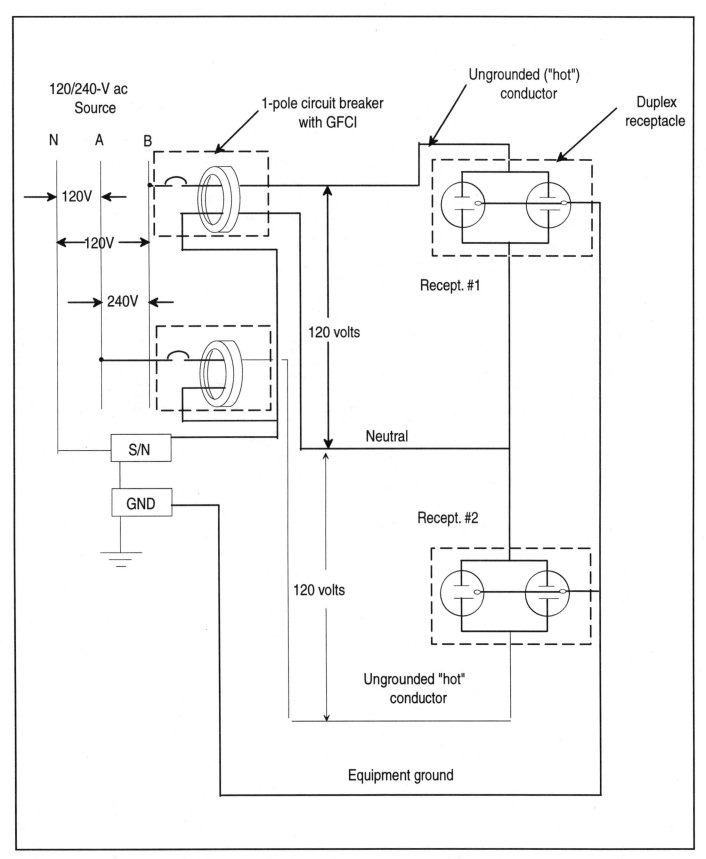

Figure 2-155: GFCI circuit depicting common load neutral.

They are available in both 15- and 20-A arrangements. The 15-A unit has a NEMA 5-15R receptacle configuration for use with 15-A plugs only. The 20-A device has a NEMA 5-20R receptacle configuration for use with 15- or 20-A plugs. Both 15- and 20-A units have a 120-V ac, 20- A circuit rating. This is to comply with *NEC* Table 210-24 which requires that 15-A circuits use 15-A rated receptacles but permits the use of either 15 or 20-A rated receptacles on 20-A circuits. Therefore, the GFCI receptacle units which contain a 15-A receptacle may be used on 20-A circuits.

These receptacles have terminals for the "hot" neutral and ground wires. In addition, they have feed-through terminals which can be used to provide ground-fault protection for other receptacles electrically "downstream" on the same branch circuit. All terminals will accept #14-#10 AWG copper wire.

GFCI receptacles have a two-pole tripping mechanism which breaks both the "hot" and the neutral load connections.

When tripped, the "RESET" button extends, making visible a red indicating band. The unit is reset by pushing this button.

GFCI receptacles have the additional benefit of being noise-suppressed. Noise suppression minimizes false tripping due to spurious line voltages or RF signals between 10 and 500 Mhz.

The GFCI receptacle can be mounted, without adapters, in wall outlet boxes that are at least 1½ in deep.

Plug-In GFCI Receptacle

The plug-in GFCI receptacle is a plug-in ground-fault protection adapter for use in either two- or three-wire 120-V ac receptacles. This device has a unique retractable ground pin which makes it possible to provide ground-fault protection at existing two-wire polarized receptacles as well as on three-wire receptacles. This unit provides two Class A GFCI-protected receptacles.

To use the unit on three-wire receptacles, lock the ground pin on the back of the unit. For two-wire receptacles, unlock the ground pin. The ground pin will retract automatically as the unit stabs are inserted into the receptacle. A yellow indicator pin on the front shows when the ground pin has been retracted.

When tripped, the plug-in GFCI receptacle has a red fault light which illuminates. To reset the unit, just push the blue reset button.

Figure 2-156 shows typical applications of GFCIs utilizing panelboards to segregate the GFCI-protected loads from normal loads.

RECEPTACLES

A receptacle is a contact device installed at the outlet for the connection of a single attachment plug. Several types and configurations are available for use with many different attachment plug caps — each designed for a specific application. For example, receptacles are available for two-wire, 120-V, 15- and 20-A circuits; others

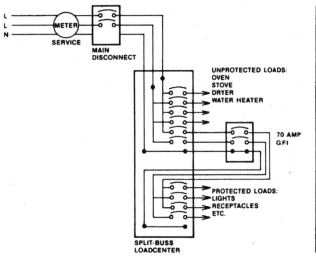

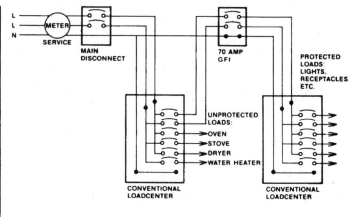

Figure 2-156: Typical applications of GFCI systems.

are designed for use on two- and three-wire, 240-V, 20-, 30-, 40-, and 50-A circuits.

Receptacles are rated according to their voltage and amperage capacity. This rating, in turn, determines the number and configuration of the contacts — both on the receptacle and the receptacle's mating plug. Figure 2-157 on the next page shows the most common configurations, along with their applications. This chart was developed by the Wiring Device Section of NEMA and illustrates 75 various configurations, which cover 38 voltage and current ratings. The configurations represent existing devices as well as suggested standards (shown with an asterisk in the chart) for future design. Note that all configurations in Figure 2-157 are for general-purpose nonlocking devices.

As indicated in the chart, unsafe interchangeability has been eliminated by assigning a unique configuration to each voltage and current rating. All dual ratings have been eliminated, and interchangeability exists only where it does not present an unsafe condition.

Each configuration is designated by a number composed of the chart line number, the amperage, and either "R" for receptacle or "P" for plug cap. For example, a 5-15R is found in line 5 and represents a 15-A receptacle.

A clear distinction is made in the configurations between "system grounds" and "equipment grounds." System grounds, referred to as grounded conductors, normally carry current at ground potential, and terminals for such conductors are marked "W" for "White" in the chart. Equipment grounds, referred to as grounding conductors, carry current only during ground-fault conditions, and terminals for such conductors are marked "G" for "grounding" in the chart.

Receptacle Characteristics

Receptacles have various symbols and information inscribed on them that help to determine their proper use and ratings. For example, a standard duplex receptacle usually contains the following printed inscriptions:

- The testing laboratory label
- The CSA (Canadian Standards Association) label
- Type of conductor for which the terminals are designed
- Current and voltage ratings, listed by maximum amperage, maximum voltage, and current restrictions

The testing laboratory label is an indication that the device has undergone extensive testing by a nationally recognized testing lab and has met with the minimum safety requirements. The label does not indicate any type of quality rating. A receptacle may be marked with the "UL" label which indicates that the device type was tested by Underwriters' Laboratories, Inc. of Northbrook, IL. ETL Testing Laboratories, Inc. of Cortland, NY is another nationally recognized testing laboratory. They provide a labeling, listing and follow-up service for the safety testing of electrical products to nationally recognized safety standards or specifically designated requirements of jurisdictional authorities.

The CSA (Canadian Standards Association) label is an indication that the material or device has undergone a similar testing procedure by the Canadian Standards Association and is acceptable for use in Canada.

Current and voltage ratings are listed by maximum amperage, maximum voltage and current restriction.

Conductor markings are also usually found on duplex receptacles. Receptacles with quick-connect wire clips will be marked "Use #12 or #14 solid wire only." If the inscription "CO/ALR" is marked on the receptacle, either copper, aluminum, or copper-clad aluminum wire may be used. The letters "ALR" stand for "aluminum revised." Receptacles marked with the inscription "CU/AL" should be used for copper only, although they were originally intended for use with aluminum also. However, such devices frequently failed when connected to 15- or 20-A circuits. Consequently, devices marked with "CU/AL" are no longer acceptable for use with aluminum conductors.

The remaining markings on duplex receptacles may include the manufacturer's name or logo, "Wire Release" inscribed under the wire-release slots, and the letters "GR." beneath or beside of the green grounding screw.

The screw terminals on receptacles are color-coded. For example, the terminal with the green screw head is the equipment ground connection and is connected to the U-shaped slots on the receptacle. The silver-colored terminal screws are for connecting the grounded or neutral conductors and are associated with the longer of the two vertical slots on the receptacle. The brass-colored terminal screws are for connecting the ungrounded or "hot" conductors and are associated with the shorter vertical slots on the receptacle.

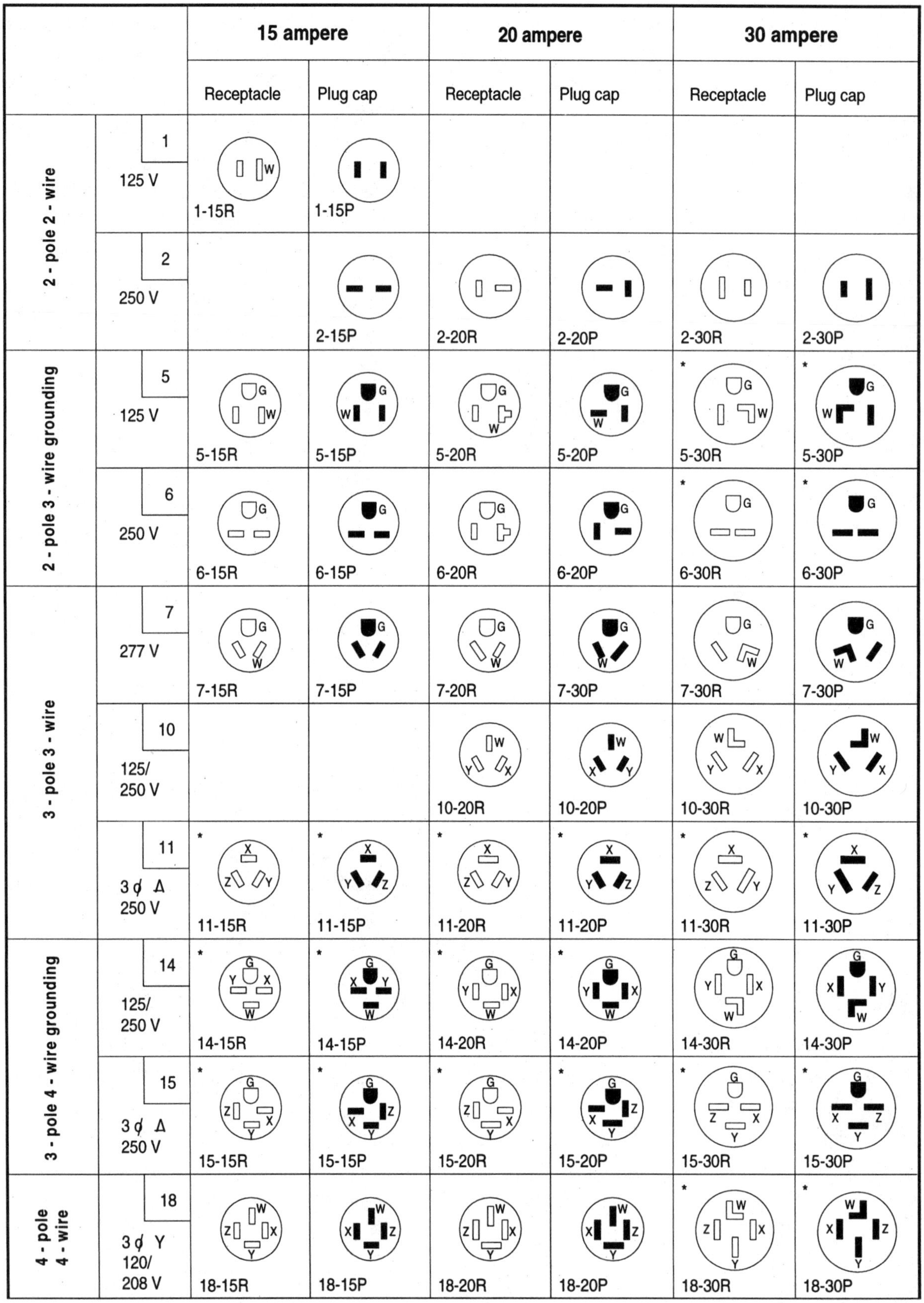

			15 ampere		20 ampere		30 ampere	
			Receptacle	Plug cap	Receptacle	Plug cap	Receptacle	Plug cap
2 - pole 2 - wire	125 V	1	1-15R	1-15P				
	250 V	2		2-15P	2-20R	2-20P	2-30R	2-30P
2 - pole 3 - wire grounding	125 V	5	5-15R	5-15P	5-20R	5-20P	5-30R	5-30P
	250 V	6	6-15R	6-15P	6-20R	6-20P	6-30R	6-30P
3 - pole 3 - wire	277 V	7	7-15R	7-15P	7-20R	7-30P	7-30R	7-30P
	125/ 250 V	10			10-20R	10-20P	10-30R	10-30P
	3 ∅ Δ 250 V	11	11-15R	11-15P	11-20R	11-20P	11-30R	11-30P
3 - pole 4 - wire grounding	125/ 250 V	14	14-15R	14-15P	14-20R	14-20P	14-30R	14-30P
	3 ∅ Δ 250 V	15	15-15R	15-15P	15-20R	15-20P	15-30R	15-30P
4 - pole 4 - wire	3 ∅ Y 120/ 208 V	18	18-15R	18-15P	18-20R	18-20P	18-30R	18-30P

Figure 2-157: NEMA configurations for general-purpose nonlocking receptacles and plug caps.

50 ampere		60 ampere	
Receptacle	Plug cap	Receptacle	Plug cap
5-50R	5-50P		
6-50R	6-50P		
7-50R	7-50P		
10-50R	10-50R		
11-50R	11-50P		
14-50R	14-50P	14-60R	14-60P
15-50R	15-50P	15-60R	15-60P
18-50R	18-50P	18-60R	18-60P

Figure 2-157: NEMA configurations *(Cont.)*.

Dimensions and characteristics of many types of receptacles are shown in Figure 2-158 beginning on page 316.

SWITCHES

Two basic terms used to identify the characteristics of switches are:

- Pole or poles
- Throw

The term *pole* refers to the number of conductors that the switch will control in the circuit. For example, a single-pole switch breaks the connection on only one conductor in the circuit. A double-pole switch breaks the connection to two conductors, and so forth.

The term *throw* refers to the number of internal operations that a switch can perform. For example, a single-pole, single-throw switch will "make" one conductor when thrown in one direction (the ON position) and "break" the circuit when thrown in the opposite direction; that is, the OFF position. The commonly used ON/OFF toggle switch is an SPST switch (single-pole, single-throw). A two-pole, single-throw switch opens or closes two conductors at the same time. Both conductors are either open or closed; that is, in the ON or OFF position.

A two-pole, double-throw switch is used to direct a two-wire circuit through one of two different paths. One application of a two-pole, double-throw switch is in an electrical transfer switch where certain circuits may be energized from either the main service, or from an emergency source such as a standby generator, battery packs, or some other alternate power source. The double-throw switch "makes" the circuit from one or the other and prevents the circuits from being energized from both sources at the same time. Three-pole, double-throw switches are used on three-phase services, but operate in the same general way as two-pole switches. Two types of transfer switches are in common use:

- Manual
- Automatic

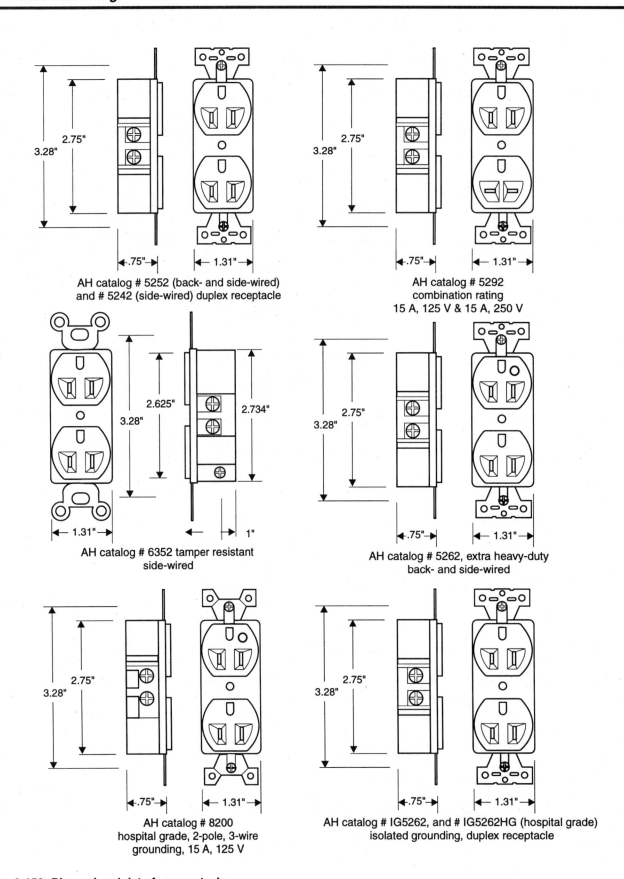

AH catalog # 5252 (back- and side-wired)
and # 5242 (side-wired) duplex receptacle

AH catalog # 5292
combination rating
15 A, 125 V & 15 A, 250 V

AH catalog # 6352 tamper resistant
side-wired

AH catalog # 5262, extra heavy-duty
back- and side-wired

AH catalog # 8200
hospital grade, 2-pole, 3-wire
grounding, 15 A, 125 V

AH catalog # IG5262, and # IG5262HG (hospital grade)
isolated grounding, duplex receptacle

Figure 2-158: Dimensional data for receptacles.

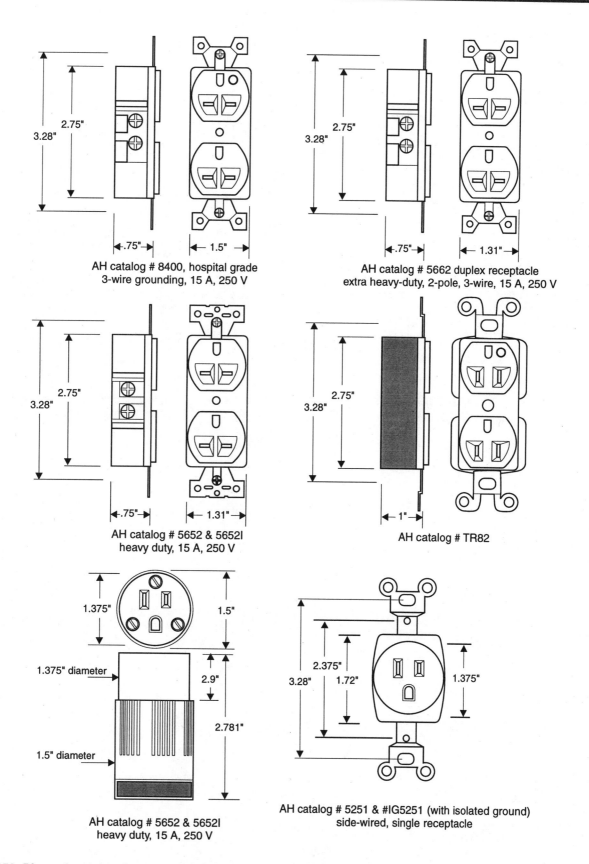

2.75"

3.28"

.75"

1.5"

AH catalog # 8400, hospital grade
3-wire grounding, 15 A, 250 V

2.75"

3.28"

.75"

1.31"

AH catalog # 5662 duplex receptacle
extra heavy-duty, 2-pole, 3-wire, 15 A, 250 V

2.75"

3.28"

.75"

1.31"

AH catalog # 5652 & 5652I
heavy duty, 15 A, 250 V

2.75"

3.28"

1"

AH catalog # TR82

1.375"

1.5"

1.375" diameter

2.9"

2.781"

1.5" diameter

AH catalog # 5652 & 5652I
heavy duty, 15 A, 250 V

3.28"

2.375"

1.72"

1.375"

AH catalog # 5251 & #IG5251 (with isolated ground)
side-wired, single receptacle

Figure 2-158: Dimensional data for receptacles. *(Cont.)*

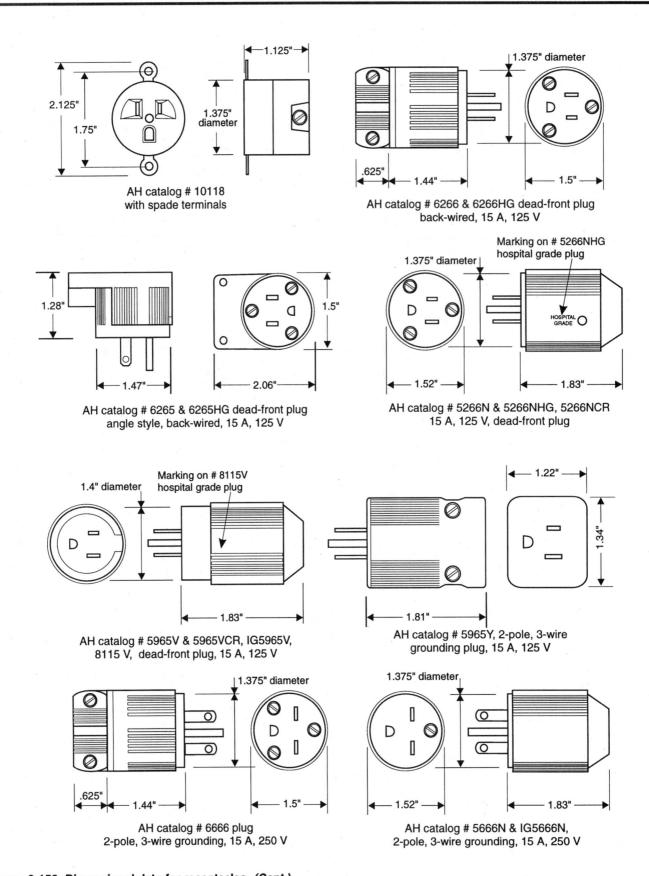

AH catalog # 10118
with spade terminals

AH catalog # 6266 & 6266HG dead-front plug
back-wired, 15 A, 125 V

AH catalog # 6265 & 6265HG dead-front plug
angle style, back-wired, 15 A, 125 V

AH catalog # 5266N & 5266NHG, 5266NCR
15 A, 125 V, dead-front plug

AH catalog # 5965V & 5965VCR, IG5965V,
8115 V, dead-front plug, 15 A, 125 V

AH catalog # 5965Y, 2-pole, 3-wire
grounding plug, 15 A, 125 V

AH catalog # 6666 plug
2-pole, 3-wire grounding, 15 A, 250 V

AH catalog # 5666N & IG5666N,
2-pole, 3-wire grounding, 15 A, 250 V

Figure 2-158: Dimensional data for receptacles. *(Cont.)*

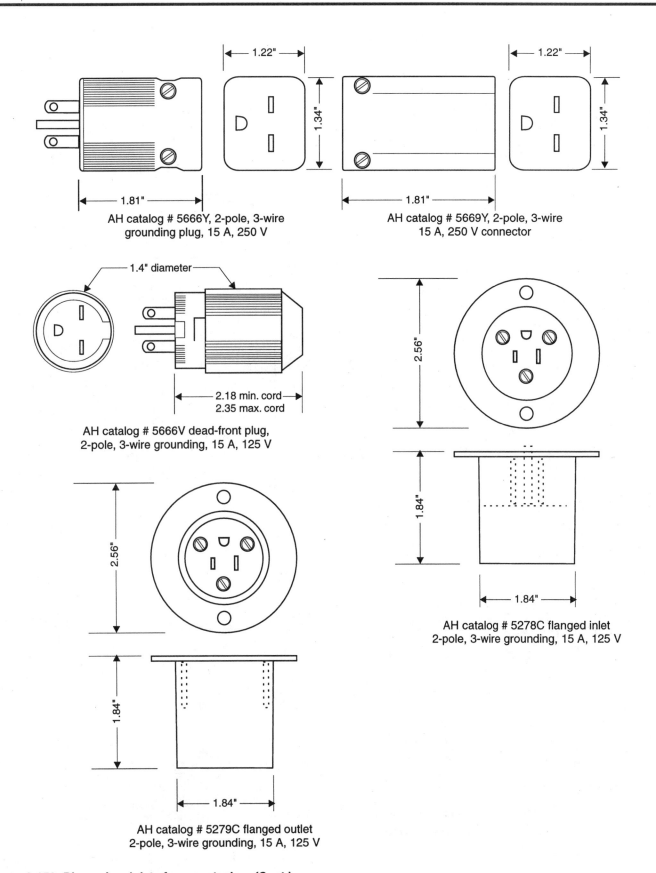

AH catalog # 5666Y, 2-pole, 3-wire
grounding plug, 15 A, 250 V

AH catalog # 5669Y, 2-pole, 3-wire
15 A, 250 V connector

AH catalog # 5666V dead-front plug,
2-pole, 3-wire grounding, 15 A, 125 V

AH catalog # 5278C flanged inlet
2-pole, 3-wire grounding, 15 A, 125 V

AH catalog # 5279C flanged outlet
2-pole, 3-wire grounding, 15 A, 125 V

Figure 2-158: Dimensional data for receptacles. *(Cont.)*

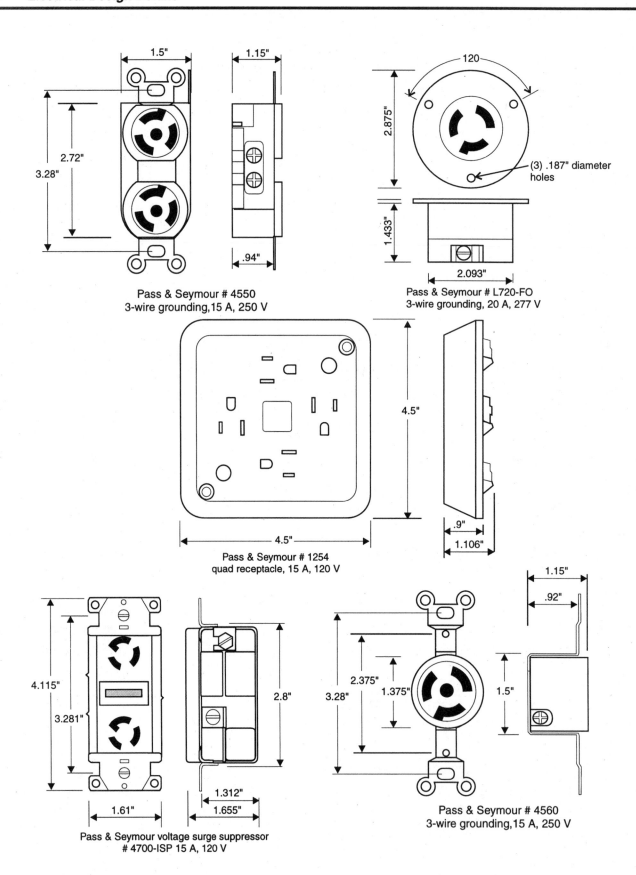

Pass & Seymour # 4550
3-wire grounding, 15 A, 250 V

Pass & Seymour # L720-FO
3-wire grounding, 20 A, 277 V

(3) .187" diameter holes

Pass & Seymour # 1254
quad receptacle, 15 A, 120 V

Pass & Seymour voltage surge suppressor
4700-ISP 15 A, 120 V

Pass & Seymour # 4560
3-wire grounding, 15 A, 250 V

Figure 2-158: Dimensional data for receptacles. *(Cont.)*

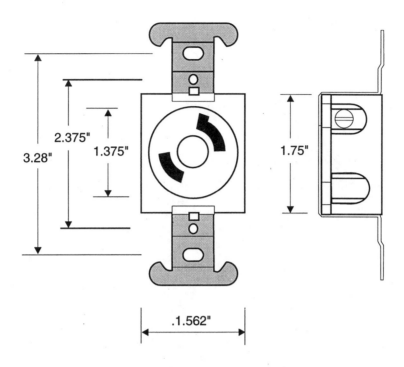

Pass & Seymour # 7535
2-wire, 15 A, 125 V

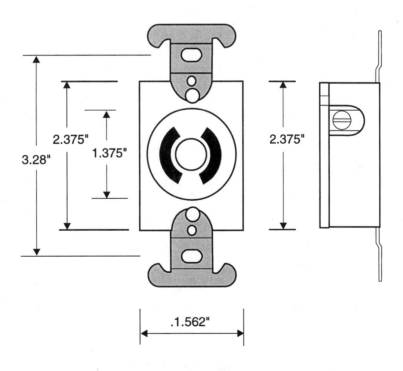

Pass & Seymour # 7210
2-wire, 20 A, 250 V

Figure 2-158: Dimensional data for receptacles. *(Cont.)*

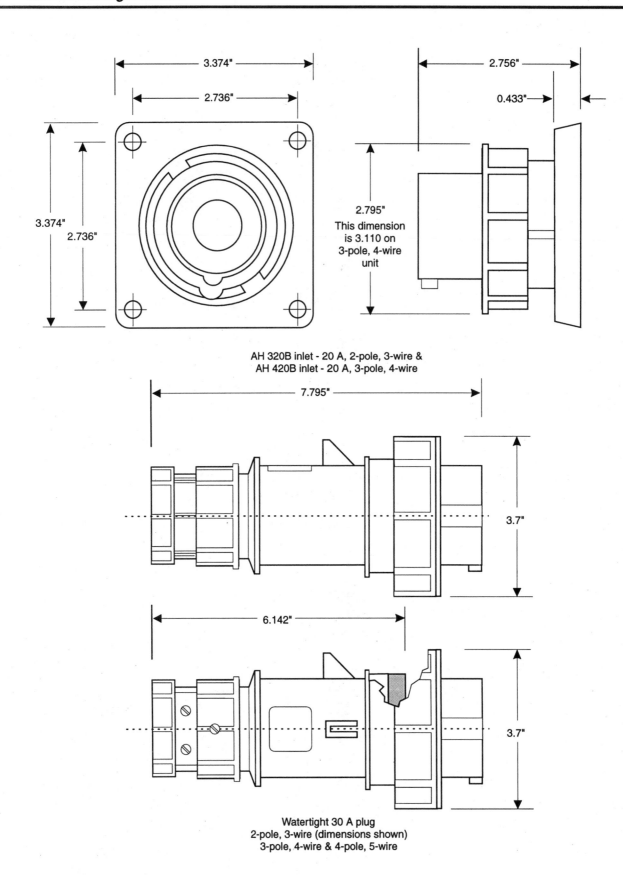

AH 320B inlet - 20 A, 2-pole, 3-wire &
AH 420B inlet - 20 A, 3-pole, 4-wire

Watertight 30 A plug
2-pole, 3-wire (dimensions shown)
3-pole, 4-wire & 4-pole, 5-wire

Figure 2-158: Dimensional data for receptacles. *(Cont.)*

Watertight 30 A Connector

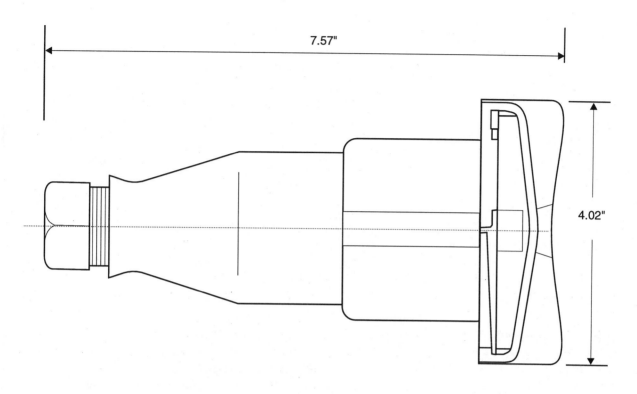

7.57"

4.02"

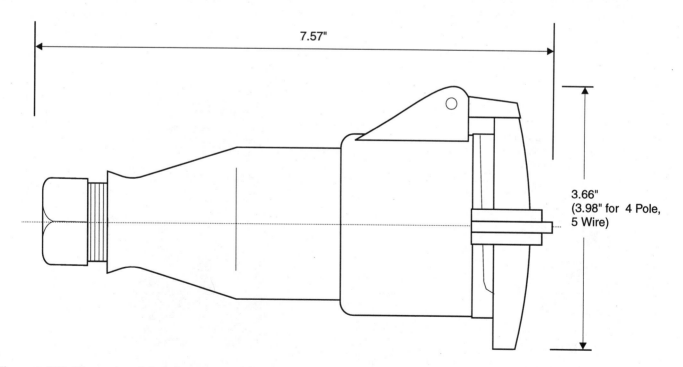

7.57"

3.66"
(3.98" for 4 Pole,
5 Wire)

Figure 2-158: Dimensional data for receptacles. *(Cont.)*

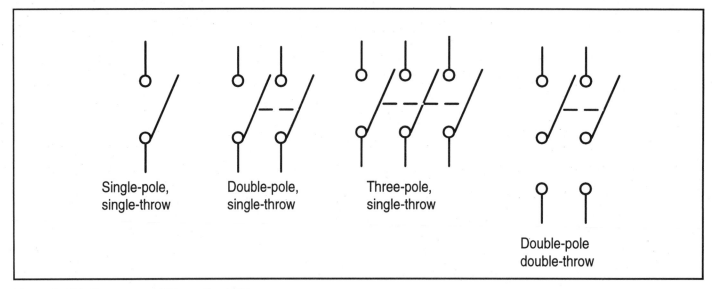

Figure 2-159: Common switch configurations.

Figure 2-159 shows common switch configurations. Other configurations are usually a combination of those show.

Where transfer switches are used in conjunction with standby emergency systems, it may be desired to bypass and isolate the transfer switch and its related equipment. A two-way bypass isolation system is shown in Figure 2-160. This is a manually operated device used in conjunction with a transfer switch to provide a means of directly connecting load conductors to a power source, and of disconnecting the transfer switch.

Limit Switches

Limit switches are pilot devices for electrically controlling or limiting the movement or function of mechanical apparatus. They may be either slow-make, slow-break or quick-make, quick-break type. Such switches may be used to limit the travel of a machine, to stop a machine at a given point, to change the direction of a machine during various phases of its cycle, and to limit the overtragel of a machine. Outline dimensions (in inches) of typical limit switches are shown in Figure 2-161, while dimensions and arm movement appear in Figure 2-162.

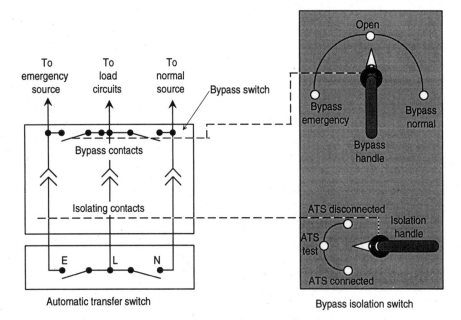

Figure 2-160: Two-way bypass isolation system.

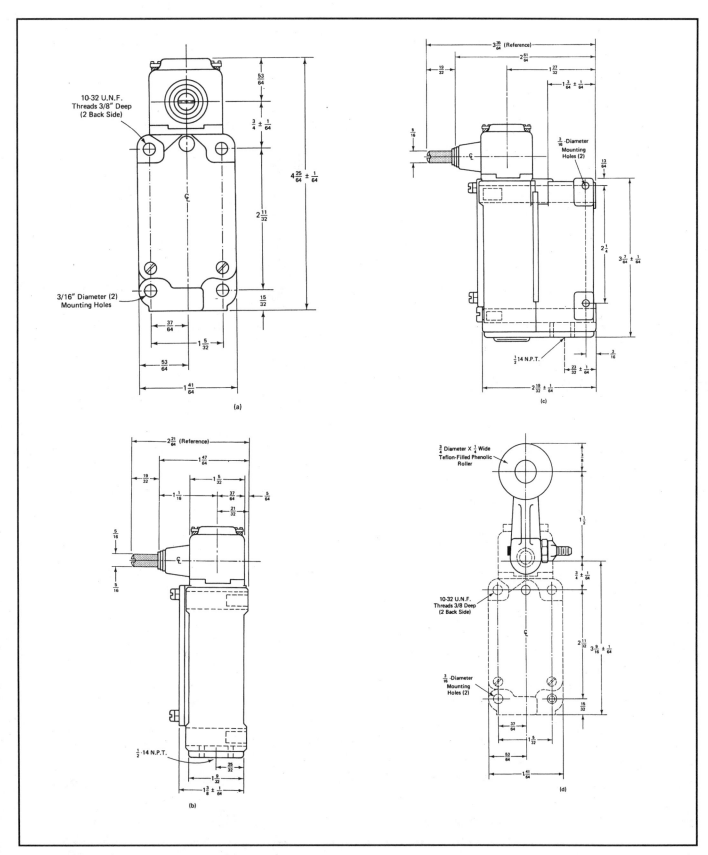

Figure 2-161: Dimensions of limit switches.

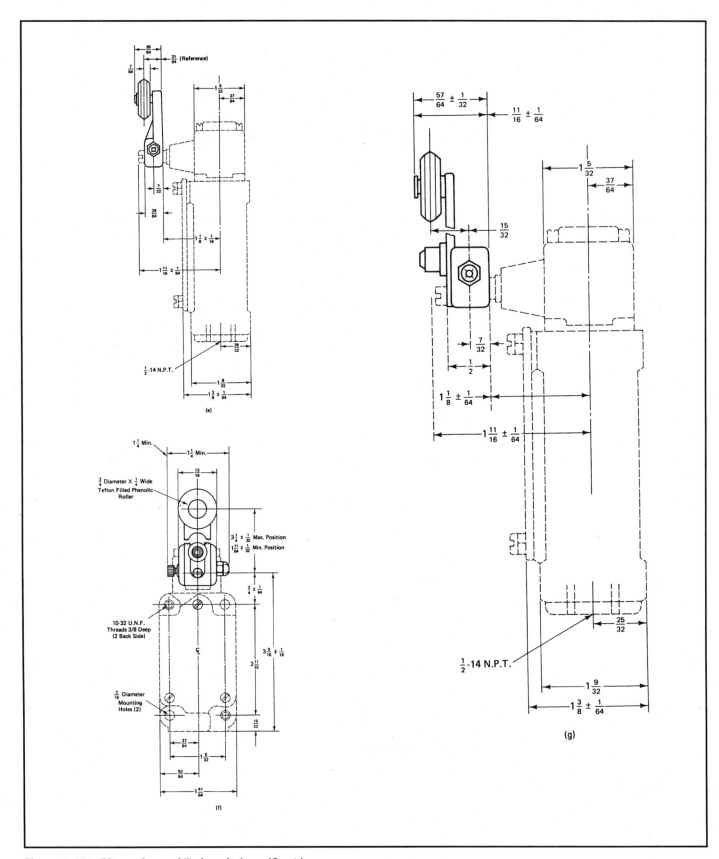

Figure 2-161: Dimensions of limit switches. (Cont.)

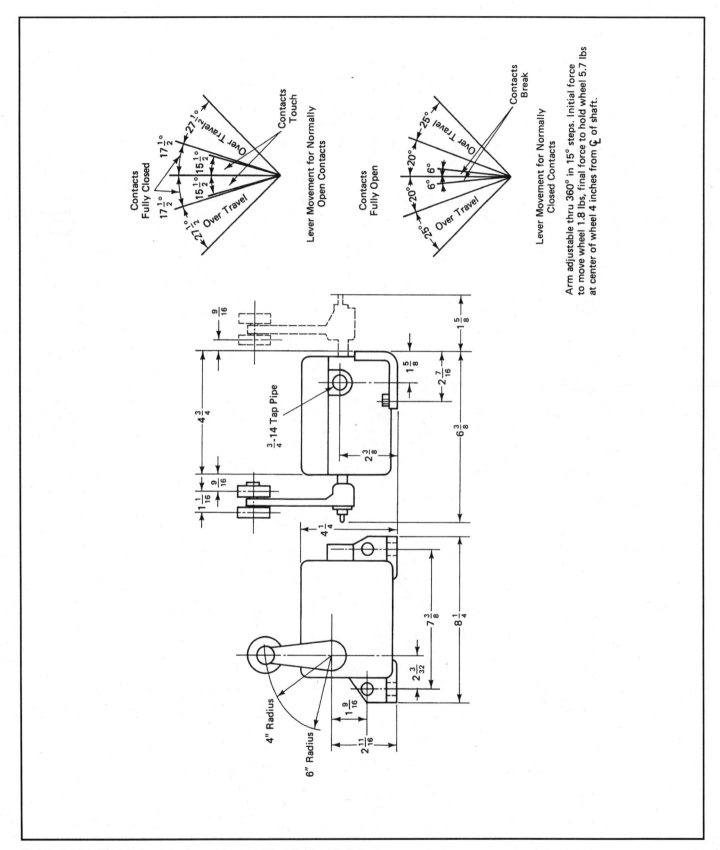

Figure 2-162: Dimensions and arm movement of limit switches.

Safety Switches

Enclosed single-throw safety switches are manufactured to meet industrial, commercial, and residential requirements. See Figures 2-163 and 2-164. The two basic types of safety switches are:

- General duty
- Heavy duty

Double-throw switches are also manufactured with enclosures and features similar to the general and heavy-duty single-throw designs.

The majority of safety switches have visible blades and safety handles. The switch blades are in full view when the enclosure door is open and there is visually no doubt when the switch is OFF. The only exception is Type 7 and 9 enclosures; these do not have visible blades. Switch handles, on all types of enclosures are an integral part of the box,

	GENERAL DUTY INDOOR NEMA 1 ENCLOSURES							
Ampere Rating	2 Pole, 2 Fuse or 3 Pole, 2 Fuse and S/N	Dimensions					Horsepower Ratings 240 ac	
		Height A	Width B	Rainshed E	W/Handle C	Depth w/cover D	Standard	Maximum
30		8.63 in	6.38 in	—	7 in	4.13 in	3	7.5
60		12.38 in	8 in	—	8.75 in	5 in	7.5	15
100		16.50 in	8.50 in	—	8.5 in	6.25 in	15	30
200		24.50 in	13 in	—	13 in	9.5 in	25	60
400		40.75 in	17.75 in	—	20.13 in	12.75 in	50	100
600		48.75 in	17.75 in	—	20.13 in	13.25 in	75	100
	3 Pole, 3 Fuse							
30		8.63 in	6.38 in	—	7 in	4.63 in	3	7.5
60		12.38 in	8 in	—	8.75 in	5 in	7.5	15
100		16.50 in	8.50 in	—	8.5 in	6.25 in	15	30
200		24.50 in	13 in	—	13 in	9.13 in	25	60
400		40.75 in	23 in	—	25.38 in	12.25 in	50	100
600		48.75 in	23 in	—	25.38 in	12.25 in	75	100

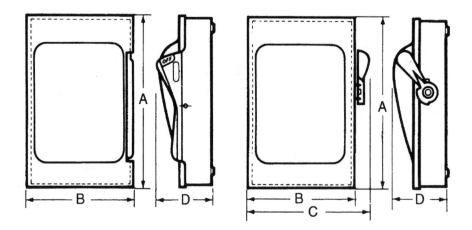

Figure 2-163: Specifications of 240-volt, nonfusible safety switches with NEMA 1 enclosures.

not the cover, so that the handle is in control of the switch blades under normal conditions.

Heavy Duty Switches

Heavy duty switches are intended for applications where ease of maintenance, rugged construction, and continued performance are primary concerns. They can be used in atmospheres where general duty switches would be unsuitable, and are therefore widely used in industrial applications. Heavy duty switches are rated 30 through 1200 amperes and 240 to 600 V ac or dc. Switches with horsepower ratings are capable of opening a circuit up to six times the rated current of the switch. When equipped with Class J or Class R fuses for 30 through 600 ampere switches, or Class L fuses in 800 and 1200 A switches, many heavy duty safety switches are UL listed for use on systems with up to 200,000 RMS symmetrical amperes available fault current. This, however, is about the highest short-circuit rating available for any heavy-duty safety

Ampere Rating	GENERAL DUTY RAINTIGHT NEMA 3R ENCLOSURES								
		Dimensions					Horsepower Ratings 240 ac	
	Poles	Height A	Width B	Rainshed E	W/Handle C	Depth w/cover D	1Ø, 2W	3Ø, 3W
30	2 w/SN or 3	10.25 in	7.75 in	7.75	9.5 in	5.63 in	3	7.5
60	2 w/SN or 3	12.25 in	8.38 in	8.75	10.5 in	6 in	10	15
100	2 w/SN or 3	18 in	11.38 in	11.75	13.75 in	7.5 in	15	30
200	2 w/SN or 3	24.25 in	13 in	13.38	15.63 in	10 in	25	60
400	2 w/SN or 3	40.63 in	22.88 in	26.75 in	27 in	12.15 in	50	100
600	2 w/SN or 3	49.63 in	22.88 in	26.75 in	27 in	12.25 in	75	100
30	3 w/SN	10.25 in	7.38 in	7.75	9.5 in	5.63 in	—	7.5
60	3 w/SN	12.25 in	8.38 in	8.75	10.5 in	6 in	—	15
100	3 w/SN	18 in	11.28 in	11.75	13.75 in	7.5 in	—	30
200	3 w/SN	24.25 in	13 in in	13.38	15.63 in	10 in	—	100

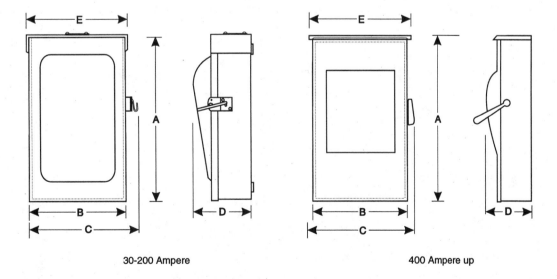

30-200 Ampere   400 Ampere up

Figure 2-164: Specifications of 240-volt, nonfusible safety switches with NEMA 3R enclosures.

switch. Applications include use where the required enclosure is NEMA Type 1, 3R, 4, 4X, 5, 7, 9, 12 or 12K.

Switch Blade and Jaws

Two types of switch contacts are used by the industry in today's safety switches. One is the "butt" contact; the other is a knife-blade and jaw type. On switches with knife-blade construction, the jaws distribute a uniform clamping pressure on both sides of the blade contact surface. In the event of a high-current fault, the electromagnetic forces which develop tend to squeeze the jaws tightly against the blade. In the butt type contact, only one side of the blade's contact surface is held in tension against the conducting path. Electromagnetic forces due to high current faults tend to force the contacts apart, causing them to burn severely. Consequently, the knife blade and jaw type construction is the preferred type for use on all heavy duty switches. The action of the blades moving in and out of the jaws aids in cleaning the contact surfaces. All current-carrying parts of these switches are plated to reduce heating by keeping oxidation at a minimum. Switch blades and jaws are made of copper for high conductivity. Spring-clamped blade hinges are another feature that help assure good contact surfaces and cool operations. "Visible blades" are utilized to provide visual evidence that the circuit has been opened.

Fuse Clips

Fuse clips are plated to control corrosion and to keep heating to a minimum. All fuse clips on heavy duty switches have steel reinforcing springs for increased mechanical strength and firmer contact pressure.

Terminal Lugs

Most heavy duty switches have front removable, screw-type terminal lugs. Most switch lugs are suitable for copper or aluminum wire except NEMA TYPES 4, 4X, 5 stainless and TYPES 12 and 12K switches which have all copper current-carrying parts and lugs designated for use with copper wire only. Heavy duty switches are suitable for the wire sizes and number of wires per pole as listed in Figure 2-165.

Insulating Material

As the voltage rating of switches is increased, arc suppression becomes more difficult and the choice of insulation material becomes more critical. Arc suppressers are usually made of insulation material and magnetic suppresser plates when required. All arc suppresser materials must provide proper control and extinguishing of arcs.

Operating Mechanism and Cover Latching

Most heavy duty safety switches have a spring driven quick-make, quick-break mechanism. A quick-breaking action is necessary if the switch is to be safely switched OFF under a heavy load.

The spring action, in addition to making the operation quick-make, quick-break, firmly holds the switch blades in the ON or OFF position. The operating handle is an integral part of the switching mechanism and is in direct control of the switch blades under normal conditions.

A one-piece cross bar, connected to all switch blades, should be provided which adds to the overall stability and integrity of the switching assembly by promoting proper alignment and uniform switch blade operation.

Dual cover interlocks are standard on most heavy duty switches where the NEMA enclosure permits. However, NEMA Types 7 and 9 have bolted covers and obviously cannot contain dual cover interlocks. The purpose of dual interlock is to prevent the enclosure door from being opened when the switch handle is in the ON position and prevents the switch from being turned ON while the door is open. A means of bypassing the interlock is provided to allow the switch to be inspected in the ON position by qualified personnel. However, this practice should be avoided if at all possible. Heavy duty switches can be padlocked in the OFF position with up to three padlocks.

Accessories

Accessories available for field installation include Class R fuse kits, fuse pullers, insulated neutrals with grounding provisions, equipment grounding kits, watertight hubs for use with Type 4, 4X, 5 stainless or Type 12 switches, and interchangeable bolt-on hubs for Type 3R switches.

Electrical interlock consists of auxiliary contacts for use where control or monitoring circuits need to be switches in conjunction with the switch operation.

TYPE 1 AND 3R HEAVY DUTY TERMINAL LUG DATA			
Ampere Rating	Conductors Per Phase	Wire Range Wire Bending Space Per NEC Table 373-6	Lug Wire Range
30	1	#12-6 AWG (Al) or #14-6 AWG (Cu)	#12-2 AWG (Al) or #14-2 AWG (Cu)
60	1	#12-3 AWG (Al) or #14-3 AWG (Cu)	#12-2 AWG (Al) or #14-2 AWG (Cu)
100	1	#12-1 /0 AWG (Al) or #14-1 /0 AWG (Cu)	#12-1 /0 AWG (Al) or #14-1 /0 AWG (Cu)
200	1	#6 AWG-250 kcmil (Al/Cu)	#6 AWG-300 kcmil (Al/Cu)
400	1 or 2	#3/0 AWG-750 kcmil (Al/Cu) or #6 AWG-300 kcmil (Al/Cu)	#3/0 AWG-750 kcmil (Al/Cu) and #6 AWG-300 kcmil (Al/Cu)
600	2	#3/0 AWG-500 kcmil (Al/Cu)	#3/0 AWG-500 kcmil (Al/Cu)
800	3	#3/0 AWG-750 kcmil (Al/Cu)	#3/0 AWG-750 kcmil (Al/Cu)
1200	4	#3/0 AWG-750 kcmil (Al/Cu)	#3/0 AWG-750 kcmil (Al/Cu)
TYPE 4, 4X, 5 STAINLESS, AND TYPE 12 AND 12K HEAVY DUTY TERMINAL LUG DATA			
Ampere Rating	Conductors Per Phase	Wire Range Wire Bending Space Per NEC Table 373-6	Lug Wire Range
30	1	#14-6 AWG (Cu)	#14-2 AWG (Cu)
60	1	#14-4 AWG (Cu)	#12-2 AWG (Cu)
100	1	#14-1 AWG (Cu)	#14-1 AWG (Cu)
200	1	#6 AWG-250 kcmil(Cu)	#6 AWG-250 kcmil(Cu)
400	1 or 2	#1 /0 AWG-600 kcmil(Cu) or #6 AWG-250 kcmil(Cu)	#1 /0 AWG-600 kcmil(Cu) and #6AWG-250 kcmil(Cu)
600	2	#4 AWG-350 kcmil(Cu)	#4 AWG-350 kcmil(Cu)

Figure 2-165: Safety switch lug specifications.

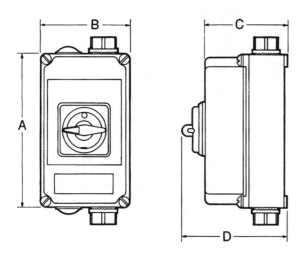

DIMENSIONS IN INCHES				
Enclosure Type	**A**	**B**	**C**	**D**
Heavy Duty	6.7	4.6	3.8	5.4
Extra Heavy Duty	10.2	6.3	5.0	6.7

Figure 2-166: Arrow Hart manual motor controller.

Motor-Circuit Switches

Motor-circuit switches are rated in horsepower (hp) and are capable of interrupting the maximum operating overload current of a motor of the same hp rating as the switch at its rated voltage. *See* Figure 2-166.

Bolted Pressure Switch

Bolted pressure switches are load-break disconnects. They are suitable for service-entrance equipment and are usually furnished in a general-purpose, dead-front enclosure. They have a fuse access door with lock. Incoming line terminals are located at the top, with load terminals at the bottom. All neutrals are full capacity. The front and side views of a typical bolted pressure switch, with dimensions, appear in Figure 2-167.

Toggle Wall Switches

General-use snap switches are so constructed that they can be installed in flush device boxes or on outlet box covers, or otherwise used in conjunction with wiring systems

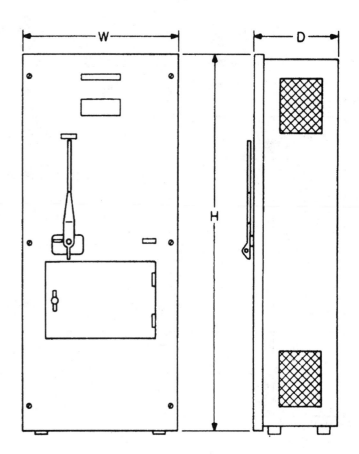

Ampere Rating	Dimensions, Inches		
	H	**W**	**D**
1200	69	48	17
1600	69	48	17
2000	69	48	17
2500	73	52	18
3000	81	49	20½

Figure 2-167: Front and side views of a bolted pressure switch.

recognized by the *NEC*. Such switches vary in grade, capacity, and purpose. It is very important that proper types are selected for the given application. In general, three types are in common use: conventional snap, quiet, and mercury.

Dimensional data for wall switches are shown in Figure 2-168, while their corresponding wall face plates are shown in Figure 2-169.

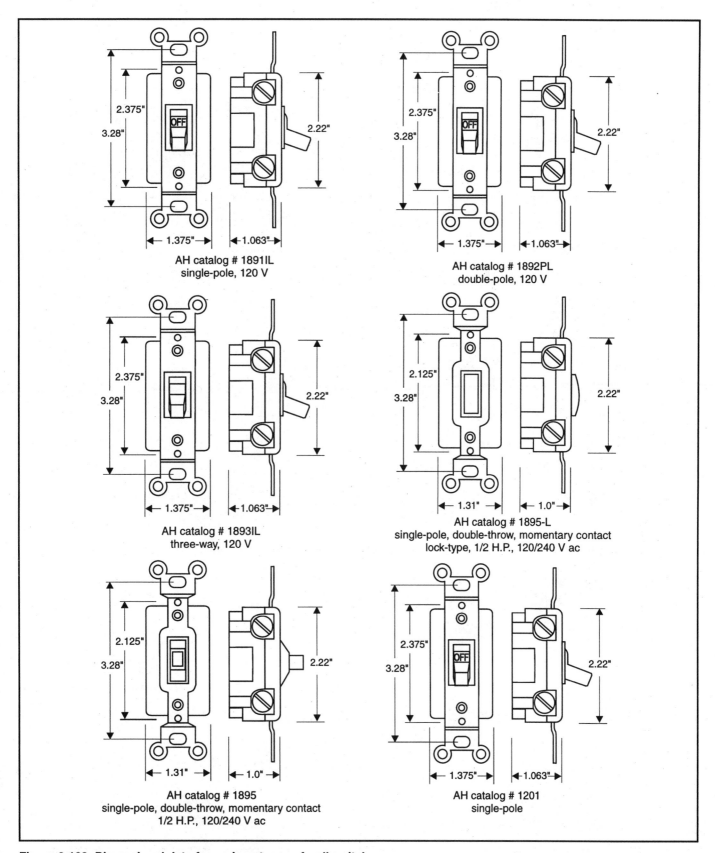

Figure 2-168: Dimensional data for various types of wall switches.

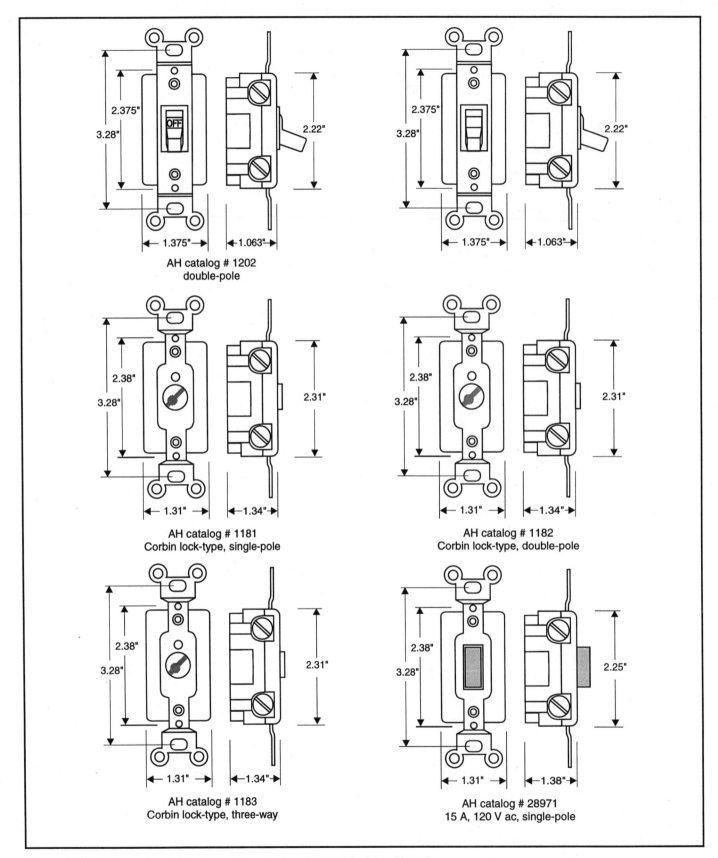

Figure 2-168: Dimensional data for various types of wall switches. *(Cont.)*

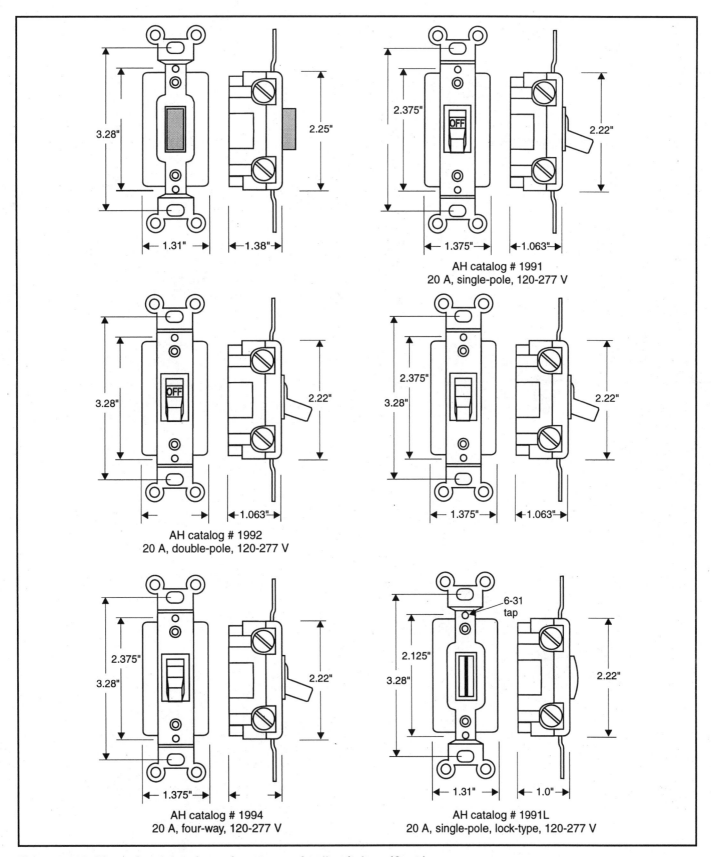

Figure 2-168: Dimensional data for various types of wall switches. *(Cont.)*

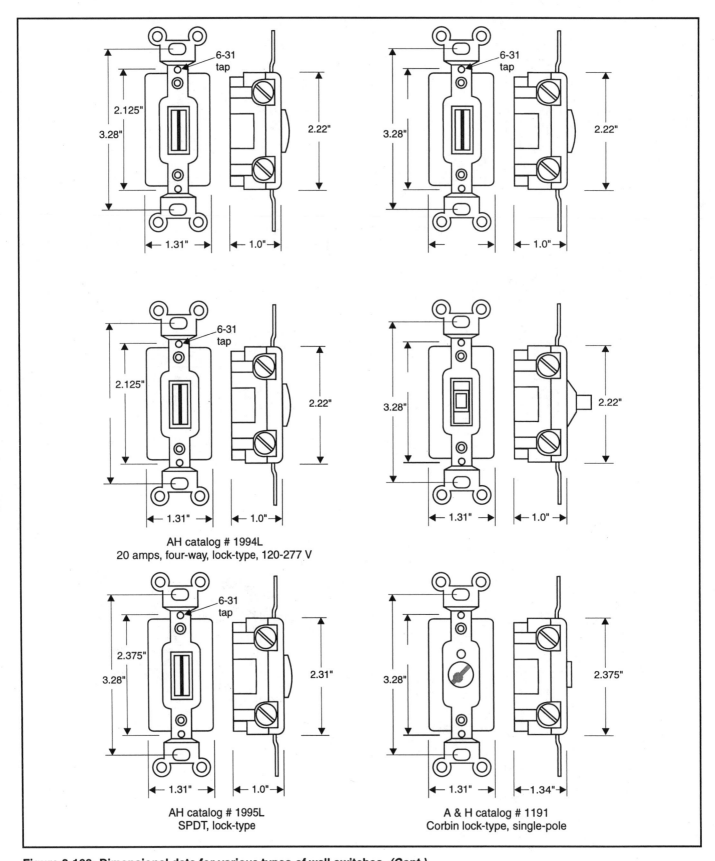

Figure 2-168: Dimensional data for various types of wall switches. *(Cont.)*

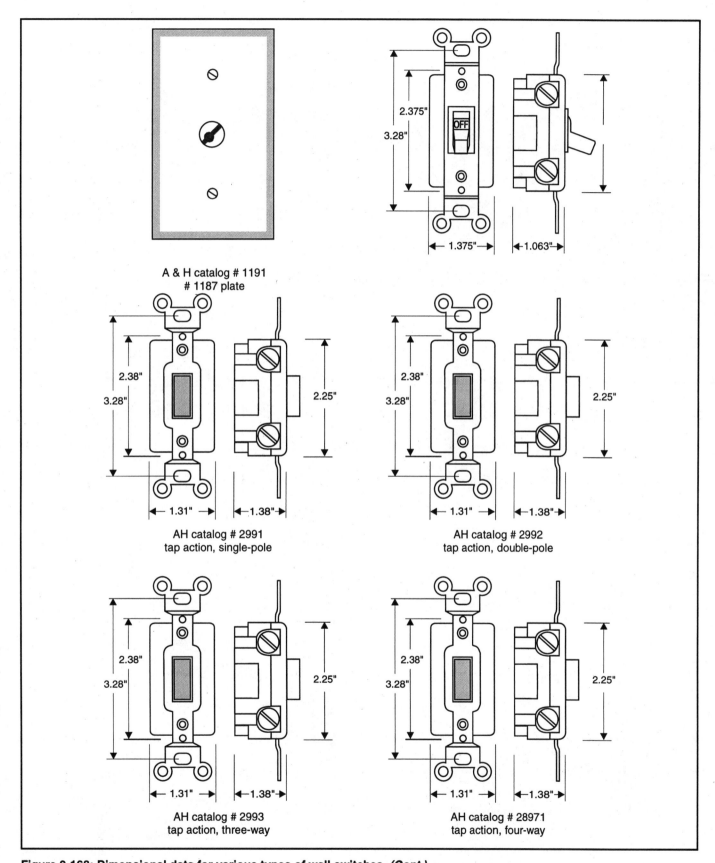

Figure 2-168: Dimensional data for various types of wall switches. *(Cont.)*

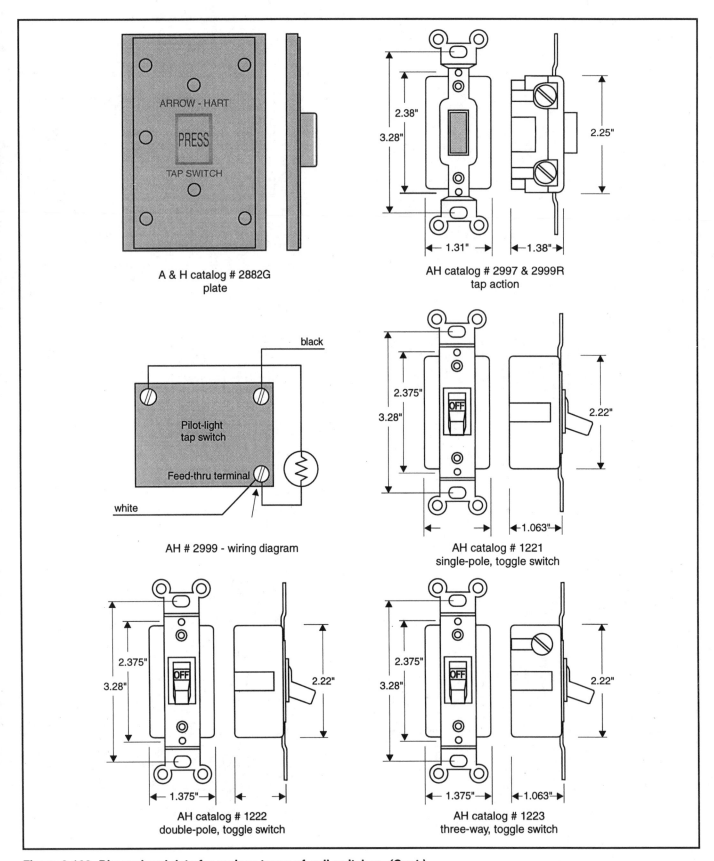

Figure 2-168: Dimensional data for various types of wall switches. *(Cont.)*

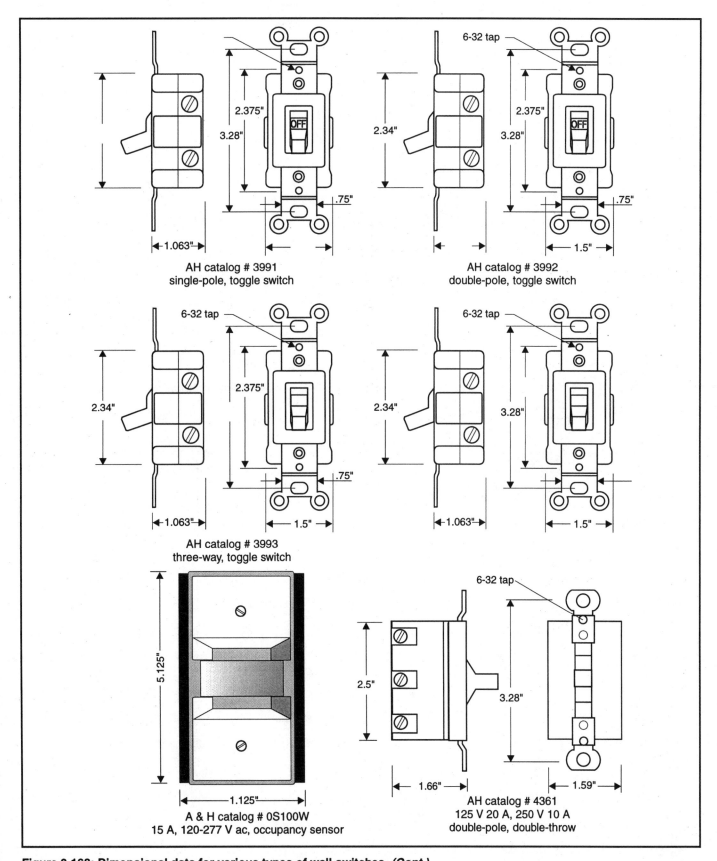

Figure 2-168: Dimensional data for various types of wall switches. *(Cont.)*

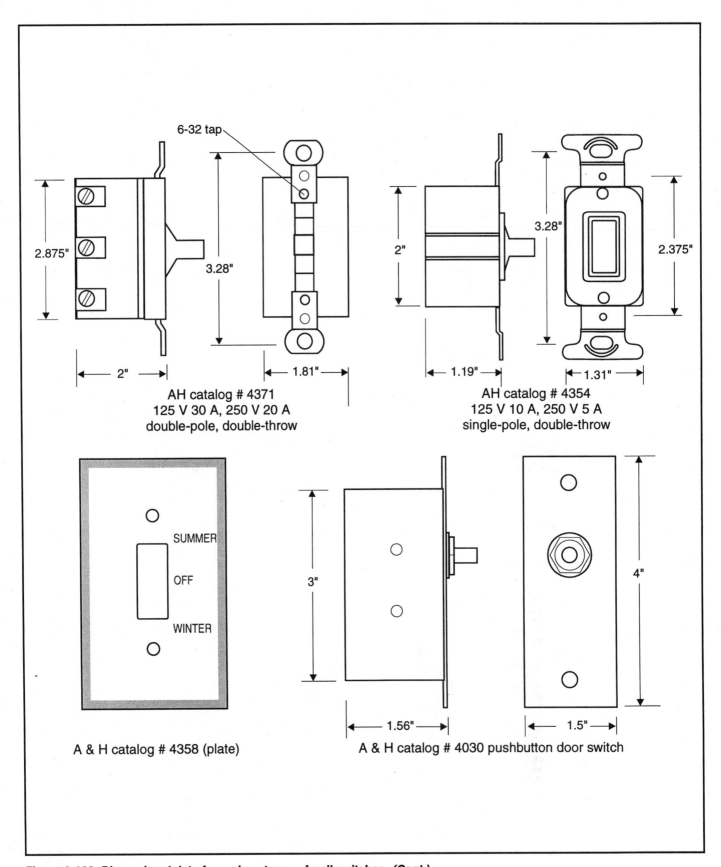

Figure 2-168: Dimensional data for various types of wall switches. *(Cont.)*

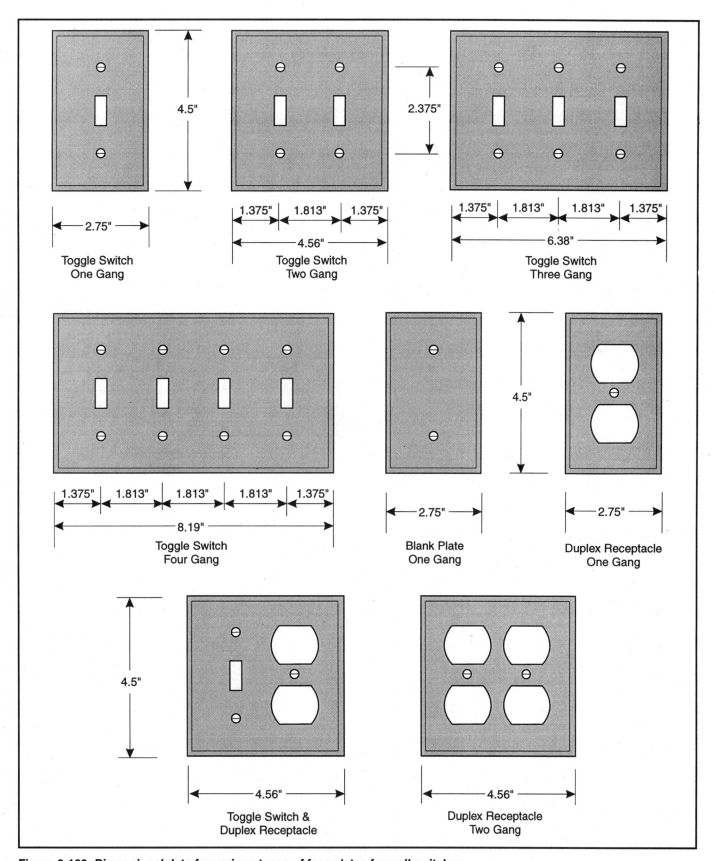

Figure 2-169: Dimensional data for various types of face plates for wall switches.

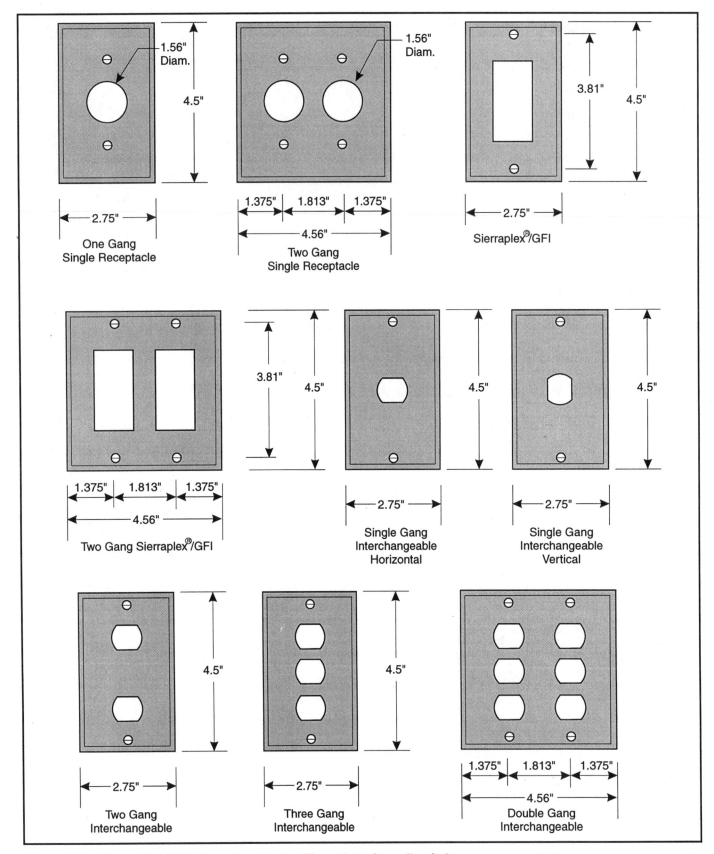

Figure 2-169: Dimensional data for various types of face plates for wall switches.

TRANSFORMERS

The electric power produced by alternators in a generating station is transmitted to locations where it is utilized and distributed to users. Many different types of transformers play an important role in the distribution of electricity. Power transformers are located at generating stations to step up the voltage for more economical transmission. Substations with additional power transformers and distribution equipment are installed along the transmission line. Finally, distribution transformers are used to step down the voltage to a level suitable for utilization.

Transformers are also used quite extensively in all types of control work, to raise and lower ac voltage on control circuits. They are also used in 480Y/277 systems to reduce the voltage for operating 208Y/120-volt lighting and other electrically-operated equipment. Buck-and-boost transformers are used for maintaining appropriate voltage levels in certain electrical systems.

It is important for anyone working in the electrical field to become familiar with transformer operation; that is, how they work, how they are connected into circuits, their practical applications and precautions to take during the installation and while working on them.

Transformer Connections — Dry Type

Electrical designs for commercial and industrial installations will more often involve dry-type transformers as opposed to oil-filled ones. Dry-type transformers are available in both single- and three-phase with a wide range of sizes from the small control transformers to those rated at 500 kVA or more. Such transformers have wide application in electrical systems of all types.

NEC Section 450-11 requires that each transformer must be provided with a nameplate giving the manufacturer; rated kVA; frequency; primary and secondary voltage; impedance of transformers 25 kVA and larger;

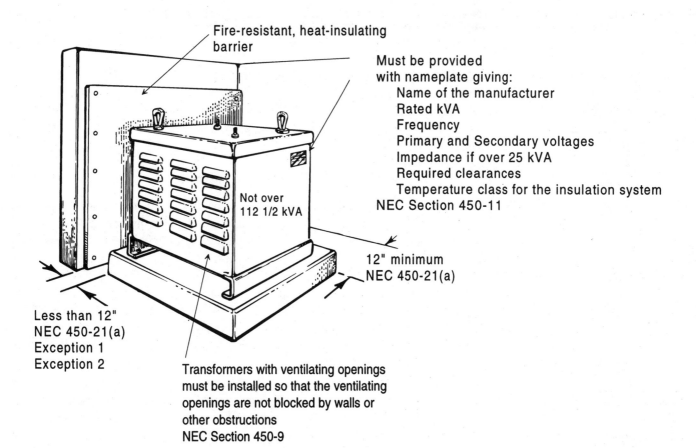

Fire-resistant, heat-insulating barrier

Must be provided
with nameplate giving:
    Name of the manufacturer
    Rated kVA
    Frequency
    Primary and Secondary voltages
    Impedance if over 25 kVA
    Required clearances
    Temperature class for the insulation system
NEC Section 450-11

Not over 112 1/2 kVA

12" minimum
NEC 450-21(a)

Less than 12"
NEC 450-21(a)
Exception 1
Exception 2

Transformers with ventilating openings must be installed so that the ventilating openings are not blocked by walls or other obstructions
NEC Section 450-9

Figure 2-170: NEC installation requirements for dry-type transformers.

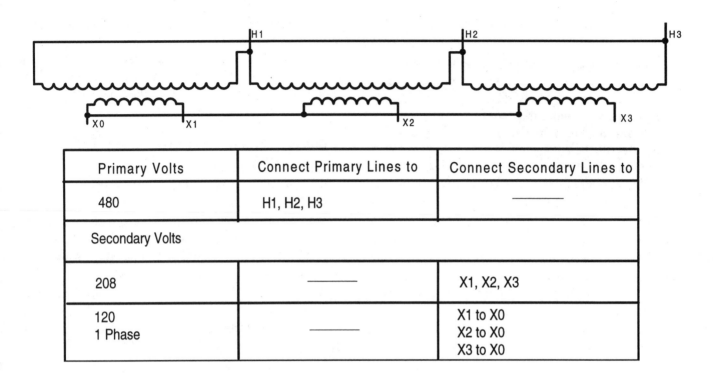

Primary Volts	Connect Primary Lines to	Connect Secondary Lines to
480	H1, H2, H3	———
Secondary Volts		
208	———	X1, X2, X3
120 1 Phase	———	X1 to X0 X2 to X0 X3 to X0

Figure 2-171: Typical transformer manufacturer's wiring diagram for delta/wye dry-type transformer.

required clearances for transformers with ventilating openings; and the amount and kind of insulating liquid where used. In addition, the nameplate of each dry-type transformer must include the temperature class for the insulation system. See Figure 2-170 on the preceding page.

In addition, most manufacturers include a wiring diagram and a connection chart as shown in Figure 2-171 for a 480-volt delta primary to 208Y/120-volt secondary. It is recommended that all transformers be connected as shown on the manufacturer's nameplate.

In general, this wiring diagram and accompanying table indicates that the 480-volt, 3-phase, 3-wire primary conductors are connected to terminals H_1, H_2, and H_3, respectively — regardless of the desired voltage on the primary. A neutral conductor, if required, is carried from the primary through the transformer to the secondary. Two variations are possible on the secondary side of this transformer: 208-volt, 3-phase, 3- or 4-wire or 120-volt, 1-phase, 2-wire. To connect the secondary side of the transformer as a 208-volt, 3-phase, 3-wire system, the secondary conductors are connected to terminals X_1, X_2, and X_3; the neutral is carried through with conductors usually terminating at a solid-neutral bus in the transformer.

Another popular dry-type transformer connection is the 480-volt primary to 240-volts delta/120 volts secondary.

This configuration is shown in Figure 2-172. Again, the primary conductors are connected to transformer terminals H_1, H_2, and H_3. The secondary connections for the desired voltages are made as indicated in the table.

Zig-Zag Connections

There are many occasions where it is desirable to upgrade a building's lighting system from 120-volt fixtures to 277-volt fluorescent lighting fixtures. Oftentimes these buildings have a 480/240-volt, three-phase, four-wire delta system. One way to obtain 277 volts from a 480/240-volt system is to connect 480/240-volt transformers in a zig-zag fashion as shown in Figure 2-173 on page 346. In doing so, the secondary of one phase is connected in series with the primary of another phase, thus changing the phase angle.

The zig-zag connection may also be used as a grounding transformer where its function is to obtain a neutral point from an ungrounded system. With a neutral being available, the system may then be grounded. When the system is grounded through the zig-zag transformer, its sole function is to pass ground current. A zig-zag transformer is essentially six impedances connected in a zig-zag configuration.

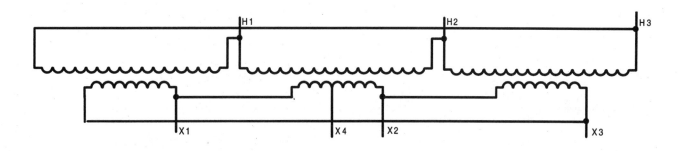

Primary Volts	Connect Primary Lines to	Connect Secondary Lines to
480	H1, H2, H3	————
Secondary Volts		
240	————	X1, X2, X3
120	————	X1, X4 or X2, X4

Figure 2-172: Typical transformer manufacturer's wiring diagram for 480 V delta to 240 V delta transformer connection.

The operation of a zig-zag transformer is slightly different from that of the conventional transformer. We will consider current rather than voltage. While a voltage rating is necessary for the connection to function, this is actually line voltage and is not transformed. It provides only exciting current for the core. The dynamic portion of the zig-zag grounding system is the fault current. To understand its function, the system must also be viewed backward; that is, the fault current will flow into the transformer through the neutral as shown in Figure 2-174.

The zero sequence currents are all in phase in each line; that is, they all hit the peak at the same time. In reviewing Figure 2-174, note that the current leaves the motor, goes to ground, flows up the neutral, and splits three ways. It then flows back down the line to the motor through the fuses which then open — shutting down the motor.

The neutral conductor will carry full fault current and must be sized accordingly. It is also time rated (0-60 seconds) and can therefore be reduced in size. This should be coordinated with the manufacturer's time/current curves for the fuse.

To determine the size of a zig-zag grounding transformer, proceed as follows:

Step 1. Calculate the system line-to-ground asymmetrical fault current.

Step 2. If relaying is present, consider reducing the fault current by installing a resistor in the neutral.

Step 3. If fuses or circuit breakers are the protective device, you may need all the fault current to quickly open the overcurrent protective devices.

Step 4. Obtain time/current curves of relay, fuses, or circuit breakers.

Step 5. Select zig-zag transformer for:

a. Fault current — the line-to-ground

b. Line-to-line voltage

c. Duration of fault (determined from time/current curves)

d. Impedance per phase at 100%; for any other, contact manufacturer

Buck-and-Boost Transformers

The buck-and-boost transformer is a very versatile unit for which a multitude of applications exist. Buck-and-

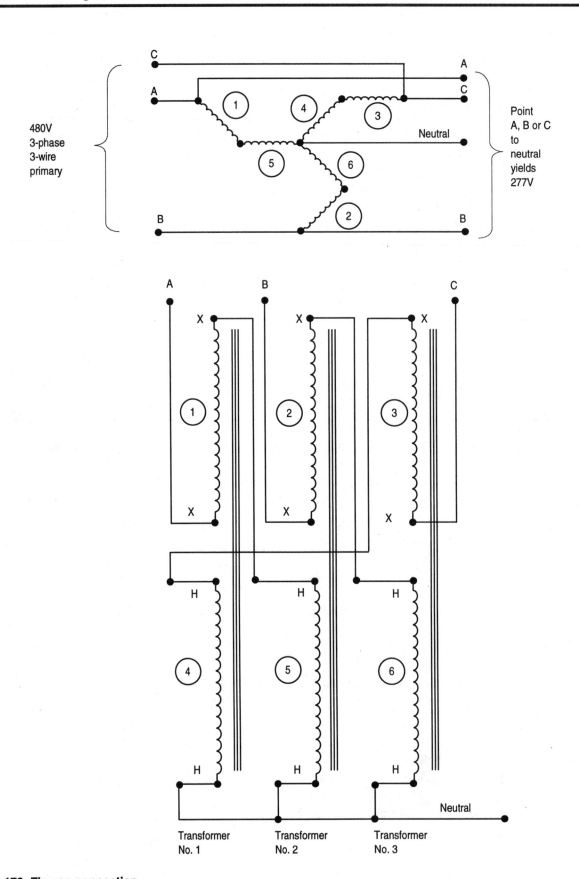

Figure 2-173: Zig-zag connection.

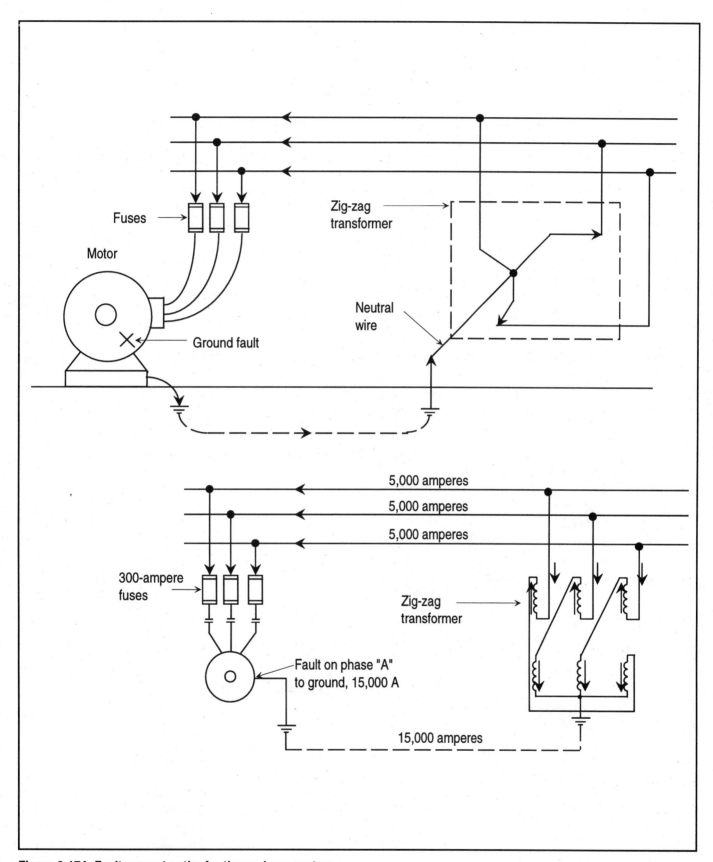

Figure 2-174: Fault-current paths for three-phase system.

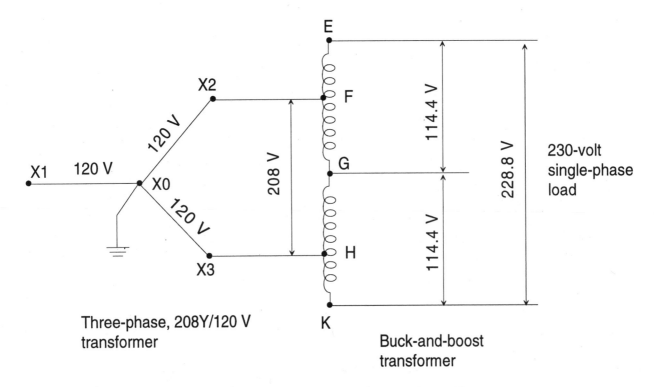

Figure 2-175: Buck-and-boost transformer connected to 208-V system to obtain 230 V.

boost transformers, as the name implies, is designed to raise (boost) or lower (buck) the voltage in an electrical system or circuit. In their simplest form, these insulated units will deliver 12 or 24 volts when the primaries are energized at 120 or 240 volts respectively. Their prime use and value, however, lies in the fact that the primaries and the secondaries can be interconnected — permitting their use as an autotransformer.

Let's assume that an installation is supplied with 208Y/120V service, but one piece of equipment in the installation is rated for 230 volts. A buck-and-boost transformer may be used on the 230-volt circuit to increase the voltage from 208 volts to 230 volts. See Figure 2-175. With this connection, the transformer is in the "boost" mode and delivers 228.8 volts at the load. This is close enough to 230 volts that the load equipment will function properly.

If the connections were reversed, this would also reverse the polarity of the secondary with the result that a voltage would be 208 volts minus 20.8 volts = 187.2 volts. The transformer is now operating in the "buck" mode.

Transformer connections for typical three-phase buck-and-boost open-delta transformers are shown in Figure 2-

176. The connections shown are in the "boost" mode; to convert to "buck" mode, reverse the input and output.

Another three-phase buck-and-boost transformer connection is shown in Figure 2-177; this time wye-connected. While the open-delta transformers (Figure 2-176) can be converted from buck to boost or vice-versa by reversing the input/output connections, this is not the case with the three-phase, wye-connected transformer. The connection shown (Figure 2-177) is for the boost mode only.

Several typical single-phase buck-and-boost transformer connections are shown in Figure 178. Other diagrams may be found on the transformer's nameplate or with packing instructions that come with each new transformer.

Control Transformers

Control transformers are available in numerous types, but most control transformers are dry-type step-down units with the secondary control circuit isolated from the primary line circuit to assure maximum safety. *See* Figure 2-179. These transformers and other components are usually mounted within an enclosed control box or control

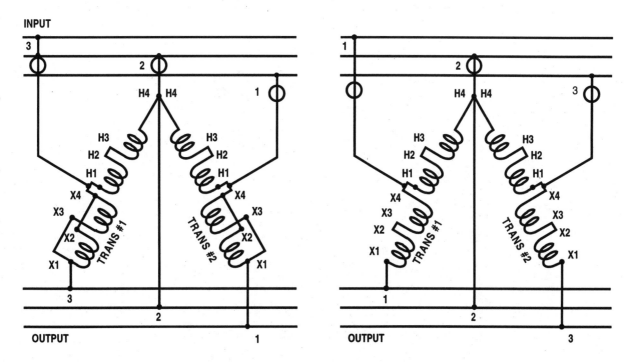

Figure 2-176: Open delta, three-phase, buck-and-boost transformer connections.

panel, which has a pushbutton station or stations independently grounded as recommended by the *NEC*. Industrial control transformers are especially designed to

INPUT ONLY

NEUTRAL

TRANS #1   TRANS #2

OUTPUT ONLY

Figure 2-177: Three-phase, wye-connected buck-and-boost transformer in the "boost" mode.

accommodate the momentary current inrush caused when electromagnetic components are energized, without sacrificing secondary voltage stability beyond practical limits. *See* NEC Section 470-32.

Other types of control transformers, sometimes referred to as control and signal transformers, normally do not have the required industrial control transformer regulation characteristics. Rather, they are constant-potential, self-air-cooled transformers used for the purpose of supplying the proper reduced voltage for control circuits of electrically operated switches or other equipment and, of course, for signal circuits. Some are of the open type with no protective casing over the winding, while others are enclosed with a metal casing over the winding.

In seeking control transformers for any application, the loads must be calculated and completely analyzed before the proper transformer selection can be made. This analysis involves every electrically energized component in the control circuit. To select an appropriate control transformer, first determine the voltage and frequency of the supply circuit. Then determine the total inrush volt-amperes (watts) of the control circuit. In doing so, do not neglect the current requirements of indicating lights and timing devices that do not have inrush volt-amperes, but are energized at the same time as the other components in

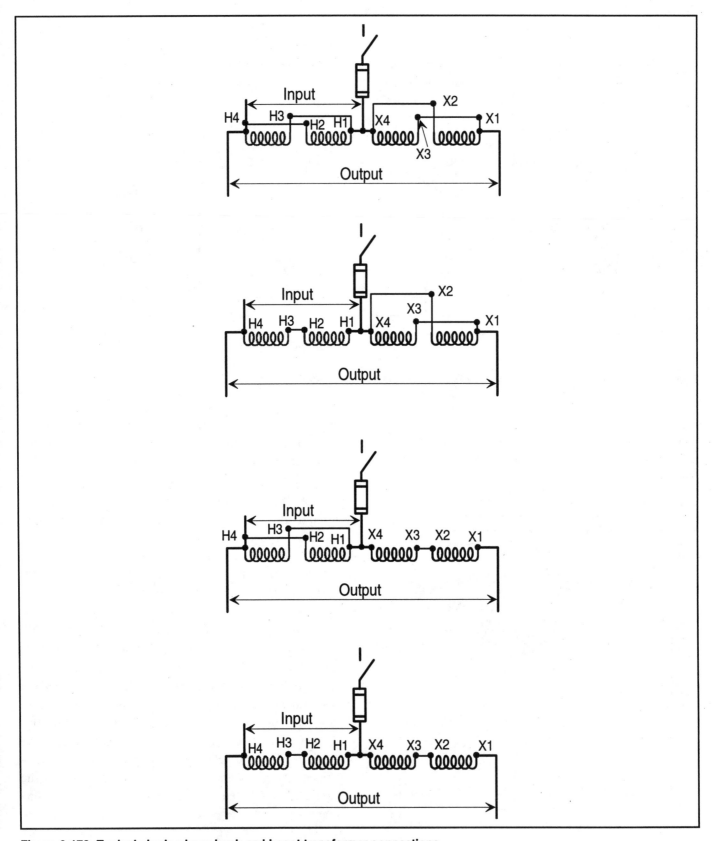

Figure 2-178: Typical single-phase buck-and-boost transformer connections.

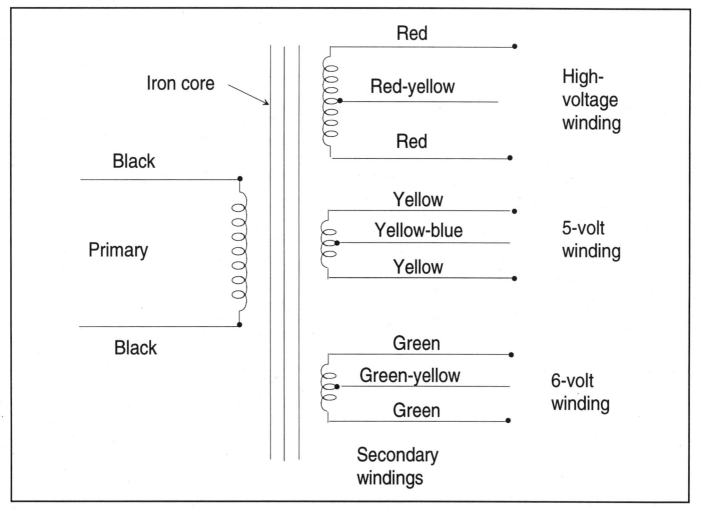

Figure 2-179: Typical control-transformer wiring diagram.

the circuit. Their total volt-amperes should be added to the total inrush volt-amperes.

Practical applications of control tranformers are shown in Figure 2-180 through 2-183.

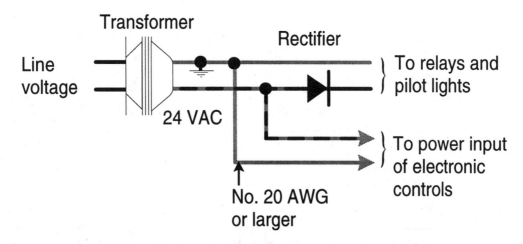

Figure 2-180: Power supply wiring diagram showing connection of 24 Vac control transformer.

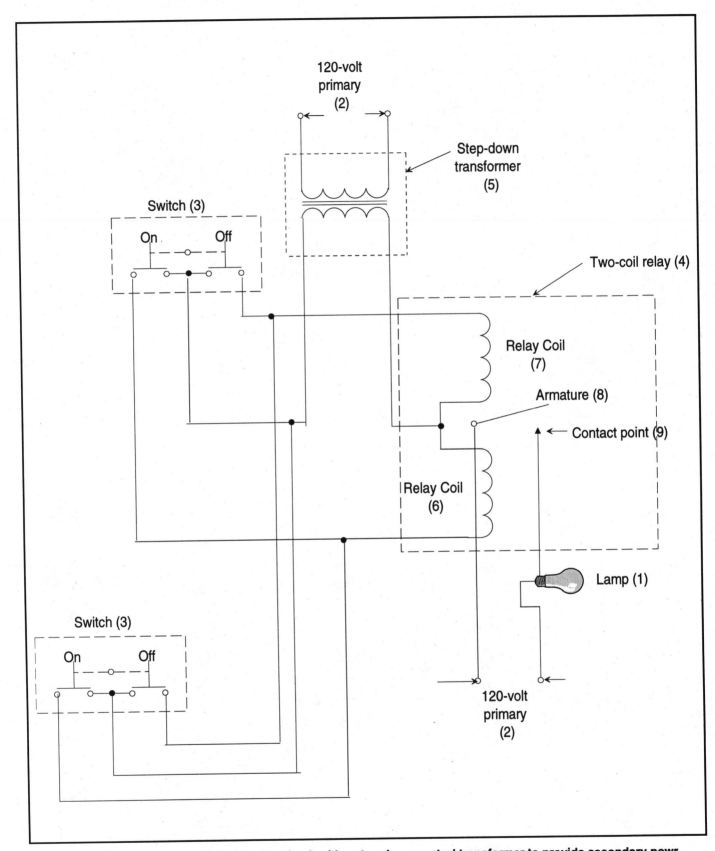

Figure 2-181: Basic remote-control switching circuit with a step-down control transformer to provide secondary powr.

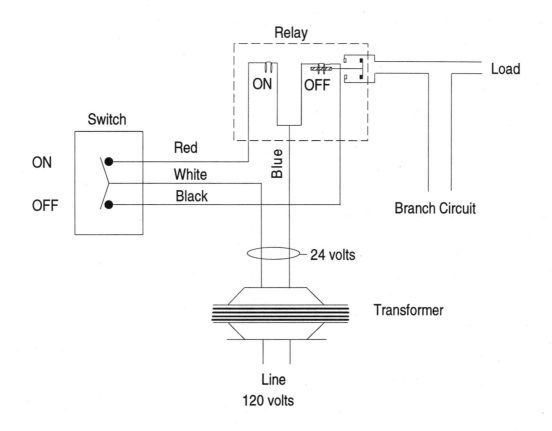

Figure 2-182: Wiring details of a remote-control relay with its related control transformer connection.

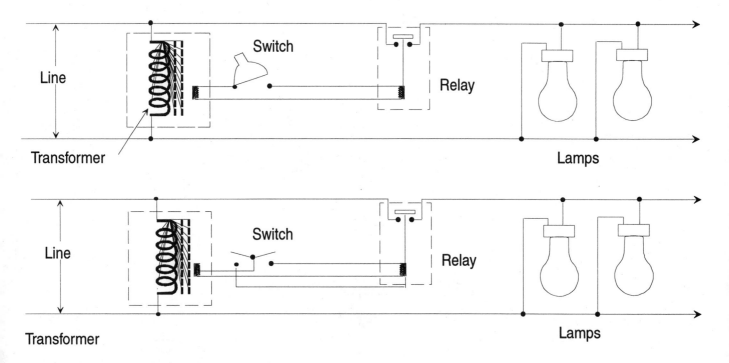

Figure 2-183: Two types of remote-control circuits and related components, including control transformers.

WIRE SPLICES AND CONNECTIONS

Splices and connections that are properly made will often last as long as the original insulation on the wire itself, while poorly made connections will always be a source of trouble; that is, the joints will overheat under load and cause a higher resistance than normal in the the circuit.

The basic requirements for a good electrical connection include the following:

- It should be mechanically and electrically secure.
- It should be insulated as well as or better than the existing insulation on conductors.
- The above characteristics should last as long as the conductor is in service.

There are many different types of electrical joints for different purposes, and the selection of the proper type for a given application will depend to a great extent on how and where the splice or connection is used.

Electrical joints are normally made with solderless pressure connectors or lugs to save time, a knowledge of the traditional splices will enable designers to approach their work with more authority.

Stripping and Cleaning

Before any connection or splice can be made, the ends of the conductors must be properly stripped and cleaned. Poorly stripped wire can result in nicks, scrapes, or burnishes. Any of these can lead to a stress concentration at the damaged cross section. Heat, rapid temperature change, mechanical vibration, and oscillatory motion can aggravate the damage, causing faults in the circuitry or even total failure. Consequently, electrical designers should clearly specify the methods used to strip and clean conductors.

Wire Sizes

It is a common mistake to believe that a certain gauge of stranded conductor has the same diameter as a solid conductor. This is a very important consideration in selecting proper blades for strippers. The table in Figure 2-184 shows the nominal sizes referenced for the different wire gauges.

Heat-Shrink Insulators

Heat-shrinkable insulators for small connectors provide skintight insulation protection and are fast and easy to use. They are designed to slip over wires, taper pins, con-

DIMENSIONS OF COMMON WIRE SIZES			
Size, AWG/kcmil	Area Circular Mils	Overall Diameter in Inches	
		Solid	Stranded
18	1620	0.040	0.046
16	2580	0.051	0.058
14	4130	0.064	0.073
12	6530	0.081	0.092
10	10380	0.102	0.116
8	16510	0.128	0.146
6	26240	-	0.184
4	41740	-	0.232
3	52620	-	0.260
2	66360	-	0.292
1	83690	-	0.332
1/0	105600	-	0.373
2/0	133100	-	0.419
3/0	167800	-	0.470

Figure 2-184: Dimensions of common wire sizes.

DIMENSIONS OF COMMON WIRE SIZES *(Cont.)*			
Size, AWG/kcmil	**Area Circular Mils**	**Overall Diameter in Inches**	
		Solid	**Stranded**
4/0	211,600	—	0.528
250	250,000	—	0.575
300	300,000	—	0.630
350	350,000	—	0.681
400	400,00	—	0.728
500	500,000	—	0.813
600	600,000	—	0.893
700	700,000	—	0.964
750	750,000	—	0.998
800	800,000	—	1.03
900	900,000	—	1.09
1000	1,000,000	—	1.15
1250	1,250,000	—	1.29
1500	1,500,000	—	1.41
1750	1,750,000	—	1.52
2000	2,000,000	—	1.63

Figure 2-184: Dimensions of common wire sizes. *(Cont.)*

nectors, terminals, and splices. When heat is applied, the insulation becomes semirigid and will provide positive strain relief at the flex point of the conductor. A vapor-proof bands seals and protects the conductor from abrasion, chemicals, dust gasoline, oil, and moisture. Extreme temperatures, both hot and cold, will not affect the performance of these insulators. A tubing selector guide appears in Figure 2-185. The types of materials used for these insulators are as follows:

PVC: This type is a general-purpose, economical tubing that is widely used in the electronics industry. The PVC compound is irradiated by being bombarded with high-velocity electrons. This results in a denser, cross-linked material with superior electrical and mechanical properties. It also assures that the tubing will resist cracking and splitting.

Polyolefin: Polyolefin tubing has a wide range of uses for wire bundling, harnessing, strain relief, and other applications where cables and components require addi-

tional insulation. It is irradiated, flame retarding, flexible, and comes in a wide variety of colors.

Double wall: This type is available and designed for outstanding protective characteristics. It is semirigid tubing that has an inner wall that actually melts and an outer wall that shrinks to conform to the melted area.

Teflon®: This type is considered by many users to be the best overall heat-shrinkable tubing — physically, electrically, and chemically. Its high temperature rating of 250°C resists brittleness or loss of translucency from extended exposure to high heat and will not support combustion.

Neoprene: Components that warrant extra protection from abrasion require a highly durable, yet flexible, tubing. The irradiated neoprene tubing offers this optimal coverage.

Kynar®: Irradiated Kynar is a thin-wall, semirigid tubsparent tubing enables easy inspection of components that are covered and retains its properties at its rated temperature.

		HEAT-SHRINK INSULATOR TUBING SELECTOR GUIDE						
Manhattan Number	Type	Material	Temperature Range, °C	Shrink Ratio	Max. Long. Shrinkage %	Tensile Strength, psi	Colors	Dielectric Strength, V/mil
MT-105	Nonshrinkable	PVC	+105	—	—	2700	White, red, clear, black	800
MT-150	Shrinkable	PVC	-35 to +105	2:1	10	2700	Clear, black	750
MT-200	Nonshrinkable	Teflon®	-65 to +260	—	—	2700	Clear	1400
MT-221	Shrinkable	Flexible polyolefin	-55 to +135	2:1	5	2500	Black, white, red, yellow, blue, clear	1300
MT-250	Nonshrinkable	Teflon®	-65 to +260	—	—	7500	Clear	1400
MT-300	Shrinkable	Polyolefin double wall	-55 to +110	6:1	5	2500	Black	1100
MT-350	Shrinkable	Kynar®	-55 to +175	2:1	10	8000	Clear	1500
MT-400	Shrinkable	Teflon®	+250	1.2:2	10	6000	Clear	1500
MT-500	Shrinkable	Teflon®	+250	11/2:1	10	6000	Clear	1500
MT-600	Shrinkable	Neoprene	+120	2:1	10	1500	Black	300

Figure 2-185: Tubing selector guide.

Most tubing is available in a wide variety of colors and put-ups. Tubing selector guides are available from manufacturers than can help in the selection of the best tubing for any given application.

When specifying heat-shrinkable tubing, the manufacturer, and material substance are usually all that is necessary. Installers will choose the correct size for the wire.

Wire Nuts

Since their invention in 1927, wire nuts (Figure 2-186) have been a favorite wire connector for use on residential and commercial branch-circuit applications. Several varieties of wire nuts are available, but the following are the most common.

- Those designed for use on wiring systems 300 V and under.
- Those designed for use on wiring systems 600 V and under (1000 V in lighting fixtures and signs).

Most brands are UL listed for aluminum to copper in dry locations only; aluminum to aluminum only; or copper to copper only; 600 V maximum; 1000 V in lighting fix-

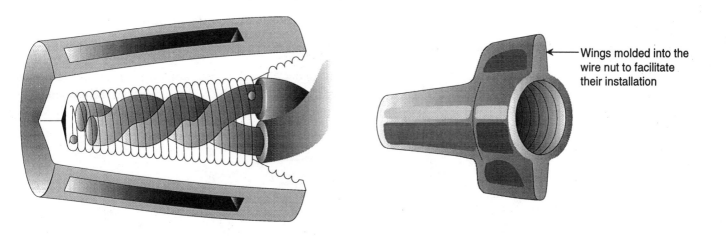

Figure 2-186: Typical wire nuts for connecting conductors up to 600 V (1000 V in lighting fixtures and signs).

tures and neon sighs. The maximum temperature rating is 105°C (221°F).

Wire nuts are frequently used for all types of splices in residential and commercial applications and are considered to be the fastest connectors on the market for this type of work. They are made in sizes to accommodate conductors as small as No. 22 AWG up to as large as No. 10 AWG, with practically any combination of thoses sizes in between.

The model numbers of wire nuts will vary with the manufacturer, but Ideal set a standard years ago with their 71B through 76B series. The table in Figure 2-187 gives the dimensions of this series.

High Voltage Terminations

High voltage is normally considered to be any voltage over 600 V nominal. Consequently, voltage ranges can vary from 601 V to 170,000 V or more.

Although any conductor splice or termination must be done with care, extreme caution must be exercised with the higher voltages. Figure 2-188 on the next pageshows the basic steps for making a basic high-voltage splice.

In general, the higher the voltage, the more fine-tuning required of the splice or termination. Every splice must provide mechanical and electrical integrity at least equal to the original conductor insulation. Furthermore, all high-

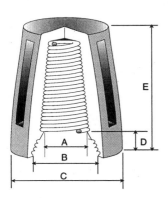

Model	A	B	C	D	E
71B	$\frac{1}{8}$	$\frac{1}{4}$	$\frac{21}{64}$	$\frac{11}{64}$	$\frac{37}{64}$
72B	$\frac{9}{64}$	$\frac{9}{32}$	$\frac{25}{64}$	$\frac{11}{64}$	$\frac{45}{64}$
73B	$\frac{9}{64}$	$\frac{11}{32}$	$\frac{7}{16}$	$\frac{5}{16}$	$\frac{55}{64}$
74B	$\frac{11}{64}$	$\frac{7}{16}$	$\frac{35}{64}$	$\frac{17}{64}$	$\frac{61}{64}$
76B	$\frac{1}{4}$	$\frac{17}{32}$	$\frac{21}{32}$	$\frac{17}{64}$	$1\frac{1}{16}$

Figure 2-187: Dimensions of wire nuts.

voltage terminations and splices must be made under the best environmental conditions possible on the jobsite. The slightest amount of moisture, dirt, and similar foreign material can render a high-voltage termination or splice useless.

High Voltage Cable Components

Despite the visible differences, all power cables are essentially the same, consisting of the components shown in Figure 2-189.

Conductors used with modern solid dielectric cables are available in four basic configurations; all are shown in Figure 2-190. High-voltage splicing details are shown in Figure 2-191 through 2-195.

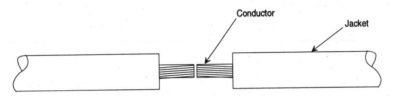

A. Typical straight splice for a single (unshielded) conductor.

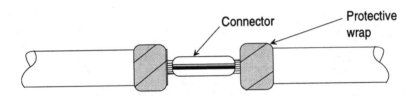

B. Apply a protective wrap of friction tape to the jacket ends to protect the jacket and insulation while cleaning and securing the connector.

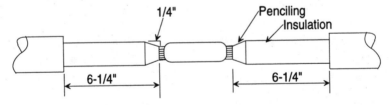

C. Remove additional jacket and insulation an amount equal to 25 times the thickness of the overall insulation.

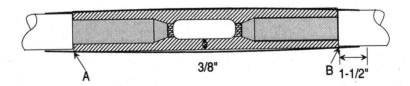

D. Method of insulating and taping the splice.

Figure 2-188: Basic steps for making a high-voltage splice.

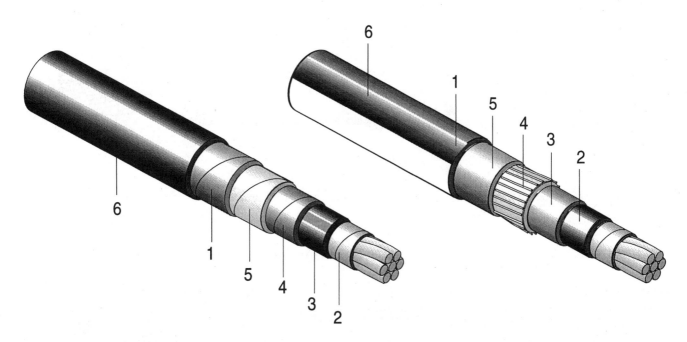

1. STRAND SHIELDING   Semi-conductive material
2. PRIMARY INSULATION   Butyl rubber, Ethylene Propylene, rubber, low and high
                                  density Polyethylene, cross-linked Polyethylene
3. SHIELD BEDDING   Semi-conductive material
4. CABLE SHIELDING   Metallic or wire
5. CABLE BEDDING   Tape or fibrous material
6. JACKET   Butyl rubber, Neoprene, PVC, low and high density
                                  Polyethylene, cross-linked Polyethylene

Figure 2-189: High-voltage cable specifications.

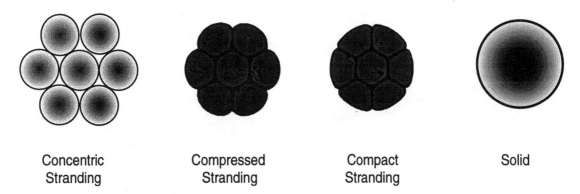

Concentric Stranding   Compressed Stranding   Compact Stranding   Solid

Figure 2-190: Four basic high-voltage cable configurations.

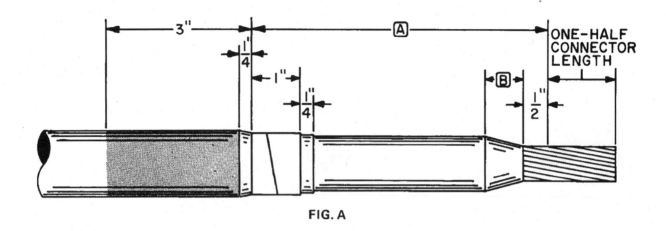

FIG. A

Figure 2-191: Cable preparation detail for inline splice. Dimensions A and B will vary with the cable size.

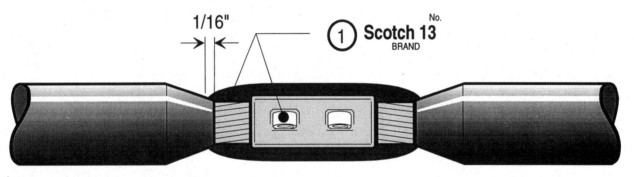

Figure 2-192: Connection detail for inline splice.

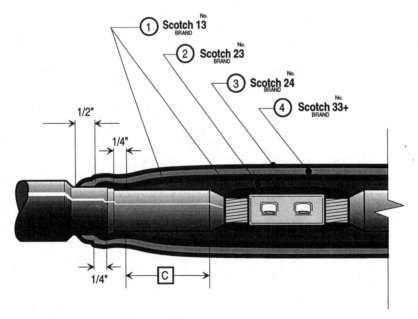

Figure 2-193: Primary Insulation detail for inline splice.

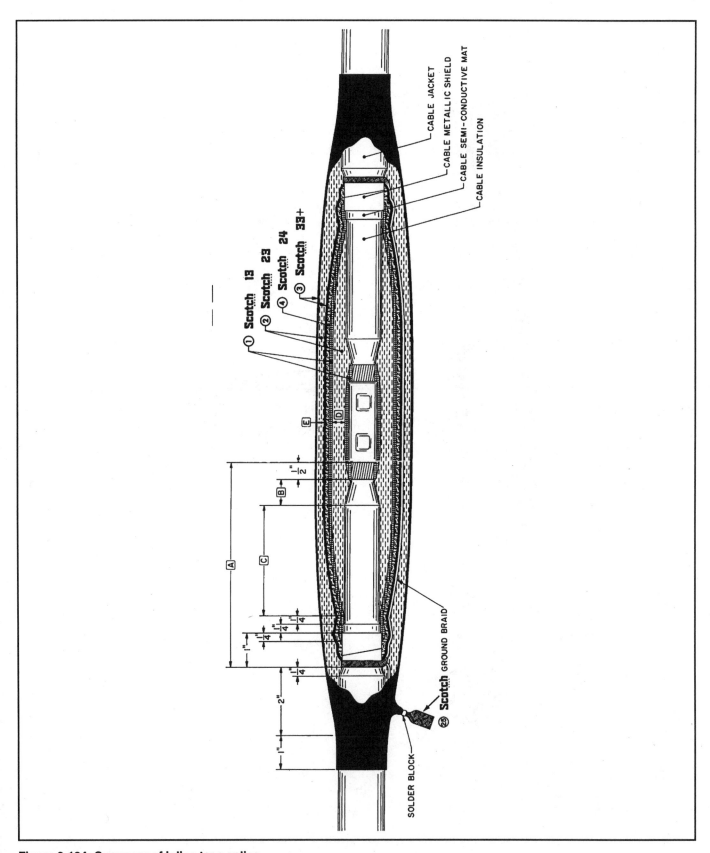

Figure 2-194: Summary of inline tape splice.

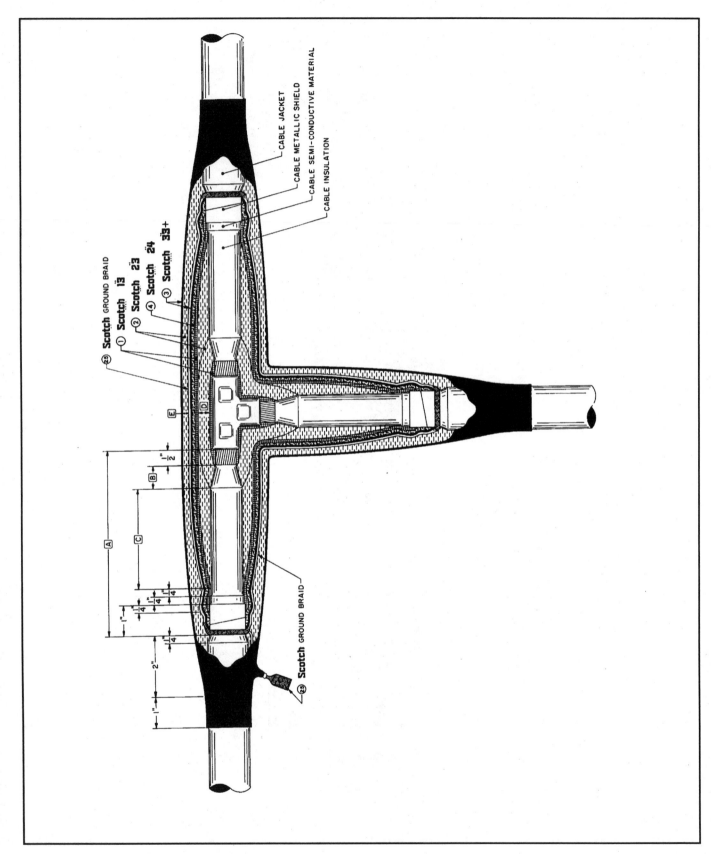

Figure 2-195: Summary of tee tape splice.

ENGINEERING DATA

Mechanical Characteristics

To Find	Use Equation
Torque in Pound Feet	$\dfrac{HP \times 5250}{RPM}$
Horsepower	$\dfrac{Torque \times RPM}{5250}$
RPM	$\dfrac{120 \times Frequency}{Number\ of\ Poles}$

I = amperes
Eff = efficiency
PF = power factor
RPM = revolutions per minute

E = volts
kW = kilowatts
HP = horsepower
kVA = kilovolt amperes

Electrical Characteristics

To Find:	Single-Phase Equation	Three-Phase Equation
Amperes when hp is known	$\dfrac{hp \times 746}{E \times Eff \times PF}$	$\dfrac{hp \times 746}{1.73 \times E \times Eff \times PF}$
Amperes when kW is known	$\dfrac{kW \times 1000}{E \times PF}$	$\dfrac{kW \times 1000}{1.73 \times E \times PF}$
Amperes when kVA is known	$\dfrac{kVA \times 1000}{E}$	$\dfrac{kVA \times 1000}{1.73 \times E}$
Kilowatts	$\dfrac{I \times E \times PF}{1000}$	$\dfrac{1.73 \times I \times E \times PF}{1000}$
kVA	$\dfrac{I \times E}{1000}$	$\dfrac{1.73 \times I \times E}{1000}$
HP (output)	$\dfrac{I \times E \times Eff \times PF}{746}$	$\dfrac{1.73 \times I \times E \times Eff \times PF}{746}$

CONVERSION FACTORS

Multiply	by	To Obtain
acres	43,560	square feet
acres	4047	square meters
acres	1.562×10^{-3}	square miles
acres	5645.38	square varas
acres	4840	square yards
amperes	0.11	abamperes
atmospheres	76.0	cm of mercury
atmospheres	29.92	inches of mercury
atmospheres	33.90	feet of water
atmospheres	10.333	kg per sq. meter
atmospheres	14.70	pounds per sq. inch
atmospheres	1.058	tons per sq. foot
British thermal units	0.2520	kilogram-calories
British thermal units	777.5	foot-pounds
British thermal units	3.927×10^{-4}	horsepower-hours
British thermal units	1054	joules
British thermal units	107.5	kilogram-meters
British thermal units	2.928×10^{-4}	kilowatt-hours

Multiply	by	To Obtain
Btu per min.	12.96	foot-pounds per sec.
Btu per min.	0.02356	horsepower
Btu per min.	0.01757	kilowatts
Btu per min.	17.57	watts
Btu per sq. ft./Min.	0.1220	watts per sq. inch
bushels	1.244	cubic feet
bushels	2150	cubic inches
bushels	0.03524	cubic meters
bushels	4	pecks
bushels	64	pints(dry)
bushels	32	quarts(dry)
centimeters	0.3397	inches
centimeters	0.01	meters
centimeters	393.7	mils
centimeters	10	millimeters
centimeter-grams	980.7	centimeter-dynes
centimeter-grams	10^{-5}	meter-kilograms
centimeter-grams	7.233×10^{-5}	pound-feet
centimeters of mercury	0.01316	atmospheres
centimeters of mercury	0.4461	feet of water
centimeters of mercury	136.0	kg per sq. meter
centimeters of mercury	27.85	pounds per sq. meter
centimeters of mercury	0.1934	pounds per sq. inch
centimeters per second	1.969	feet per minute
centimeters per second	0.03281	feet per second
centimeters per second	0.036	kilometers per hour
centimeters per second	0.6	meters per minute
centimeters per second	0.02237	miles per hour
centimeters per second	3.728×10^{-4}	miles per minute
cubic centimeters	3.531×10^{-5}	cubic feet
cubic centimeters	6.102×10^{-2}	cubic inches
cubic centimeters	10	cubic meters
cubic centimeters	1.308×10^{-6}	cubic yards
cubic centimeters	2.642×10^{-4}	gallons

Multiply	by	To Obtain
cubic centimeters	10	liters
cubic centimeters	2.113×10^{-3}	pints(liq)
cubic centimeters	1.057×10^{-3}	quarts(liq)
cubic feet	62.43	pounds of water
cubic feet	2.832×10^{4}	cubic cm
cubic feet	1728	cubic inches
cubic feet	0.02832	cubic meters
cubic feet	0.03704	cubic yards
cubic feet	7.481	gallons
cubic feet	28.32	liters
cubic feet	59.84	pints (liq)
cubic feet	29.92	quarts (liq)
cubic feet per minute	472.0	cubic cm per sec.
cubic feet per minute	0.1247	gallons per sec.
cubic feet per minute	0.4720	liters per sec.
cubic feet per minute	62.4	lb of water per min.
cubic inches	16.39	cubic centimeters
cubic inches	5.787×10^{-4}	cubic feet
cubic inches	1.639×10^{-5}	cubic meters
cubic inches	2.143×10^{-5}	cubic yards
cubic inches	4.329×10^{-3}	gallons
cubic inches	1.639×10^{-2}	liters
cubic inches	0.03463	pints(liq)
cubic inches	0.01732	quarts(liq)
cubic yards	7.646×10^{5}	cubic centimeters
cubic yards	27	cubic feet
cubic yards	46,656	cubic inches
cubic yards	0.7646	cubic meters
cubic yards	202.0	gallons
cubic yards	764.6	liters
cubic yards	1616	pints(liq)
cubic yards	807.9	quarts(liq)
cubic yards per minute	0.45	cubic feet per sec.
cubic yards per minute	3.367	gallons per second
cubic yards per minute	12.74	liters per second

Multiply	by	To Obtain
degrees (angle)	60	minutes
degrees (angle)	0.01745	radians
degrees (angle)	3600	seconds
dynes	1.020×10^{-3}	grams
dynes	7.233×10^{-5}	poundals
dynes	2.248×10^{-6}	pounds
ergs	9.486×10^{-11}	kilograms
ergs	1	dyne-centimeters
ergs	7.376×10^{-8}	foot-pounds
ergs	1.020×10^{-3}	gram-centimeters
ergs	10^{-7}	joules
ergs	2.390×10^{-11}	kilogram-calories
ergs	1.020×10^{-8}	kilogram-meters
feet	12	inches
feet	0.3048	meters
feet	0.36	varas
feet	1/3	yards
feet of water	0.02950	atmospheres
feet of water	0.8826	inches of mercury
feet of water	304.8	kg per sq. meter
feet of water	62.43	pounds per sq. ft.
feet of water	0.4335	pounds per sq. inch
foot-pounds	1.286×10^{-3}	British thermal units
foot-pounds	1.356×10^{7}	ergs
foot-pounds	5.050×10^{-7}	horsepower hours
foot-pounds	1.356	joules
foot-pounds	3.241×10^{-4}	kilogram-calories
foot-pounds	0.1383	kilogram-meters
foot-pounds	3.766×10^{-7}	kilowatt-hours
foot-pounds per minute	1.286×10^{-3}	Btu per minute
foot-pounds per minute	0.01667	foot pounds per sec.
foot-pounds per minute	3.030×10^{-5}	horsepower
foot-pounds per minute	3.241×10^{-4}	kg-calories per min.
foot-pounds per minute	2.260×10^{-5}	kilowatts
foot-pounds per sec.	7.717×10^{-2}	Btu per minute

Multiply	by	To Obtain
foot-pounds per sec.	1.818×10^{-3}	horsepower
foot-pounds per sec.	1.945×10^{-2}	kg-calories per min.
foot-pounds per sec.	1.356×10^{-3}	kilowatts
gallons	8.345	pounds of water
gallons	3785	cubic centimeters
gallons	0.1337	cubic feet
gallons	231	cubic inches
gallons	3.785×10^{-3}	cubic meters
gallons	4.951×10^{-3}	cubic yards
gallons	3.785	liters
gallons	8	pints (liq)
gallons	4	quarts (liq)
gallons per minute	2.228×10^{-3}	cubic ft per sec.
gallons per minute	0.06308	liters per second
grains (troy)	1	grains (av)
grains (troy)	0.06480	grams
grains (troy)	0.04167	pennyweights (troy)
grams	980.7	dynes
grams	15.43	grains (troy)
grams	10^{-3}	kilograms
grams	10^{3}	milligrams
grams	0.03527	ounces
grams	0.03215	ounces (troy)
grams	0.07093	poundals
grams	2.205×10^{-3}	pounds
horsepower	42.44	Btu per min
horsepower	33,000	foot-pounds per min.
horsepower	550	foot-pounds per sec.
horsepower	1.014	horsepower (metric)
horsepower	10.70	kg-calories per min.
horsepower	0.7457	kilowatts
horsepower	7.457	watts
horsepower (boiler)	33,520	Btu per hour
horsepower (boiler)	9,804	kilowatts
horsepower-hours	2547	British thermal units

Multiply	by	To Obtain
horsepower-hours	1.98×10^6	foot-pounds
horsepower-hours	2.684×10^6	joules
horsepower-hours	641.7	kilogram-calories
horsepower-hours	2.737×10^5	kilogram-meters
horsepower-hours	0.7457	kilowatt-hours
inches	2.540	centimeters
inches	10^3	mils
inches	0.03	varas
inches of mercury	0.03342	atmospheres
inches of mercury	1.133	feet of water
inches of mercury	345.3	kg per sq meter
inches of mercury	70.73	pounds per sq ft.
inches of mercury	0.4912	pounds per sq in.
inches of water	0.07355	inches of mercury
inches of water	25.40	kg per sq. meter
inches of water	0.5781	ounces per sq in.
inches of water	5.204	pounds per sq ft.
inches of water	0.03613	pounds per sq in.
kilograms	980,665	dynes
kilograms	10^3	grams
kilograms	70.93	poundals
kilograms	2.2046	pounds
kilograms	1.102×10^{-3}	tons (short)
kilogram-calories	3.968	British thermal units
kilogram-calories	3086	foot-pounds
kilogram-calories	1.558×10^{-3}	horsepower-hours
kilogram-calories	4183	joules
kilogram-calories	426.6	kilogram-meters
kilogram-calories	1.162×10^{-3}	kilowatt-hours
kg-calories per min.	51.43	foot pounds per sec.
kg-calories per min.	0.09351	horsepower
kg-calories per min.	0.06972	kilowatts
kilometers	10^5	centimeters
kilometers	3281	feet
kilometers	10^3	meters

Multiply	by	To Obtain
kilometers	0.6214	miles
kilometers	1093.6	yards
kilowatt-hours	3415	British thermal units
kilowatt-hours	2.655×10^6	joules
kilowatt-hours	1.341	horsepower-hours
kilowatt-hours	3.6×10^6	joules
kilowatt-hours	860.5	kilogram-calories
kilowatt-hours	3.671×10^5	kilogram-meters
kilowatts	56.92	Btu per min.
kilowatts	4.425×10^4	foot-pounds per min.
kilowatts	737.6	foot-pounds per sec.
kilowatts	1.341	horsepower
kilowatts	14.34	kg-calories per min.
kilowatts	10^3	watts
log10N	2.303	logEN or ln N
logEN or lnN	0.4343	log10N
meters	100	centimeters
meters	3.2808	feet
meters	39.37	inches
meters	10^{-3}	kilometers
meters	10^3	millimeters
meters	1.0936	yards
miles	1.609×10^5	centimeters
miles	5280	feet
miles	1.6093	kilometers
miles	1760	yards
miles	1900.8	varas
miles per hour	44.70	centimeters per sec.
miles per hour	88	feet per minute
miles per hour	1.467	feet per second
miles per hour	1.6093	kilometers per hour
miles per hour	0.8684	knots per hour
miles per hour	0.4470	M per sec.
months	30.42	days
months	730	hours

Multiply	by	To Obtain
months	43,800	minutes
months	2.628×10^6	seconds
ounces	8	drams
ounces	437.5	grains
ounces	28.35	grams
ounces	0.0625	pounds
ounces per sq. inch	0.0625	pounds per sq. inch
pints (dry)	33.60	cubic inches
pints (liq)	28.87	cubic inches
pounds	444,823	dynes
pounds	7000	grains
pounds	453.6	grams
pounds	16	ounces
pounds	32.17	poundals
pounds of water	0.01602	cubic feet
pounds of water	27.68	cubic inches
pounds of water	0.1198	gallons
pounds of water per min.	2.669×10^{-4}	cubic feet per sec.
pounds per cubic foot	0.01602	grams per cubic cm.
pounds per cubic foot	16.02	kg per cubic meter
pounds per cubic foot	5.786×10^{-4}	pounds per cubic inch
pounds per cubic foot	5.456×10^{-9}	pounds per mil foot
pounds per square foot	0.01602	feet of water
pounds per square foot	4.882	kg per sq. meter
pounds per square foot	6.944×10^{-3}	pounds per sq. inch
pounds per square inch	0.06804	atmospheres
pounds per square inch	2.307	feet of water
pounds per square inch	2.036	inches of mercury
pounds per square inch	703.1	kg per sq. meter
pounds per square inch	144	pounds per sq. foot
quarts	32	fluid ounces
quarts (dry)	67.20	cubic inches
quarts (liq)	57.75	cubic inches
rods	16.5	feet
square centimeters	1.973×10^5	circular mils

Multiply	by	To Obtain
square centimeters	1.076×10^{-3}	square feet
square centimeters	0.1550	square inches
square centimeters	10^{-6}	square meters
square centimeters	100	square millimeters
square feet	2.296×10^{-5}	acres
square feet	929.0	square centimeters
square feet	144	square inches
square feet	0.09290	square meters
square feet	3.587×10^{-8}	square miles
square feet	0.1296	square varas
square feet	1/9	square yards
square inches	1.273×10^{6}	circular mils
square inches	6.452	square centimeters
square inches	6.944×10^{-3}	square feet
square inches	10^{6}	square mils
square inches	645.2	square millimeters
square miles	640	acres
square miles	27.88×10^{6}	square feet
square miles	2.590	square kilometers
square miles	3,613,040.45	square varas
square miles	3.098×10^{6}	square yards
square yards	2.066×10^{-4}	acres
square yards	9	square feet
square yards	0.8361	square meters
square yards	3.228×10^{-7}	square miles
square yards	1.1664	square varas
temp.(degs.C)+17.8	1.8	temp.(degs.F)
temp.(degs.F)-32	5/9	temp.(degs.C)
tons (long)	2240	pounds
tons(short)	2000	pounds
yards	0.9144	meters

DECIMAL/MILLIMETER EQUIVALENTS OF COMMON FRACTIONS		
Fraction	Decimal	MM
$\frac{1}{64}$	.015625	.397
$\frac{1}{32}$	.03125	.794
$\frac{3}{64}$	.046875	1.191
$\frac{1}{16}$	.0625	1.588
$\frac{5}{64}$	.078125	1.984
$\frac{3}{32}$	.09375	2.381
$\frac{7}{64}$	.109375	2.778
$\frac{1}{8}$	.125	3.175
$\frac{9}{64}$	.140625	3.572
$\frac{5}{32}$	.15625	3.969
$\frac{11}{64}$	.171875	4.366
$\frac{3}{16}$	.1875	4.763
$\frac{13}{64}$	.203125	5.159
$\frac{7}{32}$	.21875	5.556
$\frac{15}{64}$	.234375	5.953
$\frac{1}{4}$	.25	6.350
$\frac{17}{64}$	.265625	6.747
$\frac{9}{32}$	.28125	7.144
$\frac{19}{64}$	.296875	7.540
$\frac{5}{16}$	.3125	7.938
$\frac{21}{64}$	.328125	8.334

Fraction	Decimal	MM
$\frac{11}{32}$	.34375	8.731
$\frac{23}{64}$	.359375	9.128
$\frac{3}{8}$	.375	9.525
$\frac{25}{64}$	.390625	9.922
$\frac{13}{32}$	.40625	10.319
$\frac{27}{64}$	.421875	10.716
$\frac{7}{16}$	.4375	11.113
$\frac{29}{64}$	.453125	11.509
$\frac{15}{32}$	.46875	11.906
$\frac{31}{64}$	.484375	12.303
$\frac{1}{2}$	.5	12.700
$\frac{33}{64}$	.515625	13.097
$\frac{17}{32}$	.53125	13.494
$\frac{35}{64}$	.546875	13.891
$\frac{9}{16}$	.5625	14.288
$\frac{37}{64}$	.578125	14.684
$\frac{19}{32}$	.59375	15.081
$\frac{39}{64}$	.609375	14.478
$\frac{5}{8}$	.625	15.875
$\frac{41}{64}$	.640625	16.272
$\frac{21}{32}$	.65625	16.669
$\frac{43}{64}$	.671875	17.066

Fraction	Decimal	MM
$\frac{11}{16}$	.6875	17.463
$\frac{45}{64}$	.703125	17.859
$\frac{23}{32}$	.71875	18.256
$\frac{47}{64}$	.734375	18.653
$\frac{3}{4}$	.75	19.050
$\frac{49}{64}$	.765625	19.447
$\frac{25}{32}$	.78125	19.844
$\frac{51}{64}$	.796875	20.241
$\frac{13}{16}$	.8125	20.638
$\frac{53}{64}$	.828125	21.034
$\frac{27}{32}$	.84375	21.431
$\frac{55}{64}$	.859375	21.828
$\frac{7}{8}$	.875	22.225
$\frac{57}{64}$	.890625	22.622
$\frac{29}{32}$	.90625	23.019
$\frac{59}{64}$	.921875	23.416
$\frac{15}{16}$	.9375	23.813
$\frac{61}{64}$	.953125	24.209
$\frac{31}{32}$	.96875	24.606
$\frac{63}{64}$	.984375	25.003
1	1.	25.400

## Chapter 3

Power Generation

An electric power system consists of several systems all of which contribute to the power available for utilization. The main parts of the power system are:

- Generating system
- Transmission system
- Distribution system

The flow of power from generation to utilization can be illustrated in a simple way by a single-line diagram, such as shown in Figure 3-1 on the next page.

The voltage is generated in one or several generators, such as the generator located in the generating station. Then the voltage is stepped up by the transmission transformer to a very high value suitable for transmission. The generator circuit breaker is used with the transformer. The transmission line carries the very high voltage from the generator bus in the station to the high-voltage bus in a substation. The line is protected by the circuit breakers. A substation which steps the voltage down for a subtransmission system is a bulk-power substation. The substation is protected by the circuit breakers, and has a step-down transformer which provides by its secondary a lower voltage for the substation bus. Subtransmission circuits branch off to the distribution substations, such as the substation with the distribution transformer. Primary distribution feeders take the still lower distribution voltage through the circuit breakers to the utilization substations, which contain the transformers. From these transformers, secondary feeders provide utilization voltage to lighting and power circuits at the customer's premises. All of these items are covered in this chapter.

PARTS OF AN ELECTRICAL DISTRIBUTION SYSTEM

The simple single-line diagram in Figure 3-1 points out only the essential parts of the power system. In each system there are many devices necessary for protection, regulation, control, and measuring. There are also deviations in connections and various combinations of generated-transmission, and distribution voltage, as well as deviations in the length of the transmission line and in the area covered by the system of substations and distribution systems.

Generating Station

A generating station contains one or more ac generators, or alternators, which generate ac power at a certain voltage. The main power system in a generating station starts with the water, steam, or nuclear energy input into a prime mover, which may be a steam or water turbine or a steam engine. The prime mover turns the alternator shaft and the voltage is generated in the armature windings, which cut across the lines of force produced by the field. The field electromagnets are excited by a dc generator called an exciter. The main power system continues to low-voltage switching, a step-up power transformer, and high-voltage switching. The transmission line conveys the high-voltage power from the station. Within the generat-

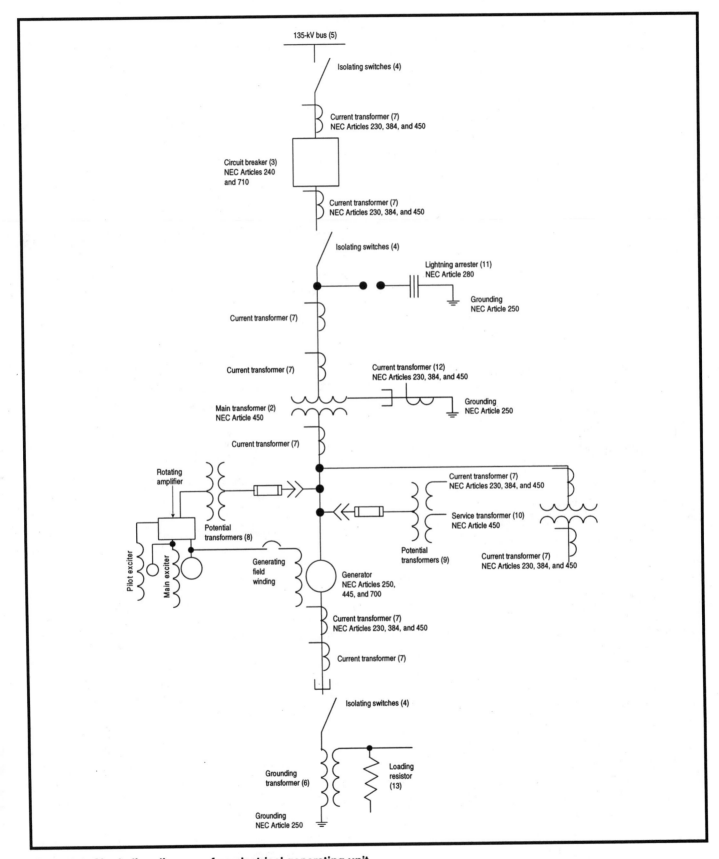

Figure 3-1: Single-line diagram of an electrical generating unit.

ing station, however, there are many other devices which make station operation possible. There is, first, the control-feedback system, which controls the prime mover through the governor, and the excitation system, which controls the generator voltage by controlling the field excitation. Another important system is the power-feedback system, which provides the power for the equipment used for station-service lighting, heating, cooling, and communication within the generating station.

Electrical Components of a Generating Unit

The single-line diagram shown in Figure 3-1 indicates all the essential electrical components of one generator, or generating unit. The voltage is generated at 13,800 volts in the generator and stepped up to 138 kV (kilovolts) in the main transformer, and through the 138-kV circuit breaker and the isolating switches, it is impressed on the 138-kV bus. The generator is grounded through the grounding transformer. The current transformers provide power for the measuring instruments and relays, and the potential transformer provides voltage for the voltage regulation. Another potential transformer serves for relays and measuring. The station-service transformer serves as a source of voltage for the station equipment. Generator protection is obtained by the lightning arrester and the

main-transformer neutral point is grounded through the current transformer. The grounding transformer is provided with a loading resistor. The generator field is energized by the main exciter and the exciter field is energized by the pilot exciter. The field is controlled by a rotating amplifier, which is energized by an auxiliary power source.

When several generating units are used in a generating station, they may be connected in a unit system shown in Figure 3-2, or in a multiple-supply system, shown in Figure 3-3 on the next page. In the unit system, each unit has its own main transformer, and in the multiple-supply system, one main transformer receives power from all units. The main components are identified by the callouts in Figures 3-2 and 3-3.

Synchronization Of High-Voltage Alternator

When an alternator is connected to the bus of an ac system, it is of the utmost importance that the incoming generator run in synchronism with the frequency of the ac voltage in the bus. A synchronizer, or synchroscope, indicates whether the generator and the bus are in synchronism. A vertical position of the indicator means that the generator and the bus are running in phase and at the same

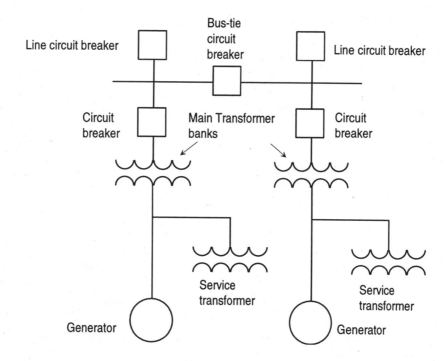

Figure 3-2: Two-unit generating station.

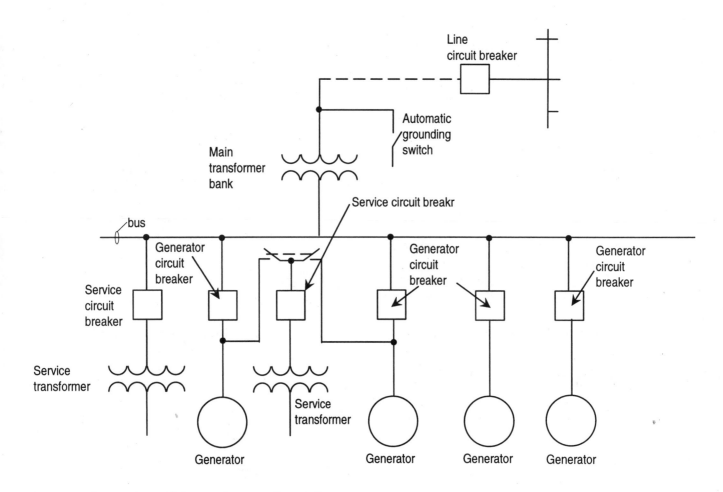

Figure 3-3: Four-unit, multiple-supply generating station.

frequency. The connections of a synchroscope are shown in Figure 3-4. The incoming high-voltage three-phase alternator is connected to the high-voltage three-phase bus through the disconnecting switches and the circuit breaker. One potential transformer is connected to one phase of the alternator and another one to the same phase of the bus. The secondaries of the potential transformers are connected to the synchronizing bus. A voltmeter is connected through the double receptacle to the synchronizing bus and then through the receptacle to the synchronizing lamps, along with the synchroscope. The voltmeter can indicate the voltage of the bus or that of the generator, depending on the way it is plugged into the double receptacle. Similarly, the generator voltage is applied through plugging to one element of the synchroscope through the reactor and the resistor. The bus voltage is applied directly to the other element of the synchroscope. If the two voltages are not in phase, the indicator of the synchroscope moves and indicates the need for synchronization.

Station-Service Supply

The electric power necessary for proper operation of station-service equipment is up to 5 to 10 percent of the power generated by the generators. Transformer banks of 10,000 kVA or more are in common use to supply power for lighting and equipment at the generating station.

TRANSMISSION SYSTEMS

From the generating station, electric power is transmitted by high-voltage transmission lines to the areas of distribution. The system of transmission lines covers great distances and provides large areas with electric power. The required reliability, security, and stability of transmission systems are insured by transmission substations which supply the switching, voltage-transformation, and control facilities. High-voltage switchgear gives protection against line faults, contains disconnecting means for maintenance purposes, and may have equipment for tying two transmission lines and synchronizing their voltages.

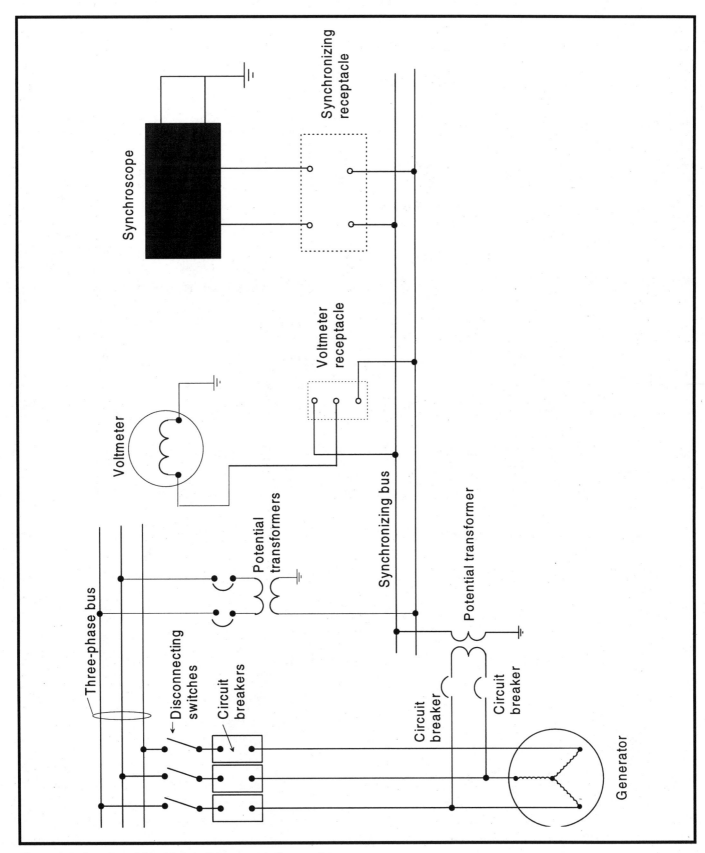

Figure 3-4: Synchronizing high-voltage generating system.

Large power transformers, mostly three-phase, and rated up to more than 100,000 kVA, and autotransformers rated as high as 215,000 kVA are located in most transmission substations in order to step down the high transmission voltage for subtransmission and distribution systems.

Voltage control in transmission lines may be obtained by line-drop compensation or by improving the power factor with synchronous capacitors.

The equipment of substations is rated first according to the nominal voltage. Some common nominal voltages are, for example, 240 or 120 kV for transmission voltage and 13.8 or 4.16 kV for distribution voltage. The substation equipment is further rated according to the high-potential test voltage and according to the basic installation level (BIL), which is the resistance to sudden stresses, such as caused by lightning. The BIL rating is normally 4 to 15 times the normal voltage rating.

Substations in the transmission system may he operated automatically, manually, or by remote control from a distance by an attendant at a supervising control substation.

Substation Switching System

Circuit breakers and other switching equipment in a substation may be arranged to separate a bus, a part of a transformer, or a control device from other equipment. The common arrangements are single-bus, double-bus, transfer-bus, ring-bus, or mesh systems of switching.

Switching systems are shown in Figure 3-5. The simple single-bus switching system in (a) has the bus protected by the circuit breakers on the incoming and outgoing lines. The double-bus switching system in (b) has two main buses, but only one is normally in operation; the other is a reserve bus. The ring-bus system in (c) has the bus arranged in a loop with breakers placed so that the opening

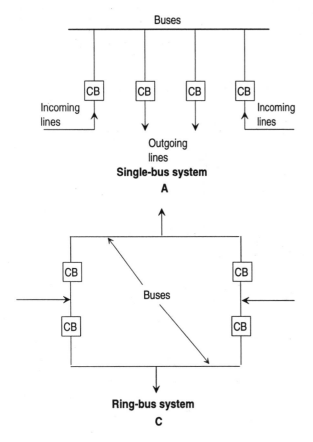

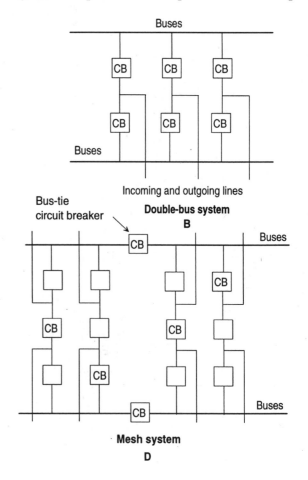

Figure 3-5: Switching systems in substation.

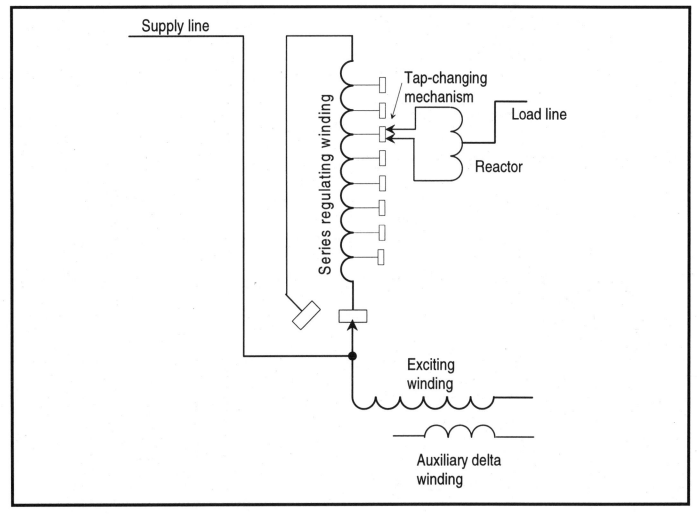

Figure 3-6: One phase of a single-core step-voltage regulator.

of one breaker does not interrupt the power through the substation. The greatest security and flexibility of switching is obtained by the mesh system shown in (d). Bus-tie breakers are added and placed between pairs of main buses. Assuming any circuit breaker opens, you may verify that the circuits will not be interrupted by tracing the circuits.

Voltage Regulation in Substations

One way of obtaining voltage control uses on-load changers in transformers. The taps may be changed automatically and in many small steps according to the need.

Another method of voltage control is possible with a step-voltage regulator. One phase of such a regulator is shown in Figure 3-6. The supply line is connected to the exciting winding of the regulator. The auxiliary delta

winding provides a path for circulating currents. The regulating winding in series with the exciting winding has taps which are connected to the tap-changing mechanism, which in turn is connected through the reactor to the load line.

A system of voltage regulation in a transmission line which uses a synchronous capacitor is shown by the block diagram in Figure 3-7 on the next page. The synchronous capacitor provides reactive power for the line and so improves the power factor of the power system. The regulation is obtained automatically by changing the strength of the exciter field of the synchronous capacitor. The changes in voltage are detected by potential transformers on the three-phase lines and are transferred to the static control devices which consist of the voltage-adjusting device, the voltage comparator, and the reactive-cur-

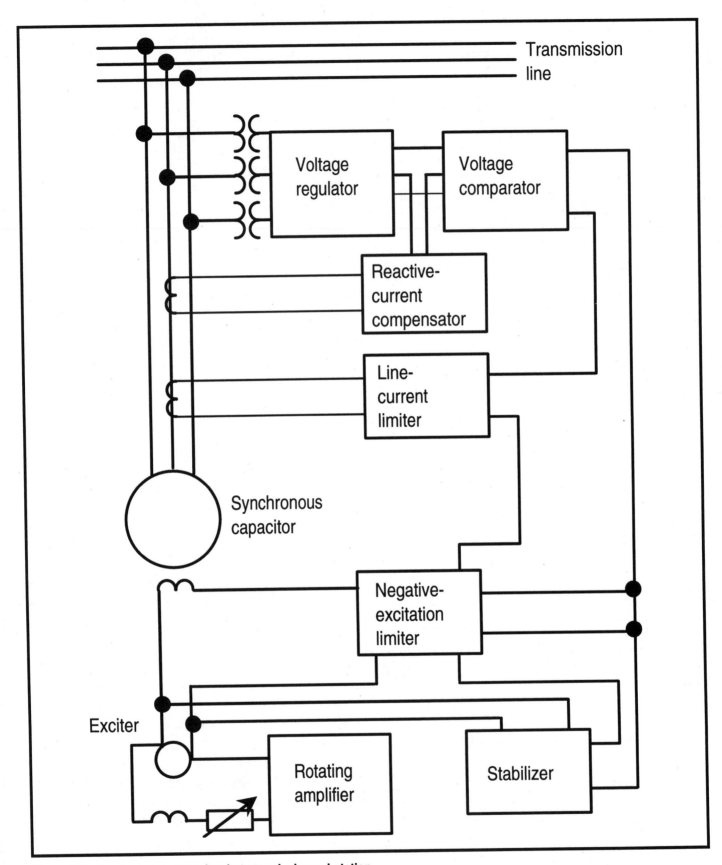

Figure 3-7: Synchronous capacitor in transmission substation.

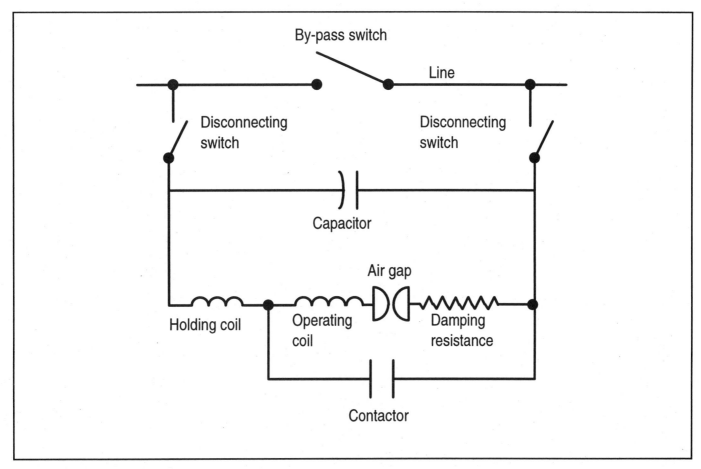

Figure 3-8: Capacitor unit in series with the line.

rent compensator. The compensator receives the current from the current transformer on one phase.

The voltage comparator sends a boosting or a bucking signal to the stabilizer and to the rotating amplifier, which in turn increase or decrease the exciter field. These changes in the exciter field are reflected in the synchronous-condenser field and finally in the output of the condenser. The line-current limiter is used in automatic operation to guard against overload of the condenser. The negative-excitation limiter prevents too high a leading current in the condenser field, which might increase the line voltage dangerously.

The power factor of the system may be improved by installation of series and shunt banks of capacitors in the transmission system. One single-phase capacitor unit is shown in Figure 3-8. The capacitor is connected in series with the line through two disconnecting switches. A normally open bypass switch is in the main line. The holding coil and the operating coil are connected in series with an air gap and a damping resistance. When the voltage across

the capacitor rises over a certain value, a spark jumps across the gap and causes a current in the circuit parallel with the capacitor. This current closes the contactor and so short-circuits the capacitor. The holding coil keeps the contacts closed as long as the current is higher than normal.

High-voltage Lines in Substations

The incoming transmission or subtransmission lines and the outgoing distribution lines of a distribution substation are most frequently arranged in a radial system or in a network system. The system selected for either incoming or outgoing lines depends on the purpose of the substation.

According to the degree of security required of the substation, the incoming, or the high-voltage, lines may be arranged in a radial, double radial, or loop system. The most commonly-used high-voltage switchgear arrangement is shown in Figure 3-9 on the next page.

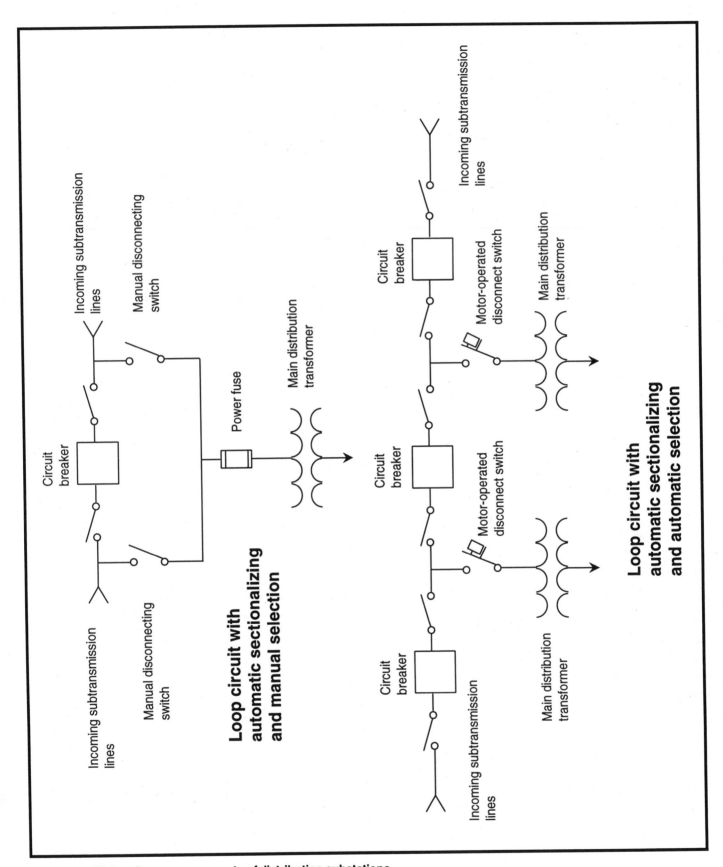

Figure 3-9: High-voltage arrangements of distribution substations.

STEEL TOWER INSTALLATIONS

Steel tower construction is used mainly for high-voltage transmission lines over long distances and over mountainous terrain. Such systems are used primarily for the support of conductors and not the mounting of transformers, switches, and other equipment. The latter equipment is normally located at substations or switchyards.

Steel tower construction is also used for the supporting structures of overhead trolley systems for electrified railroads, and modified steel tower structures are sometimes used for sports lighting.

The basic components of a steel tower consist of a base or footing, the main tower structure, bridge or cross-arm members, insulators and vibration dampeners, and tower grounding.

The size and type of footing depend upon the type and height of the tower and the nature of the ground and terrain. Footings are usually completely detailed in the contract drawings, and specifications and details such as the ones in Figure 3-10 are common on working drawings.

POWER TRANSMISSION SPECIFICATIONS

The following specification is for a 230 kV transmission line and covers items that the design consultant must address in the written specifications and on the construction working drawings.

(A) Right of way shall be cleared a distance of 62.5 feet each side of the center line of the transmission line unless otherwise directed. All trees, brush, and stumps within a radius of 25 feet of any tower leg of the finished transmission line shall be cut off as close to the ground as practicable as determined by the contracting officer, and in no case shall they be cut off at a height of more than 12 inches above the ground. All trees, brush, and stumps more than five feet in height in other areas to be cleared shall be cut off at not more than 18 inches above the ground. If directed by the contracting officer, the clearing shall also include the cutting or trimming of all trees outside of the right of way if such trees upon falling would come within 10 feet of the nearest conductor of the line. The cleared material shall be burned or otherwise disposed of as approved in writing by the contracting officer. All materials to be burned shall be piled and when in suitable condition shall be completely burned. Piling for burning shall be done in such a manner and such locations as will cause the least fire risk, and all materials which cannot be completely burned as the work proceeds shall be piled in locations approved by the contracting officer and thereafter completely disposed of by burning within the period of time covered by the contract. Payment for clearing land and right of way will be made at the lump sum price bid.

(B) Where the right of way is through well-developed areas such as orchards, clearing will be confined to the tower sites, except that the contracting officer may require trimming or removal of all trees or obstruction that interfere with operation of the transmission line. All trees, brush, or stumps within ten feet of any tower member shall be cut off as close to the ground as practicable.

(C) Blasting will be permitted only when proper precautions are taken for the protection of persons, the work, and public or private property, and any damage done to the work or public or private property by blasting shall be repaired by the contractor at the contractor's expense. Caps or other exploders or fuses shall be in no case stored, transported, or kept in the same place in which dynamite or other explosives are stored, transported or kept. The location and design of powder magazines, methods of transporting explosives, and in general, the precautions taken to prevent accidents shall be subject to the specifications, but the contractor shall be liable for all injuries to or deaths of persons or damage to property caused by blasts or explosives.

(D) The conductors will be furnished by others. The conductors for the transmission line will be 795,000-circular mil aluminum conductor, steel reinforced (ACSR), having an outside diameter of 1.108 inches, an ultimate strength of 31,200 pounds and a weight of approximately 1,098 pounds per 1,000 feet. Unless otherwise shown on the drawings or directed by the contracting officer, all clearances, measured in still air at 60°F, shall conform to provisions of the local safety standards. The equipment and methods used for stringing the conductors shall be such that the conductors will not be damaged or injured, and shall be subject to the approval of the contracting officer. Particular care shall be taken at all times to insure that the conductors do not become kinked, twisted, or abraded in any manner. If the conductors are damaged in the contractor's operations, the contractor shall repair or replace the damaged sections, including the furnishing of the necessary additional material, in a manner satisfactory to the contracting officer and at no additional cost. All sections of the conductors damaged by the application of gripping attachments shall be repaired or replaced before the conductors are sagged in place. The conductors shall be laid along the ground from moving reels and then raised

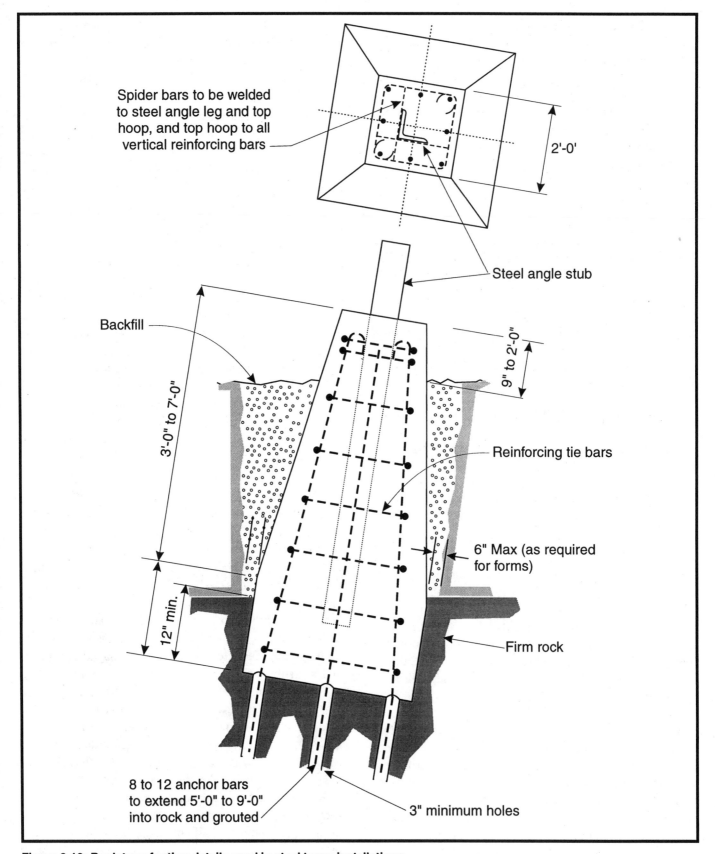

Figure 3-10: Rock type footing details used in steel tower installations.

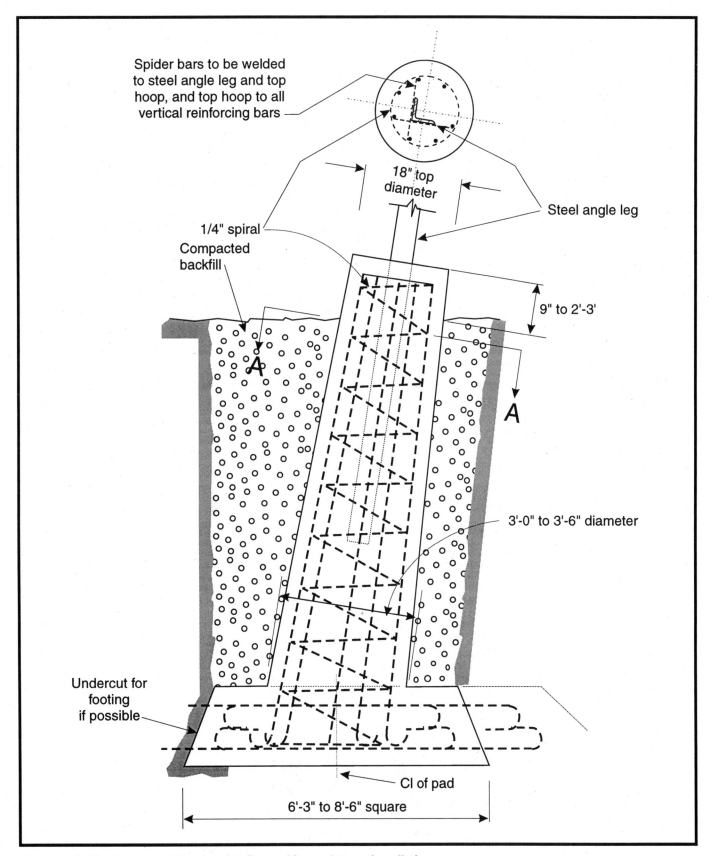

Figure 3-11: Hand excavated footing details used in steel tower installations.

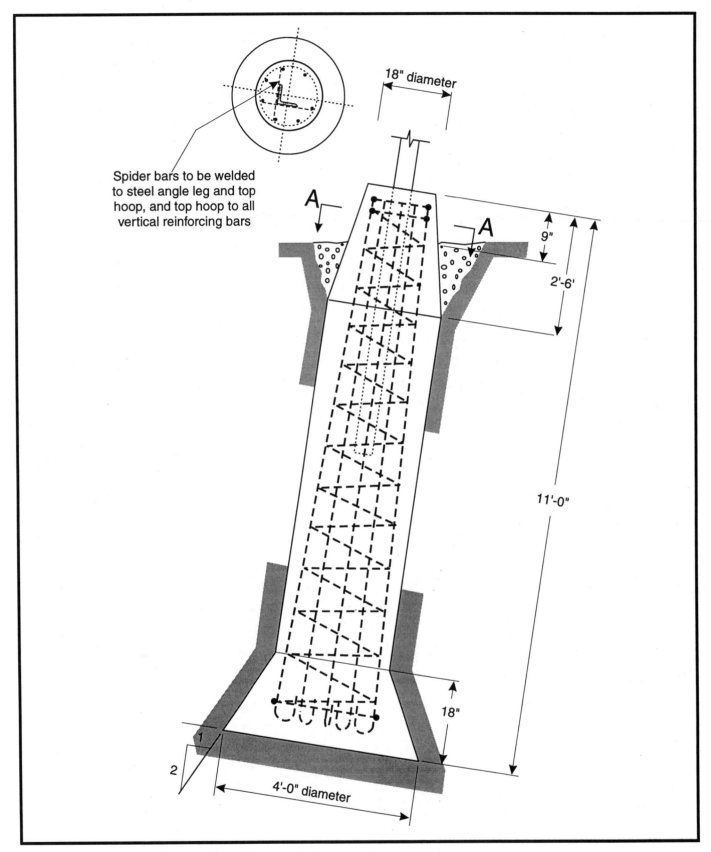

18" diameter

Spider bars to be welded to steel angle leg and top hoop, and top hoop to all vertical reinforcing bars

A

A

9"

2'-6'

11'-0"

18"

1

2

4'-0" diameter

Figure 3-12: Auger excavated footing details used in steel tower installations.

into position in the stringing blocks by means of running lines, at the direction of the contracting officer. The running lines shall be of sufficient length to avoid applying undue strain to the insulators and structures. The running lines shall be connected to the conductors with swivel connectors and stocking type grips as directed by the contracting officer. The end of the grips shall be taped to the conductor so that the grips will run freely in the sheaves. Conductor splices shall not be passed through a sheave except as specifically permitted by the contracting officer. Reel stands shall be heavily constructed and provision shall be made for braking the reels. The conductor shall not be dragged over the ground, rock, fence wires, or any object which may damage the conductor. Suitable guards or sheaves protect the conductor from damage in places where it would otherwise be impossible to keep the conductor from coming in contact with objects which may injure the conductor. Guards shall consist of material over which the conductor may slide without injury and shall be subject to the approval of the contracting officer. The minimum diameter of the sheaves shall be 14 inches at the bottom of the groove and the size and shape of the groove shall conform to the conductor manufacturer's recommended to be equipped with high-quality ball or roller bearings to reduce friction to a minimum. The conductor shall not be allowed to hang in the stringing blocks more than 18 hours before being pulled to the specified sag. After being sagged, the conductor shall be allowed to hang in the stringing blocks for not less than two hours before being clipped in, to permit the conductor tension to equalize. The total time which the conductor is allowed to remain in the stringing blocks before being clipped in shall be not more than 36 hours. The conductors shall not be prestretched, and shall be sagged in accordance with sag tables furnished by the contracting officer. The sag tables will be furnished to the contractor after notice to proceed has been issued. The length of conductor sagged in one operation shall be limited to the length that can be sagged satisfactorily. In sagging lengths of more than one reel, the sag of three or more spans near each end and the middle of the length being sagged shall be checked. The length of the spans used for checking shall be approximately equal to the ruling span. The sag of all spans more than 1,500 feet in length shall be checked, and at sharp vertical angles the sag shall be checked on both sides of the angle. The sag of spans on both sides of all horizontal angles of more than 10° shall be checked. After the conductors have been pulled to the required sag, intermediate spans shall be inspected to determine whether the sags are uniform and correct. Sagging operations shall not be carried on when, in the opinion of the contracting officer, wind prevents satisfactory sagging. A tolerance of plus or minus one-half inch of sag per hundred feet of span length, but not to exceed six inches in any one span, will be permitted, provided, that all conductors in the span assume the same sag and the necessary ground clearance is obtained; provided further, that the conductor tension between successive sagging operations is equalized so that the suspension insulator assemblies will assume the proper in. The contracting officer will check the sag at all points to be checked, and the contractor shall furnish the necessary personnel for signaling and climbing purposes. At all suspension or tension structures, the conductors shall be attached to the insulator assemblies by suspension or strain clamps as shown on the drawings, and all nuts shall be tightened adequately. Payment for stringing conductors will be made at the unit price per mile of line bid in Schedule No. _____ for stringing six 795,000-circular mil, steel-reinforced aluminum conductors, which unit price shall include the cost of stringing, splicing, connecting, armoring, and sagging the conductors, as described in this paragraph and the following paragraphs.

(E) Unless otherwise directed by the contracting officer, all joints or splices and deadends in the conductors shall be made in accordance with the recommendations of the conductor manufacturer using the special compressor tool recommended for this purpose. The contractor shall furnish all necessary tools, including compressors, required for making joints and splices. Compression-type connectors and compression-type deadends will be furnished by others. All joints or splices shall be located at least 50 feet away from the structures, and no joints or splices shall be made in spans crossing over or adjoining important highways, railroads, or other utility lines without express permission of the appropriate authorities or approval of the contracting officer. At all deadend and large angle structures, the jumper connections shall be made with compression-type jumper terminals which shall be bolted to the compression-type deadends. The cost of making all joints, splices, deadends, and installing jumpers shall be included in the unit price bid in Schedule No. for stringing conductors.

(F) At each suspension insulator assembly on the transmission line, armor rods shall be attached to the aluminum conductor in accordance with this paragraph or as directed by the contracting officer. If it becomes necessary to change the point of attachment by more than 2.5 inches either way from the midpoint of the armor rods after the armor rods are attached, a new set of armor rods shall be furnished and attached by the contractor without additional cost. Compression repair sleeves or compression joints may be used on the jumper suspension insulator assemblies in lieu of armor~rod sets. The cost of attaching armor rods and compression joints or compression repair sleeves on jumper insulator assemblies shall be included in the unit price bid in Schedule No. for strinqinq conductors.

(G) Vibration dampers shall be attached to the aluminum conductor at the ends of those spans so designated on the plan-profile drawings or on the structure-list sheets to be furnished by the contracting officer, and at other points required by the contracting officer. The vibration dampers shall be attached in accordance with Drawing No. , and fastened securely so that they will hang in vertical planes. The contractor shall ascertain that the drain holes are open after the vibration dampers are attached. Payment for attaching vibration dampers will be made at the unit price bid therefor in Schedule No. _____

(H) Two overhead ground wires shall be strung for the entire length of the transmission line, and shall be attached to the towers by the contractor in accordance with the details shown on Drawing No. . For the northern half of the line, the overhead ground wires will be $\frac{1}{2}$-inch diameter high strength double-galvanized, stranded steel wire. For the southern one-half of the line, the overhead ground wires will be $\frac{7}{16}$-inch, extra high-strength copperweld stranded wire. The ground wire, together with appurtenant material, will be furnished by others. The equipment and methods used for stringing the overhead ground wires shall be the same as for stringing the conductors as described elsewhere, and the same degree of care shall be exercised to avoid damage or injury to the ground wires. If damaged, they shall be repaired by the contractor or the damaged sections replaced, including the furnishing of the necessary additional material, in a manner satisfactory to the contracting officer and at no additional cost, before the wires are finally sagged in place. Joint or splices in the overhead ground wires shall be located at least 50 feet away from the structures, and no joints or splices shall be made in spans crossing over or adjoining important highways, railroads, or other public utility lines, unless approved by the contracting officer. The cost of making all joints and splices in the ground wires shall be included in the unit prices bid on Schedule No. for stringing the overhead ground wires. The overhead ground wires shall be sagged in place in accordance with the sag tables furnished by the contracting officer. The methods used in checking the sag of the overhead ground wires shall be the same as those outlined previously for checking the sag of the conductors. Payment for stringing the two overhead ground wires will be made at the unit prices per mile of line bid therefor in Schedule No. , which unit prices shall include the cost of installing all fittings required for the installation of the ground wires, and the stringing, splicing, connecting, deadening, sagging, and clipping of the overhead ground wires as stated in this paragraph.

(I) All metal and wire fences which cross under the transmission line or which are located in the proximity of and parallel to the transmission line shall be grounded with ground posts which will be furnished by others. The ground posts will be ten feet long, with attached tongues for holding the wires of the fence. For each one-quarter of a mile of fence that the fence is within 75 feet of the center line of the transmission line, and on each side of the right of way where a fence crosses under the transmission line, one ground post shall be driven to a depth of not less than five feet, or as directed by the contracting officer. The fence shall be fastened securely to the ground posts by the tongues provided therefor on the posts. Payment for placing fence ground posts and for grounding fences will be made at the unit price per post bid therefor in Schedule No._.

(J) The contractor shall install the ground rods and connect the ground rods to the tower. Bare stranded copper conductor, connectors and ground rods will be furnished by others. Conductor will be #4/0 AWG. Ground rods will be $\frac{7}{8}$-inch by 10 foot copperweld. Four ground rods for each tower shall be installed as shown Drawing No. . Ground rod resistance tests will not be required. Payment for placing tower ground rods and grounding towers will be made at the unit price per rod bid therefor in Schedule No._____.

Substations

(A) The contractor shall furnish and install a complete substation of open framework construction complete with transformers and oil circuit breakers as shown in the working drawings and the supplemental detailed drawings, schedule, etc.

(B) All concrete pads and foundations shown in the drawings or called for elsewhere in the specifications shall also be furnished and installed by the electrical contractor.

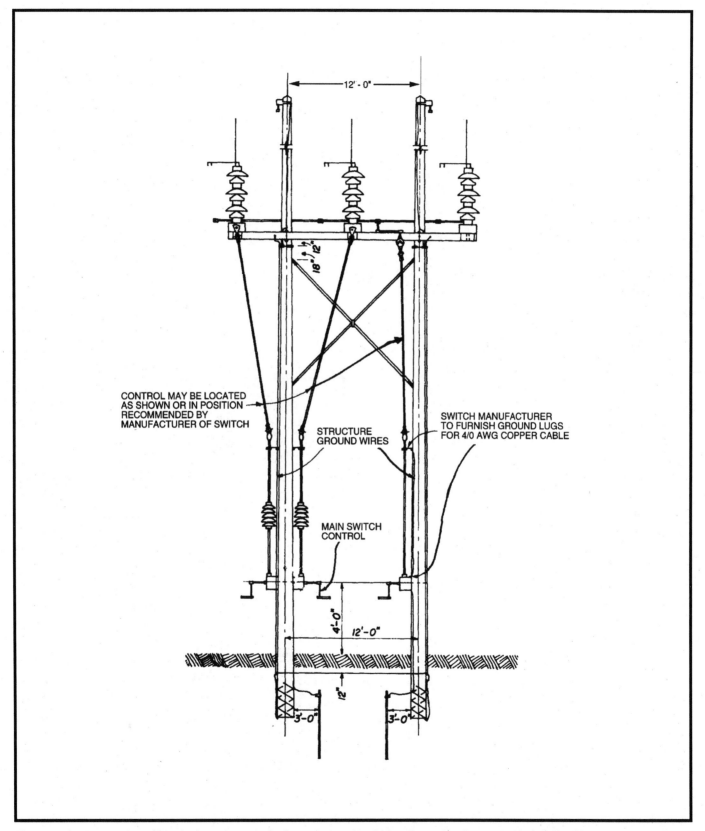

Figure 3-13: A complete description of transmission towers should be shown on the construction drawings, along with large-scale detail drawings where appropriate. This wood-pole detail was used on a project that extended an existing system.

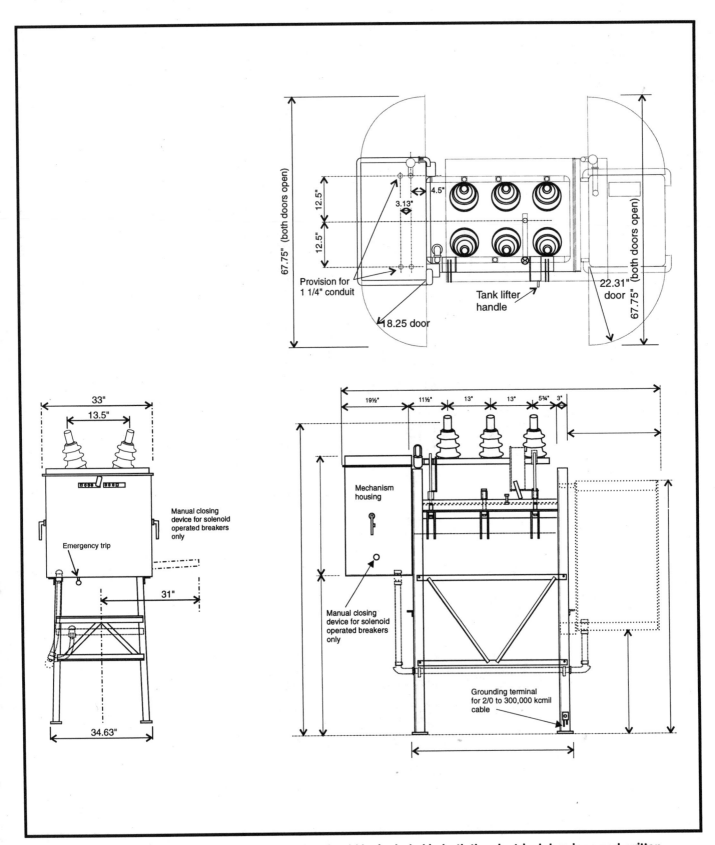

Figure 3-14: Details of outdoor power circuit breakers should be included in both the electrical drawings and written specifications.

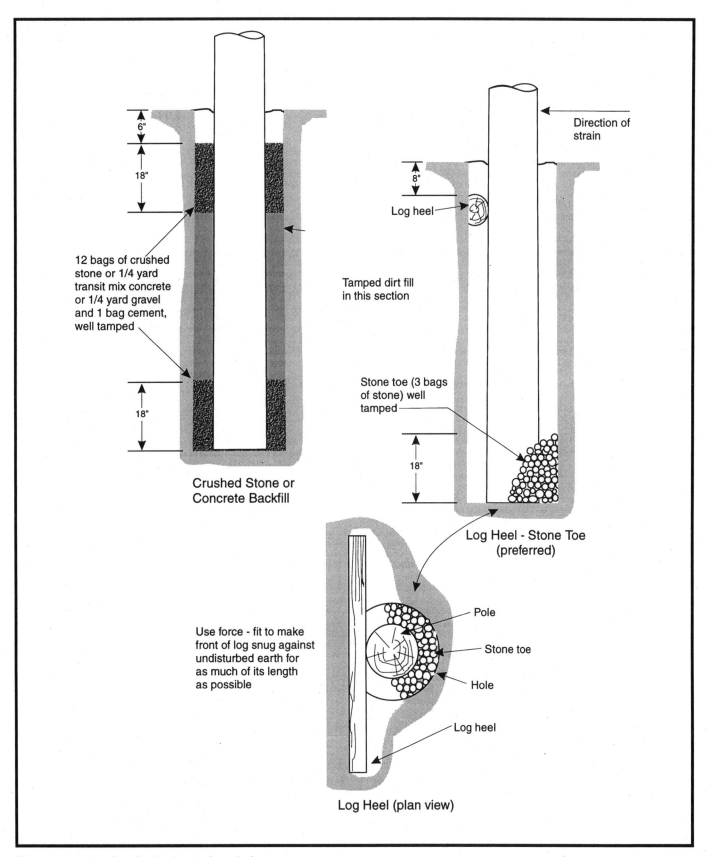

6"

18"

12 bags of crushed
stone or 1/4 yard
transit mix concrete
or 1/4 yard gravel
and 1 bag cement,
well tamped

18"

Crushed Stone or
Concrete Backfill

Direction of
strain

8"

Log heel

Tamped dirt fill
in this section

Stone toe (3 bags
of stone) well
tamped

18"

Log Heel - Stone Toe
(preferred)

Use force - fit to make
front of log snug against
undisturbed earth for
as much of its length
as possible

Pole

Stone toe

Hole

Log heel

Log Heel (plan view)

Figure 3-15: Details of typical pole foundations.

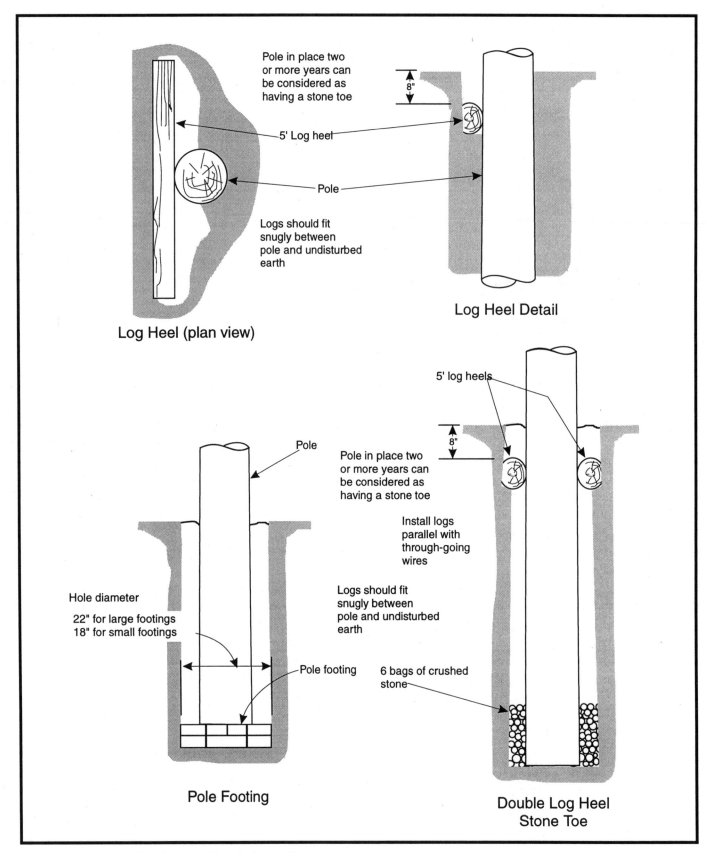

Figure 3-15: Details of typical pole foundations. *(Cont.)*

PRIMARY DISTRIBUTION SYSTEMS

The source of most commercial electric energy is a generator or a combination of generators. The generator is driven either by engines, hydraulics, or steam — although wind power is once again being experimented with. Steam can be developed by coal (the most popular means at present), oil, gas, or nuclear fission.

Most generating stations utilize an outdoor substation, which contains step-up transformers to transform the generator voltage of, say, 14 kV to the transmission voltage, which may be as high as 750 kV. A three-phase transformer is normally used for each generator.

Other equipment contained in the outdoor substation includes disconnect switches for isolating circuit breakers, line-grounding disconnect switches, and lightning arresters. Current and potential instrument transformers are also usually located in the outdoor substation to supply operating power for protective relays and metering devices.

The main transmission lines between the generating station and the bulk power station are usually protected by differential protective relays which measure the incoming and outgoing currents. The relays operate to open the line circuit breakers, usually placed at each end of the line, when a fault occurs and the incoming and outgoing currents do not balance. This way, any faults may be quickly isolated to main system stability and preserve service to the unfaulted portion of the system.

The outdoor bulk Dower substation. mentioned in the preceding paragraph, is very similar to the generating station substation in that it contains circuit breakers, the outdoor substation, and other related switchgear equipment. At the generating station substation, the generated voltage is stepped up to the transmission voltage. However, at the bulk power substation, this voltage must again be stepped down, or reduced, by the use of step-down transformers. From this point, power is supplied to distribution substations, and each of these subtransmission circuits are protected by some form of overcurrent protection.

Subtransmission and primary feeder circuits have many variations and voltages. The actual voltage used is governed by a large number of factors — mainly economic but also including such practical matters as the area to be served, load densities, estimated future growth, terrain, and availability of rights-of-way and substation sites.

Various designs of utilization substations are also needed to provide the many types of services that are in demand. Typical loads include residential services, which are usually 240/120 V, single phase, and commercial services, which may be 240/120 V, single phase, or perhaps 208/120 V, Y, three phase. Some larger installations are utilizing 480/208Y, three-phase services, while some industrial applications may require 13-kV service.

The local utility companies may require synchronous condenser or capacitor substations to maintain better system voltage, or some feeders may supply a low-voltage network where network protectors are used. Such circuits are usually equipped with some form of fault detector and with an automatic-reclosing switching scheme for automatic restoration of service after an outage.

Conversion substations are used in some areas to change alternating current to direct current for services such as 600, 1500, or 3000 V for railways such as used on the Pennsylvania Railway. Certain types of mills and electrochemical plants also require direct current for their manufacturing processes. Sometimes the plant itself provides this conversion equipment (see Chapter 5), while the local power companies provide the conversion in other cases.

Protection of Power System

A variety of situations may interfere with the normal operation of a power system. The predominant abnormal conditions on distribution circuits are line faults, system overloads, and equipment failures. Atmospheric disturbances and both animal and human interference with the system are generally the underlying causes of these conditions.

Line faults can be caused by strong winds which whip phase conductors together or which blow tree branches on the lines. In winter, freezing rain can produce a gradual buildup of ice on a circuit, and eventually one or more conductors may break and fall to the ground. Squirrels and other animals have been known to place themselves between an energized portion of the circuit and ground, producing sad results for the animal as well as the circuit.

On underground systems, cables severed by earth-moving equipment are a prevalent cause of faults. Lightning strokes can fault a system by opening lines or initiating arcs between conductors. Unforeseen load growth is the primary cause of overloads. Equipment failure can be caused by lightning; insulation deterioration; improper design, manufacture, installation, or application; and system faults.

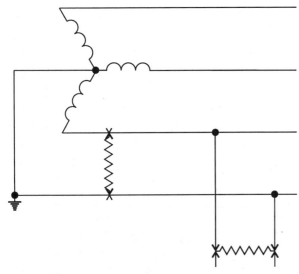

Figure 3-16: Points of a distribution system where line-to-line ground faults are possible.

Types Of Faults

Line-to-ground, line-to-line, and double line-to-ground are faults common to single- and three-phase systems. The three-phase fault, naturally, is a characteristic only of the three-phase system.

Line-to-ground faults occur when one conductor falls to ground or contacts the neutral wire. Possible points along a distribution system where this type of fault can result are shown in Figure 3-16.

Line-to-line faults can happen when conductors or a single- or three-phase system are short-circuited, as shown in Figure 3-17. They can occur anywhere along a three-phase wye or delta system or along a two-phase branch.

Double line-to-ground faults occur when two conductors fall and are connected through ground or when two conductors contact the neutral of a three-phase or single-phase grounded system. Figure 3-18 shows a typical faulted circuit.

The various faults illustrated in Figures 3-16 through 3-18 are all unsymmetrical. Faults such as these and other unbalanced conditions on polyphase systems are traditionally analyzed by the application of symmetrical component theory. This theory is adequately covered in a number of books on electrical theory, and due to the length required to cover it adequately, we shall omit it here. Protective relays and their connections are many and varied. Adequate protection for a small distribution system connected to a few small generators may consist of time-cascaded overcurrent relays. This means that the most remote load may be protected by a straight solenoid-operated

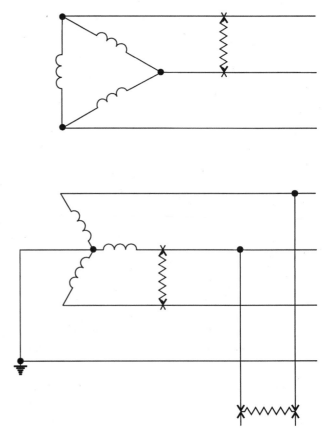

Figure 3-17: Points of a distribution system where line-to-line faults sometime occur.

overcurrent relay that has no intentional time delay. When the current through the relay reaches overcurrent values, the relay operates to trip the breaker as fast as possible. The relays and breakers nearer the power source may operate on currents of fault magnitudes but with enough time delay to permit the more remote breaker to operate if it is going to. The relays still nearer the source of generation will operate on the same principle but with longer and longer time delays. Time separations of one-quarter of a second are adequate in most cases.

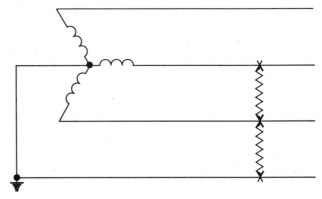

Figure 3-18: Typical double line-to-ground faults.

Large power systems fed from large generating units require a different type of protecting system. Large systems denote large inertial which cannot be changed rapidly without seriously endangering the stability of the system. High-speed relays that can quickly detect the location of the fault and breakers that can open the fault current in about one-twentieth of a second are now a standard combination.

Transfer Schemes

Recent developments in distribution protective equipment, primarily the advent of electronically controlled recloser and sectionalizer controls, have allowed the usage of more complex, automated loop-type distribution systems. This development, along with more advanced load transfer schemes and supervisory equipment, has led to substantial improvements in the reliability of modern distribution systems.

Switched Load Transfer Schemes: A line schematic of a simple load transfer scheme using a McGraw-Edison Type S transfer control and electrically operated oil switches is shown in Figure 3-19.

In this load transfer scheme, power is normally supplied from a preferred source and automatically switched to an alternate, standby source if the preferred service is lost for any reason. Upon restoration of preferred source voltage, the load is switched back automatically or manually. Return switching can be either in a closed-transition (parallel return) mode (the preferred source closes before the alternate source opens) or in an open-transition mode (nonparallel return; the alternate source opens before the preferred source closes).

The type S control is designed for use with three single-phase switches or one three-phase switch in each line; either single-phase or three-phase sensing can be employed.

Load Transfer — Manual Return: This scheme (Figure 3-20) uses electronically controlled reclosers in both the

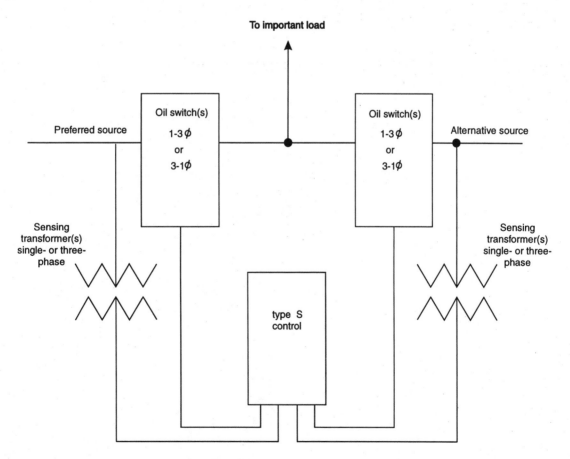

Figure 3-19: Line schematic of a single-load transfer scheme.

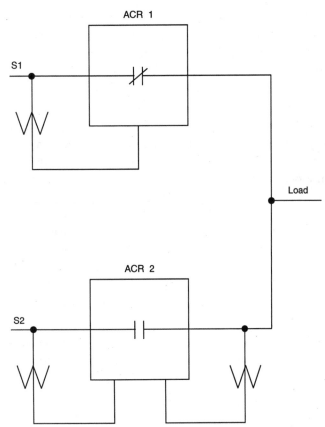

Figure 3-20: Load transfer scheme with manual return. Note that recloser ACR 1 is closed to receive normal power while ACR 2 is open and on standby.

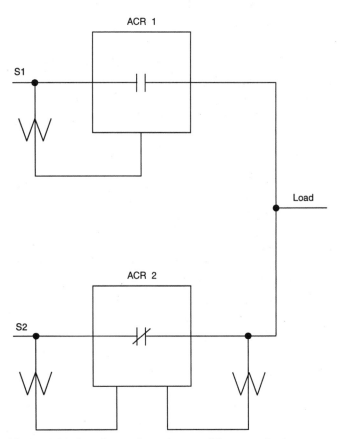

Figure 3-21: Load transfer scheme with manual return. Note that recloser ACR 1 is now open and ACR 2 is closed to allow the alternate power source to feed the load.

preferred source and alternate source lines. In these examples, reclosers are equipped with McGraw-Edison Type LS controls. Darker lines on illustrations represent energized lines. Load is normally fed from the preferred source, S1.

Recloser ACR 1 is normally closed and senses voltage (one or three phases) on its source side. It opens after a time delay upon loss of S1 voltage.

Recloser ACR 2 is normally open and senses voltage (one or three phases) on its load side. It closes after a time delay (longer than ACR 1) upon loss of load voltage. In addition, ACR 2 is equipped with a block of-reclose accessory energized from the alternate source (S2) which prevents any attempt to close ACR 2 if S2 is not energized. When preferred source voltage (S1) is lost, the controls of both reclosers sense the loss of voltage. If voltage is not restored within the time delay selected, ACR 1 opens.

After a longer time delay, ACR 2 closes to restore service to the load (Figure 3-21). A momentary cold-load pickup accessory can be provided for ACR 2 to prevent tripping on cold-load inrush.

When preferred source voltage (S1) is restored, transfer back to the preferred source is accomplished manually.

If a permanent fault occurs on the load side of the system as shown in Figure 3-22, the preferred source recloser (ACR 1) operates to lockout. The alternate source recloser (ACR 2) senses the loss of load voltage, and after a time delay, ACR 2 closes into the fault and also operates to lockout, as shown in Figure 3-23. (A momentary non-reclose accessory is available for one-shot lockout of ACR 2 to minimize load disturbances.) After the fault is cleared, service from the preferred source is restored manually.

Load Transfer — Automatic Return: Upon loss of preferred source voltage the load is automatically transferred to an alternate source. When preferred source voltage is restored, the load is automatically transferred back to the preferred source. The scheme shown in Figure 3-24 on page 400 uses electronically controlled reclosers in both the preferred source and alternate source lines. Both reclosers are equipped with type LS controls or with a McGraw-Edison Type S control. A requirement of this scheme is that the reclosers must be near enough to

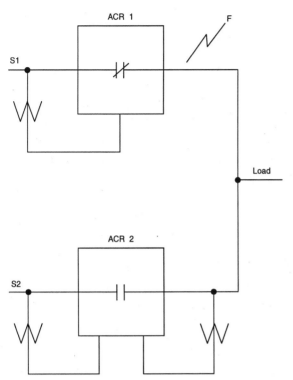

Figure 3-22: Recloser ACR 1 operates to lockout any faults that occur on the load side of the system.

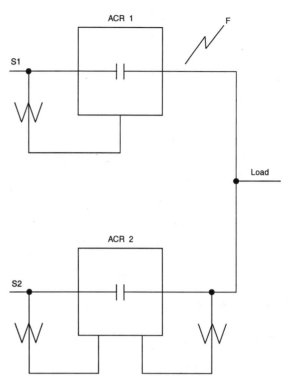

Figure 3-23: Recloser ACR 2 senses the loss of load voltage, and after a time delay, also operates to lockout the fault.

establish a communication link between them. Load is normally fed from the preferred source, S1. Recloser ACR 1 is normally closed and senses voltage (one or three phases) on its source side. It opens after a time delay upon loss of S1 voltage. The control of the normally open alternate source recloser (ACR 2) is connected to the control of ACR 1. In addition, ACR 2 is equipped with a block-of-reclose accessory energized from the alternate source (S2) which prevents any attempt to close ACR 2 if S2 is not energized.

When the preferred source voltage (S1) is lost, the control of ACR 1 senses the loss of voltage. If voltage is not restored within the time delay selected, ACR 1 opens and signals ACR 2 to close which restores service to the load, provided the alternate source (S2) is energized, as shown in Figure 3-25 on the next page. (A momentary cold-load pickup accessory is sometimes provided for the alternate source recloser to prevent tripping on a cold-load inrush).

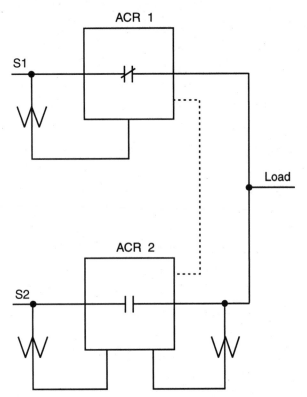

Figure 3-24: Electronically controlled reclosers in both the preferred source and alternate source lines.

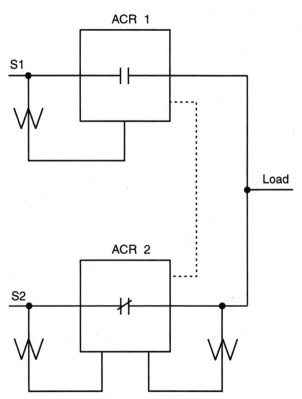

Figure 3-25: If voltage S1 is not restored with the time delay selected, ACR 1 opens and signals ACR 2 to close which restores service to the load.

## Chapter 4

Medium Voltage Distribution

Details of equipment and installation methods related to the transmission, distribution, and control of electric power ranging from 601 to 35,000 V are covered in this chapter. Such equipment includes the following:

- Medium Voltage Substations
- Medium Voltage Transformers
- Medium Voltage Power Factor Correction
- Medium Voltage Insulators and Lightning Arresters
- Medium Voltage Switchboards
- Medium Voltage Circuit Breakers
- Medium Voltage Reclosers
- Medium Voltage Interrupter Switches
- Medium Voltage Fuses
- Medium Voltage Power Distribution
- Medium Voltage Underground Power Distribution
- Medium Voltage Converters
    Frequency Chargers
    Voltage Rectifiers
- Medium Voltage Primary Grounding

Most medium-voltage electrical power installations are handled by the utility or power companies on all residential and small commercial projects. However, some electrical contracts — especially those involving large industrial establishments or government projects — will include the installation of medium-voltage power systems. In some cases, outside area lighting, sports lighting, street lighting, etc., may also fall under the heading of medium-voltage equipment.

Clear and detailed drawings and specifications for medium-voltage power distribution systems are necessary to decrease the hazards of omission or misjudgment of labor and variable factors affecting the labor cost and to ensure an accurate installation.

The construction documents should also include a profile drawing of the system showing the varying elevations of terrain — if other than level ground — and a schedule of ground conditions where possible. If the latter is not practical, the specifications should call for the bidding contractors to make an actual field survey of the area.

ELECTRICAL DISTRIBUTION

The essential elements of an ac electrical system capable of producing useful power include generating stations, transformers, substations, transmission lines and distribution lines. Figure 4-1 on the next page shows these elements and their relationships.

Generation

Electricity is produced at the generating plant at voltages varying from 2,400 to 13,200 V. Transformers are also located at the generating plant to step up the voltage to hundreds of thousands of volts for transmission — a kind of wholesale block technique for economically moving massive amounts of power from the generation point to key locations.

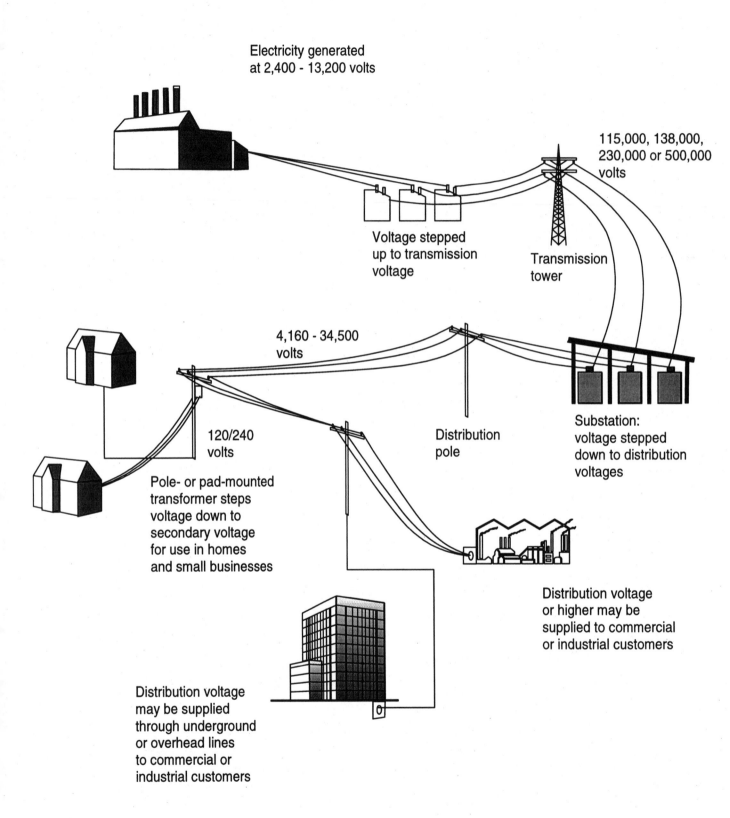

Electricity generated
at 2,400 - 13,200 volts

115,000, 138,000,
230,000 or 500,000
volts

Voltage stepped
up to transmission
voltage

Transmission
tower

4,160 - 34,500
volts

Substation:
voltage stepped
down to distribution
voltages

120/240
volts

Distribution
pole

Pole- or pad-mounted
transformer steps
voltage down to
secondary voltage
for use in homes
and small businesses

Distribution voltage
or higher may be
supplied to commercial
or industrial customers

Distribution voltage
may be supplied
through underground
or overhead lines
to commercial or
industrial customers

Figure 4-1: Parts of a typical electrical distribution system.

Electricity is transported from one part of the system to another by metal conductors, cables made up of many strands of wire. A continuous system of conductors through which electricity flows is called the distribution circuit.

Transmission

The system for moving high-voltage electricity is called the transmission system. Transmission lines are interconnected to form a network of lines. Should one line fail, another will take over the load. Such interconnections provide a reliable system for transporting power from generating plants to communities for use in residential, commercial, and industrial establishments.

Most transmission lines installed by power companies utilize three-phase current — three separate streams of electricity, traveling on separate conductors. This is an efficient way to transport large quantities of electricity. At various points along the way, transformers step down the transmission voltage at facilities known as substations. This is usually the first step in conditioning the voltage for utilization by consumers.

Substations

Substations can be small buildings or fenced-in yards containing switches, transformers, and other electrical equipment and structures. Substations are convenient places to monitor the system and adjust circuits. Devices called regulators, which maintain system voltage as the demand for electricity changes, are also installed in substations. Another device, which momentarily stores energy, is called a capacitor, and is sometimes installed in substations; this device reduces energy losses and improves voltage regulation. Within the substation, rigid tubular or rectangular bars, called busbars or buses, are used as conductors.

At the substation, the transmission voltage is stepped down to voltages below 69,000 V which feed into the distribution system.

The distribution system delivers electrical energy to user's energy-consuming equipment — such as lighting, motors, machines and appliances from residential to industrial establishments.

Conductors called feeders, radiating in all directions from the substation, carry the power from the substation to various distribution centers. At key locations in the distribution system, the voltage is stepped down by transformers to the level needed by the customer. Distribution conductors on the high-voltage side of a transformer are called primary conductors (primaries); those on the low-voltage side are called secondary conductors (secondaries).

Transformers are actually smaller versions of substation distribution transformers that are installed on poles, on concrete pads, or in transformer vaults throughout the distribution system.

Distribution lines carry either three-phase or single-phase current. Single-phase power is normally used for residential and small commercial occupancies, while three-phase power serves most of the other users.

Underground

Most power companies now utilize transmission systems that include both overhead and underground installations. In general, the terms and the devices are the same for both. In the case of the underground system, distribution transformers are installed at or below ground level. Those mounted on concrete slabs are called padmounts (Figure 4-2), while those installed in underground vaults are called submersibles.

Buried conductors (cables) are insulated to protect them from soil chemicals and moisture. Many overhead conductors do not require such protective insulation.

When underground transmission or distribution cables terminate and connect with overhead conductors at buses or on the tops of poles, special devices called potheads or cable terminators are employed. These devices prevent moisture from entering the insulation of the cable and also serve to separate the conductors sufficiently to prevent arcing between them. The cable installation along the length of the pole is known as the cable riser. *See* Figure 4-3.

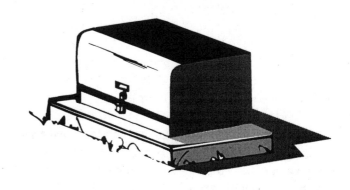

Figure 4-2: Padmount transformer.

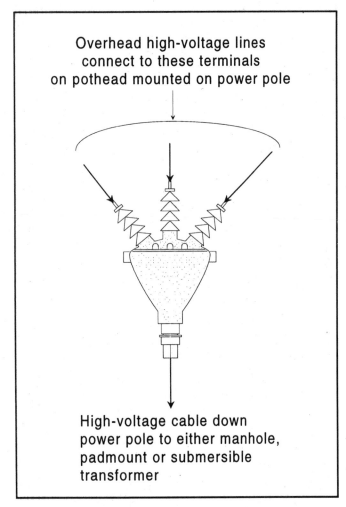

Overhead high-voltage lines
connect to these terminals
on pothead mounted on power pole

High-voltage cable down
power pole to either manhole,
padmount or submersible
transformer

Figure 4-3: Typical pole-mounted pothead.

Secondary Systems

From a practical standpoint, most of the electricians' work will be with the power supply on the secondary (usage) side of the transformer.

Two general arrangements of transformers and secondaries are in common use. The first arrangement is the sectional form, in which a unit of load, such as one city street or city block, is served by secondary conductors, with the transformer located in the middle. The second arrangement is the continuous form in which the primary is installed in one long continuous run, with transformers spaced along it at the most suitable points to form the secondaries. As the load grows or shifts, the transformers spaced along it can be moved or rearranged, if desired. In sectional arrangement, such a load can be cared for only by changing to a larger size of transformer or installing an additional unit in the same section.

One of the greatest advantages of the secondary bank is that the starting currents of motors are divided among transformers, reducing voltage drop and also diminishing the resulting lamp flicker at the various outlets.

Power companies all over the United States and Canada are now trying to incorporate networks into their secondary power systems, especially in areas where a high degree of service reliability is necessary. Around cities and industrial applications, most secondary circuits are three-phase, either 120/208 V or 277/480 V and wye-connected. Usually, two to four primary feeders are run into the area, and transformers are connected alternately to them. The feeders are interconnected in a grid, or network, so that if any feeder goes out of service the load is still carried by the remaining feeders.

The primary feeders supplying networks are run from substations at the usual primary voltage for the system, such as 4160, 4800, 6900, or 13,200 V. Higher voltages are practicable if the loads are large enough to warrant them.

Common Power Supplies

The most common power supply used for residential and small commercial applications is the 120/240-V, single-phase service; it is used primarily for light and power, including single-phase motors up to about $7\frac{1}{2}$ horsepower (hp). A diagram of this service is shown in Figure 4-4, depicting all of the major components.

Four-wire, wye-connected secondaries (Figure 4-5) and four-wire, delta-connected secondaries (Figure 4-6) are common for commercial and industrial applications.

The characteristics of the electric service and the equipment connected to the service must match; also, the characteristics of an electric service will often dictate those for the electrical equipment or vice versa.

Referring again to the three-phase, wye-connected service in Figure 4-5, note that the voltage between any one of the three-phase conductors and the grounded (neutral) conductor is 120 V. Consequently, one would probably assume that the voltage between any two of the phase conductors would be 240 V. However, this is not the case. When dealing with any three-phase system, a factor — the square root of 3 ($\sqrt{3}$) — enters the picture. Therefore, to find the voltage between any two-phase conductors, multiply the voltage of one phase conductor to ground (120 V)

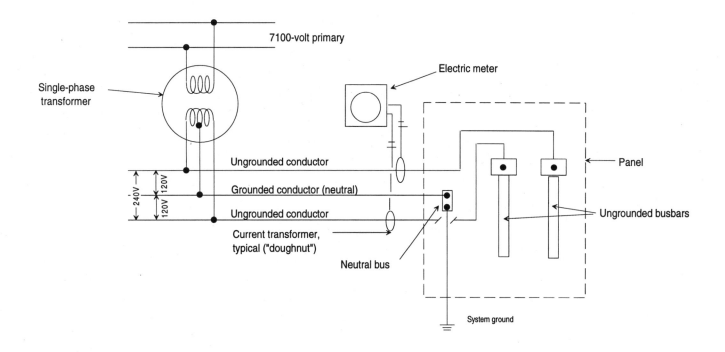

Figure 4-4: Single-phase, 3-wire, 120/240-V electric service.

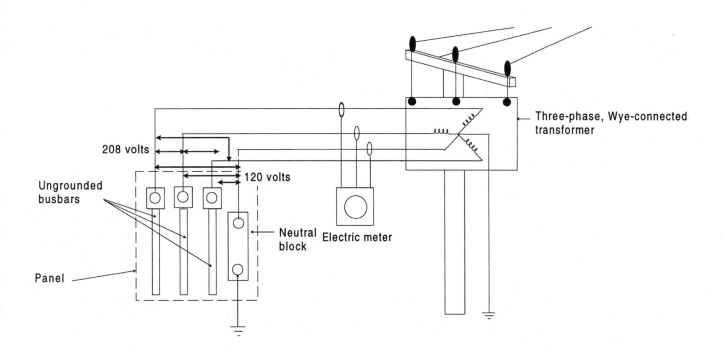

Figure 4-5: Three-phase, 4-wire, 120/208-V wye-connected electric service.

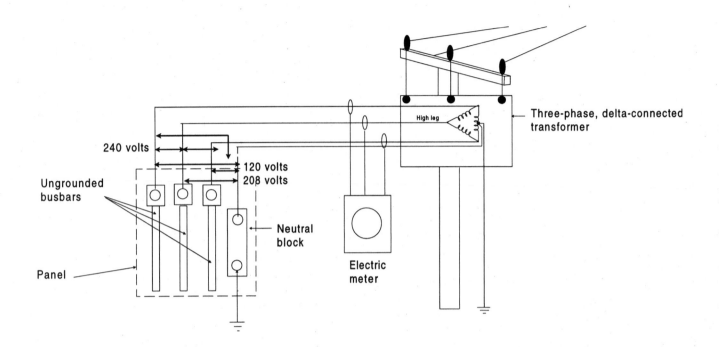

Figure 4-6: Three-phase, 4-wire, 120/240-V delta-connected electric service.

by the square root of 3. Thus,

$$120 \times \sqrt{3} =$$
$$120 \times 1.73(2050808) = 207.84$$
$$\text{Rounds off to } 208 \text{ V}$$

Therefore, feeder and branch circuits connected to 120/208-V, three-phase, four-wire systems can supply the following loads:

- 120-V, single-phase, two-wire.
- 208-V, single-phase, two-wire.
- 208-V, three-phase, three-wire.
- 120/208-V, three-phase, four-wire.

The 120/208-V, three-phase, four-wire system yields an electrical supply for loads rated 120/208-V requiring only three wires. These loads usually include such items as HVAC equipment, cooking units, washers and dryers used in high-rise apartments.

Another popular wye-connected system is the three-phase, four-wire, 277/480-V system.

Feeder and branch circuits connected to the 277/480-V, three-phase, four-wire systems can supply loads rated:

- 277-V, single-phase, two-wire.
- 480-V, single-phase, two-wire
- 480-V, three-phase, three-wire.
- 277/480-V, three-phase, four-wire.

The delta-connected system in Figure 4-6 operates a little differently. While the wye-connected system is formed by connecting one terminal from three equal voltage transformer windings together to make a common terminal, the delta-connected system has its windings connected in series, forming a triangle or the Greek delta symbol Δ. Note in Figure 4-7 that a center-tap terminal is used on one winding to ground the system. On a 240/120-V system, there are 120 V between the center-tap terminal and each ungrounded terminal on either side. There are 240 V across the full winding of each phase.

Refer again to Figure 4-7. Note that a high leg results at point "B." This is known in the trade as the "high leg,"

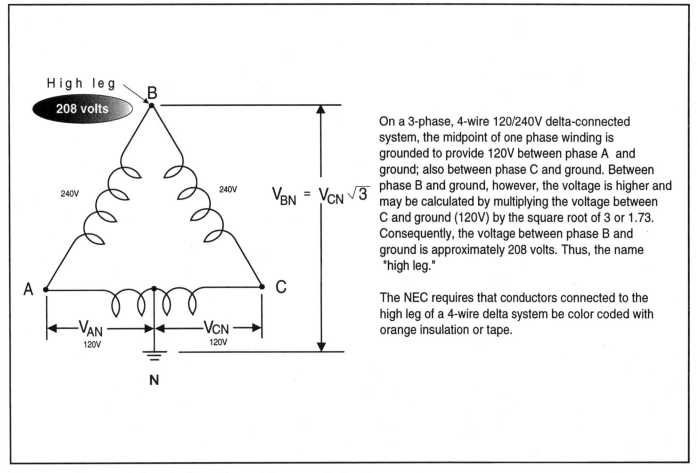

On a 3-phase, 4-wire 120/240V delta-connected system, the midpoint of one phase winding is grounded to provide 120V between phase A and ground; also between phase C and ground. Between phase B and ground, however, the voltage is higher and may be calculated by multiplying the voltage between C and ground (120V) by the square root of 3 or 1.73. Consequently, the voltage between phase B and ground is approximately 208 volts. Thus, the name "high leg."

The NEC requires that conductors connected to the high leg of a 4-wire delta system be color coded with orange insulation or tape.

Figure 4-7: Characteristics of a three-phase delta-connected system.

"red leg," or "wild leg." This high leg has a higher voltage to ground than the other two phases. The voltage of the high leg can be determined by multiplying the voltage to ground of either of the other two legs by the square root of 3. Therefore, if the voltage between phase A to ground is 120 V, the voltage between phase B to ground may be determined as follows:

$$120 \times \sqrt{3} \; = \; 207.84 = 208 \text{ V}$$

From this, it should be obvious that no single-pole breakers should be connected to the high leg of a center-tapped, four-wire, delta-connected system. In fact, *NEC* Section 215-8 requires that the phase busbar or conductor having the higher voltage to ground be permanently marked by an outer finish that is orange in color. By doing so, this will prevent future workers from connecting 120-V single-phase loads to this high leg which will probably result in damaging any equipment connected to the circuit. Re-

member the color *orange*; no 120-V loads are to be connected to this phase.

ONE-LINE DIAGRAMS

The one-line diagram shown in Figure 4-8 (beginning on the next page) is typical of drawings used on medium voltage electrical systems. In general, a one-line diagram is never drawn to scale. Rather, these diagrams show the major components in an electrical system and then utilize only one drawing line to indicate the connections between the various system components. Even though only one line is used between components, this single line may indicate a raceway of two, three, four, or more conductors. Notes, symbols, tables, and detailed drawings are used to supplement and clarify a one-line diagram.

Referring again to Figure 4-8, this drawing shows installation details for a 2000 kVA substation, utilizing a 13.8 kV primary and a 4.16 kV, 3-phase, 3-wire, 60 Hz

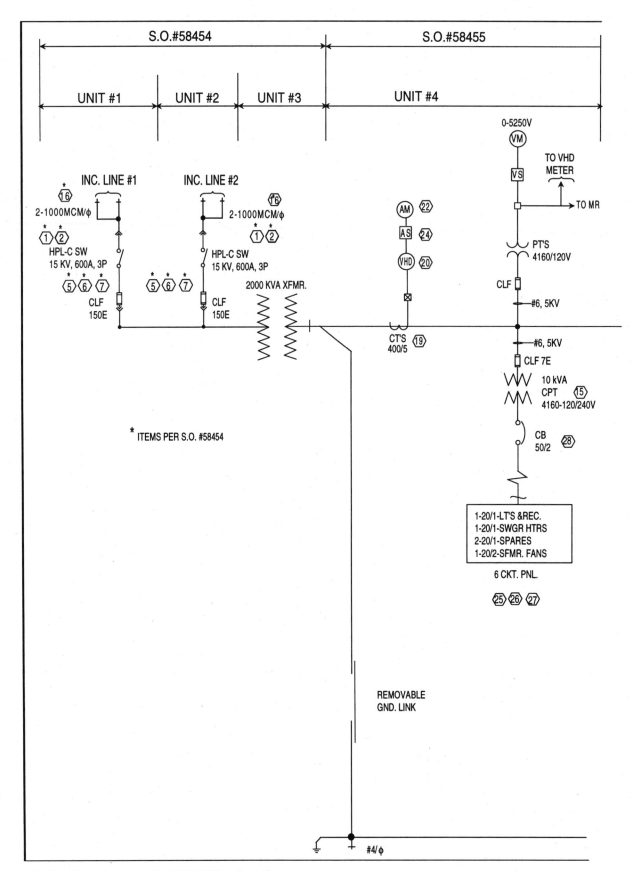

Figure 4-8: One-line diagram of a 2000-kVA substation.

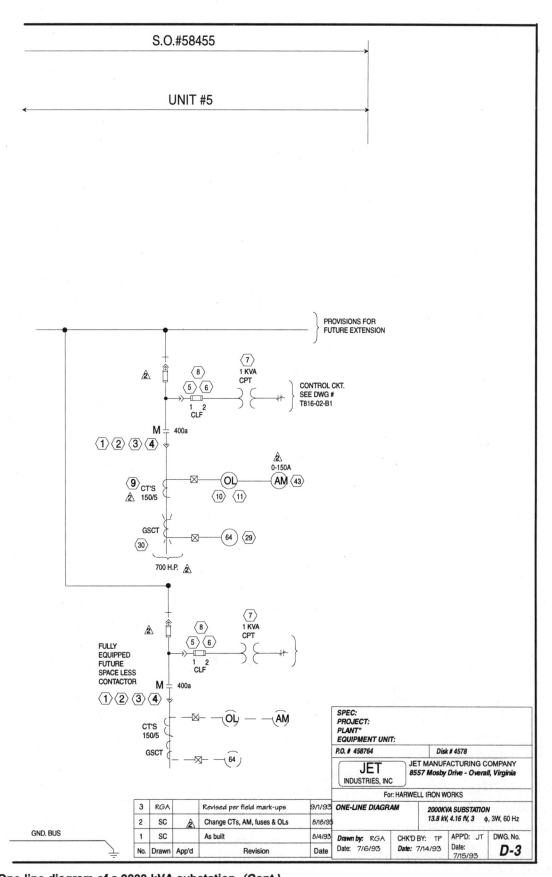

Figure 4-8: One-line diagram of a 2000-kVA substation. *(Cont.)*

secondary. Note that this drawing sheet is divided into the following sections:

- Service order numbers
- Unit numbers
- One-line diagram
- Title block
- Revision notes

Service order numbers: These numbers are arranged at the top of the drawing sheet in a time-sequence, bar-chart type arrangement. For example, S.O. #58454 deals with the primary side of the 2000-kVA transformer, including the transformer itself. This section includes a high-voltage switchgear, with an in-outdoor enclosure. The switchgear itself consists of two HPL-C interrupter switches, each rated at 15 kV, 600 A, and 150E current-limiting fuses (CLF).

Service order #58455 deals with the wiring and related components on the secondary side of the transformer and begins with a low-voltage switchgear with an in-outdoor enclosure.

Unit numbers: Service order #58454 is further subdivided into three units which are indicated as such on the drawing immediately under the S.O. number. Unit #1 deals with incoming line #1; Unit #2 deals with incoming line #2, and Unit #3 covers the 2000-kVA transformer and its related connections and components.

Service order #58455 is further subdivided into two units — Units # 4 and #5. Basically, Unit 4 covers grounding, the installation of current transformers, various meters, potential transformers, a 10-kVA 4160/240-V transformer, and a 6-circuit panel — all derived from a 600-A, 4160-V, 3-wire, 60 Hz main bus.

Unit #5 continues with the main bus and covers the installation and connection of a complete motor-control center, along with another fully-equipped future space, less contactors.

One-line diagram: The one-line diagram takes up most of the drawing sheet and gives an overview of the entire installation. Let's begin at the left side of the drawing where incoming line #1 is indicated. For clarification, this section of the drawing is enlarged in Figure 4-9.

Incoming line #1 (partially abbreviated on the drawings as "INC. LINE #1") consists of two, 1000 MCM (kcmil) conductors per phase as indicated by note "2–1000 MCM/f." In other words, incoming line #1 consists of par-

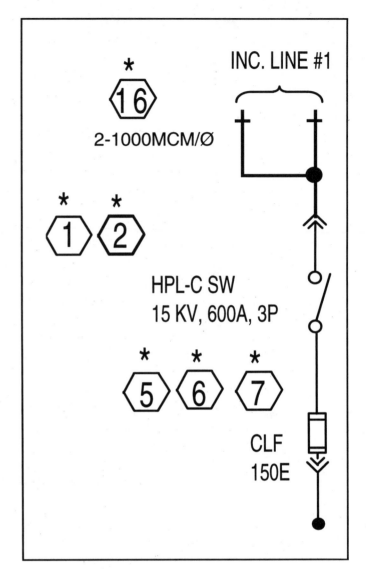

Figure 4-9: Incoming 13.8 kV high-voltage line.

allel 1000 kcmil conductors. Since this is a three-phase system, a total of six 1000 kcmil conductors are utilized.

The single-line continues to engage separable connectors at the single-throw 15-kV, 600-A, 3-pole switch. Overcurrent protection is provided by current-limiting fuses as indicated by the fuse symbol combined with a note. The single-line continues to the high-voltage bus which connects to the primary side of the 2000-kVA transformer.

Incoming line #2 (partially abbreviated on the drawings as "INC. LINE #2") has identical components as line #1. This line also connects to the high-voltage bus which connects to the primary side of the 2000-kVA transformer.

Notice the numerals, each enclosed by a hexagon, placed near various components in these two high-voltage primaries. Note also that an asterisk is placed above each of these marks. A note on the drawing indicates the following:

* ITEMS PER S.O. #58454

These marks appear in a supplemental schedule titled "Bill of Materials" which describes the marked items, lists the number required, manufacturer, catalog number, and a brief description of each. Such schedules are extremely useful to estimators, job superintendents, and workers on the job to ensure that each required item is accounted for and installed.

Title block: Every electrical drawing should have a title block, and it is normally located in the lower right-hand corner of the drawing sheet; the size of the block varies with the size of the drawing and also with the information required.

The title block for the project in question is shown in Figure 4-10 and contains the following:

- Name of the project
- Address of the project
- Name of the owner or client
- Name of the person or firm who prepared the drawing
- Date drawing was made
- Scale(s), if any
- Initials of the drafter, checker, designer, and engineer, with dates under each
- Job number
- Drawing sheet number
- General description of the drawing

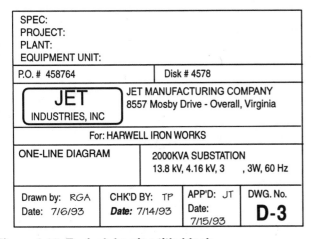

Figure 4-10: Typical drawing title block.

3	RGA		Revised per field mark-ups	9/1/93
2	SC		Change CTs, AM, fuses & OLs	8/18/93
1	SC		As built	8/4/93
No.	Drawn	App'd	Revision	Date

Figure 4-11: Typical drawing revision block.

Revision block: Sometimes electrical drawings will have to be partially redrawn or modified during the planning or construction of a project. It is extremely important that such modifications are noted and dated on the drawings to ensure that the workers have an up-to-date set of drawings to work from. In some situations, sufficient space is left near the title block for dates and description of revisions as shown in Figure 4-11.

CAUTION

When a set of electrical drawings has been revised, always make certain that the most up-to-date set is used for all future layout work. Either destroy the obsolete set of drawings, or else clearly mark on the affected drawing sheets, "Obsolete Drawing — Do Not Use." Also, when working with a set of working drawings and written specifications for the first time, thoroughly check each page to see if any revisions or modifications have been made to the originals. Doing so can save much time and expense to all concerned with the project.

Secondary One-Line Diagrams

Referring again to Figure 4-8, note that a 600-A, 4160-V, 3-phase, 3-wire, 60 Hz aluminum main bus is used to feed the remaining secondary elements. This bus consists of busbars with dimensions of $1\frac{5}{16} \times 2$ in for each phase.

A grounding conductor is shown immediately at the secondary side of the 2000-kVA transformer. This grounding conductor angles off the main line to a vertical line that proceeds toward the bottom of the drawing where it connects to the ground bus of the system which, in turn, is bonded to all qualifying grounding electrodes on the premises. Note the removable ground link between the transformer and the grounding bus connection. The drawing shows this conductor to be No. 4/0 AWG.

Looking back at the 2000-kVA transformer, note that the main bus continues in a horizontal line to the right of the transformer symbol. The first group of equipment encountered is the metering section. Note the current transformers (CTs) are designated by both symbol and note. The "400/5" note indicates that the CTs have a ratio of 400 to 5; that is, if 400 A are flowing in the main bus, only 5 A will flow to the meters. Again, numerals enclosed in hexagons are placed at each component in this section. Referring to the "Bill of Materials" schedule in Figure 4-12, we see a description of Item #19 to be "CT's 400/5 Type JAF-0;" two are required; catalog number is 750X10G304 and is manufactured by GE (General Electric). Continuing from the CTs up to Item #20, the schedule describes this as a three-phase, three-wire, watthour meter with a 15 minute demand and is designed to register with CTs with a ratio of 400/5 and PTs with primary/secondary at 4160/120 V.

Mark	Req'd	Cat. No.	Mfg.	Description
1	2	IC2957B103C	GE	Disc. Handle & Elec. Interlock ASM. (400A) (CAT#116C9928G1)
2	2	IC2957B108E	GE	Vert. Bus (CAT#195B4010G1)(400A) Shutter ASM. (CAT#116C9927G1) (400A)
3	2	1C2957B10BF	GE	Coil Finger ASM. (CAT#194A6949G1) (400A) Safety Catch (CAT#194A6994G1) (400A) Stab Fingers (CAT#232A6635G) (400A)
4	1		Toshiba	5 kV, 300 A, 3P, Vacuum Contactor 120 Vac Rectified Control Type CV461J-GAT2
5	4	2033A73G03	W	5kV Fuse MTG (2/CPT)
6	4	677C592G09	W	5kV, CLF, 2E Fuses Type CLE-PT
7	2	HN1K0EG15	Micron	1kVA, 4160-120 CPT
8	3	9F60LJD809	GE	CLF Size 9R (170A) Type EJ-2 (600HP)
9	3	615X3	GE	CT'S 150/5A Type JCH-0
10	1	CR224C610A	GE	200 Line Block O.L. Rly. 3 Elements Ambient Compensated W/INC. Contact
11	0	CR123C3.56A	GE	O.L. HTR (600HP)

Figure 4-12: Bill of materials schedule.

Mark	Req'd	Cat. No.	Mfg.	Description
11A	3	CR123C3.26A	GE	O.L. HTR. (2.79A) (700HP)
12	1	7022AB	AG	Off Delay R.Y .5-5 SEC.
13	0	CR2810A14A	GE	Machine Tool RLY. INO&INC 120VAC (MR)
14	1	CR294OUM301	GE	Emergency Stop PB (Push to Stop; Pull to Reset) W/NP
15	1	9T28Y5611	GE	10kVA CPT. 4160-120/240V
16	2	643X92	GE	PT'S 4160/120V Type JVM-3/2FU
17	2	9F60CED007	GE	CLF 7E, 4.8kV Type EJ-1
18	2	9F61BNW451	GE	Fuse Clips Size C
19	2	750X10G304	GE	CT'S 400/5 Type JAF-0
20	1	700X64G885	GE	DWH-Meter 3φ, 3W, 60HZ, Type DSM-63 W/15MIN. Demand Register CT'S Ratio 400/5 & PT'S 4160-120V
21	1	50-103021P	GE	VM Scale 0-5250V Type AB-40
22	1	50-103131L	GE	AM Scale 0-400A Type AB-40
23	1	10AA004	GE	VS Type SBM
24	1	10AA012	GE	AS Type SBM
25	1	TL612FL	GE	6 CKT. PNL.
26	4	TQL1120	GE	20/1 C/B Type TQL.
27	1	TQL2120	GE	20/2 C/B Type TQL.
28	1	TEB12050WL	GE	50/2 C/B Type TEB
29	1	3512C12H02	W	Type GR Groundgard RLY. Solid State
30	1	3512C13H03	W	GRD. Sensor
31	2	H	Smout Hollman	1/2 LT. REC.

Figure 4-12: Bill of materials schedule. *(Cont.)*

Mark	Req'd	Cat. No.	Mfg.	Description
32	2	7604-1	GE	LT. SW. & Receptacle
33	2	4D846G20	GE	120VAC, 250W HTR
34	1		Econo	Econo Lift for Contactor
35	11	Lot	Cook	NP/Schedule DWG. 58455-A1
36	3	Hold	T & B	Lug
37	0	50250440LSPK	GE	AM Scale 0-100A PNL. Type 2%
				ACC. Type 250 4-1/2 Case
38	1	NON10	Bus	10A, 250V Fuse
39	1	CP232	AH	2P, 250V Pull-Apart Fuse Block
40	1		Cook	SWGR NP S.O.#58455

Figure 4-12: Bill of materials schedule. *(Cont.)*

The next group to the right of this first group of metering equipment is a second group of metering equipment, connected to the vertical line above the main bus. Notes on the drawing indicate #6 AWG, 5 kV conductors are connected to the main bus and are protected with current-limiting fuses. These conductors continue to two 4160/120-V PTs and then the 120-V conductors continue to a junction box, a voltage-sensing device (voltage synchroscope), and finally to a voltmeter. Also note the branches "To VHD Meter" (varhours demand meter) and "To MR" (meter recorder). This latter device is normally referred to as a "recording demand meter."

At this point on the main bus, note also that a vertical line extends below the main bus. Again, #6 AWG, 5 kV conductors tap onto the main bus and are protected with current-limiting fuses. These high-voltage conductors terminate at a 10 kVA 4160-120/240-V transformer. The secondary side of this transformer has its conductors protected by means of a 50-A, 2-pole circuit breaker which feeds a 6-circuit panel. Note that this panel contains four, 20-A, 1-pole circuit breakers and one 20-A, 2-pole circuit breaker. The circuits for the 1-pole breakers are also indi-

cated on the drawing as follows:

- 1-20/1-LTs & REC.
- 1-20/1-SWGR HTRS
- 2-20/1-SPARES
- 1-20/2-XFMR FANS

The interpretation of the abbreviations is as follows:

LTS	=	Lights
REC	=	Receptacles
SWGR HTRS	=	Switchgear heaters
XFMR FANS	=	Transformer fans

The remaining two taps from the main bus in the drawing in question are for feeding two motor-control centers (MCC); one to be put into use immediately while the other is a fully-equipped MCC, less contactors, for future use. Let's look at the complete MCC first. This is the last tap from the main bus in Figure 4-8. An enlarged view of this section is shown in Figure 4-13 for clarification.

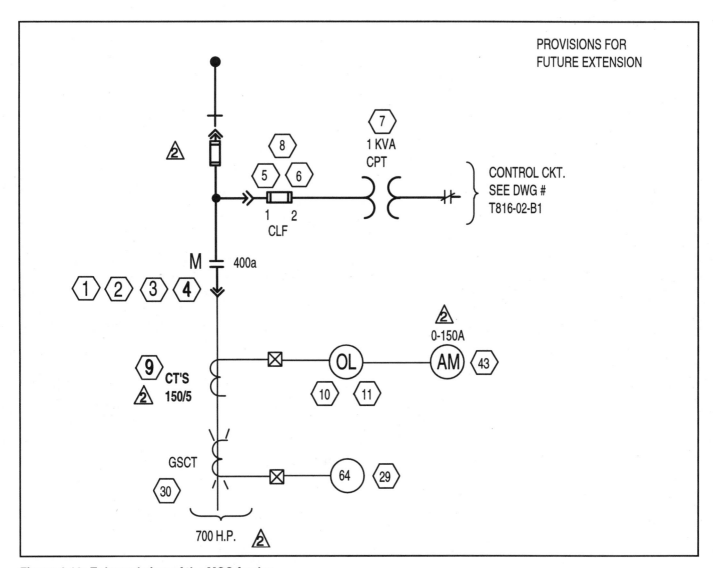

Figure 4-13: Enlarged view of the MCC feeder.

This feeder is provided with overcurrent protection by means of current-limiting fuses (CLF 9R), which are fuse type EJ-2, rated at 170 A. Immediately beneath this device, note that a tap is taken from the main line, fused with 5 kV MTG fuses and also 5 kV, CLF, 2E fuses before terminating at a 1 kVA, 4160/120-V CPT transformer. This transformer is provided to accommodate the 120-V control circuit shown in Figure 4-14 on the next page.

Now let's backtrack to the main feeder and continue downward to a contactor before another group of current transformers (CTs) are installed in the circuit. These CTs are accompanied by notes and Mark No. 9. Referring to the schedule in Figure 4-12 for a description of Mark #9, we see that these three CTs have a ratio of 150/5; that is, when the circuit is drawing 150 A, the metering devices will receive only 5 A, but the meter itself will indicate 150 A.

This circuit continues to a 200 line block overload relay with three elements, and then on to an ammeter with a range of 0-150 A.

The next item on this main vertical feeder is a ground sensor which is connected to a solid-state groundguard relay. The feeder then enters, and connects to the busbars, in a motor-control center (MCC). The remaining feeder in the one-line wiring diagram under consideration is for future use and is similar to the circuit just described.

Shop Drawings

When large pieces of electrical equipment are needed, such as high-voltage switchgear and motor control cen-

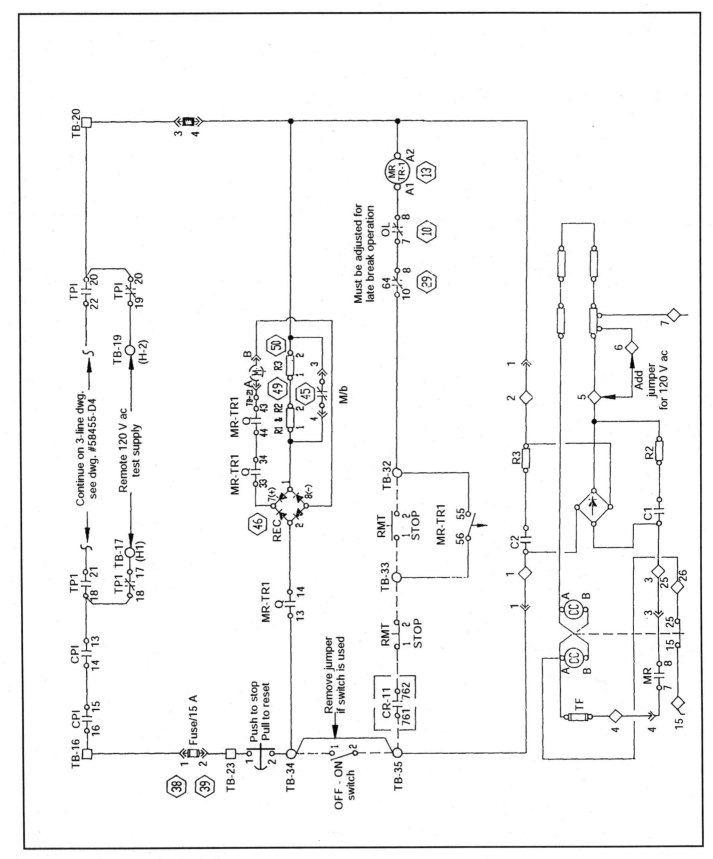

Figure 4-14: Motor-control circuit diagram.

ters, most are custom built for each individual project. In doing so, *shop drawings* are normally furnished by the equipment manufacturer — prior to shipment — to ensure that the equipment will fit the location at the shop site, and also to instruct workers on the job how to prepare for the equipment; that is, rough-in conduits, cable tray, etc.

Shop drawings will usually include connection diagrams for all components that must be "field wired" or connected. As-built drawings, including detailed factory-wired connection diagrams are also included to assist workers and maintenance personnel in making the final

connections, and then to assist them in troubleshooting problems once the system is in operation. These same drawings are also essential for additions.

Details of this combined switchgear and transformer equipment for the 2000 kVA substation are shown in Figures 4-15, 4-16, and 4-17. Figure 4-18 shows a cross-sectional view of the switchgear. Such detail drawings are provided in this situation to show the anchoring arrangement (including dimensions) of the switchgear equipment. These drawings are also useful for providing unloading and access space.

ANCHORING DETAIL

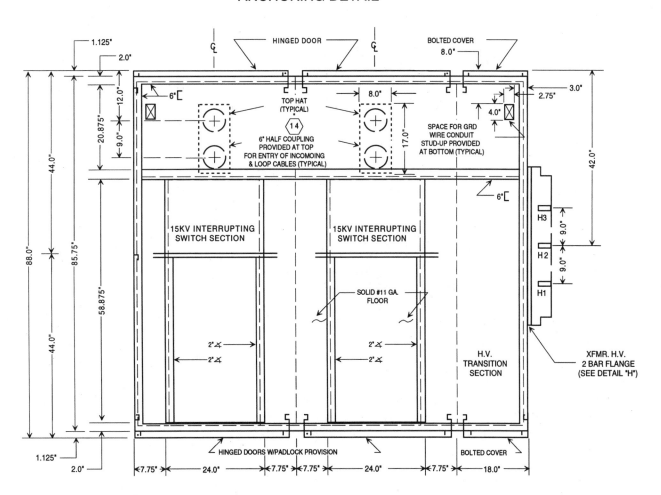

BASE PLAN

Figure 4-15: Base plan of the switchgear and transformer arrangement.

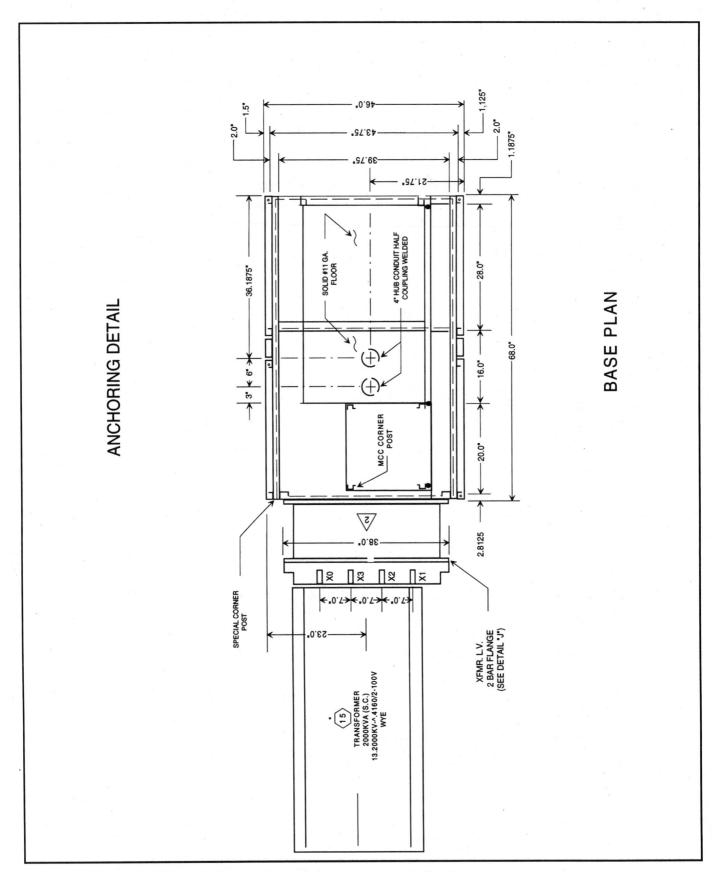

Figure 4-15: Base plan of the switchgear and transformer arrangement. *(Cont.)*

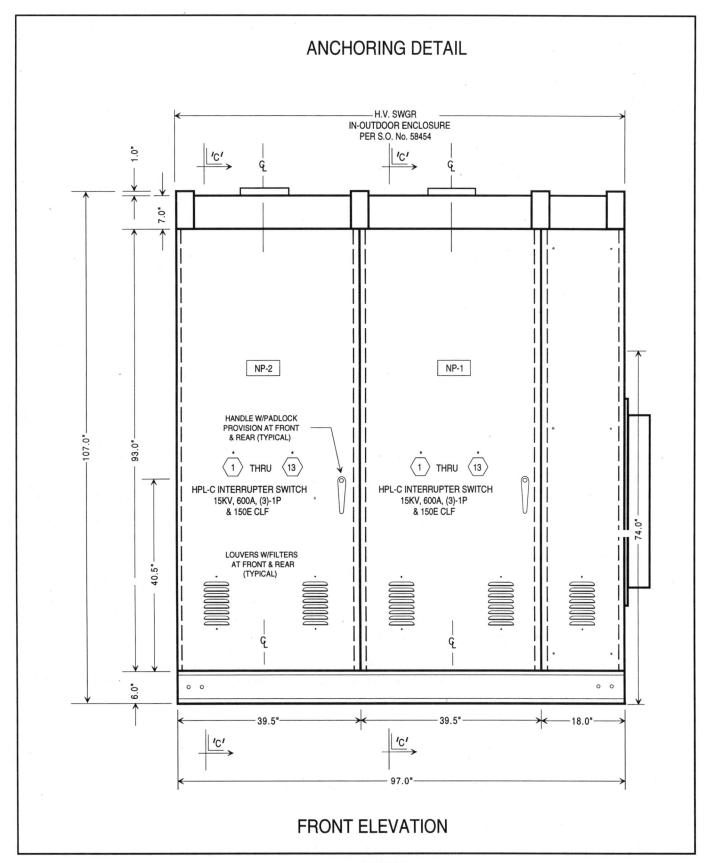

Figure 4-16: Front elevation of the switchgear and transformer arrangement.

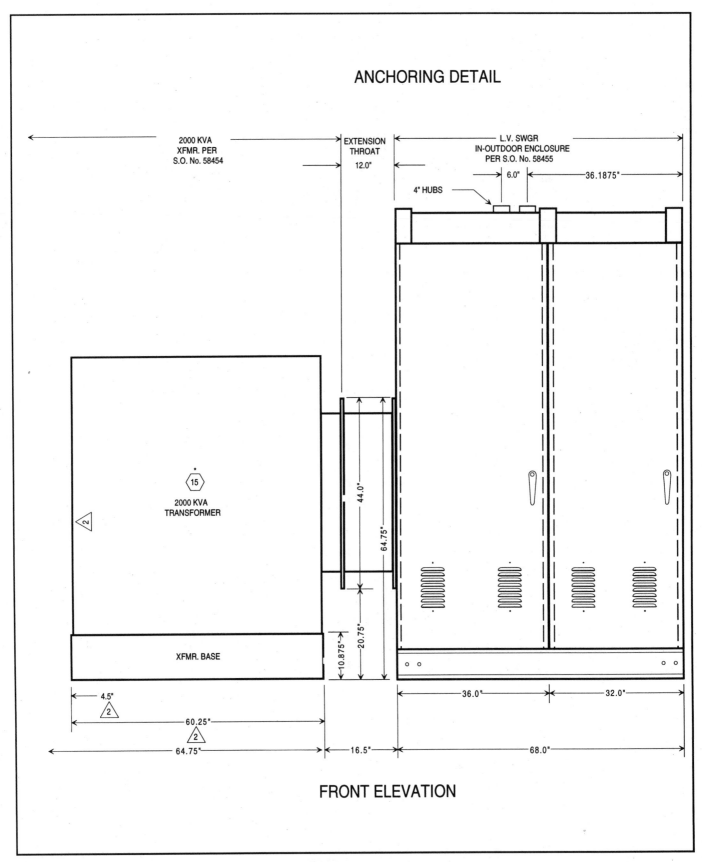

Figure 4-16: Front elevation of the switchgear and transformer arrangement. (Cont.)

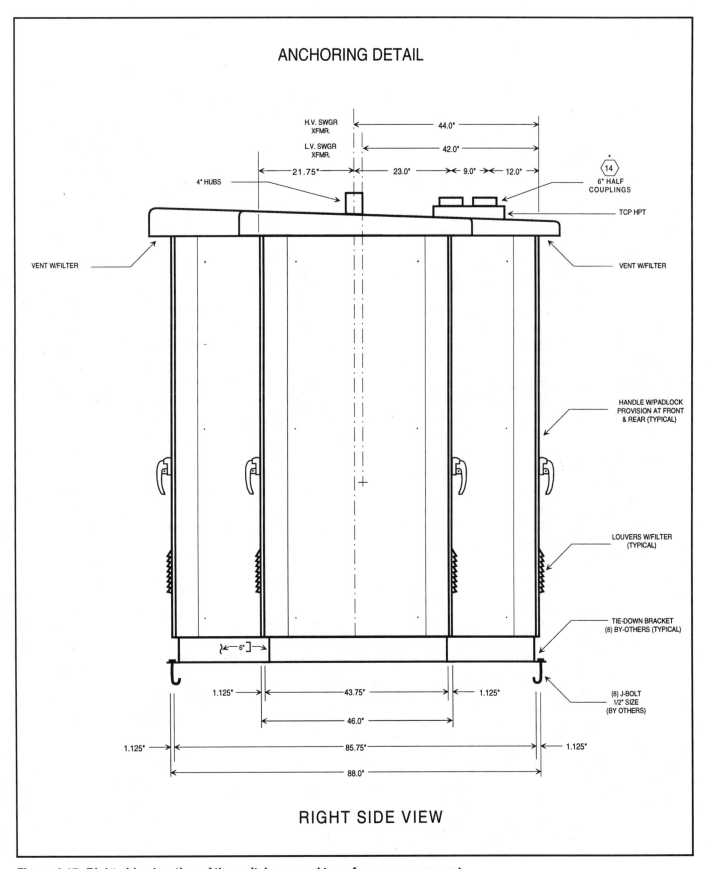

Figure 4-17: Right side elevation of the switchgear and transformer arrangement.

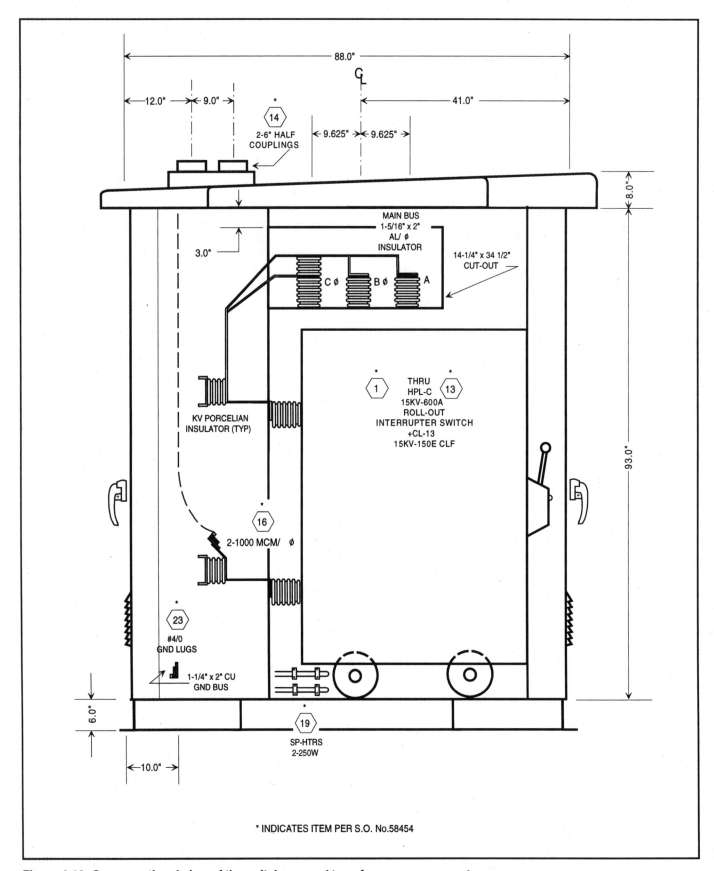

Figure 4-18: Cross-sectional view of the switchgear and transformer arrangement.

Distribution Systems

In the flow of power from the generating station to the utilization point, distribution systems provide the link between the high-voltage transmission or subtransmission systems and the consumer of electric power. Key points of distribution systems are the distribution substations, which contain transformers and the switchgear necessary to deliver low-voltage power to the distribution-system primary feeders. A common voltage transformation is from 33 kV to the standard distribution voltage of 4.16 kV.

Voltage control is obtained economically with on-load tap changers on the transformers in distribution substations, but some installations require separate voltage-regulating transformers, individual feeder-voltage regulators, or induction-voltage regulators.

Metering, relaying, and automatic controlling are functions included in most distribution substations. There is rarely a distribution substation which does not operate automatically. The main items to be controlled are the feeder circuit breakers, which makes the control installation relatively simple. Metering is required for statistical purposes or for billing if the power-utility company providing the power does not own the distribution system.

Rural Substations

The main types of substations, classified according to their purposes, are the rural and the industrial substation. The rural substation is very simple and is designed with the lowest possible cost in mind. A single-line diagram of a rural substation is shown in Figure 4-19. The subtransmission line connects through the disconnecting switch and the high-voltage fuse to the substation transformer. No circuit breaker is used. The low voltage from the transformer is supplied to the distribution bus and from there to several feeders. Each feeder has a load-disconnecting switch and a fuse. Meters are connected to the current transformer and the potential transformer, which has its own disconnecting switch.

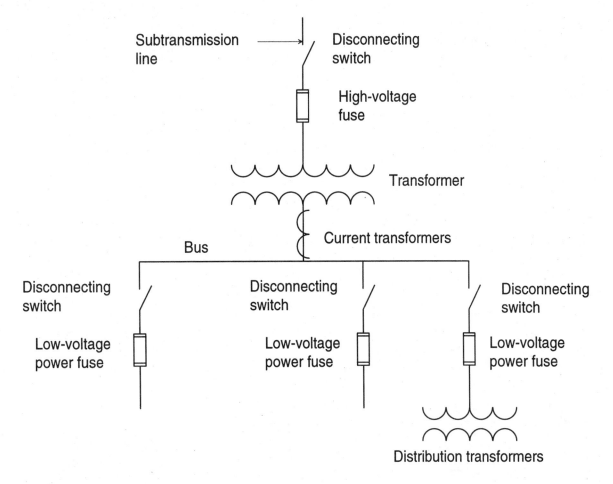

Figure 4-19: Single-line diagram of a substation.

Industrial Substation

An industrial substation could be located near an industrial plant or in the plant building. It must supply a low utilization voltage of 120 V for lighting and small motors, and voltages of 6.9 or 13.8 kV for large motors, and should provide a high security of service. Industrial substations very often include capacitors or a synchronous condenser to control the power factor. An industrial substation built as a unit is shown diagrammatically in Figure 4-20. It is a structure containing five sections 1 to 5. Section 1 is the supply section, which houses the pothead and the disconnecting switch. The transformer section 2 houses the 4160/600-V transformer and the 600-V bus. The feeder sections 3 to 5 each contain a bus, an air circuit breaker, and a feeder cable which connects to the motor load. One section includes the 600-120/208-V lighting transformer and the secondary feeder, which connects to the distribution panel for lighting and similar 120-V circuits.

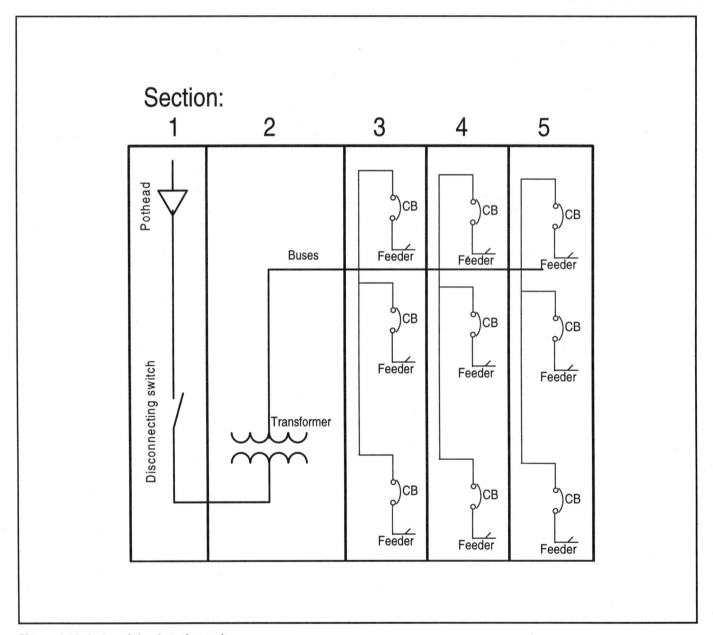

Figure 4-20: Industrial substation unit.

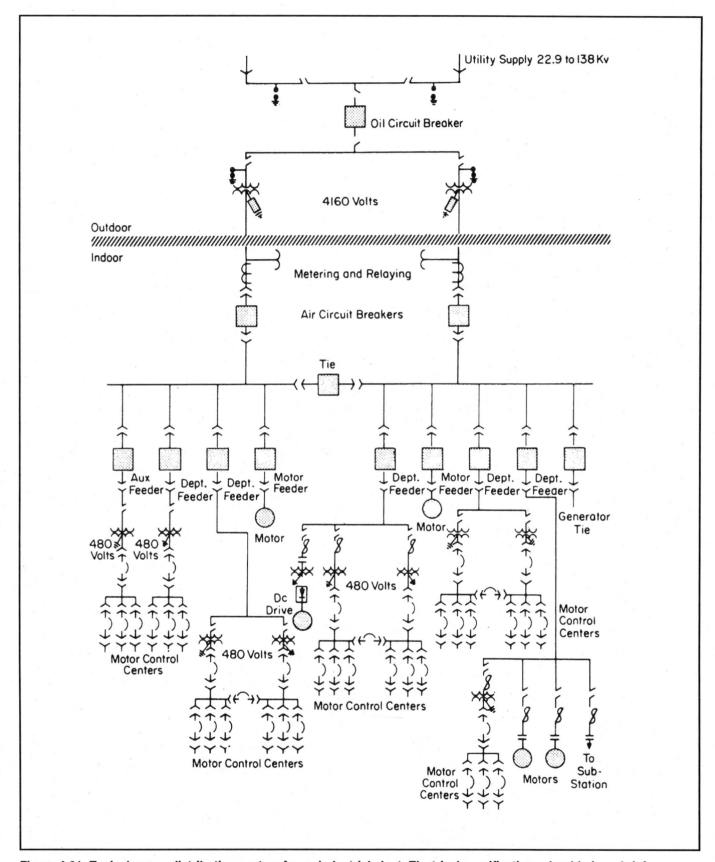

Figure 4-21: Typical power distribution system for an industrial plant. Electrical specifications should give a brief description of the overall electrical distribution system, complete with electrical diagrams as shown here.

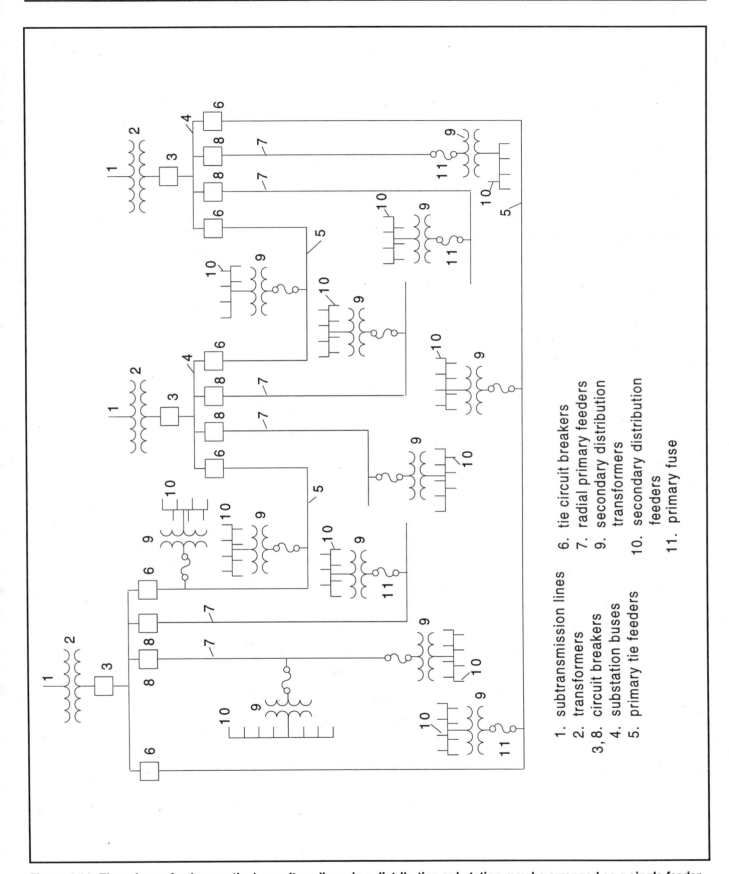

Figure 4-22: The primary feeders, or the low-voltage lines, in a distribution substation may be arranged as a single feeder, double feeder, or a network as shown in this drawing.

INTERRUPTER SWITCHES

As a result of electrical loads on distribution lines, utilities have moved to higher and higher operating voltages — 13.8 kV, 23 kV, 34.5 kV and higher. This higher voltage has led to the need for more and more sectionalizing points to minimize the impact of outage downtime. Compounding the problem, the probability of fault exposure has increased as operating voltages have moved to higher levels. Not only are there more miles of line, but the higher operating voltages are generally more prone to faults from tree limbs, transformer bushing flashovers, and similar problems.

At one time, conventional disconnects met the need for sectionalizing devices, but this is no longer true. The switching capability of a disconnect, while marginal at 2.4 to 4.8 kV, is completely inadequate at 13.8 kV and higher. To isolate a section of line by opening a disconnect thus requires dropping the entire feeder as a preliminary operation, aggravates the situation still further. And under emergency conditions, the chances of operator error increase proportionally.

A wide variety of modern switch styles is now available, each style geared to the economics and the switching duty of the specific application. The single-pole styles and the side-break styles are intended for pole-top installation on distribution feeders; the vertical-break styles are intended for distribution substations or feeders. These switches perform all their switching duties with no external arc and also provide foolproof isolation of a visible air gap. Here are a few examples of their use.

- During emergency situations requiring quick action, a modern interrupter switch can drop the load without complicated breaker-and-switch sequencing.
- There is no need to drop individual loads as a preliminary operation, since the switch can drop the entire load.
- Lines may be extended and additional load accommodated (within the range of the switch) without affecting switching ability.
- A loaded circuit may be dropped inadvertently (through an error or misunderstanding), with no hazard to the operator or to the system.
- Interlocking may be dispensed with: for example, in transformer operation or applications, between the primary switch and the secondary breaker.

Because of the no-external-arc feature of most modern interrupter switches, phase spacing may be much less than those established for the older horn-gap switches. Figure 4-23 gives minimum switch phase spacings on feeders or in substations compared with industry standard phase spacing.

On the secondary side of the substation there are more feeders, more heavily loaded lines, and longer lines, all

Interrupter Switch	Minimum Phase Spacing in Inches				
	ANSI Standards				
Style	Rating (kV) Nominal	Alduti Interrupter Switches	Vertical Break Disconnects and Bus Supports	Side-Break Disconnects	Horn-Gap Switches
Single-pole or three-pole vertical-break	7.2	18	18	30	36
	14.4	24	24	30	36
	23	30	30	35	48
Three-pole side-break formed-channel-base	7.2	27	18	30	36
	14.4	32	24	30	36
	23	39	30	36	48

Figure 4-23: Minimum switch phase spacings on feeders or in substations.

operating at voltages of 13.8 kV and up. Even under these adverse conditions, modern interrupter switches are ideal devices for splitting load, dropping lines or cables, or even dropping the entire load in an emergency. A wiring dia-

gram of a substation application of interrupter switches is shown in Figure 4-24, while typical applications of interrupter switches on distribution feeders are shown in Figure 4-25.

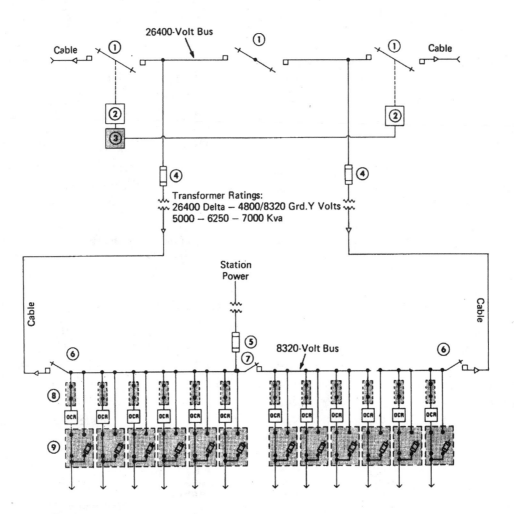

① S&C Interrupter Switches — Three-Pole, Double-Break Style
② S&C Switch Operators
③ S&C Standard Control Panel — For Automatic Transfer
④ S&C Power Fuses — Type SMD-1A, Right-Angle Style
⑤ S&C Power Fuse — Type SMD-20, Vertical-Offset Style
⑥ S&C Interrupter Switches — Single-Pole — Station Style
⑦ S&C Interrupter Switch — Three-Pole, Vertical-Break Style
⑧ S&C Loadbuster Disconnects
⑨ S&C Loadbuster Bypass Disconnects

Figure 4-24: Wiring diagram of a substation application utilizing interrupter switches.

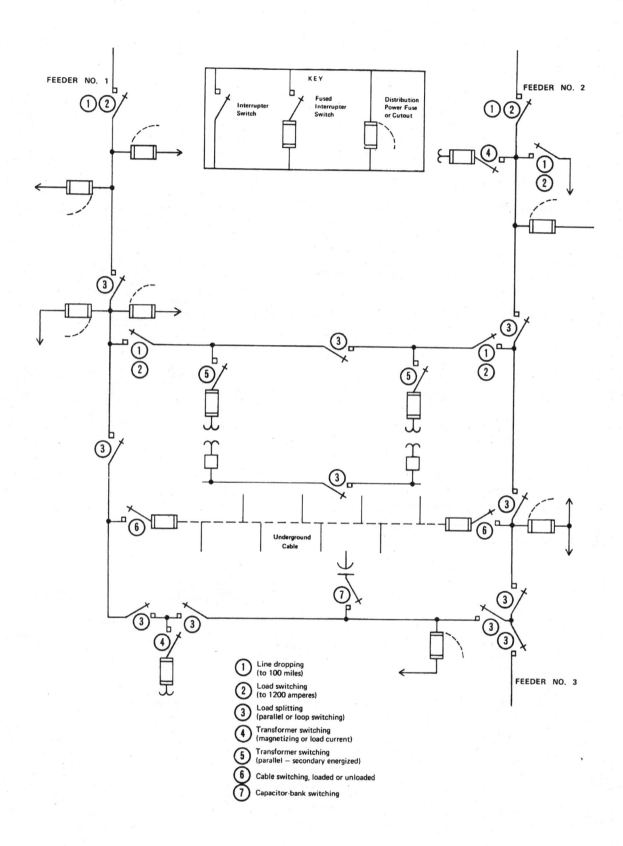

Figure 4-25: Typical application of interrupter switches used on distribution feeders.

UNDERGROUND SYSTEMS

There are several methods used to install underground wiring, but the most common include direct-burial cables and the use of duct lines or duct banks.

The method used depends on the type of wiring, soil conditions, allotted budget for the work, etc.

Direct-burial installations will range from small, single-conductor wires to multiconductor cables for power or communications or alarm systems. In any case, the conductors are installed in the ground either by placing them in an excavated trench, which is later backfilled, or by burying them directly by means of some form of cable plow, which opens a furrow, feeds the conductors into the furrow, and closes the furrow over the conductor.

Sometimes it becomes necessary to use lengths of conduit in conjunction with direct-burial installations, especially where the cables emerge on the surface of the ground or terminate at an outlet or junction box. Also, where the cables cross a roadway or concrete pavement, it is best to install a length of conduit under these areas in case the cable must be removed at a later date. By doing so, the road or concrete pad will not have to be disturbed.

Figure 4-26 shows a cross section of a trench with direct-burial cable installed. Note the sand base on which the conductors lie to protect them from sharp stones and such. A treated board is placed over the conductors in the trench to offer protection during any digging that might occur in the future. Also, a continuous warning ribbon is laid in the trench, some distance above the board, to warn future diggers that electrical conductors are present in the area.

Conductor Types

Type USE (Underground service-entrance cable): This type of cable is approved for underground use since it has a moisture-resistant covering.

Cabled single-conductor Type USE constructions may have a bare copper conductor cabled with the assembly. Type USE single, parallel, or cabled conductor assemblies may have a bare copper concentric conductor applied. These constructions do not require an outer overall covering.

Type USE cable may be used for underground services, feeders, and branch circuits.

Type UF (Underground feeder and branch-circuit cable): This type of cable is manufactured in sizes from No. 14 AWG copper through No. 4/0. In general, the overall covering of Type UF cable is flame-retardant, moisture-, fungus-, and corrosion-resistant, and is suitable for direct burial in the ground.

Type UF cable may be used for direct-burial, underground installations as feeders or branch circuits when provided with overcurrent protection of the rated ampacity as required by the *NEC*.

Where single-conductor cables are installed, all cables of the feeder circuit, subfeeder circuit, or branch circuit (including the neutral and equipment grounding conductor, if any) must be run together in the same trench or raceway.

Nonmetallic-armored cable: This type of cable is also used for underground installations. The interlocking armor consists of a single strip of interlocking tape that extends for the length of the cable. The surface of the cable is rounded, which allows it to deflect blows from picks and shovels much better than flat-bend armor. The cable must have an outer covering that will not corrode or rot. An asphalt-jute finish may be placed over the cable if it is to be subjected to particularly harsh corrosive environments.

Minimum Cover Requirements

The *NEC* specifies minimum cover requirements for direct-buried cable. Furthermore, all underground installations must be grounded and bonded in accordance with *NEC* Article 250.

Where direct-buried cables emerge from the ground, they must be protected by enclosures or raceways extending from the minimum cover distance as specified in *NEC* tables. However, in no cases will the protection be required to exceed 18 in below the finished grade.

Practical Application

Methods of installing direct-burial underground cable vary according to the length of the installation, the size of cable being installed, and the soil conditions. For short runs, from, say, a residential basement to a garage located 20 or so feet away, the excavation is often done by hand. For longer runs, power equipment is almost always used.

In general, the trench is opened to the correct depth with an entrenching tractor or backhoe. All sharp rocks, roots, and similar items are then removed from the trench to prevent these objects from damaging the direct-burial cable. If soil conditions dictate, a 3-in layer of clean sand is

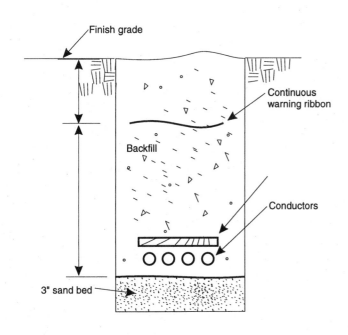

Figure 4-26: Cross-sectional view of a trench with a direct-burial cable installed.

poured into the bottom of the trench to further protect the direct-burial cable.

Once installed, another layer of sand may be placed over the cable for protection against sharp rocks; the trench is then backfilled. A treated wooden plank may also be used for cable protection; a yellow warning ribbon — designed for the purpose — is also a good idea; both are shown in Figure 4-26.

Duct Systems

By definition, a *duct* is a single enclosed raceway through which conductors or cables are pulled. One or more ducts in a single trench is usually referred to as a *duct bank*. A duct system provides a safe passageway for power lines, communication cables, or both.

Depending upon the wiring system and the soil conditions, a duct bank may be placed in a trench and covered with earth or enclosed in concrete. Underground duct systems also include manholes, handholes, transformer vaults, and risers.

Manholes are set at various intervals in an underground duct system to facilitate pulling conductors or cables when first installed, and to allow for testing and maintenance later on. Access to manholes are provided through *throats*

extending from the manhole compartment to the surface (ground level). At ground level, a manhole cover closes off the manhole area tightly.

In general, underground cable runs normally terminate at a manhole, where they are spliced to another length of cable. Manholes are sometimes constructed of brick and concrete. Most, however, are prefabricated, reinforced concrete, made in two parts — the base and the throat — for quicker installation. Their design provides room for workers to carry out all appropriate activities inside them, and they are also provided with a means for drainage. *See* Figure 4-27.

There are three basic designs of manholes: two-way, three-way, and four-way. In a two-way manhole, ducts and cables enter and leave in two directions. A three-way manhole is similar to a two-way manhole, except that one additional duct/cable run leaves the manhole. Four duct/cable runs are installed in a four-way manhole. *See* Figure 4-28. Also see Figure 4-29 for specifications of a typical manhole.

Transformer vaults house power transformers, voltage regulators, network protectors, meters and circuit breakers. Other cables end at a substation or terminate as risers — connecting to overhead lines by means of a pothead.

Types Of Ducts

Ducts for use in underground electrical systems are made of fiber, vitrified tile, metal conduit, plastic or poured concrete. In some existing installations, the worker may find that asbestos/cement ducts have been used. In most areas, a contractor must be certified before removing or disturbing asbestos ductwork, and then extreme caution must be practiced at all times.

The inside diameter of ducts for specific installations is determined by the size of the cable that the ducts will house. Sizes from 2 to 6 in are common.

Fiber duct: Fiber duct is made with wood pulp and various chemicals to provide a lightweight raceway that will resist rotting. It can be used enclosed in a concrete envelope with at least 3 in of concrete on all sides. The extremely smooth interior walls of this type of duct facilitates cable pulling through them.

Vitrified clay duct: Vitrified clay duct is sometimes called *hollow brick*. Its main use is in underground systems for low-voltage and communication cables and is especially useful where the duct run must be routed around underground obstacles, because the individual pieces of duct are shorter than other types.

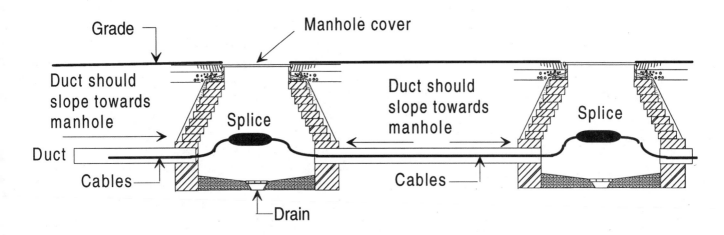

Figure 4-27: Cross-sectional view of typical underground duct system.

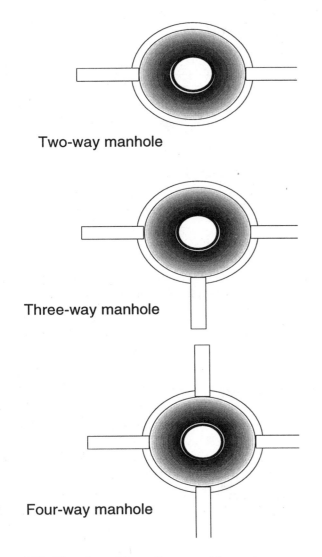

Two-way manhole

Three-way manhole

Four-way manhole

Figure 4-28: Plan view of two-, three-, and four-way manholes.

The four-way multiple duct is the type most often encountered. However, vitrified duct is available in sizes up to 16 ducts in one bank. The square ducts are usually $3\frac{1}{2}$ in in diameter, while round ducts vary from $3\frac{1}{2}$ to $4\frac{1}{2}$ in.

When vitrified clay ducts are installed, their joints should be staggered to prevent a flame or spark from a defective cable in one duct from damaging cable in an adjacent duct.

Metal conduit: Metal conduit, such as iron, rigid metal conduit, intermediate metal conduit, etc., is relatively more expensive to install than other kinds of underground ductwork. However, it provides better protection than most other types, especially against the hazards caused by future excavation.

Plastic conduit: Plastic conduit is made of polyvinyl-chloride (PVC), polyethylene (PE), or styrene. Since they are available in lengths up to 30 ft, fewer couplings are needed than with many types of duct systems. PVC conduit is currently very popular for underground electrical systems since it is light in weight, relatively inexpensive, and requires less labor to install.

Monolithic concrete ducts: This type of system is poured at the job site. Multiple duct lines can be formed using tubing cores or spacers. The cores may be removed after the concrete has set. Although relatively expensive, this system has the advantage of creating a very clean duct interior with no residue that can decay. It is also useful when curves or bends in duct systems are necessary.

Cable-in-duct: This is another popular duct type that offers a reduction in labor cost when installed. It is manufactured with cables already installed. Both the duct and the cable it contains is shipped on a reel to facilitate installing

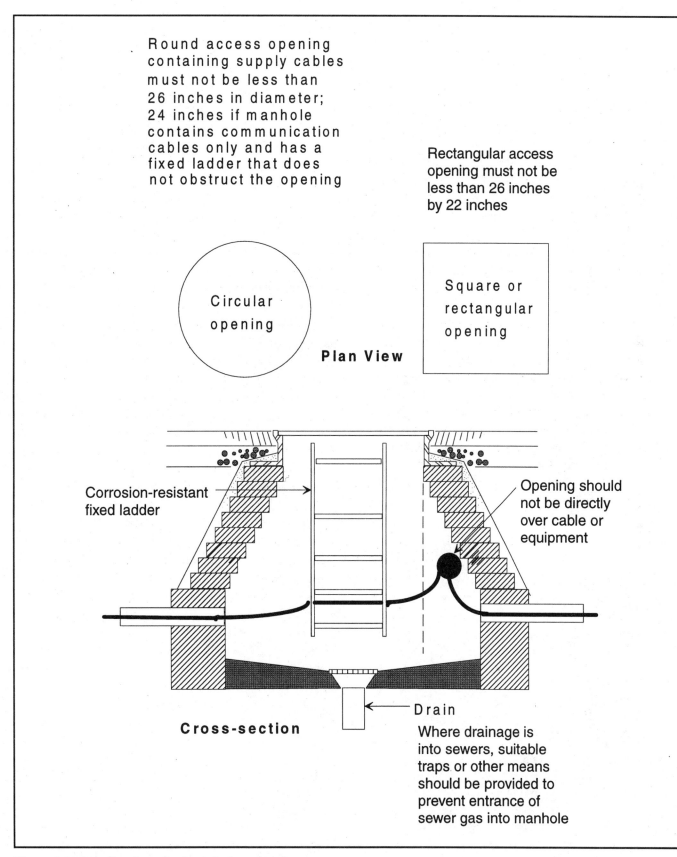

Round access opening containing supply cables must not be less than 26 inches in diameter; 24 inches if manhole contains communication cables only and has a fixed ladder that does not obstruct the opening

Rectangular access opening must not be less than 26 inches by 22 inches

Circular opening

Square or rectangular opening

Plan View

Corrosion-resistant fixed ladder

Opening should not be directly over cable or equipment

Cross-section

Drain

Where drainage is into sewers, suitable traps or other means should be provided to prevent entrance of sewer gas into manhole

Figure 4-29: Details of a typical manhole.

the entire system with ease. Once installed, the cables can be withdrawn or replaced in the future if it should become necessary.

Duct Installation

Soil conditions will dictate whether concrete encasement is required. If the soil is not firm, concrete encasement is mandatory. Concrete is also required with certain types of duct lines that are not able to withstand the pressure of an earth covering. If the soil is firm, and concrete encasement is still desired (or specified), the trench need only be wide enough for the ducts and concrete encasement. The concrete is then poured between the conduit and the earth wall. If the soil is not very firm and concrete is required, 3 additional inches should be allowed on each side of the duct banks to permit the use of concrete forms.

Ducts may be grouped in any of several different ways, but for power distribution, each duct should have at least one side exposed to the earth or the outside of the concrete envelope. This means that the pattern of ducts for power distribution should be restricted to either a two-conduit width or a two-conduit depth. This permits the heat generated by power transmission to dissipate into the surrounding earth. In other words, ducts for power distribution should not be completely surrounded by other ducts. When this type of situation exists, the inner ducts may be referred to as *dead ducts*. The heat that these ducts radiate is not dissipated as fast as from the ducts surrounding them. While not suited for power cable, these dead ducts may be used for street lighting, control cable, or communication cable. The heat generated by these types of cables is relatively low, so the ducts can be arranged in any convenient configuration.

Figures 4-30 through 4-33 show duct-bank configurations and dimensions given in the *NEC*, and should serve as a guide for design details.

Part 341A of the *National Electric Safety Code* covers control of bending, pulling, tensions and sidewall pressures during the installation of cable. This section also covers cleaning foreign material from ducts.

SECONDARY DISTRIBUTION SYSTEMS

Secondary distribution systems are installed and protected in a similar fashion to the higher-voltage primary distribution systems. The more important lines in the secondary system will require differential protection, while the less important ones may need nothing more than conventional overcurrent protection. The same rules apply, however, in that a fault must be quickly removed from the system by tripping the minimum number of circuit breakers or blowing the minimum number of fuses, leaving the balance of the system in operation.

There are two general arrangements of transformers and secondaries used. The first arrangement is the sectional form, in which a unit of load, such as one city street or city block, is served by a fixed length of secondary, with the transformer located in the middle. The second arrangement is the continuous form where the secondary is installed in one long continuous run — with transformers spaced along it at the most suitable points. As the load grows or shifts, the transformers spaced along it can be moved or rearranged, if desired. In sectional arrangement, such a load can be cared for only by changing to a larger size of transformer or installing an additional unit in the same section.

One of the greatest advantages of the secondary bank is that the starting currents of motors are divided between transformers, reducing voltage drops and also diminishing the resulting lamp flicker at the various outlets.

In the sectional arrangement, each transformer feeds a section of the secondary and is separate from any other. If a transformer becomes overloaded, it is not helped by adjacent transformers; rather, each transformer acts as a unit by itself. Therefore, if a transformer fails, there is an interruption in the distribution service of the section of the secondary distribution system that it feeds. This is the most often used layout for secondary distribution systems at the present time.

Power companies all over the United States are now trying to incorporate networks into their secondary power systems, especially in areas where a high degree of service reliability is necessary. Around cities and industrial applications, most secondary circuits are three-phase — either 120/208 V or 480/208 V, wye connected. Usually, two to four primary feeders are run into the area, and transformers are connected alternately to them. The feeders are interconnected in a grid, or network, so that if any feeder goes out of service, the load is still carried by the remaining feeders.

To protect a grid-type power system, a network protector is usually installed between the transformer and the secondary mains. This protector consists of a low-voltage circuit breaker controlled by relays, which cause it to open when reverse current flows from the secondaries into the transformer and to close again when normal conditions are restored. If a short circuit or ground fault should occur on a primary feeder or on any transformer connected to it, the

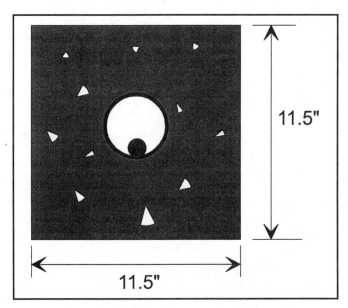

Figure 4-30: Electrical duct bank for one electrical duct.

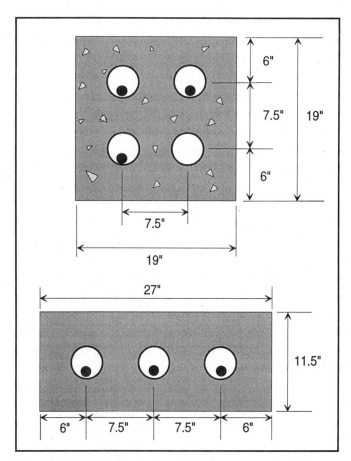

Figure 4-32: Two arrangements for three electrical ducts.

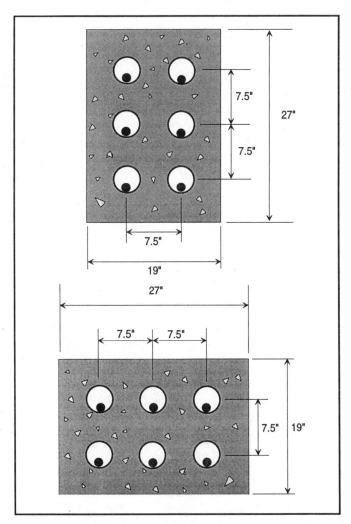

Figure 4-31: Two arrangements for duct banks containing six electrical ducts.

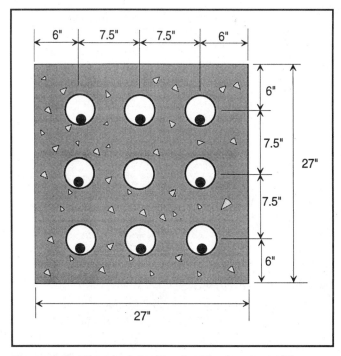

Figure 4-33: Electrical duct bank with nine ducts. The center duct, however, should not be used for power transmission; only control or communication cables.

feedback of current from the network into the fault through all the transformers on that feeder will cause the protective switches to open, disconnecting all ties between that feeder and the secondary mains. When the trouble is repaired and normal voltage conditions are restored on that feeder, the switches will reclose, putting the transformers back into service.

Designers of a network are always cautious about the placement of transformers. The transformers should be large enough and close enough together to be able to burn off a ground fault on the cable at any point. If not, such a fault might continue to burn for a long time.

The primary feeders supplying networks are run from substations at the usual primary voltage for the system, such as 4160, 4800, 6900, or 13,200 V. Higher voltages are practicable if the loads are large enough to warrant them.

Network power systems are usually installed underground with primaries, secondaries, and transformers all underground. The transformers and secondaries may, however, be overhead, or they may use a combination of overhead and underground construction.

Secondary Services

The 4160-V transformer came about through the 4160Y-V connection on 2400-V transformers. In some cases, it was advantageous to connect transformers between phase wires on a 2400/4160Y-V system, and this required a transformer having a winding voltage of 4160 V. These 4160-V transformers are now used in several ways. First, they are used in three-phase delta banks connected to 2400/4160Y-V systems.

Another application is on 4160-V single-phase lines taken off of a 2400/4160Y-V three-phase system, necessitating the use of 4160-V transformers.

In some instances, the 4160-V transformers are used for rural systems rated 4160/7200Y. With this system, 4160-V transformers can be used between phase wire and neutral of a three-phase, four-wire system, and 7200-V transformers can be used between phase wires.

The 4800-V transformers are frequently used in some sections of the United States where distribution circuits run through thickly populated rural and suburban areas. Distribution lines in these localities are necessarily much longer than in cities, and therefore the 2400-V system is not high enough voltage to be economical. On the other hand, 4800-V distribution systems in these areas have proved to be quite logical and satisfactory.

Again, the systems originally were 4800-V delta, three-phase systems with 4800-V, single-phase branch lines. These delta systems, however, are now being converted, in many cases, to 4800/8320Y, giving a higher system voltage but using the same equipment that was used on 4800-V delta systems.

Rural electrification in thinly populated areas required still higher voltage for good performance and economy. Therefore, for rural power systems in certain sections of the United States, 7200-V distribution systems have been used quite extensively and successfully. The early rural systems were 7200-V delta, three phase in most cases, with 7200-V branch lines. These systems are now giving way to 7200/12470Y-V, three-phase, four-wire systems. In fact, this system is probably the most popular in use today.

Although less popular than the 7200-V class of transformers, 7620-V systems are sometimes used for rural electrification. Most of these systems are actually 7620/13,200Y, three-phase, four-wire systems. On this type of system, 7620-V single-phase transformers can be used between phase wire and neutral of the three-phase system, or 13,200-V transformers can be used between phase wires. This type of system works out very economically for power companies that have both 7620- and 13,200-V distribution systems. In this situation, transformers can be used on either system, thereby making stocks of transformers flexible.

There are some 12,000-V, three-phase delta systems that were installed some time ago for transmission and power over greater distances than were feasible in lower voltages. There are now two applications for 12,000-V transformers. The first is for use on 12,000-V delta systems and the second for use on 7200/12,470Y-V systems.

The 13,200-V transformers also have two applications. First, they can be used on distribution systems that are 13,200-V delta, three-phase, which were built to distribute electrical energy at considerable distance. The second application has already been mentioned in connection with the 7620/13,200-V, three-phase, four-wire system. On this system, the 13,200-V standard transformer can be used between phase wires of the three-phase four-wire system. This connection is made quite often when it is necessary to connect a three-phase bank of transformers to the 7620/13,200Y-V systems.

In addition to use in rural areas, the 12,000- and the 13,200-V distribution systems are quite often used in urban areas. In relatively large cities having considerable industrial loads, 13,200- and 12,000-V lines are quite often run to serve industrial loads, while the 2400/4160Y-V system is used for the residential and commercial loads.

The 24,940-Grd-Y/14,400-V units have one end of the high-voltage winding grounded to the tank wall and are suitable only for use on systems having the neutral grounded throughout its length.

System voltages to 68 kV have been designated as distribution, although transmission lines also operate at these same voltages. The trend is to convert these lines to four-wire distribution systems and use transformers with primary windings connected in wye. As an example, a multi-grounded neutral is added to a 34,500-V system, and 20,000-V transformers are used to supply the customers. The system is also in use for new construction in high-load-density areas.

Use of Capacitors

A capacitor is a device that will accept an electrical charge, store it, and again release the charge when desired. In its simplest form, it consists of two metallic plates separated by an air gap. The larger the surface of the plates and the closer they are together, the greater the capacity will be.

If the space between the plates is filled with various insulating materials, such as kraft paper, linen paper, or oil, the capacity will be greater than with air. The increase in this capacity for any specific material as compared to air is called the dielectric constant and is expressed by the letter *K*. If plate area and spacing remain unchanged but a certain grade of paper gives a capacitor twice the capacity it had with air, then the value for *K* would be 2. All insulating materials have a value of *K* which merely expresses its dielectric effectiveness as compared to air. Thus, only three factors govern capacity:

1. How big are the plates?
2. How close are they?
3. What material separates them?

Capacitors operate on both direct- and alternating-current circuits. If connected to the terminals of a dc circuit, a pair of plates without a charge on them will accept a static charge. Current will rush into the capacitor until each plate is at the potential of the line to which it is connected. Once this potential is reached (a very rapid process), no further current, other than leakage current, will flow. If removed from the line, the capacitor plates will maintain their charge until it is dissipated by leakage between plates or by deliberate contact between plate terminals. This principle is used in surge generators. In the ac application, the plates are alternately charged and discharged by the voltage changes of the circuit to which they are connected. It is this condition which makes possible the use of capacitors for power-factor correction.

Power Factor

Power factor is a ratio of useful working current to total current in the line. As power is the product of current and voltage, the power factor can also be described as a ratio of real power to apparent power and be expressed as

$$\text{Power factor} = \frac{\text{kW}}{\text{kVA}}$$

Apparent power is made up of two components, namely, real power (expressed in kilowatts) and the reactive component (expressed in kilovars or kvar). This relationship is shown in Figure 4-34. The horizontal line AB represents the useful real power (or kilowatts) in the circuit. The line BC represents the reactive component (kvar) as drawn in a downward direction. Then a line from A to C represents apparent power (kilovolt-amperes). To the uninitiated, the use of lines representing quantity and direction is often confusing. Lines so utilized are called *vectors*. Imagine yourself at point A desiring to reach point C, but, due to obstructions, you must first walk to B, and then turn a right angle and walk to C. The energy you dissipated in

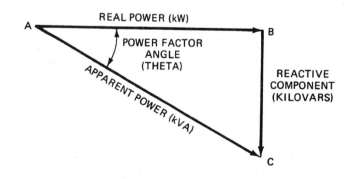

Figure 4-34: Power-factor triangle used to show angle between real power and apparent power.

reaching C was increased because you could not take the direct course, but, in the final analysis, you ended up at a point the direction and distance of which can be represented by the straight line AC. The power-factor angle shown is called theta.

There is interest in power factor because of the peculiarity of certain ac electrical equipment requiring power lines to carry more current than is actually needed to do a specific job unless we utilize a principle that has long been understood but not until recent years given the attention it deserves. This principle utilizes the application of capacitors.

Lead and Lag

We cannot add real and reactive components arithmetically. To understand why, consider the characteristics of electrical circuits. In a pure resistance circuit, the alternating voltage and current curves have the same shape, and the changes occur in perfect step, or phase, with each other. Both are at zero with maximum positive peaks and maximum negative peaks at identical instants. Compare this with a circuit having magnetic characteristics involving units such as induction motors, transformers, fluorescent lights, and welding machines. It is typical that the current needed to establish a magnetic field lags the voltage by 90°.

Visualize a tube of toothpaste. Pressure must be exerted on the tube before the contents oozes out. In other words, there is no flow until pressure is exerted, or, analogously, the flow (current) lags the pressure (voltage). In like manner, the magnetic part of a circuit resists, or opposes, the flow of current through it. In a magnetic circuit, the pressure precedes or leads the current flow, or, conversely, the current lags the voltage.

Peculiarly, in a capacitor, we have the exact opposite: current leads the voltage by 90°. Visualize an empty tank to which a high-pressure air line is attached by means of a valve. At the instant the valve is opened, a tremendous rush of air enters the tank, gradually reducing in rate of flow as the tank pressure approaches the air line pressure. When the tank is up to full pressure, no further flow exists. Accordingly, you must first have a flow of air into the tank before it develops an internal pressure. Consider the tank to be a capacitor and the air line to be the electrical system. In like fashion, current rushes into the capacitor before it builds up a voltage, or, in a sense, the current leads the voltage.

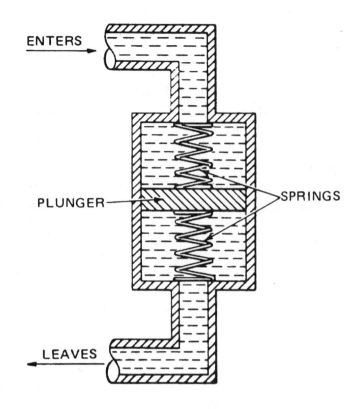

Figure 4-35: Water system used to show comparison of real power to apparent power.

A water system provides a better comparison. The system shown in Figure 4-35 is connected to the inlet and outlet of a pump. For every gallon that enters the upper section, a gallon must flow from the lower section as the plunger is forced down. If the pressure on the upper half is removed, the stored energy in the lower spring will return the plunger to midposition. When the direction of flow is reversed, the plunger travels upward, and if flowmeters were connected to the inlet and outlet, the inlet and outlet flows would prove to be equal. Actually, there is no flow through the cylinder but merely a displacement. The only way we could get a flow through the cylinder would be by leakage around the plunger or if excessive pressure punctured a hole in the plunger.

In a capacitor, we have a similar set of conditions. The electrical insulation can be visualized as the plunger. As we apply a higher voltage to one capacitor plate than to its companion plate, current will rush in. If we remove the pressure and provide an external conducting path, the stored energy will flow to the other plate, discharging the capacitor and bringing it to a balanced condition in much

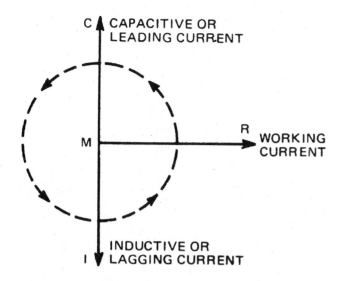

Figure 4-36: Phasor diagram used to compare leading current, working current, and lagging current.

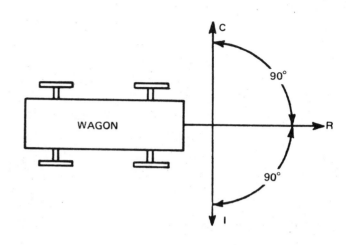

Figure 4-37: Horse and wagon comparison.

the same fashion as the spring returned the plunger to mid-position.

Since no insulator is perfect, some leakage current will flow through it when a voltage differential exists. If we raise the voltage to a point where we break down or puncture the insulation, then the capacitor is damaged beyond repair and must be replaced.

Cognizance of the three different kinds of current discussed, namely, in phase, lagging, and leading, permits the drawing of the relationship in Figure 4-36.

The industry accepts a counterclockwise rotation about point *M* as a means of determining the relative phase position of voltage and current vectors. These may be considered as hands of a clock running in reverse. *MC* is preceding *MR;* hence, it is considered leading. *MI* follows *MR;* therefore, it is lagging.

The two angles shown are right angles (90°); therefore, it becomes apparent that *MC* and *MI* are exactly opposite in direction and will cancel out each other if of equal value.

Consider a wagon to which three horses are hitched, as in Figure 4-37. If C and I pull with equal force, they merely cancel one another's effort, and the wagon will proceed in the direction of R, the working horse. If only R and I are hitched, the course will lie between these two.

Most utility lines contain quantities of working current and lagging current, and one type is just as effective in loading up the line as the other. If we know how much inductive current a line is carrying, then we can connect enough capacitors to that line to cancel out this wasteful and undesired component.

Just as a wattmeter will register the kilowatts in a line, a varmeter will register the kvar of reactive power in the line. If an inductive circuit is checked by a meter which reads 150,000 var (150 kvar), then application of 150-kvar capacitor would completely cancel out the inductive component, leaving only working current in the line.

Power-Factor Correction

In actual practice, full correction to establish the unity power factor is rarely, if ever, recommended. If a system had a constant 24-hr load at a given factor, such correction could be readily approached. Unfortunately, such is not the case, and we are faced with peaks and valleys in the load curve.

If we canceled out, by the addition of capacitors, the inductive (lagging) kvars at peak conditions, our capacitors would continuously pump into the system their full value of leading kilovars. Thus, during early morning hours when inductive kvars are much below peak conditions, a surplus of capacitive kvars would be supplied, and a leading power factor would result. Local conditions may justify such overcorrection, but, in general, overcorrection is not recommended.

A recording kilovarmeter can readily give us the condition curve, or else it can be calculated if we know the kW curve and the power factor throughout the day. From 2 to 6 a.m., refrigerators, transformer excitation, night factory loads, and the like result in relatively low readings. When the community comes to life in the morning, televisions, radios, appliances, and factory loads build up a high inductive kvar peak.

Switched Capacitors

There are several ways in which switching of capacitors can be accomplished. A large factory would arrange by manual or automatic operation to switch in a bank of capacitors at the start of the working day and disconnect them when the plant shuts down. The energizing of a circuit breaker control coil can be affected with Kvar. current, voltage, temperature, and time controls.

Making a Survey

The preparation for a survey of system conditions is not as complicated as might first be assumed. Even a very complicated system resolves itself down to a combination of individual feeder studies. Thus, the size of the system does not materially complicate the problem except to increase the amount of manual labor in arriving at the final results.

Assume that Figure 4-38 represents the kilovar curve for a given municipality or for a specific feeder on a large system. Visual inspection of the curve shows that, with the lowest kilovar value being approximately 450 kvar at 4 a.m., we could permanently install on the lines a 450-kvar bank and know that at no time during the day or night would we be operating with a leading power factor. An additional 300-kvar bank or two 300-kvar banks could readily be installed as switched capacitors to correct the power factor in the shaded area.

The larger facilities invariably have recording instruments which reveal this kvar demand directly from charts. In the case of the municipal or smaller operator who is not so elaborately equipped, the same results can be obtained from the data available on daily load sheets. Knowledge of the hourly kW demand and the hourly power-factor reading will permit the engineer to readily determine the kvar value at those corresponding times by utilizing available charts. If power-factor readings are not available, kVA can readily be determined from either the armature or the line amps and the system voltage. Once kVA and kW are known, the same chart permits ready determination of power-factor and kvar demand.

Once the total kvar of capacitors required for the system is determined, there remains only the question of proper

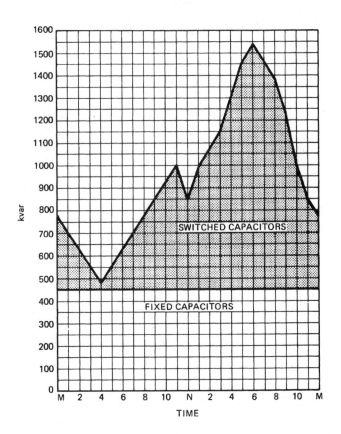

Figure 4-38: Graph depicting relation of kvar to time with fixed capacitors.

location. If a severe voltage drop is experienced on the line, the capacitors will serve their best purpose out on the distribution system. However, if voltage problems are not serious on the distribution system and the prime purpose of the installation is to relieve the generators, then the units could very readily be installed at the generating plant.

Justification for Capacitors

The application of capacitors to electrical distribution systems has been justified by the overall economy provided. Loads are supplied at reduced cost. The original loads on the first distribution systems were predominantly lighting so the power factor was high. Over the years, the character of loads has changed. Today, loads are much larger and consist of many motor-operated devices that impose greater kilovar demands upon electrical systems. Because of the kilovar demand, system power factors have been lower.

The result may be threefold:

1. Substation and transformer equipment may be taxed to full thermal capacity or overburdened.
2. High kilovar demands may, in many cases, cause excessive voltage drops.
3. A low power factor may cause an unnecessary increase in system losses.

Capacitors can alleviate these conditions by reducing the kilovar demand from the point of demand all the way back to the generators. Depending on the uncorrected power factor of the system, the installation of capacitors can increase generator and substation capability for additional load at least 30% and can increase individual circuit capability, from the standpoint of voltage regulation, 30 to 100%.

OIL-FILLED TRANSFORMER CONNECTIONS

There are numerous transformer connections for various applications. The majority of these are listed on the pages to follow. Manufacturer's data may be used for other types that may become necessary.

Single-Phase to Supply 120-V Lighting Load

The transformer connection in Figure 4-39 on the next page shows a transformer connection between the high-voltage line and load, with the 120/240-V winding connected in parallel. This connection is used where the load is comparatively small and the length of the secondary circuit is short. This type of connection is often used for isolated rural pump houses, roadside vegetable stands, and for similar applications.

Single-Phase for Light and Power

The diagram in Figure 4-40 on the next page shows a transformer connection that is used quite extensively for residential and small commercial applications. It is the most common single-phase distribution system in use today. It is known as the 240/120-V, three-wire, single-phase system, and is used where 120 and 240 V are used simultaneously.

Delta-Delta for Light and Power

The transformer connection shown in Figure 4-41 is often used to supply small single-phase lighting loads and three-phase power loads simultaneously. Note that the mid-tap of the secondary of one transformer is grounded. Consequently, the small lighting load is connected across the transformer with the mid-tap and the ground wire which is common to both 120-V circuits. The single-phase lighting load reduces the available three-phase capacity, however, since the phases will be unbalanced. This connection requires special metering and is not always available from all utilities. Before specifying or installing such systems, always check with the local power company to make certain the connection can be provided.

Open-Delta for Light and Power

Where the secondary load is a combination of single-phase lighting and three-phase power, the open-delta connected bank is frequently used. This type of connection is almost identical to the delta-delta connection except only two transformers are used. This arrangement is also used on a delta system when one of the three transformers becomes damaged. The damaged transformer is disconnected from the circuit and the remaining two transformers carry the load. However, the three-phase load carried by the open delta bank is only 86.6% of the combined rating of the remaining two equal sized units. It is only 57.7% of the normal full-load capability of a full bank of three transformers. In an emergency, however, this capability permits single- and three-phase power at a location when one unit burns out and a replacement is not readily available. The total load must obviously be curtailed to avoid another burnout. *See* figure 4-42 on page 443.

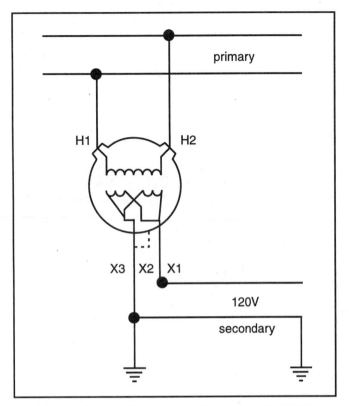

Figure 4-39: Single-phase connection to supply 120-V lighting.

Figure 4-40: Single-phase transformer connection to supply 120/240-V, 3-wire system for lighting and power.

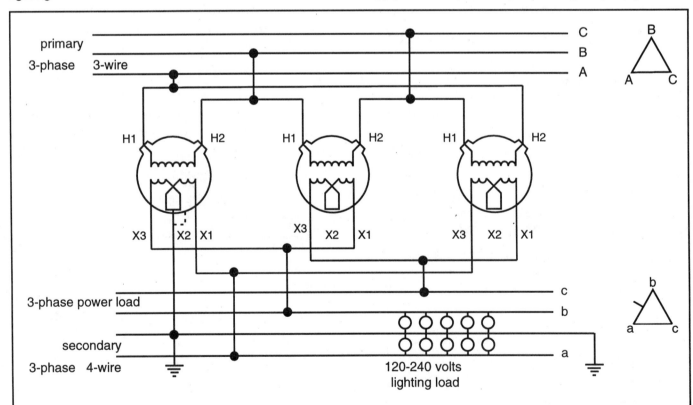

Figure 4-41: Delta-delta transformer connection for lighting and power.

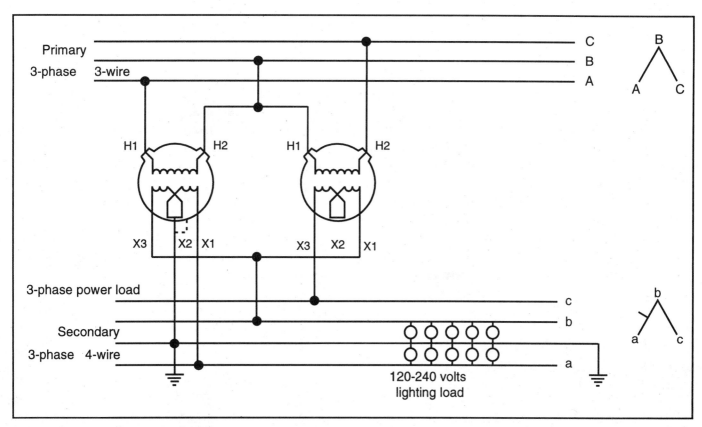

Figure 4-42: Open-delta transformer connection for lighting and power.

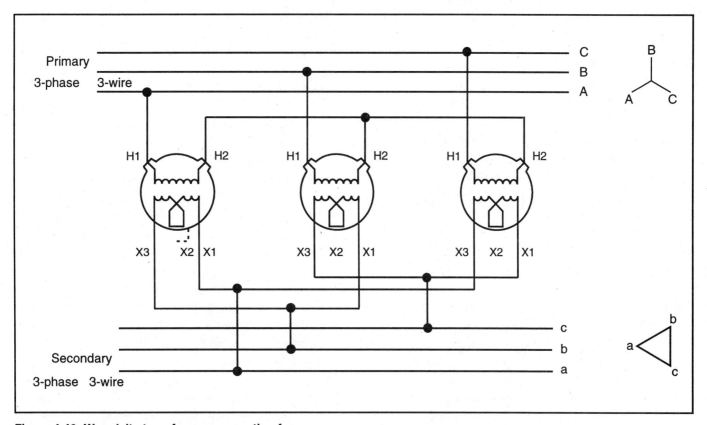

Figure 4-43: Wye-delta transformer connection for power.

Wye-Delta Connection for Power

Sometimes it is desirable to increase the voltage of a circuit from 2400 to 4160 V to increase the potential capacity. The diagram in Figure 4-43 shows such a system after it has been changed to 4160 V. The previously delta-connected distribution transformer primaries are now connected from the line to neutral so that no major change in equipment is necessary. The primary neutral should not be grounded or tied into the system neutral since a single-phase ground fault may result in extensive blowing of fuses throughout the system.

Wye-Delta Connection for Lighting and Power

The diagram in Figure 4-44 shows the connection for a wye-delta transformer bank to supply both lighting and power. This connection is similar to the delta-delta bank with only the primary connections changed. The primary neutral should not be grounded or tied into the system neutral, since a single-phase ground fault may result in extensive blowing of fuses throughout the system. The single-phase load reduces the available three-phase capacity. This connection requires special metering.

Open Wye-Delta Connection

When operating wye-delta systems, and one phase is disabled, service may be maintained at reduced load as shown in Figure 4-45. The neutral in this case must be connected to the neutral of the step-up bank through a copper conductor. The system is unbalanced, electro-statically and electro-magnetically, so that telephone interference may be expected if the neutral is connected to ground. The useful capacity of the open wye-delta bank is 87% of the capacity of the installed transformers when the two units are identical. The capacity is 57% of a three transformer bank.

Delta-Wye for Lighting and Power

The connection shown in Figure 4-46 has the neutral of the secondary three-phase system grounded, and the single-phase loads are connected between the different phase wires and the neutral, while the three-phase loads are connected to the phase wires. Consequently, the single-phase loads can be balanced on three phases in each bank, and banks may be paralled if desired.

Wye-Wye for Lighting and Power

The diagram in Figure 4-47 shows a system on which the primary voltage is increased from 2400 V to 4160 V to increase the potential capacity of the system. Note that the secondaries are connected in wye and the primary neutral is connected to the neutral of the supply voltage through a metallic conductor and carried with the phase conductor to minimize telephone interference. If the neutral of the transformer is isolated from the system neutral, an unstable condition results at the transformer neutral caused primarily by third harmonic voltages. If the transformer neutral is connected to ground, the possibility of telephone interference is greatly enhanced, and there is also a possibility of resonance between the line capacitance to ground and the magnetizing impedance of the transformer.

Wye-Wye Autotransformers

When the ratio of transformation from the primary to secondary voltage is small, the most economical way of stepping down the voltage is by using autotransformers as shown in Figure 4-48. For this application, it is necessary that the neutral of the autotransformer bank be connected to the system neutral. Branch circuits, however, must not be supplied by autotransformers in most cases.

Scott Connections

One interesting group of transformer connections are those using the Scott scheme which was originated by C. F. Scott to take care of the necessity of changing two-phase power generated at Niagara Falls to three-phase power for transmission to Buffalo, NY. *See* Figure 4-49.

This same connection that was used to change two-phase to three-phase may be reversed as shown in Figure 4-50 to change three-phase to two-phase.

When two-phase systems are changed to three-phase, it is not necessary to replace the old two-phase banks; a reconnection is all that is necessary. The teaser transformer will operate with 86.6% voltage on both the primary and secondary, so it should have the same ratio as the main transformer as shown in Figure 4-51.

The modern version of the Scott connection is in a duplex transformer called a T connection, which can simulate the various wye or delta connections except that no neutral will be made available on the primary side. *See* Figure 4-52.

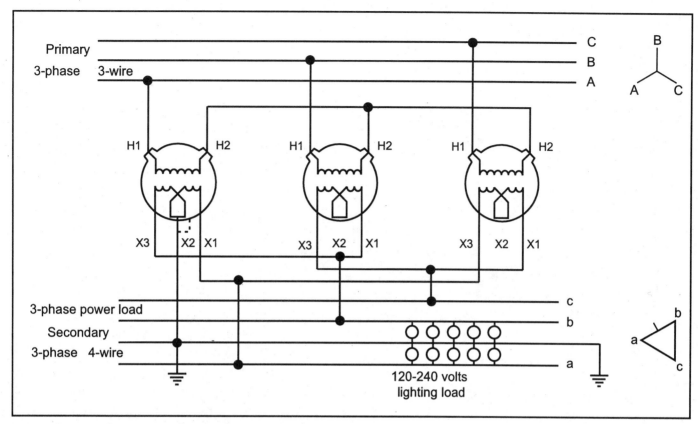

Figure 4-44: Wye-delta transformer connection for lighting and power.

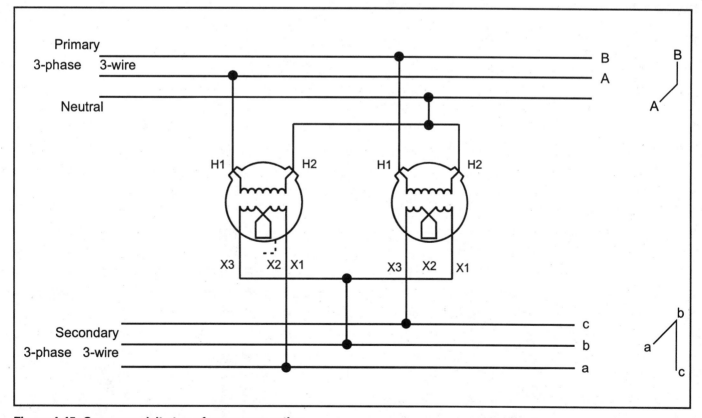

Figure 4-45: Open wye-delta transformer connection.

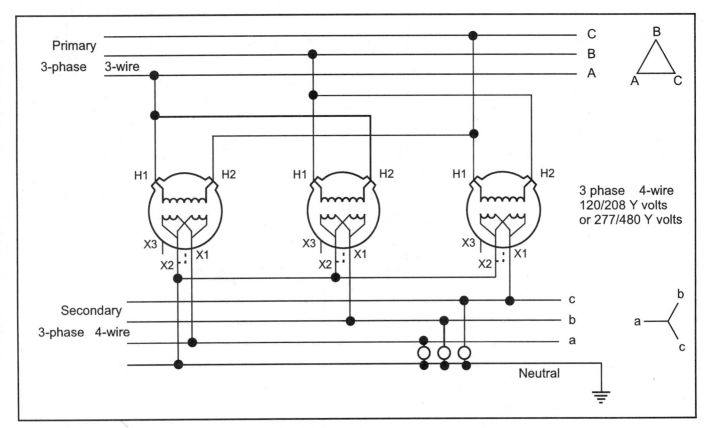

Figure 4-46: Delta-wye transformer connection for lighting and power.

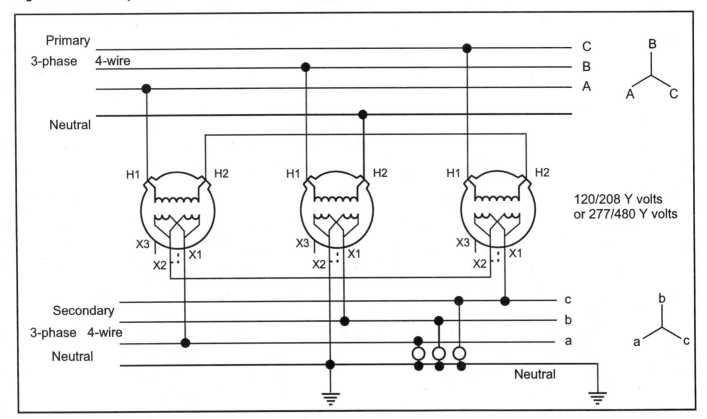

Figure 4-47: Wye-wye transformer connection for lighting and power.

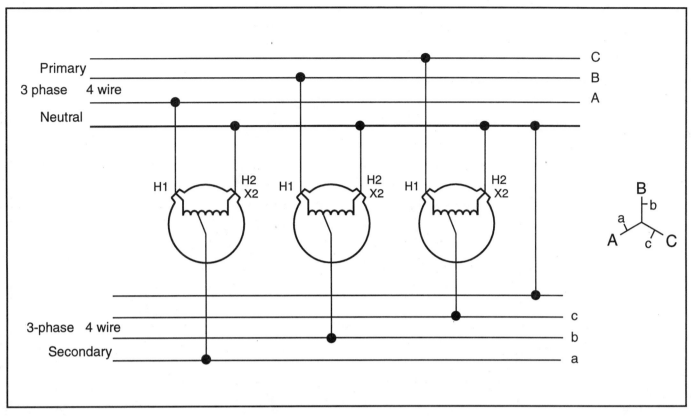

Figure 4-48: Wye-wye autotransformer connection.

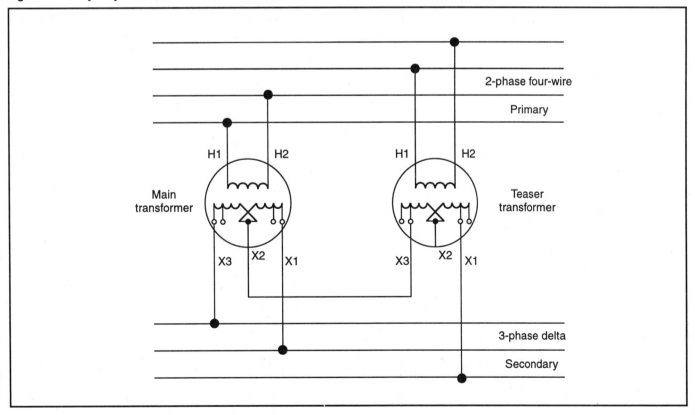

Figure 4-49: Scott connection — two-phase to three-phase.

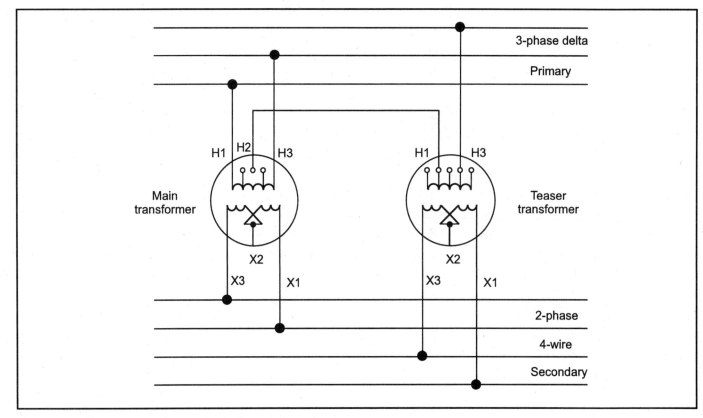

Figure 4-50: Scott connection — three-phase to two-phase.

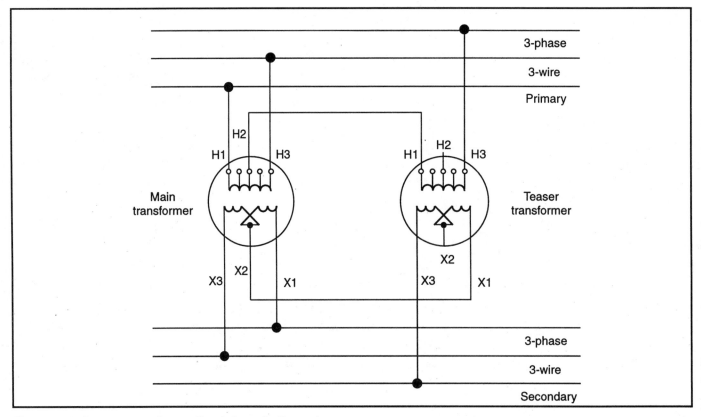

Figure 4-51: Scott connection — three-phase to three-phase.

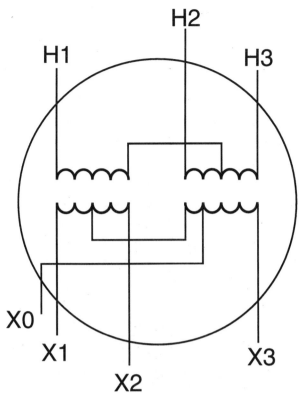

Figure 4-52: Modern version of the Scott-T connection.

PARALLEL OPERATION OF TRANSFORMERS

Transformers will operate satisfactorily in parallel on a single-phase, three-wire system if the terminals with the same relative polarity are connected together. However, the practice is not very economical because the individual cost and losses of the smaller transformers are greater than one larger unit giving the same output. Therefore, paralleling of smaller transformers is usually done only in an emergency. With large transformers, however, it is often practical to operate units in parallel as a regular practice. *See* Figure 4-53.

In connecting large transformers in parallel, especially when one of the windings is for a comparatively low voltage, the resistance of the joints and interconnecting leads must not vary materially for the different transformers, or it will cause an unequal division of load.

Two three-phase transformers may also be connected in parallel provided they have the same winding arrangement, are connected with the same polarity, and have the same phase rotation. If two transformers — or two banks of transformers — have the same voltage ratings, the same turn ratios, the same impedances, and the same ratios of reactance to resistance, they will divide the load current in proportion to their kVA ratings, with no phase difference between the currents in the two transformers. However, if any of the preceding conditions are not met, then it is possible for the load current to divide between the two transformers in proportion to their kVA ratings. There may also be a phase difference between currents in the two transformers or banks of transformers.

Some three-phase transformers cannot be operated properly in parallel. For example, a transformer having its coils connected in delta on both high-tension and low-tension sides cannot be made to parallel with one connected either in delta on the high-tension and in Y on the

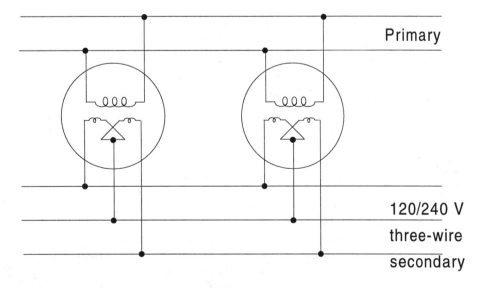

Figure 4-53: Parallel operation of single-phase transformers.

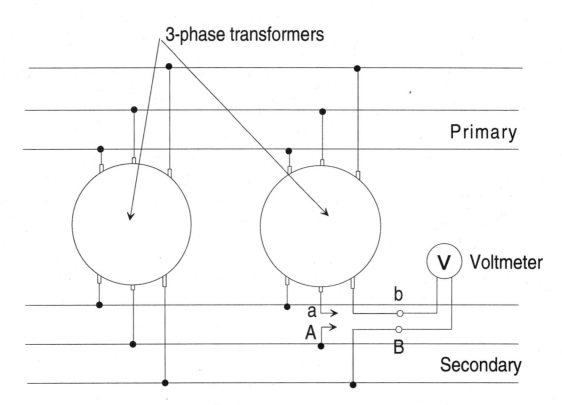

Figure 4-54: Testing three-phase transformers for parallel operation.

low-tension side. Those connected in wye on the high-tension, and in delta on the low-tension side and in wye on the low-tension side can be made to parallel with transformers having their coils joined in accordance with certain schemes; that is, connected in star or wye on the high-tension side and in delta on the low-tension side.

To determine whether or not three-phase transformers will operate in parallel, connect them as shown in Figure 4-54, leaving two leads on one of the transformers unjoined. Test with a voltmeter across the unjoined leads. If there is no voltage between the points shown in the drawing, the polarities of the two transformers are the same, and the connections may then be made and put into service.

If a reading indicates a voltage between the points indicated in the drawing (either one of the two or both), the polarity of the two transformers are different. Should this occur, disconnect transformer lead A successively to mains 1, 2, and 3 as shown in Figure 4-54 and at each connection test with the voltmeter between b and B and the legs of the main to which lead A is connected. If with any trial connection the voltmeter readings between b and B and either of the two legs is found to be zero, the transformer will operate with leads b and B connected to those two legs. If no system of connections can be discovered

that will satisfy this condition, the transformer will not operate in parallel without changes in its internal connections, and there is a possibility that it will not operate in parallel at all.

In parallel operation, the primaries of the two or more transformers involved are connected together, and the secondaries are also connected together. With the primaries so connected, the voltages in both primaries and secondaries will be in certain directions. It is necessary that the secondaries be so connected that the voltage from one secondary line to the other will be in the same direction through both transformers. Proper connections to obtain this condition for single-phase transformers of various polarities are shown in Figure 4-55. In Figure 4-55(a), both transformers A and B have additive polarity; in Figure 4-55(b), both transformers have subtractive polarity; in Figure 4-55(c), transformer A has additive polarity and B has subtractive polarity.

Transformers, even when properly connected, will not operate satisfactorily in parallel unless their transformation ratios are very close to being equal and their impedance voltage drops are also approximately equal. A difference in transformation ratios will cause a circulating current to flow, even at no load, in each winding of both

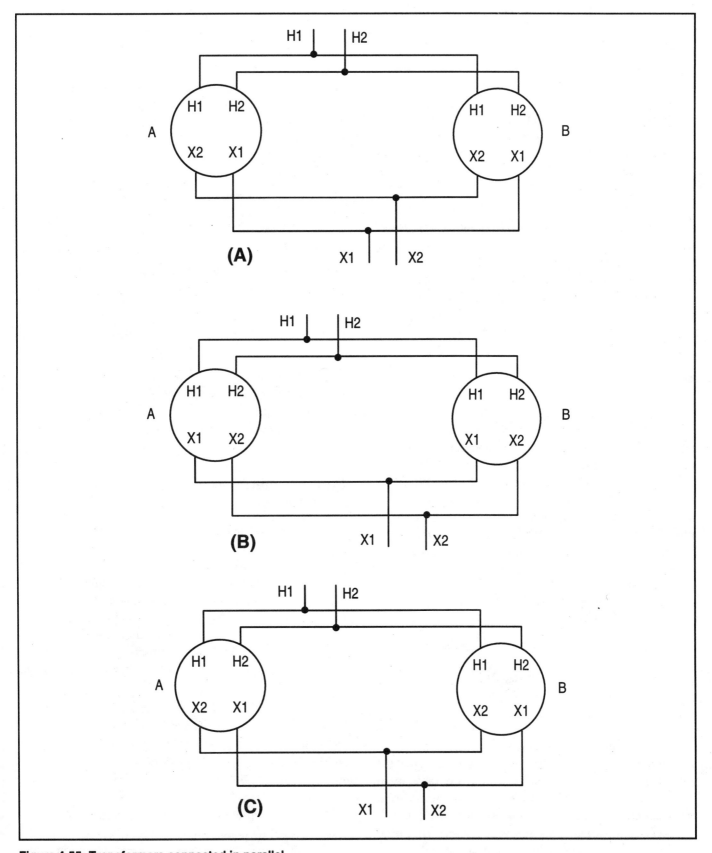

Figure 4-55: Transformers connected in parallel.

transformers. In a loaded parallel bank of two transformers of equal capacities, for example, if there is a difference in the transformation ratios, the load circuit will be superimposed on the circulating current. The result in such a case is that in one transformer the total circulating current will be added to the load current, whereas in the other transformer the actual current will be the difference between the load current and the circulating current. This may lead to unsatisfactory operation. Therefore, the transformation ratios of transformers for parallel operation must be definitely known.

When two transformers are connected in parallel, the circulating current caused by the difference in the ratios of the two is equal to the difference in open-circuit voltage divided by the sum of the transformer impedances, because the current is circulated through the windings of both transformers due to this voltage difference. To illustrate, let I represent the amount of circulating current — in percent of full-load current — and the equation will be

$$I = \frac{Percent\ voltage\ difference \times 100}{Sum\ of\ percent\ impedances}$$

Let's assume an open-circuit voltage difference of 3% between two transformers connected in parallel. If each transformer has an impedance of 5%, the circulating current, in percent of full-load current, is I = (3 × 100)/5 + 5) = 30%. A current equal to 30% full-load current therefore circulates in both the high-voltage and low-voltage windings. This current adds to the load current in the transformer having the higher induced voltage and subtracts from the load current of the other transformer. Therefore, one transformer will be overloaded, while the other may or may not be — depending on the phase-angle difference between the circulating current and the load current.

Impedance in Parallel-Operated Transformers

Impedance plays an important role in the successful operation of transformers connected in parallel. The impedance of the two or more transformers must be such that the voltage drop from no load to full load is the same in all transformer units in both magnitude and phase. In most applications, you will find that the total resistance drop is relatively small when compared with the reactance drop and that the total percent impedance drop can be taken as approximately equal to the percent reactance drop. If the

percent impedances of the given transformers at full load are the same, they will, of course, divide the load equally.

The following equation may be used to obtain the division of loads between two transformer banks operating in parallel on single-phase systems. In this equation, it can be assumed that the ratio of resistance to reactance is the same in all units since the error introduced by differences in this ratio is usually so small as to be negligible:

$$power = \frac{(kVA-1)/(Z-1)}{(kVA-1)/(Z-1)+(kVA-2)/(Z-2)}$$

$$\times\ total\ kVA\ load$$

where
kVA - 1 = kVA rating of transformer 1
kVA - 2 = kVA rating of transformer 2
Z - 1 = percent impedance of transformer 1
Z - 2 = percent impedance of transformer 2

The preceding equation may also be applied to more than two transformers operated in parallel by adding, to the denominator of the fraction, the kVA of each additional transformer divided by its percent impedance.

Parallel Operation of Three-Phase Transformers

Three-phase transformers, or banks of single-phase transformers, may be connected in parallel provided each of the three primary leads in one three-phase transformer is connected in parallel with a corresponding primary lead of the other transformer. The secondaries are then connected in the same way. The corresponding leads are the leads which have the same potential at all times and the same polarity. Furthermore, the transformers must have the same voltage ratio and the same impedance voltage drop.

When three-phase transformer banks operate in parallel and the three units in each bank are similar, the division of the load can be determined by the same method previously described for single-phase transformers connected in parallel on a single-phase system.

In addition to the requirements of polarity, ratio, and impedance, paralleling of three-phase transformers also requires that the angular displacement between the voltages in the windings be taken into consideration when they are connected together.

Phasor diagrams of three-phase transformers that are to be paralleled greatly simplify matters. With these, all that is required is to compare the two diagrams to make sure they consist of phasors that can be made to coincide; then connect together terminals corresponding to coinciding voltage phasors. If the diagram phasors can be made to coincide, leads that are connected together will have the same potential at all times. This is one of the fundamental requirements for paralleling.

CONVERSION EQUIPMENT

Conversion substations and apparatus are those that change alternating current to direct current for certain specialized needs. Manual and automatic switching controls have been standardized for the main applications:

- Electromechanical industries
- Electric railways
- Coal and ore mining
- Special industrial dc power supplies

While much of this dc power is obtained directly from dc generators, synchronous converters, and batteries, the greatest part is derived from ac sources through the use of rectifiers.

The application of rectifiers in industry ranges from very small amounts of rectified power, such as that needed for battery chargers and small dc motors, up to the hundreds of thousands of kilowatts needed for a large electrochemical process.

Other converters include equipment used to convert electrical systems from single phase to three phase.

Rectifying Devices

A rectifying device, in general, is an elementary device which has the property of effectively conducting current in only one direction. If a voltage of a certain polarity is applied to the rectifying device, the current will flow through the device in a certain direction which is called the forward direction. If a voltage of the opposite polarity is applied, the current will not flow through the rectifying device. Therefore, when an alternating current is applied to a rectifying device, the device will conduct current during one alternation of each cycle but will block, or stop, the current flow during the reversed alternation. The resulting current is an intermittent and pulsating current but one that is unidirectional, or rectified.

From the preceding paragraph, we can see how a rectifying device is used to obtain a direct current from an alternating-current source. To obtain the desired rectification, one single rectifier may be used, or a combination of the two may be used.

All rectifying devices are designed to conduct current satisfactorily in one direction only. However, some types are more effective than others; that is, some types of rectifiers block reverse current completely, while others merely have lower ratios of forward to reverse conductivity.

Types of Rectifiers

High-Vacuum Thermionic Rectifying Devices: A hot cathode tube may be used as a rectifying device and consists of a highly evacuated tube with a heated cathode and an anode; it is sometimes known as a kenotron. The objective of this type of rectifier is to withstand very high inverse voltages, and such rectifiers are best adapted to applications requiring high dc voltages and small currents, such as X-ray applications and in testing certain types of insulation. Few are ever used anymore in the generation of electric power.

Gaseous-Discharge Thermionic Rectifying Devices: Hot cathode tubes based on gaseous discharge have been filled with various gases, but mercury vapor seems to be the most popular. A small amount of mercury is introduced into the tube during manufacture and so located in the tube that it will be at a temperature near that of the ambient air. At this temperature, the mercury-vapor density is high enough to make plenty of mercury atoms available to keep the space charge down. At the same time, the mercury-vapor pressure is not high enough to break down readily when the anode voltage is negative. This type of tube has been used rather extensively for battery charging and is called a phanotron.

Another type of gaseous rectifier uses argon gas and a tungsten filament. This type of tube was once very popular for battery charging up to 100 V and 12 A.

A kenotron tube modified with the introduction of a wire-mesh screen or a metallic cylinder punched with holes and surrounding the cathode is called a pliotron. Such grid control is the basis for the operation of audio-frequency and radio-frequency amplifiers used in radio and TV in years past but has little use in the power generating field.

A gaseous-discharge tube designed with a grid is called a thyratron. Since the grid can be used to control the dc output voltage, thyratrons have been used quite extensively as a supply for small variable-speed motors and for other control functions. The usual operating voltage of thyratron tubes is under 600 V, and the current rating is 15 A or less per tube. Therefore, the practical horsepower limit for motor control is about 25 hp, but most are used on motors of less than 5 hp.

Arc-Discharge Rectifying Devices: Rectifying devices utilizing the mercury pool as the cathode are known as mercury-pool tubes, and they operate on the arc-discharge principle. Most are enclosed in high-vacuum tubes or steel tanks.

Semiconductor Rectifying Devices: One of the earliest semiconductors was the copper-oxide rectifying device, which is formed by oxidizing the surface of a copper plate. When voltage is applied between the oxidized surface and the copper, the current can be conducted readily from the oxidized surface to the copper but not in the other direction.

A later development was the selenium cell, or rectifying device. This device is formed by condensing selenium vapor on an aluminum plate in a vacuum and applying a conducting surface to the selenium. Selenium rectifying devices are built in the form of disks or plates up to about a foot square and around 16 in thick. Several devices may be mounted on a stud with insulating bushings and connected in series or parallel, or both, to form one or more circuit elements. Due to the large area of the element, such rectifiers are good at dissipating heat via their ventilating housing. In fact, with a moderate amount of forced ventilation, their current rating can be doubled.

Germanium rectifying devices are capable of increasing their current density over selenium devices more than a thousand-fold — with an accompanying reduction in bulk. This device consists of a thin wafer of very pure germanium placed between thin layers of materials such as antimony or indium. The wafer is heated to a point where the indium and antimony diffuse into the surface of the germanium and is then enclosed in a hermetically sealed housing to exclude contaminants.

The silicon rectifying device is similar in size and physical appearance to the germanium rectifying device, but its allowable temperature limit is about twice that of the germanium device, making it more desirable for many applications.

Ripple

The output voltage of a rectifier contains a ripple component whose frequency depends on the number of phases of the rectifier. The magnitude of the ripple depends on the number of rectifier phases, the amount of phase control, and the loading of the rectifier. This ripple is of no consequence in applications where the rectifier has six phases or more. However, in some applications — such as in electric railway applications — it is sometimes necessary to filter the rectifier output and reduce the ripple to prevent interference with the communications system. In applications employing small motors with single-phase rectifiers, there may be overheating of the motors due to the ripple, and in such applications oversize motors are used.

Rectifier Units

An operative assembly of rectifiers used as a group — with the necessary rectifier auxiliaries, transformer equipment, and essential switchgear — is known as a rectifier group. In most installations there are one or more rectifier auxiliary devices, such as filament transformers, excitation equipment, cooling equipment vacuum pumps, and regulating equipment.

Many rectifiers require some means to maintain constant output voltage or current. This is often accomplished by means of phase control which is obtained by varying the level of a voltage or current in the excitation circuit. One way to accomplish this is to use variable resistance or a magnetic amplifier.

When semiconductors are connected in parallel, the problem of current division arises. This is caused by the volt-amp characteristics of individual cells differing from each other, making the impedances of the parallel circuit branches unequal.

When the rectifier voltage is so high that the PRV rating of a cell would be exceeded, it becomes necessary to connect two or more cells in series. In such cases, some cells will be subject to more voltage than others because their leakage currents are unequal. This condition may be corrected, however, by shunting each cell with a resistor.

Mechanical Rectifiers

Rectifiers falling under the mechanical category mainly include the synchronous motor, which is operated so that contacts are opened and closed at approximately the same rate that a rectifying element enters and leaves conduction with the same transformer connection. Voltage control can be achieved by shifting the phase of the driving motor, simulating grid control in a mercury-arc rectifier. The same result can also be achieved by suitable excitation of the commutation reactors. Best operation is obtained by avoiding the use of phase control and, instead, using some means to change the ac voltage, such as an induction regulator or a tap-changing transformer.

Since there is no rectifier element loss, mechanical rectifiers are often more desirable in industrial applications than other types. This is particularly important in the electrochemical field where the load is applied almost continuously.

Rectifier Equipment

Besides the rectifying elements, their circuits, and auxiliary equipment, a complete rectifier unit also includes transformers, saturable reactors, and overcurrent protection.

The transformer, of course, provides proper voltage to a rectifier when installed between the ac line and the rectifier. Transformers used in this way differ from distribution transformers in several respects.

1. The secondary kVA is greater than the primary kVA.
2. The reactance is usually greater than in a distribution transformer.
3. Stronger winding bracing is required in a single-way rectifier transformer than in other types.

Transformers for double-way circuits are very similar in winding arrangement to distribution transformers, except for the value of the secondary voltage. When semiconductor rectifiers are used, no surge protection is needed, nor is extra bracing.

Three general methods of rectifier voltage regulation are available: adjustment of the ac voltage supplied to the rectifier, adjustment of the impedance in the ac circuit, and phase control of the rectifier firing.

Nearly all electrical devices — including rectifier equipment — require protection against sustained overloading and ground faults. In addition, protection may be needed against specific dangers such as overspeed, loss of field, and incorrect phase sequence. In rectifier equipment the items most in need of overload protection — outside the rectifier auxiliaries — are the transformer in a single-way rectifier circuit and the cells in a semiconductor-rectifier circuit.

The overload and arc-back protection in the ac circuits of rectifier installations is in most instances obtained by circuit breakers. When the voltage is 600 V or less, air circuit breakers are used for protection against overloads. At higher voltages, high-voltage air circuit breakers or oil circuit breakers are used in the supply circuit. At extremely high voltages, the expense of suitable circuit breakers is sometimes avoided by using fuses and disconnecting switches in the supply.

Phase Converters

It is sometimes desirable to use three-phase equipment on single-phase electric services. One such converter that will enable this connection is called the Roto-Phase manufactured by Arco Electric Products Corp. While called a phase converter, it is more accurately a phase generator because it generates one voltage which, when paralleled with the two voltages generated from a single-phase line, produces three-phase power. Induction as well as resistance three-phase loads can be operated from a single-phase supply.

A schematic diagram of a Roto-Phase connected to a single-phase source to operate two three-phase motors is shown in Figure 4-56 on the next page. In general, L_1 and L_2 are connected through a separate safety switch to Roto-Phase leads T_1 and T_2 at the junction box. The Roto-Phase lead, T_3, is then run through the third pole of the three-pole safety switch to the motor controls. L_1 and L_2, along with T_3, are then connected to the three-phase motor controls to provide three-phase service to the motors.

In using the Roto-Phase, do not connect any single-phase load or magnetic controls to T_3. This lead can be readily identified by the leg with the highest voltage to ground with the Roto-Phase running alone.

Should the motor starter have only two overload relays, do not run lead T_3 through a relay. Be sure to properly ground all electrical equipment as per the *NEC*.

Always start the Roto-Phase before energizing motors and be sure to follow wire sizings carefully. Properly maintained voltages on motor starts are very important, so wire distances, sizes, and voltage drops should be studied carefully.

SATURABLE CORE REACTORS

A saturable core reactor is a magnetic device having a laminated iron core and ac coils similar in construction to a conventional transformer. This device is uniquely effective in the control of all types of high power-factor loads. The coils in a saturable core reactor are called gate wind-

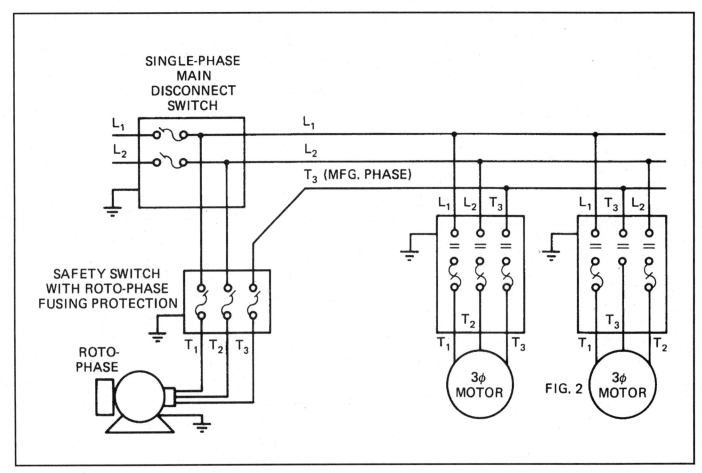

SINGLE-PHASE
MAIN
DISCONNECT
SWITCH

L_1

L_2

L_1

L_2

T_3 (MFG. PHASE)

L_1 L_2 T_3

L_1 T_3 L_2

SAFETY SWITCH
WITH ROTO-PHASE
FUSING PROTECTION

ROTO-
PHASE

T_1 T_2 T_3

T_2

T_1

T_3

T_3

T_1

T_2

3ϕ
MOTOR

FIG. 2

3ϕ
MOTOR

Figure 4-56: Wiring diagram of a Roto-Phase converter being used on a single-phase system to operate two three-phase induction motors.

ings. In addition, it is designed with an independent winding by which direct current is introduced; that is, the control winding.

When the ac coils of a saturable reactor are carrying current to the load, an ac flux (magnetism) saturates the iron core. With only ac coils functioning, the magnetism going into the iron core restricts the ac flow to the load. This restriction of current flow (impedance) causes the voltage output to the load to be about 10% of the line supply voltage. Since the iron core is always fully saturated with magnetic flux, the use of direct current from the control winding introduces dc flux, which displaces the ac flux from the iron core. This action reduces the impedance and causes the voltage output to the load to increase. By adjusting the dc flux saturation, the impedance of the ac gate windings may be infinitely varied. This provides a smooth control ranging from approximately 10 to 94% of the line voltage at the load.

Since there is practically no power loss in the control of the impedance in a saturable reactor, a relatively small amount of direct current can control large amounts of alternating current.

Application

Saturable core reactors eliminate the need for mechanical and resistance controls and are a very efficient means of proportional power control for resistance heating devices, vacuum furnaces, infrared ovens, process heaters, and other current-limiting applications.

In lighting control — especially where wattage per circuit is large — a saturable reactor eliminates the loss of dissipated power of a resistance control and provides an infinitely smooth regulated power output to the lighting load.

Wound rotor motors can be started smoothly and operate at speeds commensurate with the load when a saturable reactor, connected in series with the motor, provides the control. This, of course, completely eliminates the maintenance of costly grid resistors and drum controllers.

Types of Saturable Reactors

Although details of design will vary with each manu-facturer, usually three basic styles are available. Small-size reactors are constructed with two ac coils (gate wind-ings) on a common core with the dc control winding on the center leg of the core.

Figure 4-57 shows a medium-size reactor which is con-structed with two ac coils each mounted on an independ-ent wound core. The dc control winding is independent but encompasses both coils.

Larger-size reactors normally utilize four ac coils as the gate winding, as shown in Figure 4-58. In addition, the dc control coil constitutes a continuous winding which cir-cumscribes each pair of ac coils.

Each pair of ac coils is connected in parallel in a manner to prevent the dc windings from being affected by induced ac voltage, harmonics, or high-peak voltages when sudden load change occurs.

The matched design of the gate winding and dc control winding results in a highly efficient magnetic coupling. The ac output of a saturable reactor with matched wind-ings, connected to a unity power-factor load, can achieve in some designs as high as 97% of the input line voltage.

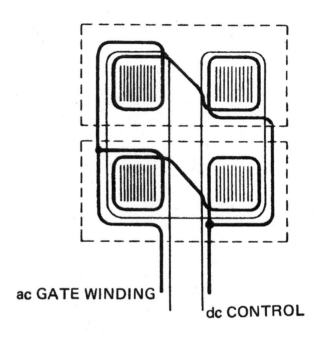

ac GATE WINDING

dc CONTROL

Figure 4-58: Larger-size reactors utilize four ac coils as the gate windings. In addition, the dc control constitutes a continuous winding which circumscribes each pair of ac coils.

Saturable reactors constructed with wound cores, in which the lamination end joints are equally distributed over a wide portion of the core's circumference, provide an unusually wide voltage adjustment for the load, par-ticularly in the lower-voltage range.

Range of Control

A saturable core reactor connected in series between the line supply and the load will provide a maximum line voltage to the load in relation to the power factor of the load. However, in no case will the available load voltage ever be 100% of the line voltage.

Since inductance is the principal electrical property of a saturable reactor, the current-impedance drop added to the normal current-resistance drop limits the percentage of voltage that will be transferred through the reactor. Be-cause of this normal voltage loss, the electrical units com-prising the load, based on their power factor, should be suitable to operate at voltages less than line voltage.

With the saturable reactor connected to the load and without dc excitation, the normal exciting current in the gate winding will supply a voltage to the load of approxi-mately 10% of the line voltage value. As the dc control is increased in power, the ac voltage to the load is increased. With the dc supply adjusted to its maximum power, the

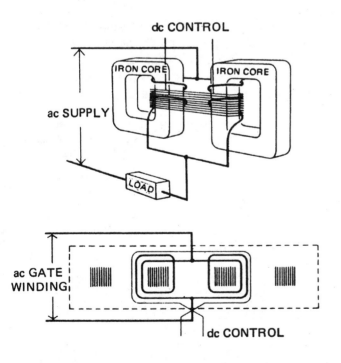

Figure 4-57: Medium-size reactors are constructed with two coils each mounted on an independent wound core. The dc control winding is independent but encompasses both coils.

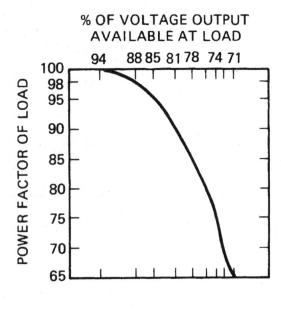

% OF VOLTAGE OUTPUT
AVAILABLE AT LOAD

MAXIMUM LOAD CHARACTERISTICS	
LOAD POWER FACTOR	MAX. VOLTAGE AVAILABLE TO LOAD IN % OF LINE VOLTAGE
1.00	94%
0.98	88%
0.95	85%
0.90	81%
0.85	78%
0.80	76%
0.75	74%
0.70	73%
0.65	71%

Figure 4-59: With the dc supply adjusted to its maximum power, the maximum voltage to the load will be slightly less than the line supply voltage, depending on the power factor and the percent of rated load.

maximum voltage to the load will be slightly less than the line supply voltage, depending on the power factor and the percent of rated load. *See* Figure 4-59.

The relationship between applied dc control and ac output can best be illustrated in the wiring diagrams in Figures 4-60, 4-61, and 4-62.

The triangles graphically illustrate the relationship of the ac voltage values of the supply line, across the reactor, and at the load.

Determining Requirements

Saturable reactors are rated in kVA equal to the unity power-factor load they control. To determine the reactor size required for a given load, use the following equation for single-phase application:

$$\text{saturable reactor kVA} = \frac{94\% \text{ of line voltage} \times \text{load current}}{1000}$$

To illustrate the use of this equation, assume a voltage of 240 V with a load current of 12 A single phase. Substituting these values in the equation, we have:

$$\frac{94\% \times 240 \times 12}{1000} = 2.71 \text{ kVA}$$

A 3-kVA saturable core reactor would be satisfactory for the preceding application provided the load is unity power factor. However, if the load is less than unity power factor, the reactor kVA required should be calculated in accordance with the power-factor constants that follow:

Load Power Factor	Power-Factor Constant Multiplier
1.00	1.00
.98	1.09
.95	1.17
.90	1.29
.85	1.42
.80	1.55
.75	1.70
.70	1.84
.65	1.89

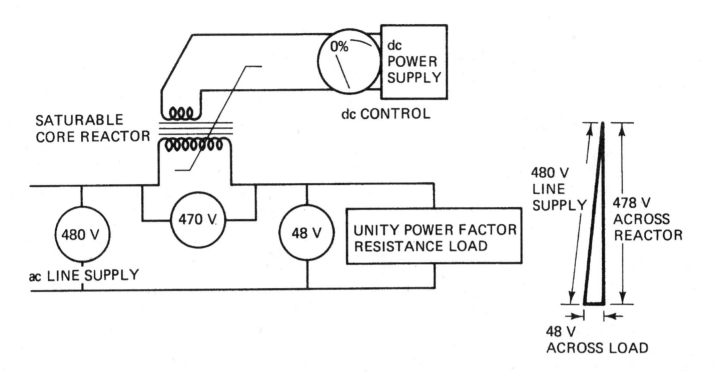

Figure 4-60: With no direct current applied, voltage across the reactor reads approximately 98% of line voltage, and the voltage at the load is approximately 10% of line voltage.

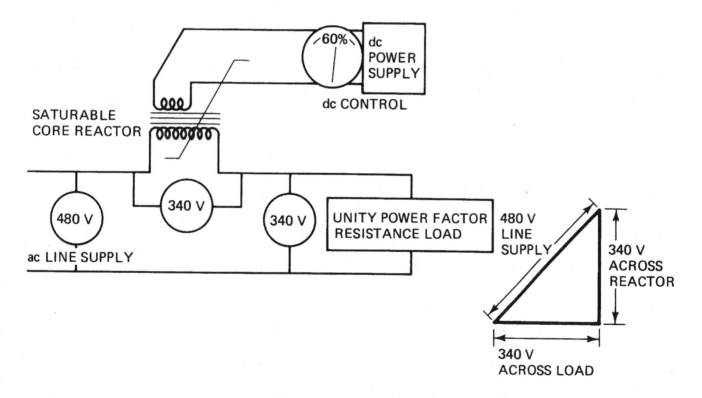

Figure 4-61: With 60% direct current applied, voltage across ... or measures 70% of line voltage, and the voltage at the load also reads 70% of line voltage.

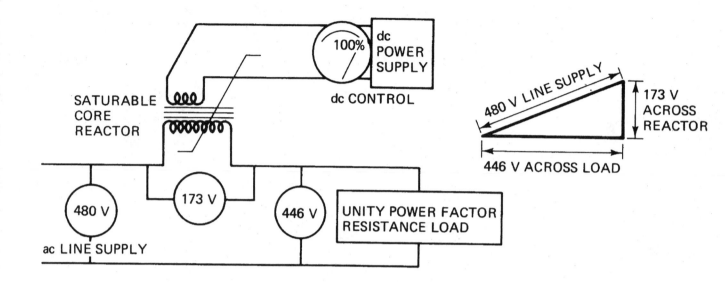

Figure 4-62: When 100% direct current is applied, the voltage across the reactor drops to 37% of line voltage, and the voltage to the load increases to 94% of line voltage.

Using this table, the load kW × power-factor constant (in the table) = kVA load of saturable reactor. For example, if the power factor is 80% (using the previous example of load), the load of 2.71 kVA × 1.55 power-factor constant = 4.2 kVA — the adjusted size needed for this application. For this application, use a 5-kVA saturable reactor, which is the nearest standard larger size.

Single-Phase Load Control

Saturable reactors can be used to control individual circuits of a lighting installation, even to control individual spotlights, or a single reactor can be used to uniformly control the light output of a multiple-circuit installation as shown in Figure 4-63.

Three-Phase Applications

Three single-phase reactors, each in series with each supply line, can be used to control a three-phase load, as three-phase reactors are not commercially feasible. Since each reactor controls a third of the total power to the load, each is rated at one-third of the total load; that is, three 2-kVA single-phase reactors will control a 6-kVA three-phase load, as shown in Figure 4-64.

Care must be exercised to make certain the proper reator voltage is used for three-phase loads. For example, for 240-V, three-phase service, use three 138-V single-phase saturable reactors. For 480-V, three-phase, use three 277-V, single-phase saturable reactors.

The dc control winding of the three reactors should be connected in series to provide symmetrical control of each phase. When this connection is used, the dc control voltage rating of each reactor is one-third of the dc supply voltage. The dc power supply used for controlling the reactors must be capable of delivering three times the wattage required to control one single reactor. Other applications are shown in Figures 4-65, 4-66, and 4-67.

Reversing Rotation of Three-Phase Motors

Saturable reactors may be used to reverse the rotation of three-phase induction motors. For this application, they should be connected as shown in Figure 4-68. Note that four saturable reactors are used but that only two reactors are saturated at any one time. Reactors 2 and 3 provide clockwise rotation. When these two are deenergized of direct current and reactors 1 and 4 are saturated, the motor shaft will reverse. This may be accomplished slowly or almost instantaneously. The latter could be used with some modification in the form of a motor brake.

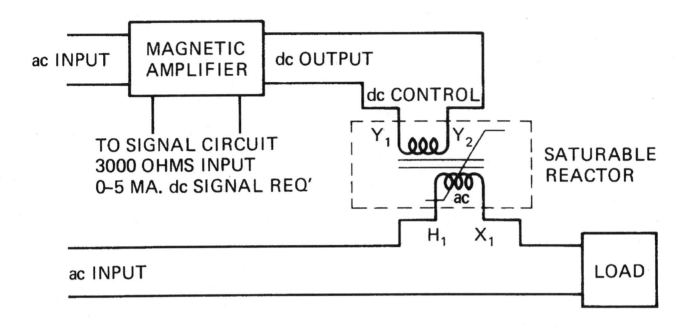

Figure 4-63: Diagram showing connections of a single-phase saturable reactor and magnetic amplifier controlling a single-phase load.

Direct-Current Requirements

The amount of dc power required for manual or automatic adjustmet of the ac load is very small — usually between .33 and 1% of the load. This dc power requirement may be expressed as gain. To illustrate, the gain of a 10-kVA reactor controlled by 60 W of direct current would be as follows:

$$\frac{10{,}000\,W}{60\,W} = \text{gain of } 167$$

The gain of a 3-kVA reactor controlled by 30 W of direct current would be as follows:

$$\frac{3000\,W}{30\,W} = \text{gain of } 100$$

The gain of a 25-kVA reactor controlled by 115 W of direct current would be as follows:

$$\frac{25{,}000\,W}{115\,W} = \text{gain of } 217$$

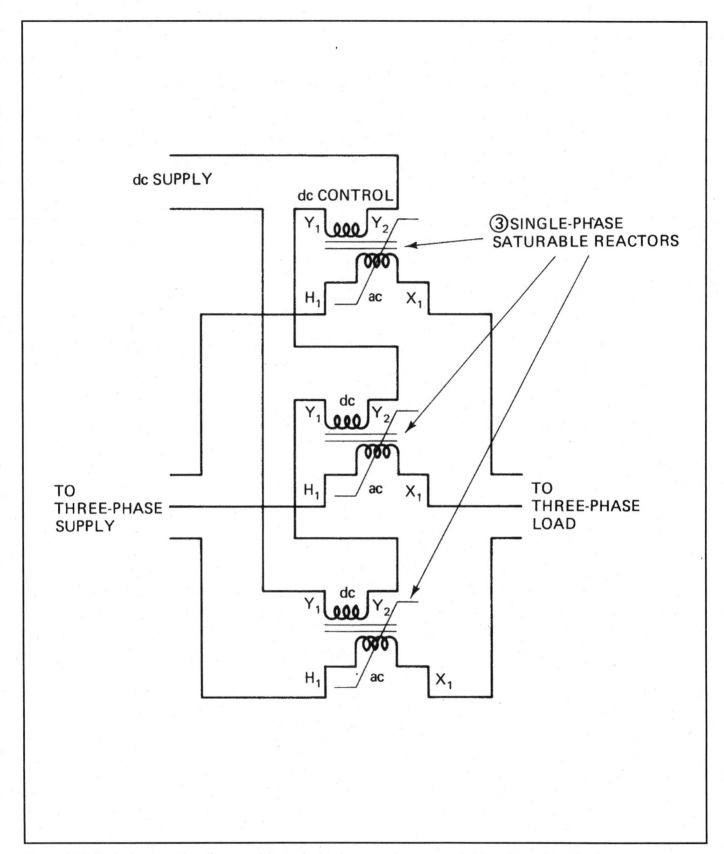

Figure 4-64: Typical connection diagram of three single-phase reactors connected to a three-phase load.

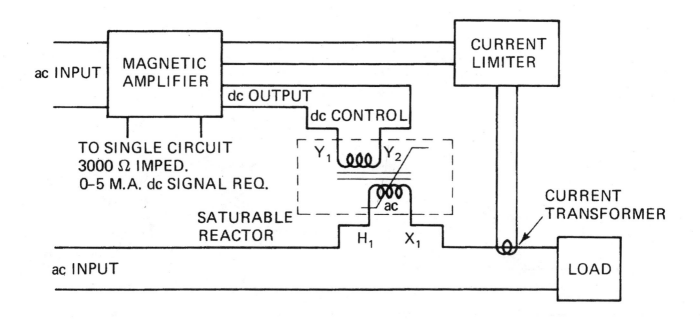

Figure 4-65: Diagram showing connection for single-phase current limiting.

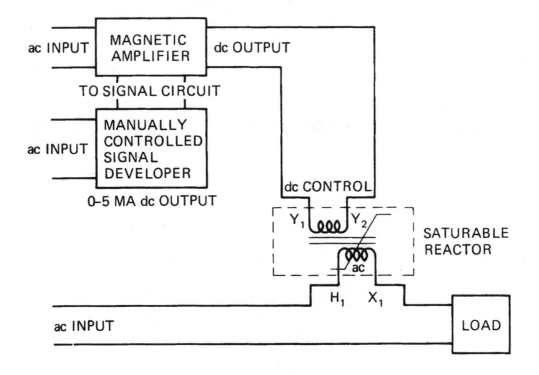

Figure 4-66: Diagram showing connections for adding a manually-controlled, signal-developer power supply to supply a signal to the magnetic amplifier.

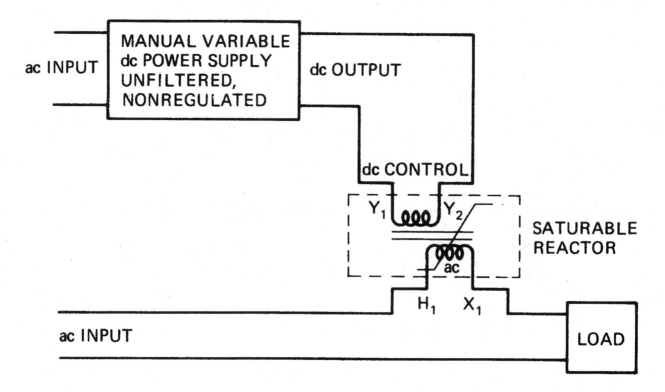

Figure 4-67: Diagram showing connections for an unfiltered, manually regulated power supply used in place of the magnetic amplifier to control the saturable reactor.

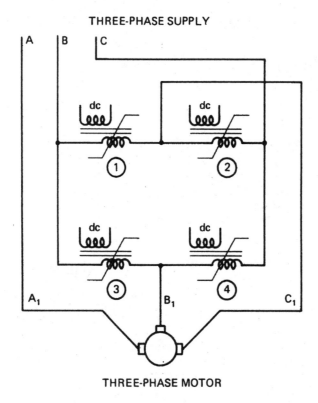

Figure 4-68: Diagram showing connections of four saturable reactors to reverse three-phase induction motors.

OVERCURRENT PROTECTION FOR TRANSFORMERS

Overcurrent protection for transformers is based on their rated current, not on the load to be served. The primary circuit may be protected by a device rated or set at not more than 125% of their rated primary current of the transformer for transformers with a rated primary current of 9 A or more.

Instead of individual protection on the primary side, the transformer may be protected only on the secondary side if all the following conditions are met:

- The overcurrent device on the secondary side is rated or set at not more that 125% of the rated secondary current.
- The primary feeder overcurrent device is rated or set at not more than 250% of the rated primary current.

For example, if a 12-kVA transformer has a primary current rating of:

$$12,000 \text{ W}/480 \text{ V} = 25 \text{ A}$$

and a secondary current rated at

$$12,000 \text{ W}/120 \text{ V} = 100 \text{ A}$$

the individual primary protection must be set at

$$1.25 \times 25 \text{ A} = 31.25 \text{ A}$$

In this case, a standard 30-A cartridge fuse rated at 600 V could be used, as could a circuit breaker approved for use on 480 V. However, if certain conditions are met, individual primary protection for the transformer is not necessary in this case if the feeder overcurrent-protective device is rated at not more than

$$2.5 \times 25 \text{ A} = 62.5 \text{ A}$$

and the protection of the secondary side is set at not more than

$$1.25 \times 100 \text{ A} = 125 \text{ A}$$

A standard 125-A circuit breaker could be used.

NOTE

The example cited previously is for the transformer only; not the secondary conductors. The secondary conductors must be provided with overcurrent protection as outlined in NEC Section 210-20.

The requirements of *NEC* Section 450-3 cover only transformer protection; in practice, other components must be considered in applying circuit overcurrent protection. For circuits with transformers, requirements for conductor protection per *NEC* Articles 240 and 310 and for panelboards per *NEC* Article 384 must be observed. Refer to *NEC* Sections 240-3, Exceptions 2 and 5; 240-21, Exceptions 2 and 8; 384-16d.

Primary Fuse Protection Only (NEC Section 450-3b1): If secondary fuse protection is not provided, then the primary fuses must not be sized larger than 125% of the transformer primary full-load A except if the transformer primary F.L.A. is that shown in *NEC* Section 450-3b1 (*see* Figure 4-69).

Individual transformer primary fuses are not necessary where the primary circuit fuse provides this protection.

Primary and Secondary Protection: In unsupervised locations, with primary over 600 V, the primary fuse can be sized at a maximum of 300%. If the secondary is also over 600 V, the secondary fuses can be sized at a maximum of 250% for transformers with impedances not greater than 6%; 225% for transformers with impedances greater than 6% and not more than 10%. If the secondary is 600 V or below, the secondary fuses can be sized at a maximum of 125%. Where these settings do not correspond to a standard fuse size, the next higher standard size is permitted.

In supervised locations, the maximum settings are as shown in Figure 4-70 except for secondary voltages of 600 V or below, where the secondary fuses can be sized at a maximum of 250%.

Primary Protection Only: In supervised locations, the primary fuses can be sized at a maximum of 250%, or the next larger standard size if 250% does not correspond to a standard fuse size.

The use of "Primary Protection Only" does not remove the requirements for compliance with *NEC* Articles 240 and 384. *See* (FPN) in *NEC* Section 450-3 which references *NEC* Sections 240-3 and 240-100 for proper protection for secondary conductors.

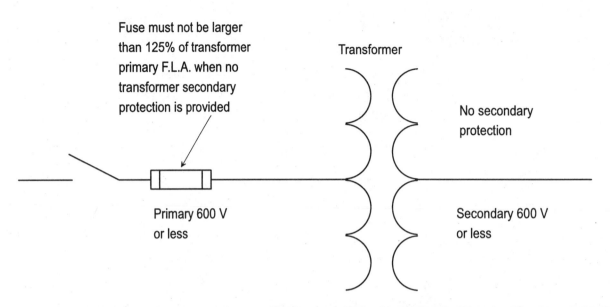

BASIC FUSE RATINGS FOR PRIMARY TRANSFORMER CIRCITS	
Primary Circuit	**Primary Fuse Rating**
9 amps or more	125% or next higher standard rating if 125% does not correspond to a standard fuse size
2 to 9 amps	167% maximum
Less than 2 amps	300% maximum

Figure 4-69: Transformer circuit with primary fuse only.

PRACTICAL APPLICATIONS

A plot plan of a partial college campus renovation project is shown in Figure 4-71. A new administration building is to be added, and the plot plan details the new service connection to the new building. Constructed details for a pad-mounted transformer are shown in Figure 4-72.

A plot plan of an underground electrical secondary distribution system for a residential subdivision is shown in Figure 4-73.

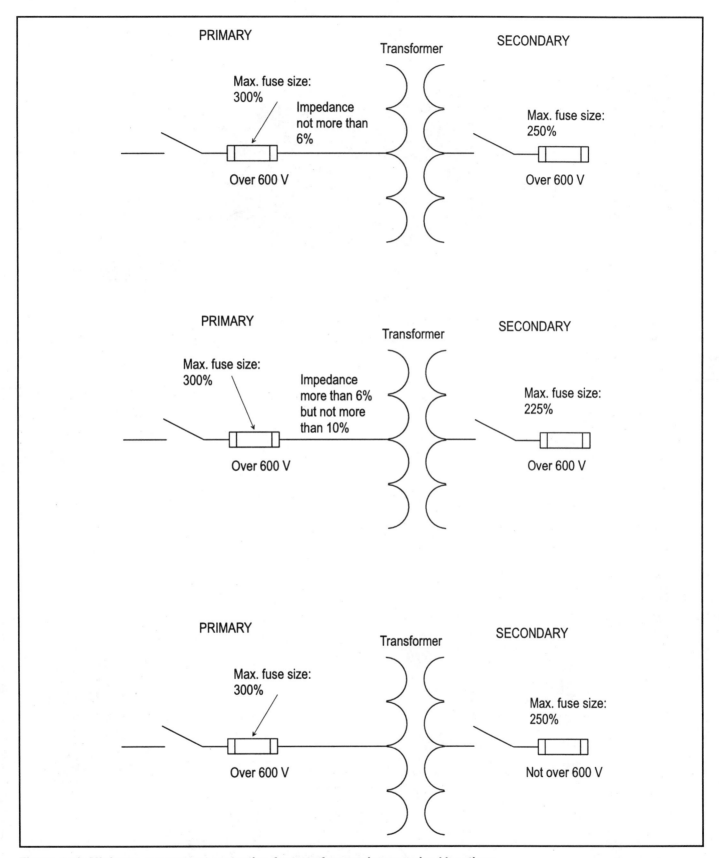

Figure 4-70: Minimum overcurrent protection for transformers in supervised locations.

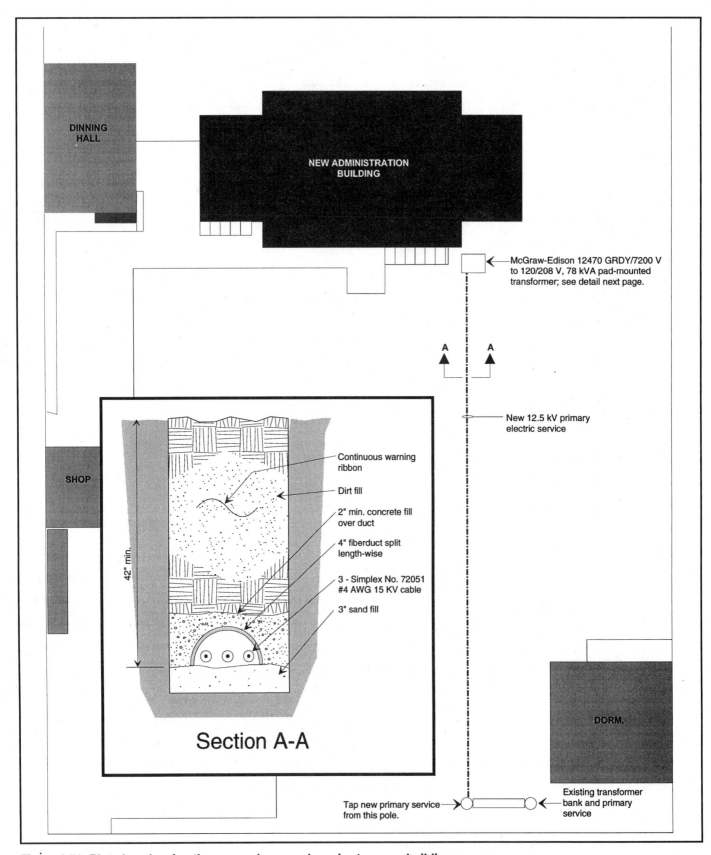

Figure 4-71: Plot plan showing the new underground service to a new building.

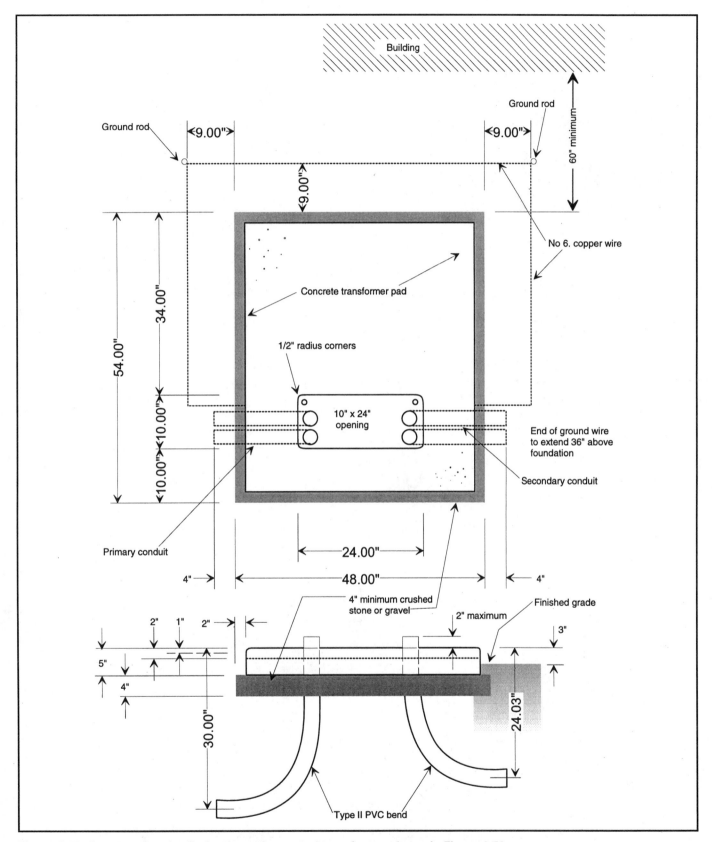

Figure 4-72: Construction details for the pad-mounted transformer shown in Figure 4-70.

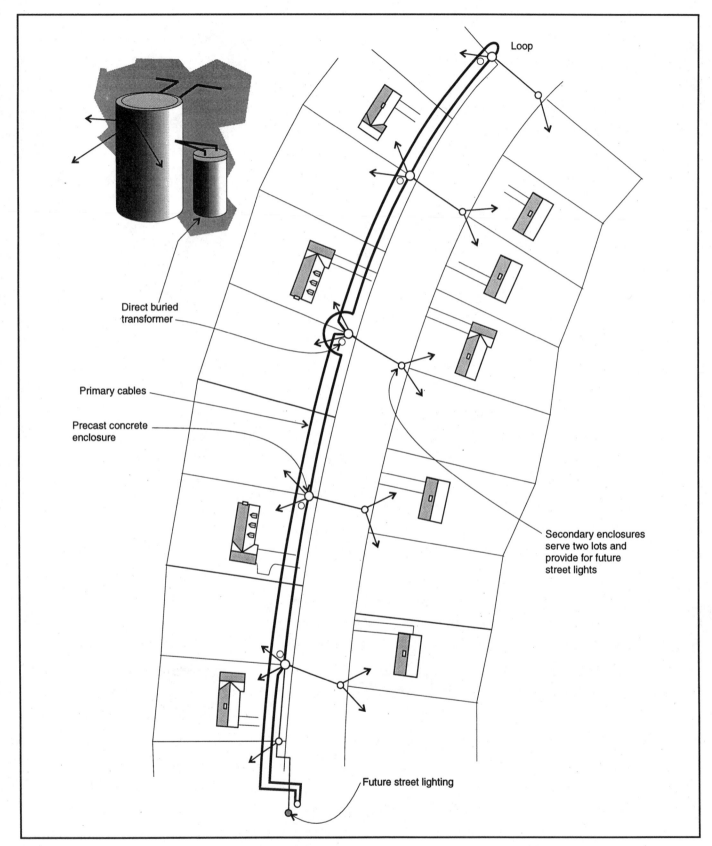

Loop

Direct buried
transformer

Primary cables

Precast concrete
enclosure

Secondary enclosures
serve two lots and
provide for future
street lights

Future street lighting

Figure 4-73: Plan view of an underground distribution system.

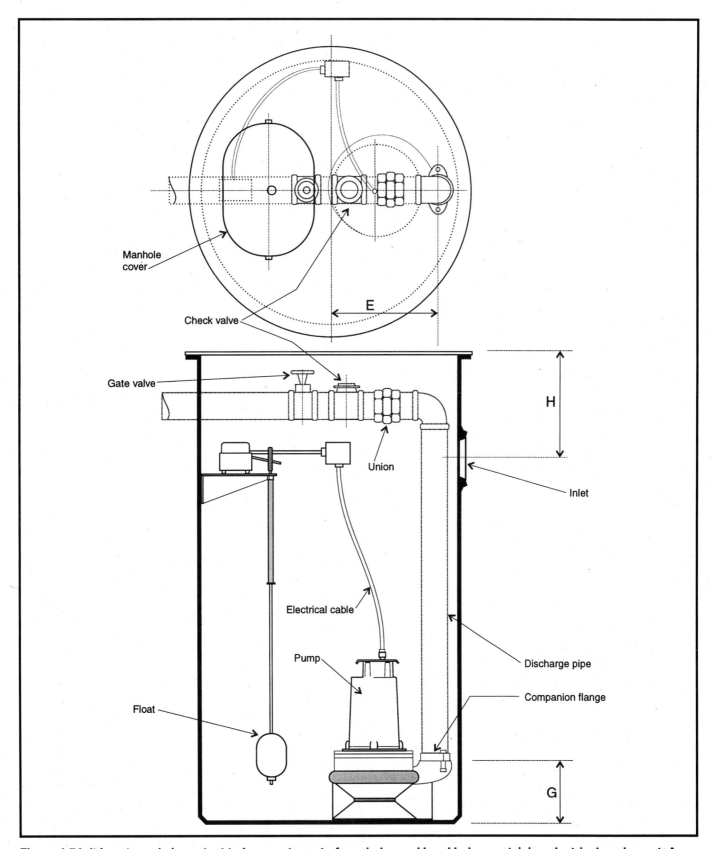

Figure 4-74: It is extremely important to keep water out of manholes and hand holes containing electrical equipment. A drain pipe is always best, but such an arrangment is not always possible; sump pumps are sometimes necessary.

Chapter 5

Service and Distribution

All buildings or premises that utilize an electrical system require an electric service (usually supplied by the local power company). An electric service may be defined as the conductors and equipment that delivers energy from the electric power distribution system to the wiring system of the premises served.

All buildings containing equipment that utilizes electricity require an electric service. An electric service will enable the passage of electrical energy from the power company's lines to points of use within the buildings. Figure 5-1 on the next page shows the basic sections of an electric service. In this illustration, note that the high-voltage lines terminate on a power pole near the building that is being served. A transformer is mounted on the pole to reduce the voltage to a usable level (120/240 V in this case). The remaining sections are described as follows:

- *Service drop:* The overhead conductors, through which electrical service is supplied, between the last power company pole and the point of their connection to the service facilities located at the building or other support used for the purpose.
- *Service entrance:* All components between the point of termination of the overhead service drop or underground service lateral and the building's main disconnecting device, except for the power company's metering equipment.
- *Service-entrance conductors:* The conductors between the point of termination of the overhead service drop or underground service lateral and the main disconnecting device in the building.

- *Service-entrance equipment:* Provides overcurrent protection to the feeder and service conductors, a means of disconnecting the feeders from energized service conductors, and a means of measuring the energy used by the use of metering equipment. *See* Figures 5-2 and 5-3.

When the service conductors to the building are routed underground, these conductors are known as the service lateral, defined as follows:

- *Service lateral:* The underground conductors through which service is supplied between the power company's distribution facilities and the first point of their connection to the building or area service facilities located at the building or other support used for the purpose.

Drawings and specifications should contain information relating to number, size, voltage rating, type of insulation of all conductors, and cables involved in the electric service and other service-entrance equipment. A wiring diagram should also appear on the drawings to show the components of the service and distribution system.

PANELBOARDS

Panelboards are normally shown and described on electrical working drawings by a combination of three methods: (1) electrical symbols, (2) schedules, and (3) power riser diagrams. In some instances, further details of panelboards and other service equipment are found in the written specifications.

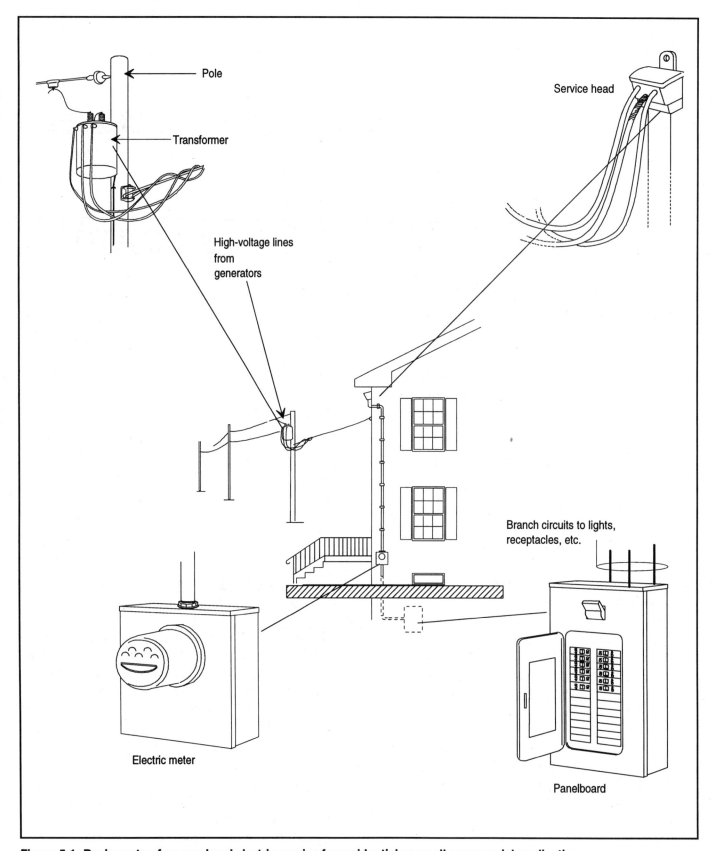

Figure 5-1: Basic parts of an overhead electric service for residential or small commercial applications.

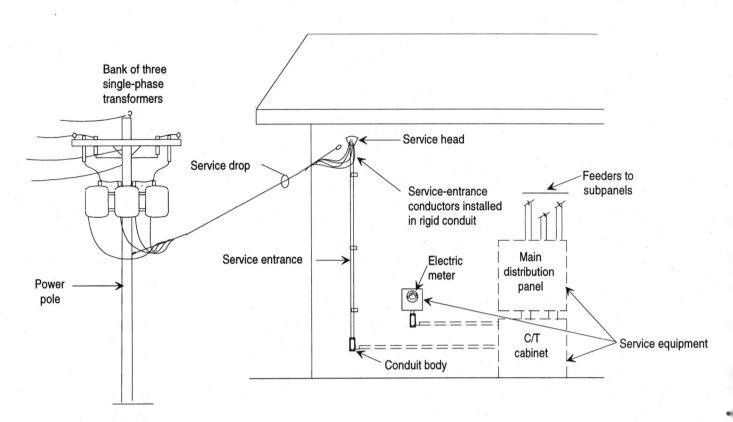

Figure 5-2: Typical three-phase overhead service.

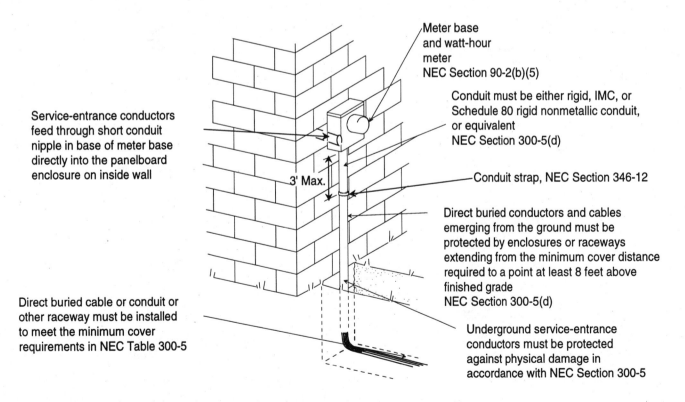

Figure 5-3: Single-phase service lateral for small commercial building.

The drawing in Figure 5-4 is a floor plan of a small renovation project. Note that the location of the service entrance and panelboards are first located on the floor plan. The panelboards are identified by two rectangular symbols, with the letters A and B enclosed in circles adjacent to the respective panelboard. Other notes are provided at this point on the drawing to give further details on the stand-by electric generator operation.

The power-riser diagram (Figure 5-5) is usually located on the same drawing sheet as the power plan. The drawing

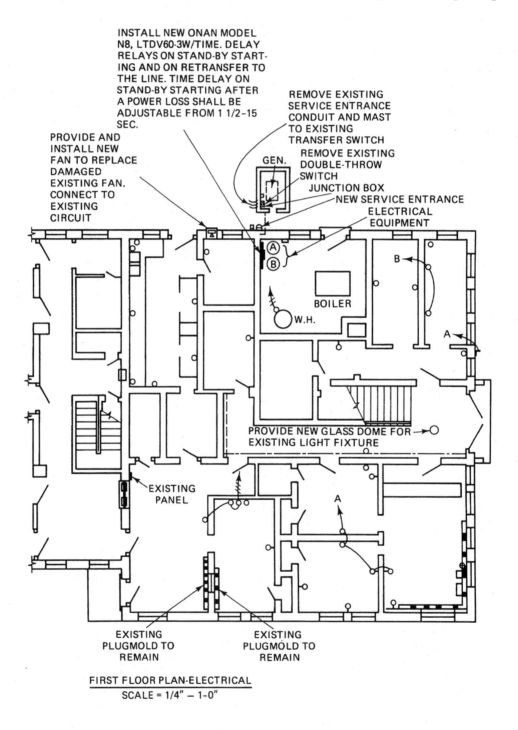

INSTALL NEW ONAN MODEL N8, LTDV60-3W/TIME. DELAY RELAYS ON STAND-BY START- ING AND ON RETRANSFER TO THE LINE. TIME DELAY ON STAND-BY STARTING AFTER A POWER LOSS SHALL BE ADJUSTABLE FROM 1 1/2-15 SEC.

PROVIDE AND INSTALL NEW FAN TO REPLACE DAMAGED EXISTING FAN. CONNECT TO EXISTING CIRCUIT

REMOVE EXISTING SERVICE ENTRANCE CONDUIT AND MAST TO EXISTING TRANSFER SWITCH

REMOVE EXISTING DOUBLE-THROW SWITCH

GEN.

JUNCTION BOX

NEW SERVICE ENTRANCE ELECTRICAL EQUIPMENT

B

BOILER

W.H.

A

PROVIDE NEW GLASS DOME FOR EXISTING LIGHT FIXTURE

EXISTING PANEL

A

EXISTING PLUGMOLD TO REMAIN

EXISTING PLUGMOLD TO REMAIN

FIRST FLOOR PLAN-ELECTRICAL
SCALE = 1/4" — 1-0"

Figure 5-4: Floor plan of renovation project showing details of service equipment.

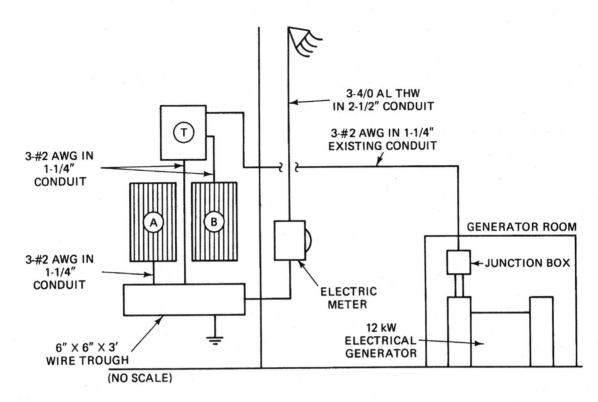

Figure 5-5: Power-riser diagram used in conjunction with the floor plan in Figure 5-4 and the panelboard schedule in Figure 5-6.

gives a diagrammatic view of the panelboards and related service wiring. The drawing on the floor plan gives the location of the equipment in relation to the building. The power-riser diagram gives further details, such as sizes of conduit and conductors feeding the panels, size of wire trough, and size of ground wire.

The panelboard schedule (Figure 5-6 on the next page) gives additional information on the individual panelboards, A and B. For example, the schedule states that panel A is surface mounted, has a 100-A capacity, is rated for 120/240 V, and is single-phase, three wire. The panel also contains a 100-A main circuit breaker. Looking to the right of this previous information, we can see that the panel is to contain seven 1-pole, 20-A circuit breakers for new circuits and thirteen 1-pole, 20-A circuit breakers for spares.

Panelboard B is the same type as A, except that it contains different circuit breakers: a total of twelve 1-pole, 20-A circuit breakers; four 2-pole, 60-A breakers; one 2-pole, 30-A breaker; and one 2-pole, 20-A circuit breaker.

With the information provided in the floor plan, power-riser diagram, and the panelboard schedule, there should

be little doubt as to how the electric panelboards should be installed and wired.

As mentioned previously, the written specifications may also contain data pertaining to the panelboards. This information is mainly provided so that substitutions for brands other than those specified can be made.

Panelboards are classified by the *NEC* in two general categories:

- Lighting and appliance panels
- Power distribution panels

A lighting and appliance branch-circuit panelboard is defined in the *NEC* as "One having more than 10% of its overcurrent devices rated 30 amp or less for which neutral connections are provided." The *NEC* also limits the number of overcurrent devices (branch-circuit poles) to a maximum of 42 in any one cabinet. When the 42 poles are exceeded, two or more separate enclosures are required.

All other panelboards not defined as lighting and appliances branch-circuit panelboards are classified as distribution panels and are restricted only to practical physical limitations such as standard enclosure heights and widths.

Panel ID	Type Cabinet	Panel Mains			Branches					Items Fed or Remarks
		Amps	Volts	Phase	1P	2P	3P	Prot.	Frame	
A	Surface	100 A	120/240 V	1ø, 3W	7	—	—	20 A	70 A	New Circuits
Sq. "D" Type NQO w/100 A Main Breaker					13	—	—	20 A	70 A	Spares
					4	—	—	—	—	Spaces
B	Surface	100 A	120/240 V	1ø, 3W	3	—	—	20 A	70 A	Recept. Cir.
Sq. "D" Type NQO w/100 A Main Breaker					6	—	—	20 A	70 A	Lighting
					3	—	—	20 A	70 A	Spares
					—	4	—	60 A	70 A	Transfer switch, range, and 2 spares
					—	1	—	30 A	70 A	Water Heater
					—	1	—	20 A	70 A	Gen. Rm. Htr.
					6	—	—	—	—	Spaces

Figure 5-6: Panelboard schedule for the project under discussion.

Panels used for service-entrance equipment must be located near the point where the service conductors enter the building. Furthermore, a panelboard having main lugs may have a maximum of six operating handles to disconnect the entire panelboard from the supply conductors. Where more than six disconnects are required, a main circuit breaker or main disconnect switch must be provided.

The use of gutter (wire trough) taps is recommended as the most efficient method of connecting each section of a multisection panelboard to the feeder. The gutter tap devices are normally not furnished with the panelboard but are available from several manufacturers. A wiring diagram of one such tap is shown in Figure 5-7.

Subfeed lugs are another means of interconnecting multisection panels. The second set of lugs are mounted directly beside the main lugs (Figure 5-8). The feeder conductors are then brought through or into the wiring gutter of the first section and connected to the main lugs. Another

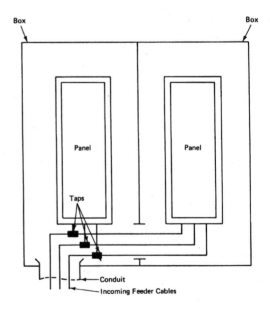

Figure 5-7: Gutter taps connecting multisection panelboards to the feeder.

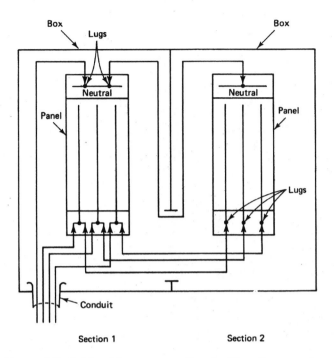

Figure 5-8: Subfeed lugs connecting multisection panels.

set of cables of the same size are connected to the subfeed lugs and are carried over to the main lugs of the adjacent panel. Cross-connection cables are normally not furnished by the panel supplier, but they are available from several manufacturers.

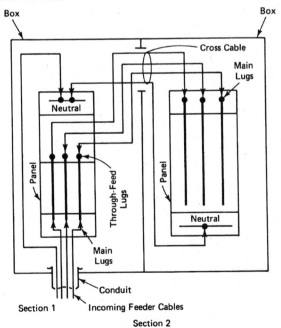

Figure 5-9: Connecting panels using through-feed lugs.

A third method of connecting panels involves the use of through-feed lugs at the bottom of the first panel section (Figure 5-9). Another set of lugs is located at the opposite end of the main bus from the main lugs. The interconnecting cables are connected to the through-feed lugs in the second section. The connection arrangement, of course, could be reversed; that is, main lugs at top, through-feed lugs at the bottom end of the panel.

This method, however, is not the most desirable to use, because all the current that flows to the second section must first flow through the bus bars of the first section. Furthermore, if the panelboard has a neutral bar, it must be moved up several inches to allow space for connecting cross cables, making the panel unnecessarily long.

Figure 5-10 shows the required clearances for overhead services.

ELECTRIC SERVICES

The following drawings show details of several electric service installations for residential and small commercial applications.

Figures 5-11 through 5-13, for example, show typical overhead service installation using Type S.E. (service-entrance) cable. The service drop point of attachment must be of sufficient height to allow for the following minimum service-drop clearances:

- Ten ft for multiplex services above sidewalks and finished grades
- Twelve ft for open wire services above sidewalks and finished grades
- Twelve ft over residential driveways
- Eighteen ft over public streets, alleys and public parking lots
- Two ft clearance from telephone and CATV wires at midspan crossings

When the situation requires a conduit mast installation, the service should be installed as shown in Figures 5-14 through 5-24. The service drop point of attachment must be of sufficient height to allow the minimum clearance as described in Figure 5-10 on the next page.

A typical 100-A underground service lateral is shown in Figure 5-25. Besides the requirements specified in the notes on the drawing, inhibitor compound should be used on all aluminum conductor terminations.

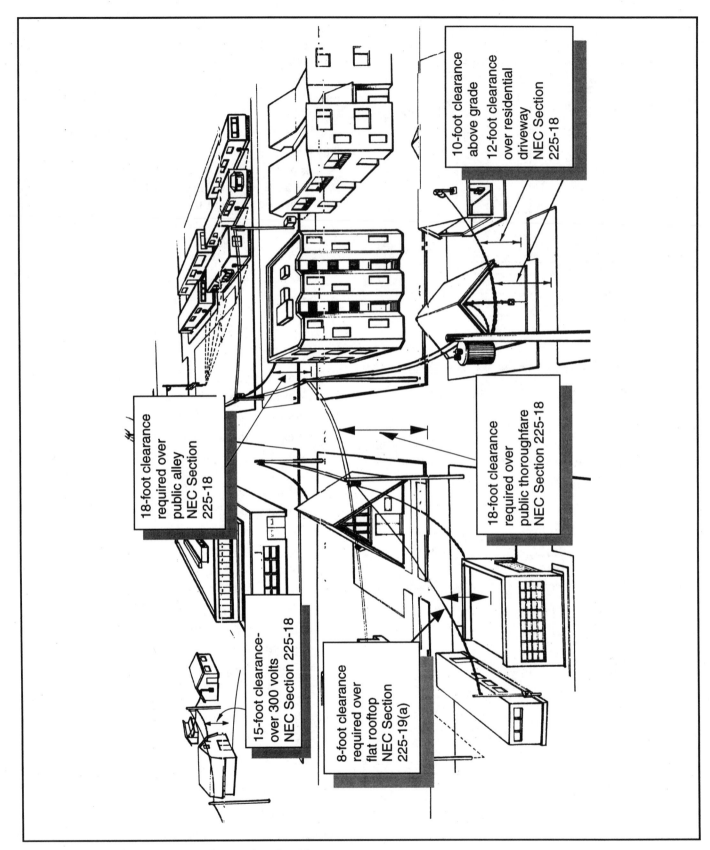

Figure 5-10: Required clearances for overhead services.

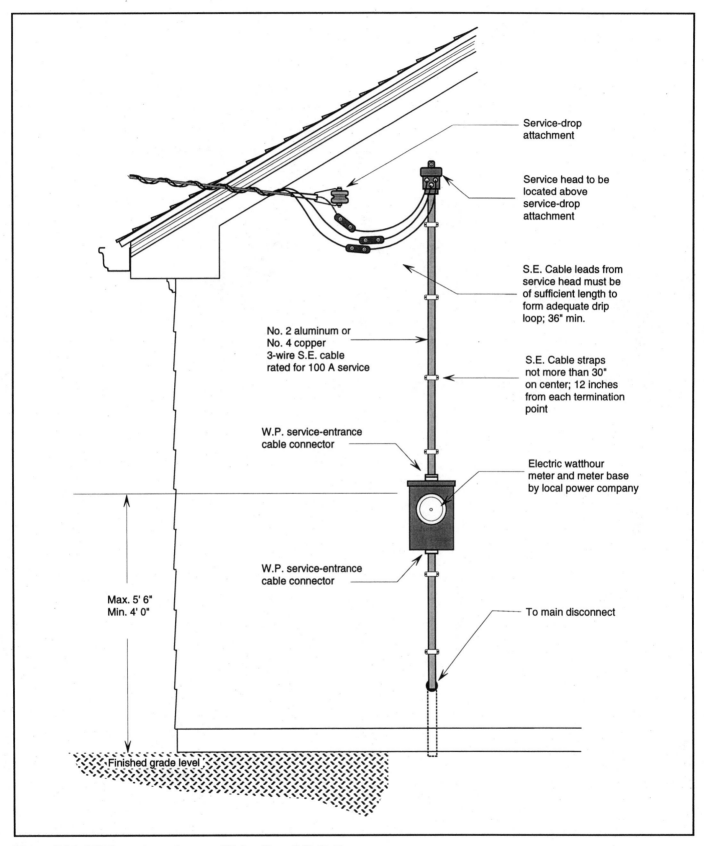

Figure 5-11: 100-A service entrance utilizing Type S.E. Cable.

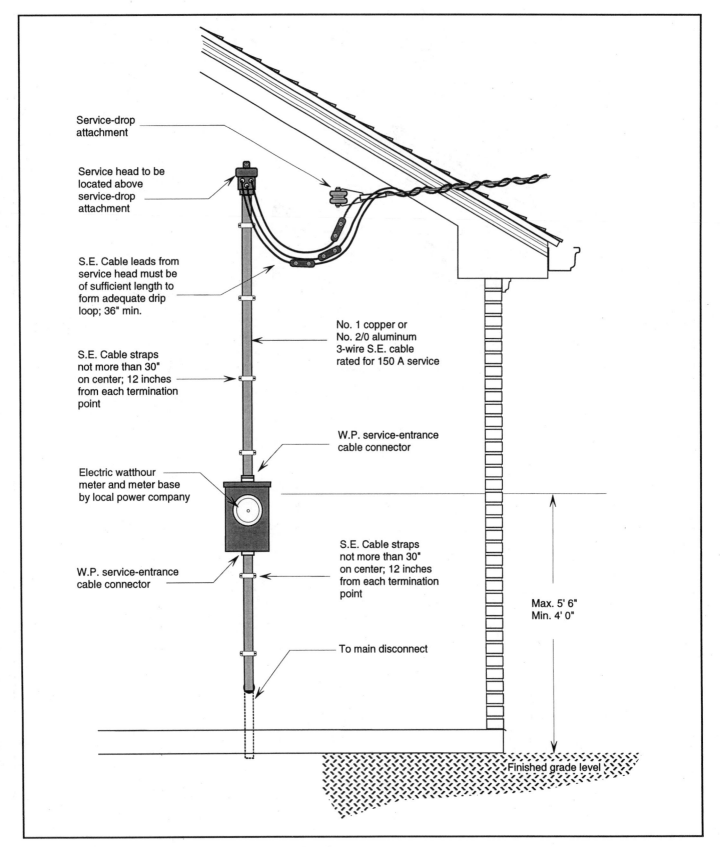

Service-drop attachment

Service head to be located above service-drop attachment

S.E. Cable leads from service head must be of sufficient length to form adequate drip loop; 36" min.

S.E. Cable straps not more than 30" on center; 12 inches from each termination point

Electric watthour meter and meter base by local power company

W.P. service-entrance cable connector

No. 1 copper or No. 2/0 aluminum 3-wire S.E. cable rated for 150 A service

W.P. service-entrance cable connector

S.E. Cable straps not more than 30" on center; 12 inches from each termination point

To main disconnect

Max. 5' 6" Min. 4' 0"

Finished grade level

Figure 5-12: 150-A service entrance utilizing Type S.E. Cable.

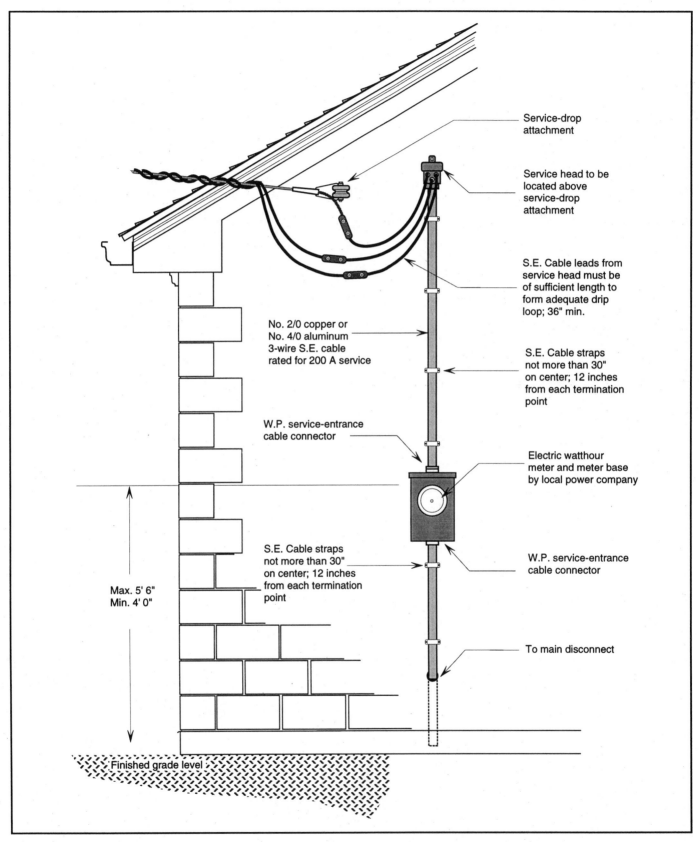

Figure 5-13: 200-A service entrance utilizing Type S.E. Cable.

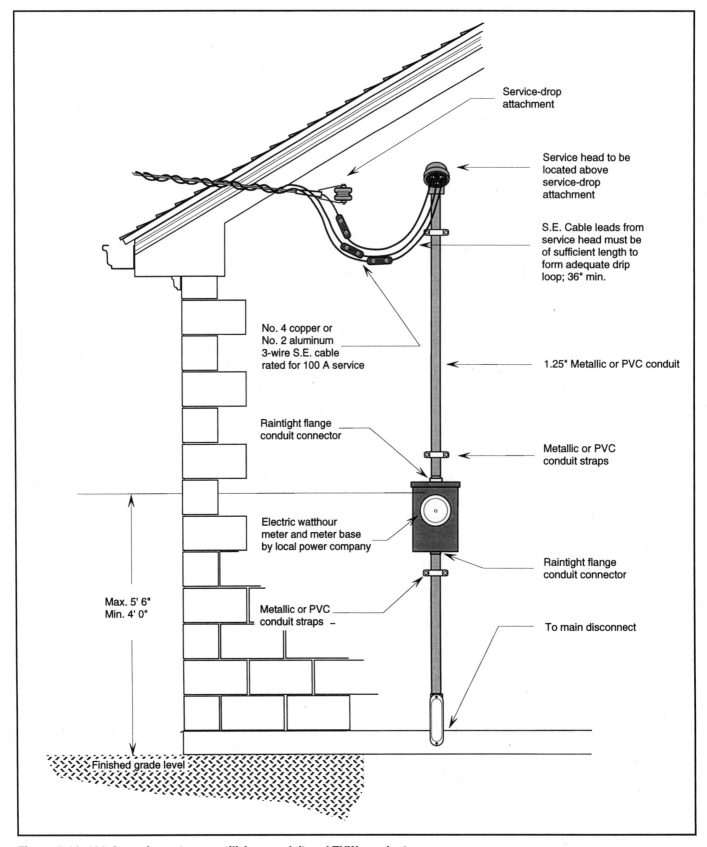

Figure 5-14: 100-A service entrance utilizing conduit and THW conductors.

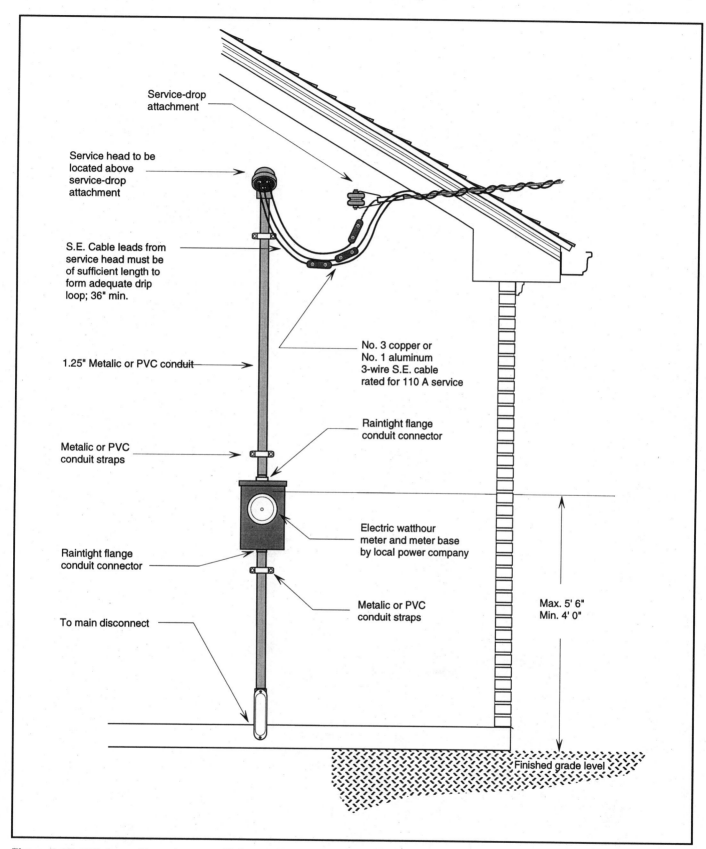

Service-drop
attachment

Service head to be
located above
service-drop
attachment

S.E. Cable leads from
service head must be
of sufficient length to
form adequate drip
loop; 36" min.

1.25" Metalic or PVC conduit

Metalic or PVC
conduit straps

Raintight flange
conduit connector

To main disconnect

No. 3 copper or
No. 1 aluminum
3-wire S.E. cable
rated for 110 A service

Raintight flange
conduit connector

Electric watthour
meter and meter base
by local power company

Metalic or PVC
conduit straps

Max. 5' 6"
Min. 4' 0"

Finished grade level

Figure 5-15: 110-A service entrance utilizing conduit and THW conductors.

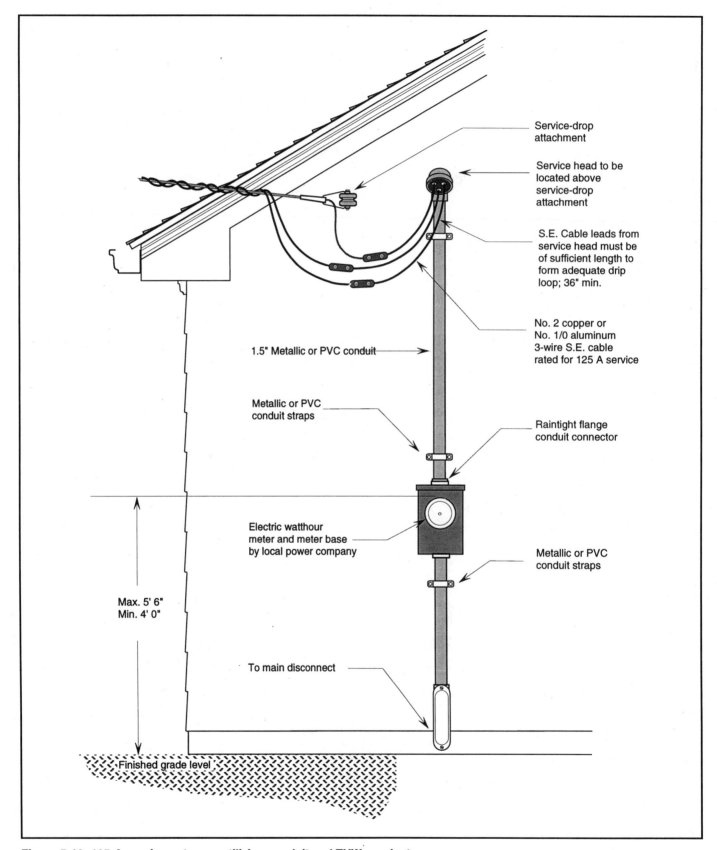

Service-drop
attachment

Service head to be
located above
service-drop
attachment

S.E. Cable leads from
service head must be
of sufficient length to
form adequate drip
loop; 36" min.

No. 2 copper or
No. 1/0 aluminum
3-wire S.E. cable
rated for 125 A service

1.5" Metallic or PVC conduit

Metallic or PVC
conduit straps

Raintight flange
conduit connector

Electric watthour
meter and meter base
by local power company

Metallic or PVC
conduit straps

Max. 5' 6"
Min. 4' 0"

To main disconnect

Finished grade level

Figure 5-16: 125-A service entrance utilizing conduit and THW conductors.

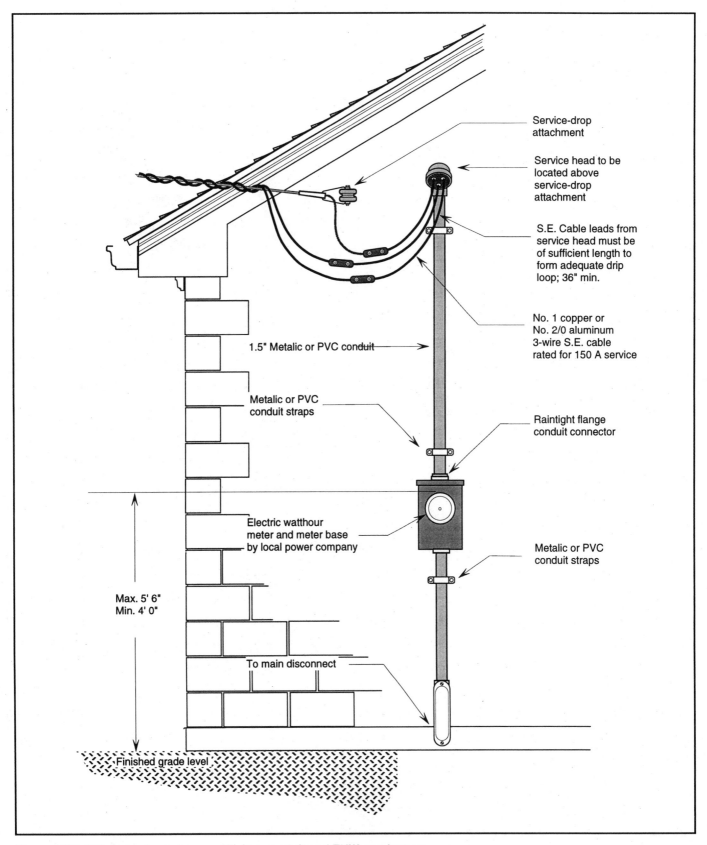

Figure 5-17: 150-A service entrance utilizing conduit and THW conductors.

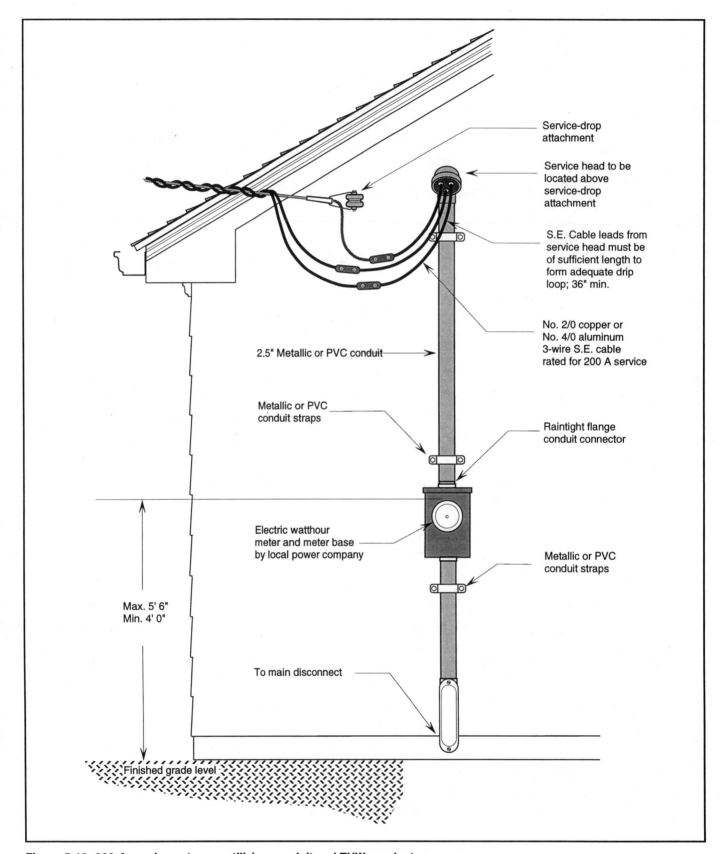

Service-drop attachment

Service head to be located above service-drop attachment

S.E. Cable leads from service head must be of sufficient length to form adequate drip loop; 36" min.

No. 2/0 copper or No. 4/0 aluminum 3-wire S.E. cable rated for 200 A service

2.5" Metallic or PVC conduit

Metallic or PVC conduit straps

Raintight flange conduit connector

Electric watthour meter and meter base by local power company

Metallic or PVC conduit straps

Max. 5' 6"
Min. 4' 0"

To main disconnect

Finished grade level

Figure 5-18: 200-A service entrance utilizing conduit and THW conductors.

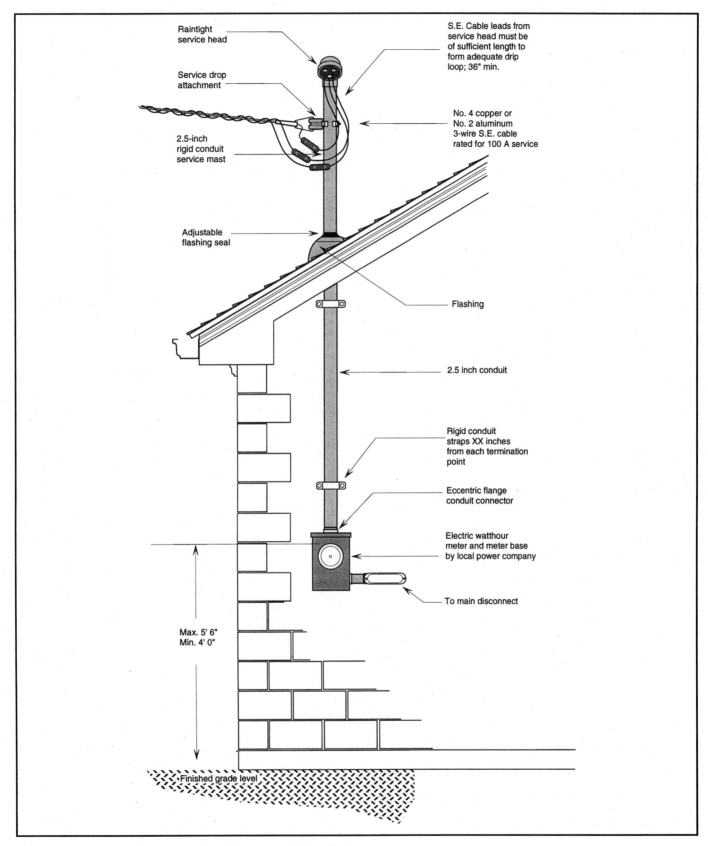

Figure 5-19: 100-A mast-thru-roof service-entrance installation.

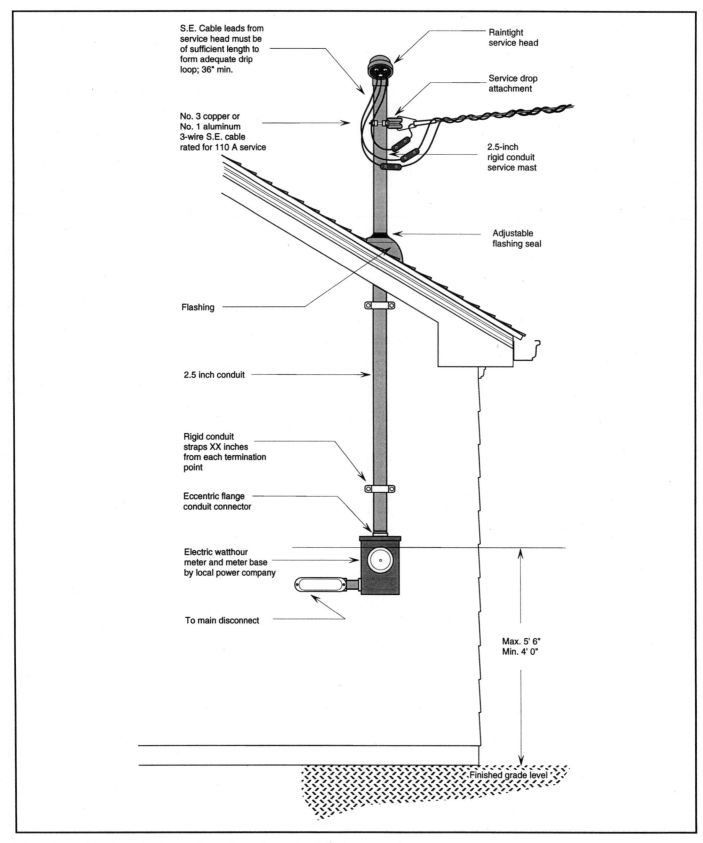

Figure 5-20: 110-A mast-thru-roof service-entrance installation.

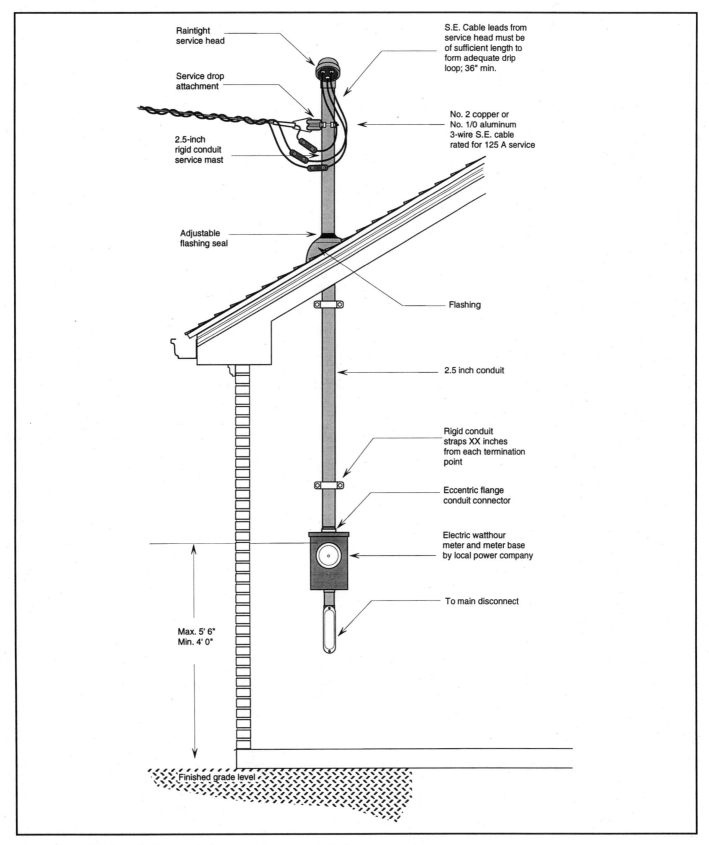

Figure 5-21: 125-A mast-thru-roof service-entrance installation.

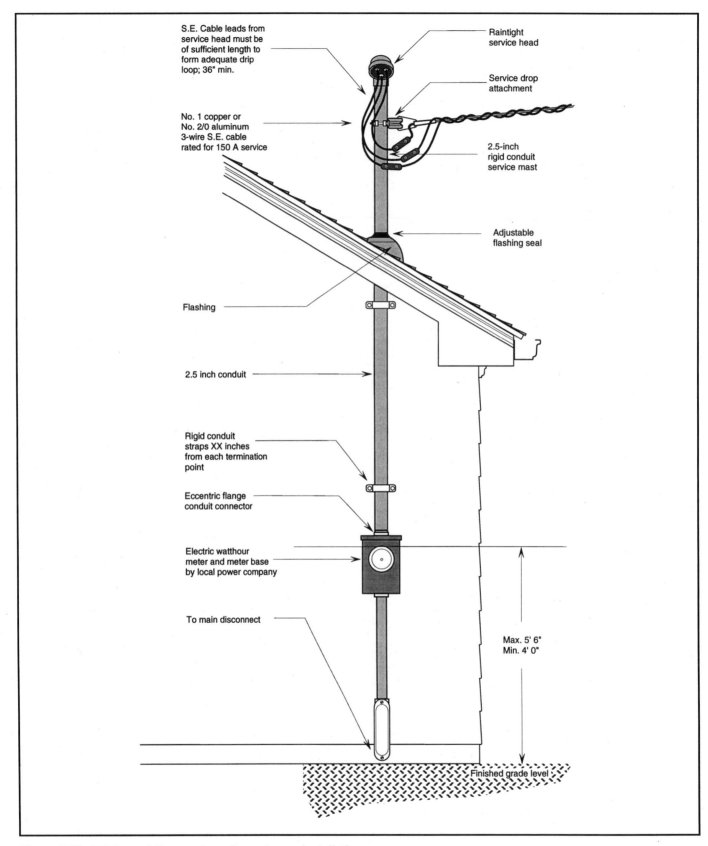

S.E. Cable leads from service head must be of sufficient length to form adequate drip loop; 36" min.

No. 1 copper or No. 2/0 aluminum 3-wire S.E. cable rated for 150 A service

Flashing

2.5 inch conduit

Rigid conduit straps XX inches from each termination point

Eccentric flange conduit connector

Electric watthour meter and meter base by local power company

To main disconnect

Raintight service head

Service drop attachment

2.5-inch rigid conduit service mast

Adjustable flashing seal

Max. 5' 6"
Min. 4' 0"

Finished grade level

Figure 5-22: 150-A mast-thru-roof service-entrance installation.

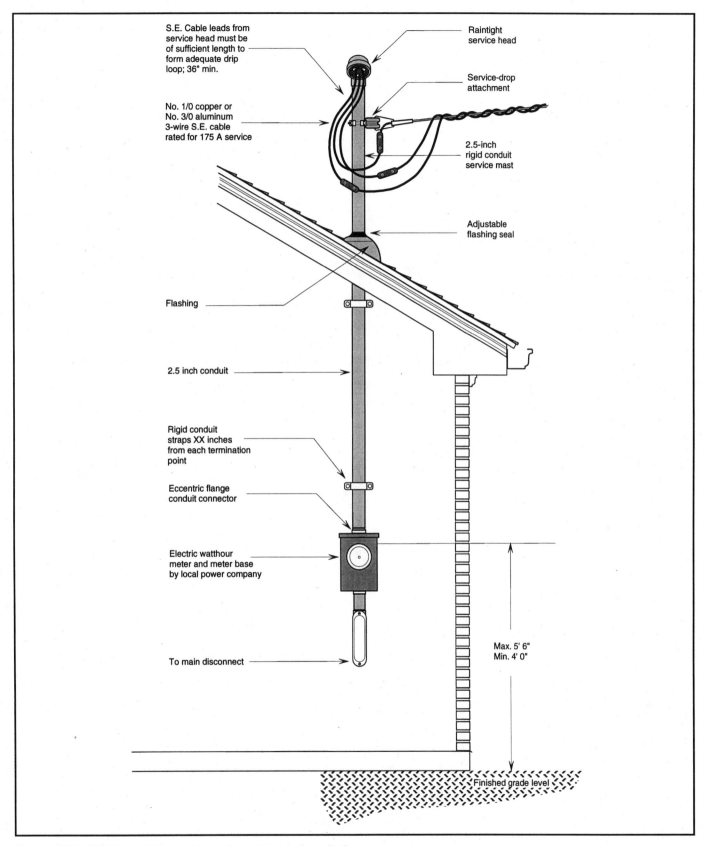

S.E. Cable leads from service head must be of sufficient length to form adequate drip loop; 36" min.

No. 1/0 copper or No. 3/0 aluminum 3-wire S.E. cable rated for 175 A service

Raintight service head

Service-drop attachment

2.5-inch rigid conduit service mast

Adjustable flashing seal

Flashing

2.5 inch conduit

Rigid conduit straps XX inches from each termination point

Eccentric flange conduit connector

Electric watthour meter and meter base by local power company

To main disconnect

Max. 5' 6"
Min. 4' 0"

Finished grade level

Figure 5-23: 175-A mast-thru-roof service-entrance installation.

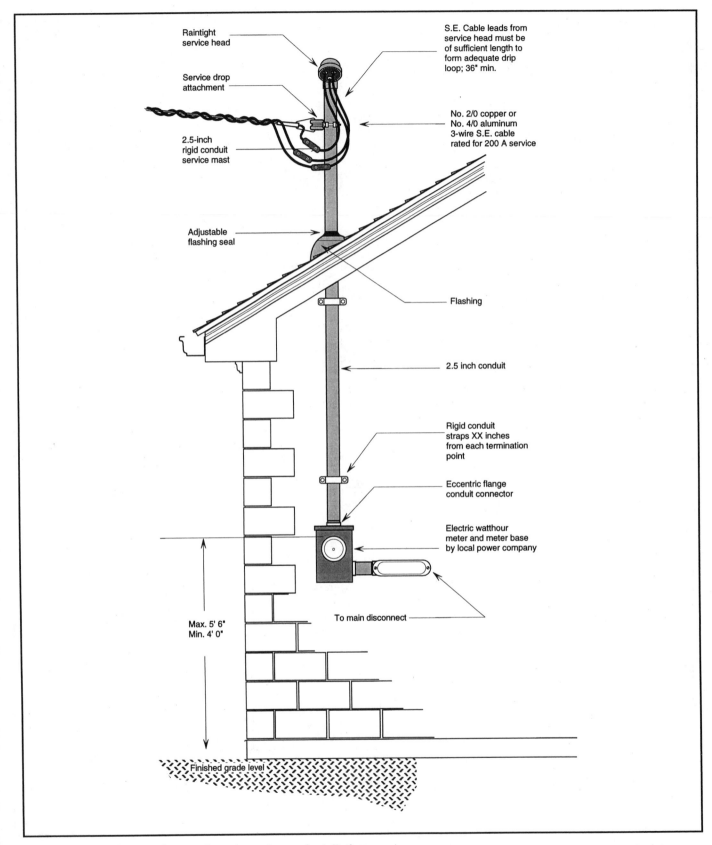

Raintight service head

Service drop attachment

2.5-inch rigid conduit service mast

Adjustable flashing seal

S.E. Cable leads from service head must be of sufficient length to form adequate drip loop; 36" min.

No. 2/0 copper or No. 4/0 aluminum 3-wire S.E. cable rated for 200 A service

Flashing

2.5 inch conduit

Rigid conduit straps XX inches from each termination point

Eccentric flange conduit connector

Electric watthour meter and meter base by local power company

To main disconnect

Max. 5' 6" Min. 4' 0"

Finished grade level

Figure 5-24: 200-A mast-thru-roof service-entrance installation.

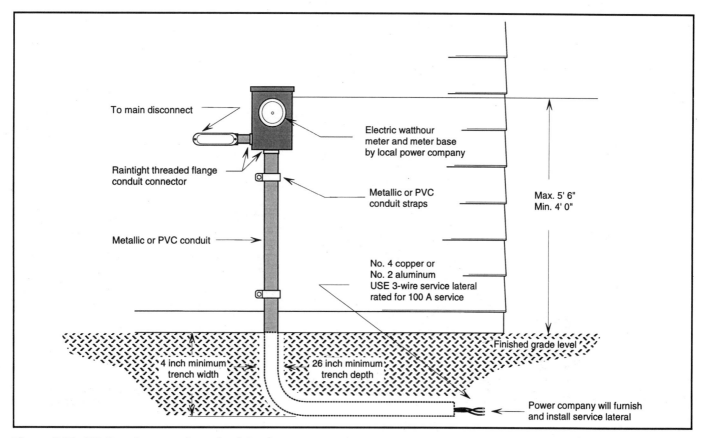

Figure 5-25: 100-A underground service lateral.

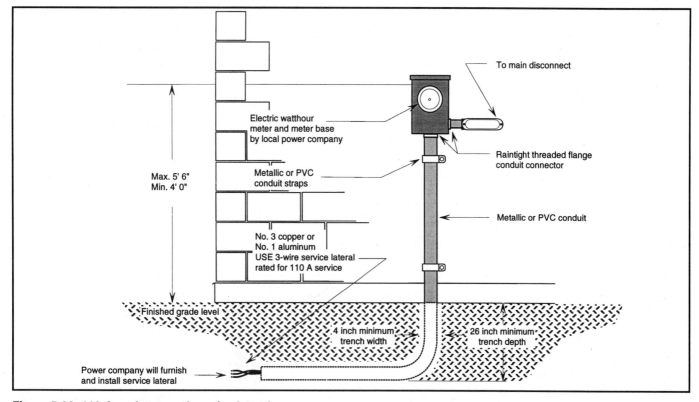

Figure 5-26: 110-A underground service lateral.

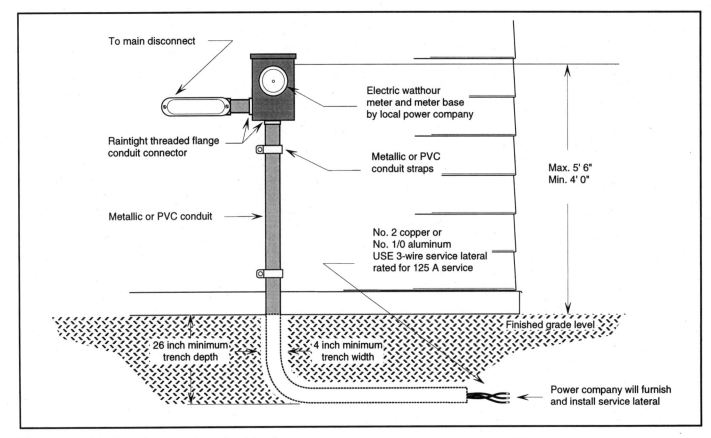

Figure 5-27: 125-A underground service lateral.

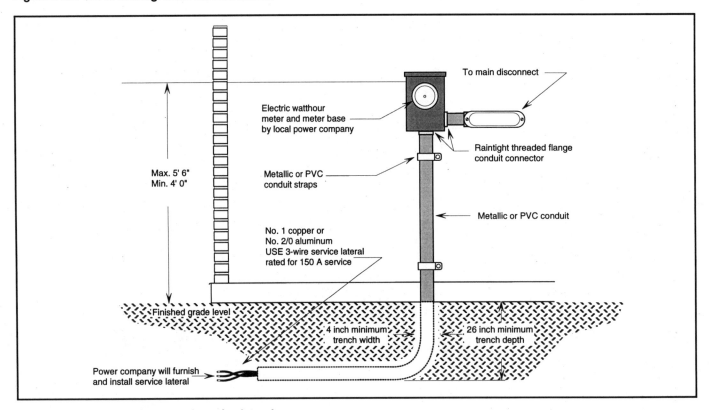

Figure 5-28: 150-A underground service lateral.

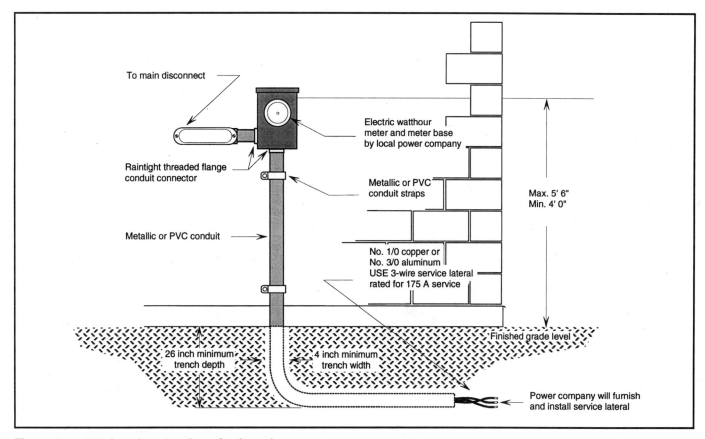

Figure 5-29: 175-A underground service lateral.

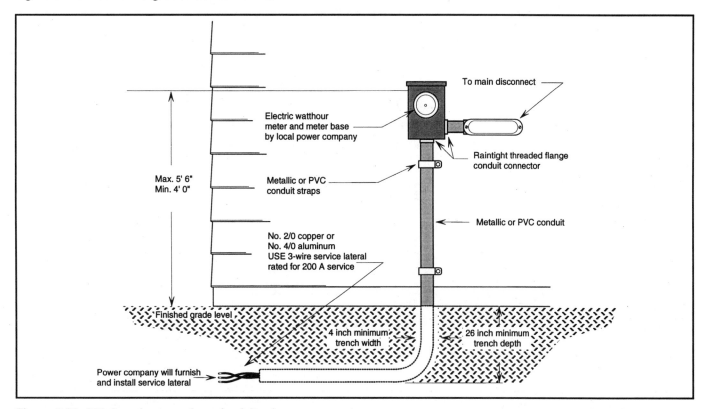

Figure 5-30: 200-A underground service lateral.

Service Equipment

In selecting a panelboard, factors such as the following must be considered: service and voltage, number and type of branch circuits, rating and type of mains, overcurrent protection, interrupting capacity, ambient temperature, and so on. An internal view of a typical 120/240-V, single-phase panelboard is shown in Figure 5-31.

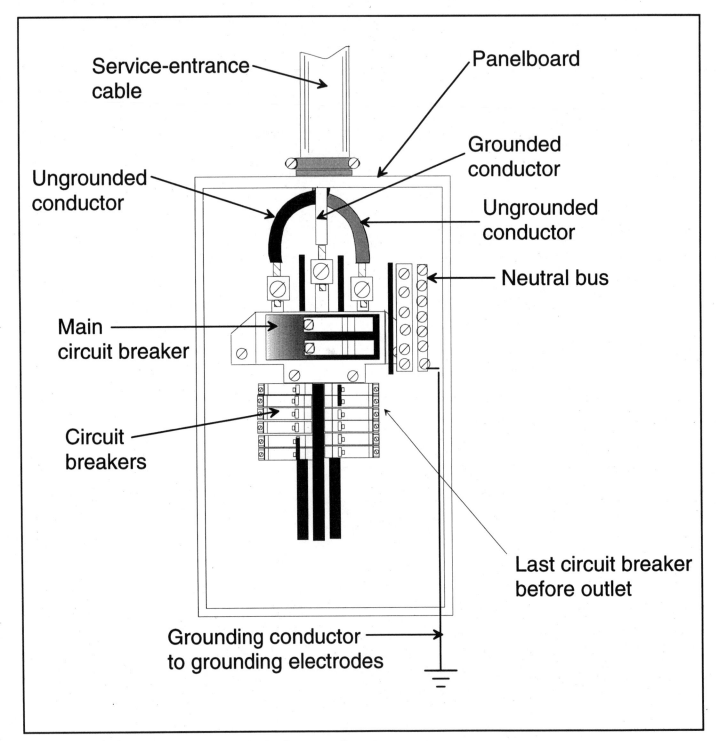

Figure 5-31: Internal view of 120/240-V, single-phase panelboard.

GROUNDING ELECTRIC SERVICES

The grounding system is a major part of the electrical system. Its purpose is to protect life and equipment against the various electrical faults that can occur. It is sometimes possible for higher-than-normal voltages to appear at certain points in an electrical system or in the electrical equipment connected to the system. Proper grounding ensures that the high electrical charges that cause these high voltages are channeled to earth or ground before damaging equipment or causing danger to human life.

When we refer to *ground*, we are talking about ground potential or earth ground. If a conductor is connected to the earth or to some conducting body that serves in place of the earth, such as a driven ground rod (electrode) or cold-water pipe, the conductor is said to be grounded. The neutral conductor in a three- or four-wire service, for example, is intentionally grounded and therefore becomes a grounded conductor. However, a wire used to connect this neutral conductor to a grounding electrode or electrodes is referred to as a grounding conductor. Note the difference in the two meanings; one is *ground***ED**, while the other provides a means of *ground***ING**.

There are two general classifications of protective grounding:

- System grounding
- Equipment grounding

The system ground relates to the service-entrance equipment and its interrelated and bonded components. That is, system and circuit conductors are grounded to limit voltages due to lighting, line surges, or unintentional contact with higher voltage lines, and to stabilize the voltage to ground during normal operation.

Equipment grounding conductors are used to connect the noncurrent-carrying metal parts of equipment, conduit, outlet boxes, and other enclosures to the system grounded conductor, the grounding electrode conductor, or both, at the service equipment or at the source of a separately derived system. Equipment grounding conductors are bonded to the system grounded conductor to provide a low impedance path for fault current that will facilitate the operation of overcurrent devices under ground-fault conditions. Equipment grounding is covered later in this book.

Article 250 of the *NEC* covers general requirements for grounding and bonding.

To better understand a complete grounding system, let's take a look at a conventional residential system beginning at the power company's high-voltage lines and transformer. A pole-mounted transformer is fed with a two-wire 7200-V system which is transformed and stepped down to a 3-wire, 120/240-V, single-phase electric service suitable for residential use. Figure 5-32 shows the voltage between phase A and phase B to be 240 V. However, by connecting a third wire (neutral) on the sec-

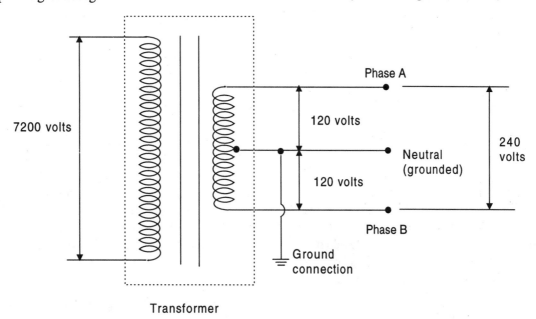

Figure 5-32: Wiring diagram of 7200-V to 120/240-V, single-phase transformer connection.

ondary winding of the transformer — between the other two — the 240 V are split in half, giving 120 V between either phase A or phase B and the neutral conductor. Consequently, 240 V are available for household appliances such as ranges, hot-water heaters, clothes dryers, and the like, while 120 V are available for lights, small appliances, tvs, and the like.

Referring again to the diagram in Figure 5-32, conductors A and B are ungrounded conductors, while the neutral is a grounded conductor. If only 240-V loads were connected, the neutral (grounded conductor) would carry no current. However, since 120-V loads are present, the neutral will carry the unbalanced load and becomes a current-carrying conductor. For example, if phase A carries 60 A and phase B carries 50 A, the neutral conductor would carry only (60 - 50 =) 10 A. This is why the *NEC* allows the neutral conductor in an electric service to be smaller than the ungrounded conductors.

The typical pole-mounted service-entrance is normally routed by messenger cable from a point on the pole to a point on the building being served, terminating in a meter housing. Another service conductor is installed between the meter housing and the main service switch or panelboard. This is the point where most systems are grounded — the neutral bus in the main panelboard. Refer back to Figure 5-30.

GROUNDING METHODS

Methods of grounding an electric service are covered in *NEC* Section 250-81. In general, all of the following (if available) and any made electrodes must be bonded together to form the grounding electrode system:

- An underground water pipe in direct contact with the earth for no less than 10 ft.
- The metal frame of a building where effectively grounded.
- An electrode encased by at least 2 in of concrete, located within and near the bottom of a concrete foundation or footing that is in direct contact with the earth. Furthermore, this electrode must be at least 20 ft long and must be made of electrically conductive coated steel reinforcing bars or rods of not less than $\frac{1}{2}$-in diameter, or consisting of at least 20 ft of bare copper conductor not smaller than No. 2 AWG wire size.

- A ground ring encircling the building or structure, in direct contact with the earth at a depth below grade not less than $2\frac{1}{2}$ ft. This ring must consist of at least 20 ft of bare copper conductor not smaller than No. 2 AWG wire size.

In most residential structures, only the water pipe will be available, and this water pipe must be supplemented by an additional electrode as specified in *NEC* Sections 250-81(a) and 250-83. With these facts in mind, let's take a look at a typical residential electric service, and the available grounding electrodes. *See* Figure 5-33.

Our sample residence has a metal underground water pipe that is in direct contact with the earth for more than 10 ft, so this is one valid grounding source. The house also has a metal underground gas-piping system, but this may not be used as a grounding electrode (*NEC* Section 250-83(a)). *NEC* Section 250-81(a) further states that the underground water pipe must be supplemented by an additional electrode of a type specified in Section 250-81 or in Section 250-83. Since a grounded metal building frame, concrete-encased electrode, or a ground ring are not normally available for most residential applications, *NEC* Section 250-83 — Made and Other Electrodes — must be used in determining the supplemental electrode. In most cases, this supplemental electrode will consist of either a driven rod or pipe electrode, specifications for which are shown in Figure 5-34.

An alternate method to the pipe or rod method is a plate electrode. Each plate electrode must expose not less than 2 ft^2 of surface to the surrounding earth. Plates made of iron or steel must be at least $\frac{1}{4}$-in thick, while plates of nonferrous metal such as copper need only be .06 in thick.

Either type of electrode must have a resistance to ground of 25 ohms or less. If not, they must be augmented by an additional electrode spaced not less than 6 ft from each other. In fact, many locations require two electrodes regardless of the resistance to ground. This, of course, is not an *NEC* requirement, but is required by some power companies and local ordinances in some cities and counties. Always check with the local inspection authority for such rules that surpass the requirements of the *NEC*.

Grounding Conductors

The grounding conductor, connecting the panelboard neutral bus to the water pipe and grounding electrodes,

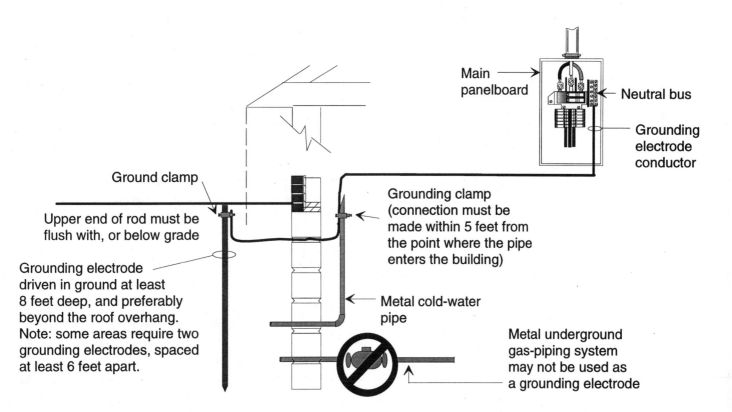

Main panelboard

Neutral bus

Grounding electrode conductor

Ground clamp

Grounding clamp (connection must be made within 5 feet from the point where the pipe enters the building)

Upper end of rod must be flush with, or below grade

Grounding electrode driven in ground at least 8 feet deep, and preferably beyond the roof overhang. Note: some areas require two grounding electrodes, spaced at least 6 feet apart.

Metal cold-water pipe

Metal underground gas-piping system may not be used as a grounding electrode

Figure 5-33: Components of a residential service grounding system.

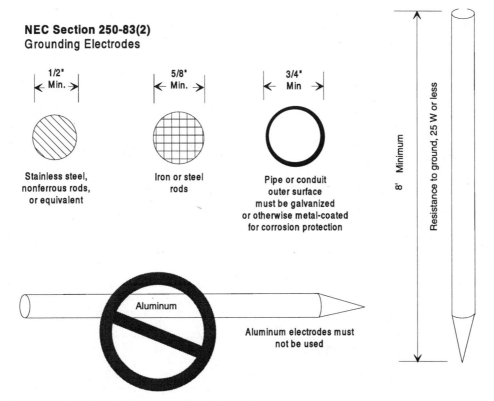

NEC Section 250-83(2)
Grounding Electrodes

1/2" Min.

Stainless steel, nonferrous rods, or equivalent

5/8" Min.

Iron or steel rods

3/4" Min

Pipe or conduit outer surface must be galvanized or otherwise metal-coated for corrosion protection

8' Minimum

Resistance to ground, 25 W or less

Aluminum

Aluminum electrodes must not be used

Figure 5-34: Specifications of rod and pipe grounding electrodes.

must be of either copper, aluminum, or copper-clad aluminum. Furthermore, the material selected must be resistant to any corrosive condition existing at the installation or it must be suitably protected against corrosion. The grounding conductor may be either solid or stranded, covered or bare, but it must be in one continuous length without a splice or joint — except for the following conditions:

- Splices in busbars are permitted.
- Where a service consists of more than one single enclosure, it is permissible to connect taps to the grounding electrode conductor provided the taps are made within the enclosures. They are not to be made outside of the enclosure.
- Grounding electrode conductors may also be spliced at any location by means of irreversible compression-type connectors listed for the purpose or the exothermic welding process. These methods prevent easy removal.

The size of grounding conductors depends on service-entrance size; that is, the size of the largest service-entrance conductor or equivalent for parallel conductors. The table in Figure 5-35 gives the proper sizes of grounding conductors for various sizes of electric services.

POWER-RISER DIAGRAMS

Power-riser diagrams are used on almost every electrical drawing for building construction. Such diagrams are great time-saving devices for giving a "picture" of the various components used in an electrical system and how they are connected in relation to one another.

A power-riser diagram used on the electrical drawings for a branch bank is shown in Figure 5-36. This diagram indicates that the electrical contractor is to provide an empty 2½-in conduit from the wall-mounted meter base (for the electric watthour meter), running underground to 3 ft beyond the building footing. The local power company will then install the service conductors underground— using direct-burial cable — and then run them through the empty conduit provided to connect them to the electric meter. The "W/BUSHING" in the note indicates that the electrical contractor is to provide a suitable bushing on the end of the electrical conduit to protect the conductor insulation as they enter the conduit from underground. Without the bushing, sharp metal edges of the conduit could cut or nick the insulation of the conductors.

The data given in the diagram in Figure 5-36 also show that the electrical contractor is responsible for installing a 2½-in conduit from the electric meter to panelboard A; this time with four 3/0 cu (copper) conductors. Connections, of course, will be made at both ends of the run — at the electric meter and at the panelboard.

Other details concerning the panelboard are given elsewhere on the drawings; that is, the location of the panelboard within the building is shown on the floor plan, while the panel's components and the circuit they feed are provided in a panelboard schedule. *See* Chapter 1 of this book.

The remaining items shown in the diagram in Figure 5-36 consist of a conduit containing three No. 12 AWG conductors to feed an air handling unit (A.H.U.), the location of which is shown on the mechanical (HVAC) drawings.

Size of Largest Service-Entrance Conductor or Equivalent Area for Parallel Conductors		Size of Grounding Electrode Conductor	
Copper	**Aluminum or Copper-Clad Aluminum**	**Copper**	**Aluminum or Copper-Clad Aluminum**
2 or smaller	1/0 or smaller	8	6
1 or 1/0	2/0 or 3/0	6	4
2/0 - 3/0	4/0 or 250 kcmil	4	2
Over 3/0 - 350 kcmil	Over 250 - 500 kcmil	2	1/0
Over 350 - 600 kcmil	Over 500 - 900 kcmil	1/0	3/0
Over 600 - 1100 kcmil	Over 900 - 1750 kcmil	2/0	4/0
Over 100 kcmil	Over 1750 kcmil	3/0	250 kcmil

Figure 5-35: Grounding conductor sizes for various electric services.

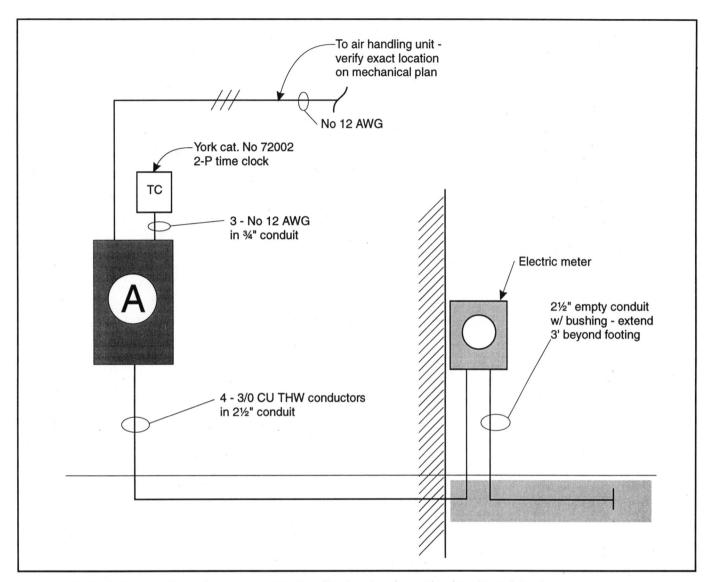

Figure 5-36: Typical power-riser diagram used to describe the electric service for a branch bank.

The York Cat. No. 72002 time clock is fed with three number 12 AWG conductors in ¾-in conduit. This time clock will be used to control outside lights; it will turn them on at dusk and off at dawn.

Note that power-riser diagrams are seldom, if ever, drawn to scale; they are diagrammatic only — showing only major parts and the connection between them.

Another power-riser diagram is shown in Figure 5-37. This is a larger system than the one shown in Figure 5-36. Note that the service in this diagram is run overhead, as opposed to the underground service on the previous example. This diagram calls for three 3½-in conduits, each containing four 500 kcmil conductors that terminate in a C/T cabinet (current transformer).

A 1-in empty conduit is called for between the C/T cabinet and the electric meter. This arrangement is for the power company's metering conductors.

Although the exact connection details are not shown in Figure 5-37, the service wires from the outside pass through the C/T cabinet and are connected to the main distribution panel (MDP). This panel contains overcurrent protection for the six feeder circuits feeding subpanels A, B, C, E, and P. Note that conduit and conductor size (as well as the number of conductors in each conduit) are indicated. One dimension line is given to remind the contractor that the top of the panelboard should be less than 6 ft above the finished floor to comply with the *NEC*. Also notice that panels D, C, and P are located on the lower level, or first floor. The note affecting panel D tells installers that this panel is mounted in the stock room adjacent to the show window; it contains overcurrent protection for the show window lights details of which appear elsewhere on the drawings or in the written specifications.

Although this power-riser diagram is not drawn to scale, the various components should be drawn and sized in reasonable proportion to each other. Furthermore, each should be identified and briefly described so there will be little doubt what is required of the workers.

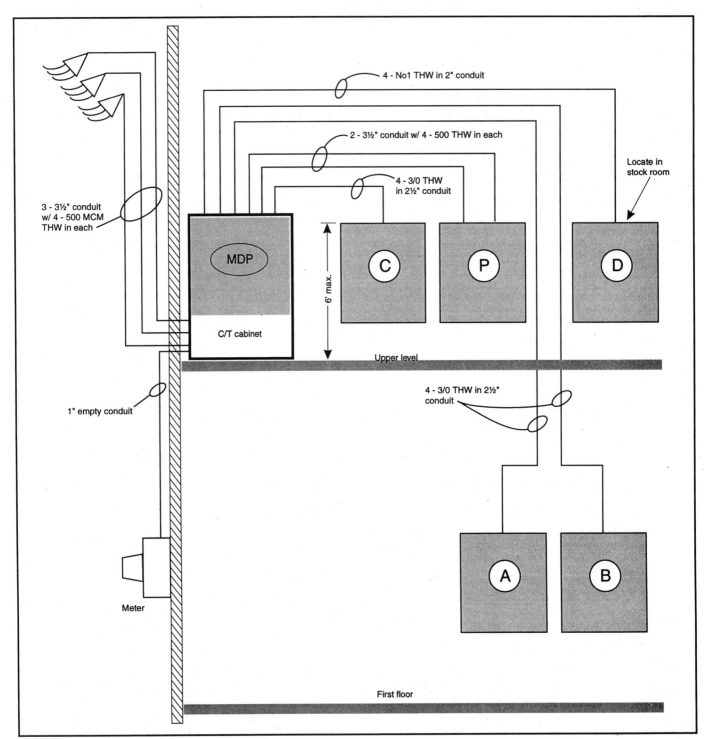

Figure 5-37: A power-riser diagram used on a department store electrical project.

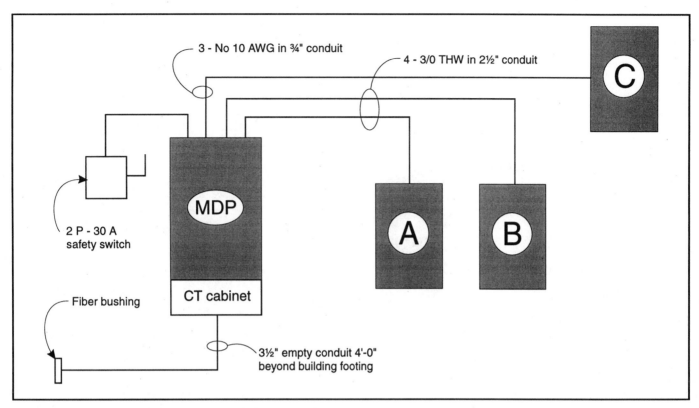

Figure 5-38: Power-riser diagram for a building extension.

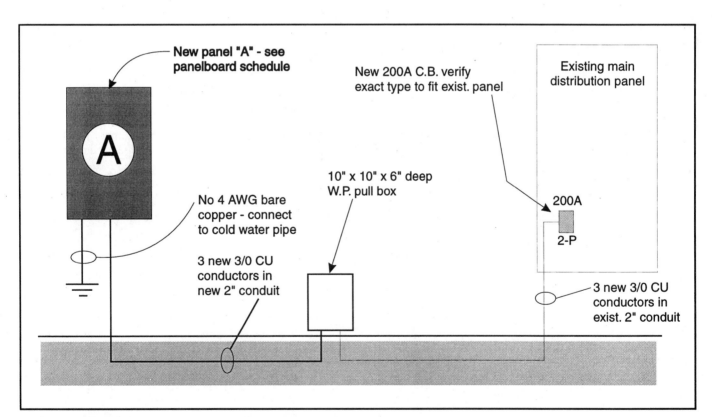

Figure 5-39: Power-riser diagram for a single-structure law office.

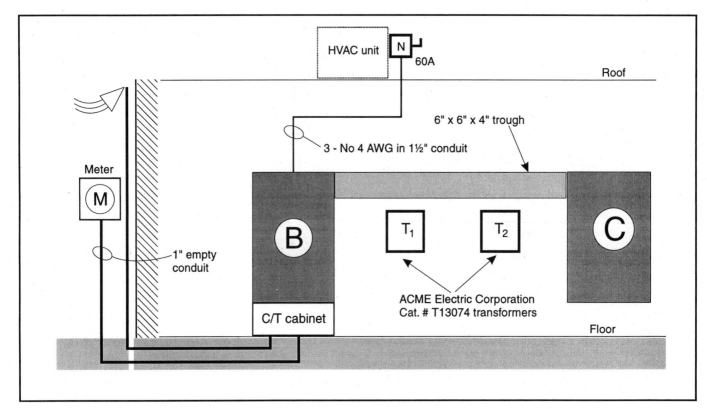

Figure 5-40: Power-riser diagram for a shopping-center store space.

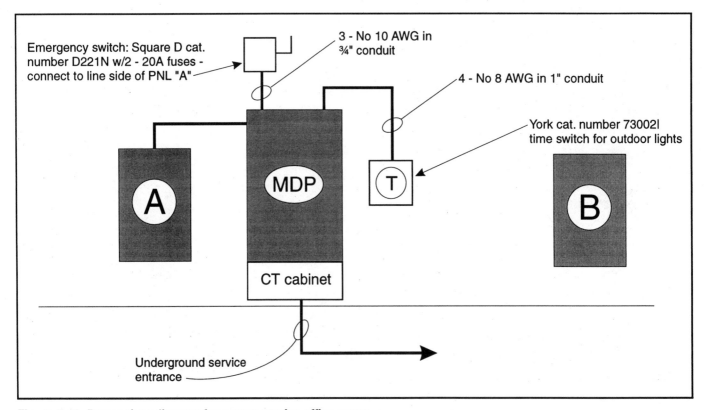

Figure 5-41: Power-riser diagram for a copy service office space.

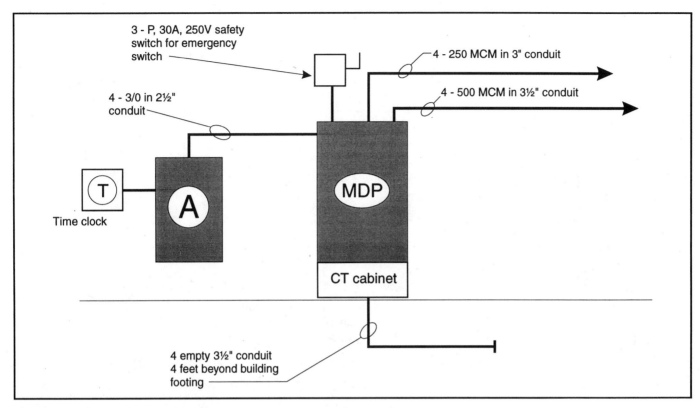

Figure 5-42: Power-riser diagram for a machine shop electrical service.

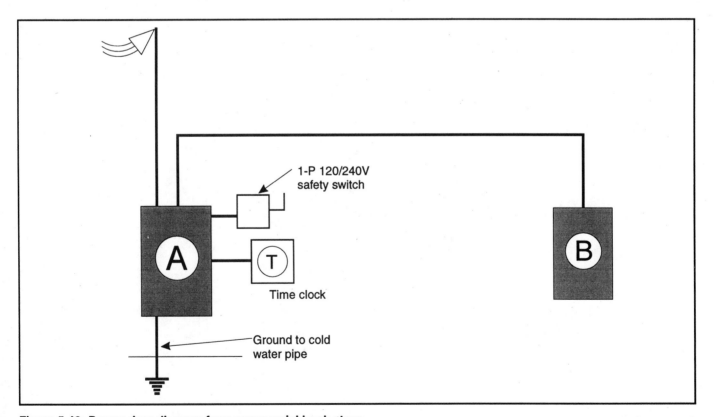

Figure 5-43: Power-riser diagram for a commercial book store.

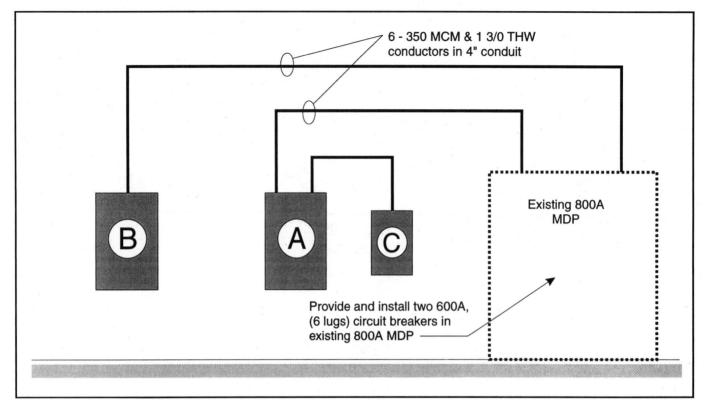

6 - 350 MCM & 1 3/0 THW
conductors in 4" conduit

Existing 800A
MDP

Provide and install two 600A,
(6 lugs) circuit breakers in
existing 800A MDP

Figure 5-44: An extension of an industrial electric service.

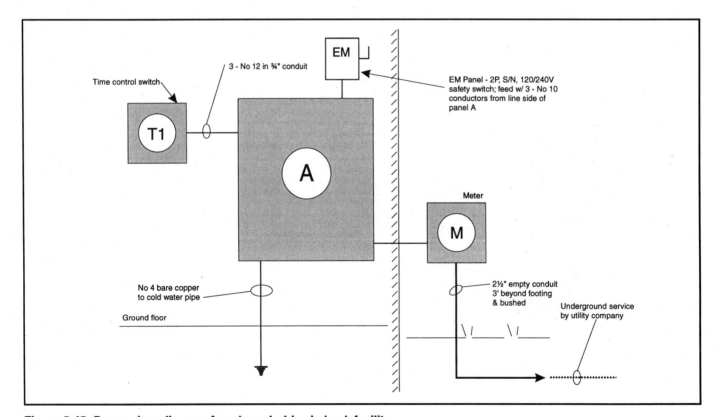

3 - No 12 in ¾" conduit

Time control switch

EM

EM Panel - 2P, S/N, 120/240V
safety switch; feed w/ 3 - No 10
conductors from line side of
panel A

T1

A

Meter

M

No 4 bare copper
to cold water pipe

Ground floor

2½" empty conduit
3' beyond footing
& bushed

Underground service
by utility company

Figure 5-45: Power-riser diagram for a branch drive-in bank facility.

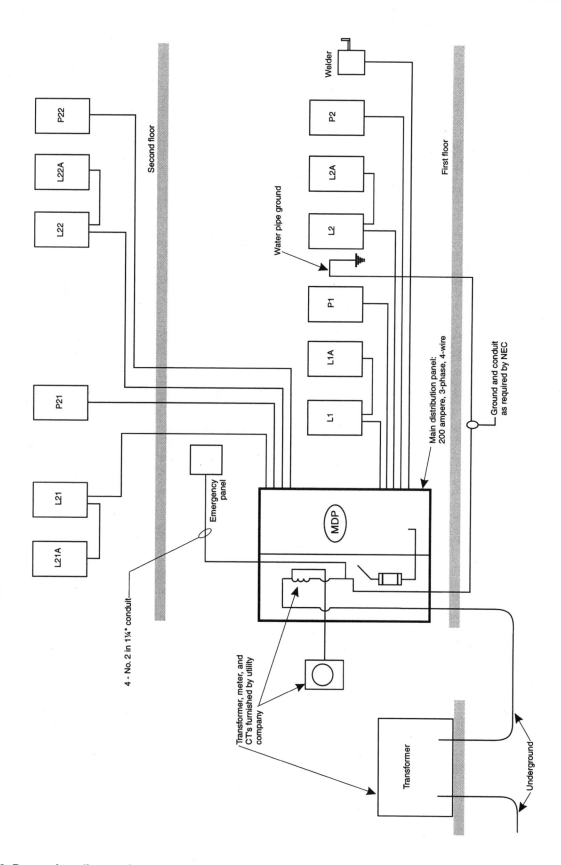

Figure 5-46: Power-riser diagram for an education administration building.

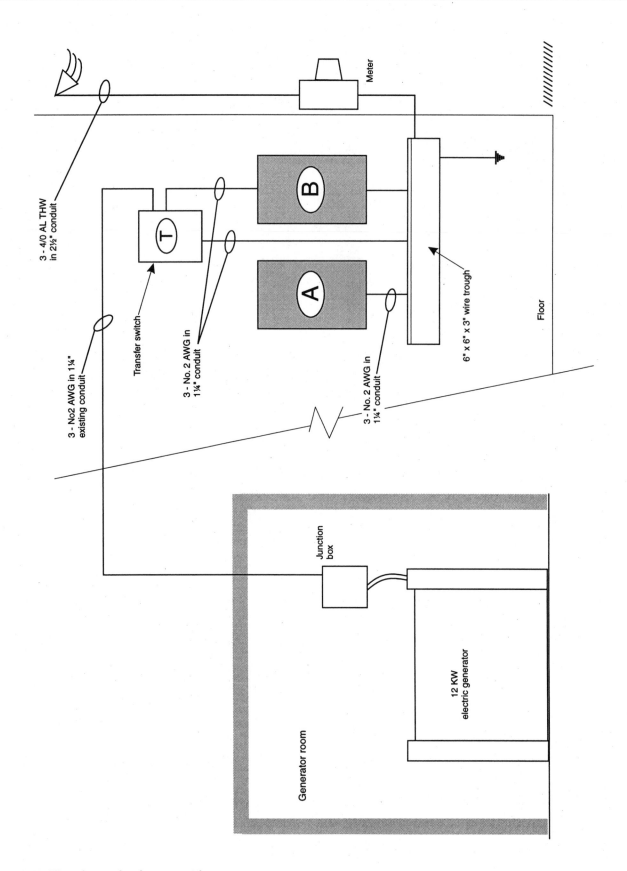

Figure 5-47: Electric service for a court house.

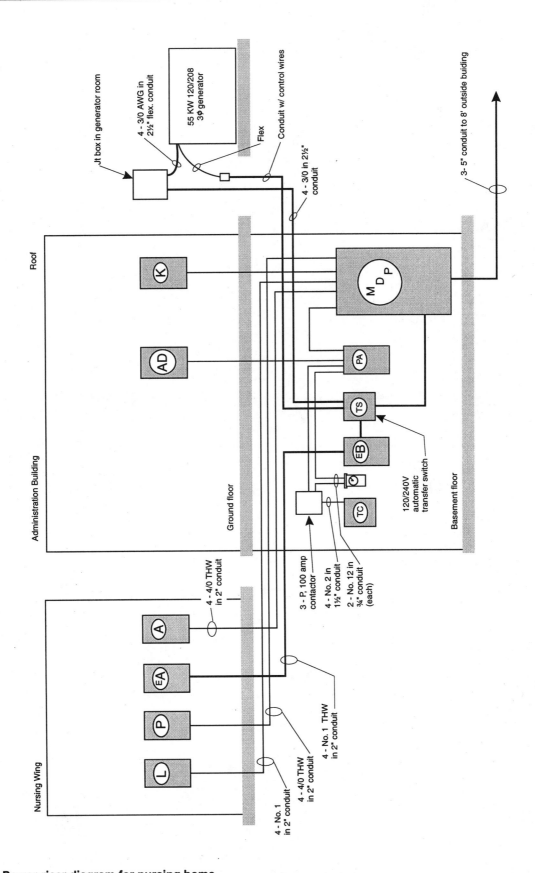

Figure 5-48: Power-riser diagram for nursing home.

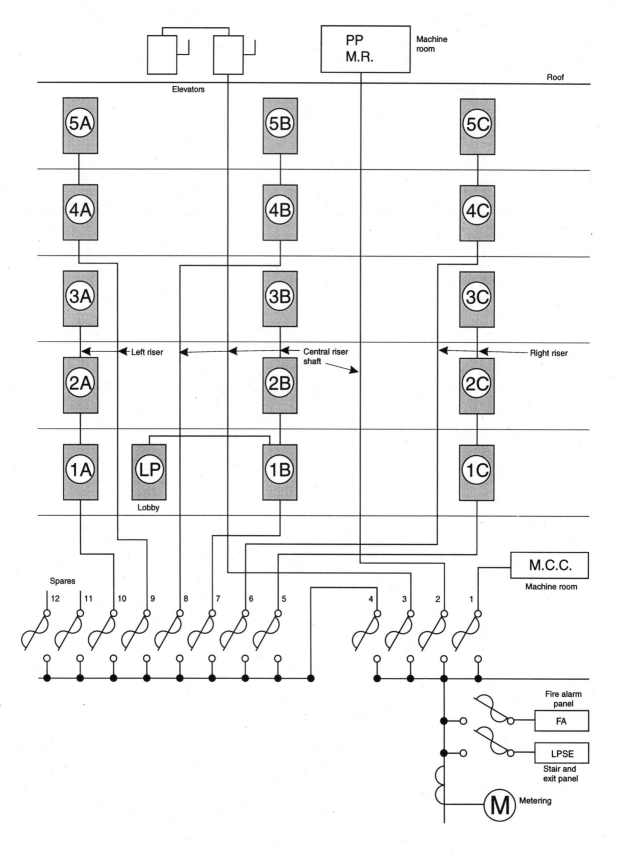

Figure 5-49: Power-riser diagram for a multistory building.

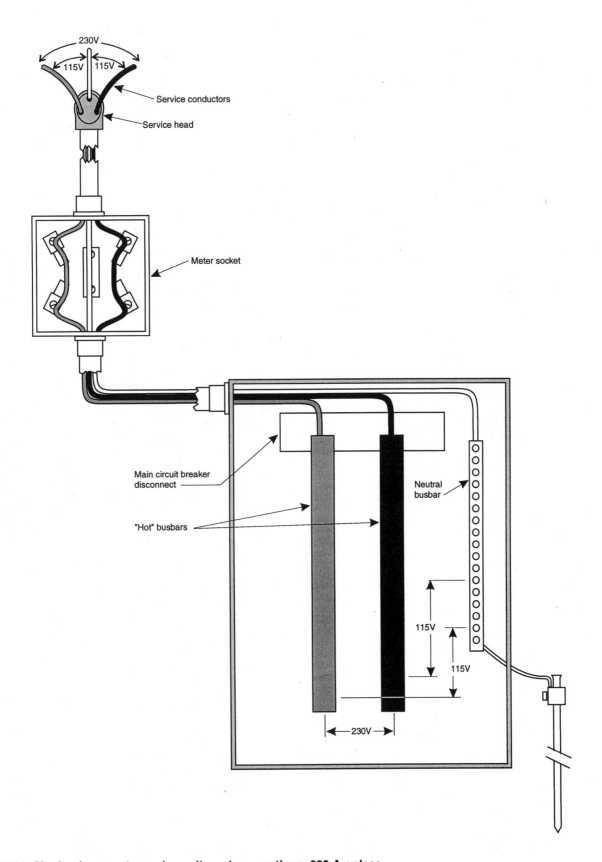

Figure 5-50: Single-phase meter and panelboard connections, 200 A or less.

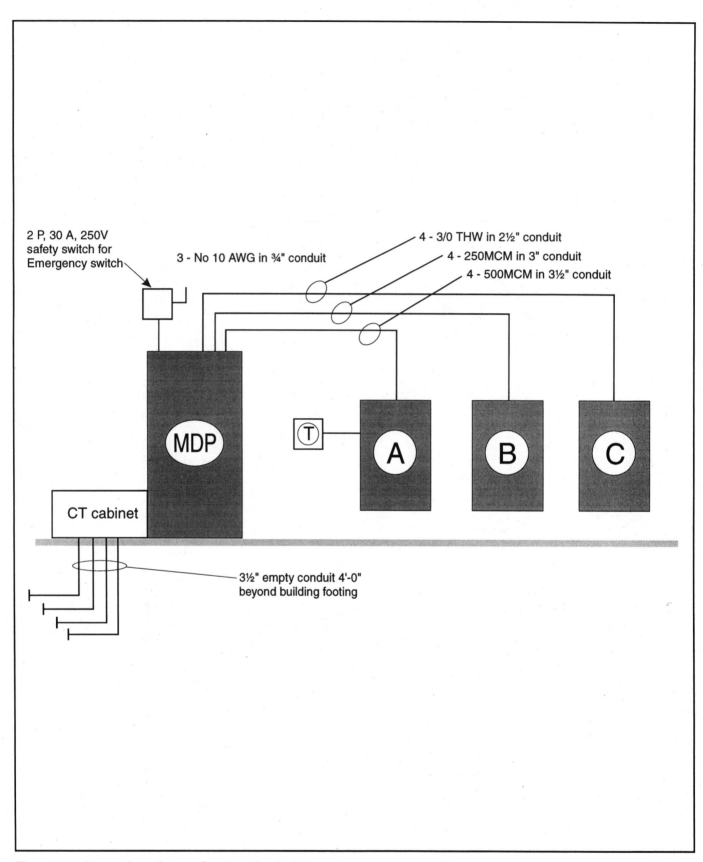

Figure 5-51: Power-riser diagram for a banking facility.

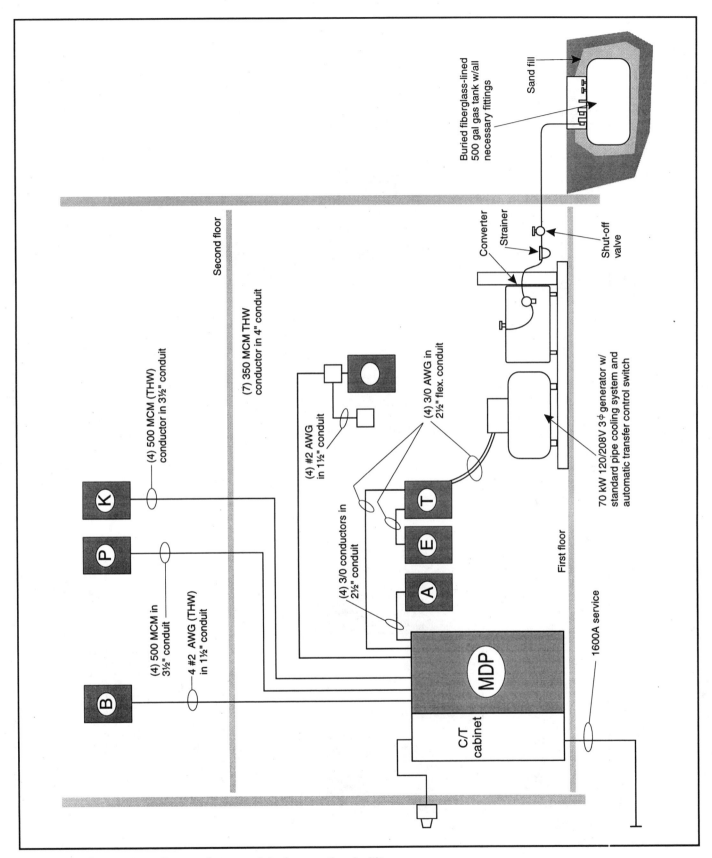

Figure 5-52: Power-riser diagram for a municipal correction facility.

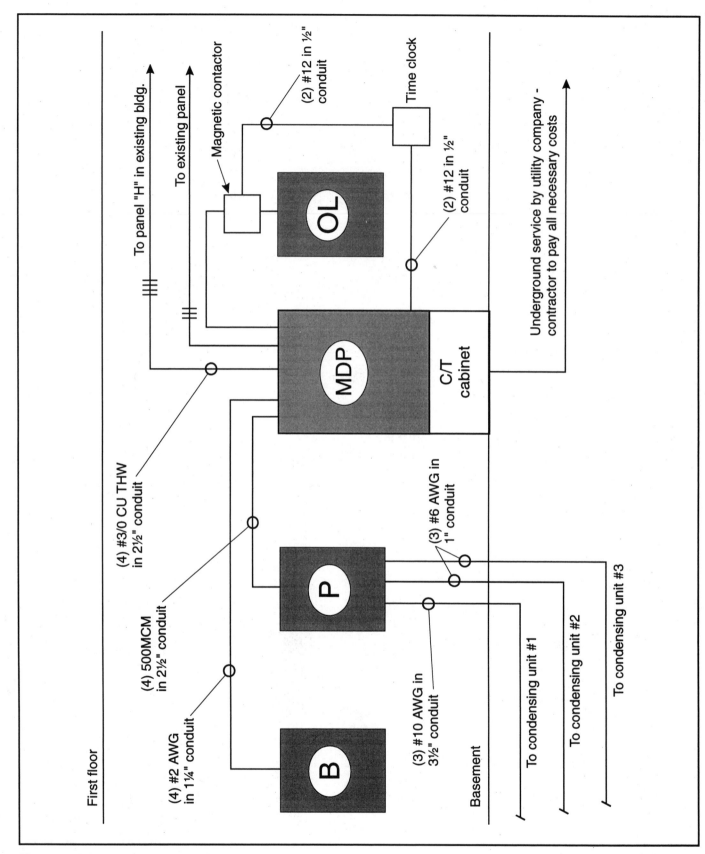

Figure 5-53: Power-riser diagram for an education wing addition.

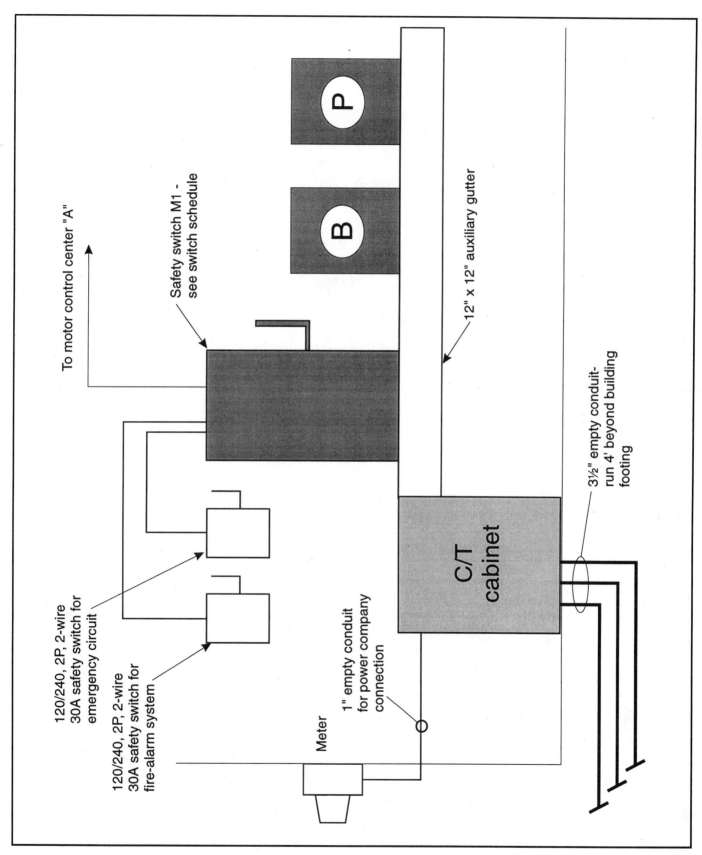

Figure 5-54: Power-riser diagram for a county clinic.

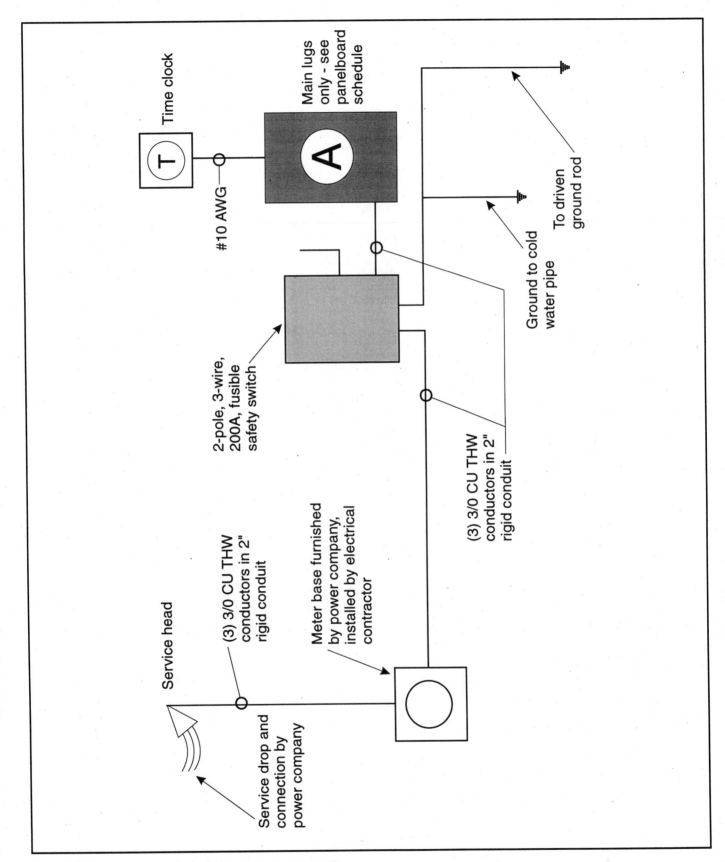

Figure 5-55: Power-riser diagram for photography studio.

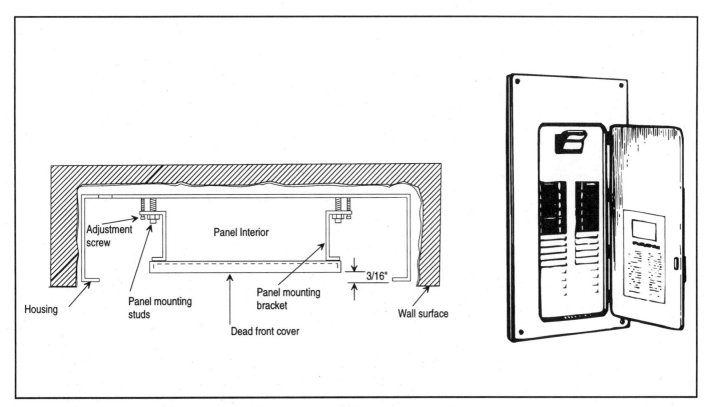

Figure 5-56: Cross-sectional plan and pictorial views of a flush-mounted panelboard.

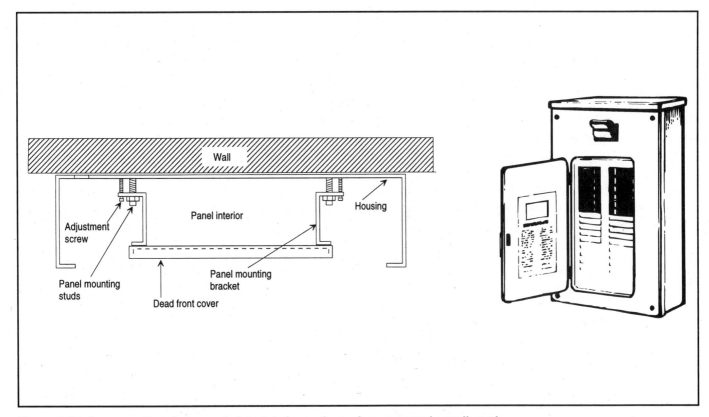

Figure 5-57: Cross-sectional plan and pictorial views of a surface-mounted panelboard.

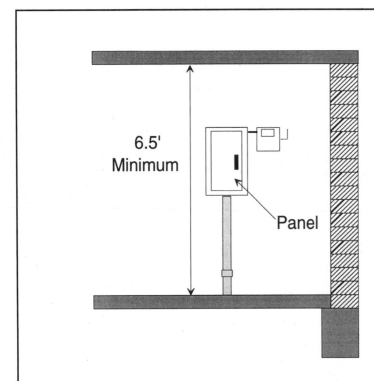

HEADROOM

The minimum headroom of working spaces about service equipment has increased from 6.25 feet to 6.5 feet NEC Section 110-16(f)

Exception No. 1 to this requirement remains the same; that is, service equipment under 200 amperes in existing dwellings does not require this much headroom

Figure 5-58: NEC headroom requirements for service equipment.

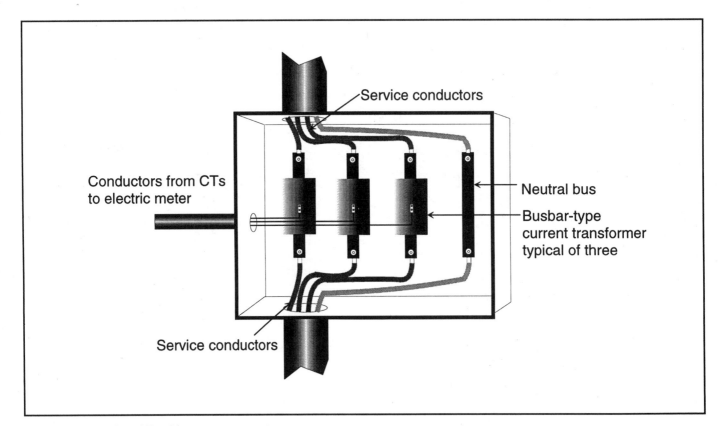

Figure 5-59: Typical CT cabinet arrangement.

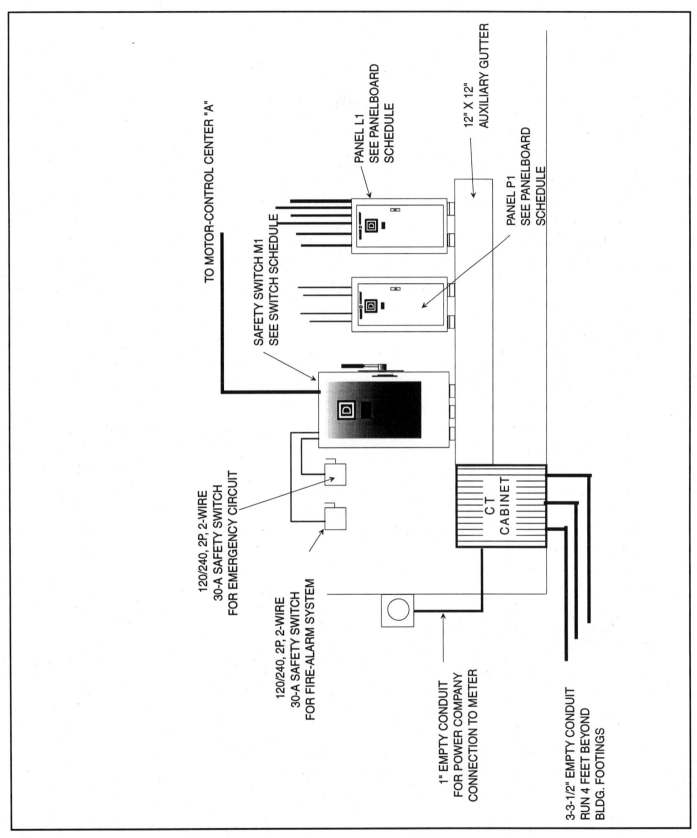

Figure 5-60: Power-riser diagram for a three-phase, 4-wire, 1200-A service.

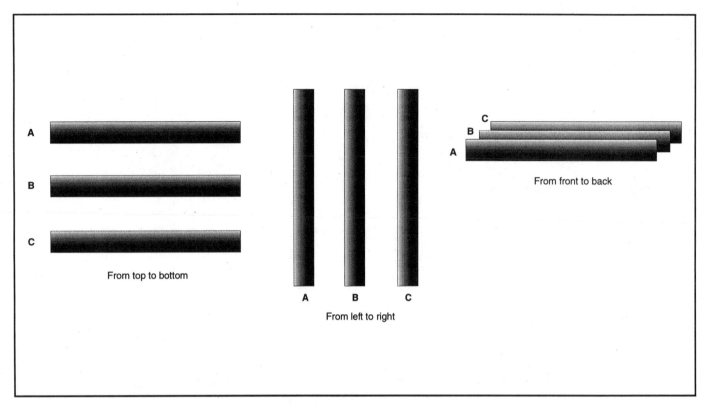

Figure 5-61: NEC approved phase arrangements.

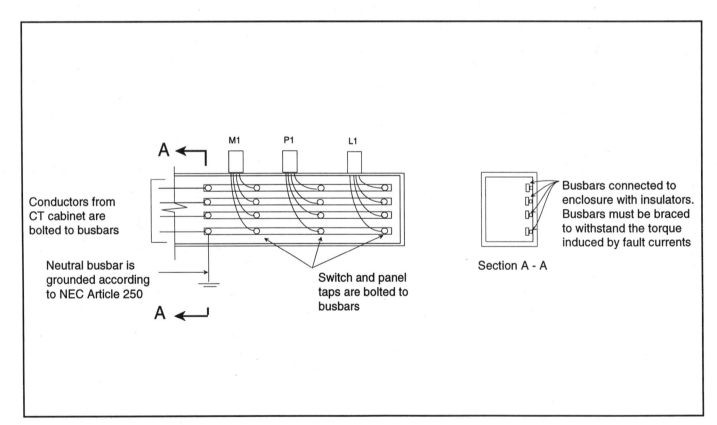

Figure 5-62: Three-phase bussed gutter.

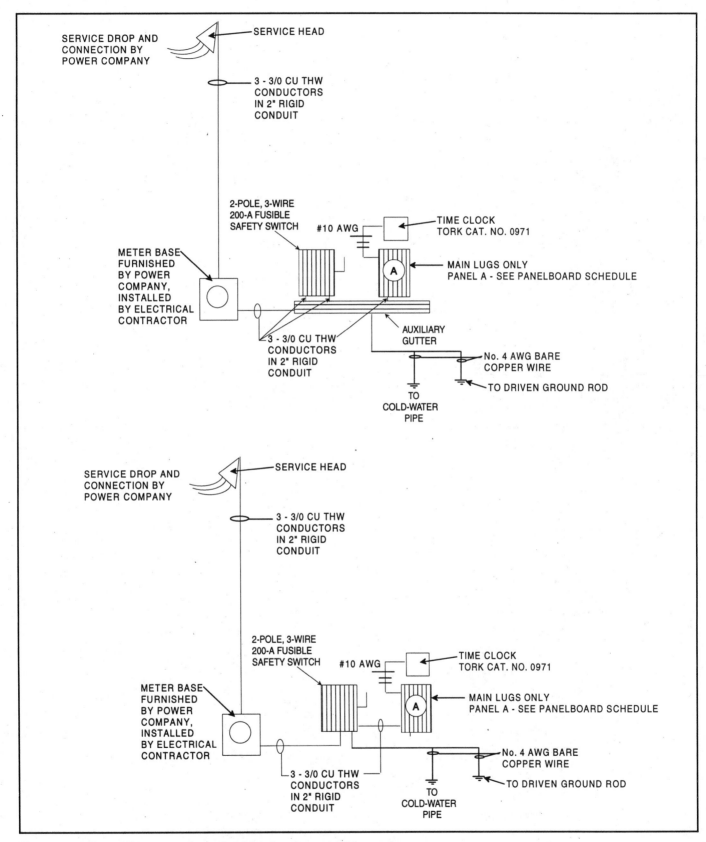

Figure 5-63: Two different configurations for the same electric service.

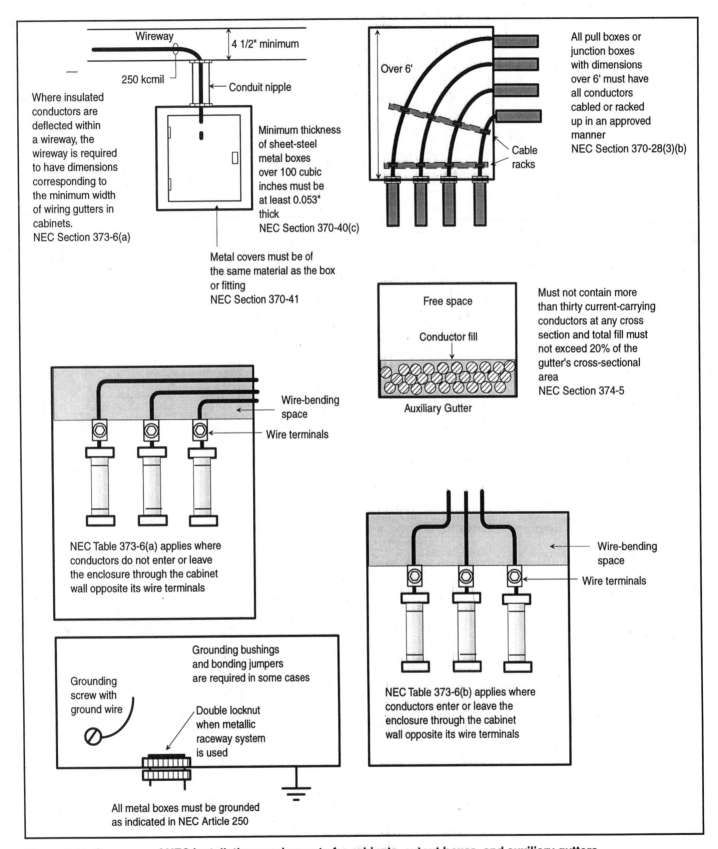

Wireway

4 1/2" minimum

250 kcmil

Conduit nipple

Where insulated conductors are deflected within a wireway, the wireway is required to have dimensions corresponding to the minimum width of wiring gutters in cabinets.
NEC Section 373-6(a)

Minimum thickness of sheet-steel metal boxes over 100 cubic inches must be at least 0.053" thick
NEC Section 370-40(c)

Metal covers must be of the same material as the box or fitting
NEC Section 370-41

Over 6'

Cable racks

All pull boxes or junction boxes with dimensions over 6' must have all conductors cabled or racked up in an approved manner
NEC Section 370-28(3)(b)

Free space

Conductor fill

Auxiliary Gutter

Must not contain more than thirty current-carrying conductors at any cross section and total fill must not exceed 20% of the gutter's cross-sectional area
NEC Section 374-5

Wire-bending space

Wire terminals

NEC Table 373-6(a) applies where conductors do not enter or leave the enclosure through the cabinet wall opposite its wire terminals

Wire-bending space

Wire terminals

NEC Table 373-6(b) applies where conductors enter or leave the enclosure through the cabinet wall opposite its wire terminals

Grounding bushings and bonding jumpers are required in some cases

Grounding screw with ground wire

Double locknut when metallic raceway system is used

All metal boxes must be grounded as indicated in NEC Article 250

Figure 5-64: Summary of NEC installation requirements for cabinets, cutout boxes, and auxiliary gutters.

Chapter 6

Lighting

The basic requirement for any lighting design is to determine the amount of light that should be provided and the best means of providing it. Lighting designers should therefore strive to select lighting equipment that will provide the highest visual comfort and performance that is consistent with the type of area to be lighted and within the budget allowed. The designer must further strive to accurately convey this information to workers on the job by means of drawings, schedules, notes, and written specifications.

A typical lighting floor plan is shown in Figure 6-1 on the next page. This scale drawing shows the location of all lighting fixtures on the level by appropriate lines, symbols, and notes. Each type of fixture has an identification number and is further identified in a lighting-fixture schedule (*See* Chapter 1 for sample lighting-fixture schedules). A lighting-fixture schedule should give the identifying symbol or mark, manufacturer, catalog number, number and types of lighting, operating voltage, and the mounting characteristics; that is, surface mounted, recessed, wall mounted, etc.

Lighting fixtures shown on floor plans, when used in conjunction with a lighting-fixture schedule, or written specifications will usually be enough information for contractors to bid and install the installation. Furthermore, most lighting fixtures are shipped with installation instructions packed within each carton.

However, there are times when it becomes necessary to provide a detail drawing of a separate item or a portion of the electrical lighting system, giving a complete and exact description of its use and the details necessary to show the

workers exactly what is required for the installation. For example, some projects require the use of special lighting fixtures or components which are not standard catalog items. But even the use of cataloged items may be so specialized that additional descriptions should be included in the specifications or large-scale details provided in the working drawings.

When a specialty lighting arrangement must be assembled on the job site, detail drawings of the assembly with help ensure that the installation is installed correctly. Such details not only save time in estimating, supervising, and installing the system, they also cut down on conference time required of the design consultant.

Take, for example, theatrical lighting around a bathroom mirror. Methods of showing such an arrangement will vary with different consulting firms, but the most common method is to show a wall-mounted junction box on the floor plan with a note giving the mounting height, fixture type, and instructions to mount the junction box over the lavatory mirror.

Let's assume that a Type 5 designation has been given this lighting arrangement. The lighting-fixture schedule will give the basic components and specifications; that is, one Lightolier (manufacturer) Catalog No. 4300 fixture, two Lightolier Catalog No. 4301 fixtures, and two Lightolier Catalog No. 4304 connectors. Still, there is a question as to exactly how these components are to be arranged. To clarify the situation, a note is placed in the "remarks" column of the lighting-fixture schedule to refer to the Type 5 lighting-fixture details on the electrical drawings. The detail appears in Figure 6-2 on page 527. When this detail is

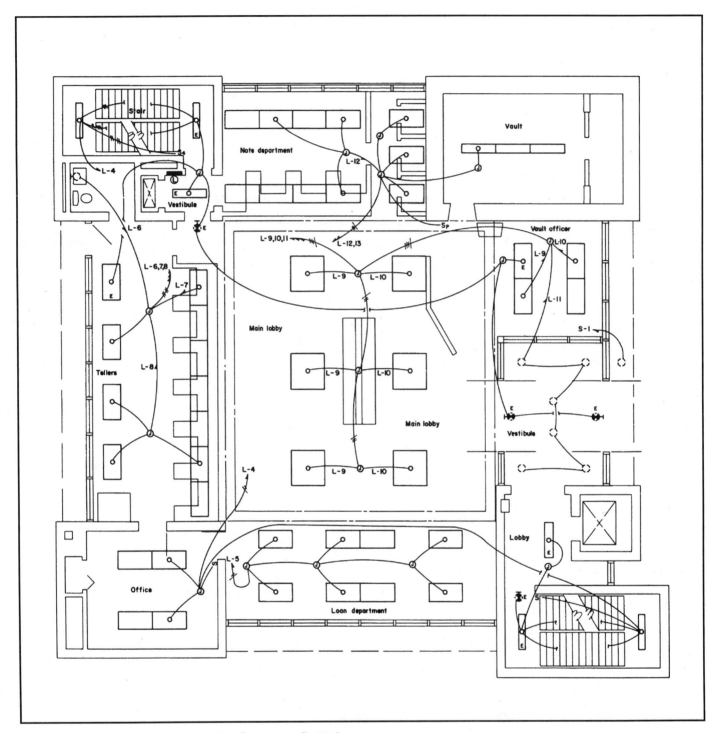

Figure 6-1: Method of showing lighting fixtures on floor plans.

used in conjunction with a lighting-fixture schedule, there is little doubt as to the installation of this lighting fixture.

LAMP CLASSIFICATIONS

In general, there are three common sources of electric light:

- Incandescent lamps
- Gaseous-discharge lamps
- Electroluminescent lamps

Despite continuous improvement, none of these light sources have a high overall efficiency. The very best

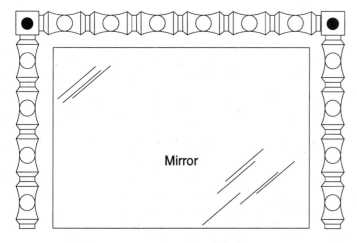

Figure 6-2: Detail drawing of a specialty lighting layout for a bathroom mirror.

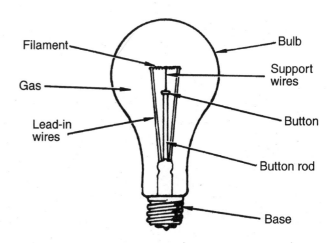

Figure 6-4: Components of a typical incandescent lamp.

light source converts only approximately ¼ of its input energy into visible light. The remaining input energy is converted to heat or invisible light. The energy distribution of a typical fluorescent lamp is shown in Figure 6-3.

Incandescent Lamps

Incandescent lamps are made in thousands of different types and colors, from a fraction of a watt to over 10,000 W each, and for practically any conceivable lighting application.

Extremely small lamps are made for instrument panels, flashlights, etc., while large incandescent lamps, over 20-in diameter, are used for spotlights and street lighting.

Regardless of the type or size, all incandescent filament lamps consist of a sealed glass envelope containing a filament as shown in Figure 6-4. The incandescent filament lamp produces light by means of a filament heated to incandescence (white light) by its resistance to a flow of electric current. Most of these elements are capable of producing 11 to 22 lumens per watt, and some produce as high as 33 lumens per watt.

The filaments of incandescent lamps were originally made of carbon. Now, tungsten is used for virtually all lamp filaments because of its higher melting point, better spectral characteristics, and strength — both hot and cold.

The sealed glass envelope enclosing the filament is used to obtain a vacuum or an atmosphere of inert gas. Without such an atmosphere, the filament would rapidly disintegrate due to oxidation.

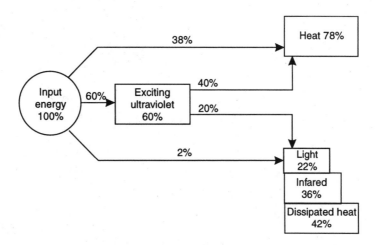

Figure 6-3: Energy distribution of a typical cool-white fluorescent lamp.

The filaments of all early incandescent lamps operated in a vacuum — all air and gas, insofar as practical, were exhausted from the space within the bulb surrounding the filament. In a vacuum lamp, the heat losses by convection and conduction are reduced, but the filament begins vaporizing at a lower temperature and therefore evaporates more rapidly than it would if pressure was applied.

The purpose of gas inside the bulb is to create pressure on the filament to retard evaporation, and this type of lamp is considered more efficient than a vacuum lamp of the same size. Since the filament can operate at a higher temperature in a gas-filled lamp, it also produces a whiter light than produced by a vacuum lamp of the same size.

Technical Descriptions of Incandescent Lamps

Incandescent lamps are made in a variety of shapes and sizes for use in many applications. Typical bulb shapes are shown in Figure 6-5.

The sizes and shapes of lamp bulbs are designated by a letter or letters followed by a number. The letter indicates the shape of the bulb, and a few designations follow:

- S: straight side
- F: flame
- G: round or globular
- T: tubular
- A: arbitrary designation applied to lamps commonly used for general lighting of 200 W or less.

The numerals in a bulb designation indicate the maximum diameter of the bulb in eighths of an inch. For example, an A-21 bulb is 21 eighths of an inch or 2⅝ in diameter at its maximum dimensions.

Bulb Finish and Color

To diffuse the light from the filament, many lamps have inside-frosted bulbs, produced by a light acid etching applied to the inner surface of the bulb. Some types of lamps are available with an inside white silica coating which provides still greater diffusion. The inside-frosted bulb absorbs no measurable amount of light, whereas the silica coating absorbs about 2 percent. With both treatments, the outer surface of the bulb is left smooth and easily cleaned. Diffusing bulbs are preferred for most general lighting purposes, but where accurate control of light is involved, as in optical systems, clear bulb lamps are necessary.

Other finishes applied to some general lighting service lamps are white bowl and silvered bowl. A white-bowl lamp has a translucent white coating on the inner surface of the bulb bowl, which serves to reduce both direct and reflected glare from open fixtures. A silvered-bowl lamp has an opaque silver coating applied to the bowl. The inner surface of this coating is a high-specula reflector which is not affected by dust or deterioration, and therefore remains efficient throughout the life of the lamp. Silvered-bowl lamps are commonly used in certain types of equipment for totally indirect lighting, and also occasionally in direct fixtures such as standard dome reflectors.

Colored light in filament lamps is produced subtractively, by means of a bulb that absorbs light colors other than that desired. Most colored bulbs are made by applying a pigmented coating to either the inner or the outer surface of a clear bulb, or by fusing an enamel into the outer surface (ceramic coating). The colors in most common use are red, blue, green, yellow, orange, ivory, flametint, and white.

Lamps with a slightly pink-colored inside silica coating (Beauty-Tone lamps) are available in the three-way design. These lamps are primarily used in residential lighting equipment. They are used where delicately tinted light is desired for a decorative effect. Ceramic coatings and in-

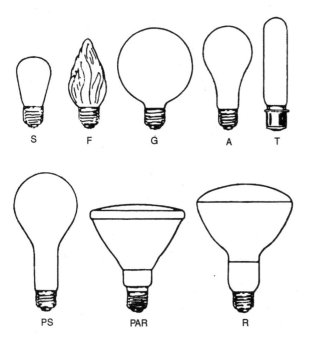

Figure 6-5: Typical incandescent bulb shapes.

side coatings are satisfactory for either outdoor or indoor use, but most outside coatings are not permanent and are recommended for use only where they are protected from the weather; that is, either indoors or in an outside lighting fixture with a weatherproof, gasketed cover.

Another type of colored lamp has a bulb of natural-colored glass, made by adding chemicals to the ingredients of the glass. Natural-colored bulbs are made in daylight blue, blue, amber, green, and ruby. They produce light of purer colors than coated bulbs and are often used for theatrical and photographic lighting purposes. Where decorative or display lighting is involved, coated lamps are preferred to natural-colored lamps because of their lower cost. If cost is no object, then the natural-colored glass bulbs are still the best choice.

The most widely used of the natural-colored lamps is the daylight blue. The characteristics of the daylight blue bulb are such as to reduce the preponderance of red and yellow light which is common to incandescent lamps, with the result that the light produced more nearly approaches daylight in color. Since this is accomplished at the expense of increased lamp cost and of some 35% absorption in light, daylight blue lamps should be used only where the lighting requirements make it necessary.

Base

The base provides a means of connecting the lamp bulb to the socket. For general lighting purposes, screw-type bases are most commonly used. Most general lighting service lamps (300 W and below) have medium screw bases. The higher wattages (300 W and above) use the mogul screw base. Some of the lower-wattage lamps, particularly the sign, indicator, and decorative types, have candelabra or intermediate screw bases. *See* Figure 6-6.

A light source (lamp filament) cannot be accurately aligned with respect to an optical system by means of a screw base. Filament orientation is provided by a number of other types of bases. The most common bases are the prefocus, bipost, bayonet, and special pin-type bases for projection lamps. *See* Figure 6-6. A bipost base, usually used on high-wattage lamps, consists of two metal pins or posts imbedded in a glass "cup" forming the end of the lamp bulb. Most screw and prefocus bases are attached to the bulb by means of a basing cement especially designed for the purpose. Other bases used on certain lamps include prong types, screw terminals, contact lugs, flexible leads, recessed single contact, and a number of other types for specific applications.

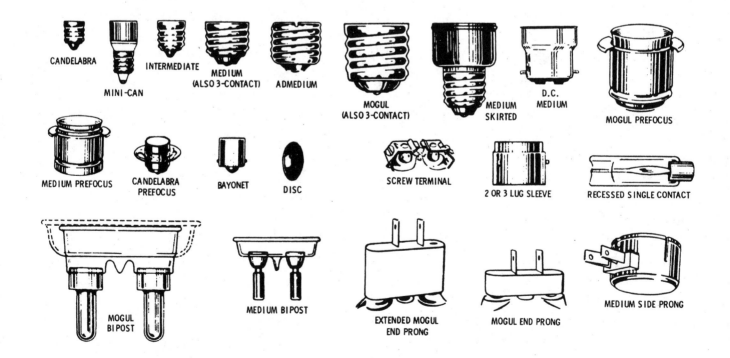

Figure 6-6: Incandescent base configurations.

Filaments

The filament (Figure 6-7) is the light-producing element of a lamp, and the primary considerations in its design are its electrical characteristics. The wattage of a filament lamp is equal to the voltage delivered at the socket times the amperes flowing through the filament. By Ohm's law, the current is determined by the voltage and by the resistance, which in turn depends on the length and the diameter of the filament wire. The higher the wattage of a lamp of a given voltage, the higher the current, and therefore the greater the diameter of the filament wire required to carry it. The higher the voltage of a lamp of a given wattage, the lower the current and the smaller the diameter of the filament wire.

The higher the operating temperature of the filament, the greater the share of the emitted energy that lies in the visible region of the radiation spectrum. Since most filament lamps radiate as light only about 10 to 12 percent of the input energy, it is important to design a lamp for as high a filament temperature as is consistent with satisfactory lamp life. Carbon, which has a higher melting point than tungsten and was one of the early filament materials, has been almost completely replaced by tungsten because carbon at high temperatures evaporates too rapidly, whereas tungsten combines the properties of a high-melting point and slow evaporation.

Since the larger the diameter of the filament wire the higher the temperature at which it can be operated without danger of excessive evaporation, high-wattage lamps are more efficient than low-wattage lamps of the same voltage and life rating. A 150-W, 120-V general lighting service lamp, for example, produces 34 percent more light than three 50-W, 120-V lamps consuming the same wattage. It also follows that low-voltage lamps, because their filament wire diameter is greater, are more efficient than higher-voltage lamps of the same wattage.

The filament forms in common use today are designated by a letter or letters indicating whether the wire is straight or coiled, a number specifying the general form of the filament, and sometimes another letter indicating arrangement on the supports. "S" as the first letter of a filament designation means a straight (uncoiled) filament

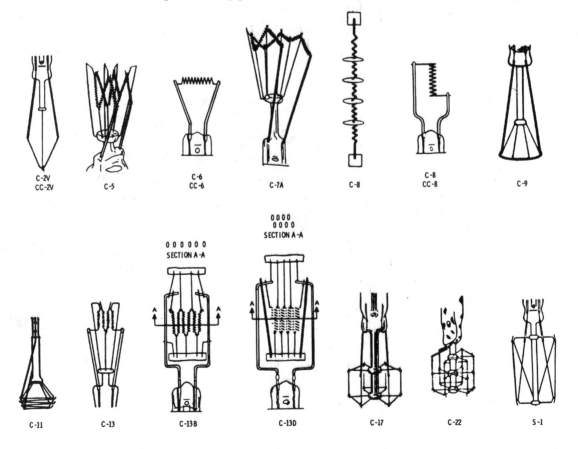

Figure 6-7: Typical incandescent filament forms.

wire, "C" a coiled wire, "CC" a coiled coil, and "R" a flat or ribbon-shaped wire. The numbers and other letters assigned to the various filament forms are purely arbitrary.

Early lamps were made with straight filaments operating in a vacuum. When inert gases were introduced into the bulb, it was found that coiling the wire decreased the effective surface exposed to the circulating gas, and therefore reduced the heat lost by conduction and convection. The coils also tend to heat each other, and the coiled filament is mechanically stronger. Today, nearly all types of lamps, both vacuum and gas-filled, have coiled filaments. The single-coil filament is formed by winding the tungsten wire on a mandrel of steel or molybdenum in a continuous process. The coil with the mandrel still in place is cut into the desired lengths and immersed in an acid bath, which dissolves the mandrel but does not attack the tungsten.

Coiled-coil, or double-coiled, filaments which provide increased efficiency and reduced light source size are at present used in various general lighting service, standard-voltage lamps in the 50- to 1000-W range, also in certain types of projection lamps. The process of making coiled-coil filaments is the same as that for making single-coil-coil filaments. With the mandrel intact, the wire is wound onto another mandrel which is later "retracted," or removed mechanically. The first mandrel is then removed from the coiled coil by dissolving.

In the general lighting service type of lamp, the arrangement of the filament coil and its supports is dictated by the limiting size of the bulb neck through which it must be inserted, and by other manufacturing considerations. Mounting a filament vertically rather than horizontally (C-8, CC-8, or 2CC-8 construction), as has recently been done in general lighting service lamps, results in a higher light output because gas convection currents raise the filament temperature and because less light is absorbed by the lamp base. Further, the bulb blackening which develops as the lamp ages is localized within a smaller area, and lumen maintenance throughout the lamp life is higher. Lamps for special purposes often require certain filament forms. For projection, searchlight, spotlight, floodlight, and similar services where accurate control of light demands a small source, the filaments are concentrated into as small a space as possible. In contrast, for showcase service where a long light source is needed, the filament may be extended along almost the full length of the bulb.

Filling Gas

Incandescent lamps were first made with evacuated bulbs, the purpose being merely to keep the filament from burning up by excluding oxygen. Later it was discovered that the pressure exerted on the filament by an inert gas introduced into the bulb retarded the evaporation of tungsten, thus making it possible to design lamps for higher filament temperatures. Vacuum lamps are now designated as "Type B" lamps, gas-filled lamps as "Type C."

The gas removes some heat from the filament, as a result of conduction and convection losses not present in the vacuum lamp. The larger the surface of the wire in proportion to its volume or mass, the greater this cooling effect becomes, until eventually it nullifies the gain achieved by using the filling gas. Filaments with a current rating of less than $\frac{1}{3}$ A have a wire diameter so small that the introduction of gas is a disadvantage. For this reason, standard-voltage general lighting service lamps of less than 40 W are of the vacuum or "Type B" construction, while lamps of 40 W and higher are gas-filled.

Nitrogen and argon are the gases most commonly used in lamp manufacture. Projection lamps use an atmosphere of 100 percent nitrogen. Most other types have a mixture of nitrogen and argon, the proportions varying with the lamp and the service for which it is designed. High-voltage lamps, for example, are filled with approximately 50 percent argon and 50 percent nitrogen, the higher wattage standard-voltage types about 90 percent argon and 10 percent nitrogen, and the lower wattage standard-voltage types and all street series lamps about 98 percent argon and 2 percent nitrogen. Some nitrogen is necessary to prevent arcing across the lead-in wires, which would occur if pure argon were used. The greater the inherent tendency of a lamp to arc, the higher the percentage of nitrogen in its gas mixture.

Krypton is a relatively rare and expensive gas which has a higher atomic weight than either argon or nitrogen, and therefore causes less energy loss by conduction and convection. It is primarily used in certain miniature lamps such as those on miners' caps, where the limited capacity of the battery power supply makes it essential to obtain the greatest possible efficiency. Hydrogen, because of its low atomic weight, is used in certain very special types of flashing signal lamps where rapid cooling of the filament is important.

Types Of Lamps

The familiar general lighting service lamps, from the 15-W A-15 to the 1500-W PS-52, designed for multiple burning on 120-, 125-, or 130-V circuits, are the most commonly used filament-type lamps. All standard general service lamps are equipped with screw bases. The larger wattages are manufactured in either clear or inside-frosted bulbs. Below 150 W, inside frosted and inside white silica-coated lamps are standard. The wattages most commonly used in the home are available in a straight-sided modified T-bulb shape, with the white silica coating.

High- And Low-Voltage Lamps

Lamps similar to those of the standard-voltage line are available for operation on 230 and 250 V. The low efficacy of these lamps, as compared to comparable lamps of standard-voltage rating, is a disadvantage. Other disadvantages, resulting from the smaller filament wire diameter of high-voltage lamps, are reduced mechanical strength and larger overall light-source size which makes them less satisfactory for use in floodlight and projection equipment. The only gain achieved by the industrial use of these higher voltages is the reduction in ampere load which results from doubling the voltage, and the consequence saving in wiring cost.

Projector and Reflector Lamps

PAR-bulb (projector) and R-bulb (reflector) lamps combine, in one unit, a light source and a highly efficient sealed-in reflector consisting of vaporized aluminum or silver applied to the inner surface of the bulb. The 100-W PAR38 and 150-W R-40 lamps are available in several colors. "PAR" bulbs are of hard glass. "PAR" lamps up to 160 W in size, as well as a few special service "R" lamps with heat-resistant-glass bulb, can be used outdoors without danger of breakage from rain or snow. Larger "PAR" lamps and all other "R" lamps are not recommended for outdoor use unless protected from the elements.

Higher-wattage R-52 and R-57 reflector lamps are designed for general lighting purposes. They are made in both wide and narrow distribution and are best adapted for high-ceilinged industrial areas where the atmosphere contains noncombustible dirt, smoke, or fumes. Where heat-resisting glass is required for protection against thermal shock, the R-60 lamps will perform similarly. These latter types are especially suited for outdoor floodlighting. In addition to flood and spotlight service, PAR-bulb lamps have found wide application in automotive, aviation, and other miscellaneous fields where compact lighting units of precise beam control are necessary.

Showcase and Lumiline Lamps

Low-wattage tubular-bulb lamps are used for showcase lighting and other applications where small bulb diameter is required. Some of these are designed to be used in reflectors, and others are provided with an internal reflecting surface extending over approximately half the bulb area, which concentrates the light to form a beam. The Lumiline lamp is a special type of tubular light source which has a filament extending the length of the lamp. The filament is connected at each end to a disc base which requires a special type of lamp holder. Lumiline lamps are considerably less efficient than conventional general lighting service lamps, but are useful where a linear source is necessary.

Spotlight, Floodlight and Projection Lamps

Characteristic features of all lamps designed for spotlight, floodlight, and projection applications include compact filaments accurately positioned with respect to the base, for purposes of light control; relatively short life, for high efficacy and luminance; comparatively small bulbs; and restricted burning position. Since spotlight lamps must produce narrower, more intense beams than floodlight lamps, they usually have smaller filaments and shorter lives. In projection lamps the light source is still more concentrated and lamp life is further reduced, with accompanying increased efficacy.

The objective in designing projection lamps is to fill the aperture of the projection system with a light source of high luminance and maximum uniformity. This is accomplished by arranging the filament coils in a single or double vertical plane and using a base which accurately locates the filament with respect to the optical system. The biplane (C-13D) filament, with coils arranged in two parallel rows so-placed that the coils of one row fill in the spaces between those of the other, has much greater uniformity and higher average luminance than the single-row monoplane (C-15) filament. Many projection lamps have such small bulbs and operate at such high temperatures that they cannot be burned without continuous forced ventilation, and some have designed lives as short as 10 hours. Lamps for use in certain types of projectors have an opaque coating on the top of the bulb to prevent the emission of stray light.

Halogen lamps

The halogen lamp is a relatively new concept in incandescent lamps. It uses a quartz envelope which is the basis for its many advantages, including the following:

- Compactness
- Thermal shock resistance
- High efficacy
- Almost perfect maintained light throughout the lamp life

Iodine is used in the lamp to create a chemical cycle with the sublimated tungsten to keep the bulb clean. The halogen lamp is used for floodlighting, aviation, photographic, special effects, photocopy, and other applications where its special features are desirable.

Infrared Lamps

Infrared lamps are essentially the same as lamps designed for illumination purposes; the principal difference between them is filament temperature. Since the production of light is not an objective, infrared lamps are designed to operate at a very low temperature, resulting in a low light output (about 7 or 8 lumens per watt) and a consequence reduction in glare. If only the advantage of low filament evaporation is desired, the life of infrared lamps is many thousands of hours; but because of the possibility of failure from shock, vibration, and other causes, the rated life is given merely as "in excess of 5000 hours."

Infrared lamps used in the home and for therapeutic purposes are commonly of the convenient self-contained 250-W R-10 bulb type with internal reflector and red bulb. Those used in industrial processes are of three types; reflector lamps (125-, 250-, and 375-W R-40), clear G-30 bulb lamps (125, 250, 375 and 500 W), and the more recently developed small linear sources in the T-3 quartz bulb. The latter are available in a number of sizes, and the effective heating length and the voltage rating increases with the wattage. Gold-plated or specular aluminum reflectors are most effective for use with unreflectorized infrared lamps.

Incandescent Lamp Specifications

General-Lighting Lamps: The most commonly used filament lamps are available from the 15 W, A-15 lamp to the 1500 W, PS-52 lamp. All are equipped with screw bases and are designed for use on 120-, 125-, and 130-V circuits.

High-voltage lamps are used primarily in commercial and industrial applications while soft-white lamps are popular in the home.

General-Purpose Lamps: This lamp type is inside-frosted and available for use on 120- to 130-V circuits.

Econ-O-Watt and Extended-Service Econ-O-Watt Lamps: These lamps provide similar lighting levels as standard incandescent lamps while consuming 15 percent less energy. They also last two to three times longer.

INSIDE-FROSTED GENERAL-LIGHTING INCANDESCENT LAMPS – 120 V					
Watts	Lamp Type	Volts	Bulb	Base	Rated Avg. Life (Hrs)
25	25A	120	A19	Med.	2500
40	40A15	115/120	A19	Med.	1000
40	40A	120	A19	Med.	1500
60	60A	120	A19	Med.	1000
75	75A	120	A19	Med.	750
100	100A	120	A19	Med.	750
150	150A	120	A21	Med.	750
200	200A	120	A23	Med.	750
300	300M IF	120	PS25	Med.	750
300	300 IF	120	PS35	Mogul	1000
500	500 IF	120	PS35	Mogul	1000

INSIDE-FROSTED GENERAL-LIGHTING LAMPS – 130 V					
Watts	Lamp Type	Volts	Bulb	Base	Rated Avg. Life (Hrs)
15	15A15	130	A15	Med.	2500
25	25A	130	A19	Med.	2500
40	40A	130	A19	Med.	1500
60	60A	130	A19	Med.	1000
75	75A	130	A19	Med.	750

INSIDE-FROSTED GENERAL-LIGHTING LAMPS – 130 V					
Watts	Lamp Type	Volts	Bulb	Base	Rated Avg. Life (Hrs)
100	100A23	120 – 130	A23	Med.	750
100	100A	130	A19	Med.	750
150	150A	130	A21	Med.	750
200	200A	125 – 130	A23	Med.	750
200	200 IF	130	PS30	Med.	750

INSIDE-FROSTED ECON-O-WATT INCANDESCENT LAMPS					
Watts	Lamp Type	Volts	Bulb	Base	Rated Avg. Life (Hrs)
34	40A-34A/EW	120	A19	Med.	1500
34	40A-34A/EW	120	A19	Med.	1500
52	60A-52A/EW	120	A19	Med.	1000
52	60A-52A/EW	120	A19	Med.	1000
67	75A-67A/EW	120	A19	Med.	750
67	75A-67A/EW	130	A19	Med.	750
90	100A-90A/EW	120	A19	Med.	750
90	100A-90A/EW	130	A19	Med.	750
135	150A-135A/EW	120	A21	Med.	750
135	150A-135A/EW	130	A21	Med.	750

INSIDE-FROSTED EXTENDED-SERVICE ECON-O-WATT INCANDESCENT LAMPS					
Watts	Lamp Type	Volts	Bulb	Base	Rated Avg. Life (Hrs)
52	60A-52A/99/EW	120	A19	Med.	2500

INSIDE-FROSTED EXTENDED-SERVICE ECON-O-WATT INCANDESCENT LAMPS *(Cont.)*					
Watts	Lamp Type	Volts	Bulb	Base	Rated Avg. Life (Hrs)
52	60A-52A/99/EW	125/130	A19	Med.	2500
67	75A-67A/99/EW	120	A19	Med.	2500
67	75A-67A/99/EW	125/130	A19	Med.	2500
90	100A-90A/99/EW	120	A19	Med.	2500
90	100A-90A/99/EW	125/130	A19	Med.	2500
135	150A-135A/99/EW	120	A21	Med.	2500
135	150A-135A/99/EW	125/130	A21	Med.	2500

Clear General-Lighting Lamps: These lamps have a crystal clear bulb and a completely visible filament, so the brightness of the light is not softened. Clear general lighting can be used in a variety of applications where brilliance and sparkle are more important than the avoidance of glare.

CLEAR GENERAL-LIGHTING LAMPS					
Watts	Lamp Type	Volts	Bulb	Base	Rated Avg. Life (Hrs)
15	15A15/CL	130	A15	Med.	2500
25	25A/CL	120	A19	Med.	2500
25	25A/CL	130	A19	Med.	2500
40	40A/CL	120	A19	Med.	1500
40	40A/CL	130	A19	Med.	1500
60	60A/CL	120	A19	Med.	1000
60	60A/CL	130	A19	Med.	1000
75	75A/CL	120	A19	Med.	750

CLEAR GENERAL-LIGHTING LAMPS *(Cont.)*

Watts	Lamp Type	Volts	Bulb	Base	Rated Avg. Life (Hrs)
75	75A/CL	130	A19	Med.	750
100	100A/CL	120	A19	Med.	750
100	100A/CL	130	A19	Med.	750
150	150A/CL	120	A21	Med.	750
150	150A/CL	130	A21	Med.	750
300	300M	120	PS25	Med.	750
300	300M	130	PS25	Med.	750

THREE-WAY SOFT-WHITE LAMPS

Watts	Lamp Type	Volts	Bulb	Base	Rated Avg. Life (Hrs)
15	15/150T/WL 12/1	120	T19	3CT-Med.	1600
30	30/100T/SW 12/1	120	T19	3CT-Med.	1200
30	30/100T/WL	120	T19	3CT-Med.	1600
50	50/150/SW	120	T19	3CT-Med.	1200
50	50/150/WL	120	T19	3CT-Med.	1600
50	50/250/WL	120	T21	3CT-Med.	1600

Long-Life Soft-White Lamps: This lamp type provides over 33 percent longer life than standard incandescent lamps. Soft-white lamps achieve maximum diffusion of light from the filament without glare or harsh shadows.

LONG-LIFE SOFT-WHITE LAMPS

Watts	Lamp Type	Volts	Bulb	Base	Rated Avg. Life (Hrs)
25	25T/WL 12/2	120	T19	Med.	1350
40	40T/WL	120	T19	Med.	1350
60	60T/WL	120	T19	Med.	1350
75	75T/WL	120	T19	Med.	1000
100	100T/WL	120	T19	Med.	1000
150	150T/WL	120	T19	Med.	1000
200	200T/WL	120	T21	Med.	1000

Three-Way Soft-White Lamps: These lamps are available in both standard and extended-life types. Both types offer maximum flexibility in lighting a room or area.

Rough Service and Industrial-Service Lamps: These lamps feature an exceptionally rugged C-9 filament construction that is supported at five points. They are suitable for places where shocks, bumps and vibrations frequently occur. The 3500-hour lamps provide a 40 percent increase in rated average life at an increased cost of less than 25 percent over comparable wattage extended-service lamps.

ROUGH SERVICE, INSIDE-FROSTED LAMPS

Watts	Lamp Type	Volts	Bulb	Base	Rated Avg. Life (Hrs)
50	50A/RS	120	A19	Med.	1000
75	75A/RS	120	A19	Med.	1000
75	75A/RS/VS	125-130	A19	Med.	1000
100	100A/RS/VS	120	A21	Med.	1000
100	100A/RS/VS	125-130	A21	Med.	1000
150	150A/RS/VS	120	A23	Med.	1000

			TUBULAR INCANDESCENTS			
Watts	Lamp Type	Volts	Description	Bulb	Base	Rated Avg. Life (Hrs.)
15	15T6	115 -125	Switchboard, clear	T6	Cand.	2000
15	15T6	140 -150	Switchboard, clear	T6	Cand.	2000
15	15T7N	115 -125	Appliance, clear	T7	Inter.	200 – 600
20	20T61/2/IF	120	Exit sign, frosted	T6-½	Inter.	200 – 600
20	20T61/2/DC	120	Exit sign, clear	T6-½	DCBay	200 – 600 or more
25	25T61/2/IF	115 -125	Frosted	T6-½	Inter.	1000
25	25T61/2/IF	130	Frosted	T6-½	Inter.	1000
25	25T6/1/2	115 -125	Clear	T6-½	Inter.	1000
25	25T6/1/2	130	Clear	T6-½	Inter.	1000
25	25T6/1/2DC	115 - 125	Clear	T6-½	DCBay	1000
25	25T6/1/2DC	130	Clear	T6-½	DCBay	1000
25	25T8DC	115 - 125	Appliance, clear	T8	DCBay	Varies
25	25T10	120	Clear	T10	Med.	1000
40	40T10	120	Clear	T10	Med.	1000

Decorative Lamps: Decorative lamps are generally low-wattage and are available in an assortment of shapes to meet a variety of decorative needs. Lamp bases are available in candelabra (cand.) and medium.

			DECORATIVE BENT-TIP LAMPS			
Watts	Lamp Type	Volts	Description	Bulb	Base	Rated Avg. Life (Hrs.)
15	BC-15BA9C/3	120	Bent tip, clear	BA9	Cand.	1500
25	BC-25BA9C/3	120	Bent tip, clear	BA9	Cand.	1500
25	BC-25BA91/2/3	120	Bent tip, clear	BA9-½	Med.	1500
40	BC-40BA9C/3	120	Bent tip, clear	BA9	Cand.	1500
40	BC-40BA91/2/3	120	Bent tip, clear	BA9-½	Med.	1500
60	BC-60BA9C/3	120	Bent tip, clear	BA9	Cand.	2000

DECORATIVE TORPEDO LAMPS

Watts	Lamp Type	Volts	Description	Bulb	Base	Rated Avg. Life (Hrs.)
25	BC-25BA91/2/3	120	Blunt tip, clear	B10	Cand.	1500
40	BC-40B10 1/2/3	120	Blunt tip, clear	B10	Cand.	1500
60	BC-60B10 1/2/3	120	Blunt tip, clear	B10	Cand.	1500

DECORATIVE FLAME LAMPS

Watts	Lamp Type	Volts	Description	Bulb	Base	Rated Avg. Life (Hrs.)
25	BC-25F15/3A	120	Flame, trans. amber	F15	Med.	1500
25	BC-25F15/3W	120	Flame, white	F15	Med.	1500
25	BC25F15/3	120	Flame, clear	F15	Med.	1500
40	BC-40F15/3	120	Flame, clear	F15	Med.	1500
40	BC-40F15/3W	120	Flame, white	F15	Med.	1500
60	BC-60F15/3	120	Flame, clear	F15	Med.	1500

DECORATIVE GLOBE LAMPS

Watts	Lamp Type	Volts	Description	Bulb	Base	Rated Avg. Life (Hrs.)
25	BC-25G16-1/2C/3LL	120	Globe clear, LL	G16-½	Cand.	2000
25	BC-25G16-1/2C/3WLL	120	Globe white, LL	G16-½	Cand.	2000
25	BC25G16-1/2C/3W	120	Globe white	G16-½	Cand.	1500
25	25G25/3	120	Globe clear	G25	Med.	1500
25	25G25/3W	120	Globe white	G25	Med.	1500
40	40G25/3	120	Globe clear	G25	Med.	1500
40	40G25/3W	120	Globe white	G25	Med.	1500
40	BC40G16-1/2C/3W	120	Globe white	G16-½	Cand.	1500

INDUSTRIAL SERVICE, INSIDE-FROSTED LAMPS

W	Lamp Type	Volts	Bulb	Base	Rated Avg. Life (Hrs)
60	60A19/35	120	A19	Med.	3500
100	100A21/35	125-130	A21	Med.	3500
150	150A25/35	125-130	A25	Med.	3500
200	200A25/35	125-130	A25	Med.	3500

Specialty Incandescent Lamps: Specialty incandescent lamps are used in a variety of applications. For example, the 6S6 and 7C7 are used in night lights, appliances, and in panelboards. The 11S14 is used in sign lighting applications and has a brass base.

SPECIALTY INCANDESCENT LAMPS

W	Lamp Type	Volts	Description	Bulb	Base	Rated Avg. Life (Hrs.)
6	6S6	120	Clear	S6	Cand.	1500
6	6S6	125-130	Clear	S6	Cand.	1500
7	7C7	130	Clear	C7	Cand.	3000
11	11S14	130	Clear	S14	Med.	3000

FLUORESCENT LAMPS

Fluorescent lamps have become the major light source for general interior lighting of commercial and institutional buildings and have challenged other sources for residential, exterior, and other lighting applications.

Fluorescent lamps are available in straight, U-shaped, or circular configurations, and in various diameters. *See* Figure 6-9.

A fluorescent lamp consists of an airtight glass tube enclosing a small drop of mercury and a small amount of argon or argon-neon gas to facilitate starting the arc. After the arc is started, the mercury vapor emits ultraviolet radiation which is invisible and does not pass through the glass. However, the inside of the glass tube is coated with a highly sensitive fluorescent powder (phosphors) which is activated by the ultraviolet radiation and in turn converts the invisible energy to visible light. By mixtures of various phosphors, a wide range of visible light colors is possible.

Colors of Fluorescent Lamps

Cool White: This lamp is often selected for offices, factories, and commercial areas where a psychologically cool working atmosphere is desirable. This is the most popular of all fluorescent lamp colors since it gives a natural outdoor lighting effect and is one of the most efficient fluorescent lamps manufactured today.

Deluxe Cool White: This lamp is used for the same general applications as the cool white, but contains more red which emphasizes pink skin tones and is therefore more flattering to the appearance of people. Deluxe cool white is also used in food display because it gives a good appearance to lean meat; keeps fats looking white; and emphasizes fresh, crisp appearance of green vegetables. This

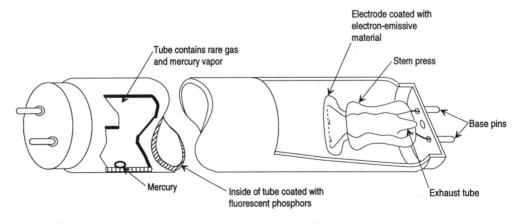

Figure 6-8: Basic components of a fluorescent lamp.

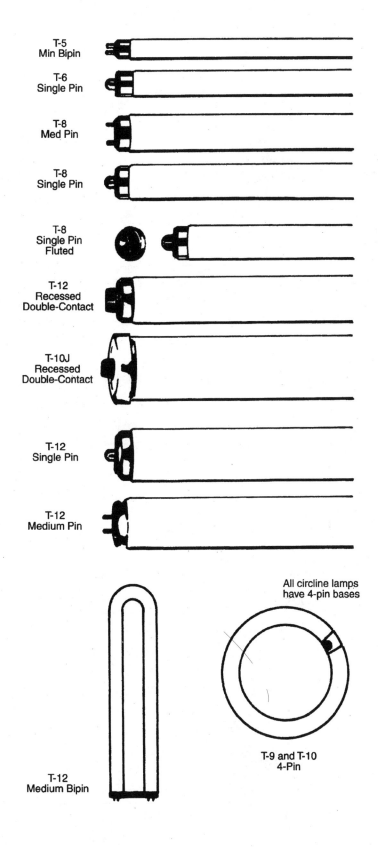

T-5 Min Bipin

T-6 Single Pin

T-8 Med Pin

T-8 Single Pin

T-8 Single Pin Fluted

T-12 Recessed Double-Contact

T-10J Recessed Double-Contact

T-12 Single Pin

T-12 Medium Pin

T-12 Medium Bipin

All circline lamps have 4-pin bases

T-9 and T-10 4-Pin

Figure 6-9: Various sizes and shapes of fluorescent lamps.

type of lamp is generally chosen wherever very uniform color rendition is desired, although it is less efficient than cool white.

Warm White: Warm-white lamps are used whenever a warm social atmosphere is desirable in areas that are not color critical. It approaches incandescence in color and is suggested whenever a mixture of fluorescent and incandescent lamps is used. While it gives an acceptable appearance to people, it has some tendency to emphasize shallowness. Yellow, orange, and tan interior finishes are emphasized by this lamp, and its beige tint gives a bright warm appearance to reds; brings out the yellow in green; and adds a warm tone to blue. It imparts a yellowish white or yellowish gray appearance to neutral surfaces.

Deluxe Warm White: Deluxe warm-white lamps are more flattering to complexions than warm white and are very similar to incandescent lamps in that they impart a ruddy or tanned hue to the skin. It is generally recommended for home or social environment applications and for commercial use where flattering effects on people and merchandise are considered important. This type of lamp enhances the appearance of poultry, cheese, and baked goods. These lamps are approximately 25 percent less efficient than warm-white lamps.

White: White lamps are used for general lighting applications in offices, schools, stores, and homes where either a cool working atmosphere or warm social atmosphere is not critical. They emphasize yellow, yellow green, and orange interior finishes. This lamp, however, is seldom used in most practical applications.

Daylight: Daylight lamps are for use in industry and work areas where the blue color associated with the "north light" of actual daylight is preferred. While it makes blue and green bright and clear, it tends to tone down red, orange, and yellow.

In general, the designations "warm" and "cool" represent the differences between artificial light and natural daylight in the appearance they give to an area. Their deluxe counterparts have a greater amount of red light, supplied by a second phosphor within the tube. The red light shows colors more naturally, but at a sacrifice in efficiency.

Other colors of fluorescent lamps are available in sizes that are interchangeable with white lamps.

These colored lamps are best used for flooding large areas with colored light; where a colored light of small area must be projected at a distant object, incandescent lamps using colored filters are best.

Classes Of Fluorescent Lamps

Preheat Lamps: Preheat, hot-cathode fluorescent lamps use a two-pin base and a starter which provides momentary current flow through the filament cathode in order to heat them. The radiation from the cathodes is possible only after the cathodes have been preheated. The time interval necessary for preheating is one drawback of this type of lamp, but this drawback is offset by the significant savings in ballast design and lamp life. The switch and starter connected across the lamp can be either automatic or manual.

Standard Preheat Rapid-Start Fluorescent Lamps: These lamps are designed to operate on both preheat and rapid-start ballast circuits. They are the most commonly used lamp type in the industry.

Instant Start: In order to overcome the slow starting of the preheat system and eliminate the need for a starter, the instant-start lamp was developed. Instant starting is accomplished by use of a specially designed ballast which delivers a high starting voltage and normal operating voltage once the lamps are started. Because no preheating is necessary with instant-start lamps, only a single pin on each end of the lamp is required. Hot-cathode lamps with single-pin bases are called slimline lamps.

Rapid-Start Lamps: A rapid-start fluorescent lamp retains the advantage of preheat starting, speeds up the starting interval, and eliminates the separate starter switch. The smooth, rapid start is accomplished by a built-in electrode heating coil in the ballast, and the lamp lights almost as quickly as instant-start lamps. These lamps are the most

MINIATURE PREHEAT T-5 FLUORESCENT LAMPS					
Watts	**Lamp Type**	**Description**	**Bulb**	**Base**	**Rated Avg. Life (Hrs.)**
4	F4T5/CW	Cool white	T5	Min. Bipin	6000
6	F6T5/CW	Cool white	T5	Min. Bipin	7500
8	F8T5/CW	Cool white	T5	Min. Bipin	7500
8	F8T5/WW	Warm white	T5	Min. Bipin	7500
13	F13T5/CW	Cool white	T5	Min. Bipin	7500
13	F13T5/27U	2700 Ultralume	T5	Min. Bipin	7500

PREHEAT T-8 FLUORESCENT LAMPS					
Watts	**Lamp Type**	**Description**	**Bulb**	**Base**	**Rated Avg. Life (Hrs.)**
15	F15T8/CW	Cool white	T8	Med. Bipin	7500
15	F15T8/D	Daylight	T8	Med. Bipin	7500
15	F15T8/C50	Colortone 50	T8	Med. Bipin	7500
15	F15T8/WW	Warm white	T8	Med. Bipin	7500
30	F30T8/CW	Cool white	T8	Med. Bipin	7500

PREHEAT T-12 FLUORESCENT LAMPS					
Watts	**Lamp Type**	**Description**	**Bulb**	**Base**	**Rated Avg. Life (Hrs.)**
14	F14T12/CW	Cool white	T12	Med. Bipin	9000
15	F15T12/WW	Warm white	T12	Med. Bipin	9000
15	F15T12/CW	Cool white	T12	Med. Bipin	9000
15	F15T12/D	Daylight	T12	Med. Bipin	9000
20	F20T12/D	Daylight	T12	Med. Bipin	9000
20	F20T12/WW	Warm white	T12	Med. Bipin	9000
20	F20T12/CW	Cool white	T12	Med. Bipin	9000
25	F25T12/CW	Cool white	T12	Med. Bipin	7500

PREHEAT RAPID-START FLUORESCENT LAMPS					
Watts	**Lamp Type**	**Description**	**Bulb**	**Base**	**Rated Avg. Life (Hrs.)**
40	F40CW	Cool white	T12	Med. Bipin	20000
40	F40WW	Warm white	T12	Med. Bipin	20000
40	F40GO	Gold, Bug-Away	T12	Med. Bipin	20000
40	F40D	Daylight	T12	Med. Bipin	20000
40	F40/C50	Colortone 50	T12	Med. Bipin	20000
40	F410W	White	T12	Med. Bipin	20000

PREHEAT RAPID-START EXTENDED-SERVICE FLUORESCENT LAMPS					
Watts	**Lamp Type**	**Description**	**Bulb**	**Base**	**Rated Avg. Life (Hrs.)**
40	F40T10/CW/99	Cool white ext.	T10	Med. Bipin	24000
40	F40T10/WW/99	Warm white ext.	T10	Med. Bipin	24000

	RAPID-START T12 FLUORESCENT LAMPS				
Watts	**Lamp Type**	**Description**	**Bulb**	**Base**	**Rated Avg. Life (Hrs.)**
30	F30T12/WWRS	Warm white	T12	Med. Bipin	18000
30	F30T12/D/RS	Daylight	T12	Med. Bipin	18000
30	F30T12/CW/RS	Cool white	T12	Med. Bipin	18000

popular and important for use in fluorescent-lighting systems.

Today, the most commonly used lamp is the rapid-start type operating at 430 mA, or approximately 10 W per ft of lamp.

Typically, the use of Econ-O-Watt fluorescent lamps cuts energy costs as much as 20 percent by consuming fewer watts. Replacing standard fluorescent lamps with Econ-O-Watt lamps in a 100,000 ft² office building can reduce energy costs as much as 10 percent per ft².

Slimline Lamps: These lamps require no starters. The ballast provides sufficient voltage to instantly light the lamp. Single-pin bases can be used on this type of fluorescent lamp.

	ECON-O-WATT ENERGY-SAVING RAPID-START FLUORESCENT LAMPS				
Watts	**Lamp Type**	**Description**	**Bulb**	**Base**	**Rated Avg. Life (Hrs.)**
25	F30T12/WW/RS/EW-11	Warm white	T12	Med. Bipin	18000
25	F30T12/CW/RS/EW-11	Cool white	T12	Med. Bipin	18000
34	F40T12/SPEC30/RS/EW-11	3000K spec	T12	Med. Bipin	20000
34	F40T12/SPEC35/RS/EW-11	3500K spec	T12	Med. Bipin	20000
34	F40T12/SPEC41/RS/EW-11	4100K spec	T12	Med. Bipin	20000
34	F40WW/RS/EW-11	Warm white	T12	Med. Bipin	20000
34	F40CW/RS/EW-11	Cool white	T12	Med. Bipin	20000
34	F40D/RS/EW-11	Daylight	T12	Med. Bipin	20000
34	F40W/RS/EW-11	White	T12	Med. Bipin	20000
34	F40LW/RS/EW-11	Lite white	T12	Med. Bipin	20000

SLIMLINE FLUORESCENT LAMPS

Watts	Lamp Type	Description	Bulb	Base	Rated Avg. Life (Hrs.)
38.5	F48T12/CW	Cool white	T12	Single pin	9000
56	F72T12/CW	Cool white	T12	Single pin	12000
75	F96T12/CW	Cool white	T12	Single pin	12000
75	F96T12/D	Daylight	T12	Single pin	12000
75	F96T12/WW	Warm white	T12	Single pin	12000
75	F96T12/C50	Colortone 50	T12	Single pin	12000

ECON-O-WATT SLIMLINE FLUORESCENT LAMPS

Watts	Lamp Type	Description	Bulb	Base	Rated Avg. Life (Hrs.)
60	F96T12/CW/EW	Cool white	T12	Single pin	12000
60	F96T12/LW/EW	Lite white	T12	Single pin	12000
60	F96T12WW/EW	Warm white	T12	Single pin	12000
60	F96T12/WW	Warm white	T12	Single pin	12000

HIGH-OUTPUT FLUORESCENT LAMPS (800ma)

Watts	Lamp Type	Description	Bulb	Base	Rated Avg. Life (Hrs.)
35	F24T12/CW/HO	Cool white (207)	T12	Recessed DC	12000
40	F30T12/CW/HO	Cool white (207)	T12	Recessed DC	12000
50	F36T12/CW/HO	Cool white (207)	T12	Recessed DC	12000
55	F42T12/CW/HO	Cool white (207)	T12	Recessed DC	9000
60	F48T12/CW/HO	Cool white (207)	T12	Recessed DC	12000
75	F60T12/CW/HO	Cool white (207)	T12	Recessed DC	12000
85	F72T12/CW/HO	Cool white (207)	T12	Recessed DC	12000
85	F72T12/D/HO	Daylight (207)	T12	Recessed DC	12000
95	F84T12/CW/HO	Cool white (207)	T12	Recessed DC	12000
95	F84T12/D/HO	Daylight (207)	T12	Recessed DC	12000

	HIGH-OUTPUT FLUORESCENT LAMPS (800ma)*(Cont.)*				
Watts	**Lamp Type**	**Description**	**Bulb**	**Base**	**Rated Avg. Life (Hrs.)**
110	F96T12/WW/HO	Warm white (207)	T12	Recessed DC	12000
110	F96T12/D/HO	Daylight (207)	T12	Recessed DC	12000
110	F96T12/CW/HO	Cool white (207)	T12	Recessed DC	12000
110	F96T12/C50/HO	Colortone 50 (207)	T12	Recessed DC	12000

	ECON-O-WATT HIGH-OUTPUT FLUORESCENT LAMPS (800ma)				
Watts	**Lamp Type**	**Description**	**Bulb**	**Base**	**Rated Avg. Life (Hrs.)**
95	F96T12/LW/HO/EW	Lite white	T12	Recessed DC	12000
95	F96T12/WW/HO/EW	Warm white	T12	Recessed DC	12000

	VERY HIGH-OUTPUT FLUORESCENT LAMPS (1500ma)				
Watts	**Lamp Type**	**Description**	**Bulb**	**Base**	**Rated Avg. Life (Hrs.)**
215	F96T12/WW/HO	Warm white	T12	Recessed DC	12000
215	F96T12/CW/HO	Cool white	T12	Recessed DC	12000

	LOW-TEMPERATURE JACKETED FLUORESCENT LAMPS (1500ma)				
Watts	**Lamp Type**	**Description**	**Bulb**	**Base**	**Rated Avg. Life (Hrs.)**
212	FJ96T12/CW/HO-O	Warm white	T12	Recessed DC	12000

All-Weather Fluorescent Lamps

Fluorescent lamps with common base sizes operate most efficiently at normal room temperatures of 70 to 80 degrees F, at which the temperature of the glass tube itself is between 100 and 120 degrees F. Where temperatures fall below this level, as in outdoor applications during winter months, a jacket placed around the outside of the lamp will maintain bulb wall temperature and will help provide reasonable light output. Rapid-start lamps with this jacket are known as all-weather fluorescent lamps.

CIRCLINE FLUORESCENT LAMPS					
Watts	Lamp Type	Description	Bulb	Base	Rated Avg. Life (Hrs.)
20	FC6T9/CW	Cool white	T9	4-pin	12000
22	FC8T9/CW	Cool white	T9	4-pin	12000
22	FC8T9/WW	Warm white	T9	4-pin	12000
32	FC12T9/CW	Cool white	T9	4-pin	12000
32	FC12T9/WW	Warm white	T9	4-pin	12000
32	FC12T9/D	Daylight	T9	4-pin	12000
40	FC16T9/CW	Cool white	T9	4-pin	12000

Cold-Cathode Fluorescent Lamps

Cold-cathode-type circuits have been used for years in neon-sign tubing because they operate at relatively low current in small-diameter tubing adaptable to bending into sign letters or luminous patterns. All lamps are instant start and require special high-voltage circuits. For the same bulb size, phosphor and current loading, the lumen output and maintenance of cold-cathode lamps are identical in performance to those with a hot cathode. These types of lamps find greatest use in sign and display lighting. For some general applications, there are standardized lengths of tubing produced in T-8 glass envelopes and four, six, and eight ft in length, but few such tubes are being installed at present.

Circline Fluorescent Lamps

All circline lamps are of the rapid-start design for operation on rapid-start ballasts. They will also operate on preheat or trigger-start ballasts — making them a universal design.

U-Bent Fluorescent Lamps

U-bent fluorescent lamps are regular 40-W lamps bent into a U-shape. This configuration allows two or three 4-ft lamps to be used in a 2-ft square fixture. It offers the advantage of allowing wiring and lampholders to be installed at one end of the fixture.

U-BENT FLUORESCENT LAMPS					
Watts	Lamp Type	Description	Bulb	Base	Rated Avg. Life (Hrs.)
40	FB40CW/3	Cool white (212)	T12	Med. bi-pin	12000
40	FB40WW/3	Warm white (212)	T12	Med. bi-pin	12000
40	FB40CW/6	Cool white (212)	T12	Med. bi-pin	12000
40	FB40WW/6	Warm white (212)	T12	Med. bi-pin	12000

SL and SLS Compact Fluorescent Lamps

These lamps incorporate a miniature fluorescent tube that is folded into a compact S shape. The fluorescent tube is enclosed in a lightweight polycarbonate housing and controlled by state-of-the-art electronics. Lamps produce light that closely resembles the appearance of incandescent lighting and has excellent color renditions.

SL lamps use only 18 W of electricity, compared to 60 or 75 W used by the incandescent lamps that the SL lamps are designed to replace. Additional sizes of 15, 20, and 23 W are also available. Electronic ballasts offer silent operation along with energy savings, and will not interfere with radio or television reception where the lamps are placed at least 3 ft away.

The SLS family of earthlight lamps offers the ultimate in high efficiency and compactness. SLS 15-, 20-, and 23-W lamps provide the light equivalent of 60-, 75-, and 90-W incandescent lights, respectively, while consuming as little as 25 percent energy.

SL Magnetic Lamp: The short overall length of this lamp type allows them to be used as direct replacements for incandescents in many fixtures which were previously unable to accommodate compact fluorescent lamps. They may also be used in total-enclosed fixtures at temperatures as low as 32 degrees F.

Earthlight SL 18: This lamp type replaces 75-W standard incandescent lamps with an energy savings of 76 percent. It will fit most incandescent lighting fixtures. Applications include wall fixtures, ceiling-mounted fixtures, auxiliary lighting, and table lamps. This lamp was primarily designed for indoor use, but will also operate outdoors in an enclosed fixture at temperatures as low as 0 degrees F. The lamp has a normal service life of approximately 10,000 hours, or about 13 times longer than the equivalent incandescent lamp.

SL 18/R40 Reflector Lamp: This lamp contains the same components as the SL18 except it has an R-40 reflector designed specifically for high-hat, recessed downlighting. However, this lamp is not designed for use on circuits with dimmers. A light output of 800 lumens is achieved using only 18 W of energy.

PL and PLC Compact Fluorescent Lamps: These low-energy, long-life lamps are available in many wattages and sizes. They offer high color rendition and energy savings in a variety of applications. These lamps are designed to replace 25 to 100-W incandescent lamps, but use only about 25 percent of the energy. Furthermore, they last up to 13 times longer than the incandescent equivalent.

High-Intensity Discharge (HID) Lamps

High-intensity discharge lamps are available in a variety of sizes, shapes and colors and offer the following important benefits:

- High efficiency
- High lumen output
- Not affected by ambient temperatures
- Long useful life
- Compact size

HID lamps are divided into three families:

- High-pressure sodium (HPS)
- Metal Halide (MH)
- Mercury vapor (MV)

HID lamps are most often used in industrial, roadway, sports, and some commercial applications. They should be used only with ballasts that match the lamp.

High-Pressure Sodium Lamps

The high-pressure sodium lamp utilizes an arc tube to enclose gases through which an electric current passes. The unique light-transmitting ceramic tube enables sodium to be operated at higher temperatures and pressure than other types of HID lamps. The result is a warm yellow light at nearly maximum theoretical efficiency — 100 to 115 lumens per watt.

This lamp type is excellent for street lighting and general outdoor lighting. Since some of all colors are present in this type of lamp, it has applications for virtually all general lighting under most conditions.

Metal-Halide Lamps

The metal-halide lamps closely resemble a regular clear mercury lamp, but the inner arc tube contains additional halide chemical compounds to increase the light output and improve lamp color.

Since the color produced by metal-halide lamps is much "warmer" than regular mercury lamps, it is suitable for many indoor applications including food displays. It has found more use, however, in outdoor floodlighting, sports-lighting, and certain general street-lighting applications.

The efficiency of the metal-halide lamp is approximately twice that of conventional mercury lamps — producing from 75 to 105 lumens per watt. However, the life of this lamp type is shorter than regular mercury lamps, which average from 6,000 to 10,000 hrs. All other operating characteristics are similar to those of the regular mercury lamp, a description of which follows. *See* Figure 6-10.

Mercury-Vapor Lamps

The mercury-vapor lamp produces light directly as a result of a current passed through gas or vapor under pressure.

While a lighted mercury-vapor lamp appears to emit white light, it actually produces light with a predominance of yellow and green rays and a small percentage of violet and blue rays. Red is absent in the basic lamp, and therefore red objects appear black or dark brown under mercury-vapor lamps. This color distortion initially prevented its use for many applications, but this disadvantage was overcome by the use of red-light-generating chemicals with the bulb.

Typical mercury-vapor lamp components include the following:

- Arc tube made to withstand the high temperatures generated as the lamp builds up to normal wattage
- Two main operating electrodes, located on opposite ends of the tube
- A starting electrode connected in series with a starting resistor and connected to the lead wire of the lower operating electrode
- Tube leads and supports
- An outer phosphor-coated glass bulb that helps stabilize the lamp operation and prevents oxidation of metal parts

The American National Standards Institute (ANSI) has developed a system of codes for mercury lamp types. This system designates a letter "H" followed by a number and two letters. The letter "H" stands for the chemical symbol "Hg" for mercury and indicates the lamp is a mercury type. The number represents the ballast type, and the two letters to follow define the physical lamp characteristics. Additional letters are used to identify the type of phosphor coating on the inside of the bulb. They are as follows:

- C — Color-improved phosphor
- W — High-efficiency phosphor
- DX — Deluxe
- Y — Yellow

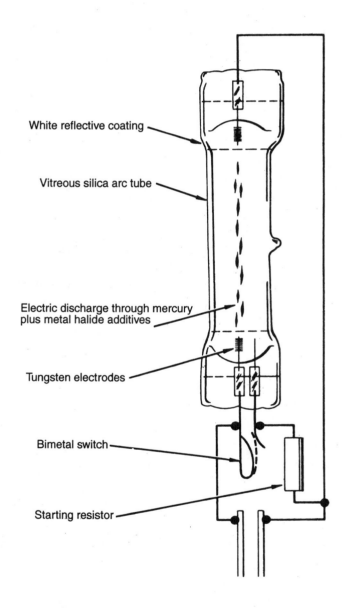

White reflective coating

Vitreous silica arc tube

Electric discharge through mercury plus metal halide additives

Tungsten electrodes

Bimetal switch

Starting resistor

Figure 6-10: Basic components of a metal-halide lamp.

RESIDENTIAL LIGHTING

A simple lighting branch circuit requires two conductors to provide a continuous path for current flow. The usual lighting branch circuit operates at 120 V; the white (grounded) circuit conductor is therefore connected to the neutral bus in the panelboard, while the black (ungrounded) circuit conductor is connected to an overcurrent protection device.

Lighting branch circuits and outlets are shown on electrical drawings by means of lines and symbols respectively; that is, a single line is drawn from outlet to outlet and then terminated with an arrowhead to indicate a home-run to the panelboard. Several methods are used to indicate the number and size of conductors, but the most common is to indicate the number of conductors in the circuit by slash marks through the circuit lines and then indicate the wire size by a notation adjacent to these slash marks. For example, two slash marks indicate two conductors; three slash marks indicate three conductors, etc. Some electrical designers omit slash marks for two-conductor circuits, stating in the symbol list or legend, "no slash marks indicate two No. 14 AWG conductors."

The circuits used to feed residential lighting must conform to standards established by the *NEC* as well as by local and state ordinances. Most of the lighting circuits should be calculated to include the total load, although at times, this is not possible because the electrician cannot be certain of the exact wattage that might be used by the homeowner. For example, an electrician may install four porcelain lampholders for the unfinished basement area, each to contain one 100-W incandescent lamp. However, in actual use the homeowners may eventually replace the original lamps with others rated at 150 W or even 200 W. Thus, if the electrician loaded the lighting circuit to full capacity initially, the circuit would probably become overloaded in the future.

From the above example, it is recommended that no residential branch circuit be loaded to more than 80% of its rated capacity. Since most circuits used for lighting are rated at 15 A, the total ampacity in volt-amperes (va) for the circuit is 15 A × 120 V = 1800 va. Therefore, if the circuit is to be loaded to only 80% of its rated capacity, the maximum initial connected load should be no more than 1440 va.

Electrical symbols are used to show the fixture types. Switches and lighting branch circuits are also shown by appropriate lines and symbols. *See* Chapter 1.

Interior Lighting Design

The basic requirement for any lighting design is to determine the amount of light that should be provided and the best means of providing it. However, since individual tastes and opinions vary greatly, there can be many suitable solutions to the same lighting problem. Some of these solutions will be dull and commonplace, while others will show imagination and resourcefulness. The lighting designer should always strive to select lighting equipment that will provide the highest visual comfort and performance that is consistent with the type of area to be lighted and the budget provided.

Properly designed lighting is one of the greatest comforts and conveniences that any homeowner can enjoy. In building new homes or remodeling old ones, the lighting should be considered equally as important as the heating/air-conditioning system, the furniture placement, and as one of the most important features of both interior and exterior decorations.

As a rule, residential lighting does not require a large quantity of elaborate calculations, as does a school or office building. However, electricians must apply their talent and ingenuity in selecting the best types of lighting fixtures for various locations in order to obtain a desirable effect, as well as the proper amount of illumination at the desired quality. Light has certain characteristics that can be used to change the apparent shape of a room, to create a feeling of separate areas within one room, or to alter architectural lines, form, color, pattern, or texture. Light also affects the mood and atmosphere within the area where used.

While calculations of any quantity are unnecessary, the designer must use some guide until he or she has gained the necessary experience to improvise. The methods described in this section are suitable for selecting the proper amount of light as well as the proper types of lighting fixtures.

Let's review two definitions before continuing.

- Lumen: A lumen is the quantity of light that will strike a surface of 1 ft^2, all points of which are a distance of 1 ft from a light source of 1 candlepower.

- Footcandle: A footcandle is a unit of measurement which represents the intensity of illumination that will be produced on a surface that is 1 ft distant from a source of 1 candlepower, and at right angles to the light rays from the source.

LIGHTING REQUIREMENTS FOR THE HOME		
Area	**Lumens Required per ft^2**	**Average Footcandles Required**
Living Room	80	70
Dining Room	45	30
Kitchen	80	70
Bathroom	65	50
Bedroom	70	30
Hallway	45	30
Laundry	70	50
Workbench	70	70

Figure 6-11: Required lumens and footcandles per ft^2 for various areas in the home.

From the preceding statements, we can say that 1 lumen per ft^2 equals 1 footcandle. Thus, the following method will be called the lumens-per-square-foot method of residential lighting design.

In using this method, it is important to remember that lighter room colors reflect light and darker colors absorb light. This method is based on rooms with light colors; therefore, if the room surfaces are dark — like one that has its walls covered with dark wood-grain paneling — the total lumens should be multiplied by a factor of at least 1.25.

The table in Figure 6-11 gives the required lumens per ft^2 for various areas in the home and also the required illumination in footcandles for those who desire to use a different method of calculating required illumination. The recommended footcandle level is "fixed" and will apply regardless of the type of lighting fixtures used. However, the recommended lumens given in this table are based on the assumption that portable table lamps, surface-mounted fixtures, or efficient structural lighting techniques will be used. If the majority of the lighting fixtures in an area will be recessed, the lumen figures in the table should be multiplied by 1.8.

Note that this method produces only approximate results; yet, quite adequate for most residential lighting applications. One of the most important considerations is to avoid shadows on work planes.

Living Room

As the living room is the social heart of most homes, lighting should emphasize special architectural features such as fireplace, bookcases, paintings, etc. The same is true of draperied walls, planters, or any other special room accents.

Dramatizing fireplaces with accent lights brings out texture of bricks, adds to overall room light level, and eliminates bright spots that cause subconscious irritation over a period of time. Use 75- to 150-W lamps in wallwash-type fixtures — either recessed or surface mounted — for this application.

While recessed downlights, cornice, or valance lighting all add life to draperies, they also supplement the general living room lighting level. Position downlights 2.5 to 3 ft apart and 8 to 10 in from the wall. Valances are always used at windows, usually with draperies (Figure 6-12 on the next page). They provide uplight, which reflects off the ceiling for general room lighting, and downlight for drapery accent. Cornices direct all their light downward to give dramatic interest to wall coverings, draperies, etc., and are good for low-ceiling rooms.

Pulldown fixtures or table lamps are used for reading areas. While the pulldown fixtures are more dramatic, the electrician must know the furniture arrangement prior to fixture placement.

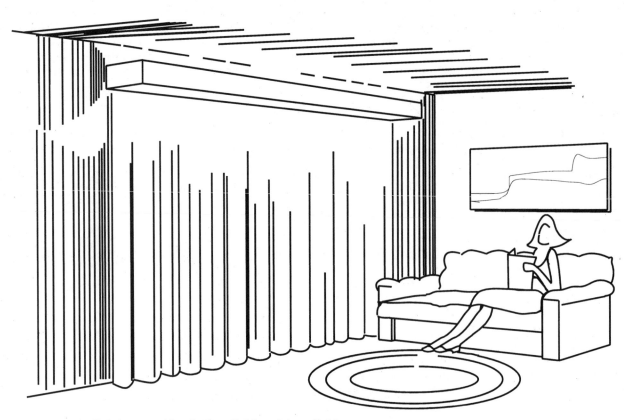

Figure 6-12: Valance lighting provides both uplight and downlight.

As a final touch, add dimmers to vary the lighting levels exactly to the living room activities — low for a relaxed mood, bright for a merry, party mood.

The floor plan in Figure 6-13 shows a living room for a small residence. Let's see how one electrician went about designing a simple, yet highly attractive and functional, lighting layout for this area. Remember that more than one solution to a lighting design is usually available.

The first step is to scale the drawings to find the dimensions of the area in question. In doing so, we find that the area is 13.75 ft wide and 19 ft long. Thus, $13.75 \times 19 = 261.25$ which is rounded off to 261 ft². The table in Figure 6-10 recommends 80 lumens per square foot for the living area. Therefore, $80 \times 261 = 20,880$ lumens required in this area.

The next step is to refer to a lamp catalog in order to select lamps that will give the required lumens. A sample sheet of lamp data is shown in Figure 6-13 on page 552. At the same time, the designer should have a good idea of the type of lighting fixtures that will be used as well as their location.

Referring again to the floor plan of the living room, note that two recessed spotlights, designated by the symbol are mounted in the ceiling above the fireplace. Each fixture contains two 75-W R-30 lamps, rated at 860 lumens each, for a total of 1720 lumens. However, since these are recessed into the ceiling (and not surface mounted), the total lumens will have to be reduced. This is accomplished by multiplying the total lumens by a factor of 0.555 to obtain the efficient lumens.

$$1720 \times 0.555 = 955 \text{ effective lumens}$$

This means that we now need 19,925 more lumens in this area to meet the recommended level of illumination.

The next section of this area will be the window area on the front side of the house. It was decided to use a drapery cornice from wall to wall which would contain four 40-W warm-white fluorescent lamps, rated at 2080 lumens each, for a total of 8320 lumens. Combining this figure with the 955 effective lumens from the recessed lamps give a total of 9275 lumens, or 11,605 more lumens to account for.

Two 3-way (100-, 200-, 300-W) bulbs in table lamps will be used on end tables, one on each side of a sofa for a total of 9460 lumens; this means that only 2145 lumens are unaccounted for. However, one 3-way (50-, 100-, 150-W) bulb will be used in a lamp on a chairside table. Since this

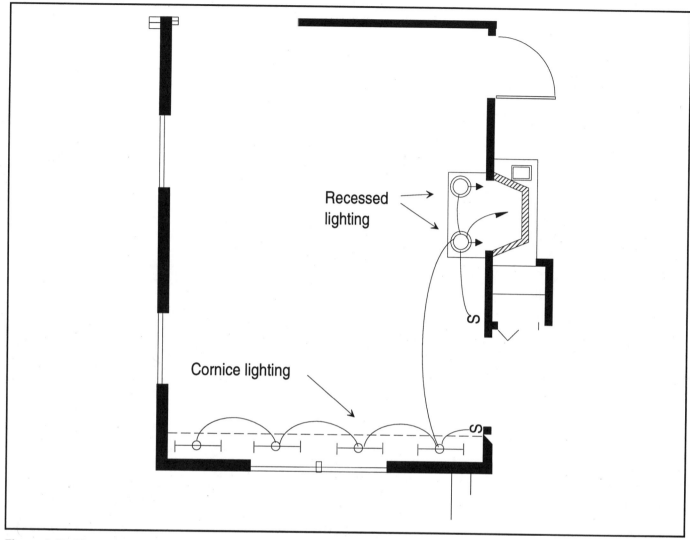

Figure 6-13: Floor plan of living room lighting layout.

lamp is rated at 2190 lumens, we now have the total lumens required for the living-room area.

It can be seen that this method of residential lighting calculation makes it possible to quickly determine the light sources needed to achieve the recommended illumination level in any area of the home.

Kitchen

The lighting layout for the kitchen must always receive careful attention since the kitchen is the area where many homemakers will spend a great amount of time. Good kitchen lighting begins with general illumination — usually one or more ceiling-mounted fixtures of a type that is close to the ceiling. The type of fixture for this general illu-

mination should be a glare-free source that will direct light to every corner of the kitchen.

If a fluorescent lamp is selected for the source of illumination, it is recommended that Deluxe cool-white lamps be used. This type of fluorescent lamp contains more red than the standard cool-white lamp. It therefore emphasizes pink skin tones and is more flattering to the appearance of people. This type of lamp also gives a good appearance to lean meat; keeps fat white; and emphasizes the fresh, crisp appearance of green vegetables. Standard cool-white fluorescent lamps are seldom recommended.

The ideal general lighting system for a residential kitchen would be a luminous ceiling, such as that shown in Figure 6-15 on page 553. This kitchen floor plan shows bare fluorescent strips, with dimming control, above ceiling panels with attractive diffuser patterns. This arrange-

INCANDESCENT GENERAL SERVICE LAMPS			
Watts	**Bulb/Base**	**Lumens (initial)**	**Life (Hours)**
60W	A-19/Med.	870	1000
75W	A-19/Med.	1190	750
100W	A-19/Med.	1750	750
100W	A-21/Med.	1710	750
150W	A-21/Med.	2880	750
150W	A-23/Med.	2780	750
150W	PS-25/Med.	2680	750
200W	A-23/Med.	4010	750
200W	PS-30/Med.	3710	750
300W	PS-25/Med.	6360	750
300W	PS-30/Med.	6110	750

FLUORESCENT TUBULAR LAMPS				
Watts	**Bulb/Base**	**Description**	**Lumens (initial)**	**Life (Hours)**
20W	T-12	CW	1300	9000
20W	T-12	CWX	850	9000
20W	T-12	WW	1300	9000
20W	T-12	WWX	820	9000
30W	T-12	CW	2300	15000
30W	T-12	CWX	1530	12000
30W	T-12	WW	2360	15000
30W	T-12	WWX	1480	12000
40W	T-12	CW	3150	18000
40W	T-12	CWX	2200	18000
40W	T-12	WW	3200	18000
40W	T-12	WWX	2150	18000
40W	T-12	Chroma 55	2020	18000
40W	T-12	Chroma 75	1990	18000
40W	T-12	Inc./Fluor.	1700	17000
40W U	T-12	CW(3⅝)	2800	12000
40W U	T-12	WW(3⅝)	2800	12000

Figure 6-14: Lamp data.

FLUORESCENT Tubular Lamps *(Cont.)*				
40W U	T-12	CW(6")	2950	12000
40W U	T-12	WW(6")	3025	12000
Circline Lamps				
Watts	**Bulb/Base**	**Description**	**Lumens (initial)**	**Life (Hours)**
22W	T-9	CW	950	7500
22W	T-9	CWX	755	7500
22W	T-9	WW	980	7500
22W	T-9	WWX	745	7500
32W	T-10	CW	1750	7500
32W	T-10	CWX	1250	7500
32W	T-10	WW	1800	7500
32W	T-10	WWX	1250	7500
40W	T-10	CW	2800	7500
40W	T-10	CWX	1780	7500
40W	T-10	WW	2350	7500
40W	T-10	WWX	1760	7500

Figure 6-14: Lamp data. *(Cont.)*

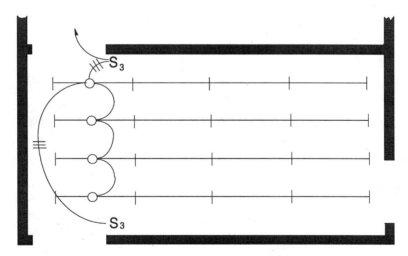

16 4-foot fluorescent
fixtures, spaced 2 feet
on center

Figure 6-15: Kitchen floor plan layout of luminous ceiling lighting.

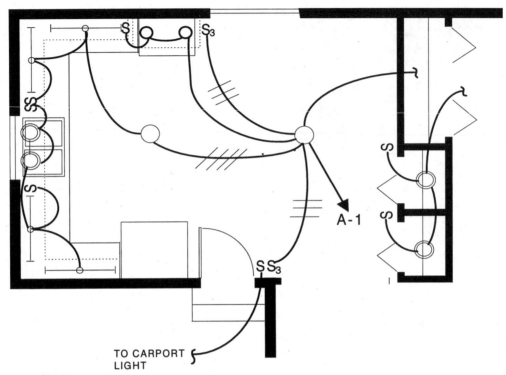

Figure 6-16: Lighting floor plan for a typical kitchen.

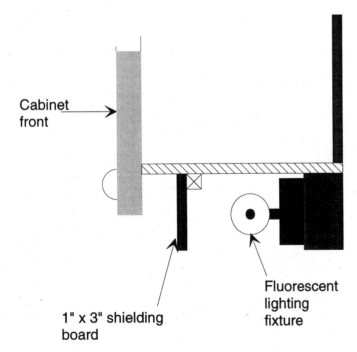

Figure 6-17: Sectional view of a fluorescent lighting fixture located behind a shielding board under kitchen cabinets.

ment, while the most expensive, provides a "skylight" effect which makes seeing easier.

Another kitchen floor-plan layout appears in Figure 6-16. First, single-tube fluorescent light fixtures were located under the wall cabinets and behind a shielding board as shown in Figure 6-17. Then warm-white fluorescent lamps were selected as the best color for lighting countertops. This shadow-free light not only accents the colorful countertops but also makes working at the counter much more pleasant.

Two 75-W R-30 floodlights installed in two recessed housings over the kitchen sink and spaced about 15 in on center offer excellent light for work at the sink. However, a two-lamp fluorescent fixture using warm-white fluorescent lamps and concealed by a faceboard, as shown in Figure 6-18, will work equally well.

The light for the electric range is taken care of by a ventilating hood with self-contained lights. The hood contains two 60-W incandescent lamps for proper illumination.

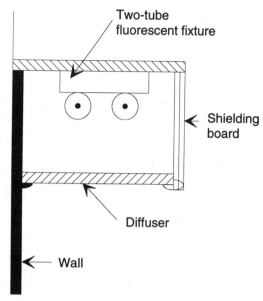

Figure 6-18: A two-tube lighting fixture located above the kitchen sink.

One surface-mounted ceiling fixture with an opal glass diffuser is used for general illumination; this fixture contains two 60-W incandescent lamps. This light source accents and enriches the wood tones of the wall cabinets. In small kitchens, concealed fluorescent strip lighting mounted in a continuous cover around the perimeter will give the effect of a larger kitchen as well as provide an excellent source of general lighting.

In this particular example, the focal point of the kitchen is the dining area. Here, a versatile pulldown light completes the lighting layout. Using 150 W of incandescent lamps, this fixture provides ample light on the table and also directs some of the light upward for a pleasing effect. This pulldown lamp is controlled by a wall-mounted dimmer for added versatility.

In homes with separate dining rooms, a chandelier mounted directly above the dining table and controlled by a dimmer switch becomes the centerpiece of the room while providing general illumination. The dimmer, of course, adds versatility since it can set the mood of the activity — low brilliance (candlelight effect) for formal dining or bright for an evening of cards. When chandeliers with exposed lamps are used, the dimmer is essential to avoid a garish and uncomfortable atmosphere. The size of the chandelier is also very important; it should be sized in proportion to the size of the dining area.

Good planning calls for supplementary lighting at the buffet and sideboard areas. For a contemporary design, use recessed accent lights in these areas. For a traditional

setting, use wall brackets to match the chandelier. Additional supplementary lighting may be achieved with a wide assortment of available fixtures, such as concealed fluorescent lighting in valances or cornices as discussed previously. Of course, there are several other possibilities. In fact, these possibilities are almost endless, limited only by the designer's knowledge, ability, and imagination.

Bathrooms

Lighting performs a wide variety of tasks in the bathroom of the modern residence. Good light is needed for good grooming and hygiene practices.

The bathroom needs as much general lighting as any other room. If the bath is small, usually the mirror and tub/shower lights will suffice for general illumination. However, in the larger baths, a bright central source is needed to transform it from the dim bath of the past to a smart, bright part of the house today.

For good grooming, the lavatory-vanity should be lighted to remove all shadows from faces and from under chins for shaving. Two wall-mounted fixtures on each side of the mirror or lighted soffits or downlights above the mirror will give the best results. Over a vanity table, pendants or downlights for concentrated light with a decorative touch may be used with equal results. For safety and health, a moistureproof, ceiling-mounted recessed fixture over the bath or shower should be included. Linen-closet lighting should also be considered.

For best results, bathroom lighting should be from three sources, like the three lighting fixtures illuminating the mirror in Figure 6-19. A theatrical effect may be obtained

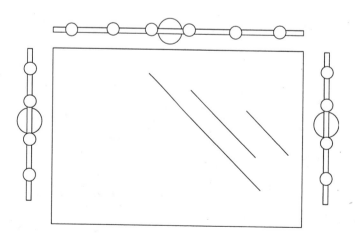

Figure 6-19: Bath mirror lighted from three different light sources.

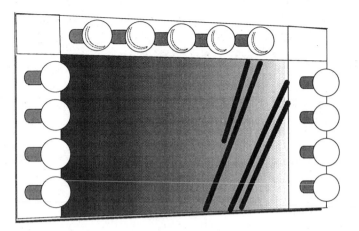

Figure 6-20: Theatrical lighting for a residential bathroom mirror.

by using exposed-lamp fixtures across the top and sides of the mirror, as shown in Figure 6-20.

A small 7- to 15-W night light is also recommended for the bath to permit occupants to see their way at night without turning on overhead lights that might disturb others.

A ceiling fixture with a sunlamp is another convenience that will be appreciated by sun worshippers. Such a lamp requires approximately two minutes to reach full ultraviolet output after it has been started and approximately three minutes to cool before it is restarted.

The sunlamp can be conveniently screwed into an ordinary household socket without the necessity of any other equipment. A good location for it in a residential bathroom is about 2 ft from the face, either over the shaving mirror or in a position where one would normally dry off after bathing.

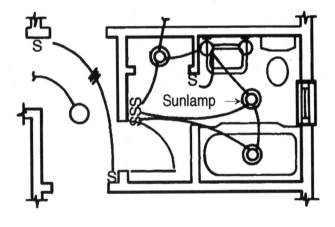

Figure 6-21: Lighting layout for a typical bath and hallway.

A typical lighting layout for a residential bathroom is shown in Figure 6-21. Note that recessed lighting fixtures are used in the linen closet and over the bath tub/shower — the latter equipped with a waterproof lens. A sunlamp is located in the center of the bathroom and is equipped with an automatic timer. Incandescent lamps are located on each side of the mirror. Also note that each lighting fixture, or group of lighting fixtures is individually controlled with a wall switch.

Bedroom Lighting

The majority of people spend at least a third of their lives in their bedroom. Still, the bedroom is often overlooked in terms of decoration and lighting as most homeowners would rather concentrate their efforts and money on areas that will be seen more by visitors. However, proper lighting is equally important in the bedroom for such activities as dressing, grooming, studying, reading, and for a relaxing environment in general.

Basically, bedroom lighting should be both decorative and functional with flexibility of control in order to create the desired lighting environment. For example, both reading and sewing (two common activities occurring in the bedroom) require good general illumination combined with supplemental light directed onto the page or fabric. Other activities, however, like casual conversation or watching television, require only general nonglaring room illumination, preferably controlled by a dimmer/switch control.

Proper lighting in and around the closet area can do much to help in the selection and appearance of clothing, and supplementary lighting around the vanity will aid in personal grooming.

One master bedroom lighting layout is shown in Figure 6-22. Cornice lighting is used to highlight a colorfully draped wall and also to create an illusion of greater depth in this small bedroom. The wall-to-wall cornice board also lowers the apparent ceiling height in the room, which makes the room seem wider.

Two wall-mounted "swing-away" lamps on each side of the bed furnish reading light, while valance-type fluorescent fixtures furnish general illumination. The owners indicated that they preferred matched vanity lamps (table lamps) for grooming. This was handled with duplex receptacles located near the vanity.

A single recessed lighting fixture in the closet provides adequate illumination for selecting clothes and identify-

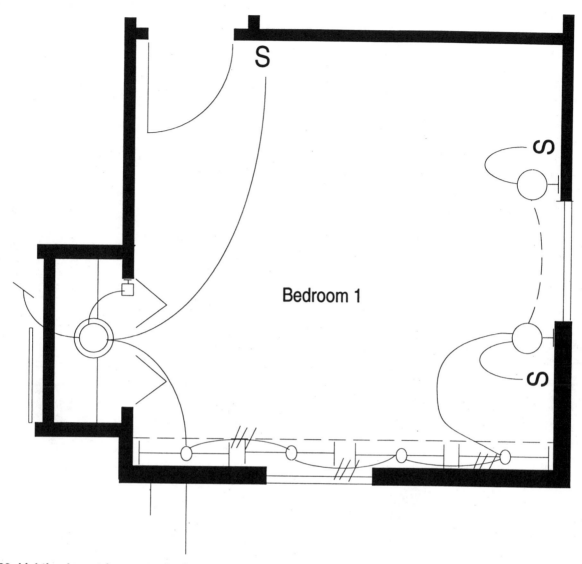

Figure 6-22: Lighting layout for master bedroom.

ing articles on the shelves. The light is controlled by a door switch which turns the light on when the door is opened and turns it out when the door is shut. This recessed lighting fixture, when combined with the general illumination of the bedroom, also illuminates a full-length mirror on the inside of the closet door.

If a closet is unusually long, two equally spaced recessed lighting fixtures may be required to provide adequate light distribution on the closet shelf; or, a full-length fluorescent fixture mounted over the closet door is another choice.

To prevent closet lamps from coming in contact with clothing hung in the closet, which would be a potential fire hazard, certain requirements have been specified by the *NEC*.

A (lighting) fixture in a clothes closet shall be installed:

- On the wall above the closet door, provided the clearance between the fixture and a storage area where combustible material may be stored within the closet is not less than 18 in, or
- On the ceiling over an area which is unobstructed to the floor, maintaining an 18-in clearance horizontally between the fixture and a storage area where combustible material may be stored within the closet . . . .
- Pendants shall not be installed in clothes closets.

The recessed fixture used in the closet of the residence in question has a solid fresnel lens and is therefore consid-

cool-white fluorescent. Each of these lamps produces ap-

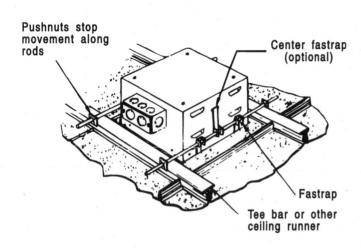

Figure 6-23: Installation details of an incandescent recessed lighting fixture housing.

ered to be located outside of the closet area. For this reason, they can be mounted anywhere on the ceiling or upper walls of the closet area.

To get a better understanding of why certain recessed lighting fixtures are considered to be located outside of the closet area, look at the drawing in Figure 6-23. This is a view of the fixture housing above the closet ceiling. Note that the make-up junction box, along with the housing for the lamp and reflector, are above the ceiling area. Consequently, if the proper lens and trim are installed in the closet area, all heat, electrical connections, and similar potential hazards (if installed within the closet) are actually outside of the closet area. Thus, the reason for the *NEC's* allowance of such fixtures almost anywhere in a clothes closet.

However, carefully review the requirements in the actual code book; that is, *NEC* Section 410-8.

Utility Rooms

The utility room in the floor plan in Figure 6-24 contains an electric water heater, clothes dryer and washer, and a pulldown ironing board. Therefore, this area will be treated as a laundry area requiring 70 lumens of illumination per ft^2. Since the dimensions of the room are 6 ft 5 in by 9 ft 2 in, the area of the room is approximately 58 ft^2 (6.41×9.16). Thus, 58×70 (lumens) = 4060 required lumens in this area.

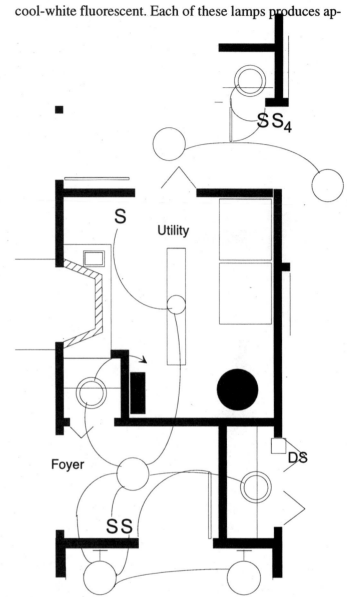

Figure 6-24: Lighting layout for utility room.

A two-lamp fluorescent lighting fixture was the interior decorator's choice for general illumination. This type of fixture was chosen for its high efficiency, long lamp life, linear source of light, and economical operation. However, since fluorescent lamps and ballasts are very sensitive to temperature and humidity, a vaporproof lens was specified on this fixture.

Warm-white lamps were used for the light source since this type of lamp approaches incandescent in color and is more flattering to complexions and clothes colors than

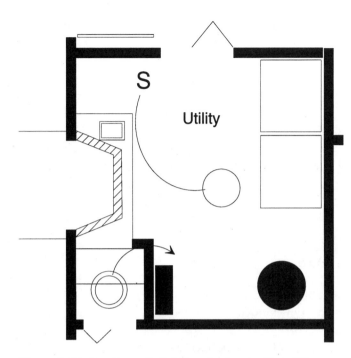

Figure 6-25: An alternate lighting design for the utility room.

proximately 2150 lumens, giving a total of 4300 lumens for the area — just about right for our calculation. An alternate lighting arrangement for the utility room is shown in Figure 6-25.

Lighting for ironing is best when a shadow-casting light is used above and in front of the operator, as shown in Figure 6-26. With this arrangement, the shadows fall toward the operator, giving the best visibility. Therefore, these supplemental fixtures were specified. Use two louvered, swivel-type bullet housing or semirecessed "eyeball" fixtures for a 150-W R-40 reflector floodlight mounted on or in the ceiling 20 to 24 in apart and 24 in ahead of the front edge of the board — aimed at its surface.

If a portable ironing board will be used in various locations throughout the house, a "pole" lamp with three louvered bullets works very well. The middle and lower housings can direct sharp light at an angle, revealing wrinkles in the clothes that are being ironed while the top lamp can be aimed at the ceiling for indirect general illumination. The middle and lower housings should contain 50-W

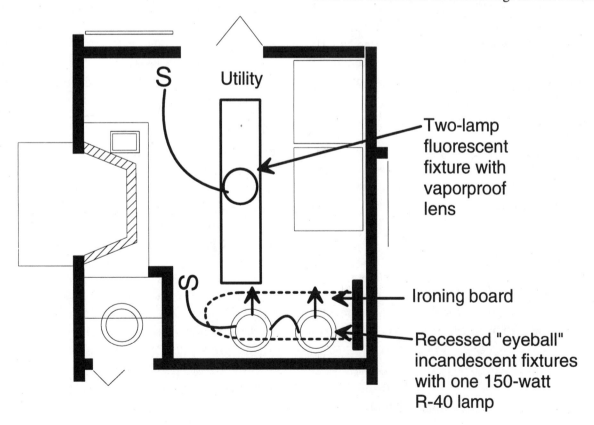

Figure 6-26: Plan view of a lighting layout for ironing.

lamps and the top housing (directed upward) should contain a 150-W reflector lamp.

Basements

If the basement area is to be unfinished and used for utility purposes only, inexpensive lighting outlets should be located to illuminate designated work areas or equipment locations. All mechanical equipment, such as the furnace, pumps, etc. should be properly illuminated for maintenance. Laundry or work areas should have general illumination as well as areas where specific tasks are performed. At least one light near the stairs should be controlled by two 3-way switches, one at the top of the stairs and one at the bottom. Other outlets may be pull-chain porcelain lampholders.

Today, many families have a portion of the basement converted into a family or recreation room. Lighting in this area should be designed for a relaxed, comfortable living atmosphere with the family's interests and activities as a starting point in design.

A typical well-designed lighting layout for a family room should include graceful blending of general lighting with supplemental lighting. For example, diffused incandescent lighting fixtures recessed in the ceiling furnish even, glare-free light throughout the room. The number of fixtures should be increased around game tables for added visual comfort.

Lamps concealed behind cornices near the ceiling enrich the natural beauty of paneled walls. This technique is particularly effective where the light shines down over books with colorful bindings. Fluorescent lamps installed end to end in a cove lighting system will not only furnish excellent general illumination for a family room, but will also give the impression of a higher ceiling; this is a very desirable effect in low-ceiling family rooms.

Light for reading can be accomplished by either table or floor lamps. Post lamps with two or three bullet fixtures are also helpful. One of the bullet housings (containing a reflector lamp) may be aimed at the proper angle for reading while the other(s) may be aimed at the ceiling for indirect general lighting.

Directional light fixtures mounted on the ceiling can be used to display mantel decorations or to illuminate a painting.

Fluorescent fixtures such as cabinet fixtures mounted on the underside of a bookcase or shelf create an excellent lighting source for displaying family portraits, collectables, hobby items, and the like.

Any family-room lighting scheme must be very flexible because most family rooms are in the scene of a variety of daily activities, and these activities require different atmospheres, which can be created by light. For instance, TV viewing requires softly lighted surroundings, while reading calls for a somewhat brighter setting with a light directed on the printed pages. Game participants feel more comfortable in a uniformly lighted room with additional glare-free light directed onto the playing area. Casual conversation flourishes amid subdued, complexion-flattering light such as incandescent or warm-white fluorescent.

Low-level lights over the bar area should be just bright enough for mixing a drink or having a late night snack.

A typical family room may be illuminated as follows. The general illumination is accomplished with recessed incandescent fixtures with fresnel lens. The electric circuit controlling these recessed fixtures is provided with a rheostat dimmer control to change the lighting level of the room as well as the atmosphere. A fluorescent lamp is installed in the center of built-in bookshelves to provide light on the counter below for writing, studying, etc. Wall-wash fixtures may also be used near the bookcase to highlight the colorful bindings of the various books.

Small recessed incandescent lamps with star-shaped lenses may be installed above a bar, while slimline fluorescent fixtures may be mounted on the inside and under the bar top to provide additional light on the work counter. If glass shelves are present behind the bar, these may be highlighted with fluorescent fixtures. With all of these lighting fixtures controlled by dimmers, many exciting effects can be achieved with this one lighting scheme.

A floor plan of a typical lighting layout for a family room is shown in Figure 6-27. In this arrangement, the homeowners desired to use table lamps for general illumination. Consequently, only one ceiling-mounted incandescent lighting fixture is shown. This fixture is designated by the symbol "7" inside of a triangle. A lighting-fixture schedule appearing on the working drawings gives the manufacturer, catalog number, number and size of lamps, and the mounting. In this case, the fixture is a surface-mounted fixture containing two 60-W incandescent lamps. A white opal wrap-around diffuser is used to curtail any glare or brightness.

Note that recessed fixtures on the drawing are designated by a circle within a circle — again with the identifying mark next to each fixture or group of fixtures. Junction boxes are provided for the two built-in medicine cabinets in the bathrooms.

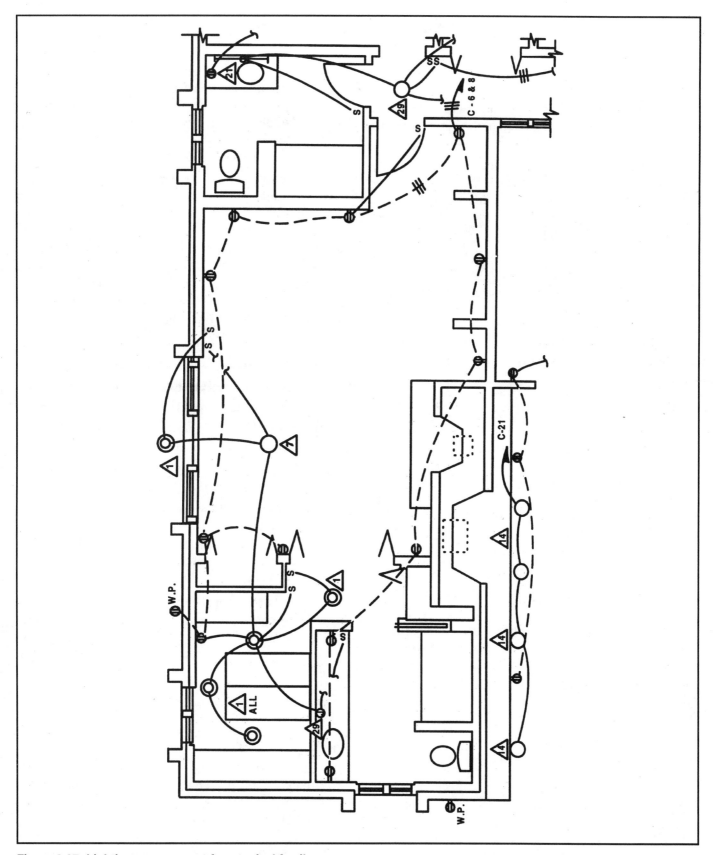

Figure 6-27: Lighting arrangement for a typical family room.

ANALYZING EXISTING LIGHTING LAYOUTS

It was previously mentioned that there is always more than one lighting solution to any given application. The final one decided upon usually depends on the homeowner's tastes and the amount allotted in the lighting-fixture budget.

One of the best ways to learn which lighting fixtures to use (and where to use them) in a residence is to analyze existing lighting layouts — those that are actually working and doing a good job for the homeowners. Such drawings follow — beginning with Figure 6-28.

When designing a lighting layout for any project, remember not to "over-kill." Selecting lighting fixtures to match the style and decor of the building should be the first consideration, while the amount of effective illumination should follow. Once this goal has been achieved, stop! All that remains is to decide on the most adequate and convenient lighting control system.

Lighting Catalogs

Manufacturers of lighting fixtures have colorful brochures and catalogs that can be extremely helpful to those designing residential lighting systems. Besides showing the types of fixtures available, and listing them by catalog number, examples of their use are also given. When a new project is encountered, a glance through a dozen or so of these brochures and catalogs will often give you some excellent ideas on how to proceed with your project.

Some of the larger lighting catalogs also provide design data for their fixtures that will prove helpful on the more sophisticated projects.

Brochures are normally available from lighting fixture manufacturers at little or no charge. Your local lighting fixture dealer will probably have free literature in their places of business. Visiting these dealers will also give you a chance to see the various types of fixtures on display.

Lighting Fixture Manufacturers

There are hundreds of lighting manufacturers in the United States. Names and addresses for most of these can be found in the CEE News Buyers' Guide, available for about $25 from Intertect Publishing Corp., 9221 Quivira Road, Overland Park, KS 66215. Some may also be found on the Internet.

UNDERWATER LIGHTING

Many buildings throughout the world incorporate water fountains into their landscaping and interior designs. This is especially true of multifamily dwellings and commercial buildings where a water fountain becomes a centerpiece for a veranda, courtyard, or lobby. Swimming pools are also used more extensively than ever before.

This section discusses the fundamentals of underwater lighting design for fountains and swimming pools. Such lighting is used mainly for decorative purposes and can be compared to a painting or other work of art. While the designer's artistic ability is very important in lighting designs of this type, there are certain basic rules that should be followed to produce good lighting layouts. *NEC* installation requirements also tightly govern the installation methods of underwater lighting.

Fountain Lighting Design

A fountain is utilized for any of the following reasons:
- Sheer fascination of visual and sound effects.
- To create product or trademark identification.
- A decorative feature piece.
- For animation.
- Air conditioning reject heat load.
- Enhance the surroundings of outstanding architectural structures.

Considerations for fountain lighting should include:
- The type of water effect to be lighted.
- Color selection.
- Maximum height to be illuminated.
- Selecting the type and number of fixtures and lamps.

Types of Water Effect to be Lighted

Single-nozzle fountains are normally lighted with two fixtures equipped with spot lamps (Figure 6-36 on page 571).

The smaller spray ring fountains may be lighted with one fixture, while the larger spray ring fountains are normally lighted with one fixture every four to five ft for illuminating heights up to 15 ft; wide-angle flood lamps are used. For heights over 15 ft, use one fixture every two to three ft with medium-spread flood lamps. This same procedure also applies to waterfalls and weirs.

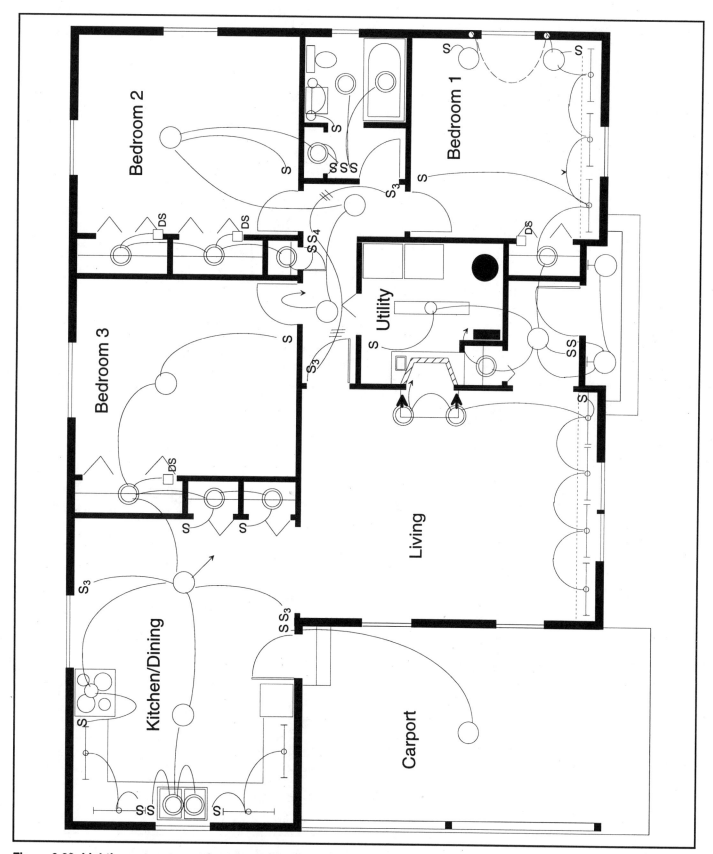

Figure 6-28: Lighting arrangement for a small, single-story residence.

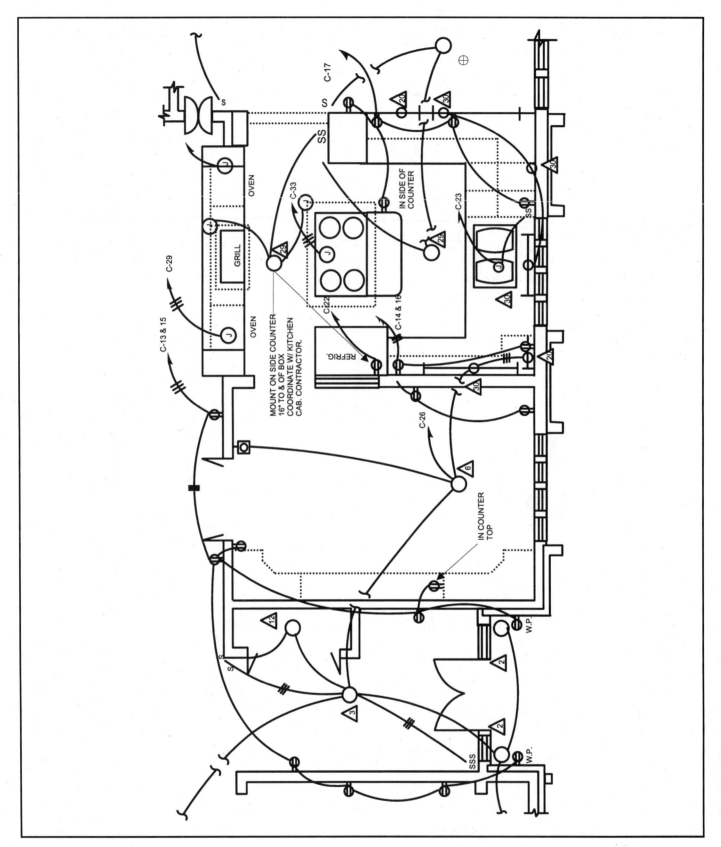

Figure 6-29: Floor plan lighting layout for a large kitchen and dining room, using both incandescent and fluorescent lighting fixtures.

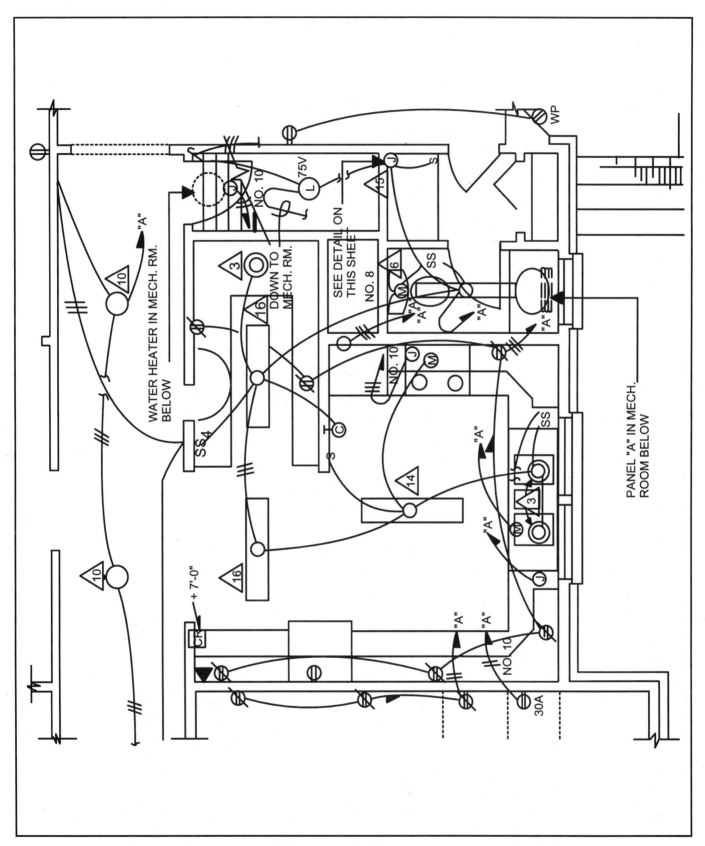

Figure 6-30: Lighting and power layout of a modern kitchen utilizing both incandescent and fluorescent fixtures.

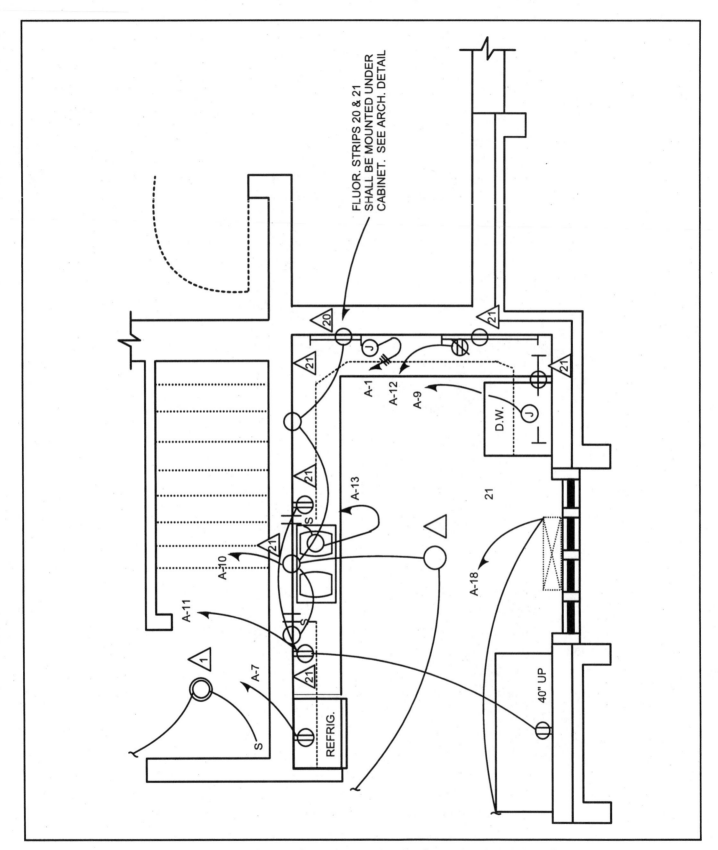

Figure 6-31: A small residential kitchen using one incandescent lighting fixture and fluorescent fixtures under the cabinets.

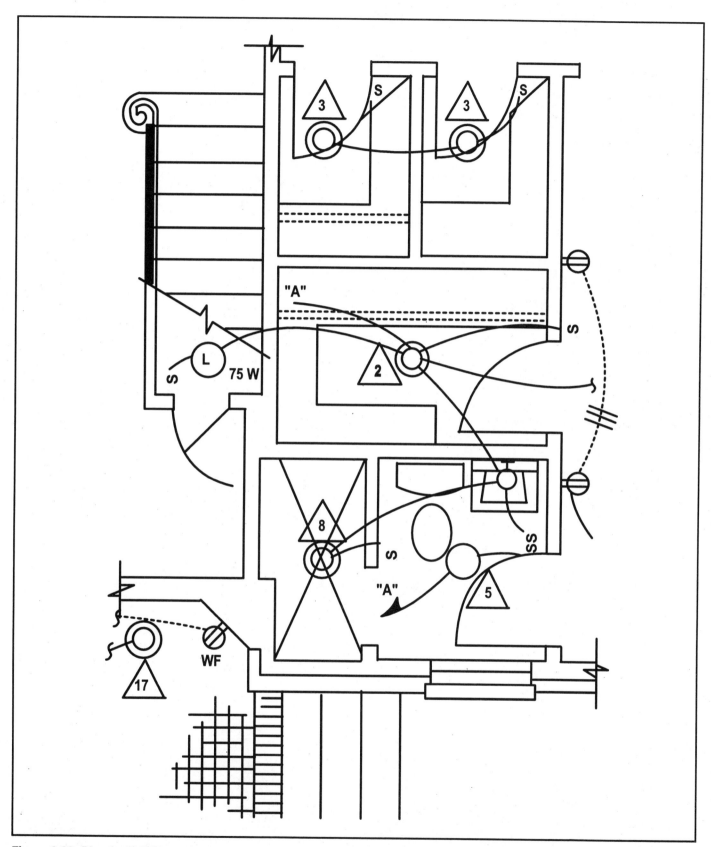

Figure 6-32: Plan for lighting a closet and a small bathroom; the junction box over the lavatory is for a built-in medicine cabinet containing its own lights.

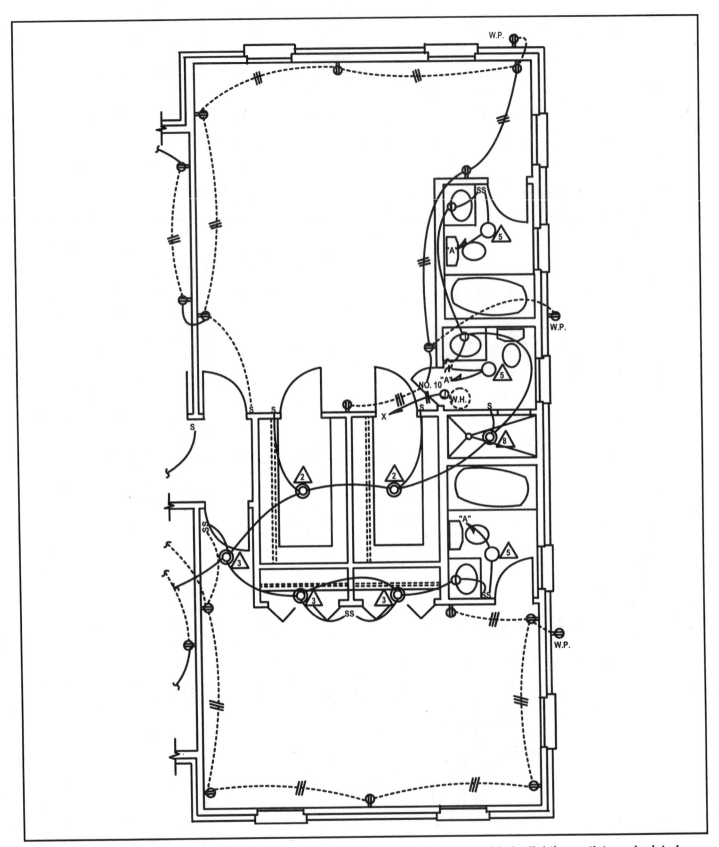

Figure 6-33: Plan view of two bedrooms, walk-in closets, and adjoining bathrooms with the lighting outlets and related circuits.

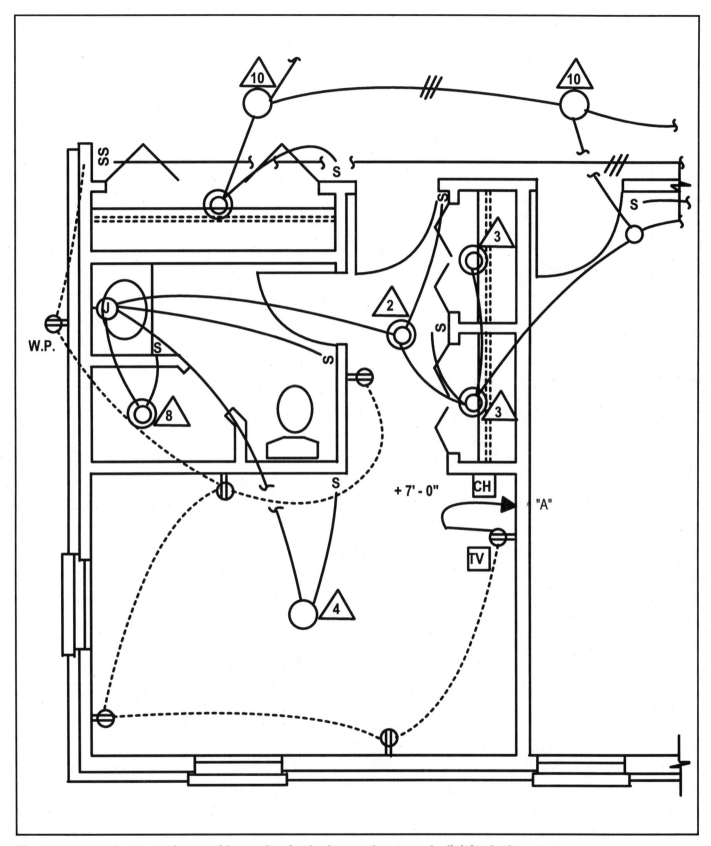

Figure 6-34: Lighting layout for a residence showing bedroom, closets, and adjoining bath.

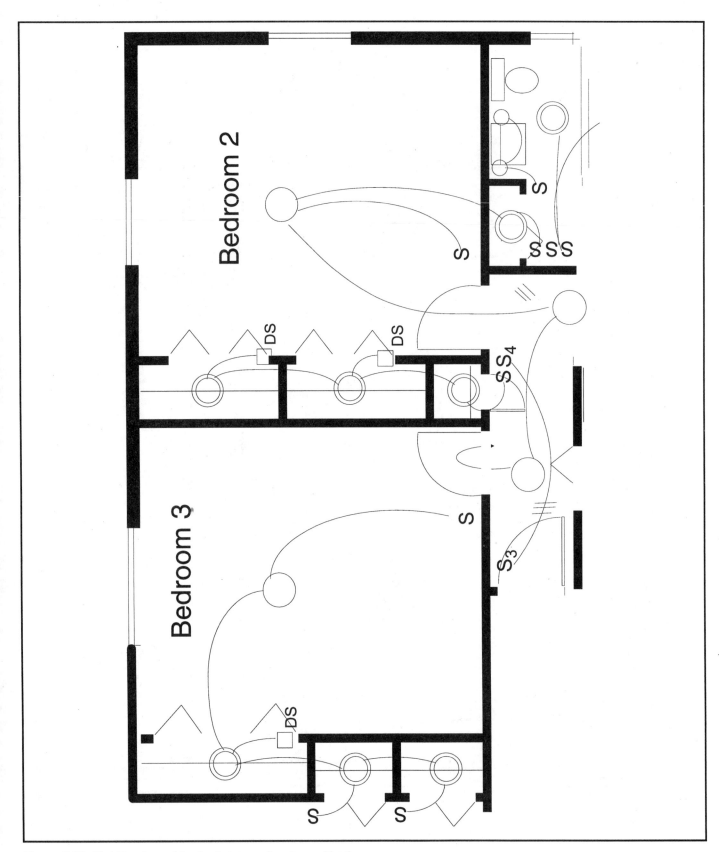

Figure 6-35: Lighting layout for bedrooms, closets, and bath.

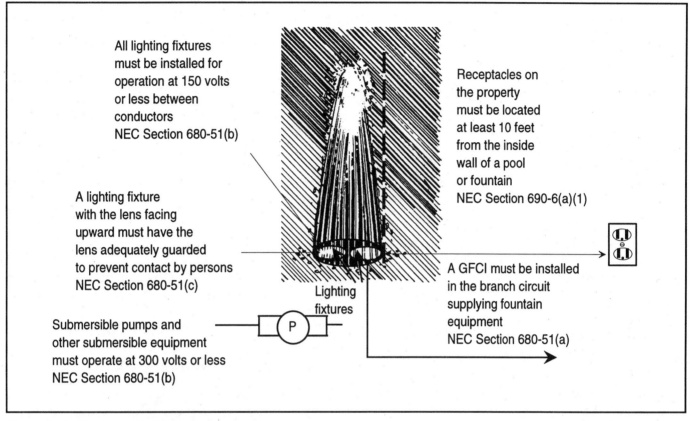

All lighting fixtures must be installed for operation at 150 volts or less between conductors NEC Section 680-51(b)

Receptacles on the property must be located at least 10 feet from the inside wall of a pool or fountain NEC Section 690-6(a)(1)

A lighting fixture with the lens facing upward must have the lens adequately guarded to prevent contact by persons NEC Section 680-51(c)

Lighting fixtures

A GFCI must be installed in the branch circuit supplying fountain equipment NEC Section 680-51(a)

Submersible pumps and other submersible equipment must operate at 300 volts or less NEC Section 680-51(b)

P

Figure 6-36: Single-nozzle fountain.

Color Selection: The selection of colors is a subjective matter and can vary as much as opinions of designers or owners of buildings. However, the following may be used as a guide in selecting colors of lamps.

- Colors directly affect the selection of fixtures and lamps, inasmuch as various colors require differing candlepower to achieve decorative effects of light.
- Amber and turquoise lamps require 50% more candlepower than a clear lens for the same level of illumination.
- Red lamps require 100% more candlepower than a clear lens for the same level of illumination.
- Blue and green lamps require 250% more candlepower than a clear lens for the same level of illumination.
- Where high levels of illumination surround the fountain, use caution when selecting colors; the surrounding light will tend to wash out the colored light. If this condition exists, it is recommended that clear, amber, or turquoise colors be used.

Heights to Be Illuminated

The table in Figure 6-37 may be used as a guide in determining the minimum beam candlepower required for a given height in order to achieve a reasonable balance of decorative color effects. The values are based upon the use of standard lenses as manufactured by Kim Lighting & Manufacturing Co., Inc.

The total footcandle requirements for a specific lighting layout would be contingent upon the perimeter of the layout.

A tabulation of standard available lamps and their rated candlepower can be found in Figure 6-38.

Typical Layout

The illustration in Figure 6-39 on page 573 shows a typical fountain layout. The twelve fountain lights surrounding the jet nozzle (inside ring) are located on a 2½-ft radius. The outside ring is lighted by 24 fountain lights located on a 6-ft radius.

Height of Water Effect (ft)	Clear	Amber & Turquoise	Red	Blue & Green
5	4,000	6,000	8,000	14.000
10	11,000	16,000	22,000	38,000
l5	21,000	31,000	42,000	73,000
20	34,000	51,000	68,000	119,000
25	50,000	75,000	100,000	175,000
30	69,000	103,000	138,000	241,000
35	91,000	136,000	182,000	378,000
40	115,000	174,000	230,000	406,000

Figure 6-37: Minimum desirable beam candlepower.

Watts	Bulb	Ordering Code	Beam Type	Beam Spread (degrees)	Initial Average Maximum Beam Candlepower	Approx. Life (hours)
150	PAR-38	150PAR/SP	Spot	30 × 30	10,500	2000
150	PAR-38	150PAR/FL	Flood	60 × 60	3500	2000
150	R-40	150R/SP	Spot	40	6300	2000
150	R-40	150R/FL	Flood	110	1300	2000
250	PAR-38	Q250 PAR38SP	Spot	26	34,000	4000
250	PAR-38	Q250PAR38FL	Flood	60	6000	4000
300	PAR-56	300PAR56 /NSP	Spot	15 × 20	70,000	2000
300	PAR-56	300PAR56/MFL	Med-Flood	20 × 35	22,000	2000
300	PAR-56	300PAR56/WFL	Wide-Flood	30 × 60	10,000	2000

Figure 6-38: Standard lamps and their rated candlepower.

Watts	Bulb	Ordering Code	Beam Type	Beam Spread (degrees)	Initial Average Maximum Beam Candlepower	Approx. Life (hours)
500	PAR-56	Q500PAR56/NSP	Spot	15 × 32	90,000	4000
500	PAR-56	Q500PAR56/MFL	Med-Flood	20 × 42	49,000	4000
500	PAR-56	Q500PAR56/WFL	Wide-Flood	34 × 66	18,000	4000
1000	PAR-64	Q1000PAR64NSP	Spot	14 × 31	160,000	4000
1000	PAR-64	Q1000PAR64MFL	Med-Flood	22 × 45	60,000	4000
1000	PAR-64	Q1000PAR64WFL	Wide-Flood	45 × 72	27,000	4000

Figure 6-38: Standard lamps and their rated candlepower. *(Cont.)*

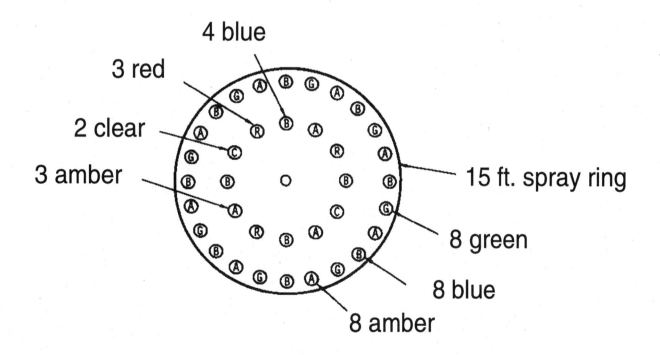

Figure 6-39: A large spray-ring fountain.

Swimming Pool Lighting

There is a large selection of underwater lights, from small 75-W units to large 1000-W units, to meet practically every underwater lighting requirement.

In selecting equipment for underwater lighting, the greatest economy is achieved through the selection of a high-quality fixture designed to give years of dependable service. By their very nature, underwater lighting fixtures are subjected to the forces most destructive to electrical fixtures — water, chemicals, and neglect.

The illustrations in Figures 6-40 and 6-41 show some recommended lighting layouts for swimming pools, wading pools, and the like.

NEC Requirements For Swimming Pools

The *NEC* recognizes the potential danger of electric shock to persons in swimming pools, wading pools, and therapeutic pools or near decorative pools or fountains. This shock could occur from electric potential in the water

itself or as a result of a person in the water or a wet area touching an enclosure that is not at ground potential. Accordingly, the *NEC* provides rules for the safe installation of electrical equipment and wiring in or adjacent to swimming pools and similar locations. *NEC* Article 680 covers the specific rules governing the installation and maintenance of swimming pools and similar installations.

The general requirements for the installation of outlets, overhead conductors and other equipment are summarized in Figure 6-42 on page 576.

Besides *NEC* Article 680, another good source for learning more about electrical installations in and around swimming pools is from manufacturers of swimming pool equipment, including those who manufacture and distribute underwater lighting fixtures. Many of these manufacturers offer pamphlets detailing the installation of their equipment with helpful illustrations, code explanations, and similar details. This literature is usually available at little or no cost to qualified personnel. You can write directly to manufacturers to request information about avail-

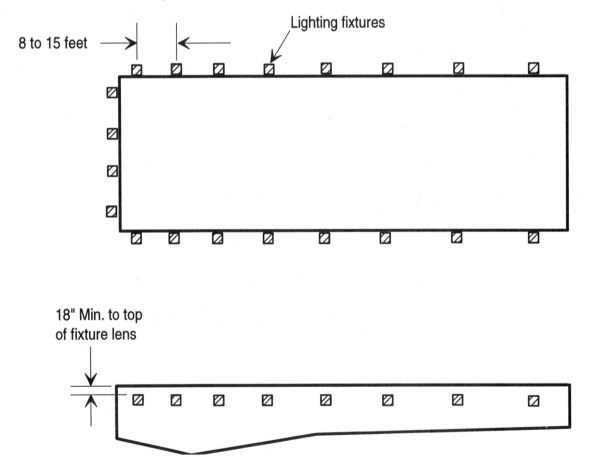

Figure 6-40: Recommended lighting spacing for commercial swimming pools.

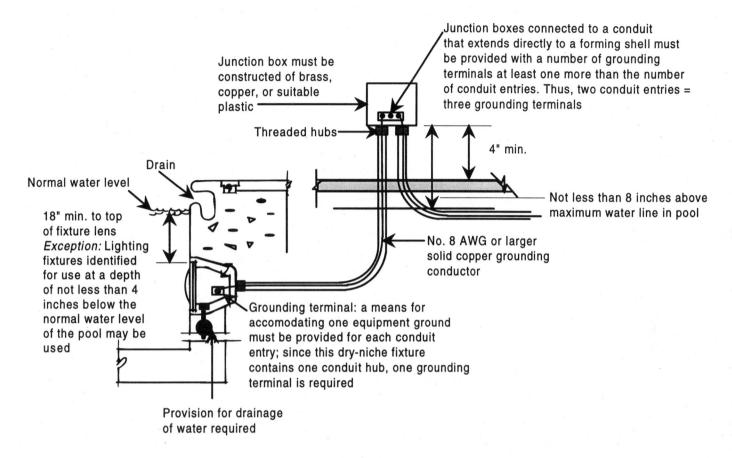

Junction box must be constructed of brass, copper, or suitable plastic

Junction boxes connected to a conduit that extends directly to a forming shell must be provided with a number of grounding terminals at least one more than the number of conduit entries. Thus, two conduit entries = three grounding terminals

Threaded hubs

4" min.

Drain

Normal water level

Not less than 8 inches above maximum water line in pool

18" min. to top of fixture lens
Exception: Lighting fixtures identified for use at a depth of not less than 4 inches below the normal water level of the pool may be used

No. 8 AWG or larger solid copper grounding conductor

Grounding terminal: a means for accomodating one equipment ground must be provided for each conduit entry; since this dry-niche fixture contains one conduit hub, one grounding terminal is required

Provision for drainage of water required

Figure 6-41: NEC installation requirements for pool lights and related components.

able literature, or contact your local electrical supplier or contractors who specialize in installing residential swimming pools.

ROADWAY LIGHTING

Street and roadway lighting have become so common that many of us fail to notice or appreciate its full value. But when one learns of the benefits derived, in the reduction of traffic accidents, crime and vagrancy, and increased business in well-lighted shopping areas, we find that it is a very important branch of illumination.

Incandescent lamps of several hundred watts were once used extensively for street and roadway lighting. However, these are being rapidly replaced by high-pressure sodium lamps because the latter have greater efficiency and longer life.

Effective roadway and street lighting can only be achieved through careful and intelligent planning and by following the American Standard Practice for Street and

Highway Lighting. The following should be taken into consideration in the order given:

- The area and roadway classification.
- The proper illumination level for the classification.
- Selection of lighting fixtures according to the requirements.
- The proper location of the lighting fixtures to provide the required quantity and quality of illumination.
- Footcandles according to roadway classification.

The average horizontal footcandles recommended in the table in Figure 6-43 on page 577 represent average illumination on the roadway pavement when the light source is at its lowest output and also when it is in its dirtiest condition.

In lighting urban roadways, the proper classification of areas and streets is important. The selection of the proper type of light source and its application is also important to a good lighting job.

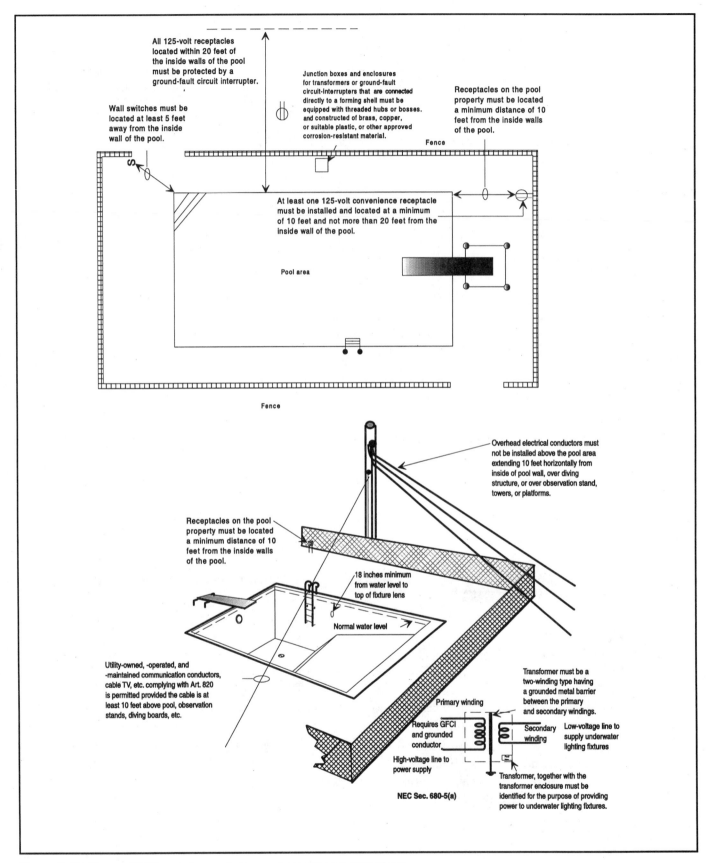

All 125-volt receptacles located within 20 feet of the inside walls of the pool must be protected by a ground-fault circuit interrupter.

Junction boxes and enclosures for transformers or ground-fault circuit-interrupters that are connected directly to a forming shell must be equipped with threaded hubs or bosses. and constructed of brass, copper, or suitable plastic, or other approved corrosion-resistant material.

Receptacles on the pool property must be located a minimum distance of 10 feet from the inside walls of the pool.

Wall switches must be located at least 5 feet away from the inside wall of the pool.

Fence

At least one 125-volt convenience receptacle must be installed and located at a minimum of 10 feet and not more than 20 feet from the inside wall of the pool.

Pool area

Fence

Overhead electrical conductors must not be installed above the pool area extending 10 feet horizontally from inside of pool wall, over diving structure, or over observation stand, towers, or platforms.

Receptacles on the pool property must be located a minimum distance of 10 feet from the inside walls of the pool.

18 inches minimum from water level to top of fixture lens

Normal water level

Utility-owned, -operated, and -maintained communication conductors, cable TV, etc. complying with Art. 820 is permitted provided the cable is at least 10 feet above pool, observation stands, diving boards, etc.

Transformer must be a two-winding type having a grounded metal barrier between the primary and secondary windings.

Primary winding

Requires GFCI and grounded conductor

Secondary winding

Low-voltage line to supply underwater lighting fixtures

High-voltage line to power supply

Transformer, together with the transformer enclosure must be identified for the purpose of providing power to underwater lighting fixtures.

NEC Sec. 680-5(a)

Figure 6-42: General NEC requirements for pool electrical installations.

AREA CLASSIFICATION			
Roadway Classification	**Downtown (footcandles)**	**Intermediate (footcandles)**	**Outlying Urban and Rural (footcandles)**
Major	2.0	1.2	0.9
Collector	1.2	0.9	0.6
Local or minor	0.9	0.6	0.2

EXPRESSWAYS AND FREEWAYS		
Roadway Classification	**Expressways (footcandles)**	**Freeways (footcandles)**
Continuous urban	1.4 – 2	0.6
Continuous rural	1.0	0.6
Interchange urban	2.0	0.6
Interchange rural	1.4	0.6

Figure 6-43: Average recommended illumination on roadway pavement.

It should be remembered that the illumination values given in Figure 6-42 are minimum maintained in service. Also, for commercial reasons, much higher levels are generally used in the commercial district than those minimum levels required for traffic safety.

The following are typical applications furnished by Westinghouse Electric Corporation and are in accordance with the ANSI-IES Standards for street and highway lighting.

Major Streets — Downtown

Figure 6-44 shows a typical layout for a 70-ft wide street using 1000-W mercury lamps in a NEMA Type III lighting fixture with a short spacing ratio and a semicutoff. They are mounted at 33 ft in an opposite arrangement and spaced 150 ft on centers for a calculated average of 2.7 footcandles. This meets ANSI-IES recommendations.

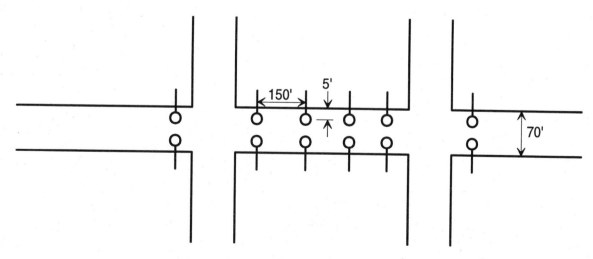

Figure 6-44: Typical lighting layout for a 70-ft street using 1000-W mercury lamps.

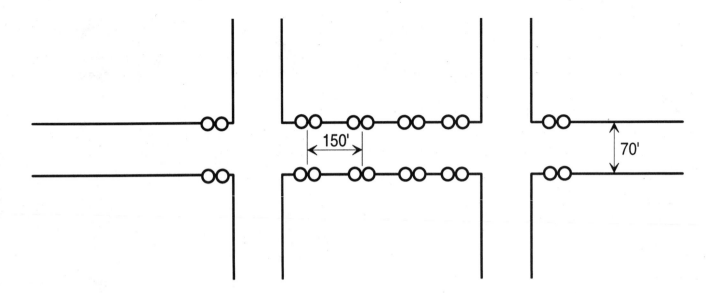

Figure 6-45: Lighting layout for a 70-ft wide street with a high level of illumination.

Figure 6-45 illustrates a typical plan for the same street but with a higher level of illumination. Two 1000-W mercury lighting fixtures are used on each pole, either side-by-side or one above the other. This arrangement gives a calculated average of 5.3 maintained footcandles.

If a higher level of illumination is required, high pressure sodium lamps may be used instead of mercury or the number of lamps per pole may be increased; increasing the number of poles is another possibility.

Intermediate Area

Figure 6-46 illustrates a layout to provide for either an intermediate area or collector street in a downtown area. Here, the 400-W clear mercury lamp is used in a Type III medium spacing ratio, semi-cutoff luminaire, or a Type IV inside-frosted phosphor coated lamp is used. Mounting height is 30 ft. The arrangement provides an average of 1 to 2 maintained footcandles throughout the traffic area.

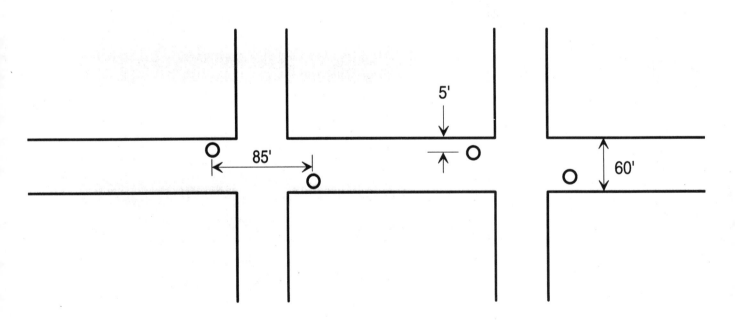

Figure 6-46: Lighting layout for an intermediate area or collector street to a downtown area.

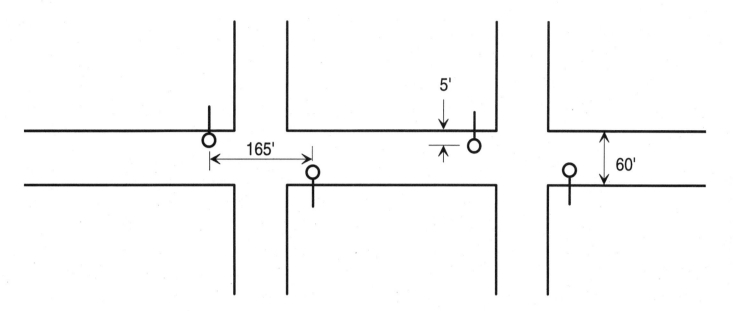

Figure 6-47: This lighting layout uses 400-W mercury lamps with medium spacing.

Outlying Area

The layout in Figure 6-47 is for an outlying area or collector street in an intermediate area. It uses a 400-W mercury lamp in a Type III, medium spacing ratio, and semi-cutoff. The mounting height is 30 ft and provides an average of 0.9 maintained footcandles.

Collector Street

The layout in Figure 6-46 is also good for a collector street in an outlying area or a local street in an intermediate area. For collector streets use 175-W mercury lamps in a Type II lighting fixtures, medium spacing ratio (115 ft on center), and semi-cutoff with a mounting height of 25 ft for an average of 0.6 maintained footcandles.

Local and Minor Street

The same basic layout in Figure 6-46 when used with 175-W mercury lamps in a Type II, long spacing ratio (250 ft on center), and semi-cutoff luminaire with a mounting height of 25 ft will maintain an average of 0.2 maintained footcandles for local and main streets 30 ft wide.

HIGHWAY, FREEWAY, AND EXPRESSWAY APPLICATIONS

Lighting fixtures on expressways or freeways should be located at a sufficient distance from the edge of the curb or traffic lane to minimize the chances of a serious accident should a car be forced from the regular traffic lane. If they are less than 30 ft from the edge of the curb or traffic lane, the poles should be protected by guard rails.

Where there is no median strip, pole arrangements should be located on the outside in either a staggered or opposite arrangement, depending upon the width of the roadway.

If the roadway is provided with a narrow median strip, and the total roadway width is not greater than 125 ft, then the entire area, including the median, should be treated as a single roadway.

If the median strip exceeds 30 ft in width, and the two roadways, including the median strip, exceed 125 ft, then the poles (standards) should be mounted in the median strip, and each roadway should be treated as a separate road.

Highway Interchange Lighting

At a highway interchange, the lighting system should be so arranged that it will aid the driver in selecting the roadway to his destination, minimizing his chances of confusion and making such roadways more comfortable and convenient for night driving (Figure 6-48).

These exit and entrance roadways are usually more narrow than the horizontal roadways they connect. They are most often constructed on a horizontal curve, but may also be on a grade curve, making the placement of lighting fixtures difficult for the maximum utilization.

The specific lighting level for interchanges should be maintained as a minimum. Preferably, it should be a little higher than that on any of the adjacent roadways.

To light interchanges, on otherwise unlighted roadways, it is desirable to start the lighting of the main through road approximately $\frac{1}{2}$ mile before the first exit road and graduate the intensities from about 0.3 footcandle for the first $\frac{1}{4}$ mile, to 0.7 footcandle for the second $\frac{1}{4}$ mile. Then use from 1 to $1\frac{1}{4}$ footcandles through the interchange, and gradually taper off in this reverse order. All values are minimum maintained.

Underpass Lighting

Medium-length underpasses, 75 to 150 ft, usually require nighttime lighting only. Generally, underpasses in excess of 150 ft will require lighting both day and night.

The horizontal illumination on the roadway for nighttime lighting should be at least twice that of the adjacent roadway illumination, and preferably more. Good recommended practice would be 5 to 10 footcandles.

Underpasses often do not have solid walls finished with high reflectance surfaces. The ceiling or overhead structure may be broken up by cross beams which make it impossible to create uniform brightness of the ceiling area. In this case, regardless of the lamp type that is chosen, the lighting fixtures should be equipped with some form of optical control that will direct the light to the roadway.

Underpass lighting fixtures may be either the fluorescent, high-pressure sodium, or mercury vapor lamp types.

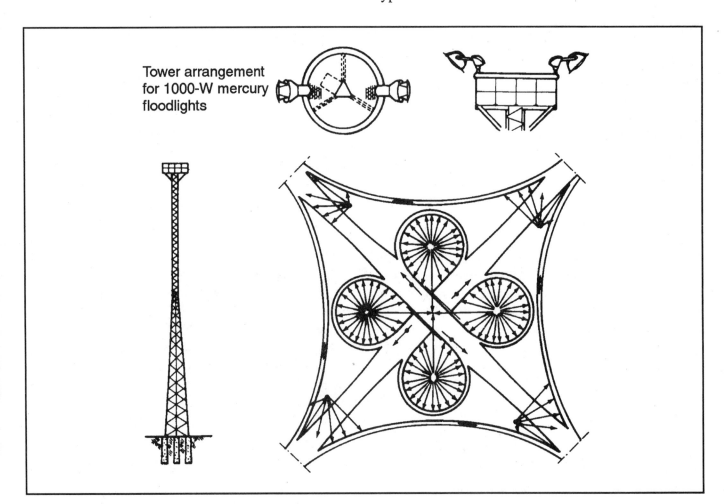

Tower arrangement for 1000-W mercury floodlights

Figure 6-48: Lighting interchange with a lighting tower.

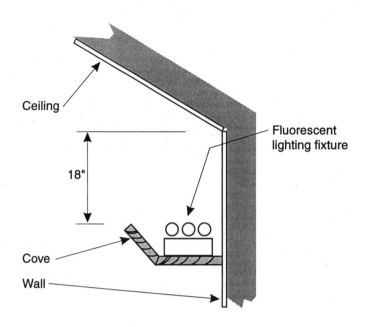

Figure 6-49: Cove lighting detail.

MISCELLANEOUS LIGHTING DETAILS

Cove Lighting Detail

The cove lighting detail in Figure 6-49 was used on a set of electrical drawings for a small church sanctuary. While the wood cove itself was detailed on the architectural drawings, a detail of the fluorescent fixture placement in relation to the cove, wall, and ceiling was considered necessary by the designer to obtain a proper installation from the contractor. The detail drawing shows bare fluorescent lighting fixtures mounted with one edge of the fixture 4 in from the wall and the other edge flush with the mounting board. Although not shown on this drawing, the cove was supported every 4 ft with steel brackets.

Deluxe warm-white fluorescent lamps were used for the church sanctuary lighting to provide a warm social atmosphere and also because warm-white lamps are more flattering to people's complexions than cool-white lamps are.

This type of lighting arrangement provided a suitable indirect lighting affect that is ideal for worship. A central chandelier controlled by a rheostat dimmer was also used in the sanctuary for decorative reasons as well as for supplementary light to be used in conjunction with the cove lighting.

Spire Lighting Detail

The church discussed in the preceding section also utilized a fiberglass spire (steeple) which required illumination to provide a glowing effect at night. Since the fiberglass spire was semitransparent, it would be best to illuminate the spire from the inside.

A narrow-beam spotlight was chosen which contained a 300-W lamp. However, the construction of the spire assembly made a detail drawing necessary to provide adequate instructions to the contractor. The detail selected is shown in Figure 6-50.

This detail shows the lighting fixture provided with a flexible cord and mounted to a 4-in-square cast-iron outlet box with 1¼-in knockouts. This box was used for support only — no electrical wiring was pulled into the box. A piece of 1¼-in rigid conduit is used to extend the lighting fixture to the proper location at the bottom of the spire. Again, this piece of 1¼-in conduit was used for support only and was secured to the building structure with two U-bolts.

The space directly under the spire did not allow enough room to provide an opening to allow complete access to

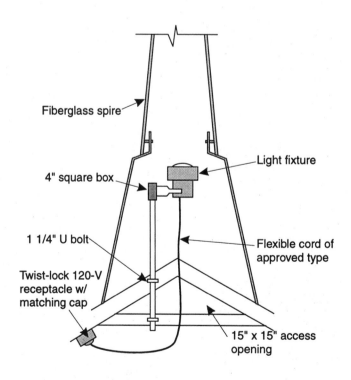

Figure 6-50: Spire lighting detail.

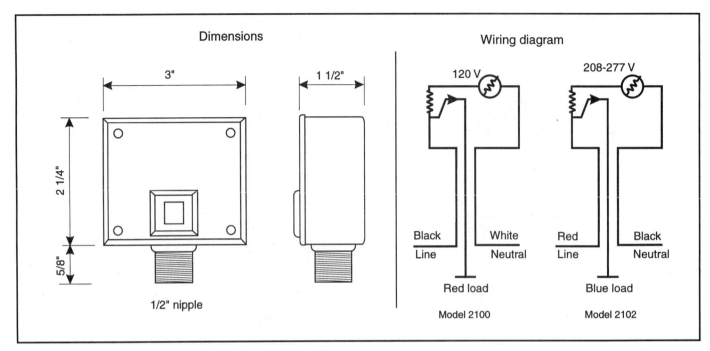

Figure 6-51: Photoelectric cell switch.

the lighting fixture. Only a 10-in × 15-in opening could be provided, and this size would not permit a person to enter in case the lamp had to be changed in the fixture. However, with the arrangement provided in the detail (Figure 6-49), the nuts on the U-bolts can be loosened and the ¼-in conduit will slip down far enough to allow the lamp to be changed. The conduit is then pushed up to its proper position, the U-bolts tightened, and the fixture is again in operation.

Since the fixture was provided with a flexible cord, an outlet box was necessary to furnish power to the fixture. This was provided with a twist-lock receptacle to prevent the plug from falling out. The circuit feeding this outlet was controlled by a photoelectric cell switch (Figure 6-51) which turned the light on at dusk and turned it off at dawn.

Pendant Mounting Detail

Figure 6-52 gives the installation details of a pendant-mounted lighting fixture used in the sales area of a hardware store with a cathedral ceiling. Note that the outlet box is mounted flush with the inside ceiling; a canopy with a swivel holding the stem or pendant is attached to the box; and the lighting fixture connects to the stem. Wires, connected to the fixture — are run inside the stem and connected to the branch circuit terminating in the outlet box.

When used on actual working drawings, the fixture type, length of stem, and size of the outlet box should be indicated on the detail; a complete description of the lighting fixture (catalog number, type and size of lamp, etc.) will normally appear in the project's lighting fixture

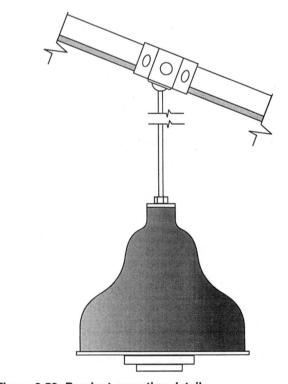

Figure 6-52: Pendant mounting detail.

schedule. If the swivel canopy does not come as a package with the lighting fixture, it should also be specified on the detail, lighting-fixture schedule, or in the written specifications by manufacturer and catalog number.

The installation is simple in that the outlet boxes, conduit, branch wiring, and so on, are roughed-in in the conventional manner before the finished ceiling is installed. Then when the ceiling is installed, each fixture is mounted to the outlet box as called for in the detail drawing. Since the stem is mounted to a swivel canopy, all fixtures will hang vertically regardless of the slope of the ceiling. However, to ensure that all lighting fixtures will be of proper height from the finished floor, dimensions should be provided, especially if the bottom of all fixtures are to be on the same plane.

Valance and Cornice Lighting

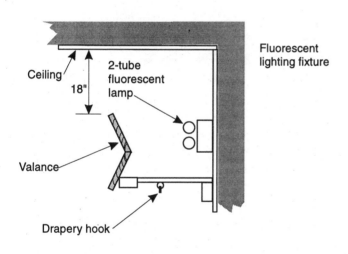

Figure 6-53: Detail of valance lighting.

While either cornice or valance lighting adds life to draperies or walls, both will also supplement the general room lighting level. Valances are nearly always used at windows, usually with draperies. They provide uplight, which reflects off the ceiling for general room lighting, and also down-light, for drapery accent. Figure 6-53 gives an example of valance lighting in a detail drawing used on an electrical working drawing.

Cornice lighting directs all light downward to give dramatic interest to wall coverings, draperies, and the like; this type of lighting is especially suited for rooms with low

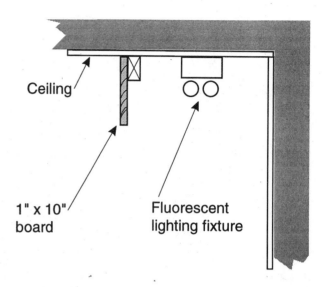

Figure 6-54: Construction detail for a typical cornice lighting arrangement.

ceilings. Figure 6-54 gives construction details for a typical cornice lighting arrangement.

Under-Cabinet Lighting Details

Bare single-tube fluorescent strip lighting installed either to the bottom of a cabinet toward the front or directly under the cabinet on the back wall is an excellent way to light kitchen or laboratory counters. Many drawing areas (Figure 6-55) and other work counters are being illuminated this way in order to use light on the task and to cut back on general illumination within an area.

Once the fixtures are installed, they may be painted to match the wall color if desired, but if they are mounted

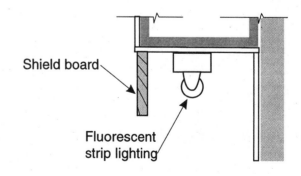

Figure 6-55: Detail of under-cabinet lighting with fluorescent lamps.

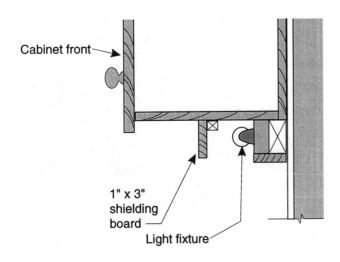

Figure 6-56: Detail of fluorescent fixture shield.

with a shield as shown in the detail in Figure 6-56, the fixture will be completely hidden from view. Another detail showing undercabinet lighting, using bare fluorescent strips, is shown in Figure 6-57.

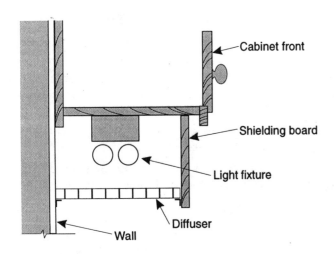

Figure 6-57: Under-cabinet lighting detail using bare fluorescent strips.

Direct-Indirect Cove Lighting

The floor plan in Figure 6-58 shows a partial plan view of a unique cove lighting arrangement which was designed to comply with the building owner's desire to have an entire lighting layout as inconspicuous as possible.

Laminated beams ran the entire width of the building over which a 2-in thick tongue-and-groove solid wood deck was installed; the bottom or interior side of the deck was painted white for light-reflection purposes. A cove was then constructed on each side of each beam, both of which ran the entire length of the beam. In this cove, alternating fluorescent and incandescent fixtures — as shown in the plan — were utilized. With this arrangement, the fluorescent lighting fixtures provided indirect general illumination by the reflected light off the ceiling. The recessed incandescent fixtures with adjustable lamps provided direct light for highlighting merchandise on the counters directly below the coves.

While the floor plan in Figure 6-57 shows the general location and arrangement of the fixtures, certain necessary construction details could not be shown in this view. Therefore, a detail drawing — like the one in Figure 6-58 on page 586 — was used to clarify the installation. When this detail is furnished with the floor plan, there should be little doubt as to how the fixtures should be installed.

Wooden Deck Installations

Many modern residential and commercial buildings are currently utilizing laminated beams in conjunction with a solid wood tongue-and-groove roof sheathing. In this type of construction, there is no practical way to conceal wiring with the solid deck. Consequently, all raceways and wiring have to be run exposed — either on the exterior or interior side of the solid-wood deck.

When run exposed on the bottom or interior side, surface-metal molding is normally the wiring method used. However, in a recent church installation that utilized laminated beams with a solid-wood cathedral ceiling, the architect did not want any wiring exposed within the sanctuary, yet relative large chandeliers were to be installed in rows on each side of the center ridge beam.

After carefully studying the architectural drawings, the electrical designer decided to specify drilled 4-in round holes through the wood deck from the top side of the roof to mount the lighting fixtures on the inside ceiling. An EMT raceway would then be used on the outside (top side) of the deck to connect the boxes and to run the conductors from outlet to outlet. Since an insulating sheathing 1-in thick was to be used over the outside wood deck, this sheathing would be channeled to accept the conduit and also to cover the tops of the outlet boxes. Once the built-up roof is installed, the entire installation would be concealed on both sides of the wood deck.

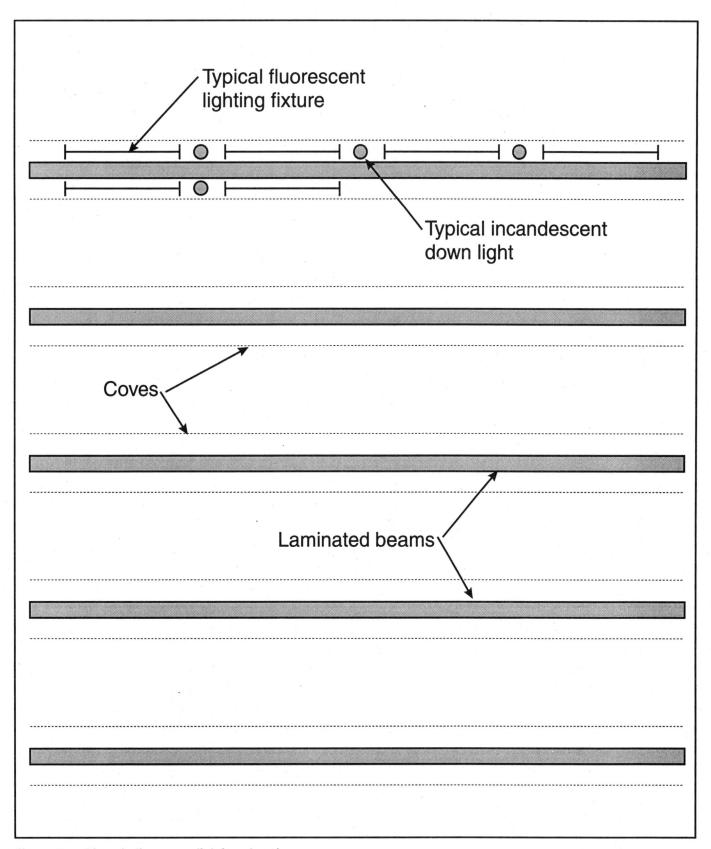

Figure 6-58: Direct-indirect cove lighting plan view.

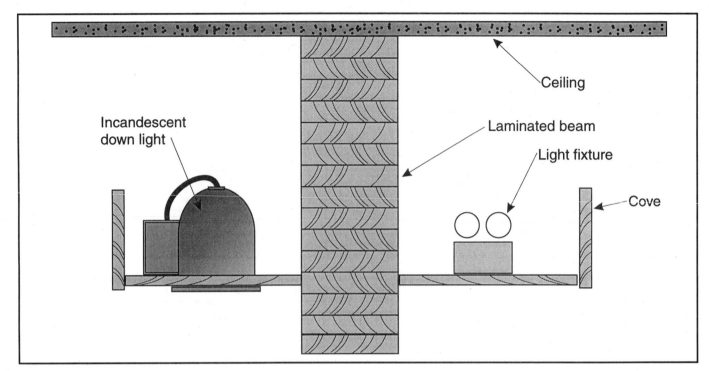

Figure 6-59: Cross-sectional detail drawing used to clarify the installation in Figure 6-57.

When preparing the drawings, the outlets were shown in a conventional manner on the floor plans. However, due to the special arrangement of the lighting outlet boxes through the wood deck, the designer used the detail in Figure 6-59 to clarify the installation. When such details are used, workers should know exactly how to install the system.

FIXTURE MOUNTING DETAILS

Most details involving lighting fixtures will be either connection diagrams or mounting details. The detail drawings to follow have been used on actual working drawings to better convey design information to workers on the job.

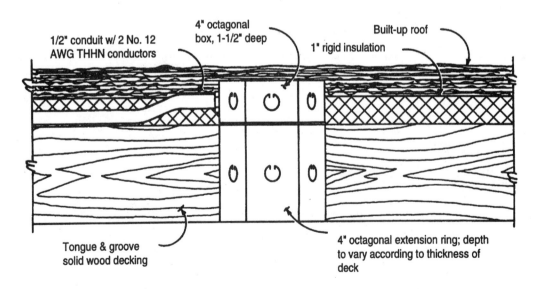

Figure 6-60: Outlet box detail for installations in a solid wood deck.

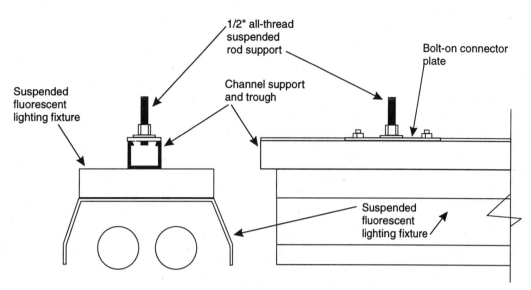

Figure 6-61: End and side view mounting details of a two-tube fluorescent lighting fixture.

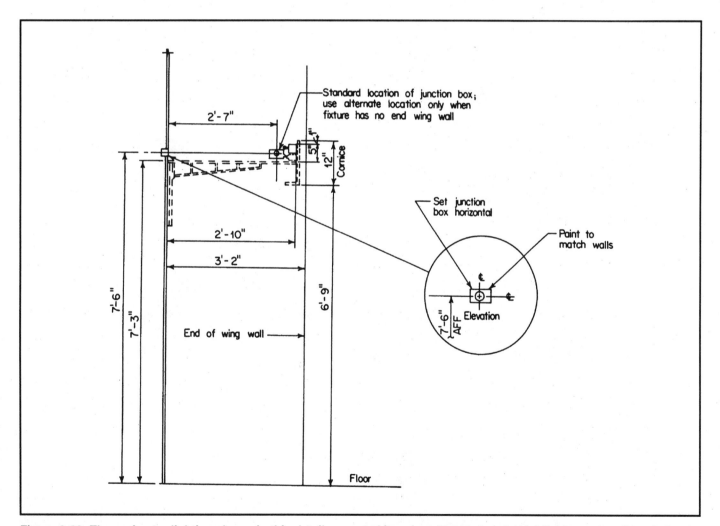

Figure 6-62: The perimeter lighting shown in this detail was used in a department store to highlight merchandise shelves around the perimeter of the store. Although all of these inexpensive fluorescent lighting fixtures utilize bare-tube lamps, the light sources are not readily visible to the public. *See* **Figure 6-63 for the floor plan layout.**

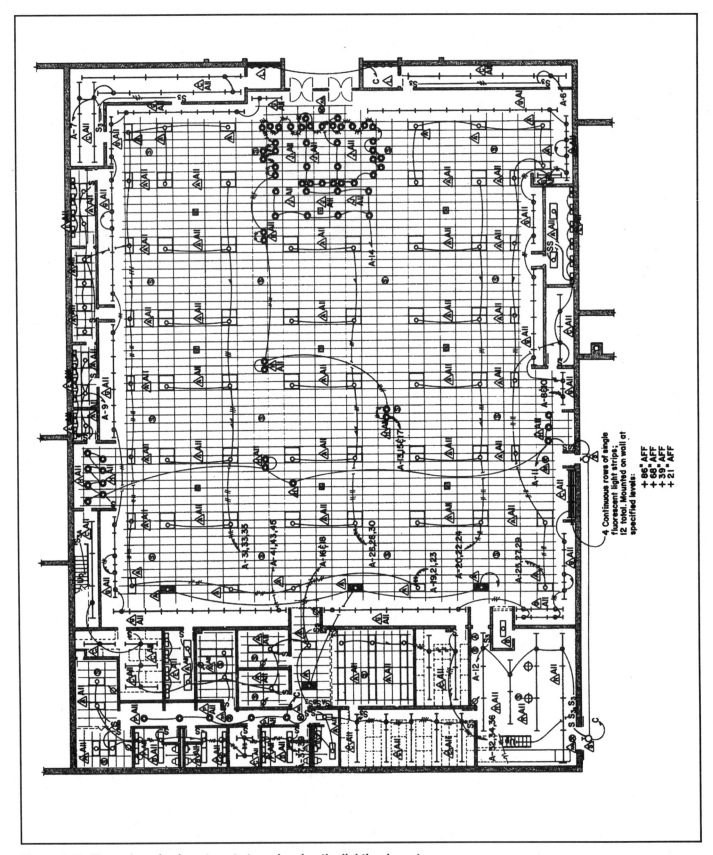

Figure 6-63: Floor plan of a department store showing the lighting layout.

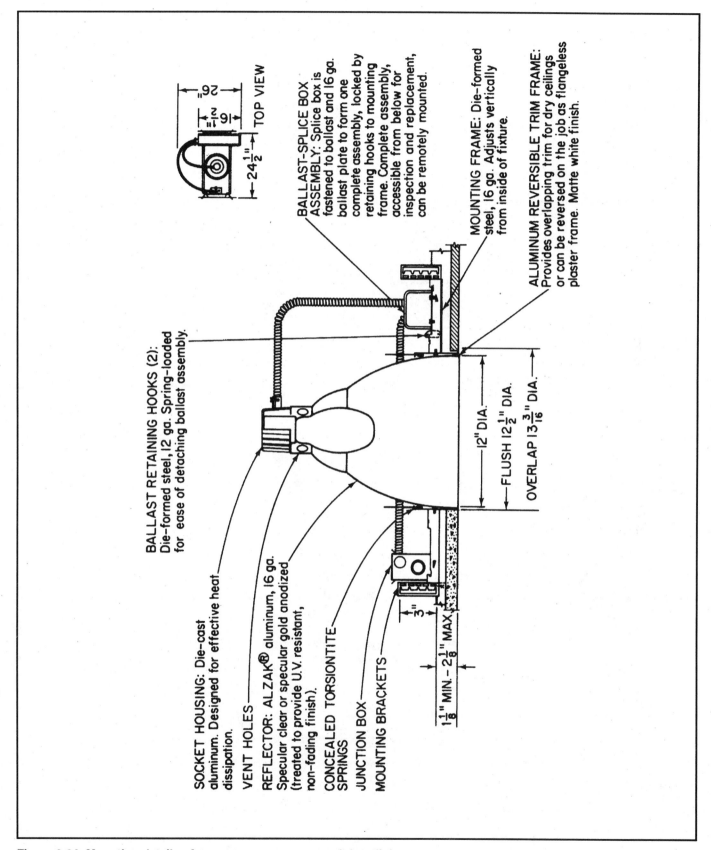

Figure 6-64: Mounting details of a mercury-vapor recessed downlight.

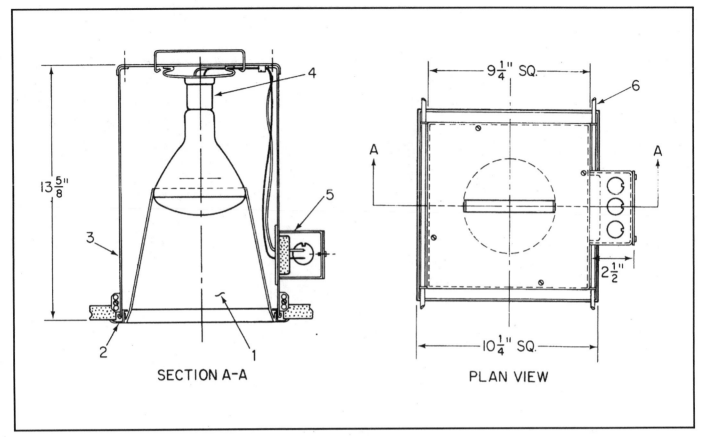

Figure 6-65: Section and plan view of a recessed spot light.

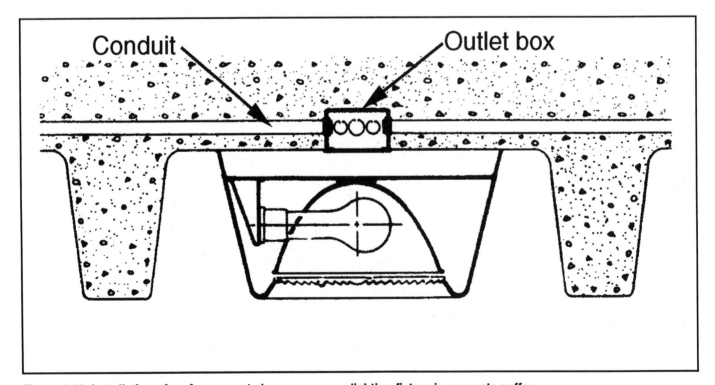

Figure 6-66: Installation of surface-mounted mercury-vapor lighting fixture in concrete coffer.

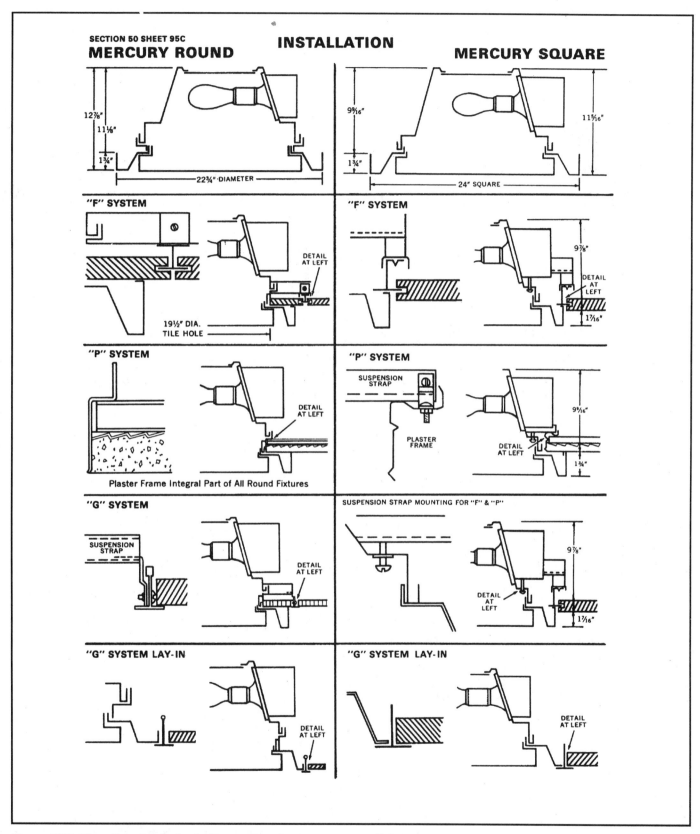

Figure 6-67: Mounting details for both round and square mercury fixtures.

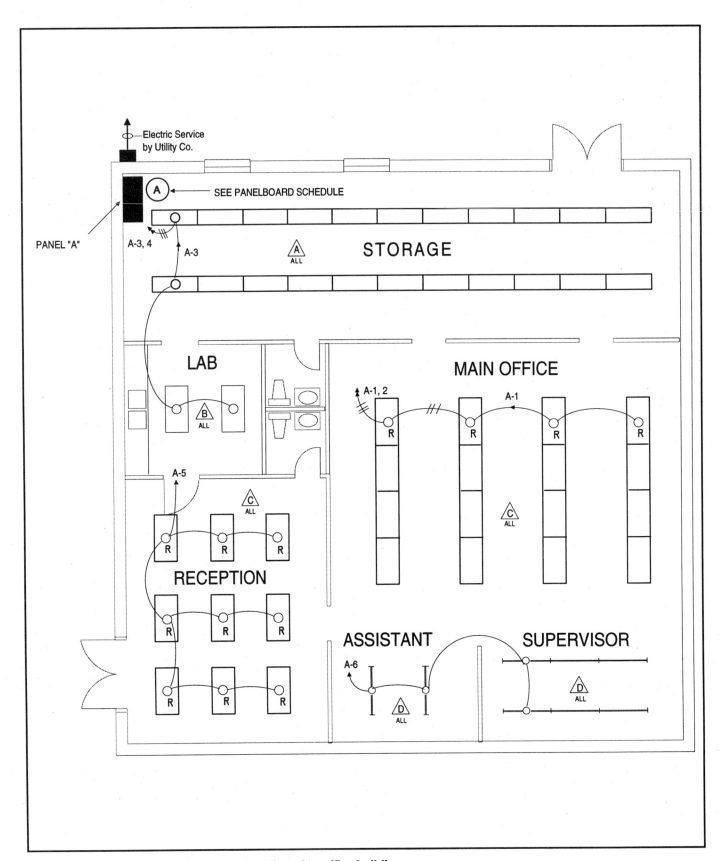

Figure 6-68: Lighting layout for a state auto licensing office building.

REMOTE-CONTROL SWITCHING

In applications where lighting must be controlled from several points, or where there is a complexity of lighting or power circuits, or where flexibility is desirable in certain systems, low-voltage remote-controlled relay systems have been applied. Basically, these systems use special low-voltage components, operated from a transformer, to switch relays which in turn control the standard line voltage circuits. Because the control wiring does not carry the line load directly, small lightweight cable can be used. It can be installed wherever and however convenient; that is, placed behind moldings, stapled to woodwork, buried in shallow plaster channels, or installed in holes bored in wall studs.

A basic remote-control switching system is shown in Figure 6-69; another is shown in Figure 6-70. In both circuits, the relay permits positive control for on and off. It can be located near the load or installed in centrally located distribution panel boxes, depending upon the application. Because no line voltage flows through the control circuits, and low voltage is used for all switch and relay wiring, it is possible to place the controls at a great distance from the source or load, thus offering many advantages.

Advantages of Remote-Control Switching

Since branch circuits go directly to the loads in this type of system, no line-voltage switch legs are required. This saves costly larger-conductor runs through all switches and saves installation time and costs if multipoint switches are used.

Protection of the low-voltage conductors is not required by the *NEC*; therefore, runs in open spaces above ceilings and through-wall partitions usually can be made without further protection. Even an outlet box is not required at the switch locations. When the electrician roughs-in the wiring, he or she merely secures a plaster ring, of the correct depth, at each switch location and usually wraps the low-voltage cable around a nail driven behind the plate. The switch and its cover are then installed after the wall is finished, leaving a neat installation.

Low-voltage switching is especially useful in rewiring existing buildings since the small cables are as easy to run as telephone wires. They are easy to hide behind baseboards or even behind quarter-round molding. The cable can be run exposed without being very noticeable because of its small size. The small, flexible wires are easily "fished" in partitions. While not recommended for perma-

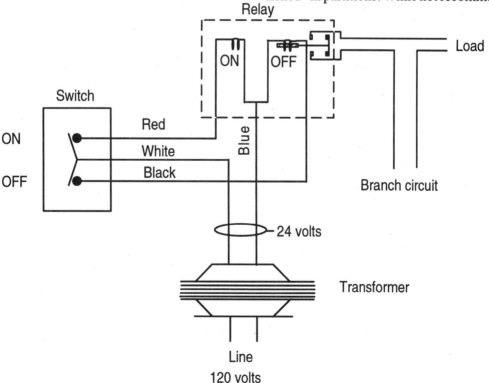

Figure 6-69: Basic layout of a remote-control switching system.

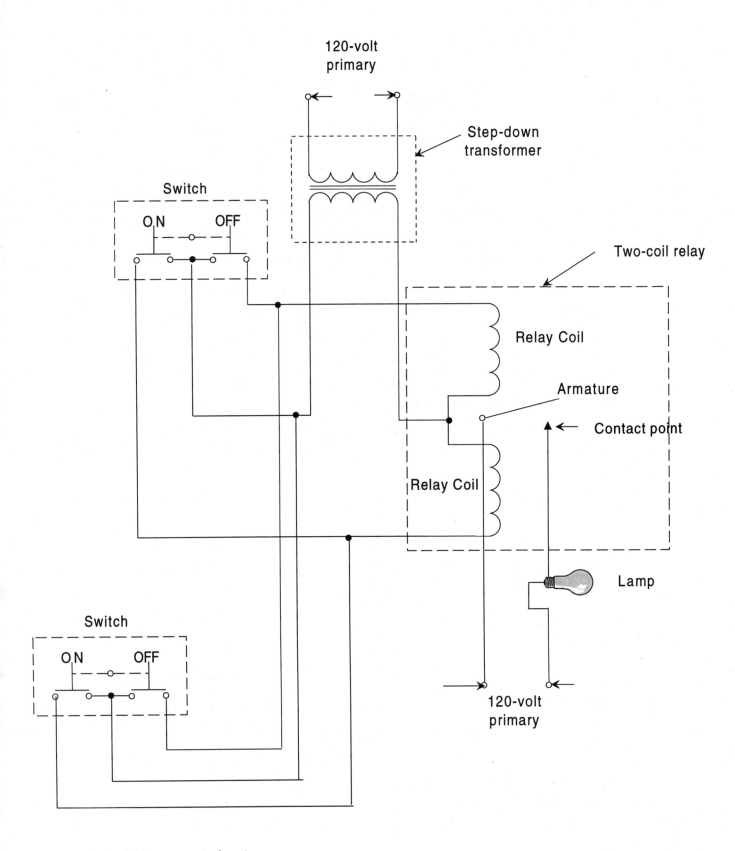

Figure 6-70: Basic remote-control system.

nent installations, low-voltage wiring can be run under rugs to switches located on a table and used for switching a relay controlling a tv outlet, for example.

Figure 6-71 shows several types of remote-control circuits. In the circuits shown, any number of on-off switches can be connected to provide control from many remote points. In a typical installation, outside lights could be controlled from any area within the building without the need to run full-size cable for three- or four-way switch legs through the building. By running a low-voltage line from each relay to a central point, such as an exit door, all relays can be operated from one spot with a selector switch; this allows quick, convenient control of all exterior and interior lighting and power circuits.

A motor-driven rotary switch is available that allows turn-on or shutdown of 25 separate circuits or more with the push of one button. Dimming can also be accom-plished by using the motorized control unit together with a modular incandescent dimming system.

System Components

The relay is the heart of a remote-control switching system. The relay employs a split low-voltage coil to move the line voltage contact armature to the ON (OFF) latched position. As illustrated in Figure 6-72 on the next page, the ON coil moves the armature to the right when a 24 VAC control signal is impressed across its leads. This is similar to a magnet attracting the handle of a standard single-pole switch to the ON position when energized. The armature (handle) latches in the ON position and will remain there until the OFF coil is energized, drawing the armature into the OFF position. Figure 6-73 on page 597 shows how a low-voltage relay is installed in an outlet box.

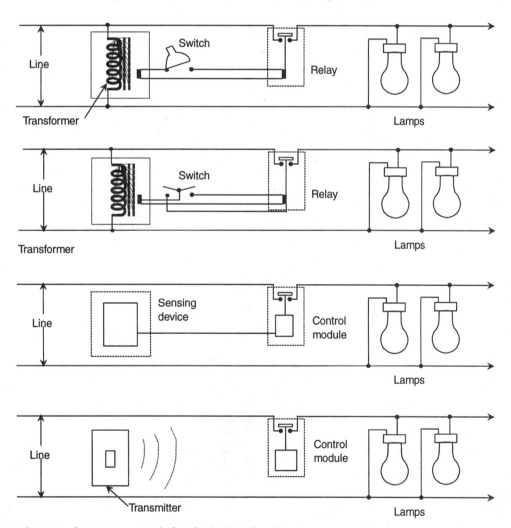

Figure 6-71: Several types of remote-control circuits and their related components.

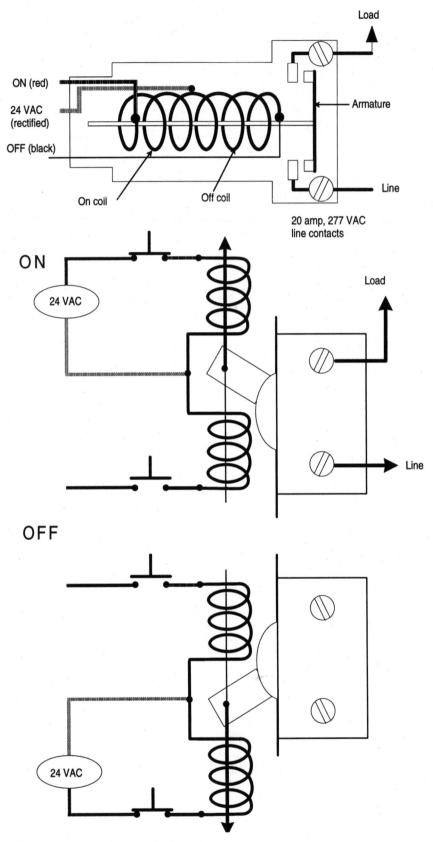

ON (red)

24 VAC
(rectified)

OFF (black)

On coil

Off coil

Load

Armature

Line

20 amp, 277 VAC
line contacts

ON

24 VAC

Load

Line

OFF

24 VAC

Figure 6-72: Relay operation.

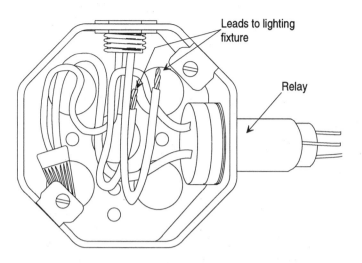

Figure 6-73: Relay installed in outlet box.

Power Supplies

Transformers designed for use on remote-control switching systems supply 24 VAC power to operate the relays and their controls. The relay and pilot switch power is rectified to extend their life; electronic control component power is not rectified. The ac input voltage to the power supply can vary by manufacturer, with the more common voltages being 120, 208, 240, and 277 V. The type of voltage used depends on the system. Figure 6-74 shows a control transformer along with connection details.

Switches

The standard low-voltage switch uses a rocker or two-button configuration to provide a momentary single-pole, double-throw action. Pushing the ON (OFF) button completes the circuit to the ON (OFF) coil of the relay, shifting the contact armature to the corresponding position. When the button is released, the relay remains in that position.

The pulse operation allows any number of switches to be wired in parallel as shown in Figure 6-75 on the next page. A group of relays could also be wired for common switch control by paralleling their control leads. Those relays would operate as a group.

Pilot switches include a lamp wired between the switch common (white) and a pilot terminal. The auxiliary contact in the relay provides power to drive this lamp when the relay is ON.

Controls

There are several variations of controls that are designed for use on low-voltage remote-control switching systems. A sampling of those available are described herein. However, there are many more variations and it is recommended that you refer to manufacturers' catalogs for a listing of all types available. Such catalogs and supplements are ideal learning "tools" and contain much valuable installation data.

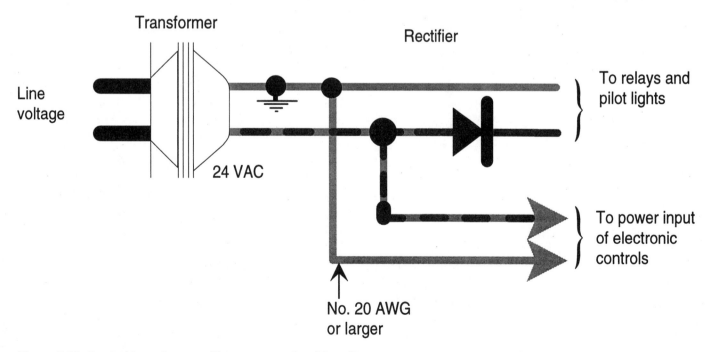

Figure 6-74: Control transformer with power-supply wiring diagram.

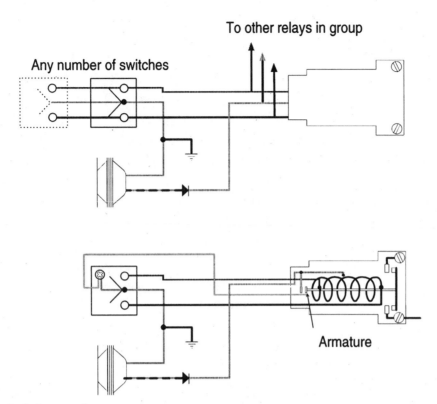

Figure 6-75: Switch operation.

Master Sequencer

The master sequencer allows relays to be controlled as a group while still allowing individual switch control for each. When the master switch is turned ON (OFF), the sequencer pulses each of its ON (OFF) relay outputs sequentially. A local switch can control an individual relay without affecting any other.

A second input channel allows timeclocks, building automation system outputs, photocells or other maintained contact devices to also control the sequencer. This provides simple automation coupled with local override of individual loads. *See* Figure 6-76.

Telephone Override Devices

Telephone override devices provide the same control functions as the sequencer; but in addition, it allows the occupant's Touchtone™ phone to be used in place of (or in addition to) hardwired switches as illustrated in Figure 6-77. Special function switches are available to allow the phone override function to be disabled or to be limited to ON overrides only.

Basic Operation Of Remote-Control Switching

Figure 6-78 shows a single-switch control of a light or a group of lights in one area. This is the basic circuit of relay control and is similar to single-pole conventional switch circuits. The schematic wiring diagram of this circuit is shown in Figure 6-79.

The addition of a timeclock and/or photoelectric cell allows applications in which any number of control circuits are turned on or off at desired intervals. For example, the dusk-to-dawn control of outdoor lighting circuits.

Multipoint switching, or the switching of a single circuit from two or more switches, is shown in Figure 6-80 along with its related wiring diagram. In conventional wiring, multipoint switching requires costly three-way and four-way switches, plus the extension of switch legs and traveler wires through all switches. One of the greatest advantages of relay switching is the low cost of adding additional switch points.

Figure 6-81 shows many lighting circuits being controlled from one location. In conventional line-voltage switches, this is accomplished by ganging the individual switches together. The same procedure can be employed with remote-control switches. When more than six

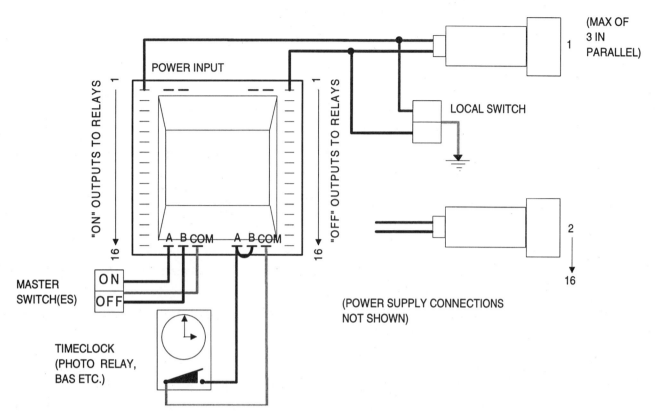

Figure 6-76: Master sequencer connection.

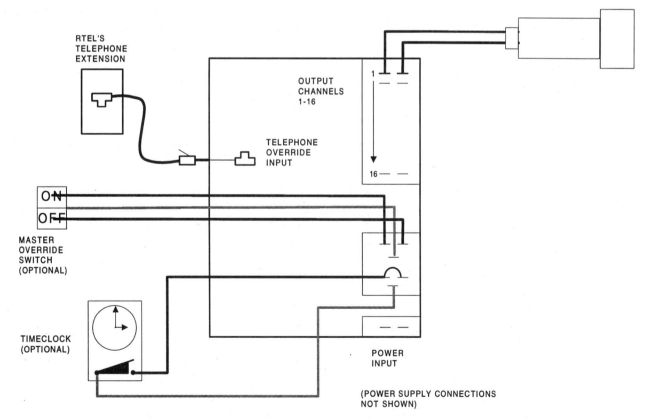

Figure 6-77: Telephone override circuit.

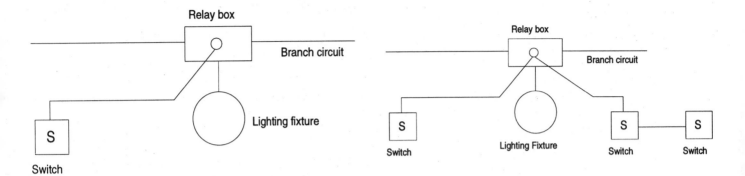

Figure 6-78: Single-switch control.

Figure 6-80: Multipoint switching.

switches are required, it is usually desirable to install a master-selector switch that permits the control of individual circuits.

Master control of many individual and isolated circuits is possible. Where more than twelve circuits are to be controlled from a single location, and where the selection of individual circuits is not required, then a motor-master control unit automatically sweeps 25 circuits from either ON or OFF at the touch of a single "master" switch, which can be located at one or more locations. When located in the master bedroom, it is possible to turn off all lights in the home with a single motion.

PLANNING A REMOTE-CONTROL SWITCHING SYSTEM

Low-voltage lighting control systems can be installed in both residential and commercial installations. The systems installed in each type of location may differ. The manufacturer's instructions and recommendations should always be consulted and adhered to, as should the blueprint specifications. The first step in the installation procedure, whether the building is residential or commercial, is to develop a plan. When laying out a remote-control switching system for a building, drawing symbols such as

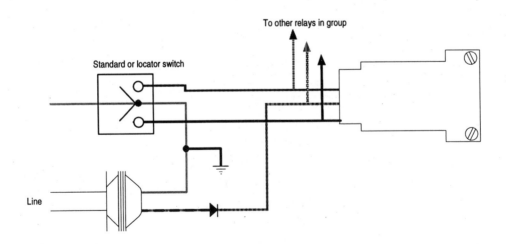

Figure 6-79: Schematic wiring diagram of single switches controlling each circuit.

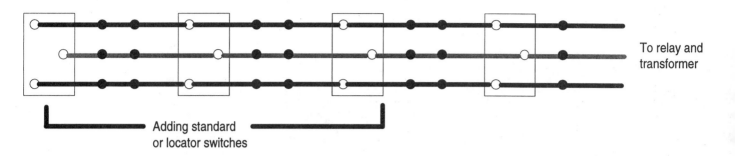

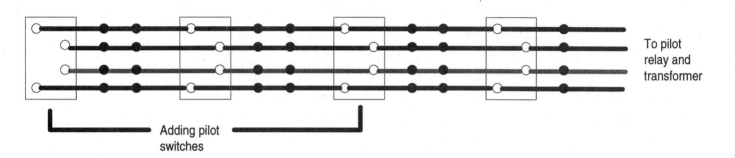

Figure 6-81: Switching circuits being controlled from one location.

the ones shown in Figure 6-82 are normally used to identify the various components. The drawing may be simple, requiring only the following:

- The location of each low-voltage control with a dashed line drawn from the switch to the outlet to be controlled.
- A schematic wiring diagram showing all components.
- The following note printed on the drawing:

"Furnish and install complete remote-control wiring system for control of lighting and other equipment as indicated on the drawings, diagrams, and schedules. System shall be complete with transformers, rectifiers, relays, switches, master-selector switches . . . wall plates, wiring, and connectors. All remote-control wiring components shall be of the same manufacture and installed in accordance with the recommendations of the manufacturer.

"Except where otherwise indicated, all remote-control wiring shall be installed in accordance with *NEC* Article 725, Class 2."

Design of Low-Voltage Switching

Many times the low-voltage lighting control system will be designed by architects, especially in new installations. However, in existing buildings, the electrician may be expected to design and install the system. If this situation occurs, the following steps, working from a set of blueprints, should be followed:

1. Assign each light or receptacle a specific relay. This can be accomplished by using the letters R1 for relay 1, R2 for relay 2, and so on.
2. Group together all the lights and outlets that are to be switched together.
3. Determine the number of relays that can be controlled by each control module.
4. Divide the number of relays by the number you computed in step 3. This is the number of control modules needed to control the relays.
5. Determine the location of the switch points on the building blueprints.
6. Indicate what lights and receptacles are being controlled by what relay. This step may be accomplished by indicating the group of lights on the blueprint. However, it is easier if a switch schedule is used.

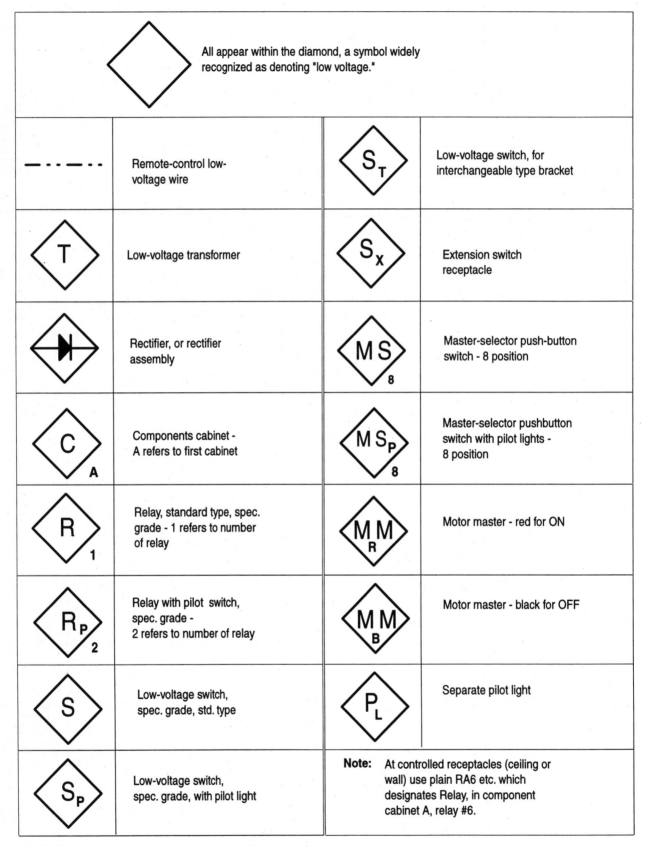

Figure 6-82: Recommended symbols for low-voltage drawings.

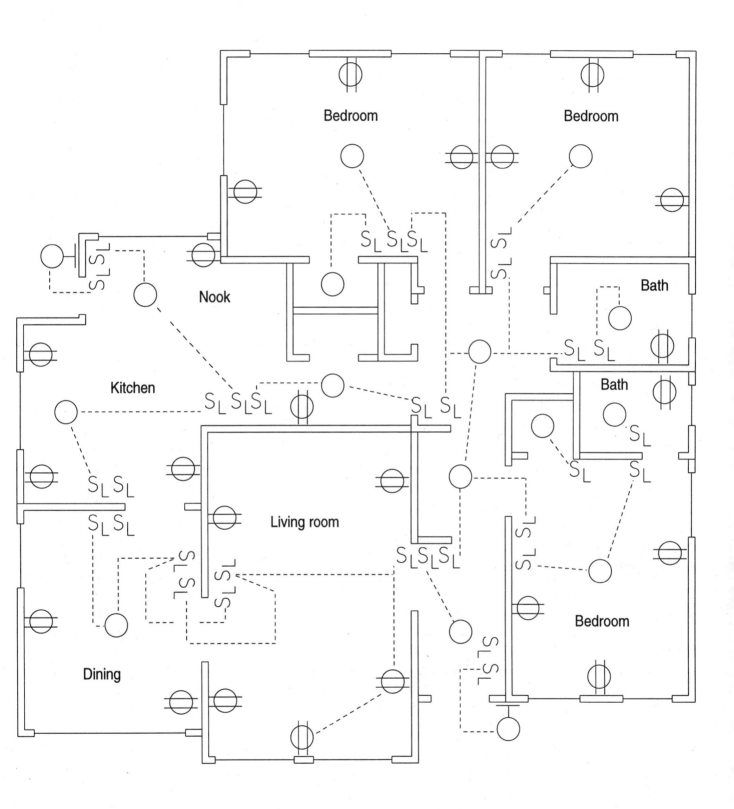

Figure 6-83: Low-voltage switching circuits laid out on a residential floor plan.

7. Review the blueprints or schedule to determine the number and size of the switching stations.

Figure 6-83 shows the floor plan of a residential building. Symbols are used to show the location of all lighting fixtures just as they are used to show the location of conventional switches. Line-voltage circuits feeding the lighting outlets are indicated in the same way (by solid lines) that they are in conventional switching. All of the remote-control switches, however, are indicated by the symbol S_L, instead of S, S_3, S_4, etc. Lines from these remote-control switches to the lighting fixtures that they control are shown by lines from the switch location to the item controlled.

Another method of showing low-voltage remote-control switching on a residential floor plan is shown in Figure 6-84 on page 605. Here the switches are shown by the manufacturer's catalog number (SF6, SF7, etc.) and each relay is numbered (R-1, R-2, etc.). When this type of layout arrangement is used a schedule is usually shown on the working drawings, or else placed in the written specifications to exactly identify each item.

A schematic wiring diagram, like the one in Figure 6-84, may sometimes be used for further details of all components and the related wiring connections. This wiring diagram aids the contractor or electrician in the installation of the system and leaves little doubt as to exactly what is required.

INSTALLATIONS

Before starting, consult both the *NEC* and local codes for installation procedures. Once the layout plan has been developed and materials are acquired, the installation of the system can begin. The rough-in instructions presented here are only an example. It is important to follow the manufacturer's recommendations and blueprint specifications.

Enclosure:. The enclosure should be placed in a location easily accessible but not hidden from sight. If at all possible, install a unit that is prewired. This will decrease installation time and also limit the possibility for mistakes. Note that, even though the panel may be prewired, the supply cables will still have to be installed. This should be done according to the manufacturer's recommendations.

Since heat will be produced by the enclosed control modules, vents should be placed around the enclosure to facilitate cooling. The location chosen for the enclosure should not hinder the cooling capabilities for which the panels are designed. The manufacturer may require that the high-voltage portion of the enclosure panel be mounted in the upper right corner. This is due to the *NEC* requirement that is designed to keep low-voltage conductors apart from high-voltage conductors. The panel itself should be mounted at eye level between two studs.

The power supply should be mounted on the enclosure and the power connections made. The number of power supplies needed per system will depend on the size of the system and the capabilities of the power supply. Some power supplies are capable of supplying up to ten modules if the average number of LEDs per channel does not exceed 2 or 2.5. Be sure to check the manufacturer's specifications to ensure that the power supply will not be overloaded.

Switch points: The first step when installing the switch points is to mark their location. As with ordinary switches, the switch points should be located as specified by the plans or specifications. Once the location has been determined, mount either a plaster ring or nonmetallic box. A box is not needed in most frame-construction installations.

A plaster ring or Romex staple should then be stapled part way in a stud, behind the plaster ring or nonmetallic box. The preparation for running the low-voltage wiring is the same as for running 120-V wiring. If the edge of the hole is closer than $1\frac{1}{4}$ in from the nearest edge of the stud, a $\frac{1}{16}$-in steel nail plate or bushing must be used to protect the cable from future damage by nails or screws.

The wires are then pulled from the switch station to the panel enclosure. They should come from the left side if the manufacturer recommends that the enclosure be installed with the high-voltage section in the upper right corner. There should be approximately 10 to 12 in of extra wire left at each switch location and 12 to 18 in left at the panel enclosure. A wire-pulling technique that reduces installation time follows.

Multiconductor cable is frequently used due to simpler

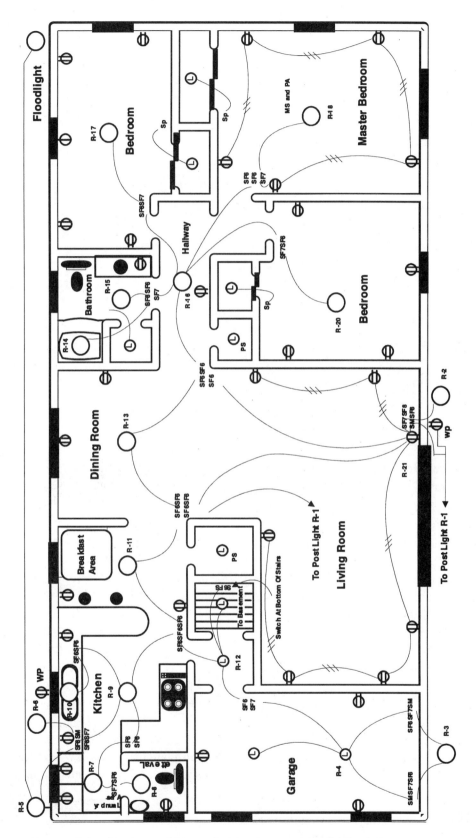

Figure 6-84: Alternate method of showing low-voltage switching on a residential floor plan.

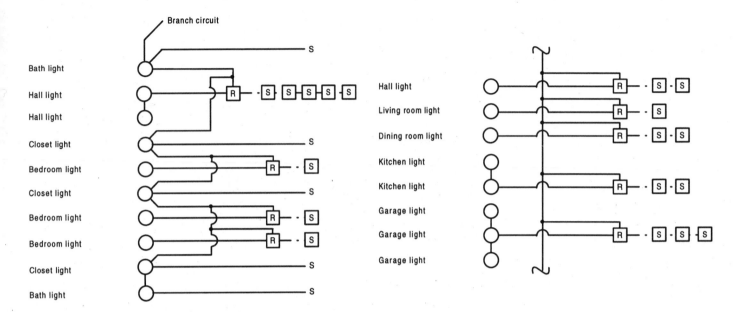

Figure 6-85: Schematic wiring diagram of a low-voltage switching system.

mon lead will have the same color throughout the system and future connection mistakes will be minimized.

The next step is to select a multiple-switch circuit. Pull a different colored wire from the enclosure to the farthest switching location in this circuit. Working back towards the panel enclosure as before, loop the wire to the other switch locations in the multiple-switch circuit. Thus, there are two leads going to each switch with one of the leads common to all the switches in the system. However, if the switches contain a pilot light, two additional No. 18 wires must be run to the LED. The wires are connected to the appropriate pilot light transformer located in the panel enclosure.

If possible, use different colored wires for each run. This will limit confusion in the future when connecting the individual circuits to power. Wire schedules should be used to limit hook up mistakes and give a clearer picture of the system.

Remember that the colors for the control wiring should not be the same as the colors for the dc power and common conductors. Additional marking methods should also be used for future reference. Note that if the wires are marked using adhesive tabs or punch cards, they should be placed

in a plastic bag and shoved behind the ring or secured into the plastic box. This will prevent the markings from being covered by paint or plaster when the walls are finished.

The low-voltage wiring should also be kept apart from the high-voltage wiring whenever possible. If the cables must be placed in parallel runs, they should be separated by at least 6 in, otherwise the noise generated from the high-voltage conductors could seriously interfere with the operation of the low-voltage system.

Power and finish trim: Connecting the circuits to power is now a fairly simple task. Each wire from the switch is connected to a relay with the other side of the relay connected to the common of the dc supply.

When the walls of the building are finished, the switches are connected to the conductors located at each switch point location. Wall plates are then added. A common type of wall plate used with low-voltage lighting control systems requires no screws since it can be snapped into place.

RELAY TROUBLESHOOTING

Electromagnetic contactors and relays of various types are used to make and break circuits that control and protect the operation of such devices as electric motors, combustion and process controls, alarms and annunciators. Successful operation of a relay is dependent upon maintaining the proper interaction between its solenoid and its mechanical elements; that is, springs, hinges, contacts, dashpots, and the like.

The operating coil is usually designed to operate the relay from 80% to 110% of rated voltage. At low voltage the magnet may not be strong enough to pull in the armature against the armature spring and gravity, while at high voltages excessive coil temperatures may develop. Increasing the armature spring force or the magnet gap will require higher pull-in voltage or current values and will result in higher drop-out values.

When sufficient voltage is applied to the operating coil, the magnetic field builds up, the armature is attracted and begins to close. The air gap is shortened, increasing the magnetic attraction and accelerating the closing action, so that the armature closes with a snap, closing normally opened contacts and opening normally closed contacts. Binding at the hinges or excessive armature spring force may cause a contactor or relay that normally has snap action to make and break sluggishly. This condition is often encountered in relays that are rarely operated. Contactors and relays should be operated by hand from time to time to make sure their parts are working freely with proper clearances and spring actions, but this procedure should be performed only by qualified personnel.

GENERAL FLOODLIGHTING

The purpose of exterior floodlighting is to extend beyond sunset the usefulness and attractiveness of any given outdoor space or object. Whether the area is a sports field for amateur or professional engagements, a parking lot, shopping center, or backyard, properly applied lighting ensures safety, convenience, and a pleasing atmosphere.

Floodlighting Calculations

Since floodlighting can encompass so many variations, and since the location of the floodlight relative to the object to be illuminated can be in any plane, any size, and any distance, standardization of design procedures is difficult. There are, however, certain fundamental laws of illumination which may be followed in designing floodlight installations.

The three most commonly used systems for floodlight calculations are the point-by-point method, the beam-lumen and the watts per square foot method. The point-by-point method, as covered later in this chapter, permits the determination of footcandles at any point and orientation on a surface. This method is valuable, since it permits a visualization of the degree of lighting uniformity realized for any given set of conditions. The beam-lumen method is very similar to the method covered in Module 01212 for interior lighting, except that it must take into consideration the fact that floodlights are not usually perpendicular to the seeing task, but instead are aimed at various angles to the surface.

Beam-Lumen Method

Beam lumens are defined as the amount of light that is contained within the beam spread of the floodlight. The lamp lumens, as found in lamp data tables, multiplied by the beam efficiency of the floodlight will give the beam lumens.

The coefficient of beam utilization (CBU), written as a decimal fraction, expresses the following ratio:

$$CBU = \frac{LA}{TBL}$$

where,

   CBU is coefficient of beam utilization

   LA is lumens reaching area

   TBL is total beam lumens

This is the percentage of the beam that falls on the area to be lighted. It can vary from 60 to 100 percent and can be accurately determined only through extensive calculations. Some guides can be established, however, which will allow a fairly accurate choice of beam utilization. In general, the larger the area to be illuminated, the higher the percentage of beam utilization.

The correct beam spread, too, is very important in obtaining the highest percentage of beam utilization. If the beam is wider than necessary, excessive light will be dispersed off the area and the beam utilization will be low. This is illustrated in Figure 6-86. Note that the beam on the left in this illustration is too wide for the surface to be illuminated. As a result, light is wasted. However, by using a narrower beam, as shown in the right illustration, most of the light beam is utilized to illuminate the surface; only a very small percentage of the beam is wasted light.

Figure 6-87 gives a comparison of approximate percentages of beam utilization factors and will serve for

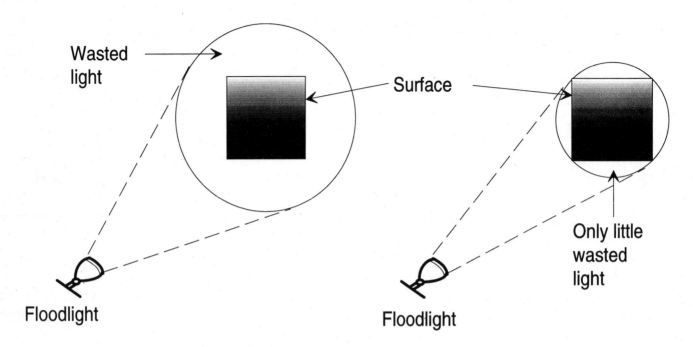

Figure 6-86: Beam spread affects the percentage of beam utilization.

most floodlight applications. Note that the beam utilization factor increases as the number of floodlights directed onto a surface increases.

In Figure 6-88, the floodlight is directed at point "C" on the sign with the following angles:

E(L)C = 20° Vertical

A(L)C = 10° Vertical

A(L)B = 40° Horizontal

C(L)D = 43° Horizontal

E(L)F = 50° Horizontal

Light distribution curves for most floodlights are available upon request. A typical curve is illustrated in Figure 6-89.

All angles found by the floodlight in question to given points on the sign are then plotted on the grid of the isocandela curve. Because of the manner in which

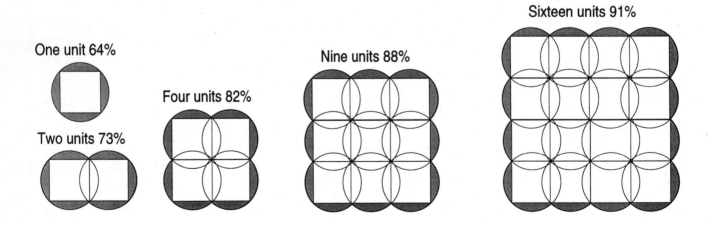

Figure 6-87: Beam utilization comparison (percentages indicate beam utilization of floodlights).

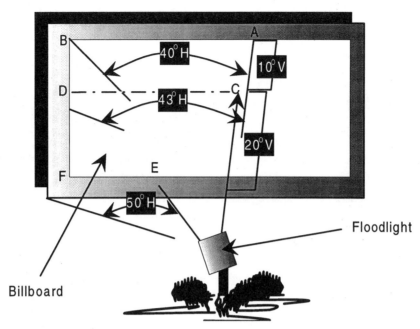

Figure 6-88: Method of calculating beam utilization.

floodlights are photometered, all horizontal lines parallel to a line perpendicular to the beam axis appear as straight horizontal lines on the grid. All vertical lines through the beam axis appear slightly curved.

Therefore, all the lumens within the solid line ABEF fall on the sign. This totals 879 lumens and when doubled, to account for the other half of the beam, gives a total of

1758 lumens falling on the sign. Since the total beam lumens is 2748, and the total falling on the sign is 1758, we substitute these values in the equation:

$$CBU = \frac{1758}{2748} = 0.64 \text{ or } 64\%$$

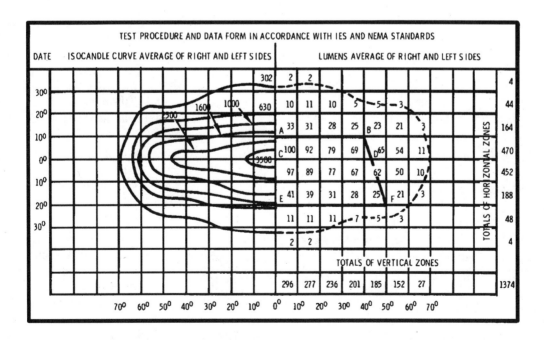

Figure 6-89: Isocandela curve of 300-W wide-beam reflector.

Type of Lamp	Maintenance Factor
Incandescent	0.75
Quartz	0.85
Clear and color-improved mercury	
175 to 700 W	0.75
1000 W	0.70
White mercury	
175 to 700 W	0.70
1000 W	0.65
Metal Halide	0.65
Lucalox (sodium)	0.75

Figure 6-90: Maintenance factor of various lamps.

The total floodlights required for a desired level of illumination may be found by the equation:

$$NF = \frac{A \times DF}{BL \times CBU \times MF}$$

Where (in the previous equations),

NF is number of footcandles

A is area

DF is desired footcandles

BL is beam lumens

CBU is coefficient of beam utilization

MF is maintenance factor

The maintenance factor figure indicates that light output from floodlights drops off as they continue operating in service. One reason is because of dirt accumulation on the lens of the floodlight. This figure, of course, varies with the atmosphere in which the units are operating. Another reason is because of the drop-off in lumen output of the lamps as they operate throughout life. The light output from some lamps drops off just slightly, while others have a higher drop-off, as may be seen in lamp-data catalogs. The table in Figure 6-90 lists the suggested total maintenance factors for various types of lamps.

Using the Beam-Lumen Method

An active outside storage area, 40 feet wide by 80 feet long, at an industrial plant is to be illuminated for an average of 2 hours per night, 5 nights a week. Here are some of the items specified by the plant owners:

- The system must be switched on and off several times during the night.
- It is adjacent to a building 30 feet high.

Here are the steps required in designing the most practical lighting system using the Beam Lumen Method:

Step 1. *Determine the Level of Illumination:* The recommended illumination levels for many floodlighting applications are listed in IES charts, available from the Illuminating Engineer's Society. These listings give a recommended illumination level for an active storage yard at an industrial plant as 20 footcandles.

Step 2. *Determine the Type of Lamps to be Used:* Due to the characteristics of the area, and especially due to the "switching" requirements, either heavy-duty incandescent or quartz lamps should be selected.

Step 3. *Determine the Type of Floodlight Fixture:* A wide-beam floodlight which contains a 1500-W quartz lamp is selected from the manufacturer's catalog. The floodlight beam lumens are also given in the catalog and are 23,338 for each fixture.

Step 4. *Determine the Coefficient of Beam Utilization and Maintenance Factor:* By referring to

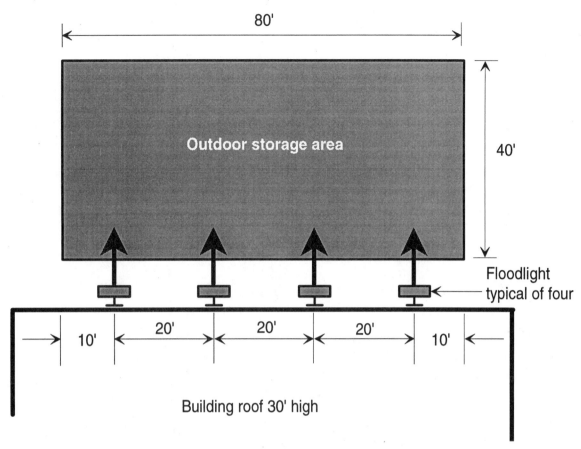

Figure 6-91: Floodlights mounted on roof of building adjacent to the area to be illuminated.

Figure 6-86, we can estimate the coefficient of beam utilization as 73% and a maintenance factor of 0.85 (from the table in Figure 6-90).

Therefore, the total number of floodlights may be found by substituting these values in the equation:

$$TFL = \frac{A \times IL}{BL \times CBU \times MF}$$

where,

TFL is total floodlights required

A is area in square feet

IL is illumination level

BL is beam lumens

CBU is coefficient of beam utilization

MF is maintenance factor

Floodlights can be mounted on the roof of the building adjacent to the area and spaced as shown on the drawing in Figure 6-91.

Watts Per Square Foot Method

For a quick and reasonably accurate way to determine the number of floodlights required for a lighting application under 10 footcandles, the "watts per square foot method" of calculation may be used. The equation is:

$$TFL = \frac{A \times IL \times WSF}{LWF}$$

where,

TFL is total floodlights required

A is area in square feet

IL is illumination level

WSF is watts per square foot factor

LWF is lamp watts per floodlight

This equation eliminates the use of technical lighting terms in the calculations. It may be used as follows:

Step 1. Determine the proper illumination level.

Step 2. Select the type of light.

Characteristics	Indandescent	Quartz	Mercury	Metallic Additive	Fluorescent
Initial cost	Low	Low	Higher	Higher	Higher
Power consumption (for equal light)	Medium to high	Medium to high	Low	Low	Low
Fixture size	Medium	Small	Medium	Medium	Large
Long burning house per year (over 1000)	Fair	Fair	Good	Good	Good
Color definition	Good	Very good	Fair	Good	Fair
Location considerations	Fair	Fair	Good	Good	Fair
Beam control	Very good	Good	Fair	Good	Poor
Cold weather operation	Very good	Very good	Good	Good	Fair
Long-range projection (narrow beam)	Best	Fair	Fair	Fair	Poor
Medium-range projection	Good	Good	Good	Good	Fair
Annual operating cost	Medium	Medium	Low	Low	Low

Figure 6-92: Light source selector table. *Courtesy Crouse-Hinds.*

Step 3. Select type of floodlight by using manufacturers' catalogs to select the proper floodlight for the given application. The table in Figure 6-92 is listed in the Crouse-Hinds Outdoor Lighting and Selector Guide and is typical of those found in manufacturers' catalogs and data.

Step 4. Determine the number and the placement of floodlights and poles.

In the watts per square foot formula, lamp lumens, floodlight beam efficiency, and lumen maintenance factors are all combined in an estimated utilization factor which will be known as the WSF factor. Four such factors are required in order to provide reasonable accuracy in different size areas. These are shown in Figure 6-93.

For example, using the watts per square foot formula, determine the quantity, wattage, and positioning of flood-

Area	WSF
Small area (1000 to 3000 ft^2)	0.16
Medium area (3000 to 20,000 ft^2)	0.11
Large area (20,000 to 80,000 ft^2)	0.08
Extra-large area (over 80,000 ft^2)	0.06

Figure 6-93: WSF factor for different areas.

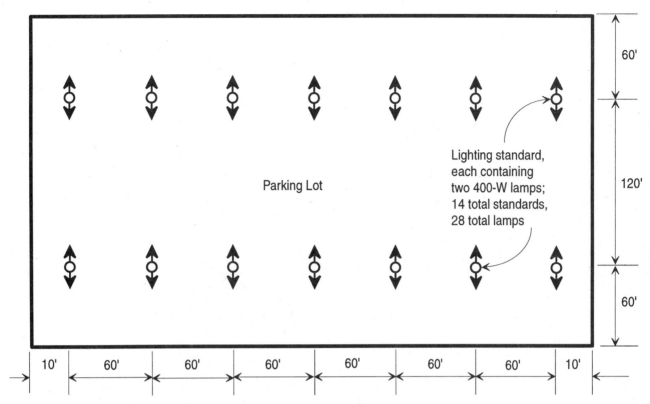

Figure 6-94: Pole and fixture layout for parking lot.

lights required to light a shopping center parking area which is 240 feet wide and 420 feet long. The floodlights will operate approximately 2000 hours per year.

Design data recommends an illumination level of two footcandles. We will use this figure for the design under consideration.

From the data in Figure 6-93 we concluded that either mercury or metallic additive lamps would be the best choice. For this application, use the metallic additive type.

We will choose a suitable lighting unit which will handle a 400-watt metallic additive type light.

By substituting our known value in the watts per square foot equation we have:

$$\text{TFL} = \frac{90,800 \times 2 \times 0.06}{400} = 27.2 \text{ floodlights}$$

Using 28 floodlights, they may be laid out as shown in Figure 6-94.

Chapter 7

Special Systems

The need for a standby source of electric power during a normal power source outage becomes increasingly evident as one considers the logic behind such an installation. Normal electric power may be interrupted at any time and for many reasons. Most of the time, the cause is nothing more than a mechanical breakdown that will last only a few minutes. Statistics show, however, that power outages are often accompanied by other dangers to our physical well being. Storms, hurricanes, floods, and explosives not only leave power lines down but a trail of property damage and personal injury as well.

It is precisely during these critical periods that we need the vital public services which rely on electric power to operate, services such as emergency lighting, power for elevators, police and fire department equipment, and life-saving equipment in hospitals.

Some states demand that all public buildings (schools, offices, etc.) have a standby electric power source. Banks rely heavily on standby electric power to ensure that their alarm systems will operate at all times and under any conditions. Some firms connect the majority of their electric lighting, office machines, and their PBX telephone switchboard to a standby electric power source — either battery banks or gasoline engine generators.

Emergency generators are also utilized extensively where ventilating or refrigerating apparatus is necessary. Hotels are required by law to have some source of emer-

gency power to operate lights, elevators, and other services essential to the safety and convenience of guests.

In high-rise office buildings, standby electric generating sets provide power to operate elevators, sump pumps, fire pumps and a certain number of emergency lights on each floor. In fact, standby electric power systems are increasingly needed in all public buildings as the public depends more and more on electrical power for work and everyday living.

NEC Articles 700 and 750 give regulations governing the design, installation, and testing of emergency and standby generating plants. It is suggested that these articles be read thoroughly before designing such a system or preparing written specifications.

The most important factor to consider in designing a standby electric power system is the size or capacity required. Therefore, the designer must be aware of the ratings and features that meet the requirements of the installation to satisfy any special needs and functions. A careful study to determine what equipment must operate during a possible outage and its total load will show exactly what degree of service the standby system should maintain.

Equipment schedules should be used where possible, listing the equipment, catalog numbers and the like, either on the drawings or in the specifications.

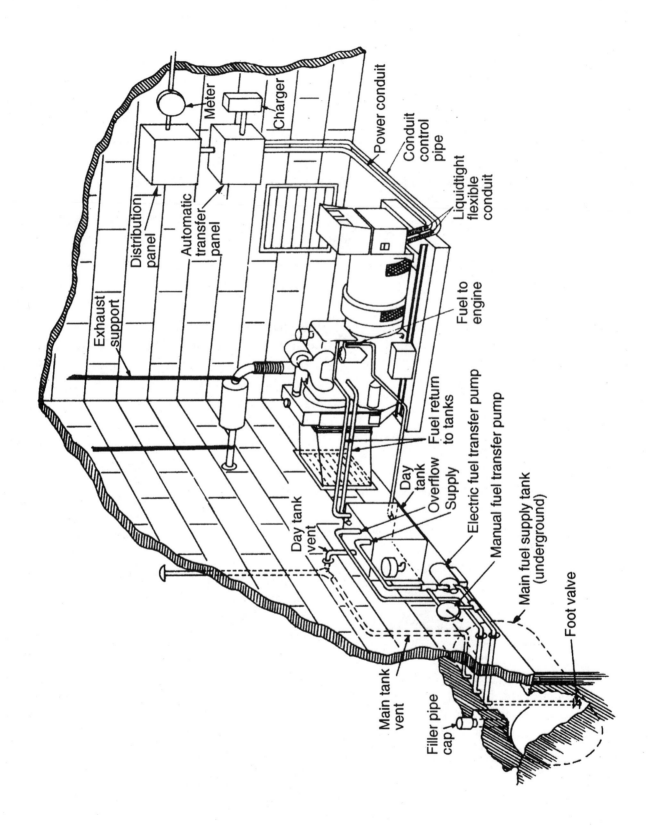

Figure 7-1: Key installation points for an emergency standby electric system.

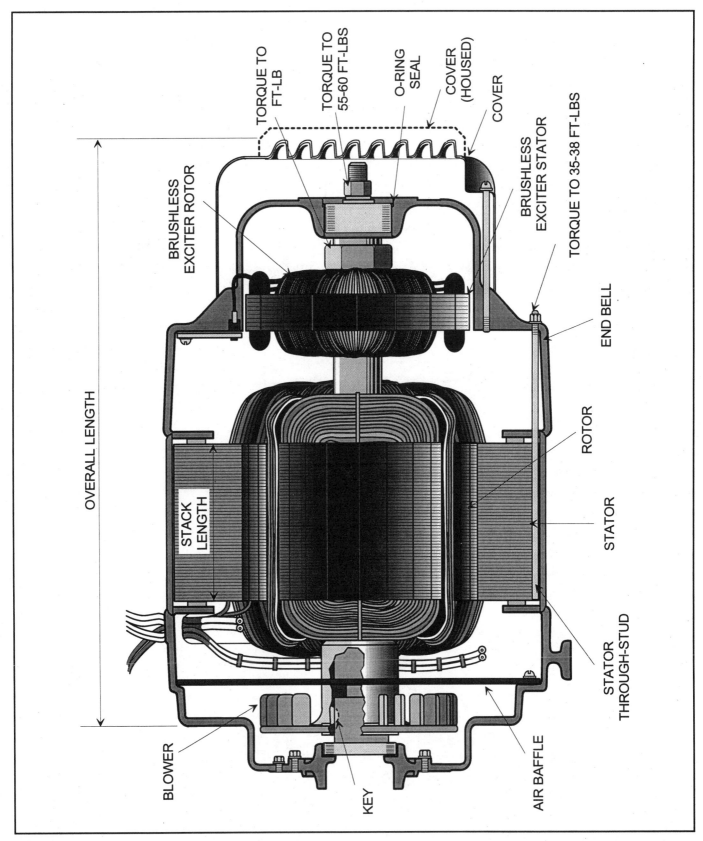

TORQUE TO FT-LB

TORQUE TO 55-60 FT-LBS

O-RING SEAL

COVER (HOUSED)

COVER

BRUSHLESS EXCITER ROTOR

BRUSHLESS EXCITER STATOR

TORQUE TO 35-38 FT-LBS

END BELL

OVERALL LENGTH

STACK LENGTH

ROTOR

STATOR

STATOR THROUGH-STUD

BLOWER

KEY

AIR BAFFLE

Figure 7-2: Cross-sectional view of a typical 4-pole, revolving field, brushless exciter alternator. Both single- and three-phase models are in common use.

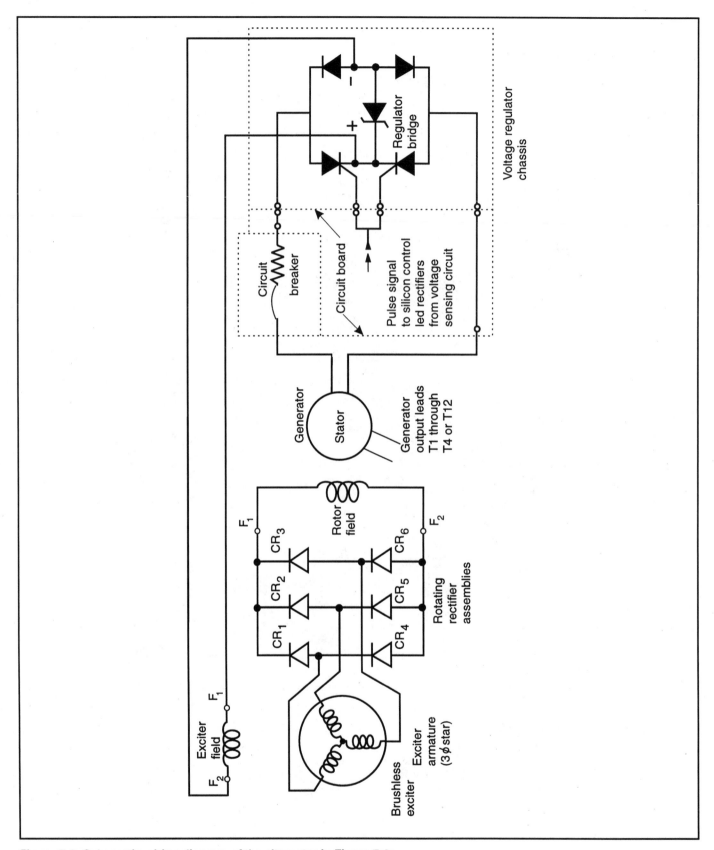

Figure 7-3: Schematic wiring diagram of the alternator in Figure 7-2.

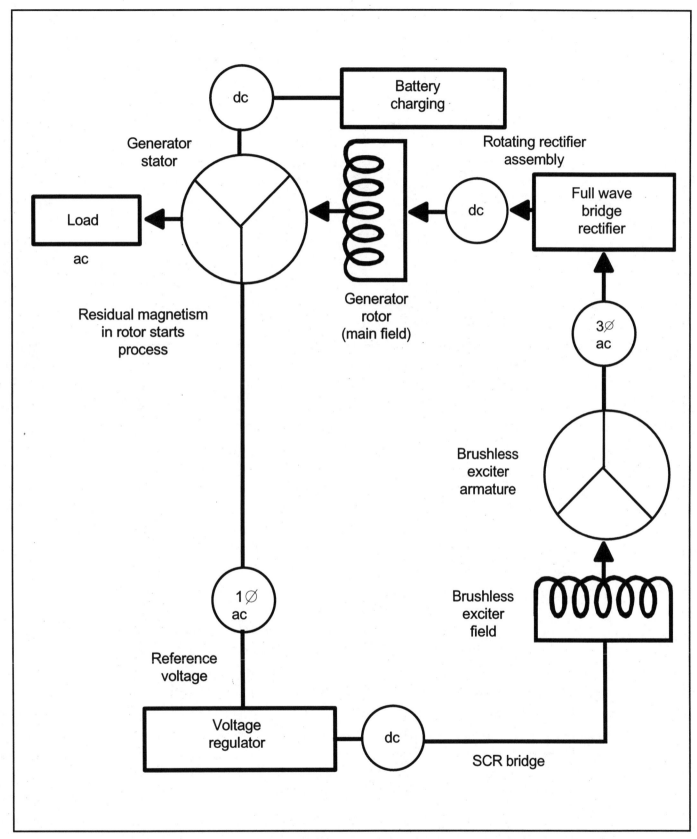

Figure 7-4: Before a generator will produce voltage, it is necessary for exciter currents to be produced. The steps of voltage buildup are shown here.

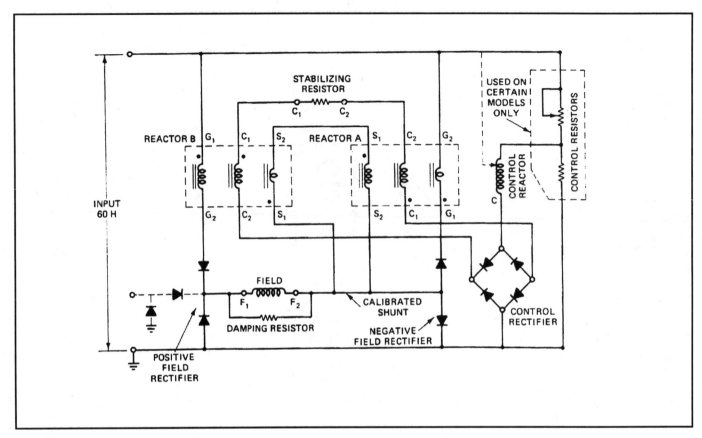

Figure 7-5: Schematic diagram of a magneciter alternator circuit.

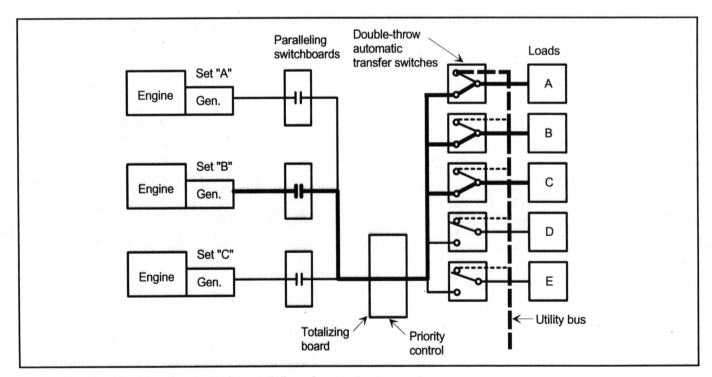

Figure 7-6: Typical automatic sequential paralleling of an emergency system.

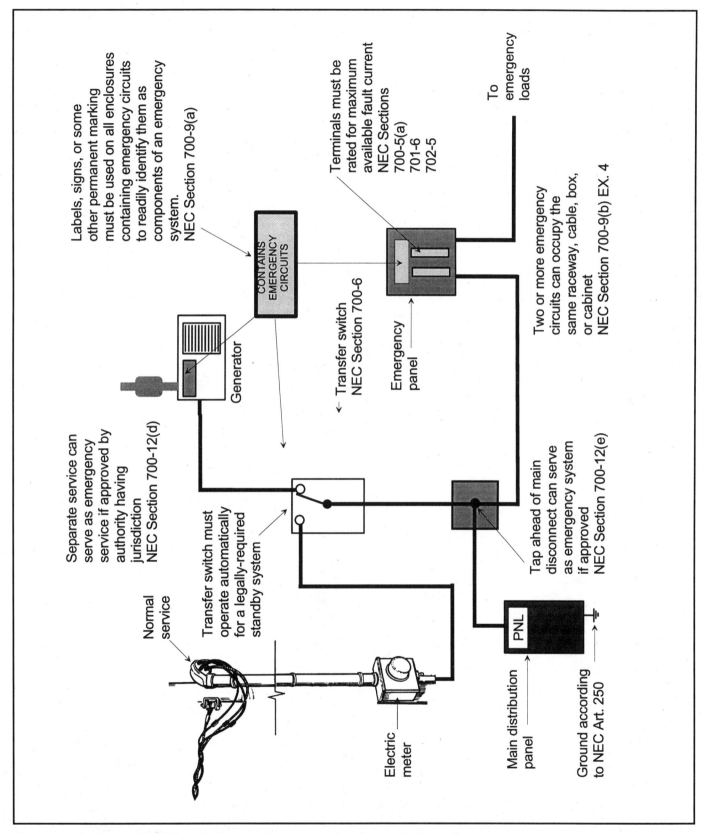

Figure 7-7: Summary of NEC installation requirements for legally-required standby systems.

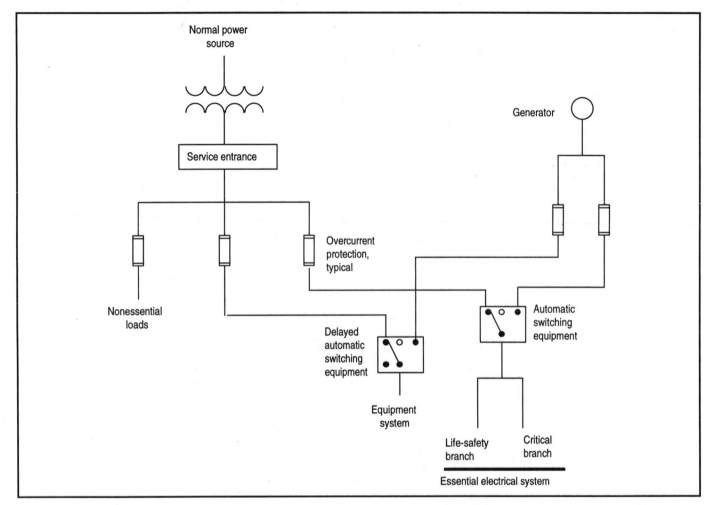

Figure 7-8: Essential electrical systems must be served by one or more transfer switches.

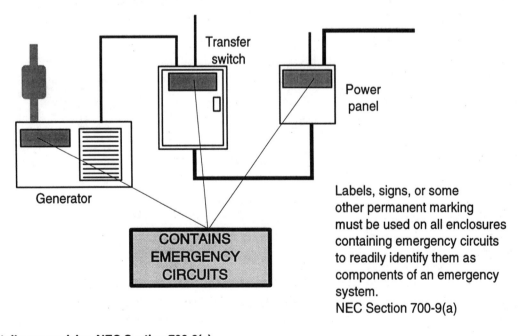

Labels, signs, or some other permanent marking must be used on all enclosures containing emergency circuits to readily identify them as components of an emergency system.
NEC Section 700-9(a)

Figure 7-9: Detail summarizing NEC Section 700-9(a).

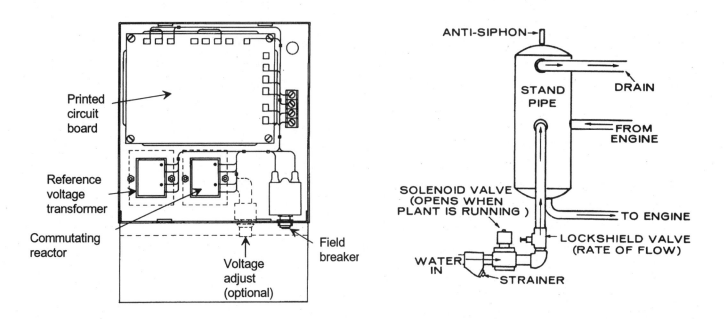

Figure 7-10: Detail of a voltage regulator assembly.

Figure 7-12: Details of a generator cooling system.

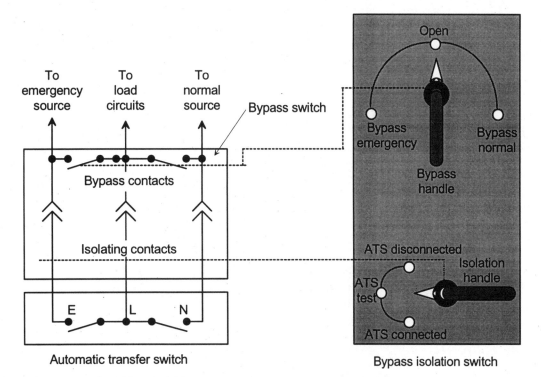

Figure 7-11: Two-way bypass isolation system.

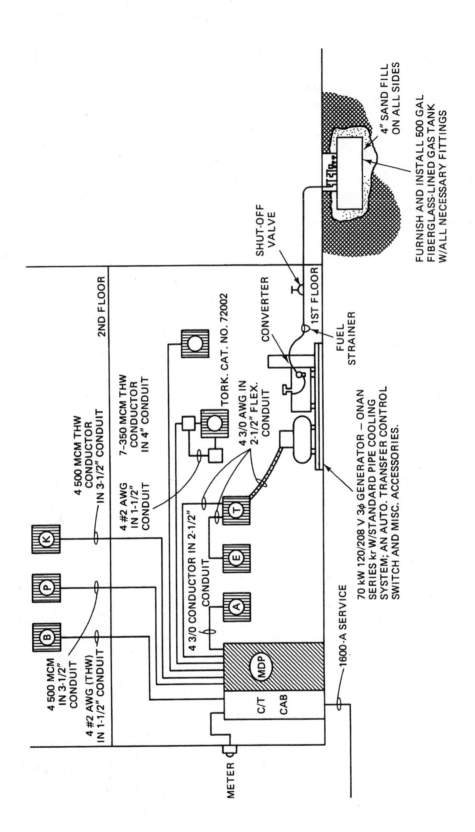

Figure 7-13: Power-riser diagram of electric service equipment with an emergency standby electrical system.

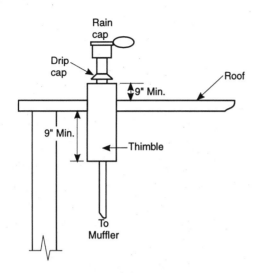

NOTES:

1. **Ventilated thimble diameter must be 6" larger than exhaust line.**

2. **Pitch exhaust lines downward or install condensation trap at bottom of riser.**

Figure 7-14: Exhaust pipe thimble detail for an emergency standby generator system.

The items in this chapter include highly specialized systems such as lightning protection, emergency light and power systems, storage batteries, battery charging equipment, cathodic protection and similar items. Also included are electrical details for wiring in hazardous (classified) areas. Most drawings for these items — because of the specialized characteristics of the items — will possibly require more detail drawings than other sections of an electrical system.

Equipment schedules should be used on the working drawings where possible, listing the equipment, catalog numbers, and the like.

WIRING IN HAZARDOUS LOCATIONS

NEC Articles 500 through 503 cover the requirements of electrical equipment and wiring for all voltages in locations where fire or explosion hazards may exist. Locations are classified according to the properties of the flammable vapors, liquids, gases, or combustible dusts and fibers that may be present, as well as the likelihood that a flammable or combustible concentration or quality is present.

Any area in which the atmosphere or a material in the area is such that the arcing of operating electrical contacts, components, and equipment may cause an explosion or fire is considered a hazardous location. In all cases, explosionproof equipment, raceways, and fittings are used to provide an explosionproof wiring system.

Hazardous locations have been classified in the *NEC* into certain class locations. Furthermore, various atmospheric groups have been established on the basis of the explosive character of the atmosphere for the testing and approval of equipment for use in the various groups.

Class I Locations

Class I locations are those in which flammable gases or vapors are or may be present in the air in quantities sufficient to produce explosive or ignitible mixtures. Examples of this type of location would include interiors of paint spray booths where volatile flammable solvents are used; inadequately ventilated pump rooms where flammable gas is pumped, and drying rooms for the evaporation of flammable solvents.

Class I atmospheric hazards are further divided into two Divisions (1 and 2) and also into four groups (A, B, C, and D). The classification involves the maximum explosion pressures, maximum safe clearance between parts of a clamped joint in an enclosure, and the minimum ignition temperature of the atmospheric mixture.

In general, the hazardous properties of the substances are greater for Group A. Group B is the next most hazardous, then Group C, with Group D the least hazardous. However, all four groups are extremely dangerous. Equipment to be used in these atmospheres must not only be approved for Class I, but also for the specific group of gases or vapors that will be present.

Class I, Division 2 covers locations where flammable gases, vapors or volatile flammable gases, vapors or volatile liquids are handled either in a closed system, or confined within suitable enclosures, or where hazardous concentrations are normally prevented by positive mechanical ventilation. Areas adjacent to Division 1 locations, into which gases might occasionally flow, would also belong in Division 2.

Class II Locations

Class II locations are those that are hazardous because of the presence of combustible dust. Class II, Division 1 locations are areas where combustible dust, under normal operating conditions, may be present in the air in quantities sufficient to produce explosive or ignitable mixtures; examples are working areas of grain-handling and storage plants and rooms containing grinders or pulverizers. Class II, Division 2 locations are areas where dangerous concentrations of suspended dust are not likely, but where dust accumulations might form.

Besides the two Divisions (1 and 2), Class II atmospheric hazards cover three groups of combustible dusts. The groupings are based on the resistivity of the dust. Group E is always Division 1. Group F, depending on the resistivity, and Group G may be either Division 1 or 2. Since the *NEC* is considered the definitive classification tool and contains explanatory data about hazardous atmospheres, refer to *NEC* Section 500-6 for exact definitions of Class II, Divisions 1 and 2.

Class III Locations

These locations are hazardous because of the presence of easily ignitable fibers or flyings, but such fibers and flyings are not likely to be in suspension in the air in these locations in quantities sufficient to produce ignitable mixtures. Such locations usually include some parts of textile mills and woodworking plants.

Once the class of an area is determined, the conditions under which the hazardous material may be present determines the division. In Class I and Class II, Division 1 locations, the hazardous gas or dust may be present in the air under normal operating conditions in dangerous concentrations. In Division 2 locations, the hazardous material is not normally in the air, but it might be released if there is an accident or faulty operation of equipment.

NEC Articles 511 and 514 regulate garages and similar locations where volatile or flammable liquids are used. While these areas are not always considered critically hazardous locations, there may be enough danger to require special precautions in the electrical installation. In these areas, the *NEC* requires that volatile gases be confined to an area not more than 4 ft above the floor. So in most cases, conventional raceway systems are permitted above this level. If the area is judged critically hazardous, explosion-proof wiring (including seal-offs) may be required.

Class	Division	Group	Typical Atmosphere/Ignition Temps.	Devices Covered	Temperature Measured	Limiting Value
I	1	A	Acetylene (305°C, 581°F	All electrical devices and wiring	Maximum external temperatures in 40°C ambient	See NEC Section 500-3
Gases, vapors	Normally hazardous	B	1,3-Butadiene (420°C, 788°F)			
			Ethylene Oxide (429°C, 804°F)			
			Hydrogen (520°C, 968°F)			
			Manufactured Gas (containing more than 30% hydrogen by volume)			
			Propylene Oxide (449°C, 840°F)			
		C	Acetaldehyde (175°C, 347°F)			
			Diethyl Ether (160°C, 320°F)			
			Ethylene (450°C, 842°F)			
			Unsymmetrical Dimethyl Hydrazine (UDMH) (249°C, 480°F)			
		D	Acetone (465°C, 869°F)			
			Acrylonitrile (481°C, 898°F)			

Figure 7-15: Summary of hazardous atmospheres.

Class	Division	Group	Typical Atmosphere/Ignition Temps.	Devices Covered	Temperature Measured	Limiting Value
I	1		Ammonia (498°C, 928°F)			
			Benzene (498°C, 928°F)			
			Butane (288°C, 550°F)			
			1-Butanol (343°C, 650°F)			
			2-Butanol (405°C, 761°F)			
			n-Butyl Acetate (421°C, 790°F)			
			Cyclopropane (503°C, 938°F)			
			Ethane (472°C, 882°F)			
			Ethanol (363°C, 685°F)			
			Ethyl Acetate (427C, 800F)			
			Ethylene Dichloride (413°C, 775°F)			
			Gasoline (280-471°C, 536-880°F)			
			Heptane (204°C, 399°F)			
			Hexane (225°C, 437°F)			
			Isoamyl Alcohol (350°C, 662°F)			
			Isoprene (220°C, 428°F)			
Gases, vapors	Normally hazardous	D	Methane (630°C, 999°F)	All electrical devices and wiring	Maximum external temperatures in 40°C ambient	See NEC Section 500-3
			Methanol (385C, 725F)			
			Methyl Ethyl Ketone (404°C, 759°F)			
			Methyl Isobutyl Ketone (449°C, 840°F)			
			2-Methyl-1-Propanol (416°C, 780°F)			
			2-Methyl-2-Propanol (478°C, 892°F)			
			Naphtha (petroleum) (288°C, 550°F)			
			Octane (206°C, 403°F)			
			Pentane (243°C, 470°F)			
			1-Pentanol (300°C, 572°F)			
			Propane (450°C, 842°F)			
			1-Propanol (413°C, 775°F)			
			2-Propanol (399°C, 750°F)			
			Propylene (455°C, 851°F)			
			Styrene (490°C, 914°F)			
			Toluene (480°C, 896°F)			

Figure 7-15: Summary of hazardous atmospheres. *(Cont.)*

Class	Division	Group	Typical Atmosphere/Ignition Temps.	Devices Covered	Temperature Measured	Limiting Value
I Gases, vapors	1 Normally hazardous	D	Vinyl Acetate (402°C, 756°F) Vinyl Chloride (472°C, 882°F) Xylenes (464-529°C, 867-984°F)	All electrical devices and wiring	Maximum external temperatures in 40°C ambient	See NEC Section 500-3
I Gases, vapors	2 Not normally hazardous	A	Same as Division 1	Lamps, resistors, coils, etc., other than arcing devices. (see Div. 1)	Max. internal or external temp. not to exceed the ignition temperature in degrees Celsius (°C) of the gas or vapor involved	See NEC Section 500-3
		B	Same as Division 1			
		C	Same as Division 1			
		D	Same as Division 1			
II Combustible dusts	1 Normally hazardous	E	Atmospheres containing combustible metal dusts regardless of resistivity, or other combustible dusts of similarly hazardous characteristics having resistivity of less than 10^2 ohm-centimeter	Devices not subject to overloads (switches, meters).	Max. external temp. in 40C ambient with a dust blanket	Shall be less than ignition temperature of dust but not more than: No overload: E—200C (392F) F—200C (392F) G—165C (329F) Possible overload in operation: Normal E—200C (392F) F—150C (302F) G—120C (248F) Abnormal E—200C (392F) F—200C (392F) G—165C (329F)
		F	Atmospheres containing carbonaceous dusts having resistivity between 10^2 and 10^8 ohm-centimeter			
		G	Atmospheres containing combustible dusts having resistivity of 10^8 ohm-centimeter or greater			
	2	F	Atmospheres containing carbonaceous dusts having resistivity of 10^5 ohm-centimeter or greater	Lighting fixtures	Max. external temp under conditions of use	Same as Division 1
	Not normally hazardous	G	Same as Division 1			
III Easily ignitible fibers and flyings	2			Lighting fixtures	Max. external temp. under conditions of use	165C (329F)

Figure 7-15: Summary of hazardous atmospheres. *(Cont.)*

NOTE

For a complete listing of flammable liquids, gases and solids, see *Classification of Gases, Vapors and Dusts for Electrical Equipment in Hazardous (Classified) Locations*, NFPA Publication No. 497M.

Class I Group	Substance	°F	°C	°F	°C	Lower	Upper	Vapor Density (Air = 1.0)
C	Acetaldehyde	347	175	-38	-39	4.0	60	1.5
D	Acetic Acid	967	464	103	39	4.0	19.9 @200°F	2.1
D	Acetic Anhydride	600	316	120	49	2.7	10.3	3.5
D	Acetone	869	465	-4	-20	2.5	13	2.0
D	Acetone Cyanohydrin	1270	688	165	74	2.2	12.0	2.9
D	Acetonitrite	975	524	42	6	3.0	16.0	1.4
A	Acetylene	581	305	gas	gas	2.5	100	0.9
B(C)	Acarolein (inhibited)	455	235	-15	-26	2.8	31.0	1.9
D	Acrylic Acid	820	438	122	50	2.4	8.0	2.5
D	Acrylonitrile	898	481	32	0	3.0	17	1.8
D	Adiponitrite	—	—	200	93	—	—	—
C	Allyl Alcohol	713	378	70	21	2.5	18.0	2.0
D	Allyl Chloride	905	485	-25	-32	2.9	11.1	2.6
B(C)	Allyl Glycidyl Ether	—	—	—	—	—	—	—
D	Ammonia	928	498	gas	gas	15	28	0.6
D	n-Amyl Acetate	680	360	60	16	1.1	7.5	4.5
D	sec-Amyl Acetate	—	—	89	32	—	—	4.5
D	Aniline	1139	615	158	70	1.3	11	3.2
D	Benzene	928	498	12	-11	1.3	7.9	2.8
D	Benzyl Chloride	1085	585	153	67	1.1	—	4.4
B(D)	1,,3-Butadiene	788	420	gas	gas	2.0	12.0	1.9
D	Butane	550	288	gas	gas	1.6	8.4	2.0
D	1-Butanol	650	343	98	37	1.4	11.2	2.6
D	2-Butanol	761	405	75	24	1.7 @ 212°F	9.8 @ 212°F	2.6
D	n-Butyl Acetate	790	421	72	22	1.7	7.6	4.0
D	iso-Butyl Acetate	790	421	—	—	—	—	—
D	sec-Butyl Acetate	—	—	88	31	1.7	9.8	4.0
D	t-Butyl Acetate	—	—	—	—	—	—	—
D	n-Butyl Acrylate (inhibited)	559	293	118	48	1.5	9.9	4.4
C	n-Butyl Formal	—	—	—	—	—	—	—
B(C)	n-Butyl Glycidyl Ether	—	—	—	—	—	—	—
C	Butyl Mercaptan	—	—	35	2	—	—	3.1
D	t-Butyl Toluene	—	—	—	—	—	—	—
D	Butylamine	594	312	10	-12	1.7	9.8	2.5
D	Butylene	725	385	gas	gas	1.6	10.0	1.9
C	n-Butyraldehyde	425	218	-8	-22	1.9	12.5	2.5
D	n-Butyric Acid	830	443	161	72	2.0	10.0	3.0

Figure 7-16: Substances used in business and industry.

Class I Group	Substance	°F	°C	°F	°C	Lower	Upper	Vapor Density (Air = 1.0)
C	Carbon Disulfide	194	90	-22	-30	1.3	50.0	2.6
C	Carbon Monoxide	1128	609	gas	gas	12.5	74.0	1.0
C	Chloroacetaldehyde	—	—	—	—	—	—	—
D	Chlorobenzene	1099	593	82	28	1.3	9.6	3.9
C	1-Chloro-1-Nitropropane	—	—	144	62	—	—	4.3
D	Chloroprene	—	—	-4	-20	4.0	20.0	3.0
D	Cresol	1038-1110	559-599	178-187	81-86	1.1-1.4	—	—
C	Crotonaldehyde	450	232	55	13	2.1	15.5	2.4
D	Cumene	795	424	96	36	0.9	6.5	4.1
D	Cyclohexane	473	245	-4	-20	1.3	8.0	2.9
D	Cyclohexanol	572	300	154	68	—	—	3.5
D	Cyclohexanone	473	245	111	44	1.1 @ 212°F	9.4	3.4
D	Cyclohexene	471	244		-7	—	—	2.8
D	Cyclopropane	938	503	gas	gas	2.4	10.4	1.5
D	p-Cymene	817	436	117	47	0.7 @ 212°F	5.6	4.6
C	C	n-Decaldehyde	—	—	—	—	—	—
D	n-Decanol	550	288	180	82	—	—	5.5
D	Decene	455	235	â	7	—	—	4.84
D	Diacetone Alcohol	1118	603	148	64	1.8	6.9	4.0
D	Dichlorobenzene	1198	647	151	66	2.2	9.2	5.1
D	1,,1-Dichloroethane	820	438	22	-6	5.6	—	—
D	1,,2-Dichloroethylene	860	460	36	2	5.6	12.8	3.4
C	1,,1-Dichloro-1-Nitroethane	—	—	168	76	—	—	5.0
D	1,,3-Dichloropropene	—	—	95	35	5.3	14.5	3.8
C	Dicyclopentadiene	937	503	90	32	—	—	—
D	Diethyl Benzene	743-842	395-450	133-135	56-57	—	—	4.6
C	Diethyl Ether	320	160	-49	-45	1.9	36.0	2.6
C	Diethylamine	594	312	-9	-23	1.8	10.1	2.5
C	Diethylaminoethanol	—	—	—	—	—	—	—
C	Diethylene Glycol Monobutyl Ether	442	228	172	78	0.85	24.6	5.6
C	Diethylene Glycol Monomethyl Ether	465	241	205	96	—	—	—
D	Di-isobutyl Ketone	745	396	120	49	0.8 @ 200°F	7.1 @ 200°F	4.9
D	Di-isobutylene	736	391	23	-5	0.8	4.8	3.9
C	Di-isopropylamine	600	316	30	-1	1.1	7.1	3.5
C	N-N-Dimethyl Aniline	700	371	145	63	—	—	4.2

Figure 7-16: Substances used in business and industry. *(Cont.)*

Class I Group	Substance	°F	°C	°F	°C	Lower	Upper	Vapor Density (Air = 1.0)
D	Dimethyl Formamide	833	455	136	58	2.2 @ 212°F	15.2	2.5
D	Dimethyl Sulfate	370	188	182	83	—	—	4.4
C	Dimethylamine	752	400	gas	gas	2.8	14.4	1.6
C	1,,4-Dioxane	356	180	54	12	2.0	22	3.0
D	Dipentene	458	237	113	45	0.7 @ 302°F	6.1 @ 302°F	4.7
C	Di-n-propylamine	570	299	63	17	—		3.5
C	Dipropylene Glycol Methyl Ether	—	—	185	85	—	—	5.11
D	Dodecene	491	255	—	—	—	—	—,
C	Epichlorohydrin	772	411	88	31	3.8	21.0	3.2
D	Ethane	882	472	gas	gas	3.0	12.5	1.0
D	Ethanol	685	363	55	1.3	3.3	19	1.6
D	Ethyl Acetate	800	427	24	-4	2.0	11.5	3.0
D	Ethyl Acrylate (inhibited)	702	372	50	10	1.4	14	3.5
D	Ethyl sec-Amyl Ketone	—	—	—	—	—	—	—
D	Ethyl Benzene	810	432	70	21	1.0	6.7	3.7
D	Ethyl Butanol	—	—	—	—	—	—	—
D	Ethyl Butyl Ketone	—	—	115	46	—	—	4.0
D	Ethyl Chloride	966	519	-58	-50	3.8	15.4	2.2
D	Ethyl Formate	851	455	-4	-20	2.8	16.0	2.6
D	2-Ethyl Hexanol	448	231	164	73	0.88	9.7	4.5
D	2-Ethyl Hexyl Acrylate	485	252	180	82	—	—	—
C	Ethyl Mercaptan	572	300		-18	2.8	18.0	2.1
C	n-Ethyl Morpholine	—	—	—	—	—	—	—
C	2-Ethyl-3-Propyl Acrolein	—	—	155	68	—	—	4.4
D	Ethyl Silicate	—	—	125	52	—	—	7.2
D	Ethylamine	725	385		-18	3.5	14.0	1.6
C	Ethylene	842	450	gas	gas	2.7	36.0	1.0
D	Ethylene Chlorohydrin	797	425	140	60	4.9	15.9	2.8
D	Ethylene Dichloride	775	413	56	13	6.2	16	3.4
C	Ethylene Glycol Monobutyl Ether	460	238	143	62	1.1 @ 200°F	12.7 @ 275°F	4.1
D	Ethylene Glycol Monobutyl Ether Acetate	645	340	160	71	0.88 @ 200°F	8.54 @ 275°F	—
C	Ethylene Glycol Monoethyl Ether	455	235	110	43	1.7 @ 200°F	15.6 @ 200°F	3.0

Figure 7-16: Substances used in business and industry. *(Cont.)*

Class I Group	Substance	°F	°C	°F	°C	Lower	Upper	Vapor Density (Air = 1.0)
C	Ethylene Glycol Monoethyl Ether Acetate	715	379	124	52	1.7	—	4.72
D	Ethylene Glycol Monomethyl Ether	545	285	102	39	1.8 @ STP	14 @ STP	2.6
B(C)	Ethylene Oxide	804	429	-20	-28	3.0	100	1.5
D	Ethylenediamine	725	385	93	34	4.2	14.4	2.1
C	Ethylenimine	608	320	12	-11	3.6	46.0	1.5
C	2-Ethylhexaldehyde	375	191	112	44	0.85 @ 200°F	7.2 @ 275°F	4.4
B	Formaldehyde (Gas)	795	429	gas	gas	7.0	73	1.0
D	Formic Acid (90%)	813	434	122	50	18	57	—
D	Fuel Oils	410-765	210-407	100-336	38-169	0.7	5	—
C	Furfual	600	316	140	60	2.1	19.3	3.3
C	Furfuryl Alcohol	915	490	167	75	1.8	16.3	3.4
D	Gasoline	536-880	280-471	-36 to -50	-38 to -46	1.2-1.5	7.1-7.6	3-4
D	Heptane	399	204	2.5	-4	1.05	6.7	3.5
D	Heptene	500	260			—	—	3.39
D	Hexane	437	225	-7	-22	1.1	7.5	3.0
D	Hexanol	—	—	145	63			3.5
D	2-Hexanone	795	424	77	25	—	8	3.5
D	Hexenes	473	245		-7	—	—	3.0
D	sec-Hexyl Acetate	—	—	—	—	—	—	—
C	Hydrazine	74-518	23-270	100	38	2.9	9.8	1.1
B	Hydrogen	968	520	gas	gas	4.0	75	0.1
C	Hydrogen Cyanide	1000	538	0	-18	5.6	40.0	0.9
C	Hydrogen Selenide	—	—	—	—	—	—	—
C	Hydrogen Sulfide	500	260	gas	gas	4.0	44.0	1.2
D	Isoamyl Acetate	680	360	77	25	1.0 @ 212°F	7.5	4.5
D	Isoamyl Alcohol	662	350	109	43	1.2	9.0 @ 212°F	3.0
D	Isobutyl Acrylate	800	427	86	30	—	—	4.42
C	Isobutyraldehyde	385	196	-1	-18	1.6	10.6	2.5
C	Isodecaldehyde	—	—	185	85	—	—	5.4
D	Iso-octyl Alcohol	—	—	180	82	—	—	—
C	Iso-octyl Aldehyde	387	197	—	—	—	—	—
D	Isophorone	860	460	184	84	0.8	3.8	—
D	Isoprene	428	220	-65	-54	1.5	8.9	2.4
D	Isopropyl Acetate,860	460	35	2	1.8 @ 100°F	8	3.5	
D	Isopropyl Ether	830	443	-18	-28	1.4	7.9	3.5

Figure 7-16: Substances used in business and industry. *(Cont.)*

Class I Group	Substance	°F	°C	°F	°C	Lower	Upper	Vapor Density (Air = 1.0)
D	Isopropylamine	756,402	-35	-37	—	—	2.0	
D	Kerosene	410	210	110-162	43-72	0.7	.5	—
D	Liquefied Petroleum Gas	761-842	405-450	—	—	—	—	—
B	Manufactured Gas (containing more than 30% H	—	—	—	—	—	—	—
D	Methyl Isobutyl Ketone	840	440	64	18	1.2 @ 200°F. 8.0 @ 200°F 3.5	—	—
D	Methyl Isocyanate	994	534	19	-7	5.3	26	1.97
C	Methyl Mercaptan	—	—	—	—	3.9	21.8	1.7
D	Methyl Methacrylate	792	422	50	10	1.7	8.2	3.6
D	2-Methyl-1-Propanol	780	416	82	28	1.7 @ 123°F	10.6 @ 202°F	2.6
D	2-Methyl-2-Propanol	892	478	52	11	2.4	8.0	2.6
D	alpha-Methyl Styrene	1066	574	129	54	1.9	6.1	—
C	Methylacetylene	—	—	gas	gas	1.7	—	1.4
C	Methylacetylene-Propadiene (stabilized)	—	—	—	—	—	—	—
D	Methylamine	806	430	gas	gas	4.9	20.7	1.0
D	Methylcyclohexane	482	250	25	-4	1.2	6.7	3.4
D	Methylcyclohexanol	565	296	149	65	—	—	3.9
D	0-Methylcyclohexanone	—	—	118	48	—	—	3.9
D	Monoethanolamine	770	410	185	85	—	—	2.1
D	Monoisopropanolamine	705	374	171	77	—	—	2.6
C	Monomethyl Aniline	900	482	185	85	—	—	3.7
C	Monomethyl Hydrazine	382	194	17	-8	2.5	92	1.6
C	Morpholine	590	310	98	37	1.4	11.2	3.0
D	Naphtha (Coal Tar)	531	277	107	42	—	—	—
D	Naphtha (Petroleum)	550	288		-18	1.1	5.9	2.5
D	Nitrobenzene	900	482	190	88	1.8 @ 200°F	—	4.3
C	Nitroethane	778	414	82	28	3.4	—	2.6
C	Nitromethane	785	418	95	35	7.3	—	2.1
C	1-Nitropropane	789	421	96	36	2.2	—	3.1
C	2-Nitropropane	802	428	75	24	2.6	11.0	3.1
D	Nonane	401	205	88	31	0.8	2.9	4.4
D	Nonene	—	—	78	26	—	—	4.35
D	Nonyl Alcohol	—	—	165	74	0.8 @ 212°F	6.1 @ 212°F	5.0

Figure 7-16: Substances used in business and industry. *(Cont.)*

Class I Group	Substance	°F	°C	°F	°C	Lower	Upper	Vapor Density (Air = 1.0)
D	Octane	403	206	56	13	1.0	6.5	3.9
D	Octene	446	230	70	21	—	—	3-9
D	n-Octyl Alcohol	—	—	178	81	—	—	4.5
Class I Group	Substance	°F	°C	°F	°C	Lower	Upper	
D	Pentane	470	243	-40	-40	1.5	7.8	2.5
D	1-Pentanol	572	300	91	33	1.2	10.0 @ 212°F	3.0
D	2-Pentanone	846	452	45	7	1.5	8.2	3.0
D	1-Pentene	527	275	0	-18	1.5	8.7	2.4
D	Phenylhydrazine	—	—	190	88	—	—	—
D	Propane	842	450	gas	gas	2.1	9.5	1.6
D	1-Propanol	775	413	74	23	2.2	13.7	2.1
D	2-Propanol	750	399	53	12	2.0	12.7 @ 200°F	2.1
D	Propiolactone	—	—	165	74	2.9	—	2.5
C	Propionaldehyde	405	207	-22	-30	2.6	17	2.0
D	Propionic Acid	870	466	126	52	2.9	12.1	2.5
D	Propionic Anhydride	545	285	145	63	1.3	9.5	4.5
D	n-Propyl Acetate	842	450	55	13	1.7 @ 100°F	8	3.5
C	n-Propyl Ether	419	215	70	21	1.3	7.0	3.53
B	Propyl Nitrate	347	175	68	20	2	100	—
D	Propylene	851	455	gas	gas	2.0	11.1	1.5
D	Propylene Dichloride	1035	557	60	16	3.4	14.5	3.9
B(C)	Propylene Oxide	840	449	-35	-37	2.3	36	2.0
D	Pyridine	900	482	68	20	1.8	12.4	2.7
D	Styrene	914	490	88	3.1	1.1	7.0	3.6
C	Tetrahydrofuran	610	321	6	-14	2.0	11.8	2.5
D	Tetrahydronaphthalene	725	385	160	71	0.8 @ 212°F	5.0 @ 302°F	4.6
C	Tetramethyl Lead	—	—	100	38	—	—	6.5
D	Toulene	896	480	40	4	1.2	7.1	3.1
D	Tridecene	—	—	—	—	—	—	—
C	Triethylamine	480	249	16	-9	1.2	7.1	3.1
D	Triethylbenzene	—	—	181	83	—	—	5.6
D	Tripropylamine	—	—	105	41	—	—	—
D	Turpentine	488	253	95	35	0.8	—	—
D	Undecene	—	—	—	—	—	—	—
C	Unsymmetrical Dimethyl Hydrazine (UDMH)	480	249	5	-15	2	95	2.0
C	Valeraldehyde	432	222	54	12	—	—	3.0
D	Vinyl Acetate	756	402	18	-8	2.6	13.4	3.0
D	Vinyl Chloride	882	472	gas	gas	3.6	33.0	2.2

Figure 7-16: Substances used in business and industry. *(Cont.)*

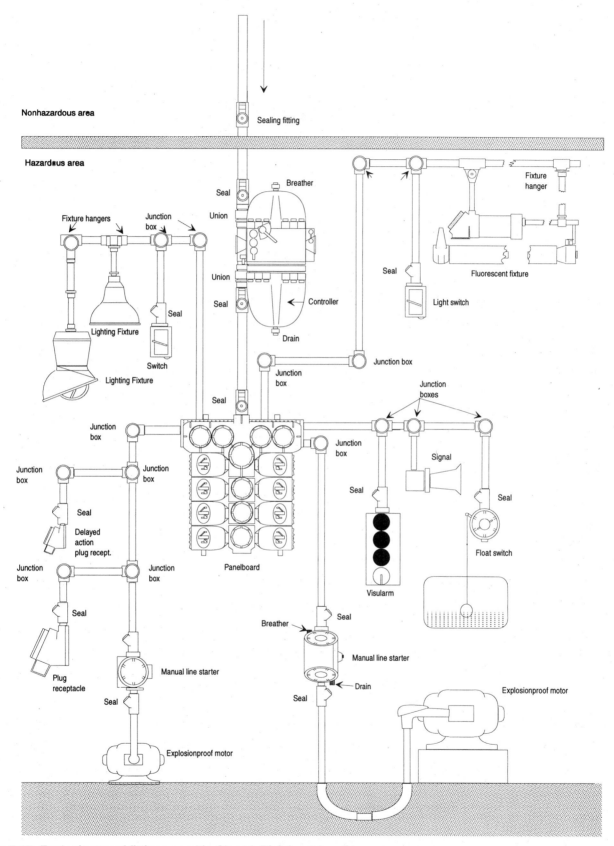

Figure 7-17: Explosionproof fittings used in Class I, Division 1 locations.

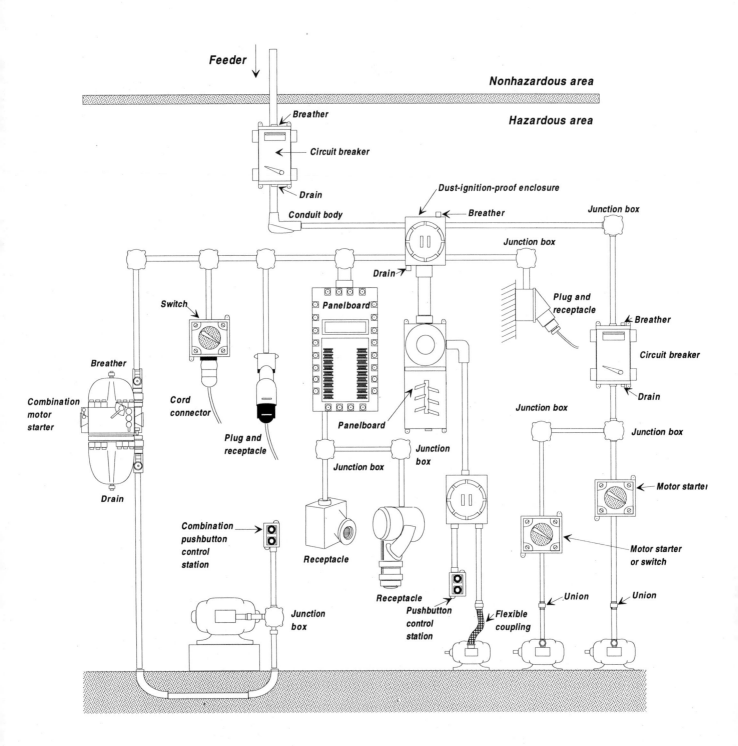

Figure 7-18: Explosionproof fittings used in Class II, Division 1 locations.

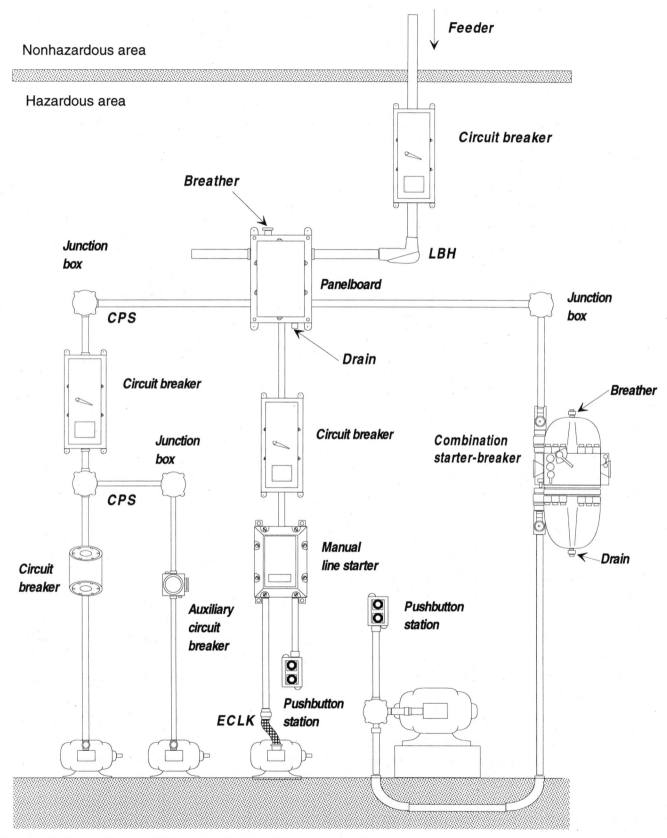

Figure 7-19: Explosionproof fittings used in Class II power installations.

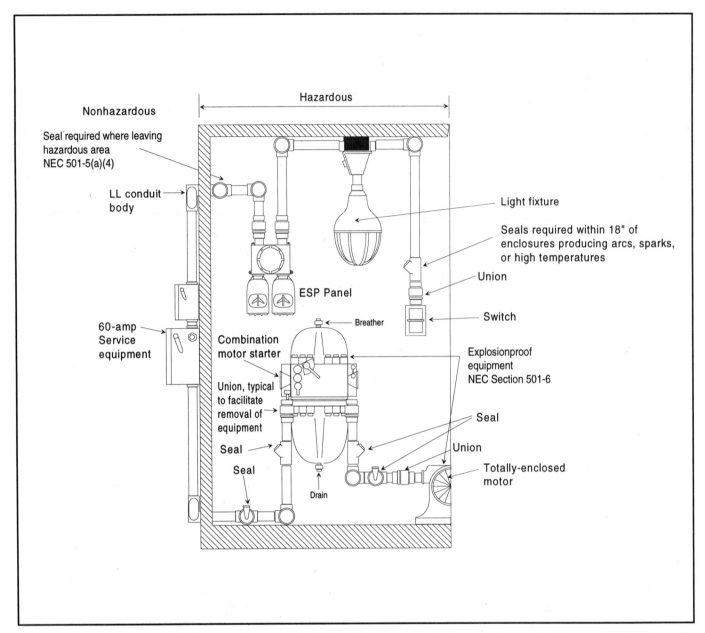

Figure 7-20: Explosionproof fittings used in Class I, Division 1 locations and showing the required locations of seals and unions.

NOTE

The components shown in Figures 7-18 through 7-25 may be ungrouped in a computer CAD program and rearranged to fit practically any hazardous situation. See Figures 7-21 and Figure 7-22 on the next page.

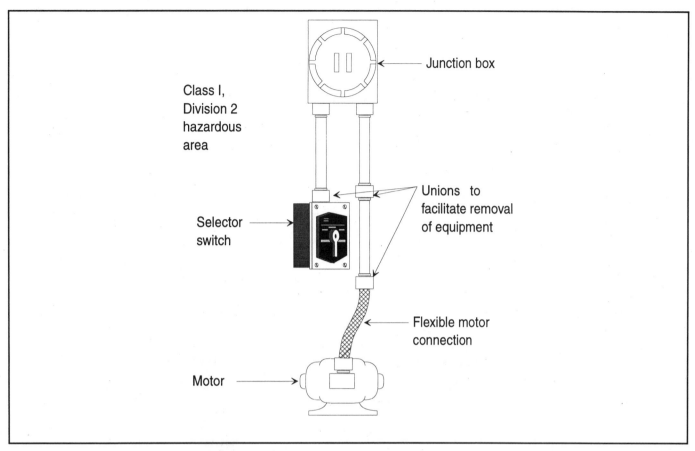

Figure 7-21: Some of the components in Figures 7-18 through 7-25 ungrouped and rearranged to form a detail drawing of explosionproof flexible connectors, used in some Class I, Division 2 locations for motor terminations.

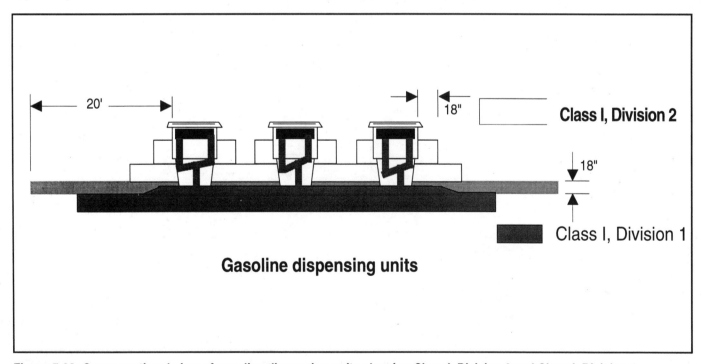

Figure 7-22: Cross-sectional view of gasoline dispensing units showing Class I, Division 2 and Class I, Division 1 locations.

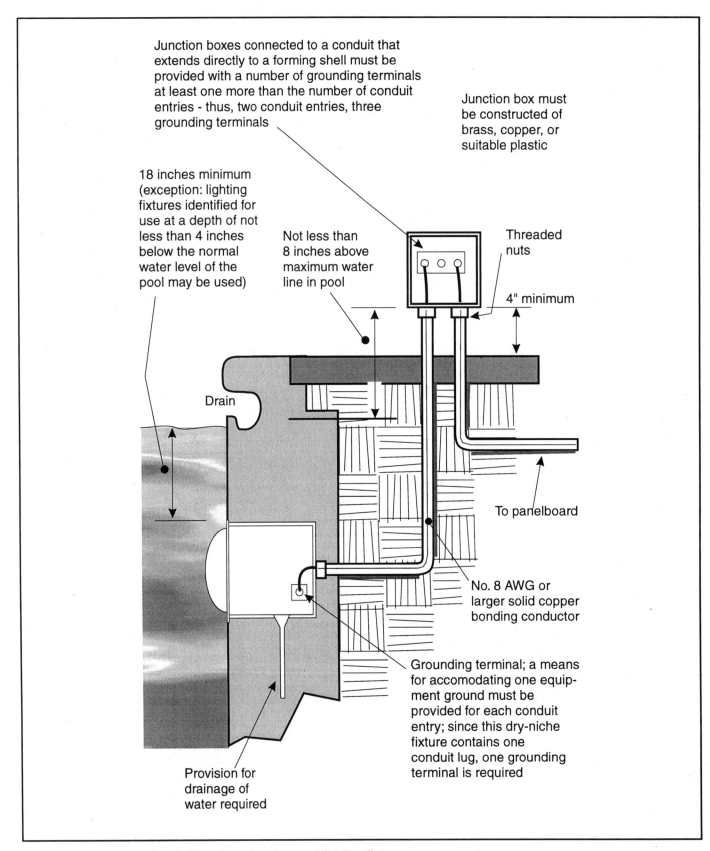

Junction boxes connected to a conduit that extends directly to a forming shell must be provided with a number of grounding terminals at least one more than the number of conduit entries - thus, two conduit entries, three grounding terminals

Junction box must be constructed of brass, copper, or suitable plastic

18 inches minimum (exception: lighting fixtures identified for use at a depth of not less than 4 inches below the normal water level of the pool may be used)

Not less than 8 inches above maximum water line in pool

Threaded nuts

4" minimum

Drain

To panelboard

No. 8 AWG or larger solid copper bonding conductor

Grounding terminal; a means for accomodating one equipment ground must be provided for each conduit entry; since this dry-niche fixture contains one conduit lug, one grounding terminal is required

Provision for drainage of water required

Figure 7-23: Cross-sectional view of a swimming pool lighting fixture.

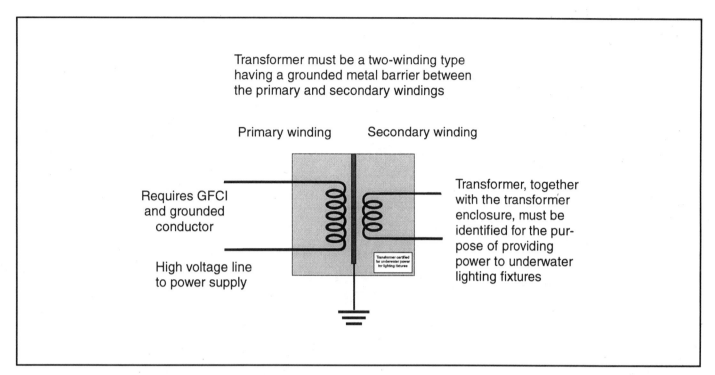

Transformer must be a two-winding type
having a grounded metal barrier between
the primary and secondary windings

Primary winding   Secondary winding

Requires GFCI
and grounded
conductor

High voltage line
to power supply

Transformer certified
for underwater power
for lighting fixtures

Transformer, together
with the transformer
enclosure, must be
identified for the pur-
pose of providing
power to underwater
lighting fixtures

Figure 7-24: Detail of low-voltage swimming pool transformer.

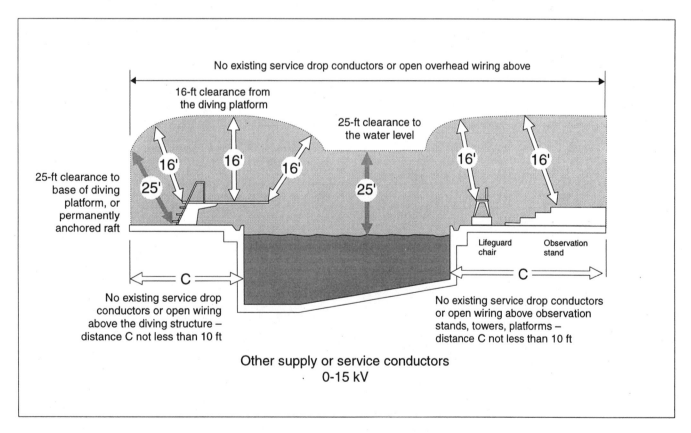

No existing service drop conductors or open overhead wiring above

16-ft clearance from
the diving platform

25-ft clearance to
the water level

25-ft clearance to
base of diving
platform, or
permanently
anchored raft

16'   16'   16'   25'   16'   16'

Lifeguard   Observation
chair   stand

C   C

No existing service drop
conductors or open wiring
above the diving structure –
distance C not less than 10 ft

No existing service drop conductors
or open wiring above observation
stands, towers, platforms –
distance C not less than 10 ft

Other supply or service conductors
0-15 kV

Figure 7-25: Spacing requirements for service conductors about swimming pools — 0 – 15 kV.

## Chapter 8

Signaling and Communication Systems

The items covered in this chapter encompass radio, shortwave, and microwave transmission; alarm and detection systems; smoke detectors; clock and program equipment; telephone and telegraph; intercommunication equipment; public address systems; television systems; and systems related to learning laboratories.

Each of the previously mentioned systems are generally designed around the equipment of a selected manufacturer. Consequently, the drawings and specifications should include the following:

- A general section in the written specifications or a schedule on the drawings that identifies the specific system and component manufacturer.
- A description of the system-operation principles.
- A list and description of the basic equipment items to be furnished.
- A description of requirements and methods for the installation and wiring of the system.

Although public telephone systems are normally installed by the utility (telephone) company, the electrical contractor frequently installs raceways, outlet boxes, and pull wires inside of the conduit runs for pulling in the telephone cable later. For all projects, the designer should contact the local telephone company to ascertain the exact requirements.

Cable television is handled in approximately the same way as telephone wires, but again, check with the local cable company for exact details and specifications.

The communication details included in this chapter are typical of those used on projects that range in size from residential to industrial and institutional establishments and cover the following:

- Fire-alarm and detection systems
- Burglar alarm systems
- Clock and program equipment
- Telephone systems
- Intercommunication equipment
- Master TV antenna equipment

SIGNALING AND COMMUNICATION SYSTEM DETAILS

When any signaling or communication system is incorporated in building construction, details of the installation are usually included in the electrical drawings and the electrical portion of the written specifications. The location of the signaling devices is normally shown on the floor plans by lines and symbols whose definition is given in a legend or symbol list. Details of the equipment (manufacturer, catalog number, etc.) are presented either in the form of schedules on the working drawings or by descriptive paragraphs in the written specifications. Certain detailed drawings are also required, giving rough-in dimensions and installation procedures as well as connection wiring diagrams.

Figure 8-1, for example, shows installation details for telephone outlets in a commercial department store that utilizes a lay-in T-bar ceiling system. Note that the telephone outlet box is mounted 18 in above the finished floor to the center line of the box and that 1-in conduit is run inside the wall and up above the finished lay-in tile ceiling. The telephone installers will then run their cable exposed in the space above the finished ceiling and "fish" their telephone cables down the conduit to each outlet box within the finished area. Some telephone companies require that a galvanized steel wire be installed in the conduit during installation to facilitate pulling the telephone cables in later.

In buildings with accessible basements or crawl spaces under the finished floor and a solid finished ceiling (preventing access to the area above the ceiling), the conduit for TV and telephone outlets may be reversed from the previous detail and installed as shown in Figure 8-2. Here, a 4-in square outlet box with a single-gang plaster ring is installed on a piece of ¾-in rigid, EMT, or PVC conduit that terminates under the floor in the crawl space of the building. The distance shown on the drawing from the center of the box to the finished floor is 16 in. Since the box is 4 in square, the electrician will probably measure up 14 in to the bottom of the box, obtaining a more exact measure on the job.

After the building is roughed in, the telephone or TV-cable company will install their telephone or coaxial cable (respectively) in the crawl space or basement and then fish the wire up to each outlet box in the finished area.

Apartment House Intercom Systems

To provide apartment house tenants with the security of a locked building entrance door to exclude undesirable

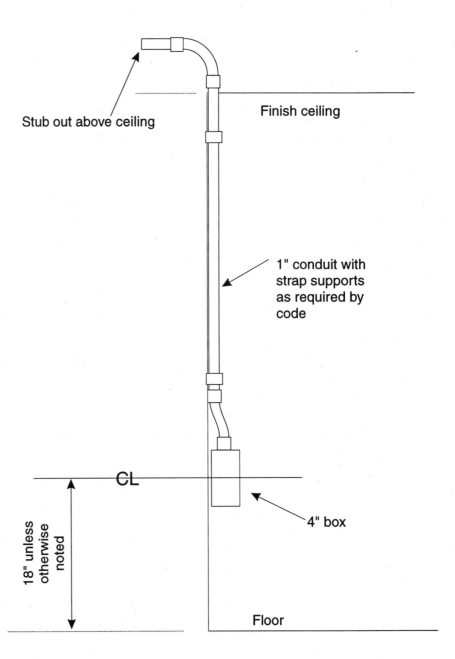

TELEPHONE OUTLET

Figure 8-1: Telephone outlet box detail used in buildings with a lay-in T-bar ceiling system.

persons from the premises, many apartment building owners are installing intercom systems. With such systems, a caller must ring the apartment he or she is visiting and announce his or her identity over the intercom system. The tenant, if satisfied with the identity of the caller, may then electrically open the building entrance door.

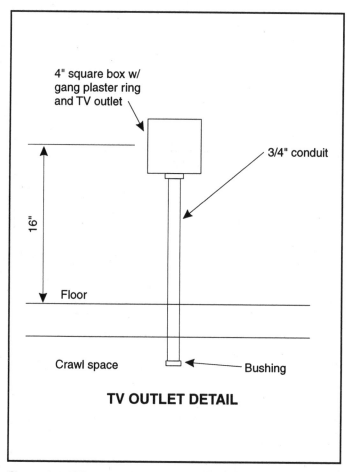

4" square box w/
gang plaster ring
and TV outlet

3/4" conduit

16"

Floor

Crawl space ← Bushing

TV OUTLET DETAIL

Figure 8-2: TV or telephone outlet detail used in buildings with a basement or crawl space.

A block wiring diagram of an apartment house intercom system is shown in Figure 8-3 on the next page. Note that a transformer is connected to a 120-V ac power supply. This transformer provides 16-V ac to the amplifier (Amplitone) unit for operation of the system. The amplifier is the heart of the system and most are solid-state and capable of providing 1 watt of undistorted power within the range of normal voice frequencies. The power is carried from the transformer to the amplifier by a pair of No. 18 AWG conductors. Another pair of No. 18 AWG conductors runs from the amplifier to the door opener.

The units, termed *Suitefones*, are equipped with three buttons:

- Talk
- Listen
- Door

Sometimes an extra button is used and marked "Rear" to open the rear or service door. When the button at the apartment entry is pressed (for the appropriate apartment)

a tone is heard at the Suitefone. The tenant then presses the "Talk" button and inquires as to who is ringing, releasing the "Talk" button and pressing the "Listen" button. If the tenant is satisfied as to the identity of the person, he or she presses the "Door" button, which activates the magnetic latch and allows the door to be opened.

Apartment House Telephone Systems

Similar to the intercom system is the apartment house telephone system. This system provides apartment dwellers with the security of a locked building entrance door. Callers must identify themselves from the building entrance, by voice, over the telephone system. The apartment occupant, if satisfied with the identity of the caller, may electrically open the building door.

These systems are intended for buildings in which entirely new telephone equipment and wiring will be installed. Apartment telephones for new buildings (under construction) are supplied with recessed wall boxes. When supplied instead with surface-mounting wall boxes, they are suitable for installation in either (a) existing buildings in which no voice communication system was previously installed, or (b) existing buildings in which a buzzer system must be rewired in accordance with the telephone system selected. A new entrance telephone and power supply cabinet must also be installed.

Existing apartment houses already equipped with an inoperative lobby-to-apartment telephone system can replace the inoperative apartment telephones with new ones. These are supplied with adapter plates to fit various makes of existing wall boxes. The existing entrance telephone should be replaced by a new entrance telephone. The wiring of the existing telephone system can be used if it is not too old, in good condition, and free of short circuits and ground faults. Naturally, when installing new replacement telephone equipment, it is advisable to install new wiring throughout. The existing power supply cabinet must be replaced. Wiring diagrams of typical apartment house telephone systems appear in Figures 8-4 and 8-5.

Nurses' Call Systems

There are several types of nurses' call systems in use in nursing homes, hospitals and other types of health-care facilities. One type is manufactured by Webster-Auth and utilizes a combination of solid-state and electromechanical components to accomplish its functions. This system is designed mainly for small hospitals and nursing homes.

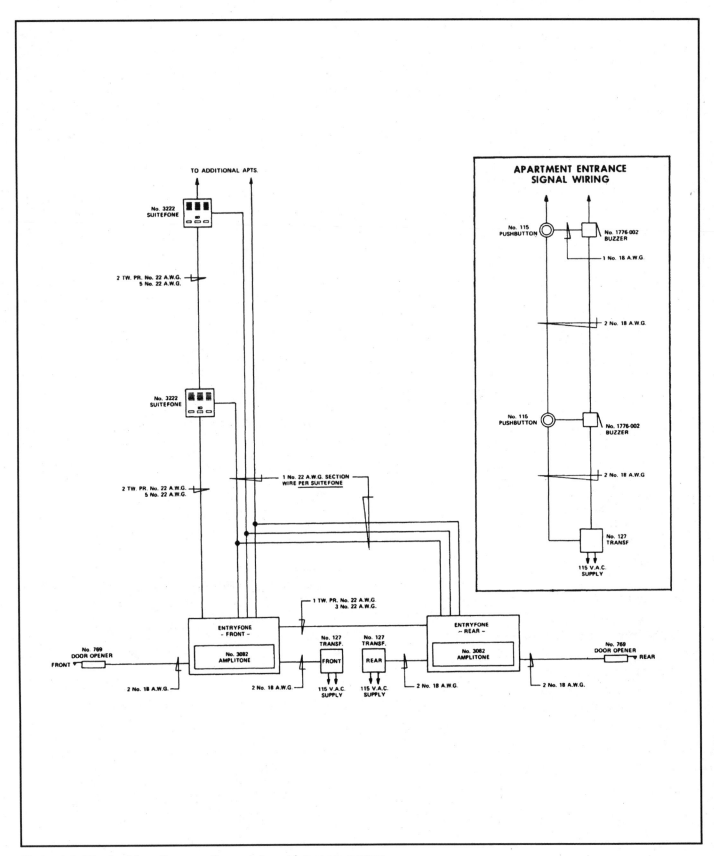

Figure 8-3: Block wiring diagram of an apartment intercom system.

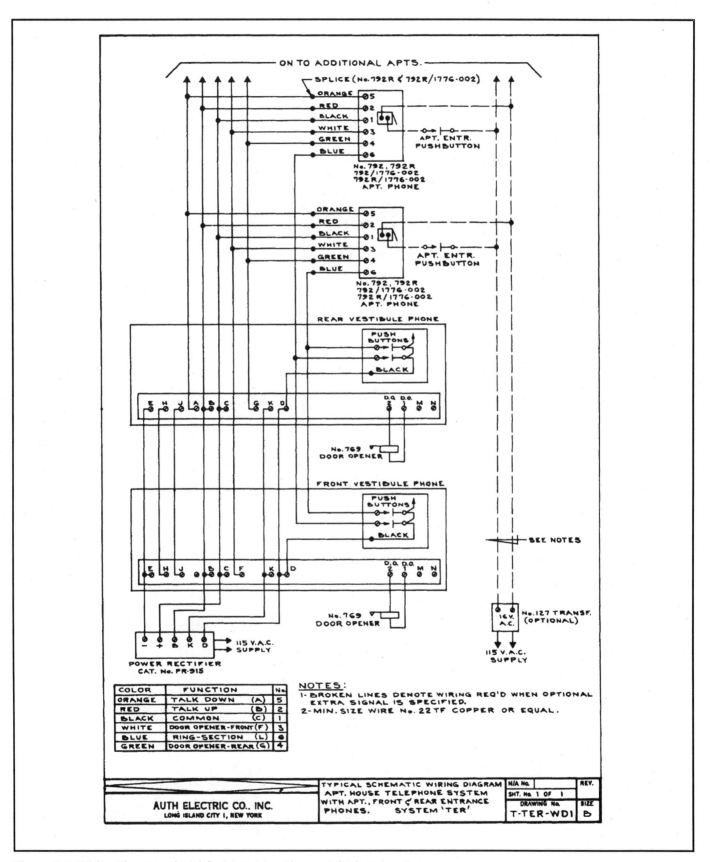

Figure 8-4: Wiring diagram of a typical apartment house telephone system.

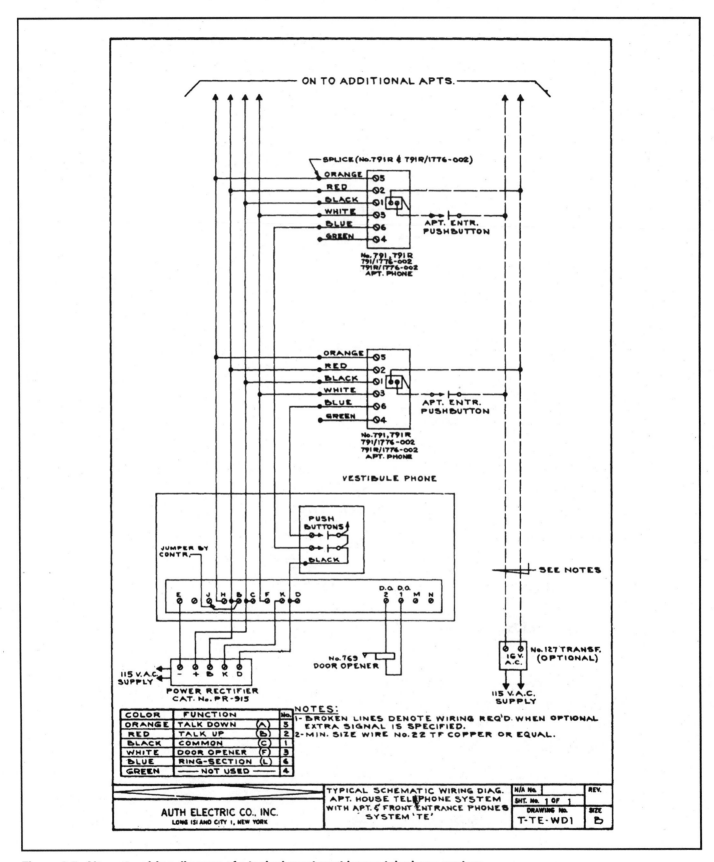

Figure 8-5: Alternate wiring diagram of a typical apartment house telephone system.

The system consists of patients' call stations, and lamp-type signal stations of various types. When a patient originates a call to the nurse, a steady lamp signal is lighted at various locations, depending upon the requirements of the system — at the bedside station (or stations) in the patient's room, in the corridor over the room door, at the nurses' station, in one or more duty rooms. In addition, a buzzer signal sounds at the nurses' station and in the duty rooms. Lamp signals are only extinguished when the nurse attends the patient's bedside and presses the reset button on the wall station. If, when attending the patient, the nurse requires emergency help, the nurse can summon it by alternately pressing and releasing the call button. This action will sound the buzzer repeatedly at the nurses' station and in the duty rooms.

When the plug of any portable cord set is accidentally removed from its receptacle, all signals are energized in exactly the same manner as when the patient calls the nurse, except that the buzzer sounds continuously.

Interchangeable cord sets are available for patients use: standard cord sets for normal usage, pressure pad for aged and incapacitated, and pull switch for oxygen patients. All plug into the same receptacles on the patient's bedside station.

Emergency-type stations in toilets, baths, solariums, and so on, enable the patient to summon assistance by pulling a pendant cord attached to a switch. This flashes a lamp signal on all associated signal stations and repeats a buzzer signal intermittently and insistently at the nurses' station and in the duty rooms. These audible and visible signals can only be canceled by throwing the switch to the "off" position on the originating station.

The circuitry employed makes possible the use of standard, momentary-contact pushbutton cord sets, and standard two-point jack-type receptacles. This lowers the cost of the original equipment and also the cost of replacement components.

A typical layout of the system just described is shown in Figure 8-6 on the next page. Note the symbols and nurses' call wire legend. The first digit of the number circling a wire run indicates the number of No. 22 AWG wires in the run, and the second number indicates the number of No. 18 AWG wires in the run. Therefore, the number "6-5" indicates that the run or conduit contains six No. 22 AWG conductors and five No. 18 AWG conductors. The remaining components are described in the symbol list; the catalog numbers correspond to those components manufactured by Auth.

Electric Clock Systems

The primary control unit of a master time and program system, such as those used in school systems and similar institutions, is commonly called the master clock. It is normally a wall-mounted panel assembly installed in the office of a responsible official in schools and institutions. It is actually a master controller, performing two major functions:

1. It is wired to a central source of unswitched power and operates all other clocks in the system. Isolated power interruptions in a building will not affect remote secondary clocks. At fixed schedule periods, which might be hourly or every 12 hours, the master controller will transmit synchronizing signals to all secondary clocks. This ensures that the time reading of all system clocks remains identical to that of the master.

2. The mechanism includes a unique multicircuit-control feature called a programmer. This permits the master controller to transmit additional signal impulses to external circuits that connect to devices such as bells, horns, chimes, and buzzers. These are referred to as program signals. Program circuits are also utilized for on-off control of building utilities.

Program signal control is established and maintained by use of continuous-run, prepunched memory tapes. Each circuit is programmed independently and uses the same principle as those used in tape control of communication transmission networks and similar to that of punch-card data processing.

In designing or installing such a system, it is necessary to obtain certain details concerning the system for the proper roughing-in of outlet boxes, conduit, cable, and so on, and certain details of construction should appear on the working drawings as prepared by the engineer for the electrical contractor. If these details are omitted on the original working drawings as received from the architect or engineer, the electrical contractor should make certain that the workers have sufficient information so that the installation may be made properly and in the least amount of time. Some of the more necessary details are described next.

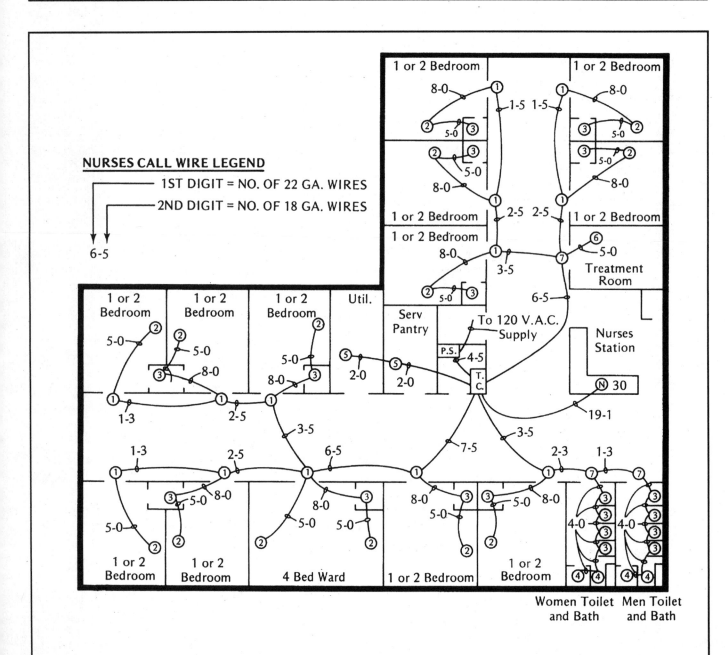

NURSES CALL WIRE LEGEND

1ST DIGIT = NO. OF 22 GA. WIRES

2ND DIGIT = NO. OF 18 GA. WIRES

6-5

SYMBOLS

① —CORRIDOR DOME LIGHTS CAT. NO. 484

② —BEDSIDE STATION CAT. NO. 481 OR CAT. NO. 482

③ —TOILET STATION CAT. NO. 257 OR CAT. No. 261

④ —SHOWER STATION CAT. NO. 257 OR CAT. NO. 261

⑤ —DUTY STATION CAT. NO. 483

⑥ —EMERGENCY STATION CAT. NO. 257 OR CAT. NO. 261

⑦ —CORRIDOR DOME CAT. NO. 302W

Figure 8-6: Typical conduit layout for nurses' call system.

Master Clock Mounting Details: The surface and flush mounting details in Figure 8-7 give such data as overall dimensions, locations and spacing of knockouts, mounting details, door hinging, and so on — all necessary for roughing-in and installation.

Wiring Details: The wiring details in Figure 8-8 show the internal wiring of a manual control station and also the connection details of secondary clocks. This type of con-

trol is used to control one or more secondary clocks when they are used without a master clock or other means of control. This type of station permits resetting and synchronizing all clocks wired to it without removing clocks from the walls.

Secondary Clock Mounting: Since the outlet boxes and raceway system for secondary clocks usually have to be roughed-in prior to finishing the building and often before

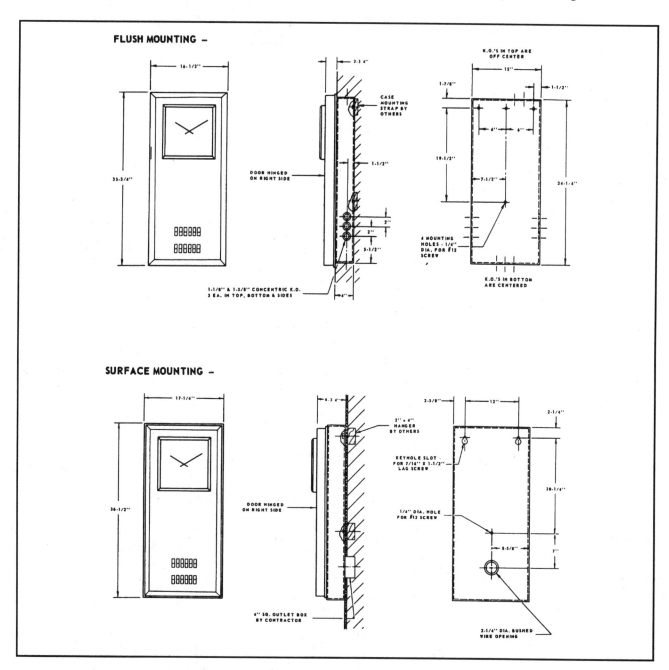

Figure 8-7: Mounting details for a master clock unit.

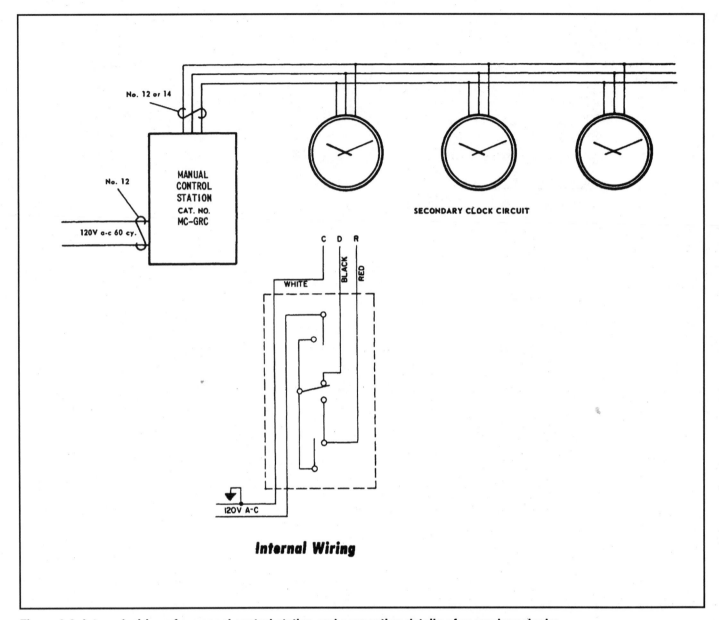

Figure 8-8: Internal wiring of a manual control station and connection details of secondary clocks.

the equipment arrives, mounting details are required to enable the workers to properly locate the equipment. The surface- and flush-mounting details in Figure 8-9 on the opposite page give the information necessary for this operation. Of course, this detail may be modified to suit existing conditions.

Telephone Mounting: Some clock systems incorporate a telephone system to communicate between classrooms and from classrooms to the main office. When such phones are used, mounting details are necessary for proper roughing-in and mounting once the walls have been finished.

Fire-Alarm Systems

The purpose of any fire-alarm system is to save lives and property. Every fire safety authority agrees that the greatest single factor contributing to low fatalities and property loss is early fire detection and warning. In fact, it is generally agreed that what happens in the first 3 minutes of a fire represents the difference between minor inconvenience and major catastrophe. To accomplish this vitally important early detection and warning, designers locate fire-alarm stations at frequent intervals in corridors, auditoriums, and other places that are usually occupied; and automatic fire detectors in boiler rooms, kitchens,

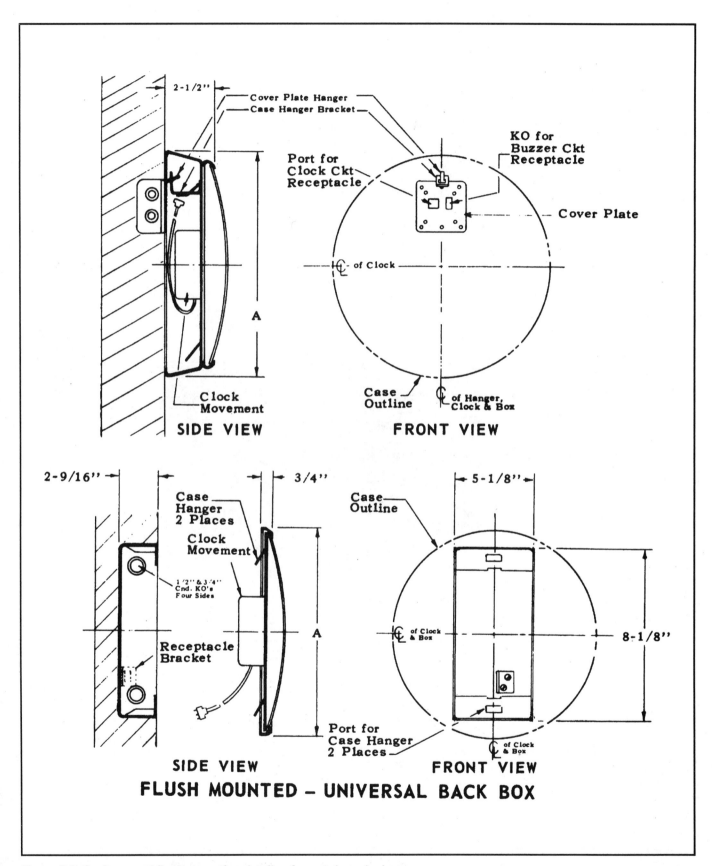

Figure 8-9: Surface- and flush-mounting details of secondary clocks.

ventilating shafts, attics, storage rooms, and so on, where a hidden fire might not otherwise be detected in its early stage. The electrical designer should also check with local ordnances and the latest recommendations of the National Fire Protection Association (NFPA).

The heart of fire alarm systems is the main or master control panel. To this panel are connected various detector and alarm circuits, as shown in Figure 8-10.

The primary power is taken from an unswitched three-wire distribution line of 120/240 (120/208) V ac (in most

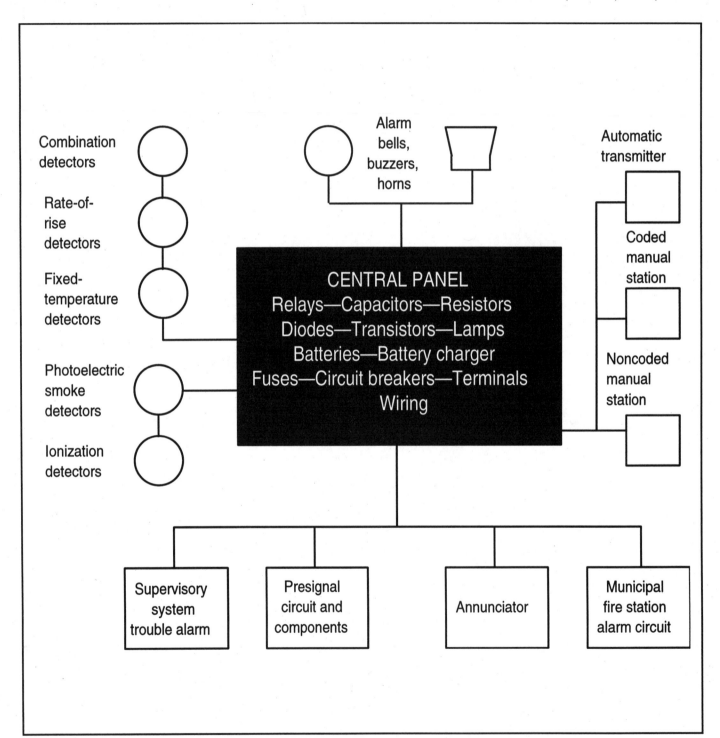

Figure 8-10: The heart of a fire-alarm system is the main or master central panel, shown here with the basic components attached.

cases). The initiating and alarm circuits are connected to the neutral ground and to one leg of the main circuit. The trouble-indicator circuits are connected to neutral ground and to the opposite leg of the circuit.

When an automatic detector or manual station is activated, the contacts close to complete a circuit path and apply 120 V to the alarm control circuits in the main panel. This includes a synchronous motor on some systems, which immediately operates cam assemblies that cause the alarm circuit switch contacts to make and break in a code sequence (if this is used). Additional cam-controlled switches stop the motor and alarm signals after, say, four complete rounds and actuate the alarm buzzer on the main panel.

Most panels contain a supplementary relay control for connection to an external auxiliary circuit providing their own electrical power. The relay usually has a single-pole double-throw contact, which operates in step with the master code signal. The circuit may be used to activate other auxiliary alarms or controls, such as a city fire-department connection, fan shutdown, or door release.

Fire-Alarm Wiring Installations

Electrical details of the complete installation should be provided in the form of working drawings, shop drawings, written specifications, or a combination of these.

The primary rule for designing fire-alarm systems, and also for showing the connections is to follow the manufacturer's instructions. To do otherwise is asking for trouble. This rule cannot be overemphasized because the require-

ment for fire-alarm circuits and their connections to initiating devices and indicating appliances makes fire-alarm system wiring very different from general wiring for light and power.

A manufacturer's installation wiring drawing routes wires and makes connections in a certain manner because of the supervision requirements. Any variance from the original layout might cause a portion of a circuit to be unsupervised and, if an open or short occurred, prevent the circuit from performing its intended function, and possibly lead to loss of life.

The rules of fire-alarm supervision are complex. Unless an installer specializes in such installations, he or she is not likely to be familiar with them. It is possible that hardware that appears to be identical in two different buildings are wired radically different. Consequently, it is the design professional's responsibility to adequately show the intent of the system, and also how it is to be wired.

Drawings supplied by the manufacturer will show how the units are connected into a system. However, it is the designer's responsibility to show how to interconnect these devices to accomplish the goals of the fire-alarm system. For example, drawings in Figure 8-11 show initiating devices correctly and incorrectly wired into a system.

Wiring and riser diagrams as shown in Figure 8-12 on the next page should be provided during the rough-in stage of construction as well as during actual connection of the system and the various components. Details of station mounting (Figure 8-13) should also be included, as should connection details (Figure 8-14) and complete schematic wiring diagrams (Figure 8-15).

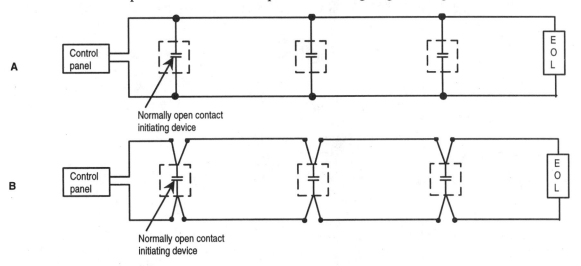

Figure 8-11: Initiating devices incorrectly wired (A) and correctly wired (B).

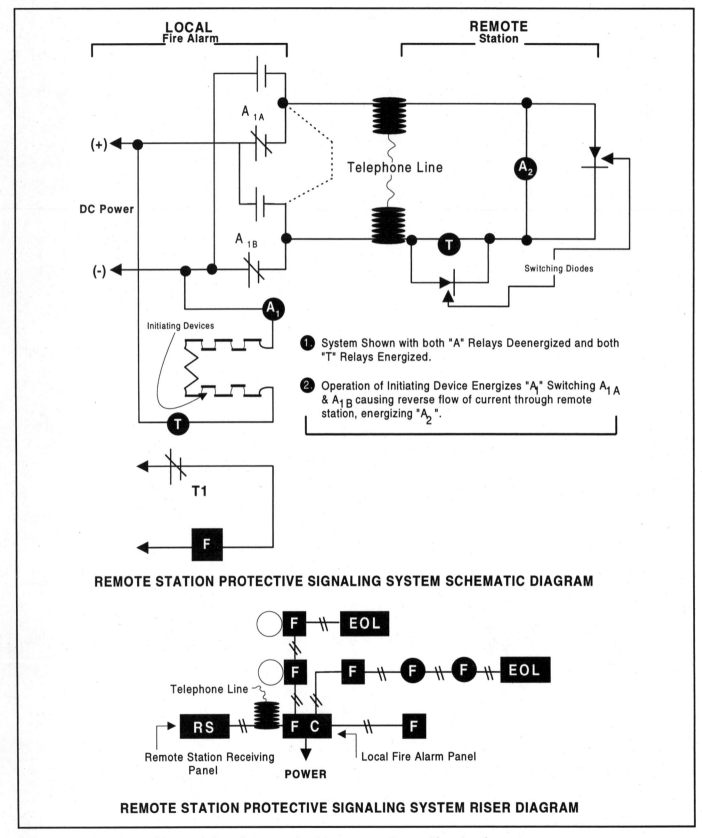

LOCAL
Fire Alarm

REMOTE
Station

A₁ₐ

(+)

Telephone Line

A₂

DC Power

A₁ᵦ

T

Switching Diodes

A₁

Initiating Devices

T

1. System Shown with both "A" Relays Deenergized and both "T" Relays Energized.

2. Operation of Initiating Device Energizes "A₁" Switching A₁ₐ & A₁ᵦ causing reverse flow of current through remote station, energizing "A₂".

T1

F

REMOTE STATION PROTECTIVE SIGNALING SYSTEM SCHEMATIC DIAGRAM

F   EOL

Telephone Line

F   F   F   F   EOL

RS   F C   F

Remote Station Receiving Panel   Local Fire Alarm Panel

POWER

REMOTE STATION PROTECTIVE SIGNALING SYSTEM RISER DIAGRAM

Figure 8-12: Fire-alarm wiring and riser diagrams should appear on the working drawings.

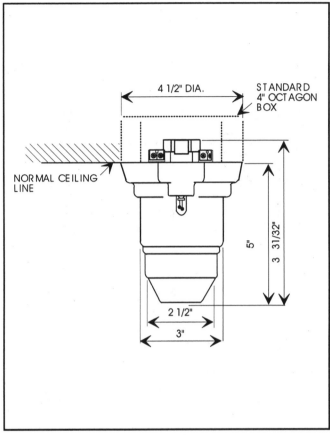

Figure 8-13: Mounting details for an ionization smoke detector.

LOW-VOLTAGE CONTROLS

Low-voltage controls offer a safe, quiet means of switching electricity that is not possible to obtain with conventional switching methods, and they have provided a new freedom in the design and use of circuit control.

With a low-voltage control system, it is practical for the designer to provide — at a nominal cost — sufficient multiple switch control of individual electrical circuits to create a new level in convenience for the user. A low-voltage control center can provide for remote control in the operation of multiple circuits from one or more location and gives visual indication of the use of each circuit. This capability in controlling lighting systems allows the electrical designer unusual latitude in providing his or her clients with the convenience desired.

Visual Aids for Schools

A variety of visual aids, such as blackboards, flip charts, maps, etc., can be illuminated by supplemental spot and flood lamps, each connected to a control center at the instructor's podium. Lights can be controlled, as required, during instruction so that the desired visual aid can be highlighted. A wiring diagram for this application is shown in Figure 8-16.

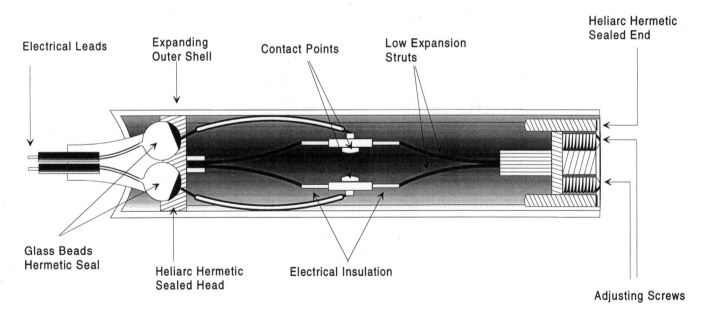

Figure 8-14: Rate compensating detector — a device that will respond when the temperature of air surrounding the device reaches a predetermined level, regardless of the rate of temperature rise.

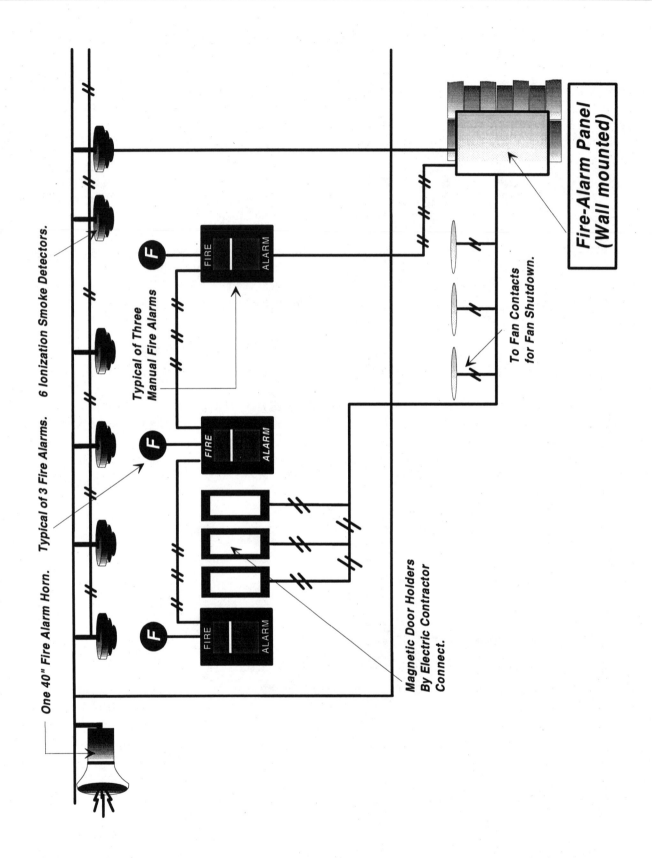

Figure 8-15: Riser diagram of a commercial fire-alarm system.

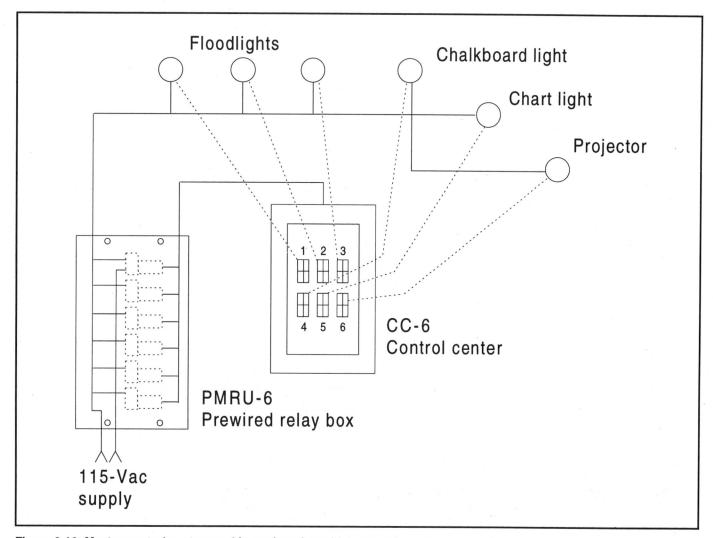

Figure 8-16: Master-control center used in conjunction with low-voltage relays.

Fail-Safe Alarm System

The purpose of this particular alarm system is to alert the bank manager or security office when any one of the teller stations is in need of assistance, without sounding a general alarm; that is, the teller may need change, an approval of a transaction, or a simular situation. Consequently, this system is not for an emergency, such as an attempted robbery.

Figure 8-17 shows the wiring diagram for a possible solution. Momentary contact switches are installed at each teller station and are wired with low-voltage wire to control centers in the offices of the manager and/or security officer. The control centers show at a glance the source of the warning. A bell, light, or buzzer may also be installed which can be turned off by key or pushbutton as action is taken.

Industrial Application

There are numerous applications of low-voltage controls in industrial establishments; the following is typical of one.

Problem: Lack of switch points owing to high ceilings and open truss construction. Attendant made rounds, turning out unused but sometimes necessary lights.

Solution: A low-voltage remote-control system was installed. Number 18 AWG wire was strung along girders, reaching previously inaccessible switching points. Larger areas could now have zone control. Control panels at main entries revealed which lights were burning and gave fingertip control, thus reducing operating cost. *See* Figure 8-18 for the wiring diagram.

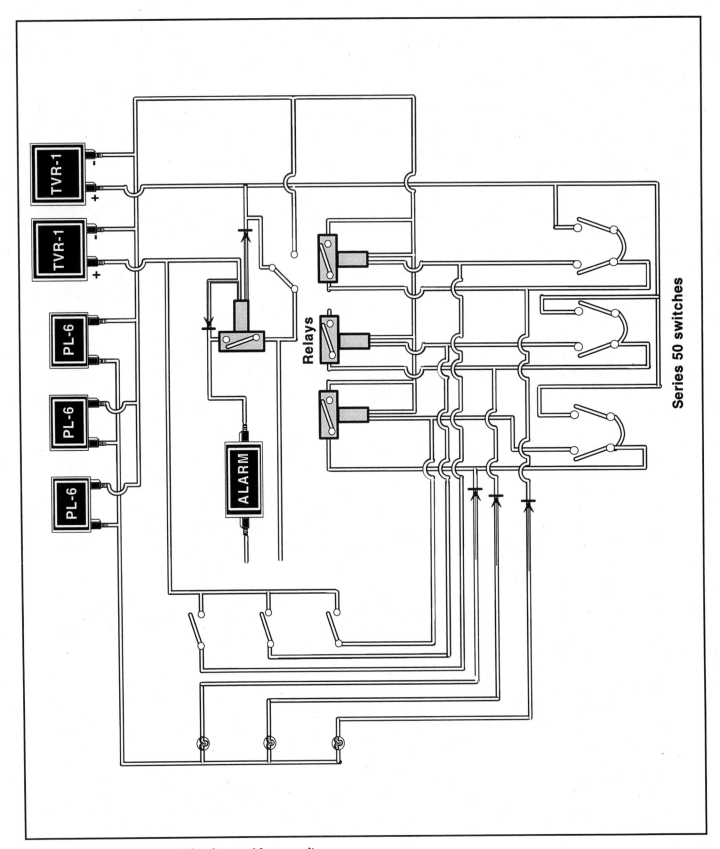

Figure 8-17: Remote-control circuits used for security purposes.

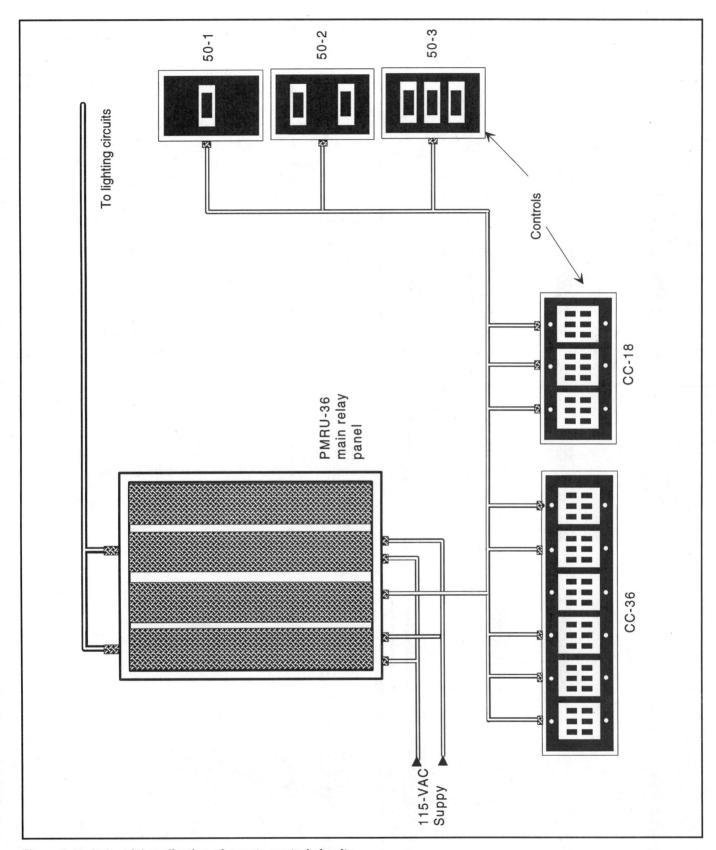

Figure 8-18: Industrial application of remote-control circuits.

Motel Management

A low-voltage master-control center installed in a motel office and also in the supply room can save much time in determining which rooms are ready for occupancy. (*See* Figure 8-19). As soon as maid service is completed, the office is signaled from the supply room. When guests check out, the manager signals for maid service, maintaining a constant readiness and flow of work.

Home Protection

Floodlights are installed at the eaves of a house. These, along with entrance lighting fixtures and three landscape lighting circuits, were connected to a master-control panel in the master bedroom. When there are sounds or visitors in the night, occupants can instantly light up the entire property. *See* Figure 8-20.

Nurse Call System

When a patient needs assistance, a low-voltage switch at the patient's bed can be used to turn on a corridor light outside the room as well as a light in the master-control center at the duty station. If it is wired to a switch in the patient's room, the light at the control center will remain on until the patient receives attention. Then the light can be turned off by the switch in the patient's room. Figure 8-21 on page 664 shows a wiring diagram of such a system.

Nursing Homes

Low-voltage switches can also provide ample switching for lights in nursing homes so that elderly patients need not walk in darkness or risk shock from wiring. In addition, master- control centers can provide zone lighting, allowing entire areas to be lit at once in case of emergencies such as fires, missing patients, and other similar situations where the ability to light the entire area is advantageous. *See* Figure 8-22.

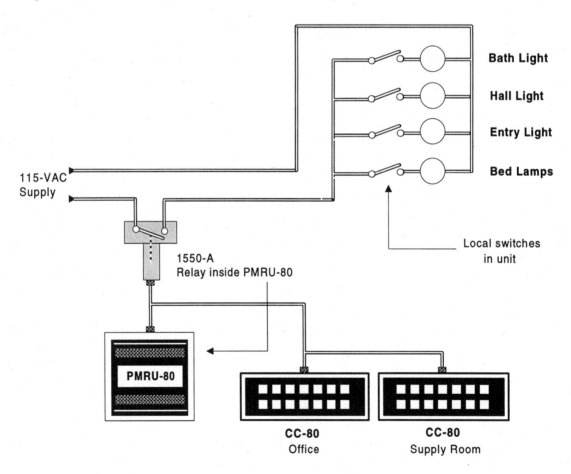

Figure 8-19: Low-voltage used in a motel.

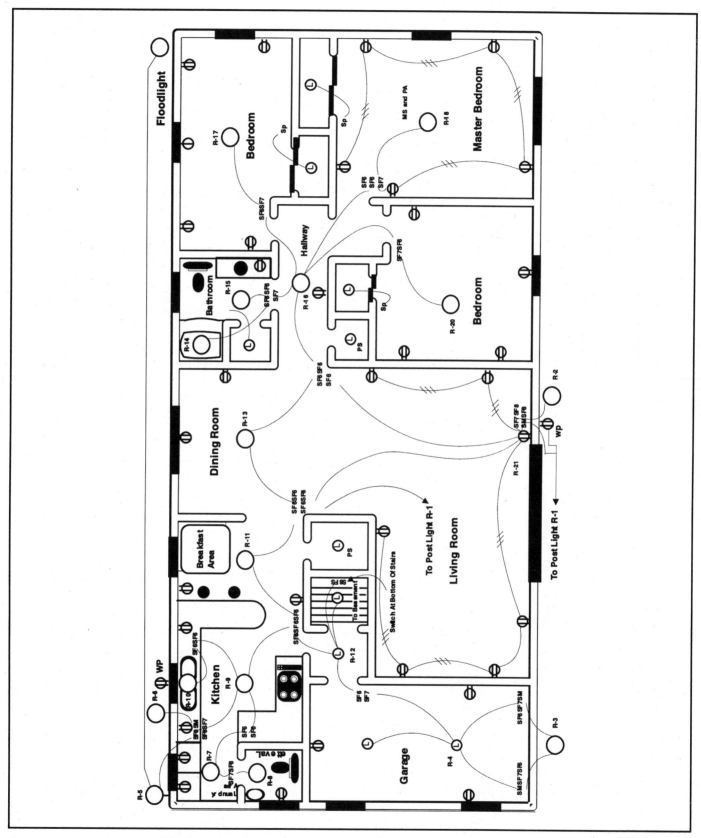

Figure 8-20: Remote-control circuits used for home protection.

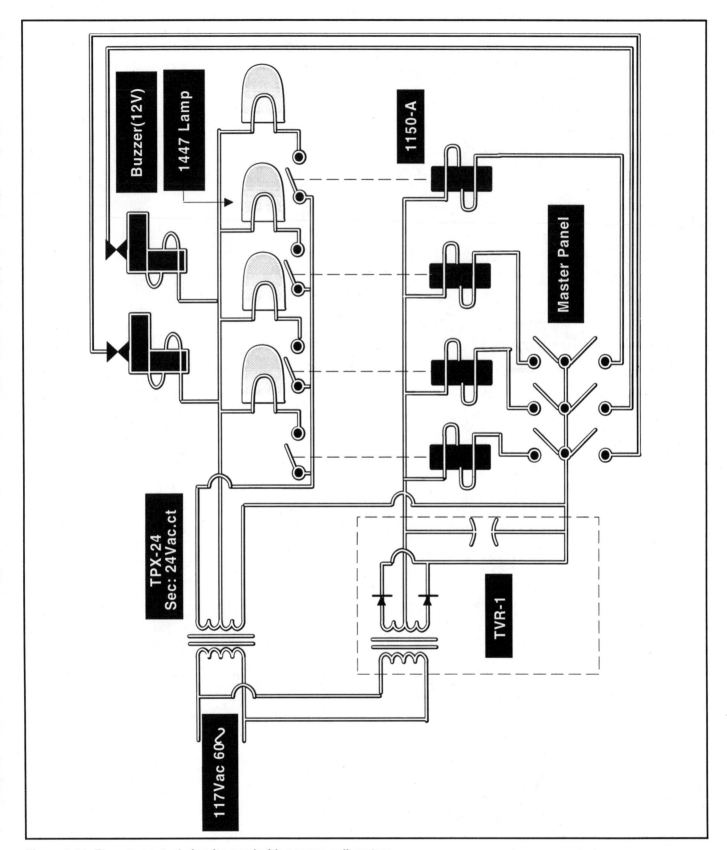

Figure 8-21: Remote-control circuits used with a nurse-call system.

Travel Trailers and Yachts

When space limitations are a problem, low-voltage switching is very practical, since the 18-gauge wire requires a minimum of space in the raceways and only a ½-in depth is necessary for the switches. Because switch legs carry low voltage, danger of fire and personal injury is kept to a minimum. Figure 8-23 on the next page shows a wiring diagram for such a system.

Office Use

Should a large firm need to know where its key personnel are at all times, low-voltage control panels installed in each office and activity area can be wired into a similar panel at the receptionist's desk. By assigning a color to each person and a number to each room, it is possible to tell at a glance where each person is at any given moment. It is also possible to know who is in a particular office. *See* Figure 8-24.

Modular Housing

Problem: Exterior walls were precast. Thin, interior wall partitions were designed to be rearranged as changing space needs dictated.

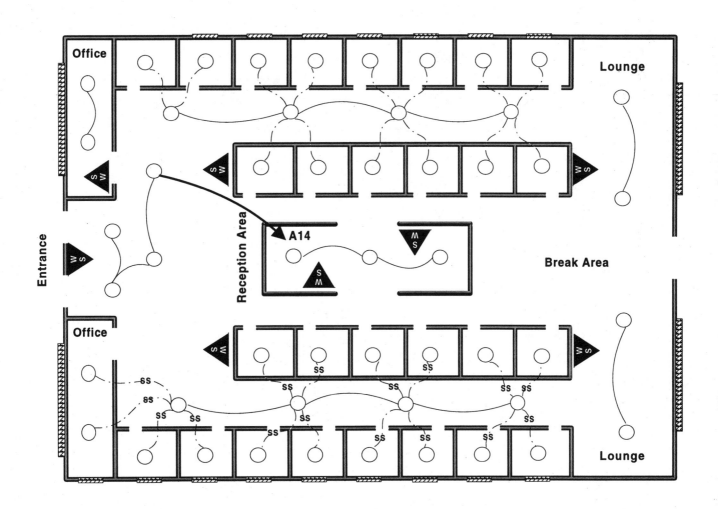

Figure 8-22: Application of a remote-control circuit used in a nursing home.

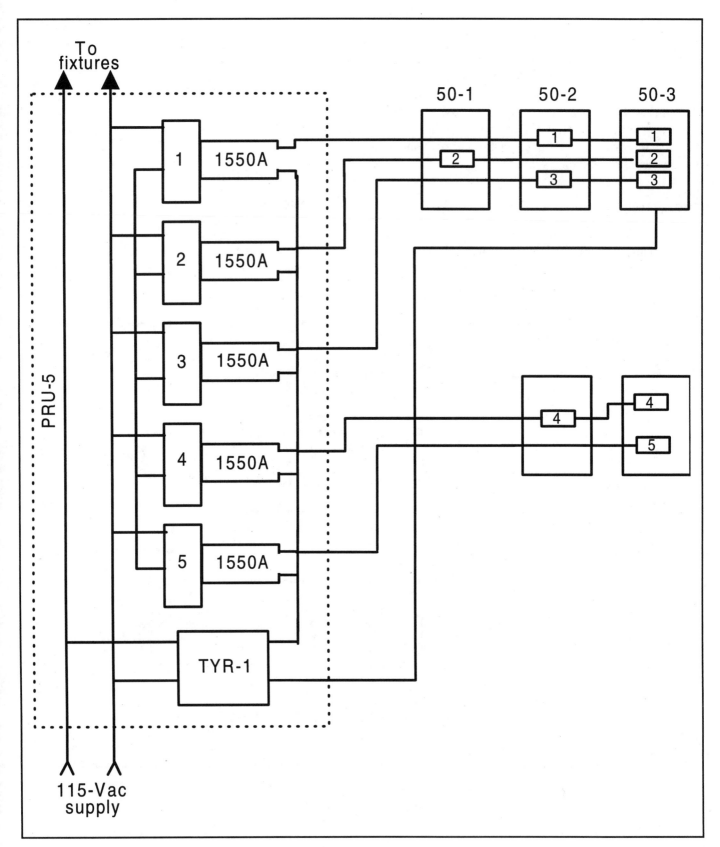

Figure 8-23: Use of remote-control circuits in travel trailers and yachts.

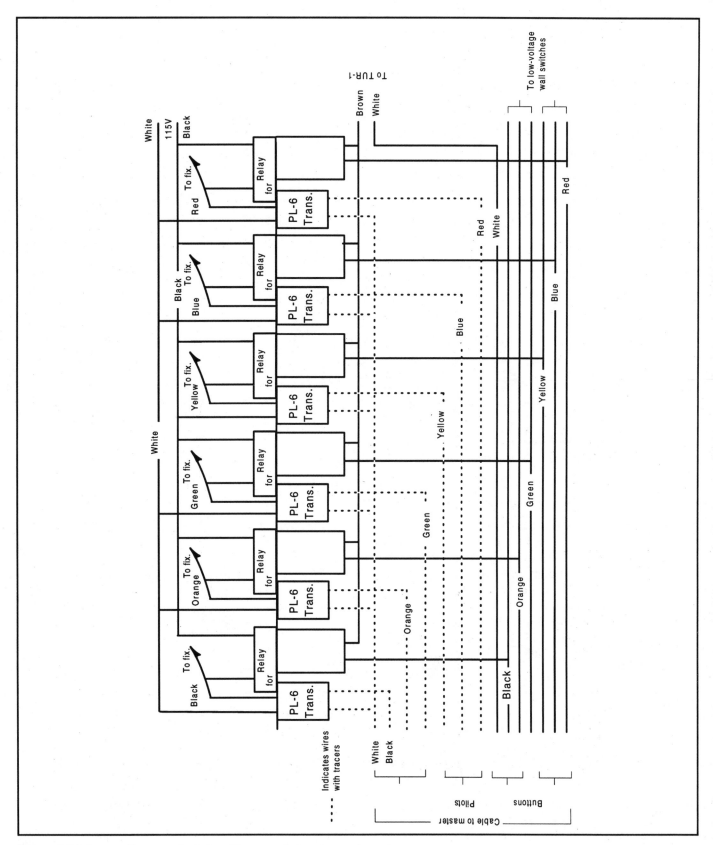

Figure 8-24: Low-voltage controls in office area to locate personnel.

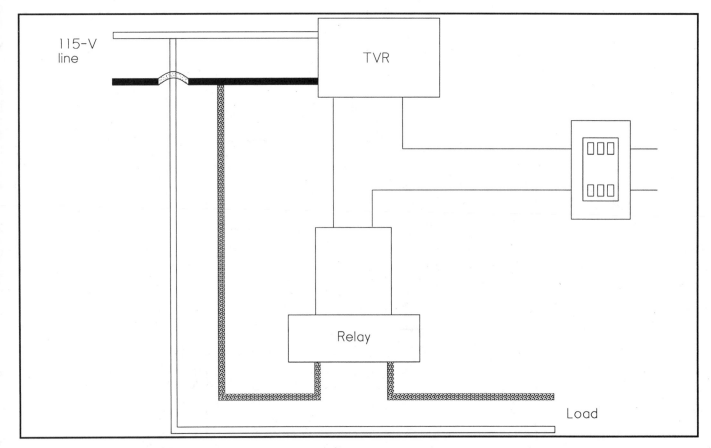

Figure 8-25: Use of low-voltage controls in modular housing.

Solution: Conduit and outlet boxes were installed in the precast sections. All interior sections were wired with double runs of low-voltage wire so that low-voltage switches could be installed at either side of the panel, however it was placed. Multibutton control centers provided maximum control of all circuits. See Figure 8-25.

Restaurants

Sometimes it is desirable for restaurant managers to know instantly when a specific table anywhere in the restaurant becomes available for seating, without a lot of leg work.

A lighted scale model of the restaurant floor plan could be wired in parallel with control centers at two serving stations. The moment a table is cleared, the waiter pushes a button on his panel, signaling the manager that the table is ready. *See* Figure 8-26.

SIGNAL AND COMMUNICATION SYSTEMS

The field of electric signaling and communications is a very broad one, covering everything from simple residential door chimes to elaborate building-alarm and detector systems.

Electric Chime Systems

One of the simplest and most common electric signal systems is the residential door-chime system. Such a system contains a low-voltage source, a pushbutton, wire and set of chimes. The quality of the chimes will range from a one-note device to those which "play" lengthy melodies.

The wiring diagram in Figure 8-27 on the next page shows a typical two-note chime controlled at two locations. One button, at the main entrance, will sound the two notes when pushed, while the other button, at the rear door, will sound only one note when pushed. Any number of pushbuttons may be added to either circuit. However,

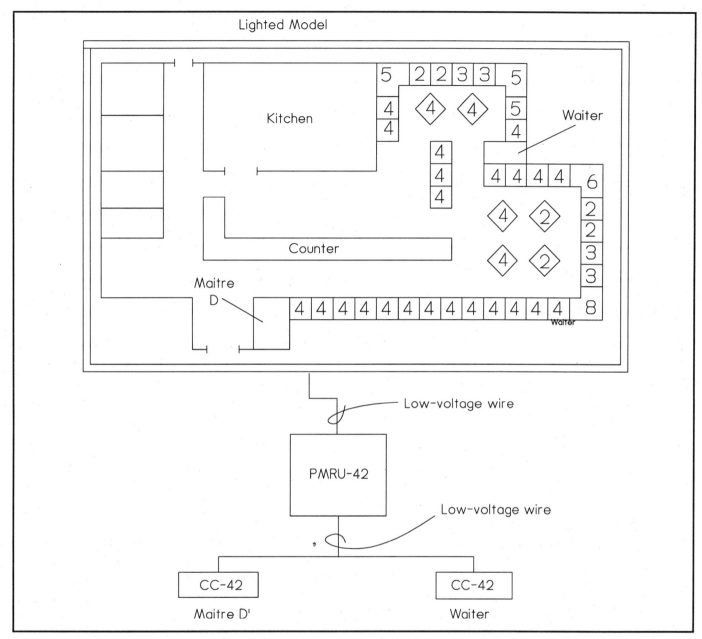

Figure 8-26: Low-voltage controls used in restaurant.

when using this particular set of chimes, only two note combinations are available; that is, either one note or two notes. When more than two doors need to be recognized, chime sets with more than two note combinations are required.

MISCELLANEOUS SYSTEMS

Intercom Systems

A local rescue squad unit recently built an addition onto their existing building. During the planning stages of this addition, it was decided to install a new intercom system throughout the building. The system was to be designed so that the dispatcher could talk or listen to the stations within the building either individually or simultaneously and so

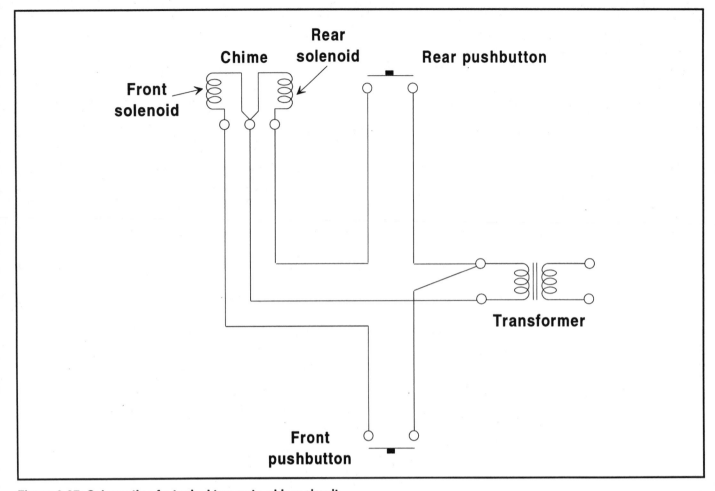

Figure 8-27: Schematic of a typical two-note chime circuit.

that any one station could call and talk with the dispatcher or any other individual station within the building.

The intercom-riser diagram in Figure 8-28 gives a schematic diagram of all the stations as well as the related wiring, conduit sizes, etc. The manufacturer and catalog number of each device are also indicated on this riser diagram along with other notes, in order to facilitate the installation of the system.

A detailed description of the system should also appear in the written specifications in case the contractor wishes to substitute another brand of equipment.

Telephone Systems

Let's assume that a branch bank utilizes several wall-mounted telephone outlets and also four floor-mounted outlets. Such outlets are normally shown on a floor-plan of the building using conventional symbols. A telephone cabinet, usually located in the mechanical room for the building, should also be shown on the floor plan. The written specifications for the project should describe the type of outlet box to be used for each of the two outlet types and a telephone-riser diagram should appear as shown in Figure 8-29. Note that this drawing shows a ¾-in. empty conduit with a pull wire in each run from each telephone outlet and terminates in the telephone cabinet. In most localities, the local telephone company will be responsible for installing the telephone cabinet as well as pulling the required cable to each telephone outlet within the building.

ALARM SYSTEMS

Bank Alarm Systems

Figure 8-30 shows riser diagram of various outlets for the bank security system such as camera junction boxes, smoke detectors, sound receivers, alarm buttons, etc. They should also be shown on a floor-plan drawing. Between these two drawings and the written specifications, suffi-

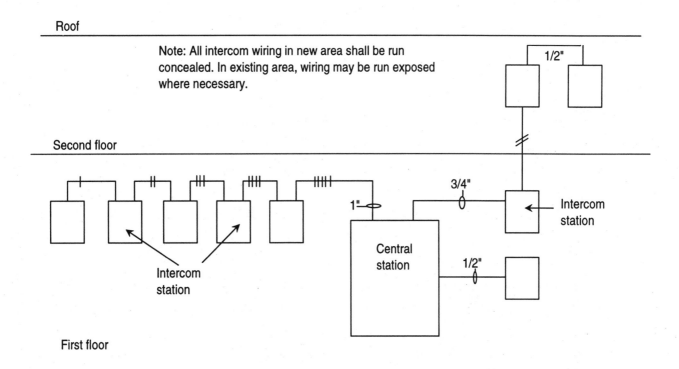

Figure 8-28: Intercom riser diagram.

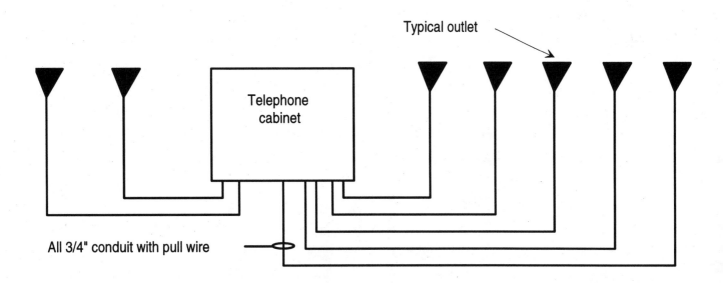

Figure 8-29: Telephone riser diagram.

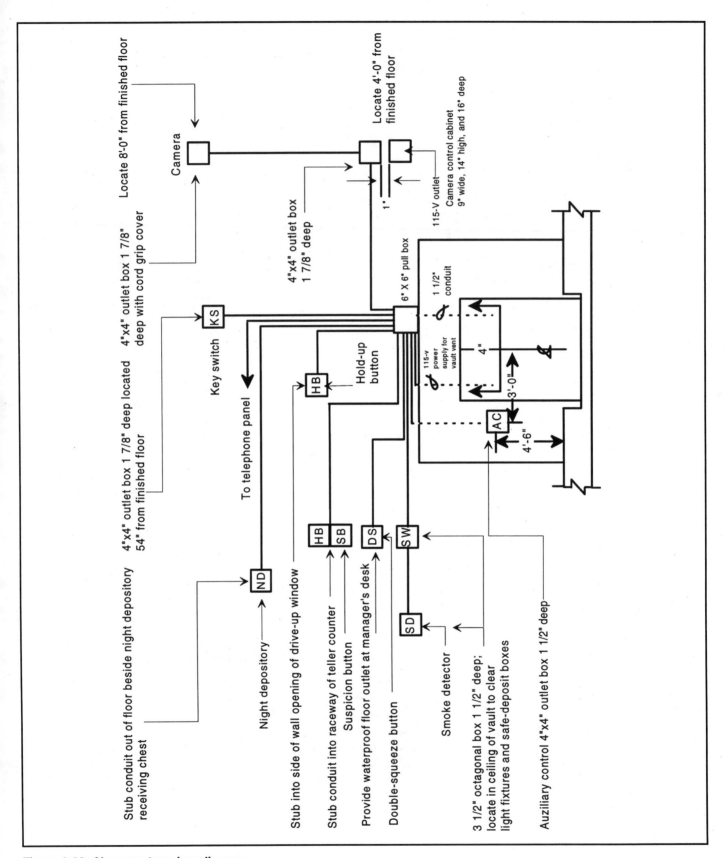

Figure 8-30: Alarm system riser diagram.

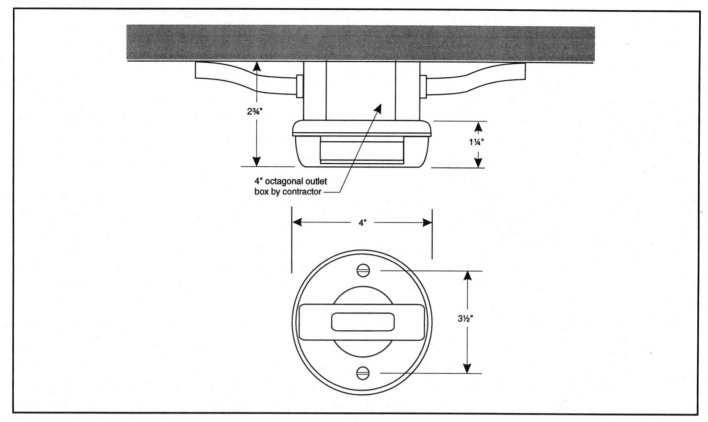

Figure 8-31: Mounting details of a surface-mounted fire detection device.

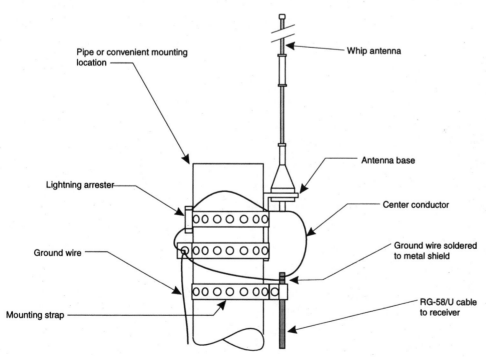

Figure 8-32: Mounting details of a whip-type antenna installation.

cient data should be supplied to indicate clearly how each outlet is to be installed.

There are many other details associated with communication and alarm systems. For example, Figure 8-31 shows the mounting arrangement of a surface-mounted fire detector; Figure 8-32 shows the mounting details of a whip-type antenna installation, while the detail in Figure 8-33 shows how a standard cast-iron floor box can be adapted for floor microphone plug-in connectors. Other details that have been used on actual working drawings begin with Figure 8-34.

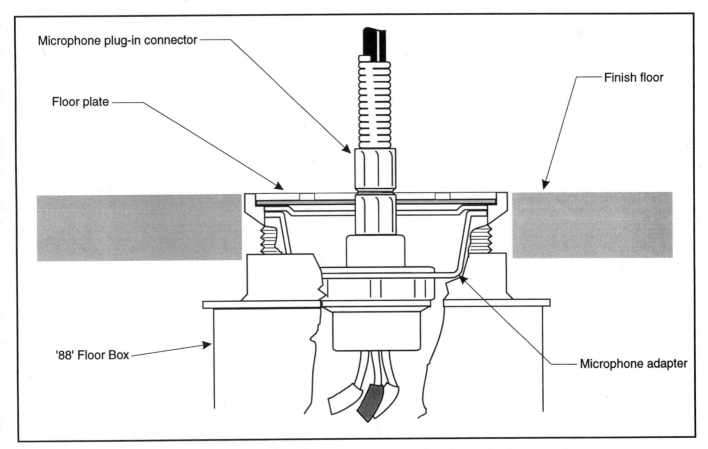

Microphone plug-in connector

Finish floor

Floor plate

'88' Floor Box

Microphone adapter

Figure 8-33: Standard cast-iron floor box shown with adapter to accept microphone plug-in connector.

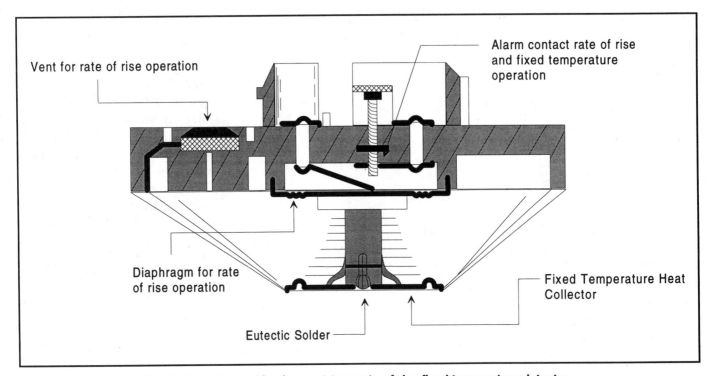

Vent for rate of rise operation

Alarm contact rate of rise and fixed temperature operation

Diaphragm for rate of rise operation

Fixed Temperature Heat Collector

Eutectic Solder

Figure 8-34: Cross-sectional view of a combination spot-type rate-of-rise fixed temperature detector.

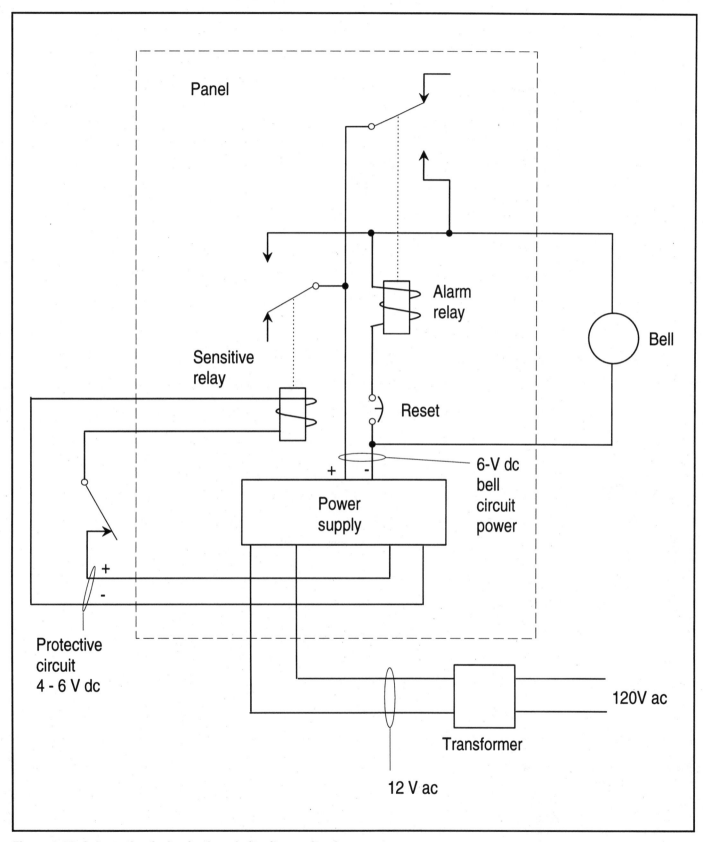

Figure 8-35: Schematic of a basic closed-circuit security alarm system.

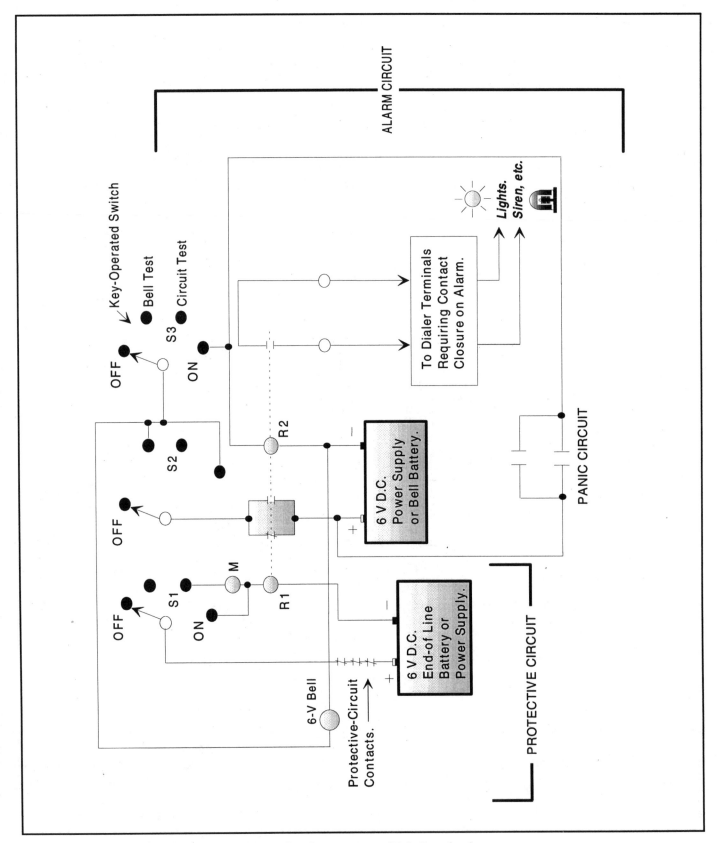

Figure 8-36: Schematic of a closed-circuit security alarm system with battery back-up.

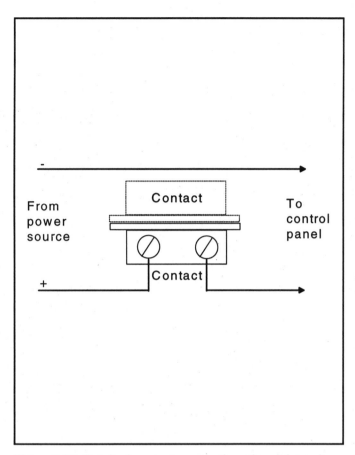

Figure 8-37: Detail of contact connections for an intrusion alarm.

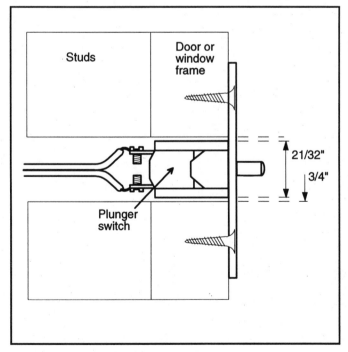

Figure 8-38: Cross-sectional view of a recess-mounted entry detector.

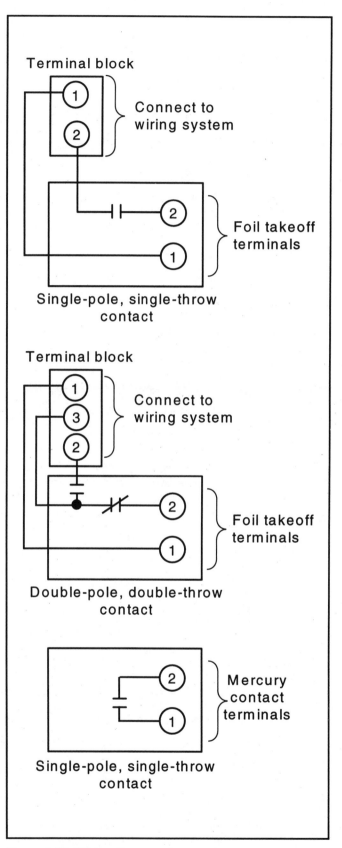

Figure 8-39: Wiring diagram of mecury contact connections.

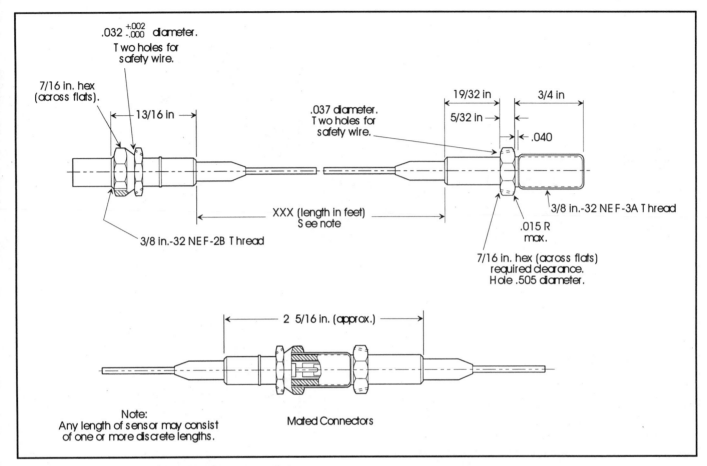

Figure 8-40: Detail of connectors used to supply the desired length of sensor cable.

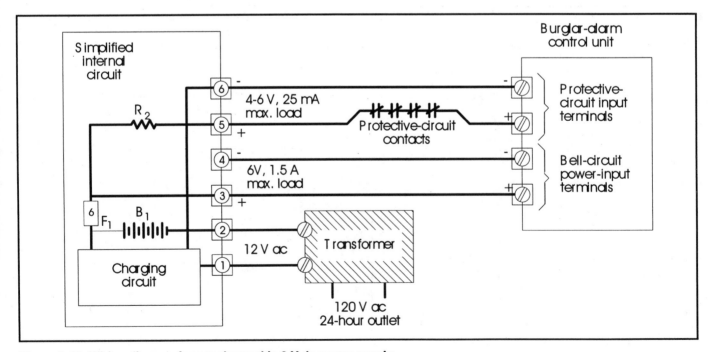

Figure 8-41: Wiring diagram for a rechargeable 6-V dc power supply.

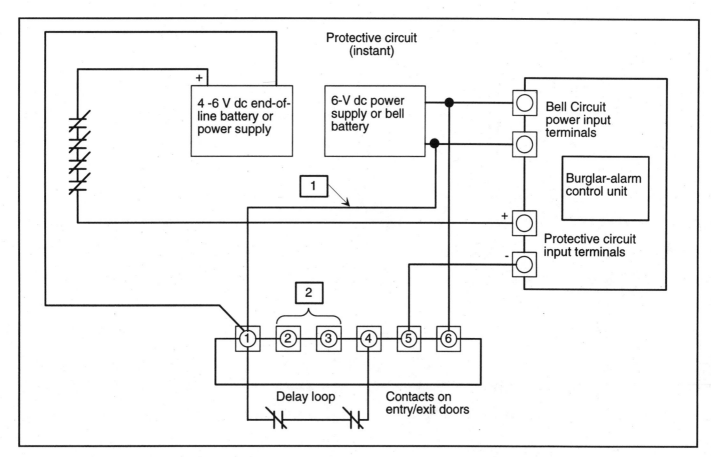

Figure 8-42: Connection detail of entry/exit delay module.

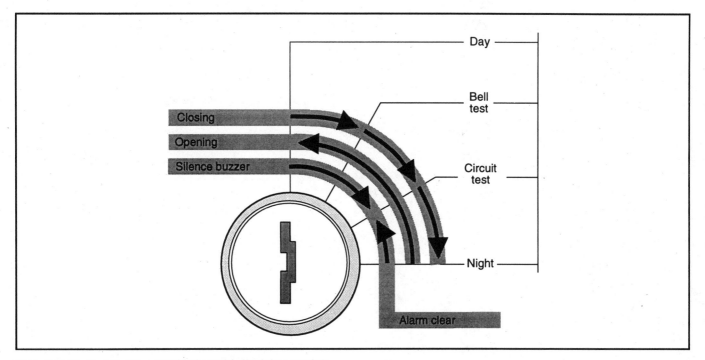

Figure 8-43: Connection detail of entry/exit delay module.

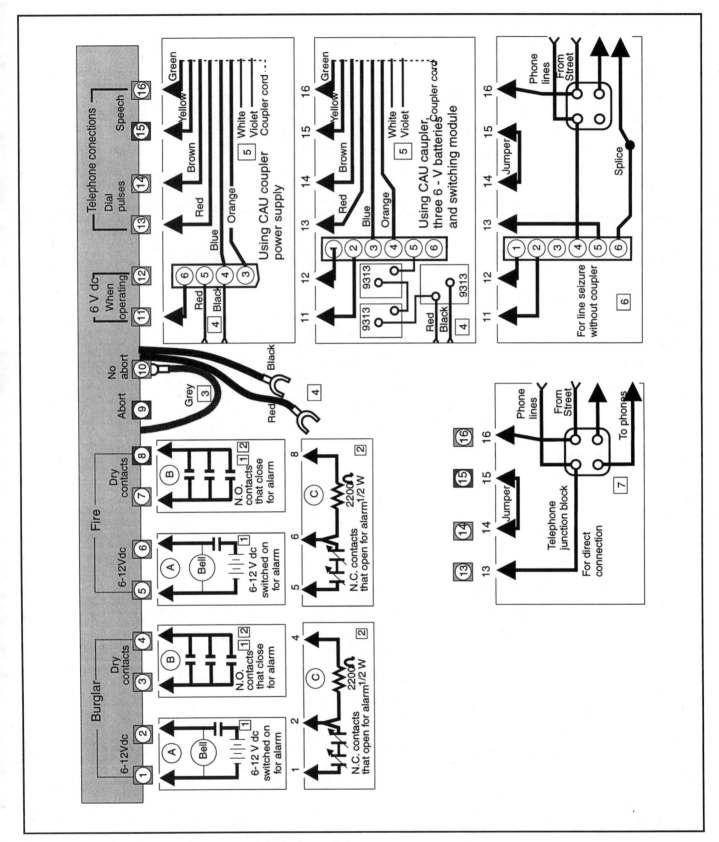

Figure 8-44: Wiring diagram of a commercial telephone dialer.

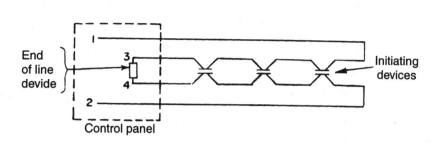

End of line device on control panel on 4-wire Class B alarm initiating circuit

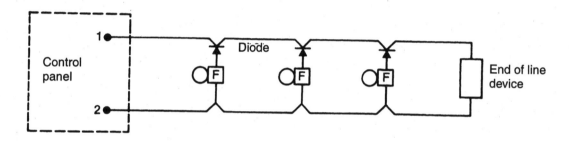

Polarized diode type alarm indicating appliances connected in parallel

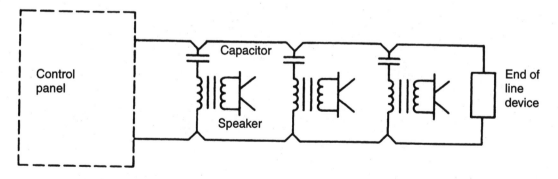

Speaker-type alarm indicating appliances

Figure 8-45: Speaker-type alarm indicating appliances.

## Chapter 9

Heating and Cooling

During the last half of the twentieth century, electric heating and cooling have found their way into thousands of homes throughout the United States, as well as in commercial and industrial establishments. The electrical contractor is always responsible for connecting branch circuits and feeders to this equipment, and in many cases, furnishes and installs the equipment and its related control wiring.

Besides electric space-heating equipment like baseboard heaters (*See* Figure 9-1) , fan (forced-air) heaters, duct heaters, heating cable, and the like, the electrical contractor is also responsible for other types of equipment,

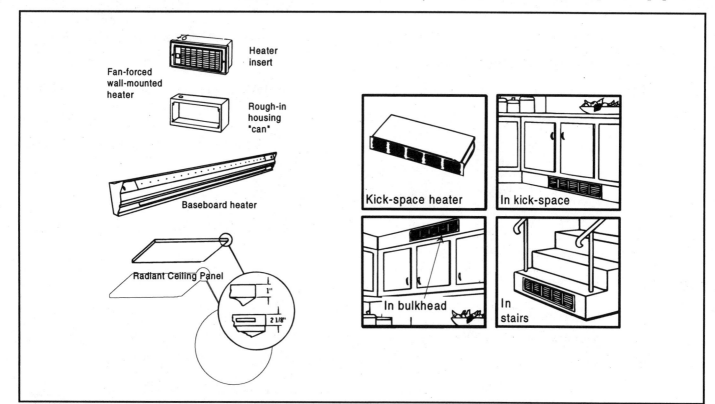

Figure 9-1: Common types of electric heaters for residential and small commercial use.

ELECTRIC HEAT SCHEDULE

HEATER TYPE	MANUFACTURER'S DESCRIPTION	DIMENSIONS	VOLTS	MOUNTING	WATTAGE/REMARKS
500 va	ElecTro-Heat Catalog No. 08531	3' x 5" x 2½"	240	Baseboard	500
1000 va	ElecTro-Heat Catalog No. 08531	6' x 5" x 2½"	240	Baseboard	1000
	ElecTro-Heat Catalog No. 08531	8' x 5" x 2½"	240	Baseboard	1500
1500 va	ElecTro-Heat Catalog No. 08531	16" x 12" x 3½"	240	Baseboard	1500 w/thermostat

Figure 9-2: Typical electric-heat schedule.

such as heat tracing, snow-melting systems, roof and gutter deicing systems, and packaged room HVAC systems.

For electric-heating units to be efficient and economical and to provide the desired comfort condition, the design of electric-heating systems must be based on precise heat-loss calculations of the areas to be served. The units must therefore be installed exactly as designed. Good electric working drawings and specifications can help to ensure the quality of such installations.

Electric-heat specifications should include the electrical characteristics of all equipment and controls. Because the location of heating units within a particular room or area is all important in obtaining a comfort condition within it, the location of the units should be shown on the project floor plans. A heating-equipment schedule is also desired and may be used to designate the equipment with symbols or other identifying marks on the working drawings. The schedule should include the manufacturer, catalog number, approximate dimensions of the units, the electrical characteristics, and other pertinent data. A typical schedule appears in Figure 9-2.

Where special conditions exists, or where manufacturer's rough-in specifications are not readily available, large-scale details (Figure 9-3) should be provided on the working drawings to ensure an accurate installation.

The following heating and cooling detail drawings have been used on several projects of various sizes. The drawings are designed so that they may be modified with very little trouble — using either manual drafting methods or electronic CAD systems — to fit exactly almost any project that is likely to be encountered by the electrical engineer, designer, or drafter.

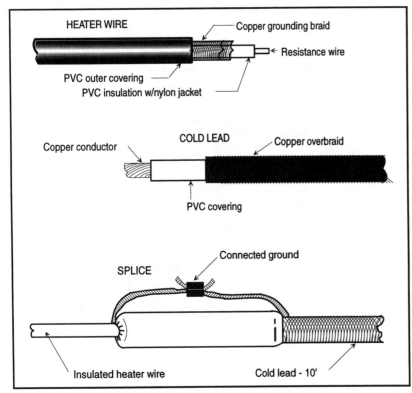

Figure 9-3: Splicing details for heat cable.

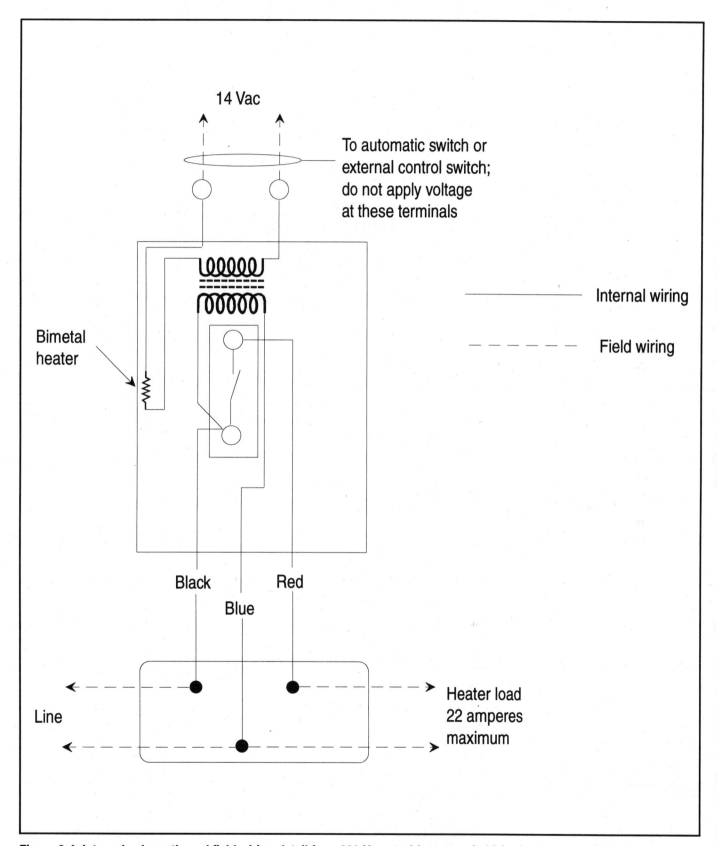

Figure 9-4: Internal schematic and field wiring detail for a 208-V control (power switch) for freeze-protection heat cable.

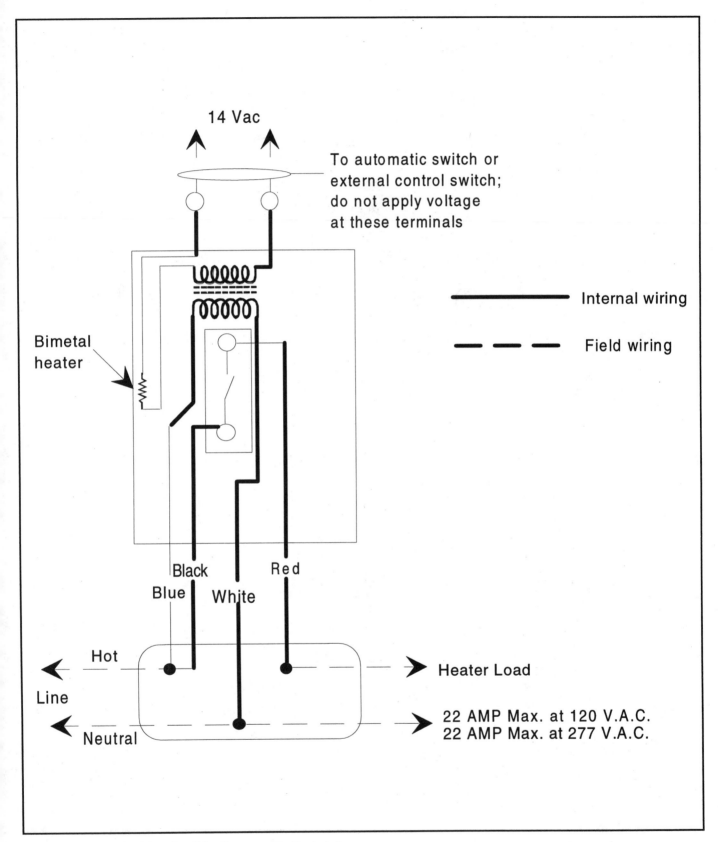

Figure 9-5: Control-wiring detail for freeze-protection systems.

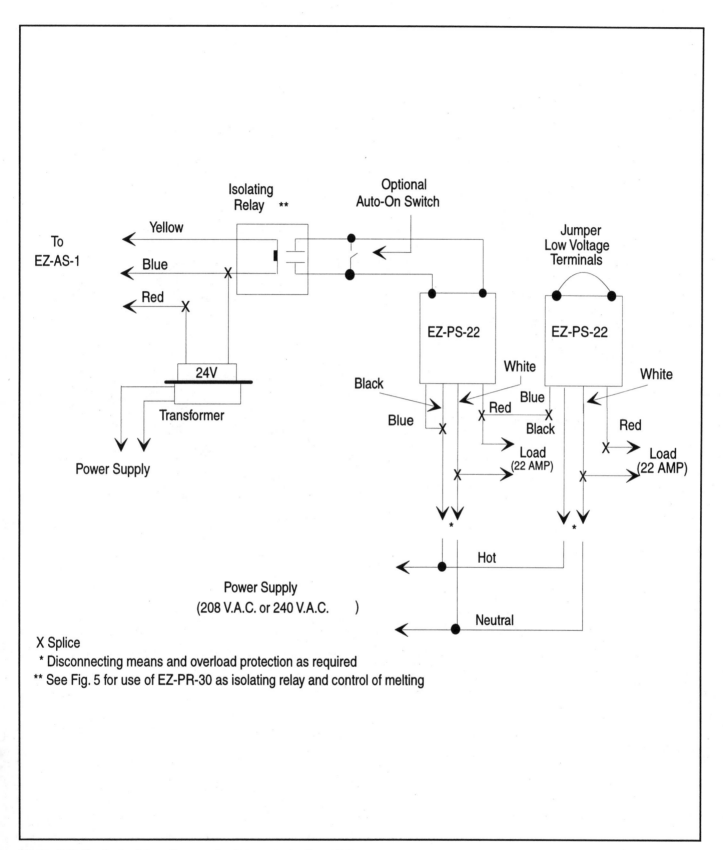

Isolating Relay **

Optional Auto-On Switch

Jumper Low Voltage Terminals

To EZ-AS-1

Yellow

Blue

Red

24V

Transformer

Power Supply

EZ-PS-22

EZ-PS-22

Black

White

White

Blue

Red Blue

Red

Black

Load (22 AMP)

Load (22 AMP)

*

*

Power Supply
(208 V.A.C. or 240 V.A.C.   )

Hot

Neutral

X Splice

 * Disconnecting means and overload protection as required

** See Fig. 5 for use of EZ-PR-30 as isolating relay and control of melting

Figure 9-6: Control-wiring diagram for freeze-protection systems.

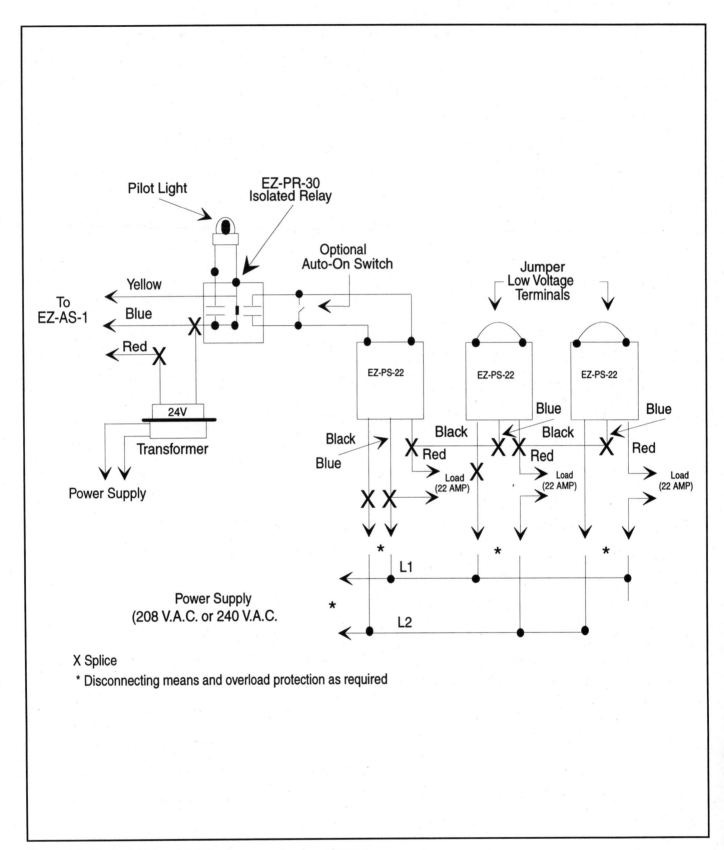

Figure 9-7: Control-wiring detail for freeze-protection systems.

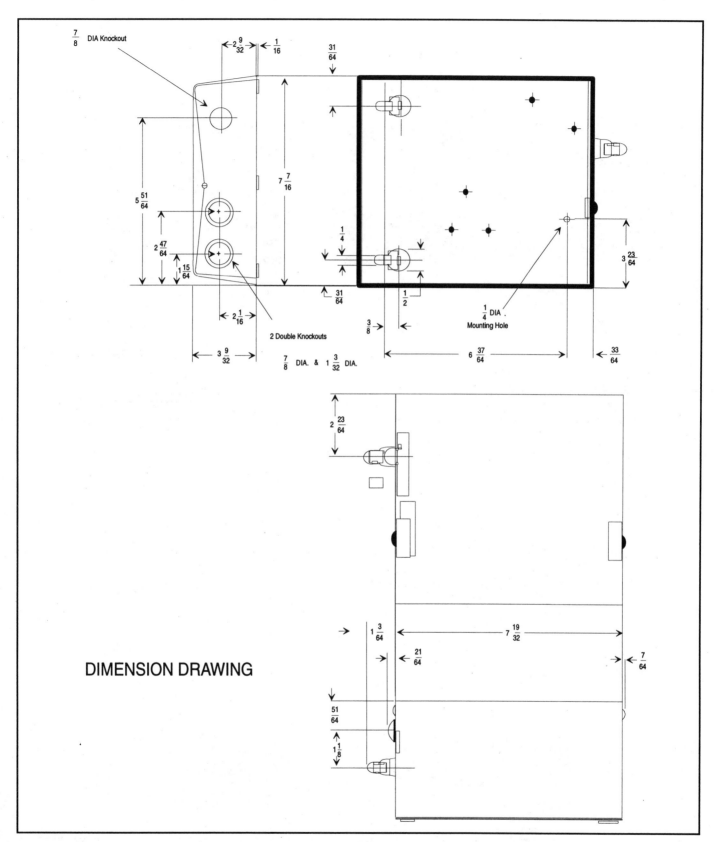

Figure 9-8: Dimensions of an electric-heating control housing. Vary dimensions to suit application.

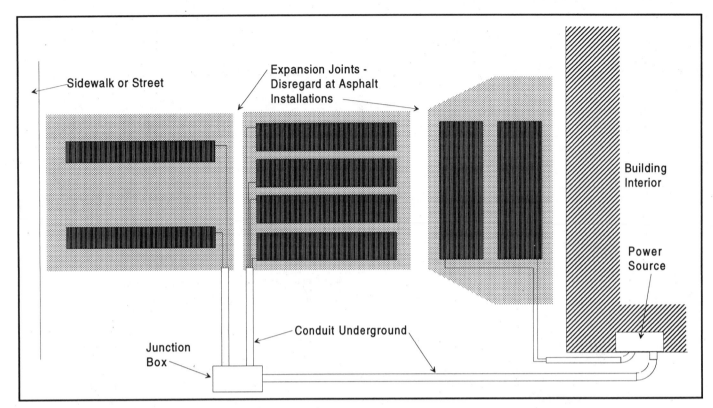

Figure 9-9: Plan view of a typical snow-melting application.

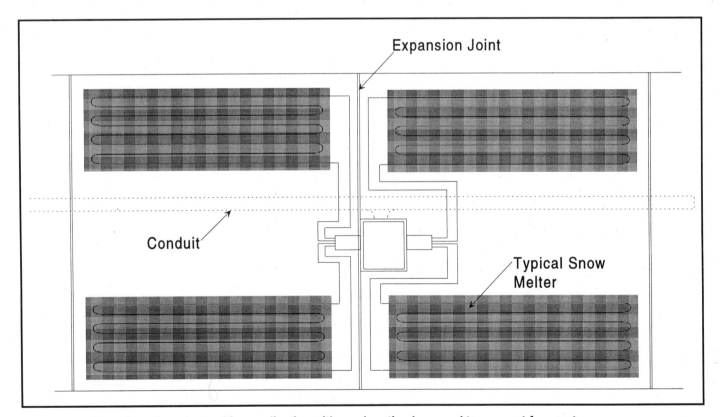

Figure 9-10: Plan view of a snow-melting application with one junction box used to connect four mats.

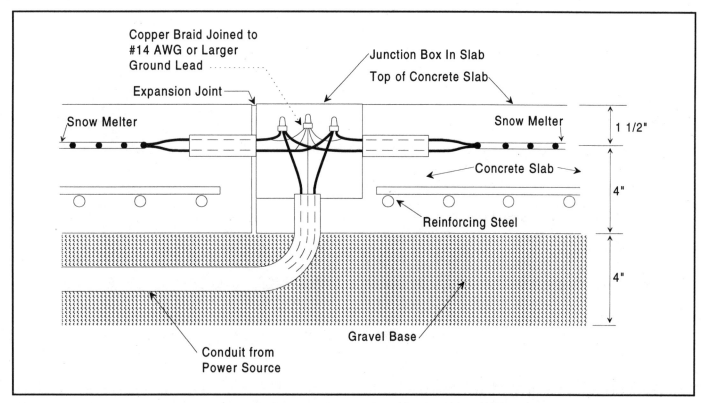

Figure 9-11: Cross-sectional view of connections for electric heater leads of snow-melting mats.

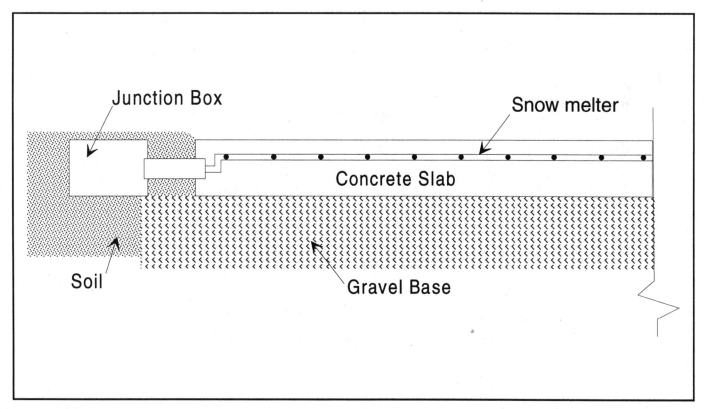

Figure 9-12: Cross-sectional view through a heated concrete slab.

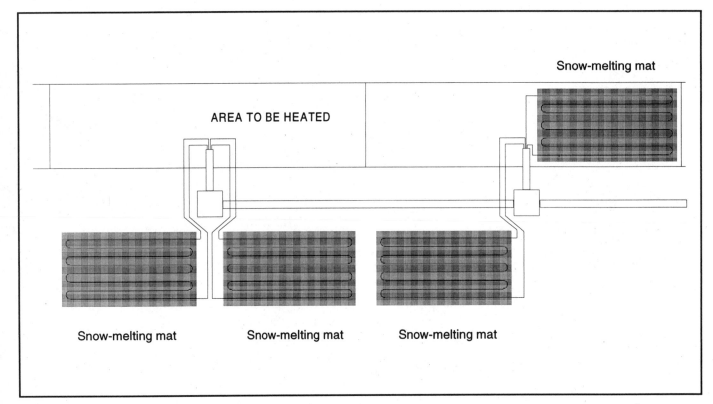

Figure 9-13: Plan view showing junction boxes with the connected heating mats.

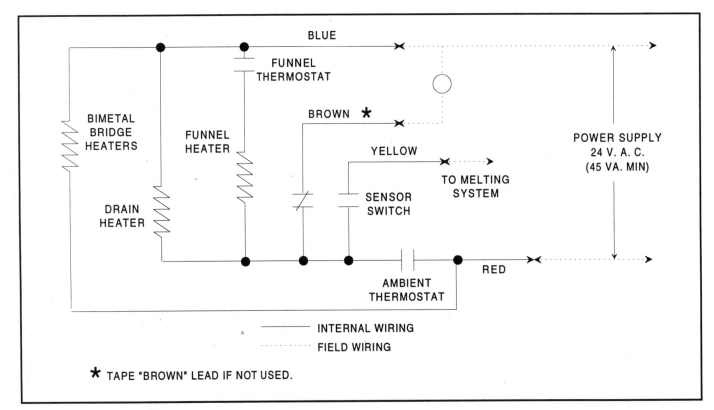

Figure 9-14: Schematic and partial wiring of an automatic snow-melting switch.

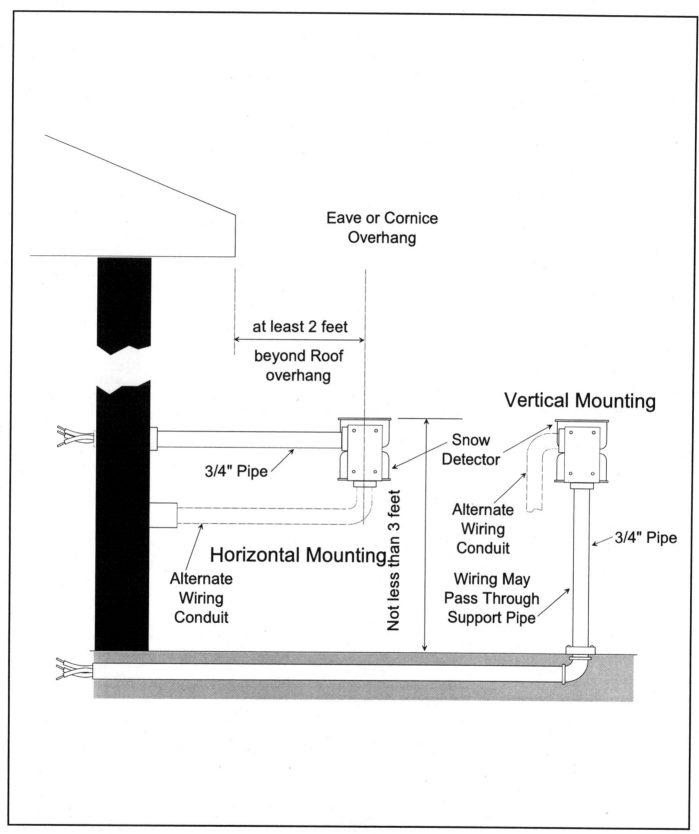

Figure 9-15: Elevation of methods used to mount automatic snow-detector switch.

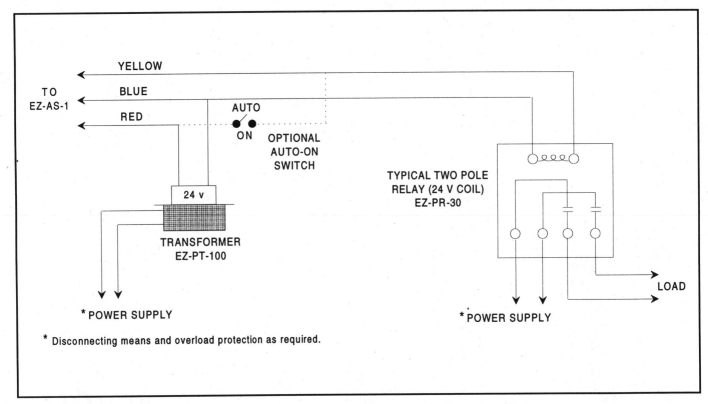

Figure 9-16: Connections of a two-pole relay with separate transformer and automatic switch.

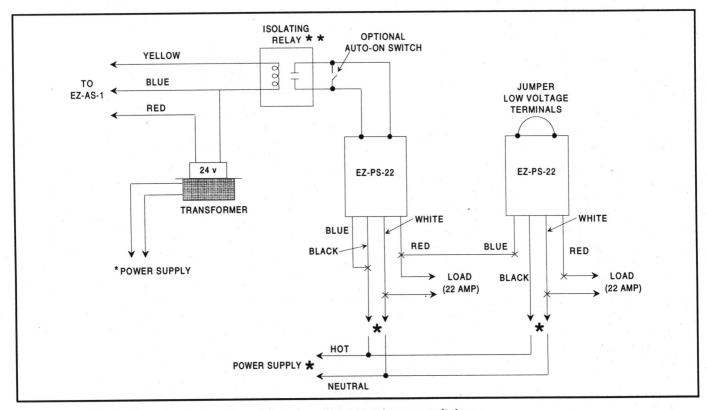

Figure 9-17: Field wiring of an automatic switch to transformer and power switches.

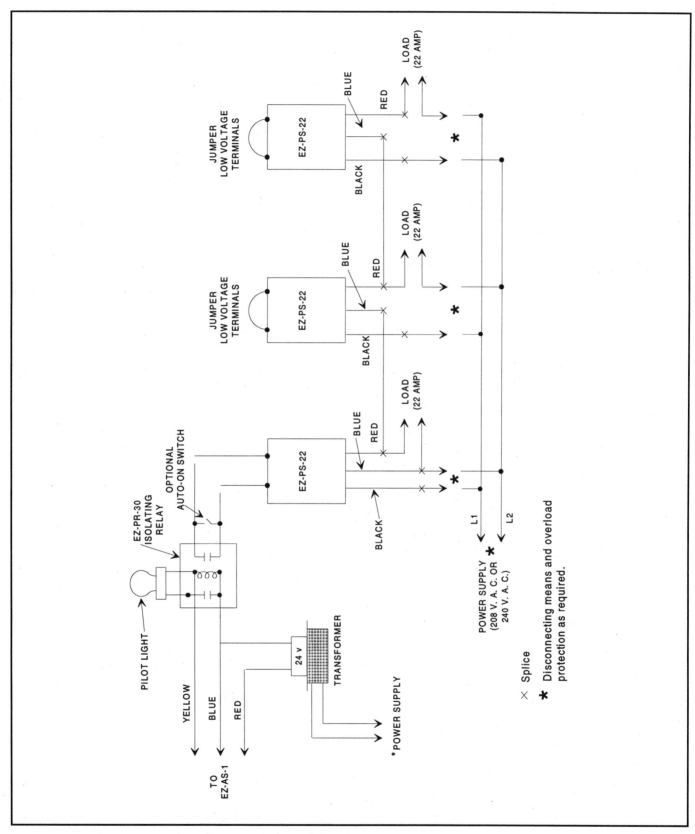

Figure 9-18: Field wiring of an automatic switch to transformer and 120-V switches.

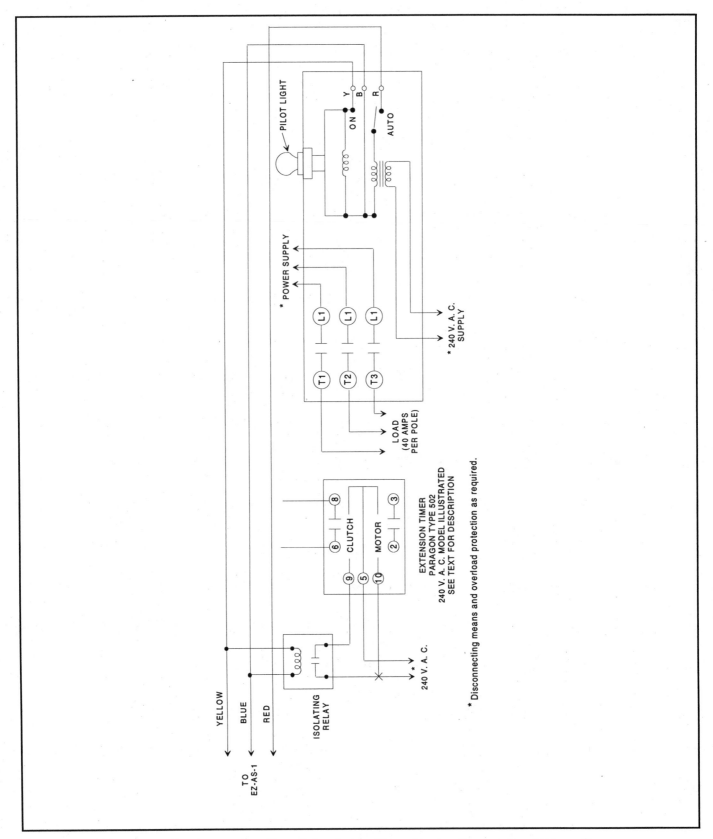

Figure 9-19: Wiring connection to an automatic switch using melting extension timer and contactor unit.

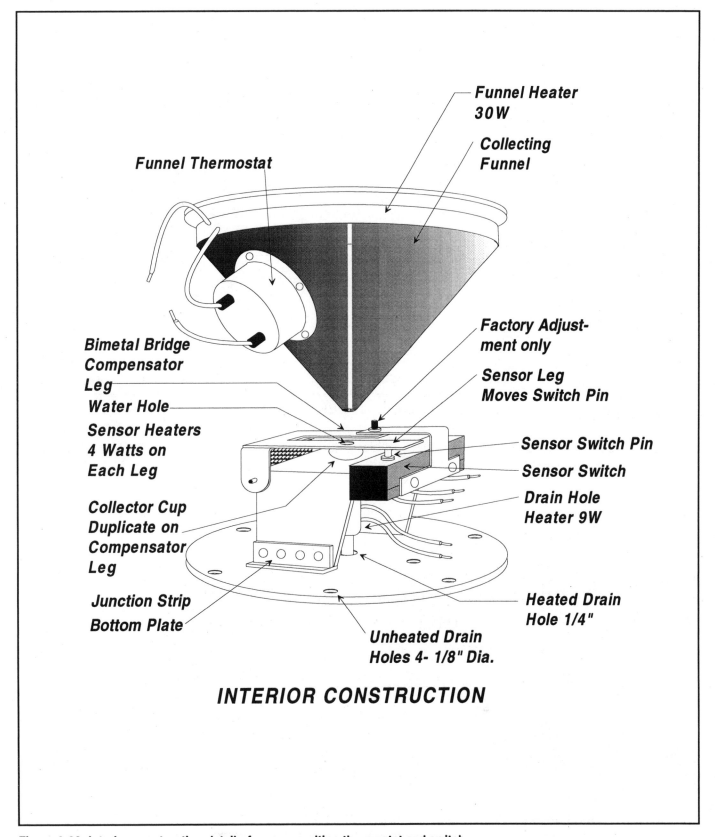

Figure 9-20: Interior construction detail of a snow-melting thermostat and switch.

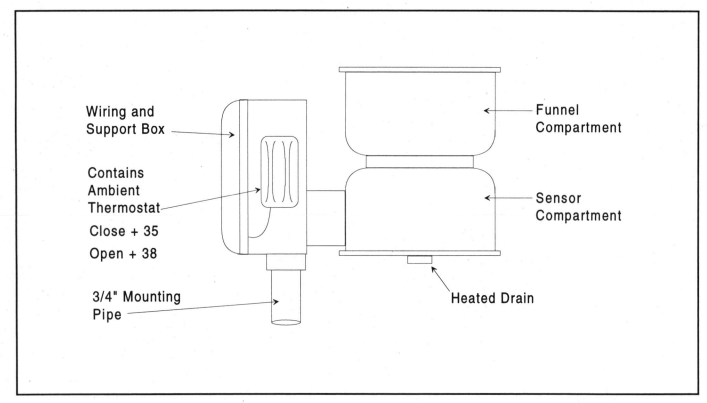

Figure 9-21: Elevation of typical snow-melting thermostat and switch.

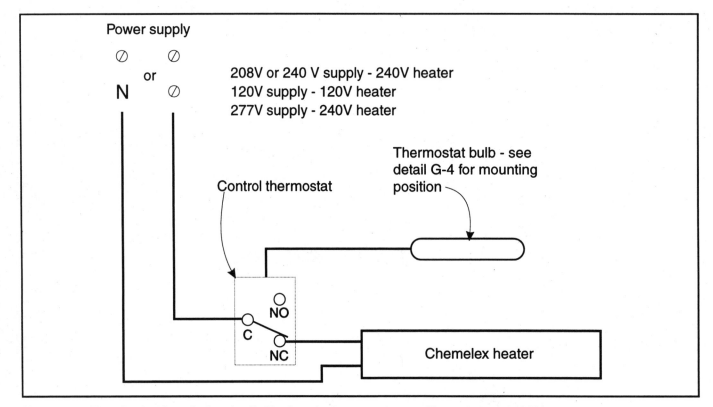

Figure 9-22: Heat-tracing installation detail: Single-pole thermostat controlling a 120- or 240-V heater.

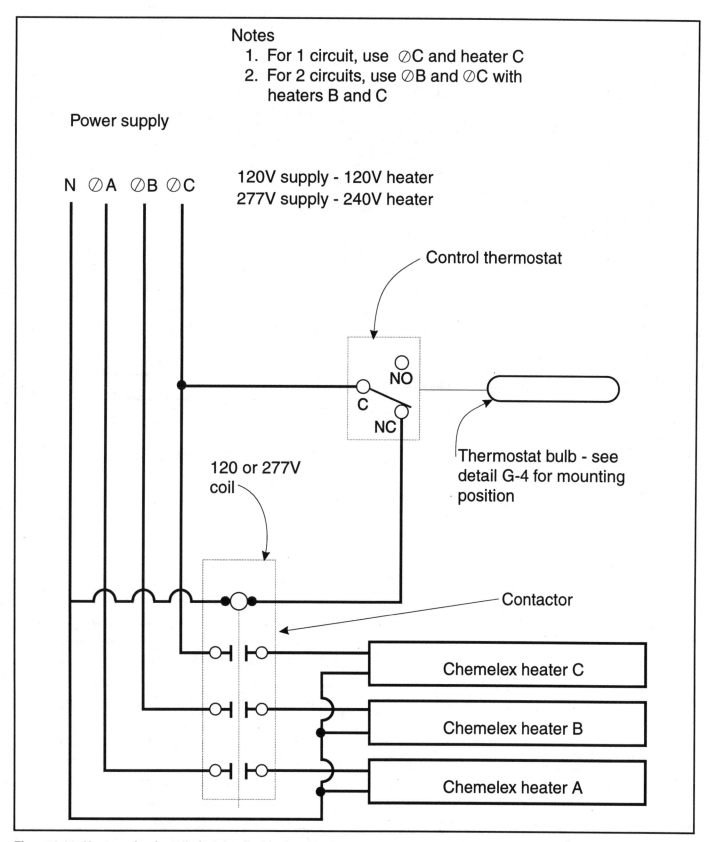

Figure 9-23: Heat-tracing installation detail: Single-pole thermostat and 3-pole contactor used to control 120- or 240-V heaters.

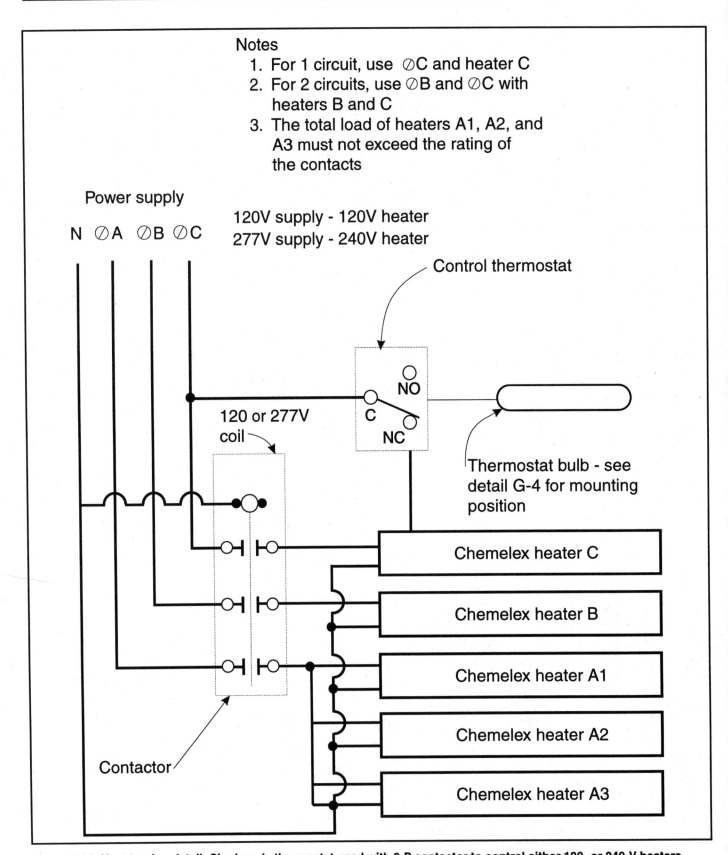

Notes
1. For 1 circuit, use ⊘C and heater C
2. For 2 circuits, use ⊘B and ⊘C with heaters B and C
3. The total load of heaters A1, A2, and A3 must not exceed the rating of the contacts

Power supply

N ⊘A ⊘B ⊘C

120V supply - 120V heater
277V supply - 240V heater

Control thermostat

NO
C
NC

120 or 277V coil

Thermostat bulb - see detail G-4 for mounting position

Chemelex heater C

Chemelex heater B

Chemelex heater A1

Chemelex heater A2

Chemelex heater A3

Contactor

Figure 9-24: Heat-tracing detail: Single-pole thermostat used with 3-P contactor to control either 120- or 240-V heaters.

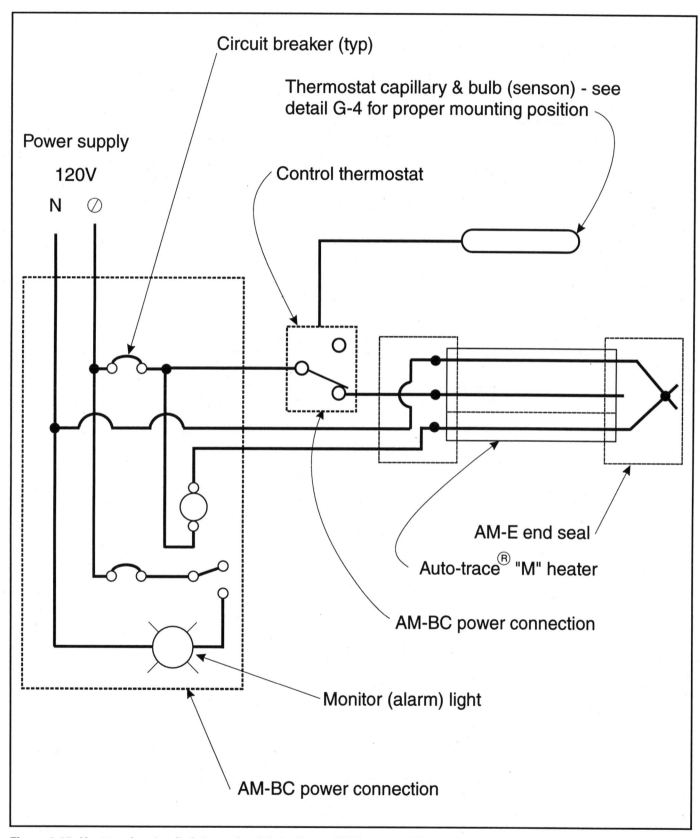

Power supply
120V
N

Circuit breaker (typ)

Thermostat capillary & bulb (senson) - see detail G-4 for proper mounting position

Control thermostat

AM-E end seal

Auto-trace® "M" heater

AM-BC power connection

Monitor (alarm) light

AM-BC power connection

Figure 9-25: Heat-tracing detail: Schematic of Auto-Tracer "M" heaters with monitor wire.

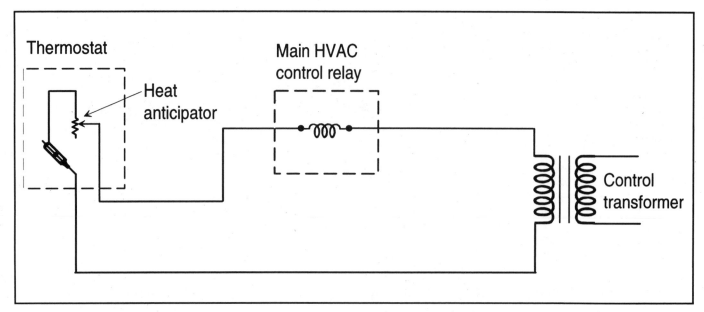

Figure 9-26: Heat-sensing control circuit with heat anticipator.

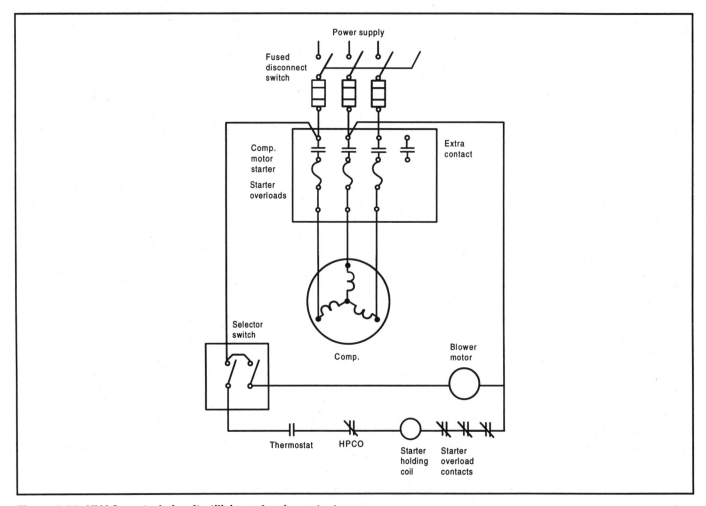

Figure 9-27: HVAC control circuit utilizing a 4-pole contactor.

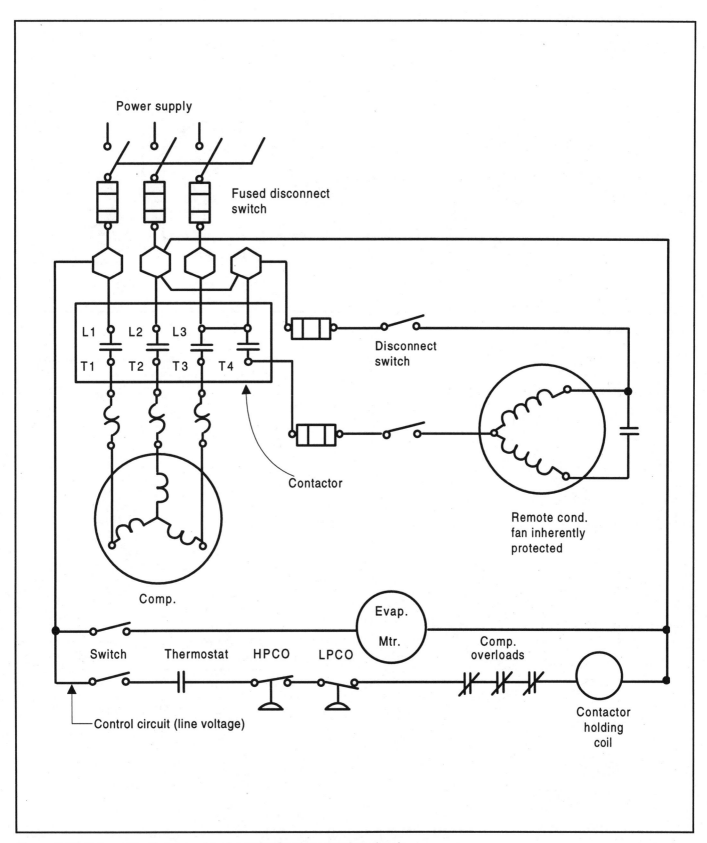

Figure 9-28: Schematic diagram of typical HVAC motor-starting circuit.

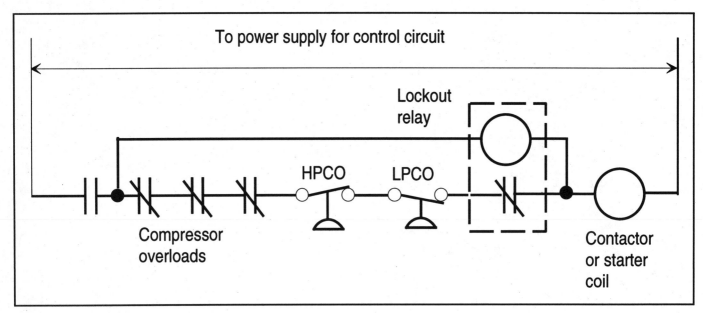

Figure 9-29: Lockout relay circuit.

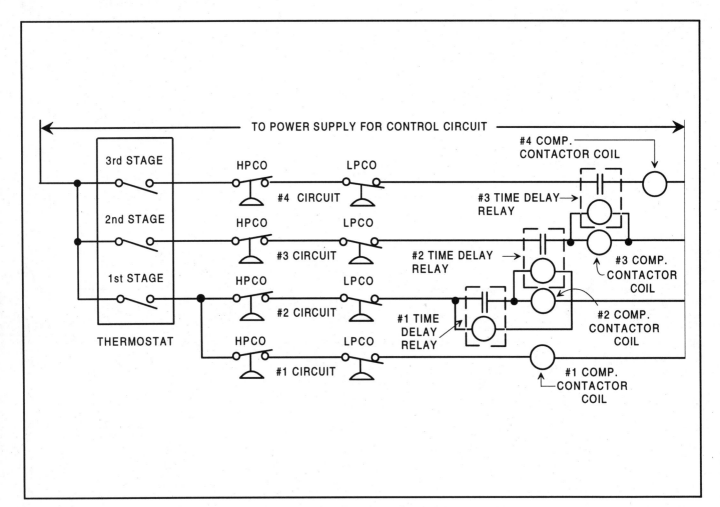

Figure 9-30: Four compressor control circuits with time-delay relay.

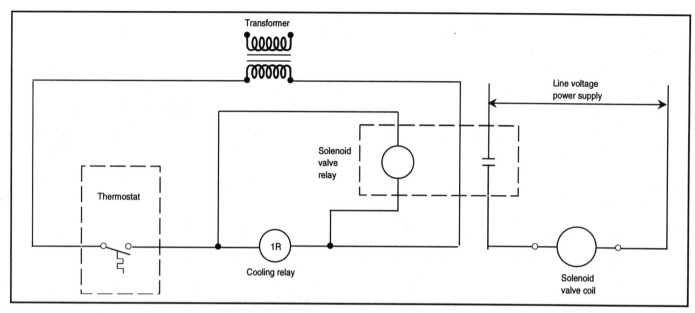

Figure 9-31: Wiring diagram utilizing a solenoid valve.

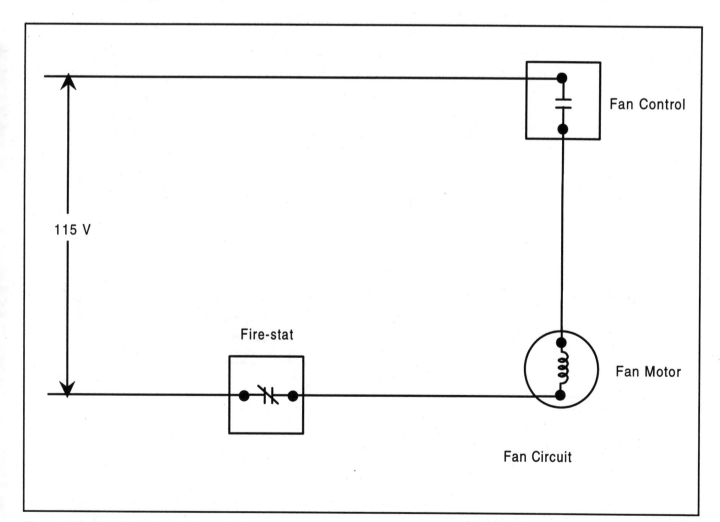

Figure 9-32: Circuit for furnace fan circuit.

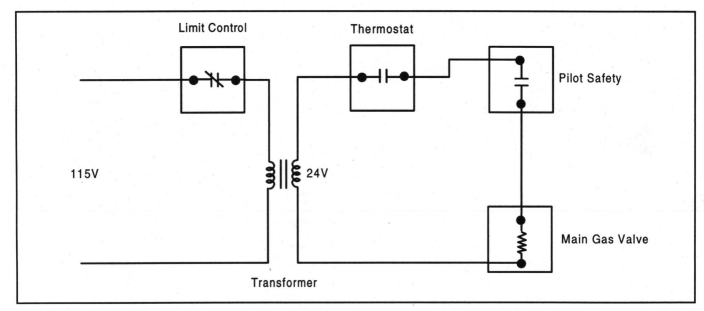

Figure 9-33: Temperature control circuit.

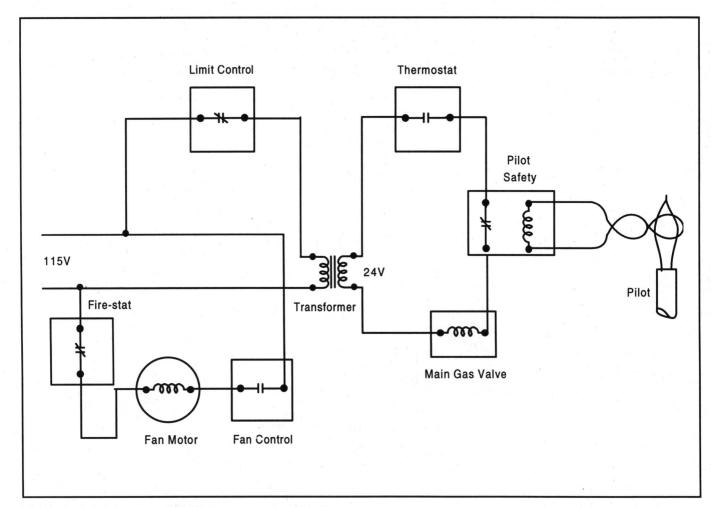

Figure 9-34: A typical 24-volt gas furnace control circuit.

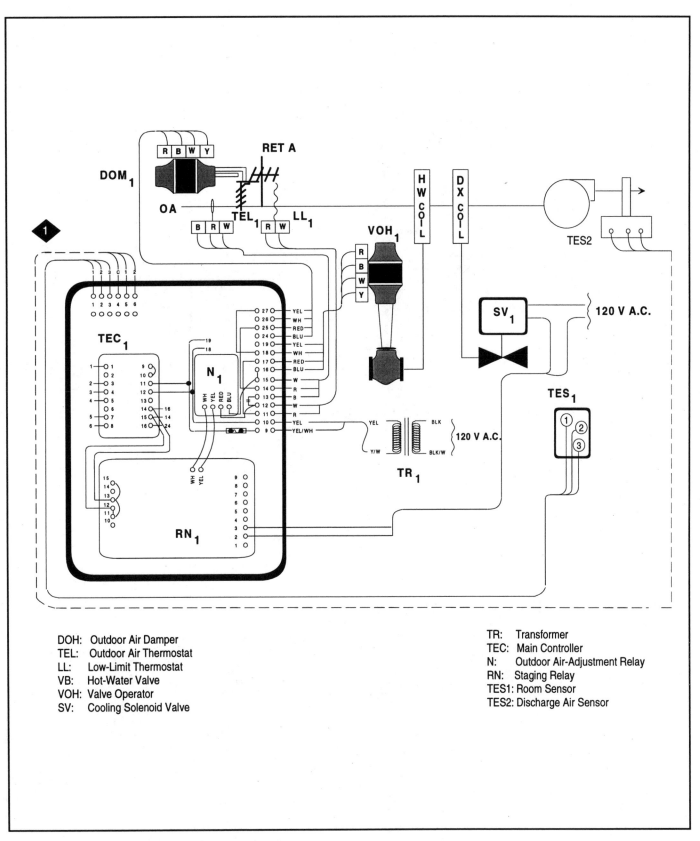

DOH: Outdoor Air Damper
TEL: Outdoor Air Thermostat
LL: Low-Limit Thermostat
VB: Hot-Water Valve
VOH: Valve Operator
SV: Cooling Solenoid Valve

TR: Transformer
TEC: Main Controller
N: Outdoor Air-Adjustment Relay
RN: Staging Relay
TES1: Room Sensor
TES2: Discharge Air Sensor

Figure 9-35: Typical electronic HVAC control system.

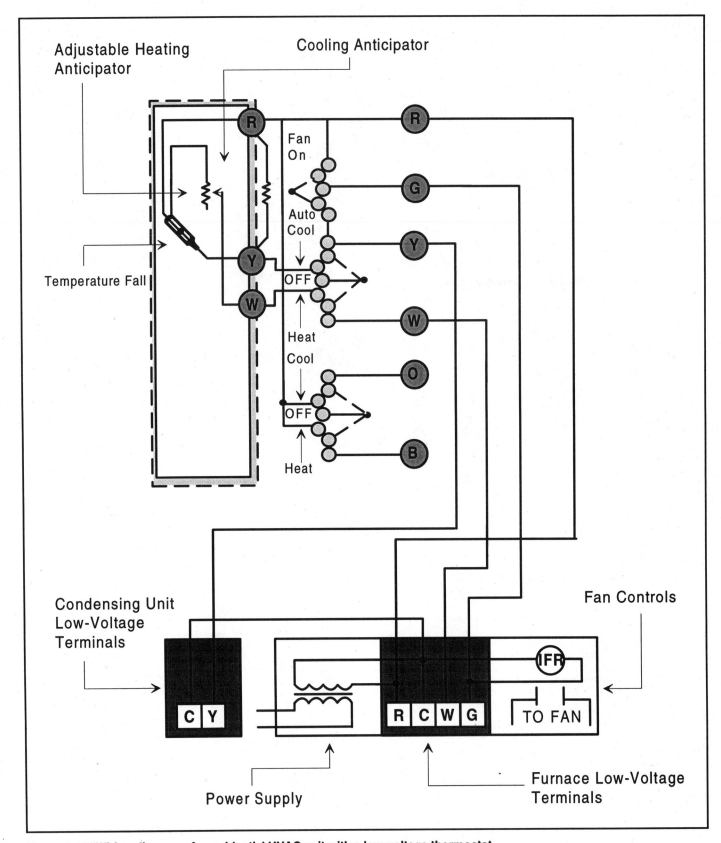

Figure 9-36: Wiring diagram of a residential HVAC unit with a low-voltage thermostat.

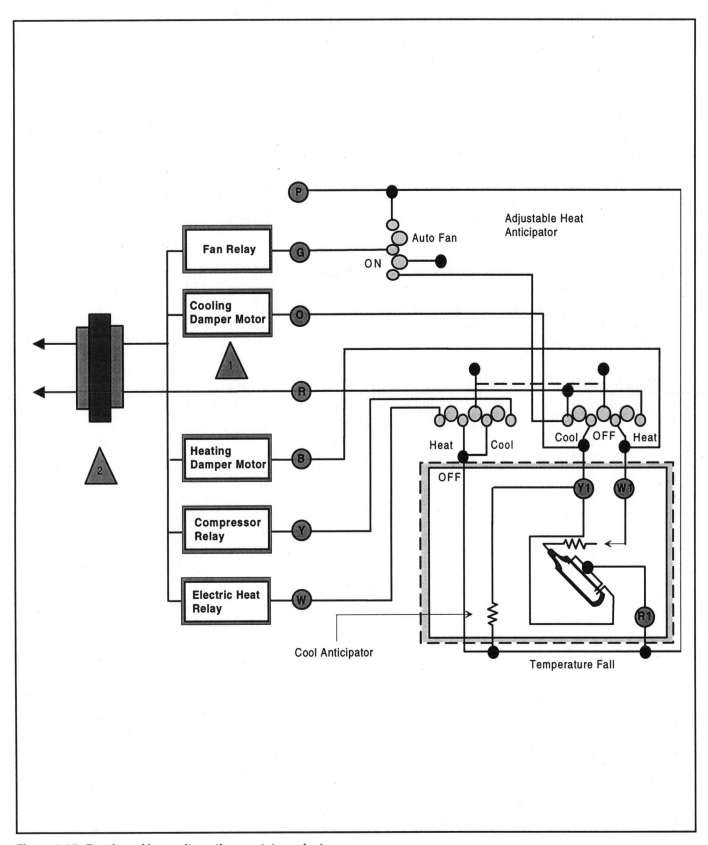

Figure 9-37: Routing of low-voltage thermostat conductors.

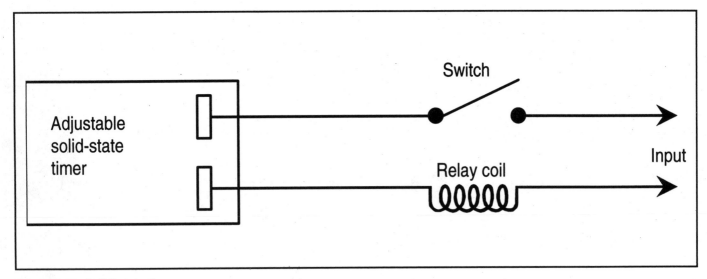

Figure 9-38: Solid-state timer.

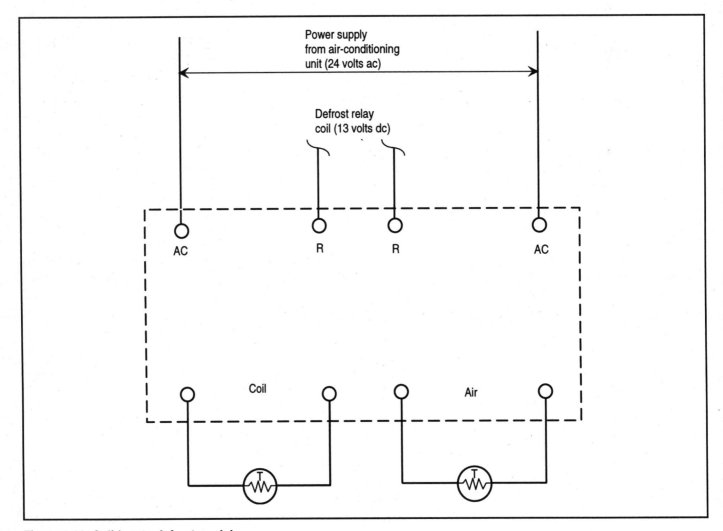

Figure 9-39: Solid-state defrost module.

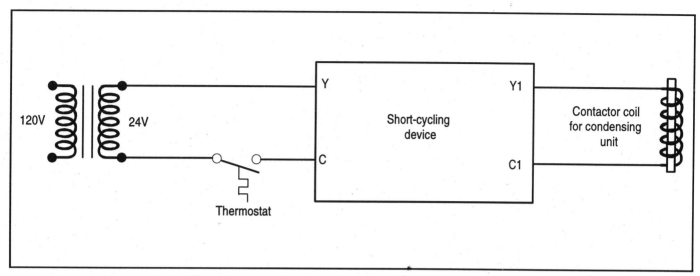

Figure 9-40: Control circuit with an anti-short-cycling device.

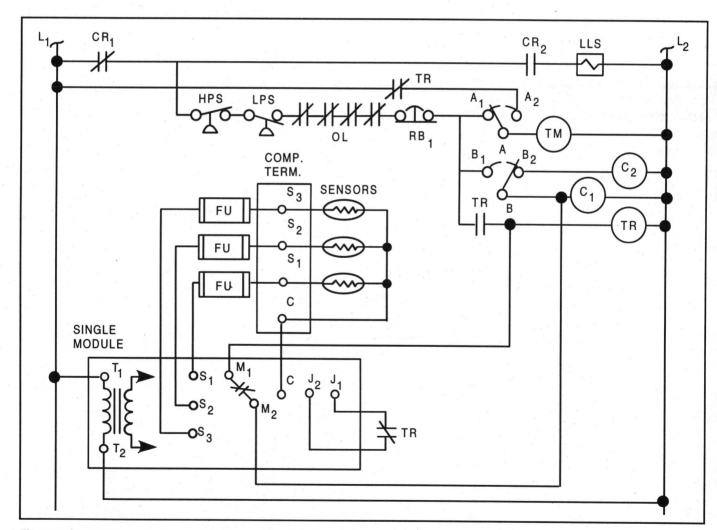

Figure 9-41: HVAC control circuit utilzing an overload module.

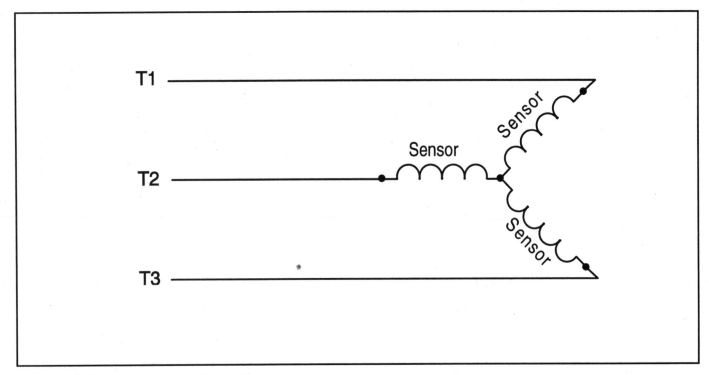

Figure 9-42: Sensor location in a three-phase motor.

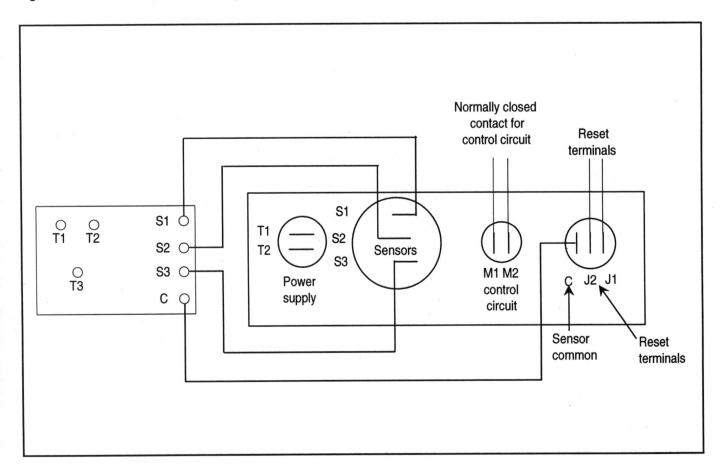

Figure 9-43: Wiring connections for an overload module.

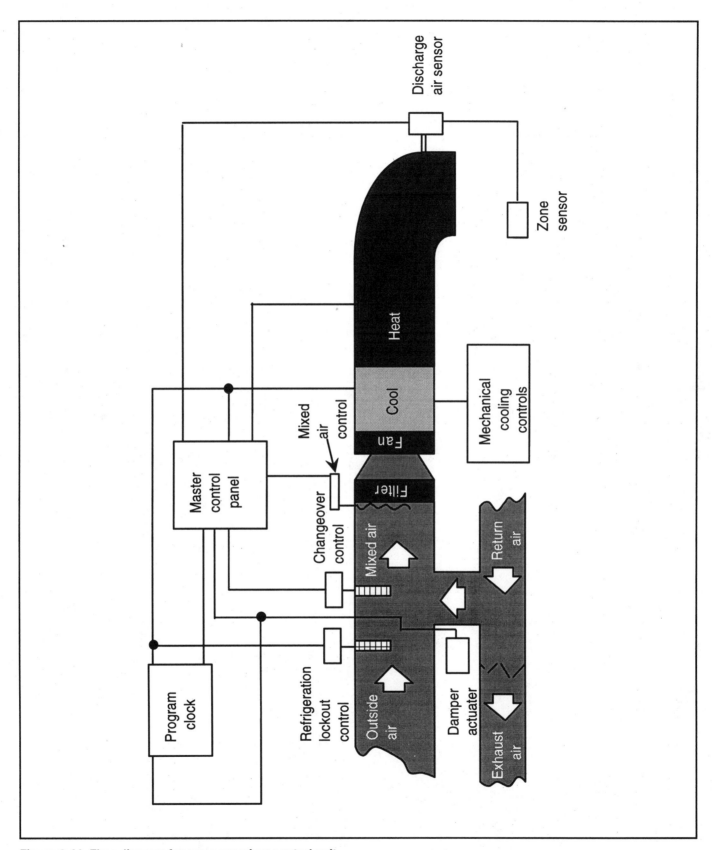

Figure 9-44: Flow diagram for an economizer control unit.

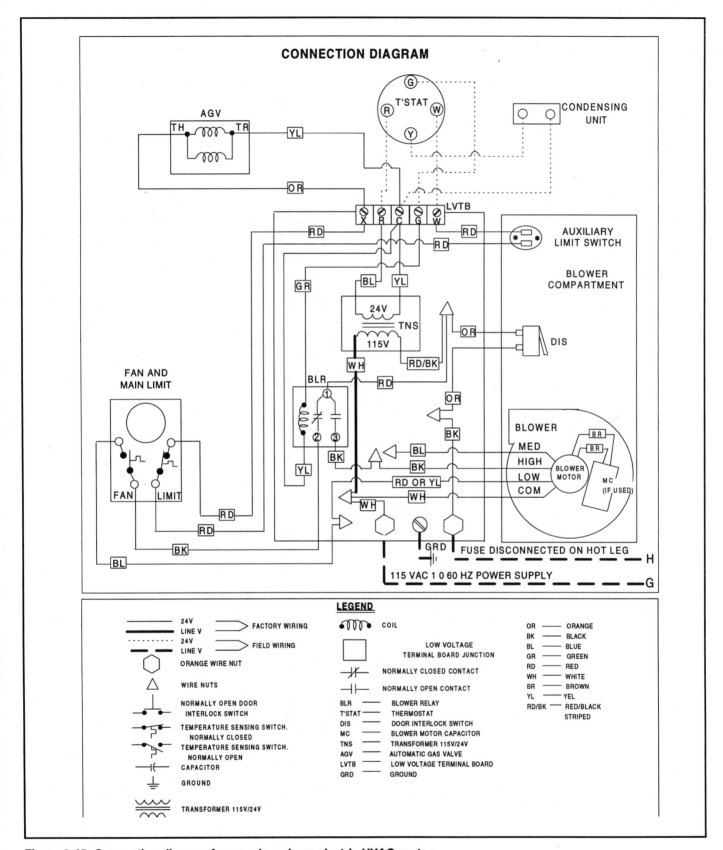

Figure 9-45: Connection diagram for a packaged gas-electric HVAC system.

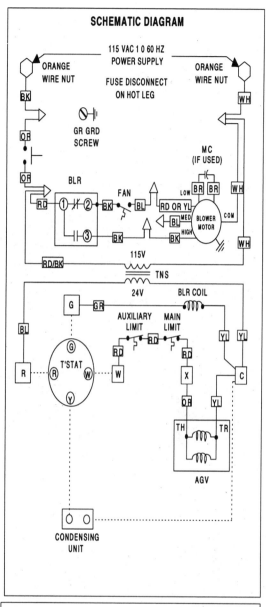

Figure 9-46: Schematic wiring diagram for a packaged gas-electric HVAC system.

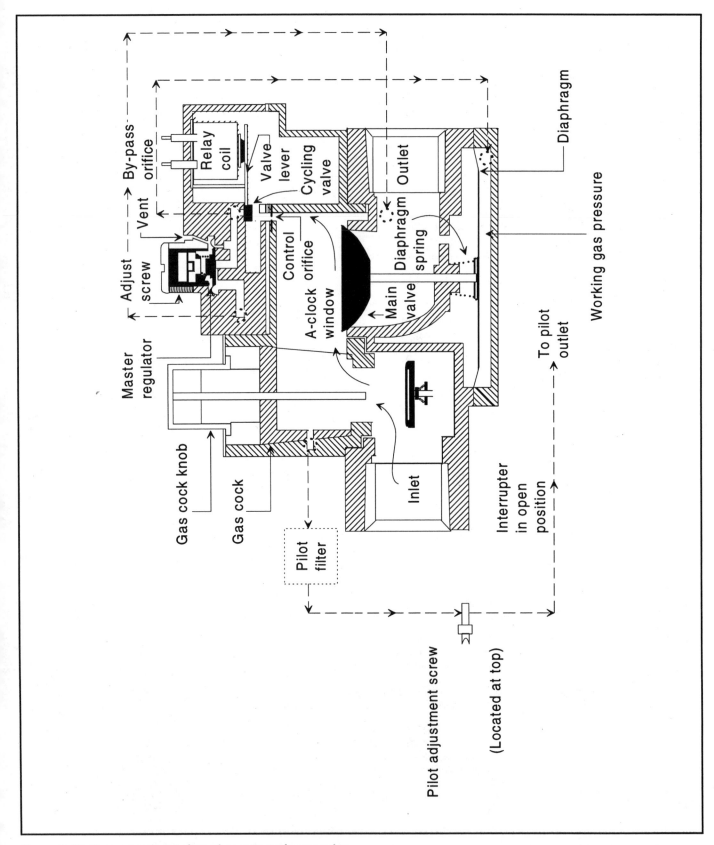

Figure 9-47: Cross-sectional view of an automatic gas valve.

Chapter 10

Controls and Instrumentation

Every piece of equipment connected to an electrical system must have some means of control, as must the system itself. Details for disconnect switches, overcurrent protection and similar control devices are covered in other chapters of this book. Items such as recording and indicating devices, metering equipment, and related equipment appear in this chapter.

METERING EQUIPMENT

Watt-Hour Meter

A watt-hour meter is an instrument that measures and registers the integral, with respect to time, of the active power in a circuit. A typical watt-hour meter consists of a combination of coils, conductors, and gears — all encased in a housing as shown in Figure 10-1. The coils are constructed on the same principle as a split-phase induction motor, in that the stationary current coil and the voltage coil are placed so that they produce a rotating magnetic field. The disc near the center of the meter is exposed to the rotating magnetic field and the torque applied to the disc is proportional to the power in the circuit. Furthermore, the braking action of the eddy currents in the disc makes the speed of rotation proportional to the rate at which the power is consumed. The disc, through a train of gears, moves the pointers on the register dials to record the amount of power used directly in kilowatt hours.

Most watt-hour meters now have five dials, as shown in Figure 10-2 on the next page. The dial farthest to the right on the meter counts the kilowatt hours singly. The second

Figure 10-1: Typical watt-hour meter.

dial from the right counts by tens, the third dial by hundreds, the fourth dial from the right by thousands, and the left-hand dial by ten-thousands. Therefore, the reading in Figure 10-2 is 2, 2, 1, 7, 9 or 22,179 kilowatt hours.

Instrument Transformers

Instrument transformers are so named because they are usually connected to an electrical instrument, such as an ammeter, a voltmeter, wattmeter, or relay. Instrument transformers fall under two general types:

- Potential transformer
- Current transformer

Figure 10-2: The reading of the five dials on this watt-hour meter shows that the total kilowatt hours is 2, 2, 1, 7, 9 or 22,179.

Potential Transformers

Potential, or voltage, transformers (Figure 10-3) are single-phase transformers used to supply voltage to instruments such as voltmeters, frequency meters, power-factor meters, and watt-hour meters. If used on three-phase systems, sets of two or three potential transformers are applied. The primary winding of a potential transformer is always connected across the main power lines. The secondary voltage is proportional to the primary voltage, but is small enough to be safe for the testing or recording instruments. The secondary of a potential transformer may be designed for several different voltages, but most are designed for 120 V. Potential transformers are primarily distribution transformers especially designed for voltage regulation so that the secondary voltage under all condi-

tions will be as nearly as possible a definite percentage of the primary voltage. Sometimes it is important to indicate the polarity of the current in the instrument and transformer. the polarity is the instantaneous direction of current at a specific moment. Wires with the same polarity usually carry a small black block or cross in the electrical diagrams as shown in Figure 10-3.

A typical connection to a single-phase, two-wire circuit is shown in Figure 10-4. The primary of the potential transformer is connected across the main line, and the secondary is connected to an instrument. Note that the secondary circuit is grounded.

All potential transformers have a wound primary and a wound secondary. Mechanically, their construction is similar to a wound current transformer. Their kVA rating

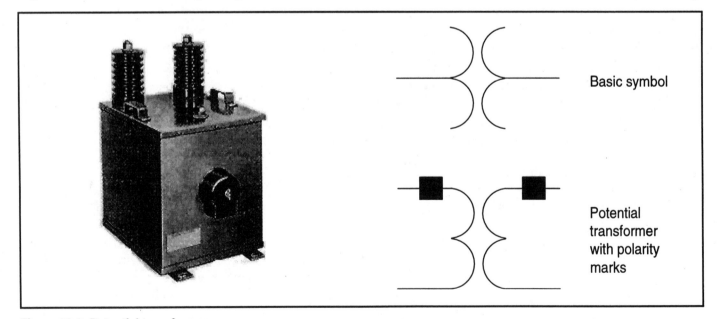

Basic symbol

Potential transformer with polarity marks

Figure 10-3: Potential transformer.

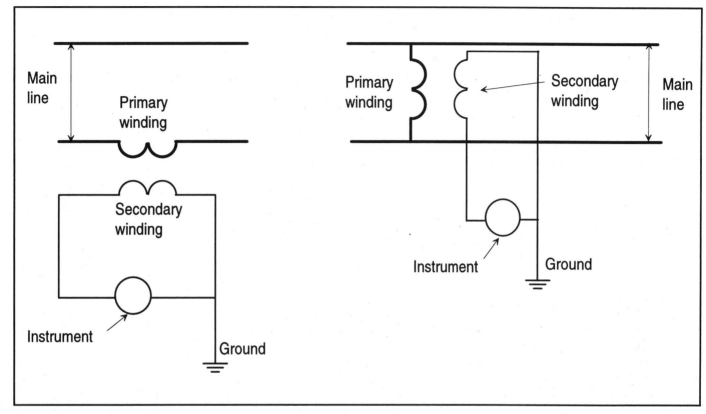

Figure 10-4: Connection of instrument transformers.

seldom exceeds 0.2 kVA, or 900 VA. Dry-type potential transformers are normally used for voltages up to 15,000 kVA, with the windings molded in epoxy.

Sometimes current and potential transformers are mounted in the same unit, called a metering outfit. The use of such units is convenient for metering in the field because the units are compact and require less work at the metering location than would be required by separate current and potential transformers.

Current Transformers

Current transformers (Figure 10-5) are used to supply current to an instrument that is connected to the transformer's secondary — the current being proportional to the primary current, but small enough to be safe for the instrument. The secondary of a current transformer is usually designed for a rated current of 5 A, although the exact ratio may vary, depending upon how the transformer is used.

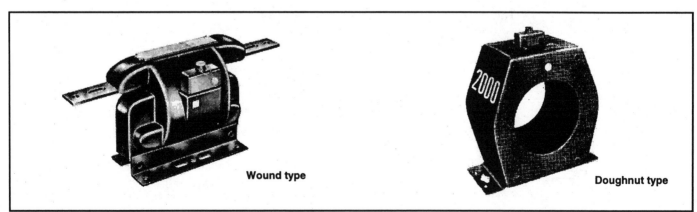

Figure 10-5: Two types of current transformers.

A current transformer operates in the same way as any other transformer in that the same relation exists between the primary and the secondary current and voltage. A current transformer is connected in series with the power lines to which it is applied so that line current flows in its primary winding. The secondary of the current transformer is connected to current devices such as ammeters, wattmeters, watt-hour meters, power-factor meters, some forms of relays, and the trip coils of some types of circuit breakers.

One current transformer can be used to operate several instruments connected in series, so that each carries the same secondary current. Current transformer and metering connection details appear in Figures 10-6 through 10-30. These details cover almost every application that will be encountered.

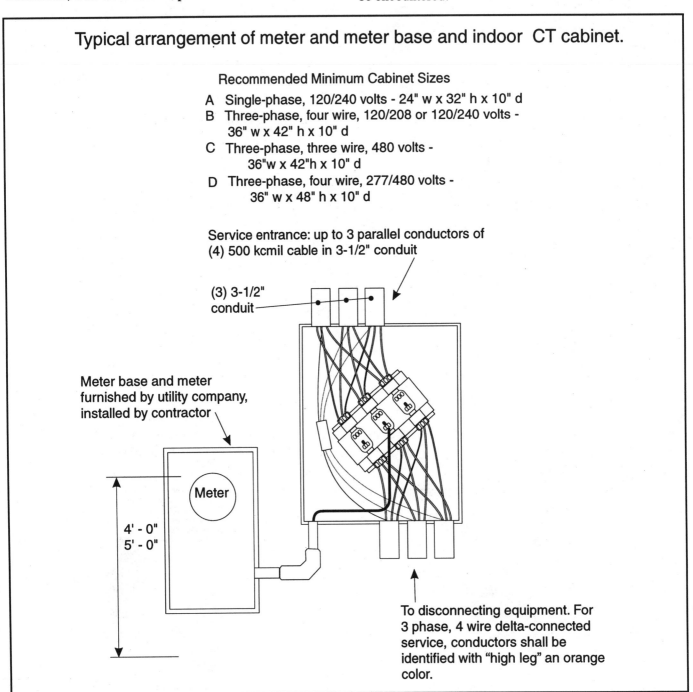

Typical arrangement of meter and meter base and indoor CT cabinet.

Recommended Minimum Cabinet Sizes

A Single-phase, 120/240 volts - 24" w x 32" h x 10" d

B Three-phase, four wire, 120/208 or 120/240 volts - 36" w x 42" h x 10" d

C Three-phase, three wire, 480 volts - 36"w x 42"h x 10" d

D Three-phase, four wire, 277/480 volts - 36" w x 48" h x 10" d

Service entrance: up to 3 parallel conductors of (4) 500 kcmil cable in 3-1/2" conduit

(3) 3-1/2" conduit

Meter base and meter furnished by utility company, installed by contractor

Meter

4' - 0"
5' - 0"

To disconnecting equipment. For 3 phase, 4 wire delta-connected service, conductors shall be identified with "high leg" an orange color.

Figure 10-6: Typical arrangement of meter, meter base, and indoor CT cabinet.

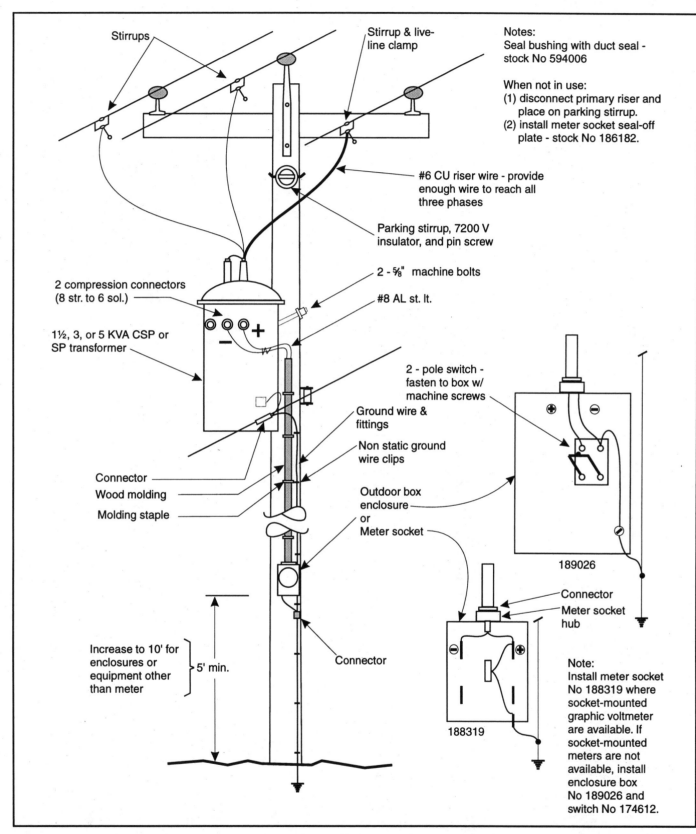

Figure 10-7: Potential transformer installation for primary voltage metering.

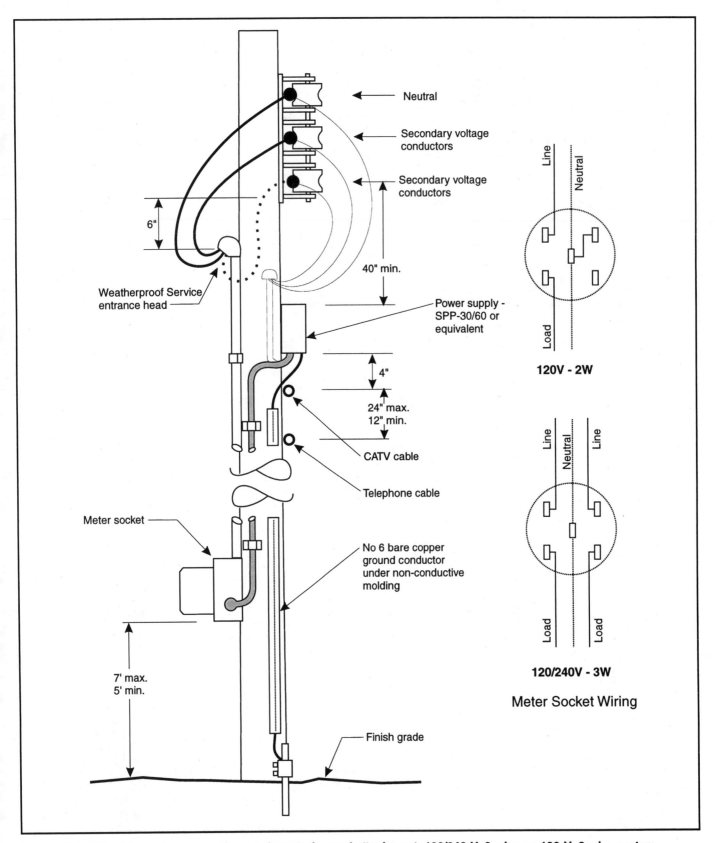

Neutral

Secondary voltage conductors

Secondary voltage conductors

6"

Weatherproof Service entrance head

40" min.

Power supply - SPP-30/60 or equivalent

4"

24" max.
12" min.

CATV cable

Telephone cable

Meter socket

No 6 bare copper ground conductor under non-conductive molding

7' max.
5' min.

Finish grade

Line
Neutral
Load
120V - 2W

Line
Neutral
Line
Load
Load
120/240V - 3W

Meter Socket Wiring

Figure 10-8: TV cable system or company pole metering and attachment; 120/240-V, 3-wire, or 120-V, 2-wire system.

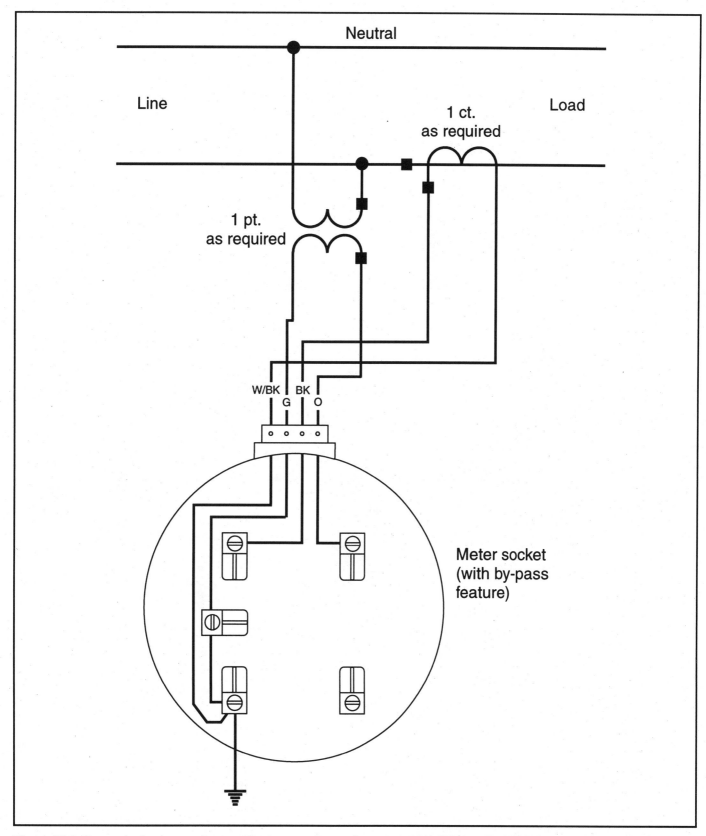

Figure 10-9: One-pole, 2-wire service transformer rated metering over 300 V. KWH or KWH/IND demand meter wiring diagram.

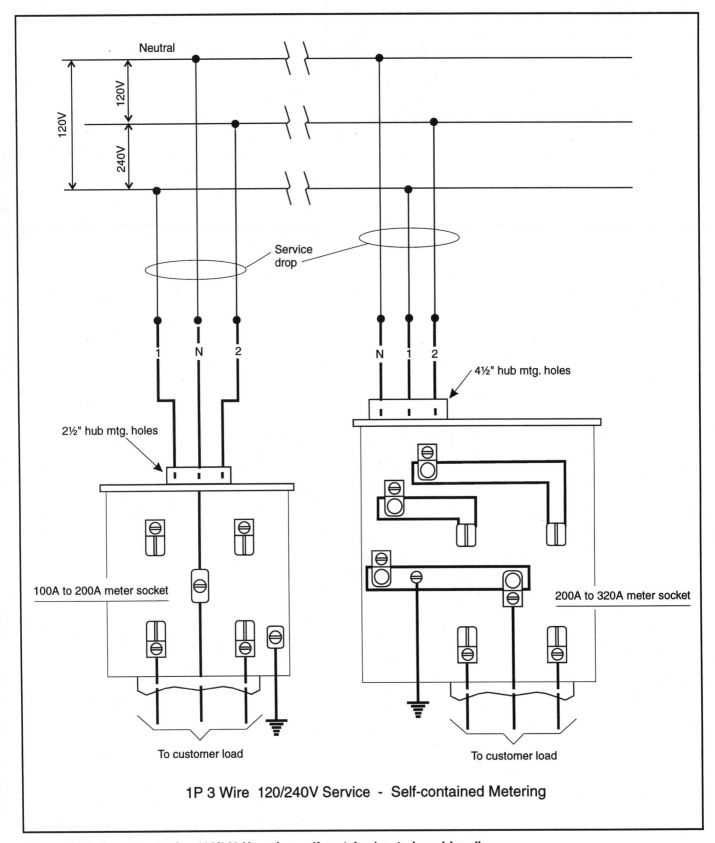

1P 3 Wire 120/240V Service - Self-contained Metering

Figure 10-10: One-pole, 3-wire, 120/240-V service; self-contained metering wiring diagram.

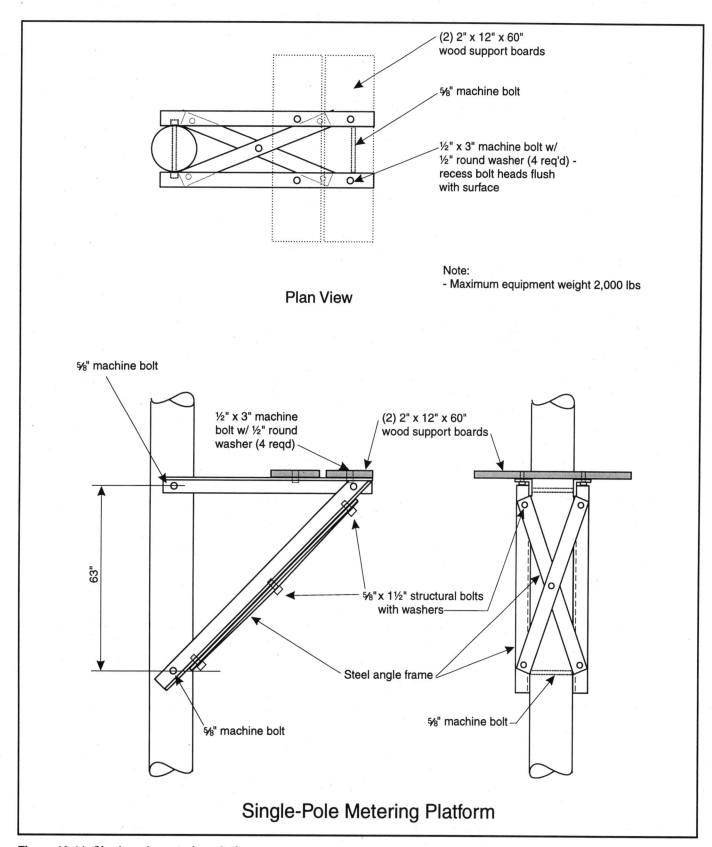

Figure 10-11: Single-pole metering platform.

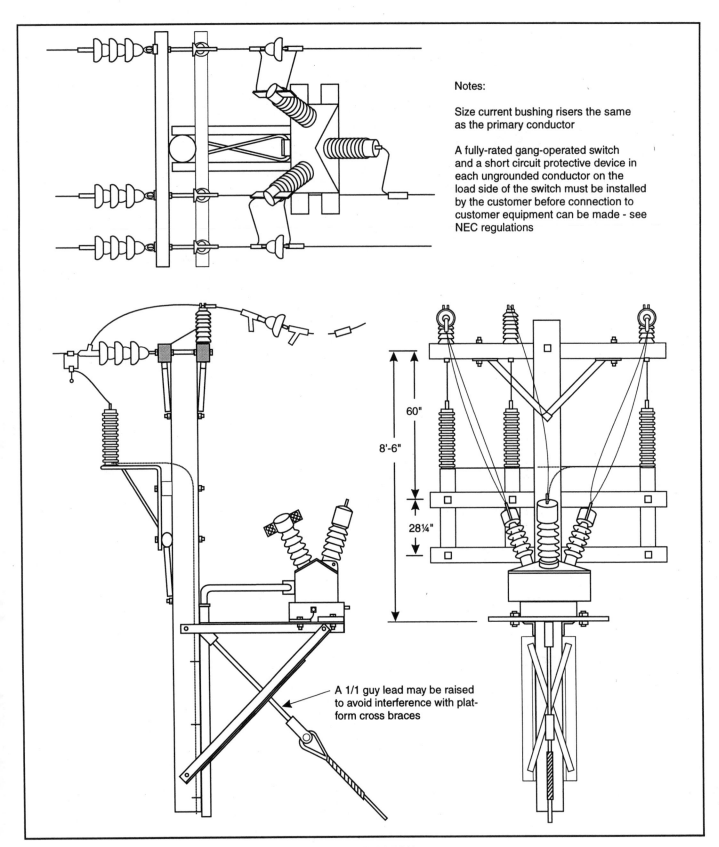

Notes:

Size current bushing risers the same
as the primary conductor

A fully-rated gang-operated switch
and a short circuit protective device in
each ungrounded conductor on the
load side of the switch must be installed
by the customer before connection to
customer equipment can be made - see
NEC regulations

A 1/1 guy lead may be raised
to avoid interference with platform cross braces

Figure 10-12: Outdoor metering for 3-phase unit, 22 kV through 34.5 kV.

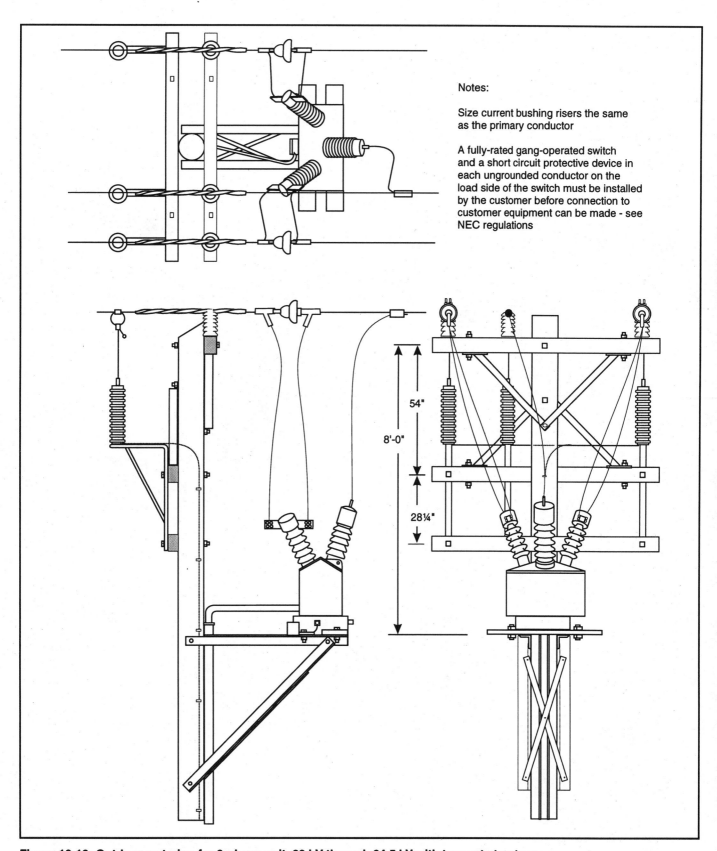

Notes:

Size current bushing risers the same as the primary conductor

A fully-rated gang-operated switch and a short circuit protective device in each ungrounded conductor on the load side of the switch must be installed by the customer before connection to customer equipment can be made - see NEC regulations

54"

8'-0"

28¼"

Figure 10-13: Outdoor metering for 3-phase unit, 22 kV through 34.5 kV with tangent structure.

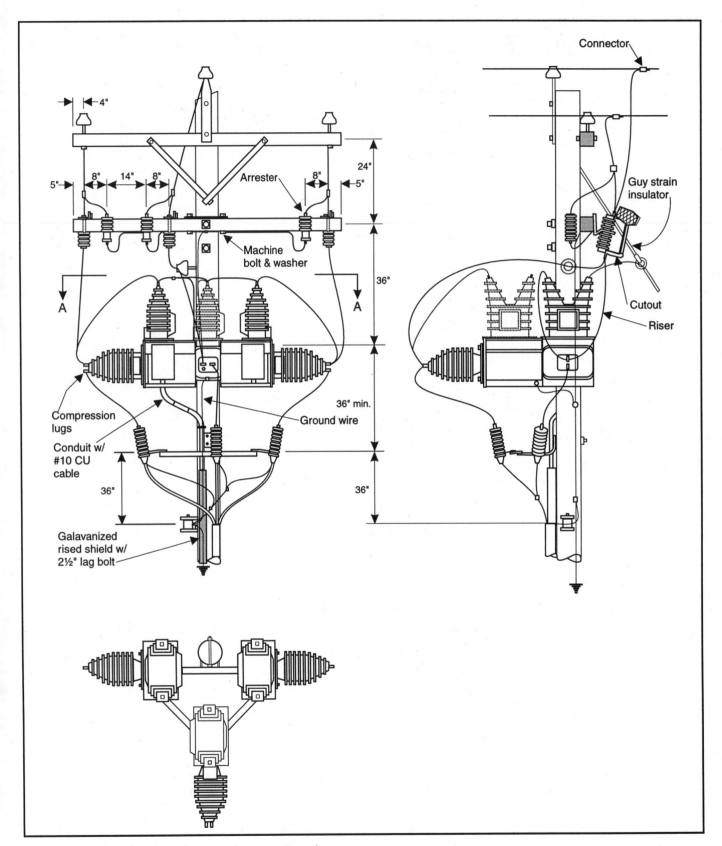

Figure 10-14: Metering for primary underground service.

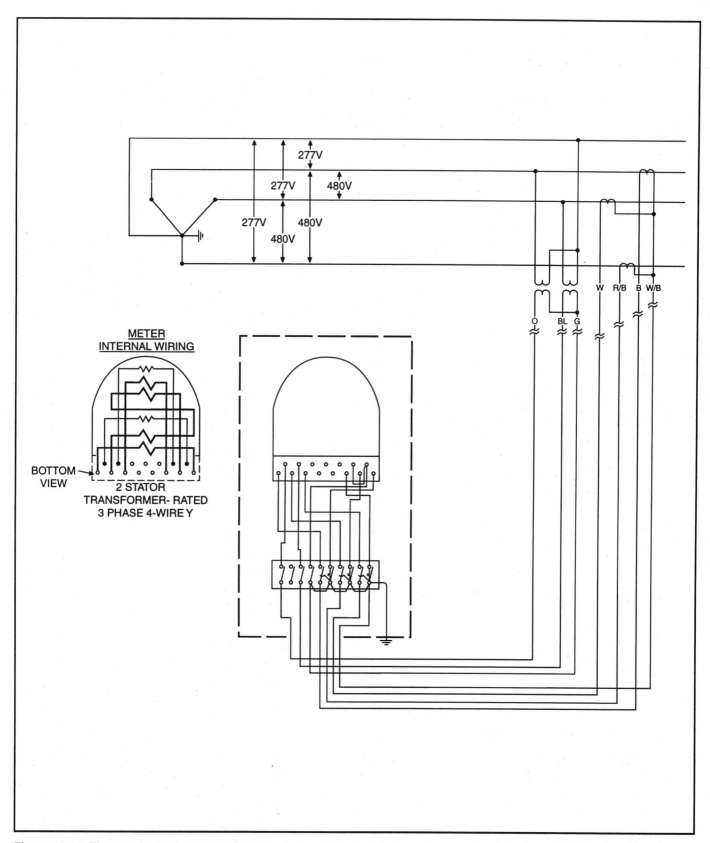

Figure 10-15: Three-pole, 4-wire, 277/480-V, wye-connected service. Transformer rated metering with A-base meter.

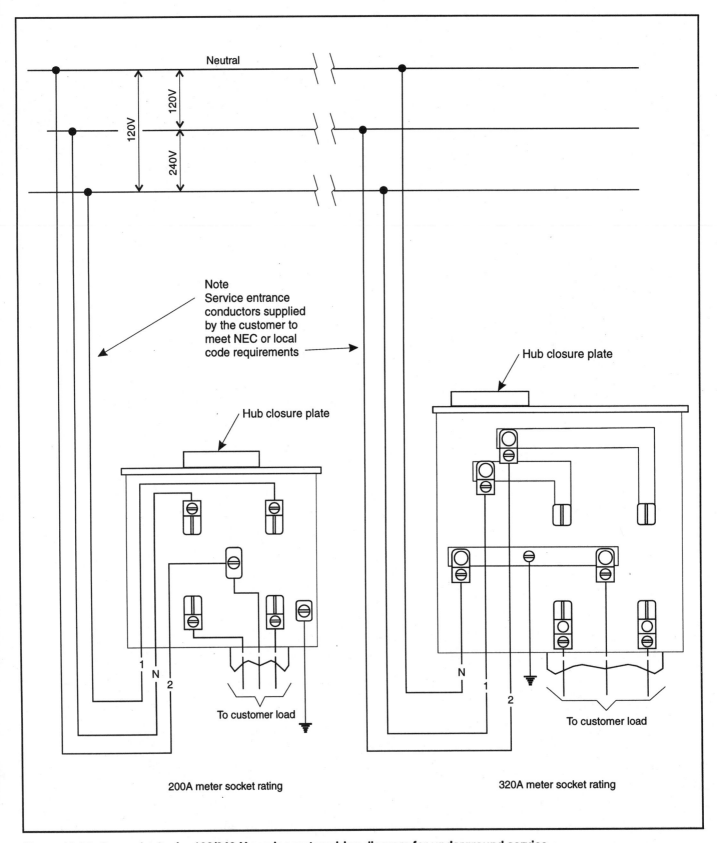

Figure 10-16: One-pole, 3-wire 120/240-V service meter wiring diagram for underground service.

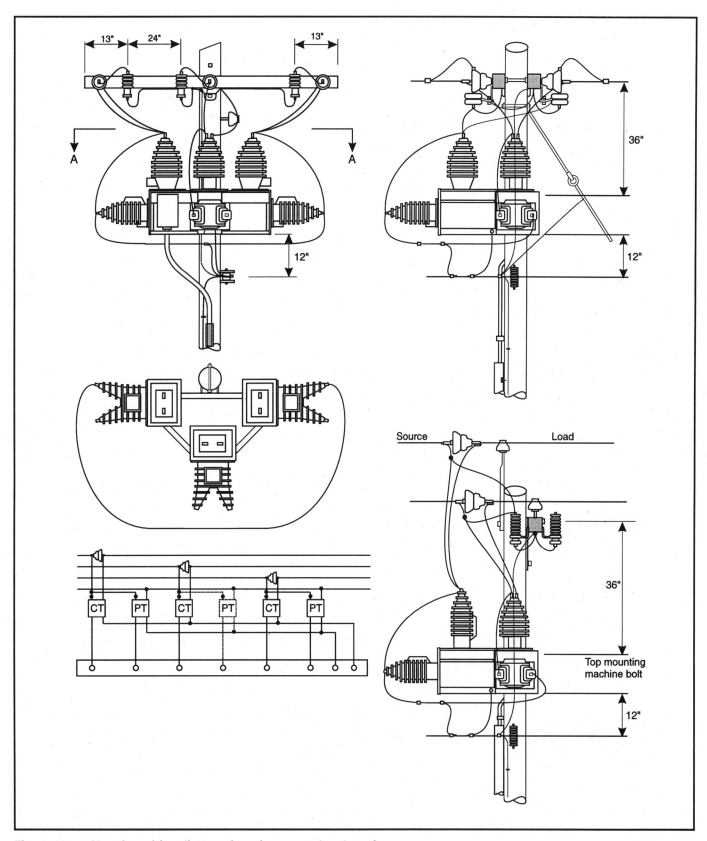

Figure 10-17: Metering wiring diagram for primary overhead service.

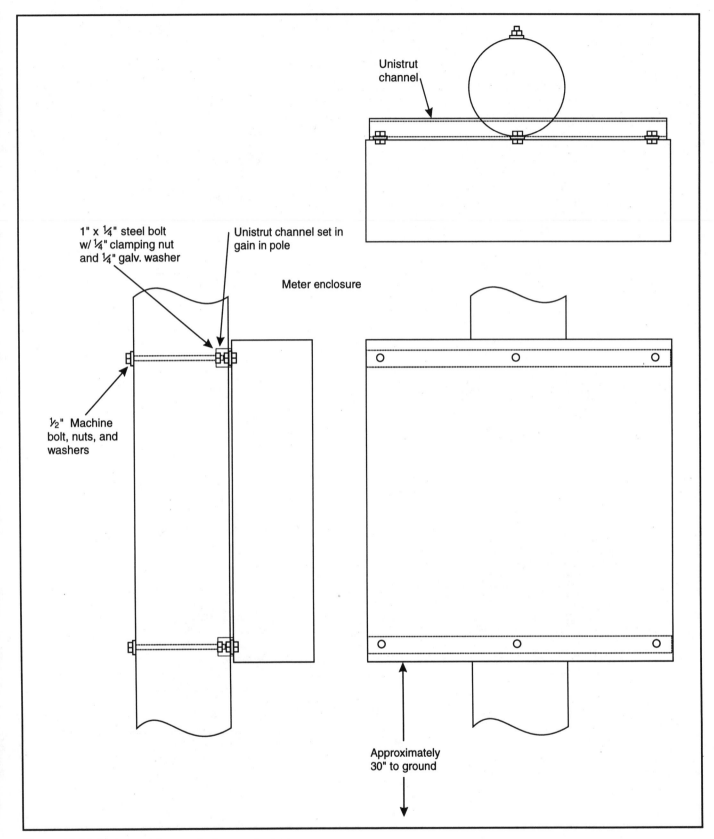

Figure 10-18: Large outdoor metering enclosure.

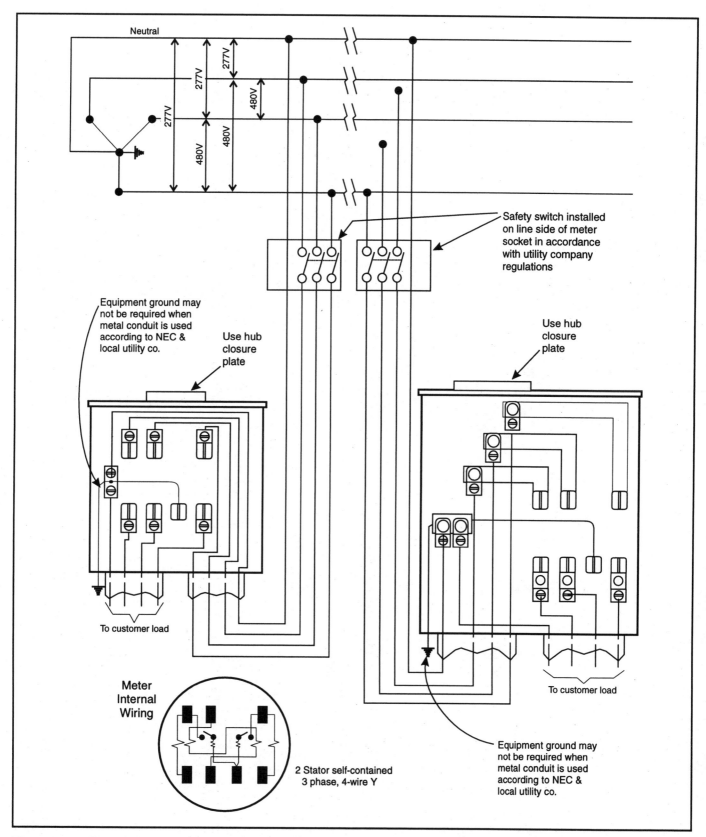

Figure 10-19: Three-pole, 4-wire, 277/480-V, wye-connected service. Self-contained metering for 200 A or less underground service.

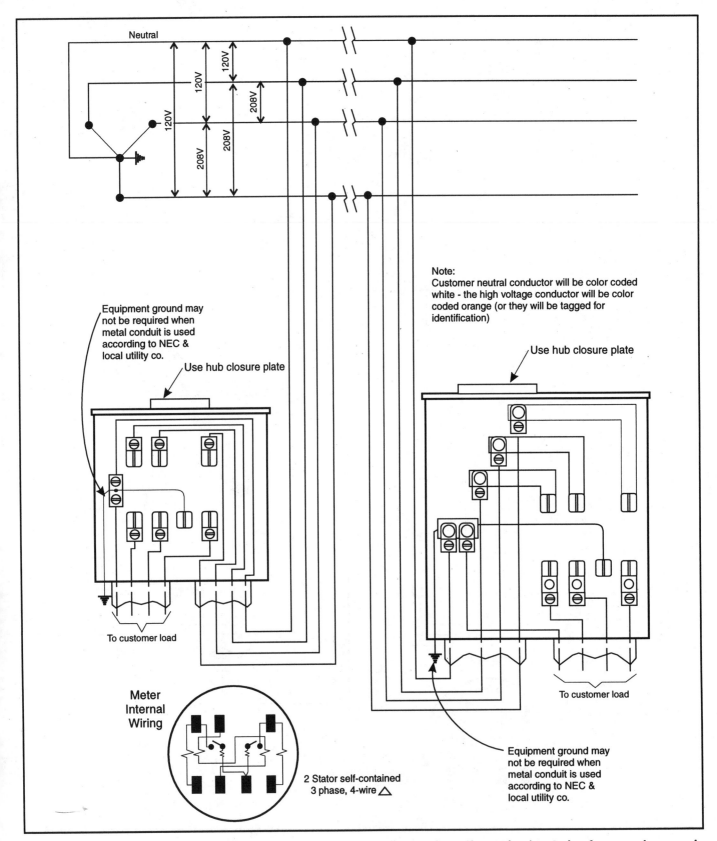

Figure 10-20: Three-pole, 4-wire, 120/208-V wye-connected wiring diagram for self-contained metering for an underground service.

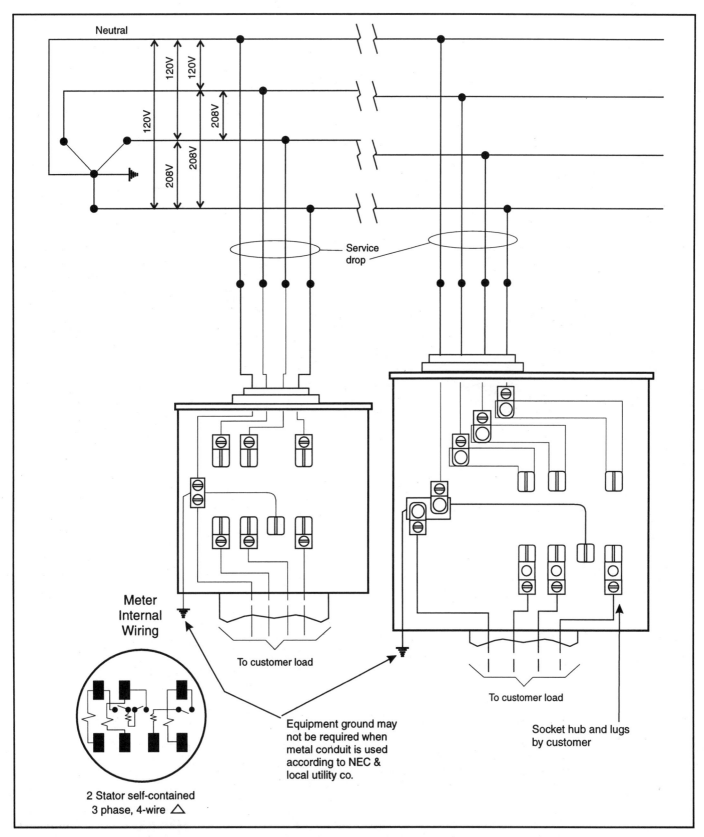

Neutral

120V

120V

120V

208V

208V

208V

208V

Service drop

Meter Internal Wiring

To customer load

Equipment ground may not be required when metal conduit is used according to NEC & local utility co.

To customer load

Socket hub and lugs by customer

2 Stator self-contained
3 phase, 4-wire △

Figure 10-21: Three-pole, 4-wire, 120/208-V service with self-contained metering for overhead service.

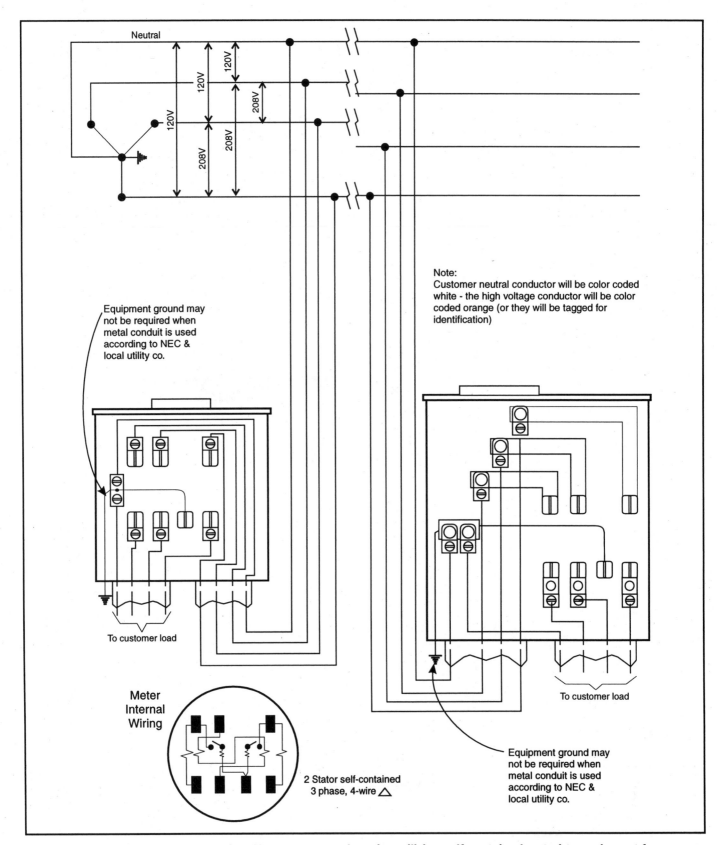

Neutral

120V
120V
120V
120V
208V
208V
208V
208V

Equipment ground may
not be required when
metal conduit is used
according to NEC &
local utility co.

Note:
Customer neutral conductor will be color coded
white - the high voltage conductor will be color
coded orange (or they will be tagged for
identification)

To customer load

Meter
Internal
Wiring

2 Stator self-contained
3 phase, 4-wire △

To customer load

Equipment ground may
not be required when
metal conduit is used
according to NEC &
local utility co.

Figure 10-22: Three-pole, 4-wire, 120/208-V, wye-connected service utilizing self-contained metering equipment for an underground service.

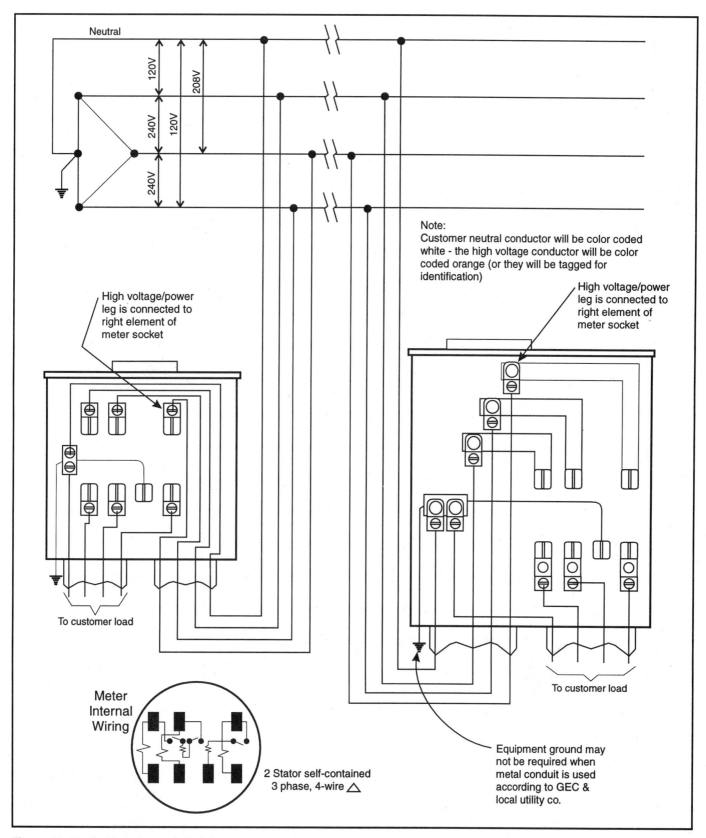

Figure 10-23: Three-pole, 4-wire, delta-connected 120/240-V service with self-contained metering for an underground service.

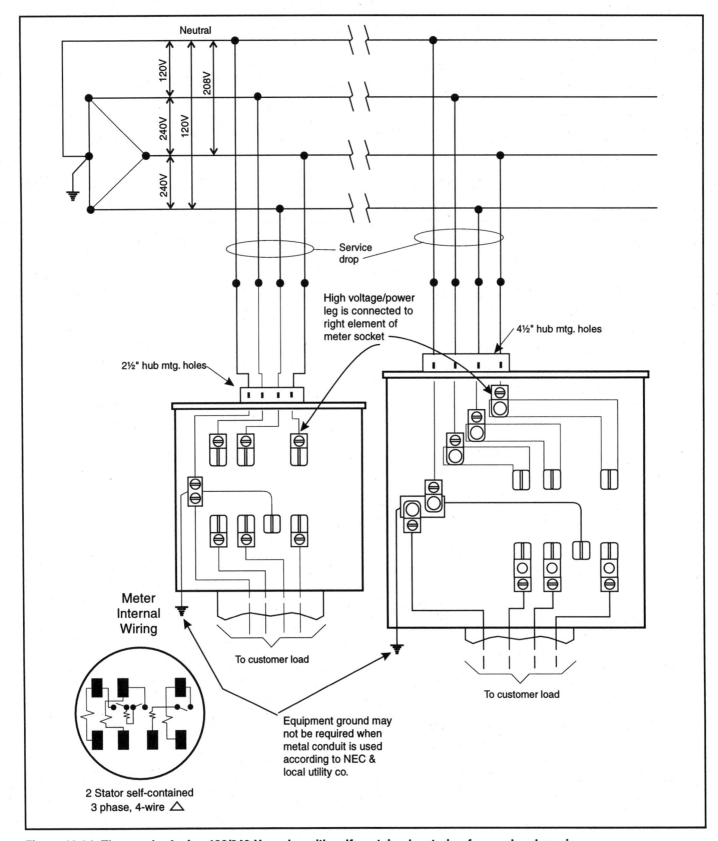

Figure 10-24: Three-pole, 4-wire, 120/240-V service with self-contained metering for overhead service.

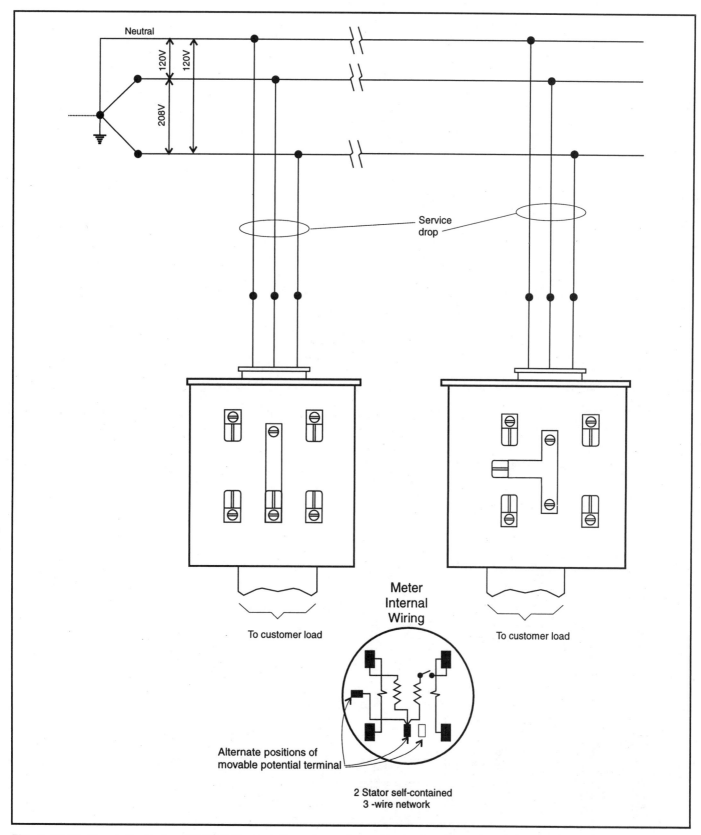

Figure 10-25: One-pole, 3-wire, 120/208-V network service using self-contained metering for an overhead service.

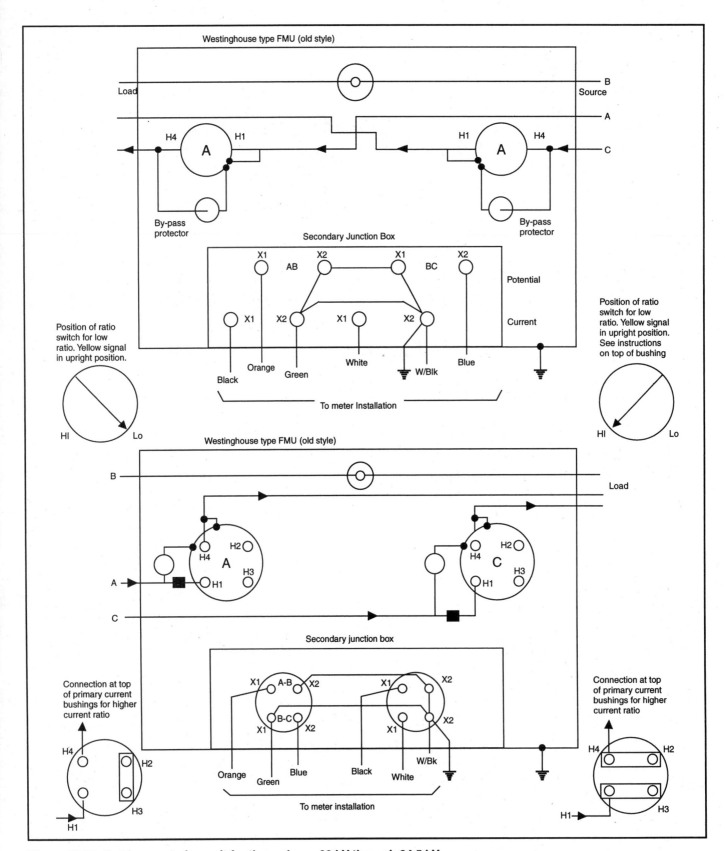

Figure 10-26: Outdoor metering unit for three-phase, 22 kV through 34.5 kV.

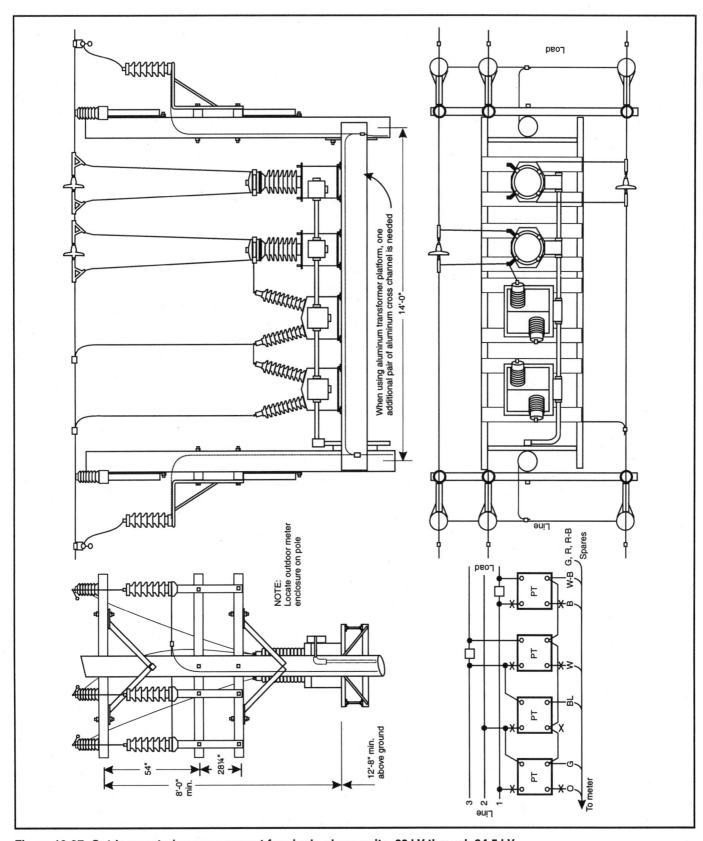

Figure 10-27: Outdoor metering arrangement for single-phase units, 22 kV through 34.5 kV.

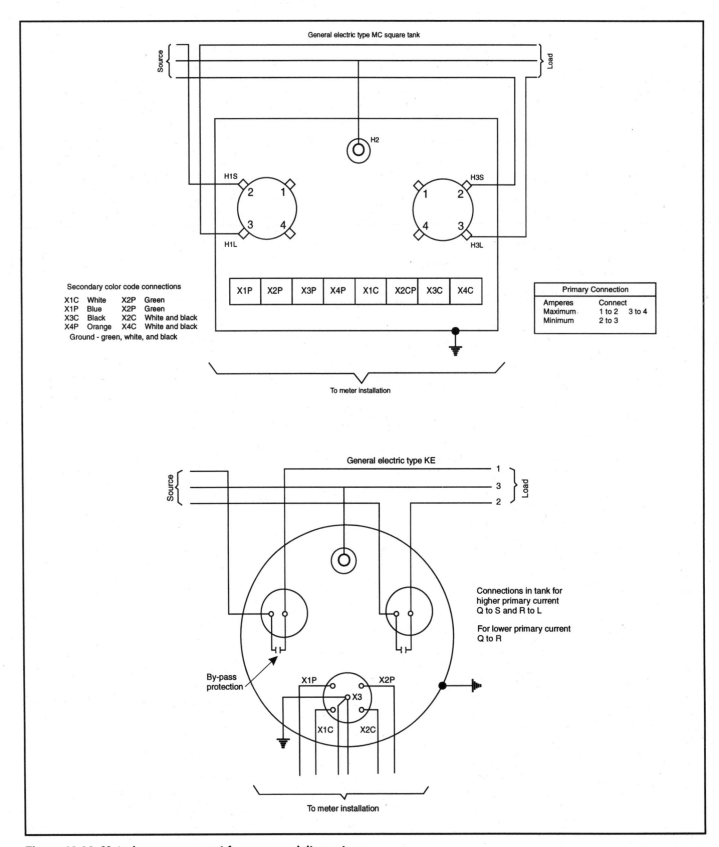

Figure 10-28: Metering arrangement for an open delta system.

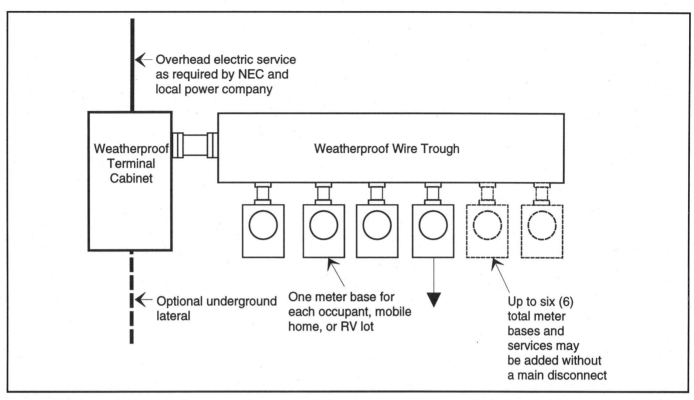

Figure 10-29: Service and meter arrangement for a small apartment complex (multifamily dwellings), mobile home parks, or RV parks.

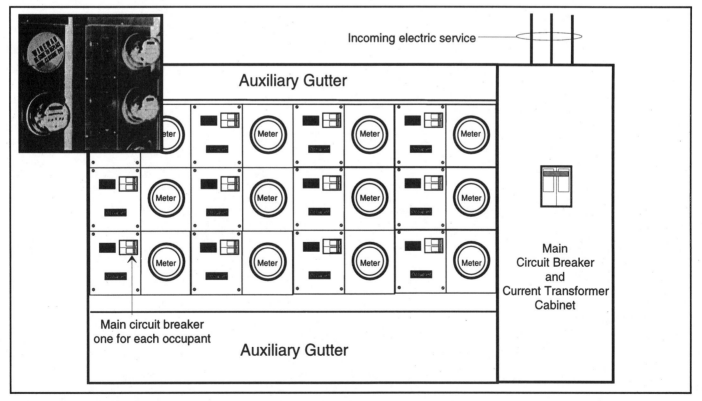

Figure 10-30: Metering arrangement for a 12-unit multifamily dwelling.

Written Specifications

The electrical specifications for a building or project are the written descriptions of work and duties required of the owner, architect, and engineer. Together with the working drawings, specifications form the basis of the contract requirements for the construction of the building or project.

Those who prepare and use construction drawings and specifications must always be alert to discrepancies between the working drawings and the written specifications. Such discrepancies occur particularly when:

1. Architects or engineers use standard or prototype specifications and attempt to apply them without any modification to specific working drawings.
2. Previously prepared standard drawings are changed or amended by reference in the specifications only and the drawings themselves are not changed.
3. Items are duplicated in both the drawings and specifications, but an item is subsequently amended in one and overlooked on the other contract document.

An example of this last case would be a power-riser diagram shown on the working drawings, which gives diagrammatic locations of all panelboards and related service equipment to be used on the project. The written specifications list all panelboards including their contents (fuses, circuit breakers, etc.) in a panelboard schedule. If another panel must be added at a later date prior to the job going out for bids, it will most often be added to the power-riser diagram on the drawings. But such a change, especially a last minute one, is often overlooked in the written specifications. For this reason, it is best not to duplicate items in the specifications and on the drawing. Rather, the information to be listed should be indicated in its proper place in one or the other of these construction documents — not in both.

In such instances, the person in charge of the project has the responsibility to ascertain whether the drawings or the specifications take precedence. Such questions must be resolved, preferably before any work is installed, to avoid added cost to either the owner, the architect/engineer, or the contractor.

Divisions 1 through 16 of the written specifications cover job requirements of a specific part of the construction work. Included in these divisions are the type and grade of materials to be used, equipment to be furnished, and the manner in which it is to be installed. Each division indicates the extent of the work covered and should be so written as to leave no doubt in anyone's mind about whether a certain part of the work is to be performed by a certain subcontractor and included in one section or another.

Division 16 of the specifications covers the electrical and related work on a given project, including the grade of materials to be used and the manner of installation for the electrical system. The following is an outline of the various sections normally included in Division 16 – Electrical of the written specifications for building construction.

16010 General Provisions: The general provisions of the electrical specifications normally consist of a selected group of considerations and regulations that apply to all sections of the division. Items covered may include the scope of the work (work included and not included) in the electrical contract, electrical reference symbols, codes and fees, tests, demonstration of the completed electrical system, and identification of the equipment and components used in the installation.

16050 Basic Electrical Materials and Methods: Definitive statements in this portion of the specifications should establish the means of identifying the type and quality of materials and equipment selected for use. This section should further establish the accepted methods of installing various materials such as raceways, conduits, bus ducts, underfloor ducts, cable trays, wires and cables, wire connections and devices, outlet boxes, floor boxes, cabinets, panelboards, switches and receptacles, motors, motor starters, disconnects, overcurrent protective devices, supporting devices, and electronic devices.

16200 Power Generation: This section normally covers items of equipment used for emergency or standby power facilities, the type used to take over essential electrical service during a normal power source outage. This section usually cites requirements for a complete installation of all emergency circuits on a given project, including emergency service or standby power in the form of a generator set or storage batteries, automatic control facilities, feeders, panelboards, disconnects, branch circuits, and outlets.

Items to be fully described in this section include the generator and its engine (reciprocating or turbine), cooling equipment, exhaust equipment, starting equipment, and automatic or manual transfer equipment.

16300 Medium Voltage Distribution: Unlike the feeders, branch circuits, and the like that carry electrical power inside a building, high-voltage (over 600 volts) power transmission is the subject of this section. Normally the specifications that require this section are for those projects constructed on government reservations and large industrial sites.

Cable and equipment specified in this section almost always is over 2.4 kV and includes such items as substations, switchgear, transformers, vaults, manholes, rectifiers, converters, and capacitors. Size and type of enclosures, as well as instrumentation, may also be included here.

16400 Service and Distribution: Power distribution facilities (under 600 volts) for the project's service entrance, metering, distribution switchboards, branch circuit panelboards, feeder circuits, and the like are described in this section by paragraphs or clauses covering selected related equipment items.

In addition to the electric service characteristics — voltage, frequency, phase, etc., the quality and capacity levels of all items involved in the service entrance and power distribution system should be clearly defined. Typical items include the size and number of conductors, installation and supporting methods, location, rating, type and circuit protection features of all main circuit breakers, and other disconnecting means. The interrupting capacity of fuses is especially important and should be a major consideration in this section of the electrical specifications.

Other items for consideration are grounding, transformers (usually dry type), underground or overhead service, primary load interrupters, converters, and rectifiers.

16500 Lighting: This section covers general conditions relating to selected lighting equipment to ensure that all such equipment is furnished and installed exactly as designated by the architect or engineer. Further clauses establish the quality and type of interior lighting fixtures, luminous ceilings, signal lighting, exterior lighting fixtures, stadium lighting, roadway lighting, accessories, lamps, ballasts and related accessories, poles, and standards. Methods of installation also are included in most sets of specifications.

Where special lighting equipment is specified, the specifications normally call for large, detailed shop drawings to be submitted to the architect or engineer for approval prior to installation.

16600 Special Systems: Items that may be covered here include a wide variety of special systems unusual to conventional electrical installation. Examples of such items are lightning protection systems, special emergency light and power systems, storage batteries, battery charging equipment, and perhaps cathodic protection. However, this section is by no means limited to these few items. Many other special systems can be described in this section of the specifications.

16700 Communications: Equipment items which are interconnected to permit audio or visual contact between two or more stations or to monitor activity and operations at remote points are covered here. Most clauses deal with a particular manufacturer's equipment and state what item will be furnished and what is expected of the system once it is in operation.

Items covered under this section include radio, shortwave and microwave transmission; alarm and detection systems, smoke detectors; clock and program equipment; telephone and telegraph equipment; intercommunication and public address equipment television systems, master TV antenna equipment, and learning laboratories.

16850 Electrical Resistance Heating: Because of working agreements among labor unions, most heating and cooling equipment is installed by workers other than electricians, and the requirements are usually covered in Division 15 – Mechanical of the written specifications. In some cases, however, the electrical contractor is responsible for installing certain pieces of heating and cooling equipment, especially on residential and apartment projects.

The main point that this section of the specifications should make is that the system installation meet with the design requirements. To do this, electric heat specifications should include pertinent data about the factors of the building insulation on which the design is based, as well as installation instructions for the selected equipment. To further aid in ensuring a proper installation, exact descriptions (manufacturer, catalog number, wattage rating, etc.) of the units normally are specified.

Items in this category of the electrical specifications include snow melting cables and mats, heating cable, electric heating coil, electric baseboard heaters, radiant heaters, duct heaters, and fan-type floor, ceiling, and wall heaters (packaged room air conditioners are listed under Division 15).

16900 Controls: As the name implies, this section covers all types of controls and instrumentation used on a given project. Examples include recording and indicating devices, motor control centers, lighting control equipment, electrical interlocking devices and applications, control of electric heating and cooling, limit switches, and numerous other such devices and systems.

Other divisions of the specifications besides Division 16, especially Division 15 – Mechanical, may involve a certain amount of electrical work. The responsibility for such work should be clearly defined. Without great care by all concerned, confusion about responsibility may result in the electrical contractor paying for work he had assumed was not under his division (Division 16).

The most common sources of confusion are in control wiring for boilers, heating, ventilating and air conditioning systems; control, signal and power wiring beyond the machine room; disconnect switches for elevator construction; automatic machinery controls; wiring on machine tools; wiring on overhead cranes and hoists; mounting and connecting of motors; connecting hospital, laundry and restaurant equipment; connecting electric signs; connecting motion picture projection and sound equipment; installing electric lighting fixtures furnished by someone other than the electrical contractor; connecting unit heaters, unit ventilators, electric fans, electric water heaters, electric water coolers, electric ranges, and other appliances when they are not furnished by the electrical contractor; connecting transformers, and similar items.

The details of discovering and resolving conflicting statements in written specifications will be covered in the paragraphs to follow. More important, however, is avoiding the conflicts altogether. Procedures for this are also given.

METHOD OF COMPILING WRITTEN SPECIFICATIONS

Writing accurate and complete electrical specifications for building construction is a serious responsibility for those who design electrical systems because the specifications, combined with the working drawings, govern practically all important decisions made during the construction span of every project.

A set of electrical specifications for a single project usually will contain hundreds of products, parts and components, and methods of installing them. No one can memorize all of the necessary items required to accurately describe the electrical system. One must rely upon reference materials; that is, manufacturer's data, catalogs, checklists, and best of all, a high-quality master electrical specification.

Sets of published master electrical specifications are available from several sources. Check with the Construction Specifications Institute, Inc. (CSI) for details. Samples of master specifications appear later in this appendix. However, before using any master electrical specifications, it is necessary to understand the format and content. The specification writer will then have to modify the master specification (by inserting or adding items and deleting others).

Selecting the Format

Although specification formats have varied over the years, the electrical work almost always falls under Division 16, and the subdivisions or sections of this division normally follow a traditional order; that is, General Provisions, Service Entrances, etc. Even so, the exact format of electrical specifications will vary.

In order to standardize specification formats, the Construction Specifications Institute, Inc. (CSI), 601 Madison St., Alexandria, VA 22192 (703) 272-0660 developed a very practical format outline for all divisions, including Division 16 — Electrical. The CSI format is currently used by most design professionals and therefore, this is the recommended format for all new projects.

If a master electrical specification is to be compiled from "scratch," necessary reference materials should first be obtained from manufacturers of electrical products. Many of the catalogs and booklets published by these companies offer a wealth of information about their products, methods of using them, and the latest materials available for use. Knowledge of these is necessary to design electrical systems and to write specifications.

Persons involved in the design and construction of electrical systems should request complete information and catalogs from all of these firms. As literature is received, study it carefully and file it under the proper headings; some manufacturers imprint the proper CSI file number on the outside of the bulletin to make this job easier. File hardcover catalogs in their proper order under the correct CSI number on bookshelves.

Most manufacturers of electrical equipment will provide a specification of their products showing how the wording should appear in a set of electrical specifications. All manufacturers want to see their products worked into electrical specifications — for obvious reasons — and will be willing to cooperate with any design professional in helping them compile a suitable specification.

Compiling Specifications

Once all reference material is relatively complete, the specification writer is ready to begin compiling a specification for any given electrical project. Most of the manufacturer's sample specifications, however, will need modification or editing to fit particular needs. Consequently, the various manufacturers' reference specifications are modified as required until all sections pertinent to the electrical project have been included. Never add more information than necessary, but make certain that all items have been adequately covered. This specification, especially if it is the first to be compiled, should be read over several times to ascertain that nothing has been left out and that everything included applies to the given project.

After compiling one such specification, the writer has a model for compiling others. Since the majority of design professionals use a computer word processing program for written specifications, save this first file for the project, and then save it as a master file, such as SpecMast.txt to use as a basic specification for the next project.

On the next electrical project, the specification writer opens the master specification file on the computer and renames it to comply with the new project. The writer reads through the newly-named specification, keeping all details of the new project in mind. If a sentence, paragraph, or perhaps a full section in the master file does not apply to the new electrical project, it is deleted. If something needs to be added, it is done as described previously; that is, use the manufacturer's sample specification as a guide. Save the computer file under the new job name, and if any new categories have been added to the original master specification file, replace the original master file with this newly revised version.

After a few such projects, the specification writer will have a good start on a master electrical specification that is suitable for most electrical projects. With each new project, items are added to the original so that it is constantly improving. As experience is gained in compiling electrical specifications in the manner described, the design professional's knowledge of the electrical industry will improve and better and better electrical specifications — taking less and less time to compile — will be the result.

The beginning specification writer also gains much knowledge by analyzing sets of existing electrical specifications that have been used on actual projects. Complete sets of construction working documents (drawings and specifications) may be reviewed at any Dodge Plan Room (McGraw-Hill Information Systems Co.) or by borrowing sets of documents from architectural and engineering firms. Such firms usually require a deposit until the documents are returned.

Using a Master Specification

The master electrical specification beginning with the next section has been used for numerous projects ranging in cost from only a few thousand dollars to several hundred thousand dollars worth of electrical work. It is the result of several years research by many electrical consulting engineering firms, electrical contractors, manufacturers of electrical equipment, and other interested parties.

No prototype specification can be compiled that will be exactly appropriate for every electrical project, but most people required to write electrical specifications will find that the material included here will form a sound basis for compiling their own master set of electrical specifications.

If, after looking over the entire set, the specification writer finds that the contents will fulfill most of his or her needs, the entire set may be keyed into a computer word processing program for immediate or future use. The entire master specification for IBM PC computers may be ordered directly from John E. Traister Associates, P.O. Box 300, Bentonville, VA 22610. Please enclose a check or money order for $20 (includes shipping charges). Also please specify the word processing format desired; otherwise the file will be sent in DOS ASCI format.

Once the complete master specification is in a computer word processing file, one is ready to customize the specification for any given project. Merely scan through the sentences and paragraphs, deleting or adding items as required. Then save your working file to a different file name to identify the project, and also so as not to change the master file.

NAVIGATING WRITTEN SPECIFICATIONS

General Conditions

The General Conditions section of written specifications consists of a selected group of regulations that apply to all subdivisions of the project. These conditions usually are the responsibility of the architect and for most projects, they are presented in the form of a standard document titled General Conditions of the Contract for Construction, which is available from the American Institute of Architects (AIA), 1735 New York Ave. NW, Washington, DC 20006, (202) 626-7300.

Because the requirements of the General Conditions concern the subcontractors (electrical, mechanical, structural, plumbing, etc.) as well as the general contractor, the electrical specification writer must make certain that they do not contain anything detrimental to Division 16 – Electrical of the specification . . . or vice versa. For example, items covering the bid, contract, performance bonds, re-

quired insurance, payment, etc., are normally covered in this division of the complete specifications. All of these will concern the electrical contractor. This division will further cover the person or trade responsible for removing rubbish, providing temporary electrical facilities, and the like; all of these items could concern the electrical contractor. For this reason, many electrical engineering firms prefer to modify the standard AIA document to state their own terms and requirements more precisely.

Specifications should call attention to these general conditions in the electrical division so that those people bidding the electrical work will be aware of them. A typical example follows:

> *The "Instructions to Bidders," "General Conditions," . . . of the architectural specifications govern work under this Division.*

In some instances, Supplementary General Conditions and Special Conditions are inserted at the beginning of the written specifications; however, these normally concern only the general contractor. Those which concern a particular trade are more frequently inserted into the division to which they directly relate. For example, supplementary conditions that define the scope of the electrical work should appear in the electrical section and not at the front of the specifications.

General Provisions

Unlike General Conditions, which gives the ground rules to *all* subdivisions of the specifications, General Provisions of Division 16 relate only to electrical work.

Depending on the size, type, and complexity of the electrical work, these provisions could involve only a few concise paragraphs or could require several pages of detailed instructions to the electrical contractor or those installing the project.

General and Special Conditions

In nearly all cases, the first few paragraphs in this specification section remind those using the specifications to refer to the architectural general and special conditions, as they will also be part of the electrical specifications.

Electrical Reference Symbols

The next subject covered should be the drawings and reference symbols. A symbol list should appear either on the drawings or in this section of the specifications. Although most engineers and designers use symbols adopted by the American National Standards Institute (ANSI), many design professionals frequently modify these standard symbols to suit their own needs. For this reason, a symbol list or legend is necessary. *See* Chapter 1 of this book.

If standard symbols must be changed, they should be easily drawn by drafters and readily interpreted by workers on the job.

Work Included

Some specifications merely give a general description or scope of the work on the assumption that this is sufficient because more detailed data will follow in subsequent pages. Other sets, however, give a more detailed explanation.

Work Not Included

Following the "work included" section, most electrical specifications also outline the work not included: what electrical equipment items are to be furnished by others, but installed by the electrical contractor, and what electrical equipment is to be furnished and installed by others, but connected by the electrical contractor.

This section of the specifications also includes information pertaining to codes and fees, tests, identification, and demonstration of a complete electrical system. If stated in the architectural General Conditions that the electrical contractor is to provide and install a temporary electrical facility, details of this service should be included in the electrical General Provision. Where the project in question is of sufficient size — requiring more than one panelboard for the temporary electric service, it is advisable to list the number, size, and location of these panels.

MASTER ELECTRICAL SPECIFICATION

DIVISION 16 — ELECTRICAL

16010 General Provisions

(A) The Architectural General and Special Conditions for the construction of this project shall be a part of the Electrical Specifications. The Electrical Contractor shall examine the general and special conditions before submitting his or her proposal.

(B) The General Contractor shall be responsible for all work included in this section and the delegation of work to the Electrical Contractor shall not relieve him of this responsibility. The Electrical Contractor and his subcontractors who perform work under this section shall be responsible to the General Contractor.

(C) Where items of the General Conditions or of the Special Conditions are repeated in this section of the specifications, it is intended to call particular attention to or qualify them; it is not intended that any other parts of the General Conditions or Special Conditions shall be assumed to be omitted if not repeated herein.

(D) The naming of a certain brand or make or manufacturer in the specifications is to establish a quality standard for the article desired. The Contractor is not restricted to the use of the specific brand of the manufacturer named unless so indicated in the specifications. However, where a substitution is requested, a substitution will be permitted only with the written approval of the Engineer. No substitute material or equipment shall be ordered, fabricated, shipped or processed in any manner prior to the approval of the Architect/Engineer. The Contractor shall assume all responsibility for additional expenses as required in any way to meet changes from the original material or equipment specified. If notice of substitution is not furnished to the Engineer within fifteen days after the General Contract is awarded, then equipment and materials named in the specifications are to be used.

(E) The Electrical Contractor shall furnish and present five (5) copies of shop drawings or brochures for all fixtures, equipment, and accessories to the Engineer for the Engineer's approval. The Electrical Contractor shall furnish and present five (5) copies of a schedule of manufacturers of all materials for which shop drawings or brochures are not presented. No equipment shall be ordered, purchased or installed prior to approval of the shop drawings, brochures, and schedules. Checking is only for general conformance with the design concept of the project and general compliance shown is subject to the requirements of the plans and specifications. Contractor is responsible for: dimensions which shall be confirmed and correlated at the job site; fabrication processes and techniques of construction; coordination of his work with that of all other trades and the satisfactory performance of his work.

(F) The Electrical Contractor shall examine drawings relating to work of all trades and become fully informed as to extent and character of work required and its relation to all other work in the project.

(G) Before submitting bid, Contractor shall visit the site and examine all adjoining existing buildings, equipment and space conditions on which his work is in any way dependent for the best workmanship and operation according to the intent of specifications and drawings. He shall report to the Architect any condition which might prevent him from installing his equipment in the manner intended.

(H) No consideration or allowance will be granted for failure to visit site, or for any alleged misunderstanding of materials to be furnished or work to be done.

(I) The Electrical Contractor shall be responsible for all arrangements and costs for providing temporary electrical metering, main switches, and distribution panels at the site as required for construction purposes. The distribution panels shall be located at a central point designated by the Architect. The General Contractor shall indicate prior to installation whether three phase or single-phase service is required.

(J) The Electrical Contractor shall furnish and install one OSHA approved pigtail socket with 150-watt lamp for every 500 square feet of floor space, evenly distributed throughout the building.

(K) The Electrical Contractor shall furnish and install power outlets to total one for every 2000 square feet or part thereof of floor area and these shall be 15-amp, single-phase receptacles for either 110 or 220 volts as directed by the General Contractor.

(L) Any light or power outlets required over the maximum quantity noted above shall be paid for by the Contractor requiring the same. The power consumption shall be paid for by the General Contractor.

16015 Electrical Drawings and Reference Symbols

(A) The Drawings are diagrammatic and indicate generally the locations of material and equipment. These Drawings shall be followed as closely as possible. The Electrical Contractor shall coordinate the work under this section with the architectural, structural, plumbing, heating and air conditioning, and the drawings of other trades for exact dimensions, clearances and roughing-in locations: This Contractor shall cooperate with all other trades in order to make minor field adjustments to accommodate the work of others.

(B) The Drawings and Specifications are complementary, each to the other, and the work required by either shall be included in the Contract as if called for by both.

(C) If directed by the Architect, the Contractor shall, without extra charge, make reasonable modifications in the layout as needed to prevent conflict with work of other trades or for proper execution of the work.

(D) Electrical symbols used on this project are shown in a Symbol List on the accompanying working drawings. This list shows standard symbols and all may not appear on the project drawings; however, wherever the symbol on project drawings occurs, the item shall be provided and installed.

16020 Work Included

(A) The scope of the work consists of the furnishing and installing of complete electrical systems — exterior and interior — including miscellaneous systems. The Electrical Contractor shall provide all supervision, labor, materials, equipment, machinery, and any and all other items necessary to complete the systems. The Electrical Contractor shall note that all items of equipment are specified in the singular; however, the Contractor shall provide and install the number of items of equipment as indicated on the drawings and as required for complete systems.

(B) It is the intention of the Specifications and Drawings to call for finished work, tested, and ready for operation.

(C) Any apparatus, appliance, material or work not shown on drawings but mentioned in the specifications, or vice versa, or any incidental accessories necessary to make the work complete and perfect in all respects and ready for operation, even if not particularly specified, shall be furnished, delivered and installed by the Contractor without additional expense to the Owner.

(D) Minor details not usually shown or specified, but necessary for proper installation and operation, shall be included in the Contractor's estimate, the same as if herein specified or shown.

(E) With submission of bid, the Electrical Contractor shall give written notice to the Architect of any materials or apparatus believed inadequate or unsuitable, in violation of laws, ordinances, rules; any necessary items or work omitted. In the absence of such written notice, it is mutually agreed the Contractor has included the cost of all required items in his proposal, and that he will be responsible for the approved satisfactory functioning of the entire system without extra compensation.

16025 Work Not Included

(A) The following equipment items and work shall be the responsibility of others.

1. Motors and controls, unless indicated otherwise, shall be furnished by others, but shall be installed and connected by the Electrical Contractor as indicated on the drawings.

2. Telephone system wires, cable, equipment, and instruments shall be furnished and installed by the telephone company.

3. Elevator signal and control wiring beyond service feeder noted on drawings shall be provided and installed by others.

4. Controls for motors on mechanical equipment unless indicated otherwise, will be furnished by others but shall be installed and wired by the Electrical Contractor.

16030 Codes and Fees

(A) All materials and workmanship shall comply with all applicable codes, specifications, local ordinances, industry standards, utility company and fire insurance carrier's requirements.

(B) In case of difference between the building codes, specifications, state laws, local ordinances, industry standards, utility company regulations, fire insurance carrier's requirements, and the contract documents, the most stringent shall govern. The Contractor shall promptly notify the Architect in writing of any such difference.

(C) Noncompliance: Should the Contractor perform any work that does not comply with the requirements of the applicable building codes, state laws, local ordinances, industry standards, fire insurance carrier's requirements, and utility company regulations, he shall bear the cost arising in correcting any such deficiency.

(D) Applicable codes and all standards shall include all state laws, local ordinances, utility company regulations and the applicable requirements of the following nationally accepted codes and standards:

1. Building Codes
   a. National Building Code
   b. Local Building Code
   c. National Electrical Code
   d. State Electrical Code
   e. Local Municipal Electrical Code

2. Industry Standards, Codes, and Specifications

    a. AMCA — Air Moving and Conditioning Association

    b. ASHRAE — American Society of Heating, Refrigeration, and Air Conditioning Engineers

    c. ASME — American Society of Mechanical Engineers

    d. ASTM — American Society for Testing and Materials

    e. EIA — Electronic Industries Association

    f. IEEE — Institute of Electrical and Electronic Engineers

    g. IPCEA — Insulated Power Cable Engineers' Association

    h. NEC — National Electrical Code (NFPA No. 70-1996)

    i. NBS — National Bureau of Standards

    j. NEMA — National Electrical Manufacturers' Association

    k. NFPA — National Fire Protection Association

    l. USASI — United States of America Standards Institute

    m. UL — Underwriters' Laboratories

3. Insurance Carriers

    a. FIA — Factory Insurance Association

    b. FMED — Factory Mutual Engineering Division

16050 BASIC ELECTRICAL MATERIALS AND METHODS

16101 General

(A) All materials and equipment shall be new, undamaged, and shall bear the UL label of approval and shall be listed for use in each specific location, unless approval does not apply.

(B) Samples of all materials proposed for use shall be presented to the Engineer for his approval when requested.

(C) Equipment Finish: All electrical equipment shall be furnished factory painted or finished with two coats of high grade enamel and in the manufacturer's standard colors unless otherwise specified.

1. Unpainted equipment and materials, except conduit, shall be cleaned and primed to be painted by the Painting Contractor in accordance with the Painting Section of these specifications.

2. The colors of all exposed electrical material and apparatus shall be as selected by the Owner.

3. Exposed cabinets and boxes (fronts and doors) shall be painted with two coats of high grade enamel as indicated in the Painting Specifications, color selected by the Architect.

16110 Raceways

(A) Where indicated on the plans, the electrical contractor shall furnish and install approved metal raceways with the necessary complement of fittings, connectors and accessory parts. Wire or raceways shall be of the "lay in" type (with, without) standard knockouts and with (hinged, screw) covers for full channel access. Wireway cross-sectional dimensions shall be as indicated on the plans or as required by the *NEC*. All sheet metal parts shall be coated with a rust inhibitor and finished in_____ baked enamel. All hardware shall be plated to prevent corrosion.

(B) Raceways shall be securely supported by approved methods at 5- foot intervals. Number of conductors per wireway shall conform to the latest edition of the *NEC*.

(C) Wireways and fittings shall be_____ type, as manufactured by, or approved equal.

(D) Wire Troughs: Horizontal wiring troughs shall be furnished and in stalled where indicated on the drawings. Troughs shall be made of_____ -gauge sheet metal steel with_____ finish with screw covers and insulated cross-brackets to support conductors at_____ foot intervals.

(E) Troughs shall be of sufficient size to accommodate feeder conduits and cables and provide ample room for installing and training the conductors.

(F) Where indicated, or considered necessary, wire troughs shall have steel barriers to separate feeder circuits, and all troughs shall be supported from the building structure independent of conduits entering them.

(G) Feeders in troughs shall be identified by fireproof tags or other approved method. Individual conductors of feeder circuits shall be tied together with cabling twine.

(H) Flexible Conduit: Connections to motors and other equipment requiring flexible connections shall be made with flexible metal conduit of the correct size. Flexible metal conduit shall be used only in making up short connections to equipment and outlets. Flexible conduit shall be galvanized, single strip, plastic-covered conduit using waterproof-type connections in areas requiring liquidtight protection. The flexible conduit shall have a copper bonding strip for equipment ground connected and inter woven into it.

(I) Surface Raceways: The Electrical Contractor shall furnish and in stall, where indicated on the plans, surface metal raceway as manufactured by _____, or approved equal. Raceway, elbows, fittings and outlets shall be of the same manufacture and designed for use together. They shall be of a size as noted, approved for the number and size of wires indicated, and shall be installed in an approved and workmanlike manner. Runs shall be parallel or at right angles to walls and partitions. Connections shall be made to other types of raceways in an approved manner with fittings manufactured for the purpose and application.

(J) Where combination metal raceways are installed for signal, lighting and power circuits, each system shall be run in separate compartments clearly identified and maintaining the same relative position throughout the system.

(K) The number of conductors installed in any raceway shall not be greater than the number for which the raceway is approved.

(L) Baseboard Types: The Electrical Contractor shall furnish and install, as indicated on the drawings, a system of metallic baseboard wireways for the service as indicated.

(M) The system shall be installed complete with junction boxes, outlet fittings, cross-connected raceways, circuit conductors, and wiring devices as indicated on the plans. The system shall be as manufactured by _____ or approved equal.

(N) At locations shown on the plans by appropriate symbol or notation, a multioutlet assembly in one or more continuous sections shall be installed. These sections shall consist of a raceway with outlets to receive standard attachment plugs spaced_____ inches apart or as otherwise noted.

16111 Conduits

(A) Rigid steel conduit shall be used for all service-entrance conduits and main feeders; and all other branch circuits and raceways unless especially excepted on the drawings and in the specifications. Rigid steel conduits shall be low carbon, hot-dipped galvanized both inside and outside, with threaded joints. Other finishes may be substituted only with the approval of the Engineer. All conduit shall be UL approved.

(B) Conduit fittings shall be cast aluminum alloy or cast ferrous alloy, galvanized, and shall be UL approved.

(C) Rigid aluminum conduit may be used for exposed wiring above grade where specifically shown or when approved by the engineer. It shall not be embedded in concrete nor otherwise used where prohibited by the *NEC*. Aluminum conduit shall be Aluminum Rigid Conduit as manufactured by_____ or approved equal.

(D) Conduit sizes shall be as indicated on the drawings, or minimum in accordance with the *NEC*, including provision for green equipment grounding conductor using inch minimum conduit except on switch legs, where $\frac{1}{2}$ conduit may be used. The use of inch conduit elsewhere may be approved if conditions warrant; verify with Engineer.

(E) Special conduit fittings shall be appropriate for each application and shall be manufactured by_____, _____, or approved equal.

(F) Conduit systems shall be installed in accordance with the latest edition of the *NEC* and shall be installed in a neat, workmanlike manner.

(G) The entire conduit system shall be installed to provide a continuous bond throughout the system.

(H) Electrical metallic tubing (EMT) may be used for branch circuits and raceways other than for service entrance and main feeders unless prohibited by the *NEC* or local ordinances. EMT shall be UL approved, galvanized inside and outside, complying with ASA C-80.3 for zinc coated EMT with fittings of the same type material and finish, and of the pressure connected type.

(I) All conduit joints shall be cut square, threaded, reamed smooth, and drawn up tight. Bends or offsets shall be made with an approved bender or hickey, or hub-type conduit fittings. Number of bends per run shall conform to the *NEC* limitations.

(J) Concealed conduits shall be run in a direct line with long sweep bends and offsets. Exposed conduits shall be parallel to and at right angles to building lines, using conduit fittings for all turns and offsets.

(K) Electrical Systems over 600 Volts: Where conduits are accessible to unauthorized personnel and contain feeders over 600 volts, conduits shall be encased in a____ -inch concrete envelope or other protective means for a distance of____ feet above floor level.

(L) Where conduits are subject to highly corrosive atmospheres, as noted on the plans, conduits shall be plastic-coated metal, all-plastic (PVC), or other corrosion-resistant type as manufactured by_____ , or approved equal. Only couplings and fittings designed specifically for the type of conduit noted shall be used. Conduit shall be supported by corrosion-resistant straps and clamps. The electrical contractor shall follow manufacturer's recommendations regarding the handling, bending, coupling, and installation of the conduit specified herein.

(M) Transitions between nonmetallic conduits and conduits of other materials shall be made with the manufacturer's standard adapters designed for such purpose.

(N) Exposed conduits shall be securely fastened in place on maximum_____ foot intervals; and hangers, supports or fasteners shall be provided at each elbow and at the end of each straight run terminating at a box or cabinet.

16113 Underfloor Ducts

(A) The electrical contractor shall furnish and install a complete system of interconnected floorducts as shown on the drawings and in accordance with the *NEC* regulations.

(B) All ducts used including connectors, fittings, and other related materials shall be of one manufacturer and shall be as manufactured by_____ or approved equal.

16114 Cable Trays

(A) The electrical contractor shall provide and install a complete cable- tray system as called for on the drawings and as called for herein. The system shall be manufactured by_____ or approved equal and shall be of the (ladder) (trough) (channel) type of construction and be (aluminum) (steel) (other).

(B) Straight sections shall be capable of supporting a cable load of_____lbs. per linear foot on a_____ foot simple span between supports where connectors are located within point of the span of tray.

(C) Fittings shall have rung spacing not to exceed four inches at maxi mum opening measured in a direction parallel to the cable for either trough or ladder type tray.

(D) The entire system shall be grounded in compliance with the *NEC* and shall not be used either as a grounded circuit conductor or as an equipment grounding conductor.

(E)   Covers and accessories supplied by the manufacturer shall be installed where indicated on the drawings and any solid bottom type tray section bottoms shall be aligned with H bars.

(F)   Supports shall be in conformance with NEMA standards and recommendations and shall be capable of carrying the required cable loads plus tray weight and any additional short time total loads not to exceed the design loads by more than one-third. On vertical runs, the cable shall be held against thrust by supports external to the tray.

16116 Manholes

(A)   The electrical contractor shall furnish and install manholes of sizes indicated on the drawings for primary electric service and telephone service. Manholes shall be constructed of reinforced concrete, complete with manhole covers and rings, sump drain, cable pulling rings, cable support racks, and other facilities as detailed on the drawings. Outside of the structures shall be waterproofed with a bituminous compound in an approved manner as recommended by the architect/owner.

16120 Wires and Cables

(A)   Wire and cable shall meet all standards and specifications applicable, and shall be in conformance with the latest edition of the *NEC*. Insulated wire and cable shall have size, type of insulation, voltage and manufacturer's name permanently marked on outer covering at regular intervals not exceeding four feet. Wire and cable shall be delivered in complete coils or reels with identifying tags, stating size, type of insulation, etc.

(B)   Wire and cable shall be suitably protected from weather and other damage during storage and handling, and shall be in first-class condition after installation.

(C)   Conductors shall be soft drawn copper, ASTM B3 for solid wire, ASTM B8 for stranded conductors.

Conductor wire sizes shall be American Wire Gauge (AWG); #6 and larger of stranded construction; #8 and smaller of solid construction.

(D)   Wire and cable shall be factory color coded with a separate color for each phase and neutral used consistently throughout the system. Color coding shall be as required by the *NEC*.

(E)   All conductors shall be rated 600 volts, unless otherwise specified or shown on the drawings, or for electronic or communication use.

(F)   Conductors for lighting, receptacle, and power branch circuits, feeders, and subfeeders size #1 and smaller shall be type THW heat and moisture-resistant thermoplastic insulated.

(G)   Conductor for feeders and subfeeders size #1/0 and larger shall be type RHW moisture- and heat-resistant rubber insulated.

(H)   Wire and cable shall be as manufactured by_____ or approved equal. Substitution of wire and cable manufacturer shall be only with the approval of the Architect/Engineer.

(I)   Branch circuits within all electric heater elements such as electric duct coils, baseboard radiation, and cabinet unit heaters shall be type THHN heat resistant, thermoplastic insulated, maximum operating temperature 90°C (194°F).

(J)   Underground feeder and branch circuit wire for direct burial in earth or in conduit shall be Type UF for use in wet or dry locations.

(K)   For any specific use not covered here above, comply with the *NEC* in conductor use.

16121 Wire Connections

(A)   Joints on branch circuits shall occur only where such circuits divide as indicated on plans and shall consist of one through circuit to which shall be spliced the branch from the circuit. In no case shall joints in branch circuits be left for the fixture hanger to make. No splices shall be made in conductor except at outlet boxes, junction boxes, or splice boxes.

(B)   All joints or splices for #10 AWG or smaller shall be made with UL approved wire nuts or compression type connectors.

(C)   All joints or splices for #8 AWG or larger shall be made with a mechanical compression connector. After the conductors have been made mechanically and electrically secure, the entire joint or splice shall be covered with Scotch #33 tape or approved equal to make the insulation of the joint or splice equal to the insulation of the conductors. The connector shall be UL approved.

16125 Pulling Cables

(A)   Install conductors in all raceways as required, unless otherwise noted, in a neat and workmanlike manner. Telephone conduits and empty conduits as noted, shall have a #14 galvanized pull wire left in place for future use.

(B)   Conductors shall be color coded in accordance with the *NEC*. Mains, feeders, subfeeders shall be tagged in all pull, junction, and outlet boxes and in the gutter of panels with approved code type wire markers.

(C)   No lubricant other than powdered soapstone or approved pulling compound may be used to pull conductors.

(D)   At least eight (8) inches of slack wire shall be left in every outlet box whether it be in use or left for future use.

(E)   All conductors and connections shall test free of grounds, shorts and opens before turning the job over to the Owner.

(F)   Pull boxes required in runs over 100 feet or when more than three 90-degree bends are used, or as indicated on the Drawings.

(G)   Feeders are to be run above ground to all power panels and lighting panels, unless indicated otherwise on drawings.

(H)   Conduit terminating inside of prestressed concrete panel voids shall be provided with necessary bushings to prevent damage to wiring run in voids.

(I)   Where motors have conduit terminal boxes, feeders shall be connected to same by flexible means.

(K)   All motors with sliding base mountings shall have not less than 18 inches nor more than 6 feet of conduit connecting rigid conduit feed to motor terminal box.

(L)   Conductor splices shall be made only in junction boxes, terminal boxes, or pull boxes.

16132 Floor Boxes

(A)   Before locating the outlet boxes, check all of the architectural drawings for type of construction and to make sure that there is no conflict with other equipment. The outlet boxes shall be symmetrically located according to room layout and shall not interfere with other work or equipment. Also note any detail of the outlets shown on the drawings.

(B)   Unless noted otherwise, floor boxes shall be manufactured by_____ catalog number_____ , or approved equal, (or terrazzo type where needed) of cast iron watertight type with aluminum cover and flange. The outlets shall be as specified under 16140, Switches and Receptacles. Each box shall have at least two threaded hubs or more when specified or called for elsewhere on the drawings. Where carpet is encountered, provide carpet flanges on the floor boxes.

(C)   Cover plates on all floor boxes, unless otherwise noted, shall be of heavy brass with permanent ring or flange and rubber gasket. Plates shall have _____-inch-diameter threaded hole in center for installation of a flat plug or fitting for receptacle or other type of outlet as indicated on the drawings.

16133 Outlet Boxes

(A) All outlet boxes for concealed wiring shall be sheet metal, galvanized or cadmium plated, at least 1½ inches deep, single or ganged, of size to accommodate devices and number of conductors noted. Boxes shall be equipped with plaster ring or cover as necessary. All outlet boxes shall be manufactured by_____ or approved equal.

(B) Boxes for exposed wiring shall be malleable iron, cadmium finish, or cast aluminum alloy, as manufactured by _____ , and shall not be less than 4 inches square by 1½ inches deep unless otherwise noted.

(C) Fixture outlet boxes shall be minimum 4 in octagonal and, where required as outlet and junction boxes, they shall be 4¹¹⁄₁₆ inches by 2⅛ inches deep.

(D) Outlet boxes for concealed telephone and signaling systems shall be of the 4 inch sq. type with plaster cover and bushed-opening cover plate.

(E) Outlet boxes for hazardous areas shall be explosionproof with appropriate fittings, sealoffs, etc.

(F) Boxes for floor outlets shall be of the cast-metal threaded-conduit-entrance, waterproof type with means for adjusting cover plate to finished floor level. Boxes shall be approximately 4 inches in diameter and 3½ inches deep, with an approved gasket or seal between adjusting ring and box.

16134 Pull and Junction Boxes

(A) The Electrical Contractor shall furnish and install junction boxes and pull boxes where indicated on the drawings, or as required by the *NEC*, or where necessary to facilitate pulling in wires and cables without damage.

(B) Boxes shall be formed from sheet steel, with corners folded in and securely welded, with inch inward flange on all four edges, with box drilled for mounting, and with flange drilled for attachment of cover. Box shall be galvanized after fabrication. Cover shall be made of one piece galvanized steel and provided with round head brass machine screws for fastening to box. Box and cover shall be made of code gauge steel, or heavier as specified. Boxes shall be a minimum of 4½ inches deep, and sized as required to meet *NEC* standards, or larger as specified, utilizing manufacturer'' standard size, or next larger to meet dimensional requirements.

(C) Pull and junction boxes shall be furnished without knockouts for field drilling and shall be manufactured by _____ or approved equal.

(D) If pull or junction box is exposed, the box shall be painted to match the finish of the building surfaces adjacent to the box, unless indicated otherwise by the engineer.

16140 Wiring Devices

(A) The wiring devices specified below with manufacturer and catalog number may also be the equivalent wiring device as manufactured by _____, _____, or _____ . All other types shall be as specified.

1. Wall Switches. Where more than one flush wall switch is indicated in the same location, the switches shall be mounted in gangs under a common plate.

Device	Manufacturer and Catalog No.
Single-pole switch	—
Three-way switch	—
Four-way switch	—
Switch w/pilot light	—
Motor switch — surface	—
— flush	—

2. Receptacles. Duplex 20A, 125V (Manufacturer and catalog number.)

(B) The Electrical Contractor shall furnish and install wall plates of appropriate type and size for all wiring and control devices, signal, and telephone outlets. Plates shall be constructed of_____ with a_____ finish and shall be in color. Special markings on the plates shall be provided as indicated on the drawings.

(C) When devices are installed in exposed outlet boxes, the plates or covers shall be of a type designed for the boxes.

(D) Where lighting branch circuits are to be controlled by a dimming device, they shall be as manufactured by_____ with rating and range of dimmer as indicated on the drawings or by the calculated load on the circuit.

(E) Time switches shall be furnished and installed where called for on the drawings and shall be manufactured by _____ , or approved equal. The switch dial shall be of the_____ type or as indicated on the drawings.

(F) Outlets and circuits indicated by the appropriate symbol on the drawings shall be controlled by a remote control low-voltage relay system as manufactured by _____ or approved equal. The system shall be complete with all relays, switches, master switches, power supply, etc.

16160 Cabinets and Enclosures

(A) Cabinets used for cable supports for service entrance, feeders, and other cables or electrical components shall be of # _____ gauge steel and shall be furnished and installed where indicated on the drawings. Boxes shall have removable screw covers fastened by corrosion-resistant machine screws and shall be of a size large enough to accommodate the feeder conduits indicated and also provide ample space to install cable supports.

(B) Wireways shall be used where indicated on the drawings and for mounting groups of switches and/or starters. Wireways shall be the standard manufactured product of a company regularly producing wireway and shall not be a local shop assembled unit. Wireways shall be of the hinged cover type, UL listed, and of sizes indicated or as

required by *NEC*. Finish shall be medium light gray enamel over rust inhibitor. Wireways shall be of raintight construction where required. Wireways shall be_____ or approved equal.

16190 Supporting Devices

(A) The Electrical Contractor shall provide and install metallic supports not more than eight feet apart or as required for the proper installation of raceway systems and all other equipment installed under this division of the contract.

(B) Conduit shall be supported on approved types of wall brackets, ceiling trapezes, strap hangers or pipe supports, secured by means of toggle bolts in hollow masonry walls or units. Expansion bolts will be used in concrete or block machine screws on metal surfaces and wood screws on wood construction.

(C) Conduit shall be securely fastened to all sheet metal outlets, junction and pull boxes with two galvanized locknuts and bushing, care being taken to see that the full number of threads project through to permit the bushing to be drawn tight against the end of the conduit, after which the locknuts shall be made tight sufficiently to draw them into firm electrical contact with the outlet box.

(D) The Electrical Contractor shall be responsible for all concrete pads, supports, piers, bases, foundations, and encasement required for the electrical equipment and conduit. The concrete pads for the electrical equipment shall be six inches larger all around than the base of the equipment unless specifically indicated otherwise.

16195 Electrical Identification

(A) The Electrical Contractor shall maintain accurate records of all deviations in work as actually installed from work indicated on the drawings. On completion of the project, two (2) complete sets of marked-up prints shall be delivered to the Architect.

(B) Provide a laminated plastic or rigid phenolic plastic nameplate with $\frac{1}{8}$-in engraved letters; and mount on each starter, disconnect switch, pushbutton station, power and lighting panel, and at each ceiling or wall access panel to electrical work. Attach name plates with self-tapping sheet metal screws.

16199 Electronic Equipment

(A) The Electrical Contractor shall be responsible for the installation and connection of a proper power supply to all electronic equipment furnished by others. He shall verify all voltage, frequency, etc., requirements prior to energizing the circuit. Those installing the equipment will be responsible for the proper operation of the equipment provided the proper power supply circuit is installed by the Electrical Contractor.

16300 MEDIUM VOLTAGE DISTRIBUTION

16301 General

(A) Right of way shall be cleared a distance of 62.5 feet each side of the center line of the transmission line unless otherwise directed. All trees, brush, and stumps within a radius of 25 feet of any tower leg of the finished transmission line shall be cut off as close to the ground as practicable as determined by the contracting officer, and in no case shall they be cut off at a height of more than 12 inches above the ground. All trees, brush, and stumps more than five feet in height in other areas to be cleared shall be cut off at not more than 18 inches above the ground. If directed by the contracting officer, the clearing shall also include the cutting or trimming of all trees outside of the right of way if such trees upon falling would come within 10 feet of the nearest conductor of the line. The cleared material shall be burned or otherwise disposed of as approved in writing by the contracting officer. All materials to be burned shall be piled and when in suitable condition shall be completely burned. Piling for burning shall be done in such a manner and such locations as will cause the least fire risk, and all materials which cannot be completely burned as the work proceeds shall be piled in locations approved by the contracting officer and thereafter completely disposed of by burning within the period of time covered by the contract. Payment for clearing land and right of way will be made at the lump sum price bid.

(B) Where the right of way is through well-developed areas such as orchards, clearing will be confined to the tower sites, except that the contracting officer may require trimming or removal of all trees or obstruction that interfere with operation of the transmission line. All trees, brush, or stumps within ten feet of any tower member shall be cut off as close to the ground as practicable.

(C) Blasting will be permitted only when proper precautions are taken for the protection of persons, the work, and public or private property, and any damage done to the work or public or private property by blasting shall be repaired by the contractor at the contractor's expense. Caps or other exploders or fuses shall be in no case stored, transported, or kept in the same place in which dynamite or other explosives are stored, transported or kept. The location and design of powder magazines, methods of transporting explosives, and in general, the precautions taken to prevent accidents shall be subject to the specifications, but the contractor shall be liable for all injuries to or deaths of persons or damage to property caused by blasts or explosives.

(D) The conductors will be furnished by others. The conductors for the transmission line will be 795,000 circular-mil aluminum conductor, steel reinforced (ACSR), having an outside diameter of 1.108 inches, an ultimate strength of 31,200 pounds and a weight of approximately 1,098 pounds per 1,000 feet. Unless otherwise shown on the drawings or directed by the contracting officer, all clearances, measured in still air at 60°F, shall conform to provisions of the local safety standards. The equipment and methods used for stringing the conductors shall be such that the conductors will not be damaged or injured, and shall be subject to the approval of the contracting officer. Particular care shall be taken at all times to insure that the conductors do not become kinked, twisted, or abraded in any manner. If the conductors are damaged in the contractor's operations, the contractor shall repair or replace the damaged sections, including the furnishing of the necessary additional material, in a manner satisfactory to the contracting officer and at no additional cost. All sections of the conductors damaged by the application of gripping attachments shall be repaired or replaced before the conductors are sagged in place. The conductors shall be laid along the ground from moving reels and then raised into position in the stringing blocks by means of running lines, at the direction of the contracting officer. The running lines shall be of sufficient length to avoid applying undue strain to the insulators and structures. The running lines shall be connected to the conductors with swivel connectors and stocking type grips as directed by the contracting officer. The end of the grips shall be taped to the conductor so that the grips will run freely in the sheaves. Conductor splices shall not be passed through a sheave except as specifically permitted by the contracting officer. Reel stands shall be heavily constructed and provision shall be made for braking the reels. The conductor shall not be dragged over the ground, rock, fence wires, or any object which may damage the conductor. Suitable guards or sheaves shall be used to protect the conductor from damage in places where it would otherwise be impossible to keep the conductor from coming in contact with objects which may injure the conductor. Guards shall consist of material over which the conductor may slide without injury and shall be subject to the approval of the contracting officer. The minimum diameter of the sheaves shall be 14 inches at the bottom of the groove and the size and shape of the groove shall conform to the conductor manufacturer's recommendation. Sheaves shall be equipped with high-quality ball or roller bearings to reduce friction to a minimum. The conductor shall not be allowed to hang in the stringing blocks more than 18 hours before being pulled to the specified sag. After being sagged, the conductor shall be allowed to hang in the stringing blocks for not less than two hours before being clipped in, to permit the conductor tension to equalize. The total time which the conductor is allowed to remain in the stringing blocks before being clipped in shall be not more than 36 hours. The conductors shall not be prestretched, and shall be sagged in accordance with sag tables furnished by the contracting officer. The sag tables will be furnished to the contractor after notice to proceed has been issued. The length of conductor sagged in one operation shall be limited to the length that can be sagged satisfactorily. In sagging lengths of more than one reel, the sag of three or more spans near each end and the middle of the length being sagged shall be checked. The length of the spans used for checking shall be approximately equal to the ruling span. The sag of all spans more than

1,500 feet in length shall be checked, and at sharp vertical angles the sag shall be checked on both sides of the angle. The sag of spans on both sides of all horizontal angles of more than 10° shall be checked. After the conductors have been pulled to the required sag, intermediate spans shall be inspected to determine whether the sags are uniform and correct. Sagging operations shall not be carried on when, in the opinion of the contracting officer, wind prevents satisfactory sagging. A tolerance of plus or minus one-half inch of sag per hundred feet of span length, but not to exceed six inches in any one span, will be permitted, provided that all conductors in the span assume the same sag and the necessary ground clearance is obtained; provided further that the conductor tension be tween successive sagging operations is equalized so that the suspension insulator assemblies will assume the proper position when the conductor is clipped in. The contracting officer will check the sag at all points to be checked, and the contractor shall furnish the necessary personnel for signaling and climbing purposes. At all suspension or tension structures, the conductors shall be attached to the insulator assemblies by suspension or strain clamps as shown on the drawings, and all nuts shall be tightened adequately. Payment for stringing conductors will be made at the unit price per mile of line bid in Schedule No._____ for stringing six 795,000 circular-mil, steel-reinforced aluminum conductors, which unit price shall include the cost of stringing, splicing, connecting, armoring, and sagging the conductors, as described in this paragraph and the following paragraphs.

(E) Unless otherwise directed by the contracting officer, all joints or splices and deadends in the conductors shall be made in accordance with the recommendations of the conductor manufacturer using the special compressor tool recommended for this purpose. The contractor shall furnish all necessary tools, including compressors, required for making joints and splices. Compression-type connectors and compression-type deadends will be furnished by others. All joints or splices shall be located at least 50 feet away from the structures, and no joints or splices shall be made in spans crossing over or adjoining important highways, railroads, or other utility lines without express permission of the appropriate authorities or approval of the contracting officer. At all deadend and large angle structures, the jumper connections shall be made with compression-type jumper terminals which shall be bolted to the compression-type deadends. The cost of making all joints, splices, deadends, and installing jumpers shall be included in the unit price bid in Schedule No._____ for stringing conductors.

(F) At each suspension insulator assembly on the transmission line, armor rods shall be attached to the aluminum conductor in accordance with this paragraph or as directed by the contracting officer. If it becomes necessary to change the point of attachment by more than $2\frac{1}{2}$ inches either way from the midpoint of the armor rods after the armor rods are attached, a new set of armor rods shall be furnished and attached by the contractor without additional cost. Compression repair sleeves or compression joints may be used on the jumper suspension insulator assemblies in lieu of armor-rod sets. The cost of attaching armor rods and compression joints or compression repair sleeves on jumper insulator assemblies shall be included in the unit price bid in Schedule No._____ for stringing conductors.

(G) Vibration dampers shall be attached to the aluminum conductor at the ends of those spans so designated on the plan-profile drawings or on the structure-list sheets to be furnished by the contracting officer, and at other points required by the contracting officer. The vibration dampers shall be attached in accordance with Drawing No. _____, and fastened securely so that they will hang in vertical planes. The contractor shall ascertain that the drain holes are open after the vibration dampers are attached. Payment for attaching vibration dampers will be made at the unit price bid therefore in Schedule No._____.

(H) Two overhead ground wires shall be strung for the entire length of the transmission line, and shall be attached to the towers by the contractor in accordance with the details shown on Drawing No. _____ . For the northern one-half of the line, the overhead ground wires will be $\frac{1}{2}$-inch diameter high strength double-galvanized, stranded steel wire. For the southern one-half of the line, the overhead ground wires will be $\frac{7}{16}$-inch, extra high-strength copperweld stranded wire. The ground wire, together with appurtenant

material, will be furnished by others. The equipment and methods used for stringing the overhead ground wires shall be the same as for stringing the conductors as described elsewhere, and the same degree of care shall be exercised to avoid damage or injury to the ground wires. If damaged, they shall be repaired by the contractor or the damaged sections replaced, including the furnishing of the necessary additional material, in a manner satisfactory to the contracting officer and at no additional cost, before the wires are finally sagged in place. Joint or splices in the overhead ground wires shall be located at least 50 feet away from the structures, and no joints or splices shall be made in spans crossing over or adjoining important highways, railroads, or other public utility lines, unless approved by the contracting officer. The cost of making all joints and splices in the ground wires shall be included in the unit prices bid on Schedule No. _____ for stringing the overhead ground wires. The overhead ground wires shall be sagged in place in accordance with the sag tables furnished by the contracting officer. The methods used in checking the sag of the overhead ground wires shall be the same as those outlined previously for checking the sag of the conductors. Payment for stringing the two overhead ground wires will be made at the unit prices per mile of line bid therefore in Schedule No. _____ , which unit prices shall include the cost of installing all fittings required for the installation of the ground wires, and the stringing, splicing, connecting, deadening, sagging, and clipping of the overhead ground wires as stated in this paragraph.

(I) All metal and wire fences which cross under the transmission line or which are located in the proximity of and parallel to the transmission line shall be grounded with ground posts which will be furnished by others. The ground posts will be ten feet long, with attached tongues for holding the wires of the fence. For each one-quarter of a mile of fence that the fence is within 75 feet of the center line of the transmission line, and on each side of the right of way where a fence crosses under the transmission line, one ground post shall be driven to a depth of not less than five feet, or as directed by the contracting officer. The fence shall be fastened securely to the ground posts by the tongues provided therefore on the posts. Payment for placing fence ground posts and for grounding fences will be made at the unit price per post bid therefore in Schedule No._____.

(J) The contractor shall install the ground rods and connect the ground rods to the tower. Bare stranded copper conductor, connectors and ground rods will be furnished by others. Conductor will be #4/0 AWG. Ground rods will be ⅝-inch by 10 foot copperweld. Four ground rods for each tower shall be installed as shown Drawing No. _____. Ground rod resistance tests will not be required. Payment for placing tower ground rods and grounding towers will be made at the unit price per rod bid therefore in Schedule No. _____.

16310 Medium Voltage Substations

(A) The contractor shall furnish and install a complete substation of open framework construction complete with transformers and oil circuit breakers as shown in the working drawings and the supplemental detailed drawings, schedule, etc.

(B) All concrete pads and foundations shown in the drawings or called for elsewhere in the specifications shall also be furnished and installed by the electrical contractor.

(C) Provide and install all ducts, terminators, and wiring for the transformers, switchgear, and other equipment concerned with the power transmission system.

16320 Medium Voltage Transformers

(A) Master station shall be designed for (outdoor or indoor) installation and shall consist of transforming and coordinating combinations of high-voltage and low-voltage switchgear, installed in accordance with the feeder diagram and at the location shown on the drawings.

(B) Transformer shall be rated for ___/____ kVA, oil insulated (non-inflammable liquid-filled, air-cooled, etc.) 3-phase, 60 cycles, _____°C temperature rise, _____volts delta primary, _____volts wye secondary, with solidly grounded neutral. High-voltage windings shall be provided with 2.5 percent taps, two above and two

below normal, externally operated manual tap-changer handle arranged for padlocking in each position. Taps may be changed only when transformer is deenergized. They provide means for adjusting to average supply voltage.

(C) Automatic tap-changing equipment for operation under load shall be installed on low-voltage side to maintain constant voltage on low-voltage terminals (or some point on the feeder).

(D Transformer shall be Type ____as manufactured by the_____Company, or approved equal.

(E) All dry-type transformers shall be ____ phase, 60-cycle, two-winding type, _____volt primary, _____/_____-volt secondary with grounded neutral. Rated capacity shall be____kVA (or as otherwise noted herein or on the plans). All transformers shall have (number) _____percent full capacity taps (below) (above) normal.

(F) Units shall be designed for quiet operation with a decibel rating not to exceed ____ dB; shall be (floor, wall, ceiling) mounted on _____ vibration eliminators. Circuit connections shall be in (rigid) (flexible) (metal) (nonmetallic) conduit in approved manner. Transformer shall be Type___ as manufactured by _____Company, or approved equal.

16321 Vaults

(A) The electrical contractor shall furnish and install vaults of sizes indicated on the drawings for primary electric and telephone services. Vaults shall be constructed of reinforced concrete, complete with cable pulling rings, cable support racks and other facilities as detailed on the drawings. Outside of the structures shall be waterproofed with a bituminous compound in an approved manner as recommended by the architect/owner.

16345 Medium Voltage Switchboards

(A) The electrical contractor shall furnish and install a ____kV outdoor, metal-clad switching center consisting of interrupter switches of capacities as noted on the drawings. The electrical contractor shall furnish and install insulated, ____kV cables, ground cable and conduit systems between the incoming _____kV sections of the metal-clad switchgear and the secondary busses of the substation, and make all connections to these busses. The electrical contractor shall furnish and install the necessary conduits and wiring between the metal-clad switchgear and the meter house as required by the owner, also the ground rods and underground cable system for equipment ground of the outdoor metal switchgear and grounding of metal fence around switchgear.

16380 Medium Voltage Converters

(A)   The electrical contractor shall furnish and install, where indicated on the plans, a ____kw, ____cycle static-frequency converter. The input shall be _____volts, _____phase, _____cycle. Output shall be _____volts, _____phase, _____cycle. The converter shall be as manufactured by_____or approved equal. The unit will be approximately ___inches high, ___inches wide, ____inches deep; and will weigh about _____pounds. Installation shall be in accordance with the manufacturer's recommendations.

(B)   The electrical contractor shall furnish and install (number) ____ kW, motor-generator frequency converter(s) which will provide _____cycle, _____volt power for the high-frequency lighting system. The cycle generator phases shall be center tapped to ground so that the voltage will not exceed _volts to ground. Each unit shall be a self-contained m-g set with associated control equipment contained in an all-metal dripproof cabinet.

16382 Capacitor

(A)   Capacitors for power-factor correction shall be type as manufactured by _____ , or approved equal. Individual units shall be rated _____kvar for operation on _____ volts, _____ phase, cycle circuits. Units shall be installed as indicated on the drawings and supported in a suitable frame or rack for required mounting. Operating losses shall not exceed percent of the kVA rating. Installation shall be

complete with necessary bus work, connectors and switching facilities. Circuit breakers or switches shall be manually operated and housed in an integral steel enclosure.

16383 Rectifiers

(A)   The electrical contractor shall furnish and install all rectifiers as called for on the drawings and shall be designed for maximum input voltage, maximum output voltage and current ratings as called for on the drawings or in the schedules.

(B)   The size and location of the rectifier electrical protective devices as well as the special installation and connection instructions shall be followed as indicated on the drawings or called for elsewhere in the written specifications.

16400 SERVICE AND DISTRIBUTION

16401 General

(A)   The Electrical Contractor shall furnish and install an electric service entrance and related distribution equipment as indicated on the floor plan, diagrams, schedules, and notes. All equipment shall be new and UL listed.

16402 Electric Service

(A)   The Electrical Contractor shall make all arrangements with the electric utility and pay all charges made by the electric utility for permanent electric service to the project. In the event that the electric utility's charges are not available at the time the project is bid, the Electrical Contractor shall qualify his bid to notify the Owner that such charges are not included.

(B)   The Electrical Contractor shall provide and install raceway, and install current transformer cabinet and/or meter trim for metering facilities as required by the electric utility serving the project. The electric utility will provide the meter installation including meter, current transformers, and connections.

(C)   The Electrical Contractor shall properly ground the electrical system as required by the *NEC*. The ground wire for the service entrance shall be run in conduit and made to the main water service and connected ahead of any valve or cutoff.

(D) The conduit used for service entrance shall be galvanized rigid steel conduit unless otherwise noted on the drawings.

(E) Conductors for the service entrance shall be copper Type RHW or THW rated at 75 degrees unless otherwise noted. The conductors indicated on the drawings are based on copper. Conductors with a size of #1/0 and larger may be aluminum, provided the size of the conductor is increased to have the same or more current carrying capacity as the copper conductors and that the terminal lugs are designed for aluminum. Also, the conduit size shall be increased accordingly.

16403 Underground Service

(A) The power company will furnish and install the underground primary service wire and the pad mounted transformer. The Contractor shall furnish and install the secondary service.

(B) Secondary service will be___ volts, _____phase, ____wire, 60 hertz ac.

(C) Metering will be by the power company. Provide one 1¼-inch empty conduit from CT cabinet to the meter cabinet. The power company will furnish the meter cabinet and CT cabinet which will be mounted by the Electrical Contractor. The power company will run control wires to the meter. The CTs will be furnished by the power company and will be installed by the Electrical Contractor.

(D) Ground service in accordance with the *NEC* and as indicated in the power-riser diagram.

16420 Service Entrance

(A) The Electrical Contractor shall provide a service-entrance system as indicated on the drawings. The Electrical Contractor shall verify the electrical requirements of all actual equipment involved prior to the installation of electrical service to same.

(B) The electrical system shall be a _____volt, _____ cycle, _____phase, _____wire service.

(C) The Electrical Contractor shall install the service-entrance conduit, the CT cabinet, conduit to meter base, and meter base as required by the power company.

16421 Emergency Service

(A) The electrical contractor shall furnish and install a complete emergency electric service as called for on the drawings and in Sections 16200 and 16620. All material and workmanship shall conform with Section 16100 of the specifications, the *NEC*, and all local codes and ordinances.

16430 Metering

(A) The Electrical Contractor shall provide and install raceway, and install current transformer cabinet and/or meter trim for metering facilities as required by the electric utility serving the project. The electric utility will provide the meter installation including meter, current transformers, and connections

(B) The electrical contractor shall verify all requirements for the metering, and furnish all miscellaneous components not provided by the utility company at no extra cost to the Owners.

16435 Converters

(A) The Electrical Contractor shall furnish and install, where indicated on the plans, a _____kW, _____ cycle static frequency converter. The input shall be _____ volts, _____ phase, _____ cycle. Output shall be _____ volts, _____ phase, _____ cycle. The converter shall be _____ as manufactured by _____ or approved equal. The unit will be approximately _____inches high, _____ inches wide, _____ inches deep; and will weigh about _____ pounds. Installation shall be in accordance with manufacturer's recommendations.

(B) The Electrical Contractor shall furnish and install (number) _____ kW, motor-generator frequency converter(s) which will provide _____ cycle _____ volt power for the high-frequency lighting system. The _____ cycle generator phases shall be center tapped to ground so that the voltage will not exceed _____ volts to ground. Each unit shall be a self-contained m-g set with associated control equipment contained in an all-metal drip-proof cabinet.

16436 Rectifiers

(A)   The Electrical Contractor shall furnish and install all rectifiers as called for on the drawings and shall be designed for maximum input voltage, maximum output voltage, and current ratings as called for on the drawings or in the schedules.

(B)   The size and location of the rectifier electrical protective devices as well as the special installation and connection instructions shall be followed as indicated on the drawings or called for elsewhere in the written specifications.

16440 Disconnect Switches

(A)   The Electrical Contractor shall furnish and install safety switches as indicated on the drawings or as required. All safety switches shall be _____ and UL listed. The switches shall be fused safety switches (FSS) or non-fused safety switches (NFSS) as shown on the drawings or required and shall be manufactured by _____.

(B)   Switches shall have a quick-make and quick-break operating handle and mechanism which shall be an integral part of the box. Padlocking provisions shall be provided for padlocking in the OFF position with at least three padlocks. Switches shall be horsepower rated for 250 volts ac or dc or 600 volts ac as required. Lugs shall be UL listed for copper and aluminum cable.

(C)   Switches shall be furnished in NEMA 1 general purpose enclosures with knockouts unless otherwise noted or required. Switches located on the exterior of the building or in "wet" locations shall have NEMA 3R enclosures (WP).

(D)   The safety switches shall be securely mounted in accordance with the *NEC*. The Contractor shall

provide all mounting materials and install fuses in the FSS. The fuses shall be dual element on motor circuits.

(E)   The Electrical Contractor shall furnish and install, as well as be responsible for the proper operation of all electronic equipment indicated thusly on the electrical drawings (see 16915 for electronic lighting controls, and 16920 for heating and cooling controls).

16450 Grounding

(A)   The conduit systems, neutral conductor for the wiring system, and the telephone system shall be securely grounded. The grounds shall be *NEC* grounds in each case. A ground shall be established and tests carried out to indicate that satisfactory ground has been established in accordance with the *NEC*. Written results of this test shall be forwarded to the Engineer before connection to the service.

16460 Transformers

(A)   The Electrical Contractor shall furnish and install all dry-type transformers shown on the drawings and indicated herein. Transformers shall be _____ phase, 60-cycle, two-winding type with Class _____ insulation; shall have _____ volt, delta-connected primaries and _____ volt wye-connected secondaries with a grounded neutral. Rated capacity shall be _____ kVA (or as otherwise noted herein or on the plans). Transformers shall have _____ percent full capacity taps below and above normal on the primary side.

(B)   Units shall have a maximum sound-level rating of _____ dB (shall not exceed decibel ratings listed in the NEMA standards for specific kVA sizes); shall be _____ mounted on _____ vibration eliminators. Circuit connections shall be in _____ conduit in approved manner.

(C)   Transformer shall be Type___ as manufactured by _____ or approved equal.

16465 Bus Ducts

(A) General: The enclosed bus bar distribution system shall be of the plug-in type as manufactured by____. It shall be constructed in accordance with NEMA recommended standards and shall bear the UL label. All necessary straight sections, fittings, take-off plugs, and duct hangers as indicated in the specifications and on the accompanying drawings are to be supplied.

(B) The bus duct shall be suitable for operation on a system rated _____volts, _____phase, _____wire, _____cycles, ( ) grounded, ( ) ungrounded.

(C) Enclosure: The enclosure shall consist of standardized, identical sheet ( ) steel ( ) aluminum casing halves bolted together to form a sturdy, self-supporting housing for the bus bars. Casing halves are to be staggered so that the ends of adjacent sections will overlap to form rigid, scarf-lap joints. A hand-hold opening shall be provided at each end of every duct section. These shall be on opposite sides of the run to provide easier access to busbar bolts. Hangers for supporting the enclosures are to be included. These shall be of a type which can be attached to any point along the duct, and shall include outboard flanges on all four sides to facilitate mounting. Two hangers shall be furnished with each duct section so that the system may be supported every five feet if required.

(D) Bus Bars: Bus bars are to be of solid rectangular design in all ratings. They shall be of round-edge copper or aluminum, completely silver-surfaced throughout their entire length. Each bar shall have one end offset by at least the thickness of the bar to provide a lapped joint free of mechanical stress. One or more ____⅜-inch spline nuts are to be swaged into the offset portion of each bar for purpose of making bolted connections to straight bars in adjacent duct sections. Loose washers and nuts shall be thereby eliminated.

(E) Bus Bar Insulators: Insulators shall be spaced no more than 12 inches from center to center along the entire length of the ducts. A felt pad shall be placed between each insulator and the metal casing to prevent mechanical contact between the two.

(F) Plug Outlets: Outlets for the attachment of bus plugs shall be available on both sides of the duct. At least 10 openings are to be provided in each 10-foot duct section. Outlets must be polarized so that the plugs may be inserted in one position only, thus assuring a continuity of phase rotation.

(G) Plug-in Units: Plugs shall be totally enclosed switches or breakers as specified. They shall be designed to make metal to metal contact between plug body and duct casing before contact stabs engage the bus bars. Stabs shall be individually enclosed by a slotted insulating sleeve to prevent accidental contact or flash-over during insertion or removal.

(H) Finish: Steel housing: All surfaces shall be ground smooth to remove all rough edges, burrs, scratches and sharp corners. All enclosing parts shall be immersed in a suitable cleaning solution, rinsed by a hot water dip, thoroughly dried, and bonderized. They shall then be sprayed with one coat of primer-surfacer oven baked, then sprayed with two coats of synthetic enamel, ASA, 61 gray color, and oven baked. Aluminum Housing: After grinding and smoothing, the parts shall be given a brush type finish, and coated with clear lacquer.

(I) Ventilated Feeder Type Bus Duct: Where ventilated bus duct is called for in the specifications or on the drawings, the duct shall be of the low impedance, ventilated ( ) indoor ( ) outdoor type. It shall be designed, manufactured, and tested in accordance with the latest applicable standards of NEMA and shall bear the UL label, and shall be manufactured by_____ or approved equal.

16470 Panelboards

(A) Furnish and install distribution and power panelboards as indicated in the panelboard schedule and where shown on the drawings. Panelboards shall be dead-front safety type, equipped with quick-make, quick-break fusible branch switches and approved for service entrance. The acceptable manufacturers of the panelboard are _____, _____, and _____ , provided they are fully equal to the type listed on the drawings. The panelboard shall be UL listed and bear the UL Label.

(B) All fusible branch switches shall be quick-make, quick-break, with visible blades and dual horsepower ratings. Switch handles shall physically indicate ON and OFF positions. Such handles shall also be able to accept three padlocks having heavy-duty industrial type shackles. Covers shall be interlocked with the switch handles to prevent opening in the ON position. A means shall be provided to allow authorized personnel to release the interlock for inspection purposes when a switch is ON. A cardholder, providing circuit identification, shall be mounted on each branch switch. Switches shall be provided with fuses or as noted on the drawings.

(C) Panelboard bus structure and main lugs or main switch shall have current ratings as shown on the panelboard schedule. The bus structure shall accommodate plug-on or bolted branch switches and motor starters as indicated in the panelboard schedule without modification to the bus assembly. Provide solid neutral assembly (S/N) when required.

(D) Switches and panelboard bus structure shall safely and without failure withstand short circuits on the systems capable of delivering up to 50,000 amperes rms symmetrical, unless otherwise noted.

(E) Panelboard assembly shall be enclosed in a steel cabinet. The rigidity and gauge of steel to be as specified in UL standard for cabinets. The size of wiring gutters shall be in accordance with UL standard. Cabinets shall be equipped with a front door and have fully concealed, self-aligning trim clamps. Fronts shall be full-finished steel with rust-inhibiting primer and baked enamel finish.

(F) Terminals for feeder conductors to the panelboard mains and neutral shall be suitable for the type of conductor specified. Terminals for branch circuit wiring, both breaker and neutral, shall be suitable for the type of conductor specified.

(G) Before installing panelboards check all of the architectural drawings for possible conflict of space and adjust the location of the panelboard to prevent such conflict with other items.

(H) The panelboards shall be mounted in accordance with the *NEC*. The Electrical Contractor shall furnish all material for mounting the panelboards.

16471 Branch Circuit Panelboard

(A) Power and lighting panels shall be of the dead-front, safety type, with thermal magnetic, quick-make, quick-break, trip free, bolted-type molded case circuit breakers. Voltage ratings, number of poles, frame size, trip ratings, main breaker or lugs, neutral bus, and ground bus are all as shown on the drawings. Bus bars shall be rectangular, solid copper, securely mounted and braced. All connections to bus bars shall be securely bolted. Cabinet boxes shall be constructed of code grade galvanized steel, sized to provide minimum 4-inch wide wiring gutters on sides, top and bottom. Fronts shall be constructed of code grade steel, adjustable indicating trim clamps and with door provided with concealed hinges and cylinder type lock and catch. Two keys per panel shall be furnished, and all locks keyed alike. Front shall be finish painted blue-gray.

(B) Power panels shall be _____ , Type _____ , _____ or _____ with branch breakers, main breakers or lugs, neutral and ground buses, etc., all as shown on the drawings.

(C) Power and lighting panel construction details shall be in accordance with UL standards and shall conform to NEMA standards. They shall bear the UL label. Panels shall meet USASI Specifications W-P-115a, Type 1, Class I.

(D) All panel directories shall be typed and terminology approved by the Owner.

(E) Molded case circuit breakers shall be _____, Type _____, _____, _____, (_____frame)_____ (_____frame), _____, (_____frame), _____ (_____frame), _____ (_____ or _____ frame), _____ (_____frame), and _____ (_____ frame).

16120 Feeder Circuit

(A) The Electrical Contractor shall provide and install a complete electrical distribution system as shown on the drawings or as required for a complete system. All materials and workmanship shall conform with Section 16.2 of the specifications, the *NEC*, and the local electrical code.

(B) Rigid conduit shall be galvanized rigid steel conduit with a minimum size of inch unless otherwise noted. Rigid steel conduit shall be installed for the following services and locations: service entrance, underground in contact with earth, in concrete slab, panel feeders, exterior of building walls, motor feeders over 10 hp electrical equipment feeders over 10 kW, wet locations, and as required by the *NEC* and local codes.

(C) Electrical metallic tubing shall be galvanized steel with a minimum size of ___ in. Electrical metallic tubing shall be used in all locations not otherwise specified for rigid or flexible conduit and where not in violation of the *NEC*.

(D) Flexible metal conduit shall be galvanized steel. Flexible metal conduit located in wet locations, shall be the liquidtight type. A short piece of flexible metal conduit shall be used for the connection to all motors and vibrating equipment, connection between recessed light fixtures and junction box, and as otherwise noted, provided the use meets the requirements of the *NEC* and local codes. The flexible metal conduit shall be the type approved for continuous grounding.

(E) The conductor material shall be as follows, unless otherwise noted:

1. Feeders shall be Type RHW or THW rated at 75° C.

2. Branch Circuits shall be Type THW rated at 75°C, except branch circuits with conductor sizes of #10 and smaller in dry locations may be Type TW rated at 60°C.

3. Special Locations. Conductors in special locations such as range hoods, lighting fixtures, etc., shall be as required by the *NEC*, local code, or as otherwise noted.

(F) No conductor shall be smaller than #12 wire, except for the control wiring and as stated in other sections of the specifications or on the drawings. Wiring to switches shall not be considered as control wiring.

(G) Conductors indicated on the drawings are based on copper. Panel, motor, and electrical equipment feeders with a size of #1/0 and larger may be aluminum, providing the size of the conductor is increased to have the same or more current-carrying capacity as the copper conductors. Also, the conduit sizes shall be increased accordingly.

(H) All conductors with the size of #8 or larger shall be stranded.

(I) All lighting and receptacle branch circuits in excess of 100 linear feet shall be increased one size to prevent excessive voltage drop.

16475 Overcurrent Protective Devices

(A) The Electrical Contractor shall furnish and install where indicated on the drawings or as required by the *NEC* molded-case circuit breakers in a NEMA type _____ enclosure. Breakers shall be manually operated, trip-free and designed so that all poles open simultaneously. Tripping mechanism shall be (thermally, magnetically) operated, shall open instantaneously on short circuits and have time delay on overloads, and have effective sealing against tampering. Breakers shall be as called for on the drawings or in the panelboard schedule and as manufactured by _____ or approved equal.

(B) Fuses, unless indicated otherwise, shall be dual element, time-lag, cartridge type as manufactured by _____. Fuses for motor circuits shall be sized in accordance with the *NEC*. Labels indicating the size and type of replacement fuses shall be glued to inside of door on all fusible switches and panelboards.

(C) All fuses shall be of the current and voltage rating as required or indicated.

(D) A duplicate set of all fuses shall be furnished to the Owner at completion of the project.

16476 Primary Load Interrupter

(A) The electrical contractor shall furnish and install a three-pole, two-position (open-close), group-operated, fused air-interrupter switch rated _____ amps with an interrupting rating of _____ amps at ___kV. Switch shall be Type____ as manufactured by _____ , or approved equal; shall be mounted in a suitable metal enclosure equipped with an interlock to permit access to fuses only when switch is in open position.

(B) Power fuses shall have an interrupting rating of _____amps rms, _____kVA symmetrical at _____ volts; shall be Type____ as manufactured by _____, or approved equal.

16481 Motors

(A) The Electrical Contractor shall furnish and install only those motors and controllers specifically indicated thusly on the electrical drawings. All motors and controls to be furnished by others shall be delivered to the job site by those furnishing the equipment. Unless otherwise noted, the electrical contractor shall then receive, handle, set, mount, and install this equipment where indicated on the drawings.

(B) All motors which are an integral part of a piece of equipment furnished by others shall not be the responsibility of the electrical contractor. However, where indicated on the electrical drawings, the electrical contractor shall provide and install disconnect switches, feeders, and control circuits for these pieces of equipment.

(C) The mechanical contractor or others furnishing motor-driven equipment will be responsible for the proper operation of the equipment, but the electrical contractor shall be responsible for the electrical work in connection therein.

16482 Motor Starters

(A) All motor controllers shall be furnished by the electrical contractor unless indicated otherwise. All manual and magnetic controllers shall contain a contact for each phase and with poles as required. All magnetic starters shall have _____ volt control circuits and holding coils.

(B) The Electrical Contractor shall provide and install fusible disconnects for all motor controllers unless nonfusible is noted.

(C) The Electrical Contractor shall install thermal overload heater elements in all switches and starters on the job whether or not the switches and starters are furnished by this subcontractor. Contractor shall assume responsibility for proper application of motor running protection for all motors in accordance with the manufacturer's recommendations and the nameplate rating of the motors actually installed. All motors will be phased to have overload protection.

(D) Standard duty pushbutton stations with maintained contact, Start/ Stop buttons and amber indicating light unless otherwise indicated will be used.

(E) The enclosures shall be surface mounted NEMA 1 unless otherwise indicated. Hand-off-automatic switches and pilots in covers will be provided as required.

(F) All controllers and accessories to be by the same manufacturer, _____ or approved equal.

16500 LIGHTING

16501 General

(A) This section of the specifications includes the furnishing and installation of lighting fixtures and lighting equipment for all outlets in the project as listed in the fixture schedule, including the connection of the fixtures and equipment to the electric wiring of the building.

(B) Lighting fixtures shall be of the types, sizes, etc., shown in the fixture schedule and notes referenced herein.

(C) Lighting fixtures described herein are intended to indicate the general fixture type, which shall be substantially as specified. It is not the intent of this specification to require the product of any particular manufacturer whose product is specified.

(D) The furnishing and installation of the lighting fixtures or lighting equipment must be executed in a manner that will insure completion coincident with the completion of the construction of the project, unless otherwise required by the contract specifications.

(E) Materials and workmanship shall be of the highest quality and in accordance with the highest commercial standards.

(F) Tubing shall be seamless drawn unless otherwise noted and if used as a supporting member of the fixture must have not less than 0.06-inch wall thickness. Curved and bent stems or other formed parts of fixtures shall be true and free from kinks, bruises and flattening. Seats between tubings, seating rings, castings, shells, and other parts shall be so closely fitted that all parts will be securely held and form inconspicuous joints. Seating rings shall be turned from solid metal. Burrs, fins, and sharp edges must be removed from wireways and all parts before they are assembled.

(G) Where aluminum members are to be fastened to dissimilar metal parts, other than steel, the aluminum shall be separated from such parts by a coat of zinc chromate, aluminum, or bituminous paint applied to the contact surfaces of the metals and allowed to dry thoroughly before assembling. The surfaces of the aluminum castings or aluminum sheet placed in contact with brick, plaster, gypsum, concrete or similar masonry construction, as in the case of recessed housings, etc., shall be back painted with zinc chromate, aluminum, or bituminous paint and allowed to dry thoroughly before installation.

(H) Unless otherwise specified, plastic or glass diffuser panels and low brightness lens panels required for fluorescent fixtures shall be contained in rigid, hinged metal frames. Diffuser

and lens panels shall be framed in a manner that will permit replacement of panels in the frame without the use of tools other than screwdriver or pliers. Panels shall be held in the frames in a neat, workmanlike, rattle-free manner that will provide proper tolerance for normal expansion and contraction without damage to the panels.

(I) Unless otherwise specified, the housing of each fluorescent lighting fixture shall be provided with a separate, factory installed grounding device.

The grounding device is to be used for connecting a separate grounding conductor to the fixture housing, and for not other purposes. For housings thinner than #18 USS gauge, the grounding device shall be a 10/32 machine screw threaded stud and nut, with a flat washer. The stud shall be of either the welded or pressure fastened type. For #18 USS gauge and heavier housings, the grounding device shall be either the above stud, etc., or a 10/32 round head machine screw, with flat grounding device that shall meet all applicable grounding requirements of the *NEC*.

(J) At the Architect-Engineer's request, the contractor shall submit for approval one each of any of the lighting fixtures required under the Contract. The fixture or fixtures shall be tagged with the name of the building or project for which the fixture is intended and be shipped, all charges prepaid, to the address specified. When fixture or fixtures have served their purpose, they will be turned over to the contractor for use in the project. In the event the submission is disapproved, the fixtures will be returned to the contractor and he shall immediately make new submission of fixture or fixtures meeting the contract requirements.

(K) Ordering of the lighting fixtures for the project shall not be commenced until the contractor has received unqualified approval of the submitted sample lighting fixtures.

16502 Lamps

(A) All lighting fixtures shall be lamped as indicated on the lighting fixture schedule. Lamps shall be those manufactured by _____, _____, or _____.

(B) At the time of acceptance of the building, the lighting fixtures shall be completely relamped as necessary and shall be thoroughly cleaned.

(C) Mercury vapor lamps are designated by USASI number. All mercury vapor lamps shall be color corrected deluxe white.

16503 Luminaire Accessories

(A) The Electrical Contractor shall furnish and install, with no additional cost to the Owner, all accessories as noted or required to install and make workable all lighting fixtures and related items contained under this section of the specification.

16505 Ballasts

(A) Each ballast shall be designed to start and satisfactorily operate the type of fluorescent lamp required in the particular fixture and shall conform to the current practice and requirements of the "Certified Ballast Manufacturers," unless the lamps are of unusual types for which "certified" ballasts are not available commercially. Ballasts shall be UL Class P of the high power factor type in sizes in which they are available of the series-sequence type. Multilamp ballasts for operating preheat and instant types of lamps shall be of the lead-lag type. Ballasts shall be securely fastened in place with mounting surface of ballast making as complete contact with surface of ballast mounting area of fixture as practical. Ballasts having four (4) mounting holes shall be attached to the mounting surface of the fixture by means of four (4) bolts and nuts of the machine screw thread type, not smaller than the #8-32 size (two at each end of ballast) and shall be fitted with lockwashers. Bolts shall be installed in a non-turning, captive manner. Nuts will not be required where bolts are shown to be threaded into sheet steel mounting plates of #18 USS gauge on Standard PPS lighting fixtures. Ballast for operating lamp sizes 30 watts or larger shall be protected against overheating by a built-in, thermally actuated automatic reclosing device sensitive to both winding temperatures and current which will prevent ballast case temperature from exceeding 110°C. After opening, the circuit shall reclose when the case temperature drops below 85°C.

(B) Ballasts for operation of fixtures for indoor use shall comply with the requirements of the UL for Class P ballasts, in lieu of the protection specified herein before. Ballast protectors shall be of the thermally actuated automatic-reset built-in type.

(C) In addition the lighting fixture manufacturer shall submit, through the contractor, a certification that the ballast, in conjunction with the fluorescent lighting fixtures proposed to be furnished on the project, will operate within the temperature limits as required by the UL Standards for Safety, Electric Lighting Fixtures, UL57.

16506 Posts and Standards

(A) The Electrical Contractor shall furnish and install all poles and standards related to exterior lighting fixtures complete with base, hardware, etc., unless otherwise indicated.

(B) The standards shall be of _____ construction and be as manufactured by _____ or as indicated in the lighting fixture schedule, drawings, and specified herein.

16510 Interior Luminaires

(A) Fixtures shall be finished as standard by the manufacturer for the fixtures specified. Finishing materials shall be applied in a manner that will produce a durable finish. Surfaces to be finished shall be thoroughly degreased and dried by adequate modern methods before finishes are applied. Metallic finishes, except chromium, shall be given an even coat of high grade lacquer, baked on. Exposed surfaces of aluminum parts, such as frames, shall be given an anodic treatment when required by drawings and/or specifications.

(B) Sheet steel fixture housings, iron and steel fixture parts for which no other finish is indicated, and steel parts on which galvanizing or cadmium plating is required by standard detail drawings shall be protected against corrosion by an even coating of zinc or cadmium on all inside and outside surfaces and edges after completion of all operations such as forming, welding, drilling. The minimum thickness of zinc coating shall be five ten-thousandths of an inch (0.0005 inch) and the minimum thickness of cadmium plating shall be

15 hundred-thousandths of an inch (0.00015 inch). Zinc coating generally may be applied by either hot dipping or electroplating. Zinc coating on alloy or heat treated steels which would be affected injuriously by the temperature of molten zinc shall be coated by electroplating only.

(C) Steel reflectors and other surfaces for which baked-on white enamel finish is required shall be made of steel of the thickness specified or noted and given a suitable primer and white color coat or coats properly applied to meet the requirements and tests described below. Reflectors shall be completely formed before application or primer or enamel color coat or coats.

(D) Lampholders, sockets and lamp receptacles for incandescent lamps shall be in accordance with Federal Specification W-L-142, of the class and style required by each fixture design. Medium screw base lampholders or sockets shall be used for incandescent lamps of 200 watts and smaller, except where the use of intermediate, candelabra, or other types of bases are required by drawings or specifications.

(E) Wiring between fluorescent lampholders and associated operating and starting equipment shall be of sizes not smaller than the sizes of the leads furnished with the approved types of ballasts and shall have equal or better insulating and heat resisting characteristics. All other wiring within fluorescent lighting fixtures or from the fixture to the splice with the building wiring shall conform to the requirements of the latest issue of the *NEC*. Unless otherwise specified or shown on drawings, all wiring in conjunction with incandescent fixtures shall also conform to the requirements of the latest issue of the *NEC* and shall not be less than #16 gauge. Wiring shall be protected with tape or tubing at all points where abrasion is liable to occur. Wiring shall be concealed within fixture construction, except where the fixture design or mounting dictates otherwise.

(F) Connection of wires to terminals of lampholders and other accessories must be made in a neat and workmanlike manner and shall be electrically and mechanically secure with no loose strands protruding. The number of wires extending to or from the terminals of a lampholder or other accessory shall not exceed the number which the accessory is designed to accommodate.

(G) Joints in wiring within lighting fixtures shall be so spliced that they will be mechanically and electrically secure and then soldered and taped to provide insulation equal to that of the conductors being joined. In lieu of solder and tape, approved types of adequately insulated solderless pressure crimped-type connectors may be furnished provided sizes used, method of application, and tools employed are in accordance with the particular manufacturer's recommendations. The use of the screwed-on type of solderless connector will not be approved for making connections in the wiring within lighting fixtures.

(H) No splice or tape will be permitted within an arm or stem. Wiring shall be continuous from the splice in the outlet box on the building wiring system to the lampholder of an incandescent fixture or to the wireway or space provided in a fluorescent lighting fixture for necessary splices between ballast and other accessories and lampholders.

(I) Each basic fluorescent fixture shall be equipped with the necessary number and type of ballasts to operate only the lamps within the particular basic unit. Single-lamp fixtures shall contain one single-lamp ballast; two-lamp fixtures shall contain one two-lamp ballasts; three-lamp fixtures shall contain one two-lamp ballast and one single-lamp ballast; and four-lamp fixtures shall contain two two-lamp ballast. Basic fluorescent fixtures containing three or four lamps shall be internally wired to have the two outer lamps operated by a common ballast and the center lamp or lamps operated by the remaining ballast.

(J) Multiple section fluorescent lighting fixtures shall be internally wired and connected to accomplish the switching arrangement required by the electrical drawings. Wiring shall conform to the requirements of the latest issue of the *NEC*.

(K) Wiring channels and wireways shall be free from projections and rough or sharp edges throughout, and all points or edges over which conductors must pass and may be subject to injury or wear shall be rounded or bushed. Insulated bushings shall be installed at points of entrance and exist of flexible wiring.

(L) Canopies, holders and similar parts shall be drawn or spun in one piece unless otherwise indicated. Spun or drawn parts shall have finished thicknesses that are not less than the undivided thickness at any point. Thickness of metal stock shall be selected to allow for all thickness as indicated; the minimum thickness of canopies, holders and similar parts shall be 0.026 inch. The dimensions of holders for reflectors or globes shall conform to the dimensions and tolerances required to suit standard fitters, i.e., those commercially referred by glassware and reflector manufacturers as $2\frac{1}{4}$, $3\frac{1}{4}$, 4, 6, and 8 inch, unless definitely specified to be otherwise.

(M) Luminous reflector, luminous side panels, diffusing panels and other luminous plastic members of fixtures shall be made of smooth, white, translucent 100 percent virgin acrylic material. Low brightness lens panels shall be made of clear, 100 percent virgin acrylic material not less than 0.125-inch average thickness smooth overall pattern of cones or hexagonal pyramidal prisms on the other. The plastic shall be nonflammable or shall have a flame spread rate of not more than 3.2 inches per minute for an $\frac{1}{2}$-inch width of the material held in a horizontal position. The plastic shall not change color materially when lighted by 4500°K fluorescent lamps. The plastic shall show no yellowing apparent to the naked eye either when subjected to the accelerated weathering test in accordance with ASTM D-795-44T (500 hours exposure, using Type S-1 lamp) or after prolonged exposure to a fluorescent lamp source under conditions identical with those existing in the fixture in which it is to be used. The plastic shall be nonelectrostatic or finished parts shall be treated with an antistatic wax.

(N) Reflectors and panels shall be formed by processes carefully controlled so that finished pieces will retain their designed contours and dimensions without change at normal operating temperatures. They shall not be subject to warping, crazing, cracking or discoloring, either in service or when stored under normal conditions in the manufacturer's standard shipping containers. The contractor shall be responsible for the development, production and furnishing of reflectors and panels meeting these requirements, controlling the surface brightness, etc.

(O) The exterior surface brightness of the plastic reflector for lighting fixtures shall be relatively uniform and shall not exceed 350 footlamberts when viewed from any position below the horizontal, nor shall it be less than 250 footlamberts when viewed crosswise at 45 from nadir, when lighted with two 40 watt, T-12, 3100 lumen, 4500°K. The interior surface of reflector shall be smooth but not necessarily polished.

(P) The exterior surface brightness of the plastic side panels for lighting fixtures shall be relatively uniform and shall not exceed 300 footlamberts nor shall they be less than 150 footlamberts when viewed on any transverse fixture axis from the horizontal to 30° below when the particular fixture is lighted with 40 watt, T-12, 3100 lumen, 4500°K.

(Q) Lampholders shall be rigidly and securely fastened to the mounting surface with necessary provisions to prevent lampholders from turning. The dimensions of lampholders shall be such as to position the lamp tube not less than 0.125 inch from mounting surface or reflector.

(R) Fluorescent lampholders and lampholder assemblies shall be of such design that lamps will be held firmly in place, electrically and mechanically secure, and shall permit easy insertion or removal of lamps. Lampholders for fixtures specified for "rapid start" operation shall be silver-plated or the equivalent thereto in electrical conductivity to insure consistently satisfactory starting and operation of the "rapid start" type of fluorescent lamp. Disconnect type of lampholder shall be used where required in connection with "rapid start" circuits.

(S) Fluorescent lighting fixtures shall be plainly marked "For Alternating Current Operation Only" unless definitely designed for operation on direct current, in which case they shall be so marked.

(T) Fixtures shall be marked for the operating voltage for which they have been designed.

(U) Markings shall be clear and so located that they will be readily visible to service personnel but shall not be conspicuous from normal viewing when fixtures are installed.

(V) The suspension length of all ceiling-mounted suspended types of lighting fixtures (fluorescent and incandescent) as listed in the fixture schedule in the project specification shall be the overall length from ceiling to the lowest point of the fixture body, reflector or glassware in its hanging position.

(W) The shop inspection shall consist of the examination of a representative sampling of the fixtures and fixture parts required for the building project.

(X) Lighting fixtures must be correctly wired for proper operation in accordance with the design circuit and shall test free from short circuits and improper grounds.

(Y) Any action taken as a result of the shop inspection must not be construed as final acceptance of the fixtures, which must fully meet all contract requirements after installation in the project and at time of final inspection.

(Z) Upon completion of the installation of the lighting fixtures and lighting equipment, they must be in first class operating order, in perfect condition as to finish and free from defects. At the time of final inspection, all fixtures and equipment must be completely lamped and be complete with the required glassware, reflectors, side panels, louvers or other components necessary to complete the fixtures. All fixtures and equipment shall be clean and free from dust, plaster or paint spots. Any reflector, glassware, side panels or other parts broken prior to the final inspection must be replaced by the contractor.

16511 Luminous Ceiling

(A) The luminous ceiling shall be as manufactured by _____ and shall be installed by the electrical contractor. Ceiling shall consist of:

1. _____strip lighting fixtures with _____ ballasts spaced in accordance with lighting layout shown on electrical plans.

2. Glass, panels (specify type) shall be installed in the grid system.

3. Supporting grid system shall consist of (_____wall angle and _____T-tracks) or (_____wall angle and_____ or _____T-tracks) installed on two or four foot centers with cross-tracks on four or two foot centers. Suspension shall be by inch threaded rod _____and rod angle bracket with_____ rod ceiling bracket.

16521 Signal Lighting

(A) The electrical contractor shall furnish and install all signal lighting equipment and related wiring as indicated on the drawings, accompanying schedules, and by notation.

16522 Exterior Luminaires

(A) The Electrical Contractor shall furnish and install all exterior lighting fixtures of the type and size indicated on the drawings, listed in the fixture schedule, or specified herein. All fixtures shall meet NEMA design and photometric standards.

(B) All lighting fixtures shall be lamped as indicated on the lighting fixture schedule. Lamps shall be those manufactured by _____.

(C) Mercury vapor lighting fixtures shall be supplied with a _____ base porcelain lamp holder which shall accommodate a_____ watt mercury vapor lamp. A _____ power factor ballast for _____ volt operation shall be supplied by _____, or approved equal.

16531 Stadium Lighting

(A) The Electrical Contractor shall furnish and install all lighting fixtures, poles and standards, ballasts, lamps, and related wiring to construct the stadium lighting system as indicated on the plans, schedules or specified herein.

(B) Concrete footings for the poles and standards shall be provided by and be the responsibility of the _____ contractor. Each footing must be constructed so that its diameter and depth take into consideration the manufacturer's wind load requirements and local soil conditions.

16532 Roadway Lighting

(A) The Electrical Contractor shall furnish and install all roadway lighting called for on the site plan, on supplemental drawings and as specified herein. The lighting fixtures shall be as manufactured by _____, and shall consist of an _____ diffuser, a _____ mounting base with all necessary accessories.

16600 SPECIAL SYSTEMS

16601 General

(A) The work under this section of the electrical specifications includes the furnishing of all labor, materials, equipment, and incidentals necessary for the installation of a complete lightning protection system as called for in the drawings, schedules, and contained herein; an emergency lighting system for exit and passageway lights including storage batteries and battery charging equipment.

(B) All work under this section shall conform in all respects to the *NEC* and local ordinances as minimum requirements. Where the plans and specifications indicate work in excess of the above minimum requirements, the plans and specifications shall be followed.

16610 Lightning Protection Systems

(A) The Electrical Contractor shall furnish and install materials in accordance with the Lightning Protection Institute Code as adopted by the American National Standards Institute (ANSI). The contractor shall use no materials nor devices which do not bear UL labels. Full compliance with the manufacturer's rules and regulations for the installation system on this particular building is also a requirement. Upon completion of the installation of the lightning protective equipment,

the contractor shall furnish the Owner with the UL Master Label Plate of Approval.

(B) Upon the request of the Architect/Engineer, the Electrical Contractor shall submit samples of terminals, anchors, conductors, and other visible parts of the system to the Architect/Engineer for his or her selection and approval at the time and place which the Architect/Engineer designates. At the same time he shall submit a typewritten schedule of the materials to be used, giving catalog numbers and complete description, to the Architect/Engineer for his or her approval.

(C) The Electrical Contractor shall employ only specially trained and thoroughly competent workers who are experienced in the installation of lightning protection equipment. The contractor shall make the entire installation in an inconspicuous manner so as not to mar the architectural design of the structure; provide an adequate number of air terminals; firmly anchor all air terminals; course the conductors properly and run them straight when they are supposed to be straight and make proper bends where bends are required; use the proper attachment for each building, or building surface; attach conductors to the building firmly so that they can't and won't come loose; see that all joints and connections are well made and will stay that way; and make all required metal work connections in a permanent and durable manner. The course of all conductors must be horizontal or downward, never upward. No branch leads may be longer than 16 feet without an additional ground.

(D) Spires, cupolas, ventilators, chimneys, high dormers, gable ends, water tanks, flagpoles, stair and elevator penthouses, and other vertical projections must be protected by air terminals.

(E) On pitched roofs the contractor shall install air terminals not more than 20 feet on centers along all ridges. There must be an air terminal within two feet of the ends of all ridges whether they occur on the main roof or on dormers. On flat roofs or pitched roofs having a slope less than 30, the contractor shall install air terminals at the corners and edges so that they are spaced not greater than 20 feet on centers. Two conductors to ground for straight ridge-line building 70 feet or less in length and one additional conductor to ground for each additional 40 feet of length will be provided. Also provided will be whatever additional conductors to ground are required to relieve low-positioned air terminals or to avoid a branch lead of over 16 feet on pitched roofs of irregular arrangement.

(F) The Electrical Contractor shall install air terminals at the corners of all flat roofs and not more than 20 feet on centers around the entire perimeter. Install an additional row of air terminals spaced 20 feet on centers for each 50 feet of roof width over 50 feet. For the first 200 feet of perimeter of flat or flat-pitched roofs, install two conductors to ground. Install one additional conductor to ground for each additional 100 feet of perimeter or fractional part thereof.

(G) The Electrical Contractor shall provide lead-covered air terminals on chimneys, so located that no chimney corner is more than 2 feet distant from an air terminal. Air terminals must extend at least 10 inches above the highest part of the chimney construction.

(H) The Electrical Contractor shall provide an adequate number of effective grounds. For the purposes of estimating and bidding, it will be assumed that the earth is permanently moist to within 3 feet from finish grade. If, during the excavation, conditions are encountered which are at variance with this assumption, an adjustment will be made between the Owner and this Contractor for the greater expense that is involved in establishing the proper ground connections for the lightning protection system, each as a rule extending into the earth to a depth of 10 feet or equivalent. The contractor shall install ground conductor guards where necessary to prevent mechanical injury.

(I) The contractor shall connect metal ventilators, metal stacks, vent pipes or other metallic objects which project above the rodded structure to the system so that they will serve as additional terminals. He shall connect metal roofing, ridge rolls, valleys, guy wires and other metal bodies of conductance, to the lightning conductor or ground them independently. Electric wires, radio wires,

and telephone wires entering buildings must be properly protected so that lightning cannot enter the building by these means.

16630 Battery Power Systems

(A) The Contractor shall furnish emergency lighting equipment, constructed in accordance with UL Classification No. 106 MO, and shall install same in locations shown in drawings in accordance with Article 700 of the *NEC*. The complete system shall consist of a wall-mounted console containing charger and battery, battery protector, six distribution circuits, solid state fuse failure alarm, automatic load contactor relay; provision for remote console monitor and failure alarm, emergency and/or exit lighting fixtures. The system shall include operating instructions and maintenance kit where required.

(B) The charger and battery console shall be a wall-mounted unit consisting of two compartments. The upper compartment shall house the charging and control chassis, with the lower compartment housing the battery. The console shall be constructed of 16 gauge steel and equipped with necessary ventilation louvers, locked access doors to both compartments, adequate electrical knockouts, and shall be finished with corrosion resistant baked enamel paint. The compartment shall have an opening large enough to expose the entire battery for servicing. The upper compartment shall house the charging controls and components which are mounted on a separate and removable chassis, completely independent of the console proper. Meters and pilot lights shall be mounted on this chassis but shall be visible from the face of the console through an opening. All switches, fuses, and controls shall be accessible only to authorized personnel by unlocking the hinged compartment door. All controls shall be properly identified. No exposed wiring shall be permitted.

(C) The charger shall be a constant voltage, fully automatic type that incorporates a solid-state charge detector and regulator to provide a continuous, infinitely variable current limited charge rate as required by the battery. The charger shall include a fully automatic surcharge device requiring no manual attention. All exposed terminal connections shall have protected coverings. The charger shall be capable of automatically recharging the battery to 100 percent of the original rated capacity. Upon the battery reaching its charge, the charger shall maintain the battery in accordance with the battery manufacturer's recommended floating voltage. The charger shall be capable of performing a minimum of 15 consecutive cycles of discharge and recharge automatically, within the specified time period and with the specified load without any loss of battery capacity.

(D) Transformers shall be of the high leakage reactance type with operating characteristics such as to limit current under simulated short circuit conditions. Rectification shall be full wave.

(E) Each dc circuit shall be fused in each polarity, and fuses and their fuse holders shall be of such design as to preclude improper fuse installation.

(F) Components shall include ammeter, voltmeter, ac ON light, high charge light, test switch, ac circuit breaker, dc charger circuit fuses, console monitor terminals, dc load fuses.

(G) One automatic, sealed, load transfer relay of the mercury plunger type shall be provided. The transformer relay shall be provided with a TEST switch to permit testing of the relay, a white pilot light to indicate ac line power to relay, and a red TROUBLE light to indicate that the emergency circuit has been energized upon failure of the normal lighting circuit power.

(H) The battery will have 27 cells, nickel cadmium plates and an alkaline electrolyte. It shall provide an estimated service life of 20 years. It shall have a sealed top, vent and fill caps and be enclosed in a translucent plastic case.

(I) Emergency lighting fixtures and/or exit lights of the types detailed or indicated on the drawings shall be furnished as part of the system. The unit shall be _____.

(J) Exit signs shall be equipped with a _____volt, sealed nickel cadmium battery

and a fully automatic solid state charger. Upon failure of normal ac power, the battery automatically provides power to illuminate the face and provide down light. Upon restoration of AC power the charger shall provide a high charge rate to recharge the battery within 12 hours, then maintain it at its floating voltage. Signs shall be equipped with two long-life lamps for ac illumination and two bayonet base lamps for emergency lighting.

(K) Battery, charger and related controls are to be recessed in the ceiling or wall.

(L) All exit signs shall be completely die cast aluminum with brushed finish with the exception of the plastic lettering.

(M) Each sign shall be as manufactured by _____, catalog number_____ or as indicated in the lighting fixture schedule.

16631 Battery Charging Equipment

(A) The electrical contractor shall furnish a solid-state fully automatic battery charger with a thermal activated surcharge control and a nickel cadmium battery.

(B) The system shall be equipped with built-in load transfer relay or may have optional remotely located relays. In addition, the console monitor shall be designed to detect and signal individual trouble conditions within the battery and charger.

(C) Upon restoration of ac electric service after a power failure, the charger immediately and automatically begins recharging the battery at a high charge rate. When the proper point in the recharge cycle is reached, the timed surcharge control is automatically activated and when the battery has been surcharged to its original rated capacity, the control mechanism ends the surcharge. The charger then reverts to its normal function of providing a continuous floating rate charge to maintain the battery at the proper floating voltage.

(D) Recharging to 100 percent of manufacturer's original rated capacity will be accomplished in 12 hours or less after discharge to 87 percent of system voltage under rated load per *NEC* Article 700. Unless otherwise specified, the unit will be furnished at _____ volts, 60 Hz, _____ watts maximum input.

16632 Storage Batteries

(A) The emergency lighting system and battery charger shall be supplied with long life nickel cadmium (alkaline electrolyte) batteries, which will provide a ten year maintenance free service interval when operated at normal ambient temperatures and at proper charging rates.

(B) Features of the batteries shall include maximum tolerance to severe discharges and overcharge; tolerance to extreme ambient temperatures; high resistance to shock and vibration; low internal resistance; low maintenance (water additions seldom required under normal conditions); no corrosion from acid fumes during recharge; negligible self discharge during periods of stand-by; unlimited shelf life.

(C) The alkaline batteries shall have nickel cadmium plates, potassium hydroxide electrolyte, seal top, vent and fill caps, and shall be enclosed in a translucent plastic case for easy viewing of electrolyte levels.

(D) The battery size, capacity and hour rate of discharge shall be selected in accordance with the connected load and specified duration of illumination.

16700 COMMUNICATIONS

16701 General

(A) The Electrical Contractor shall furnish and install all communication, alarm and detection equipment, and clock and program equipment as called for in the drawings or indicated herein.

(B) In general, the work under this section shall include the installation of all alarm and detection systems, smoke detectors, burglar alarms, clock and program equipment, telephone and telegraph equipment, intercommunication equipment, public address equipment, television systems, and learning laboratories.

16720 Fire Alarm and Detection Systems

(A) The Electrical Contractor shall furnish and install all fire alarm equipment and related wiring as indicated on the drawings or herein.

(B) The fire alarm system shall meet the requirements of NFPA Code 72 and the equipment shall be UL listed.

(C) The system shall be as manufactured by _____ and shall be wired and installed in accordance with the manufacturer's specifications and shall be in first class operating condition at the completion of the project.

(D) At each stairway, exit, and at other locations shown on plans, there shall be a noncoded fire alarm station. At each location where shown, there shall be a bell (or horn). Operating any station shall cause all sounding devices to operate continuously until the fire alarm station has been restored to normal. It shall also be possible for those in authority to transmit a test signal from any station.

(E) The stations and the sounding devices shall be connected to a control panel which shall permit a small supervisory current to pass through the entire system. A trouble bell shall also be provided and shall sound continuously in the event of failure of the main power supply source or a break or a ground fault of its installation wiring circuits.

(F) The Electrical Contractor shall furnish and install, where shown on the drawings, all conduit, wiring, connections, and other materials to make the fire alarm system workable according to the fire alarm riser diagram and related notes.

(G) All wiring shall be in strict accordance with the *NEC*, NFPA Codes 70 and 72, and all local electrical codes applying. Size and number of wires shall be in accordance with the wiring diagram furnished by the manufacturer or as shown on the drawings.

16727 Burglar Alarm

(A) The Electrical Contractor shall provide and install an intruder alarm system which shall be manufactured by _____.

(B) The activation and deactivation of the system in each apartment shall be entirely controlled by the tenant.

Leaving apartment:

1. Slide In-Out switch to Out position.

2. Open apartment door.

3. Insert key and turn key switch until amber On light illuminates.

4. Remove key, leave apartment, close door.

(C) If the apartment is illegally entered when the system is activated and the slide switch is on the Out position, a very loud audible signal will emanate from the speaker approximately 15 seconds after entry. If the slide switch is on the In position the audible signal will sound immediately.

(D) The tenant, upon reentering, shall have approximately 15 seconds to deactivate the system by using a key. This shall also extinguish the amber light. (Optional: If the system includes audible and visual signals in the lobby and/or manager's quarters, the alarms, both visual and audible, will commence 20 seconds after the apartment alarm starts. (Specification Writer—Optional Equipment. Include following paragraph, if desired.)

1. An audible alarm signal shall sound in the lobby and/or manager's quarters accompanied by a lighted indication identifying the (approximate area) (exact apartment) from which the alarm originated.

(E) Remaining in apartment:

1. Slide In-Out switch to In position.

2. Insert key and turn key switch until amber On light illuminates, then remove key.

(F) To deactivate system the tenant uses his key to operate key switch until amber On light is extinguished.

(G) When the system is activated with the slide switch in the In position, the alarm will sound immediately upon opening the door.

(H) Canceling an alarm: It shall be possible to cancel an alarm only by operating the key switch on the security panel in the apartment where the alarm originated, or by operating a master key switch on the control panel/amplifier.

(I) The security panel shall be _____ flush mounted so that the top is approximately 58 inches above the finished floor line, and located as shown on plans. Faceplate shall be satin finished stainless steel with operating instructions silk-screened thereon. Each unit shall be equipped with a five-pin tumbler lock key switch and two keys, a slide switch, an amber On lamp, printed circuit board wiring with all solid-state circuitry, and color-coded wire leads with an edge-type connector. The wall box shall be any standard outlet box with two-gang plaster cover (supplied by others).

(J) The door contact shall be _____. It shall mount recessed in the door jamb of each apartment entrance door in a special box with knockouts for conduit entrance.

(K) The control panel/amplifier shall be _____, solid-state, printed circuit board type. Each unit shall contain an amplifier and other components necessary for the required operation of the intercom and intruder alarm systems. A key-operated, system reset switch shall be mounted on the cabinet. The unit shall be surface mounted with removable cover.

(L) The power supply shall be _____ and will supply ac and dc power to the control panel/amplifier for the intercom and intruder alarm systems. One power supply is required for each 100 apartments, or fraction thereof. The unit shall be surface mounted with removable cover.

16730 Clock and Program Equipment

(A) The Electrical Contractor shall furnish and install a complete minute impulse, automatically synchronized clock and program system as shown on plans and as specified herein. The equipment shall be that of the _____ Company. Equipment of other manufacturers may be considered provided that the proposed substitute is named and an amount to be added to or deducted from the base bid for such substitution is indicated in the bid proposal. Substitute equipment will not be considered unless approval is received at least ten days prior to bid date.

(B) The system shall be controlled and synchronized by a master program clock. The secondary clocks shall be minute-impulse driven and be capable of having hour and minute hands individually and automatically synchronized once each hour for up to 25 minutes fast or slow. The correction cycle shall affect only those clocks which are "off-time."

(C) The master clock and program machine shall be a single integral unit to prevent differences between indicated time and program time. The "memory tape" shall provide direct readout of each minute and hour for a total of 1440 possible impulses during a 24-hour period. A separate channel shall be provided for six (or 12) program circuits. It shall be possible to interchange "memory tapes" without stopping, disconnecting, removing or resetting the program machine. A calendar device shall provide automatic disconnect of any program circuit for any 12-hour period of any day. Program signal duration shall be adjustable by means that prevent unauthorized changes in duration.

(D) The master program clock shall be housed in a steel case finished in baked enamel with a 10" clock dial on the outside. Exterior controls shall be provided to manually operate any or all program signals as desired. Interior controls shall include on-off switch, circuit breaker with pilot, motor switch, manual clock reset switch and program circuit on-off switches. The manual program controls shall be so arranged that manual sounding of program signals cannot result in inadvertently leaving a program circuit disconnected. The master program clock shall operate from a heavy duty synchronous motor and provide a constant torque spring reserve to keep the unit on time for up to 18 hours in case of power failure. The entire system shall operate from a common time base. No battery power supplies or supplemental time bases shall be used.

(E) System shall operate from a rectified power source provided from an adjustable transformer and full wave rectifier. Unit shall also provide 24 V ac available for low voltage signals. Unit shall be surface mounted with sufficient capacity to operate the system.

(F) The Electrical Contractor shall provide where shown on plans combination clock-speaker units. The clock and speaker baffle shall be square with identical frames mounted together. The speaker baffle shall have studs for mounting 8" speakers furnished and installed by the sound system contractor. The outlet box shall be provided with a divider and the speaker section shall contain a sound deadening material.

(G) A bell control board with total number of points required to provide individual circuit selection and control of all program signals shall be provided. Board shall be equipped with pushbuttons to ring any individual signal.

(H) Wiring shall be in general two #12 wires common to all clocks. The program wiring shall be a #12 wire common to all clocks. The program wiring shall be a #12 common wire with a #14 to all signals on one circuit. Provide two #12 wires from an unswitched source protected by a 20-amp circuit breaker to the master program clock and system power source. Provide six #14 wires from power source to the master clock. All wiring shall be color coded and shall be installed as shown on plans and as directed by manufacturers' wiring diagrams.

16741 Telephone

(A) The Electrical Contractor shall provide and install empty raceway, outlet boxes, pull boxes and associated equipment required for a complete telephone system as indicated on the drawings and specified herein. All materials and workmanship shall conform with Section 16100 of the specifications. All wiring shall be installed by the local telephone company. The entire installation shall be in accordance with the requirements of the local telephone company.

(B) Rigid conduit shall be installed in the following locations: service entrance, underground in contact with earth, in concrete slab and "wet" locations.

(C) Electric metallic tubing shall be used in all locations not otherwise specified to be rigid conduit.

(D) Telephone wall outlets shall consist of a 4" two-gang outlet box, raised device cover and a telephone device plate of the same material as the receptacle device plates. The conduit shall extend from the outlet to the designated telephone space unless otherwise noted.

(E) The Electrical Contractor shall install a #14 ga galvanized pull wire in the raceway system for future use.

(F) The wall outlets shall be mounted at approximately the following heights unless otherwise noted on the drawings or required by telephone company: desk phones, 18" AFF; wall phone, 58" AFF; telephone booth, 7'6" AFF.

16742 Intercommunication Equipment

(A) The electrical contractor shall provide and install, as shown on the plans and/or as described below, a _____ intercom system as manufactured by _____.

(B) A caller may press any pushbutton on the front entrance entry station. This shall sound a signal in the apartment suite station being called. The tenant may answer the call and converse with the caller by alternately pressing the Talk and Listen pushbuttons on the suite station. If satisfied with the caller's identity, the tenant may press the button marked Door which shall electrically operate the front entrance door opener permitting the caller to enter the building.

(C) A caller may also call and converse with any apartment on the rear entrance entry station. The rear door opener shall operate separately from a pushbutton marked Rear on apartment suite station.

(D) Tenants may ring and talk with service stations in service areas using pushbuttons on suite stations marked for the service areas to be called. Service stations may reply to apartments but cannot call apartments.

(E) Callers from the entrance(s) may ring and talk with suite stations in service areas using pushbuttons on entry station marked for the service areas to be called. Suite stations in service areas may reply to entrance and operate door opener, and they may call service stations in other service areas.

(F) Suite stations shall be _____ and shall be equipped with pushbuttons marked Talk, Listen, Door. When an extra door opener button is called for elsewhere in this specification or indicated on the drawings, it shall be marked Rear, and Door button shall be marked Front. When extra pushbuttons to ring service stations in service areas are called for, they shall be appropriately marked.

(G) Faceplates shall be vinyl bonded to steel inlaid in a moulded white cycolac frame. Finish shall be white woodgrained vinyl.

(H) Audible signals shall be electronically produced and the internal wiring shall consist of printed circuitry. Wire connections shall be either screw type terminals with edge-connector assembly; screw type terminals with edge-connector assembly plus wire leads; or wire leads skinned for wire nut connection.

(I) Suite stations shall be located as shown on the plans so that the top is approximately 60 inches above the finished floor line. Flush suite stations shall be supplied with steel backbox; surface-mounted suite stations with cast aluminum backbox.

(J) The entry station for the front entrance shall be flush mounted. It shall contain at least as many pushbuttons as there are apartments and service areas to be called, with provision to properly identify each pushbutton. It shall also contain provision for an alphabetical directory of tenants' names with apartment numbers listed alongside. The entry station shall fasten to a suitable metal wall box which shall be mounted so the top of the box is 56 inches above the finished floor line. When the entry station is combined with mail boxes in the same frame, the top of the wall opening for the combination shall be 56 inches above the finished floor line to comply with the U.S. postal regulations. The face of the entry station shall be finished _____.

(K) Multi-receptacle mail boxes, approved by the U.S. Postal Service, shall be furnished and installed as indicated on the plans and in conformance with post office regulations. They shall be combined in the same frame with the entry station. Each mail box door shall be equipped with a lock and two keys. Mail boxes shall be as manufactured by _____, catalog number _____.

(L) Door openers shall be _____ company, catalog number_____, reversible for right or left hand doors. To operate satisfactorily with intercom system, current demand shall not exceed 0.8 amps.

(M) The entry station for the rear entrance shall be cordless loudspeaking type similar in all respects to the front entrance entry station but with the addition of a press-to-talk pushbutton. It shall contain the same number of pushbuttons, and shall fasten to a suitable metal wall box which shall be mounted so the top is 56" above the finished floor line. The face of the entry station shall be finished _____. The entry station shall be as manufactured by _____, catalog number _____.

(N) The service stations for the service areas designated on the plans which shall receive calls from apartments shall consist of a telephone handset suspended on a live buckhorn hook attached to an extruded aluminum faceplate. The face of the service station shall be finished _____, and shall be manufactured by the _____, catalog number _____.

(O) In addition, each service area designated shall be equipped with a suite station identical to those in the apartments. These shall receive calls from entrances, operate door openers, and call the service stations in other service areas.

(P) The ac poser shall be derived from a _____ catalog number _____, _____watt, _____ volt transformer designed to mount on a standard outlet box. This shall provide low-voltage current to the amplifying unit.

(Q) One amplifying unit shall be supplied for the system unless indicated otherwise. It shall consist of the following individual sections enclosed in a protective steel case: a solid state dc supply; a solid state class B amplifier capable of providing one watt of undistorted power within the range of normal voice frequencies; one-eighth amp fuse; a solid-state dual-tone oscillator capable of producing a clear tone signal at a frequency of either 250 c/s or 700 c/s, as selected by the owners; a pair of audio switching relays with crossbar palladium contacts and individual nylon dust covers; a door opener relay with nylon dust cover; a volume control to adjust voice volume of system; a volume control to adjust audible signal sound volume, and screw-type terminals for connection of installation wiring.

(R) The amplifier shall be mounted in the backbox of the front entrance entry station except when the latter is combined with mail boxes, in which case it shall then be mounted in a suitable protected location. It shall be manufactured by _____, catalog number _____.

(S) The system shall be wired in strict accordance with the manufacturer's wiring diagram and recommendations. All wiring shall be free of grounds and short circuits before installation of the intercom equipment.

16781 Master TV Antenna System

(A) The master antenna TV system shall be provided to receive channels __, __, ___, and mix, amplify and split the signals to several distribution circuits around the building, as shown on the drawings. The system components and coaxial cable will be furnished and installed by others.

(B) The electrical contractor shall install the conduit and outlet boxes, and all other necessary components, together with pulling in the wires.

16850 ELECTRIC RESISTANCE HEATING

16851 General

(A) The Electrical Contractor shall furnish and install electric heating equipment as indicated on the drawings, in the electric heating schedules or noted elsewhere in the construction documents. The installation of all such equipment shall be in strict conformance to the *NEC* and applicable local ordinances.

(B) All circuits feeding the electric heating equipment shall be as indicated on the drawings and all connections to the heater junction box shall be made with an approved type of connector.

(C) Unless otherwise specified, all electric heating equipment shall be manufactured by _____, or approved equal and shall be for operation on a _____ volt, _____phase, _____wire distribution system.

(D) All equipment shall be furnished complete with required blank sections, corner and trim accessories to provide an installation as shown on the drawings.

(E) All electric heating equipment shall be automatically controlled by thermostats installed where indicated on the plans or in some cases built into the individual units as called for in the schedule.

16856 Snow Melting Cable and Mat

(A) The Electrical Contractor shall supply and install electric heating mat(s) for use with _____ volts. Each snow melting mat shall be an assembled unit consisting of nylon jacketed 105°C polyvinyl chloride insulated heating cable with a copper grounding overbraid over its entire length and spaced at a minimum of one inch centers on wire mesh. The heating mats shall be complete with ten feet of nylon jacketed TW-12 stranded cold leads, connected to the heating cables by a molded waterproof seal. Cold leads shall have a copper grounding overbraid and shall be anchored to the mat for strain relief. Each unit shall dissipate _____ watts per square foot when supplied

with _____ volts. The complete unit shall conform to UL standards. Installation is to be in accordance with manufacturer's specifications as UL listed and shall be in compliance with the *NEC* as revised.

16857 Heating Cable

(A) The Electrical Contractor shall furnish and install electric heat cable according to the drawings, schedules, and specified herein. The cable shall be as manufactured by _____.

(B) Under no circumstances should the cables be shortened. The heating cables must be installed in their complete lengths as supplied by the factory. In the event that the factory labels have been inadvertently removed from the spool, the cable can be identified by the identification tag on the nonheating leads or by connecting suitable test meters into the circuit.

(C) Ceiling lath should be gypsum board or sheet rock, fire resistant type, and cable is stapled directly to it. Before the first layer of plaster board is nailed in place, the centerline of each joist should be marked on the side walls. This shows the nailing line.

(D) If joists are run parallel to an outside wall, 1.5-inch spacings could be used for the first 2 feet from the outside wall. When crossing from one joist space to the next, cross the joist four inches from the wall. Do not have a cable urn nearer than 1.25 inches from the centerline of the joist. Nailing of the gypsum board should be in accordance with the best trade practice.

(E) All feeder circuits from the main distribution panel to the cable shall be in accordance with the *NEC*. All branch circuits wiring that will be located above heated areas should be kept above the building insulation. Where it is necessary to enclose the wires in this insulation, or between the heated ceiling and a floor above, these wires should be considered as operating in ambient temperature of 50° C (112° F). This wiring shall have its current-carrying capacity figured according to the correction factors given in the *NEC*. In no case shall the wiring be supported less than two inches above the heating ceiling.

16881 Electric Baseboard

(A) Electric baseboard heater(s) shall be _____ catalog _____ number designed for continuous operation. They shall be constructed of heavy-duty cold rolled steel. The heating element shall be encased in a cast aluminum heating grid provided with Teflon cushions at each end and have a maximum operating temperature of 400°F in an ambient of 70°F.

(B) Wattage and voltage shall be as specified and within normal manufacturing tolerances. A thermal limit control shall be supplied. The heaters shall be prewired for service connection at either end and approved wireway shall be provided. Connection boxes of equal dimensions shall be built in on both ends of the heater and both boxes shall be of sufficient size for making electrical connections as well as installing desired accessories.

16882 Packaged Room Air Conditioners

(A) Furnish permanently installed packaged terminal air conditioners of sizes and capacities as shown on drawings. Electrical circuits and the like have been designed to fit incremental conditioners. Contractor may submit similar equipment of other manufacture that satisfies these specifications provided, ten days prior to bid date, he submits detailed drawings for the correlation of other trades and also provided that he includes with his bids all additional costs accruing to other trades. Conditioners shall not exceed the following dimensions unless otherwise approved: Overall conditioner depth _____in. (min.); room cabinet height _____in, width _____in, depth _____ in (min.)—optional depths are available in 1-inch increments to 20.5 inches). Each conditioner shall consist of wall box; outside air louver; heating section; control box; cooling chassis; and room cabinet.

(B) The wall box shall be fabricated of 14 and 16 gauge zinc-coated phosphatized steel, enrobed in a continuous film of thermosetting plastic (epichlorhydrin and bis-phenol). It shall be in one piece, without concealed joints, and designed so the outside air louver may be fastened to it from within the building.

(C) The contractor shall provide with each conditioner one anodized aluminum extruded outside air louver in natural (or other) finish. (Louvers furnished by others must be approved as to free area and design by equipment manufacturer. If louver is part of panel wall construction, it should be omitted from these specifications.)

(D) For electric heat, the heat section shall contain sheath type finned electrical heating elements with safety device to turn off heaters if for any reason the heater temperature becomes excessive. Heating section shall also contain two double inlet centrifugal blowers with 7.4375-inch aluminum wheels direct-connected to a PSC motor with built-in overload protection. Centrifugal blower assembly shall be removable independent of heating elements or cooling chassis.

(E) The electrical contractor shall provide with each conditioner an easily accessible master control box of the pushbutton type for selecting off, cool, heat, high, and low to accomplish the following:

   1. For wet heat only a built-in adjustable thermostat shall regulate room air temperature through control of refrigeration compressor on cooling and through control of the automatic valve on heating.

   2. For electric heat a built in adjustable thermostat shall regulate room air temperature through control of the refrigeration compressor on cooling and through control of the electric heat elements on heating.

   3. The High-Low pushbuttons shall allow roomside blowers to operate at _____ or _____ rpm in either the heating or cooling cycle, for both wet and electric heat models. In addition, the electric heat High-Low switch shall allow for heating operation at the full connected kW (High) or one-half the connected kW (Low).

(F) The cooling chassis shall consist of a self-contained welded hermetically sealed air-cooled refrigeration system exclusive of evaporator fans and controls. Compressor shall be internally spring mounted with vibration isolators and shall include PSC motor, overload protection for all motors and fused compressor capacitor. All sheet metal shall be zinc-coated, phosphatized and enrobed in a continuous film of thermosetting plastic (epichlorhydrin and bis-phenol).

(G) Provision shall be made for easy removal and insertion of this chassis as a unit, without removal or disconnection of heating coils, heating controls or blower assembly. Stainless steel fasteners shall be used throughout. Direct connected condenser blower assemblies shall be centrifugal type with two ____-inch DWDI aluminum wheels designed to reevaporate all condensate on the condenser surface without drip, splash or spray on building exterior. Refrigerant metering shall consist of capillary restrictors, supplemented by a constant pressure expansion valve, so designed as to prevent frosting of the evaporator coil and short cycling of the compressor at outdoor temperatures down to 35°F, while providing not less than 100 percent of its full-rated capacity. Cooling chassis shall also contain quick removable washable air filter, arranged to filter both ventilating and recirculated air. Ventilation shall be two-stage positive pressure, accomplished by operating the condenser and roomside blowers in series. A motorized fail-safe damper mechanism with manual override switch shall control fresh air quantities.

(H) The Electrical Contractor shall provide a cabinet made of 16 gauge furniture steel finished with mist gray baked enamel. Cabinet to include four-way adjustable extruded aluminum discharge grilles and control access cover. The cabinet shall be equipped with tamperproof front panel, which will permit access to and removal of air filter and cooling chassis without the use of tools. Adjustable kick plate shall be provided to assure a snug fit at the floor level.

16883 Radiant Heaters

(A) Electric infrared heater(s) shall be _____ for (outdoor) installation and shall be constructed of sheet metal with minimum of 20 gauge (outdoor-zinc grip, paint grip steel) for the housing. All surfaces of the housing shall have a baked-on enamel finish providing a durable

corrosion-resistant coating or serve as a prime coat for repainting. The fixture shall be equipped with spring loaded sockets for push-pull element installation. The fixture shall include a snap-in plated wire guard for element protection. The fixture shall include an 18 gauge clear anodized aluminum reflector for (90° symmetric), (60° symmetric), (30° symmetric), (60° asymmetric), (30° asymmetric) beam control. The fixture (indoor) shall be designed for surface mounting or chain suspension and shall include a hinged mounting bracket with wire-way and convenient knockouts. The fixture (outdoor) shall be designed for stem mounting from ceilings by rigid conduit. The fixture shall be designed for use with two heating elements. The heating element shall be (select one of the following):

1. Quartz lamp consisting of a tungsten wire sealed in a gas-filled quartz tube which shall produce near infrared energy with approximately 7 lumens per watt of light.

2. Quartz tube consisting of a nickel chrome wire coiled in an unsealed tube.

3. Metal sheath consisting of a nickel chrome wire encased in a stainless steel sheath filled with a magnesium oxide insulation.

(B) Wattage and voltage shall be as specified and within normal manufacturing tolerances. Heater(s) shall be UL listed.

(C) An accessory recessed box assembly No. _____ shall be available for use with No. _____ outdoor fixture. In addition to the recessed box, the assembly package shall include trim frame, junction box with cover, flexible conduit with fittings and shall be designed for installation in any ceiling material.

16884 Duct Heaters

(A) The Electrical Contractor shall furnish and install, as indicated on the drawings, duct heaters as manufactured by _____. Voltage and other electrical characteristics, size, wattage, number of steps and accessories shall be as indicated.

(B) Heaters shall be UL listed for zero clearance and meet all the applicable requirements of the *NEC*.

(C) Heaters shall be made with galvanized steel frame. The terminal box shall be provided with solid cover in order to minimize dust infiltration, and this cover shall be hinged if built-in fuses are provided.

(D) All resistance coil terminals and nuts shall be made of stainless steel and terminal insulators and bracket bushings shall be made of high grade ceramic and securely positioned. Resistance wire shall be iron free, 80 percent nickel and 20 percent chromium. Bracket supports for the resistance wire shall be reinforced with stiffening ribs and gussets, and spaced no more than four inches apart. Heaters shall be tested dielectrically for 1000 V plus twice the rated voltage or 2000 V, whichever is higher. Heaters rated 150 kW and over shall be furnished with heavy-duty coils, aerated to 35 watts/inch of wire surface, to insure long life.

(E) Heaters shall be interchangeable for mounting in a horizontal or vertical duct and air flow may be through the heater in either direction.

(F) Electric heaters shall be of the slip-in or flanged type as scheduled. Heaters shall be furnished for _____volt, _____ phase power. Three phase heaters shall be furnished with balanced three phase steps. The control voltage shall be ____volts.

(G) Heaters shall be rated for kW and furnished for step control. Heaters shall be sized for ducts of outside duct width x duct height.

(H) Magnetic contactors shall be provided to open all ungrounded conductors. One contactor required per step. The contactors shall be UL listed and be rated for 100,000 cycles of operation. Contactors shall be constructed and listed for 600 volts operation.

(I) Mercury-type contactors shall be provided to open all ungrounded contactors. One contactor is required for each step. The contactors shall be UL listed and be rated for 100,000 cycles of operation. Contactors shall have a maximum per-pole resistive rating that matches the rating of the stages to be controlled.

(J) Automatic circuit breakers of the thermal-magnetic type shall be provided for overcurrent protection and shall be built in and prewired. The operating handle shall protrude through the control box cover. Automatic circuit breakers shall be sized for 80 percent of the circuit. One set of circuit breakers shall protect each step. An automatic circuit breaker shall be provided to protect the entire heater for those heaters rated 24 amps or less.

(K) A control transformer shall be provided within each heater control box with a VA rating that is matched to the quantity of devices it must supply power to. All ungrounded primary taps are to be provided with fuses. The primary voltage shall be as scheduled and the secondary voltage shall be 60 Hz.

(L) Heaters shall be supplied with overcurrent protection per latest edition of the *NEC*. Overcurrent protection shall consist of built-in and prewired fuses:

> 1. with one overcurrent device for each 24-ampere circuit.
>
> 2. with one overcurrent device for each heating step.
>
> 3. with one overcurrent device for entire heater for those heaters rated 24 amperes or less only.

(M) A disc type automatic reset thermal cutout, serviceable through terminal box, shall be furnished for primary protection. Heat limiters in power lines shall serve as secondary protection. In addition, a disc type manual reset thermal cutout shall be furnished. Manual reset thermal cutouts are to be in series with automatic reset thermal cutout. All three devices shall be serviceable through the terminal box, without having to remove heater from duct.

(N) Heaters intended for use with multizone equipment shall be provided with a linear cutout extending across the entire face width of the heater. A linear thermal cutout shall be supplied on all duct heaters exceeding 72" duct width.

(O) An interlock relay shall be built-in and prewired to terminal blocks and field wired by others in parallel with the fan starting circuit, to prevent heater operation when the fan starter is not energized. Fan control circuit voltage to be 120 V, 60 Hz.

(P)   A pilot light shall be installed in the control box cover, prewired and labeled to indicate the following:

1. when air flow switch has opened.

2. one light per stage to indicate when stage is energized.

(Q)   Provide a pilot switch installed in the control box cover prewired and labeled to provide the following:

1. open control circuit.

2. deenergize each step contactor (one required per step).

(R)   An air flow sensing switching of the pressure sensing type shall be built in and prewired into the control circuit, the air flow switch to have a range adjustment .05 to 8" WC.

(S)   There shall be a 10-second time delay between each step. The time delay will be provided by a rotary cam driven set of switches. Should a limit condition or power failure occur, all steps will open and the sequencer will return to its no heat position and recycle from a no-heat position.

(T)   All heaters shall be provided with a nonfused disconnect switch, interlocked with the control box door. The switch must be in the OFF position before the control box may be opened.

16885 Control of Electric Heating

(A)   The electric heating system shall be controlled by thermostats furnished by the electrical contractor and installed _____ feet above floor level at locations shown on the plans. Each thermostat shall control the number of heating units indicated.

(B)   Thermostats shall be _____, model number _____ , catalog number _____ , or approved equal. Units shall be line-voltage, single-pole, single-state, type, rated watts, volts. Operating temperature range shall be _____ °F to _____ °F, with a rated differential of _____ °F.

Units shall be for surface mounting as indicated and covers shall have a finish.

(C)   Miscellaneous electrical appurtenances such as electric relays, selector switches, On-Off switches, pilot lights, and other similar items required by the electric sequence control diagrams and not shown to be provided by the electrical contractor shall be provided as part of the mechanical contract.

1. On-Off switches shall be toggle type, 20 amp contact rating complete with engraved cover plate where required.

2. Selector switches shall be manual selector type with the indicated poles and contacts and engraved cover plate. Contact rating shall be sufficient for the connected load and appropriate for the application.

3. Relays shall be _____ , _____, or of sufficient coil rating and contacts as required for the sequence indicated.

(D)   Time clock shall be _____ or _____, seven-day model with 16-hour reserve time feature. Reserve time spring shall rewind itself after power source is restored.

(E)   Thermostats:

1. Night low-limit thermostats shall be ____, line voltage type with SP-ST switching action rated 6 amp full load and 36 amp locked rotor at 120 volts. Outdoor thermostats shall be_____.

2. Firestats shall be UL approved, manual reset type _____ with an adjustable temperature setting. Set at 125°F.

16900
CONTROLS AND INSTRUMENTATION

16901 General

(A)   All equipment and materials used in relation to control work for the project shall be new and shall bear the manufacturer's name and trade name. The equipment and material shall be essentially the standard product of a manufacturer regularly engaged in the production of the required type of equipment and shall be the manufacturer's latest approved design.

(B) The Electrical Contractor shall receive and properly store the equipment and material pertaining to the electrical work. The equipment shall be tightly covered and protected against dirt, water, chemical or mechanical injury and theft. The manufacturer's directions shall be followed completely in the delivery, storage, protection and installation of all equipment and materials.

(C) The Electrical Contractor shall provide and install all items necessary for the complete installation of the equipment as recommended or as required by the manufacturer of the equipment or required by code without additional cost to the Owner, regardless of whether the items are shown on the plans or covered in the specifications.

(D) It shall be the responsibility of the electrical contractor to clean the electrical equipment, make necessary adjustments and place the equipment into operation before turning equipment over to Owner. Any paint that was scratched during construction shall be "touched-up" with factory color paint to the satisfaction of the Architect. Any items that were damaged during construction shall be replaced.

16902 Motor Control

(A) Motor controllers or starters will be furnished as part of the equipment on which they are used. These controls shall be furnished by the respective subcontractors and shall be mounted and connected by the electrical contractor.

(B) Fractional horsepower, single phase motor starters shall provide overload protection as well as On-Off control for single phase 120 volt and 208 volt motors. Unless otherwise indicated, they shall be flush mounted in a standard single gang switch box with a standard stainless steel flush plate. Manual motor starters shall be _____ , Type _____, Class

_____, with toggle switch; one or two pole and with pilot light where shown or directed.

(C) Integral horsepower starters:
1. Motor starters shall be _____ , Class ____ , Type _____ , as required for motors shown on drawings.

2. Starters shall be line voltage, rated 600 volts, and controlled by 120 volt control circuits in all cases. Provide control power transformers, with fuse in control circuits.

3. Furnish combination starters, _____ , Class _____, where indicated or required by *NEC*.

4. Three phase starters to be provided with three overload heater elements sized on motor nameplate FLA.

5. Starter enclosures shall be NEMA Type _____, except those outdoors shall be NEMA Type _____enclosures.

(D) The Electrical Contractor shall furnish and install starter mounted and remote pushbutton stations, selector switches, etc., as required. These controls shall match starters provided by the electrical contractor, and shall be compatible with starters furnished by others.

(E) All motor starters shall be of the magnetic type unless indicated otherwise. Provide an overload relay for each phase and with poles as required. Where disconnect switches are required, provide fusible equipment unless nonfusible equipment is indicated.

16915 Lighting Control Systems

(A) The Electrical Contractor shall furnish and install complete remote control wiring system for control of lighting, receptacles and other equipment as indicated on drawings, diagrams and schedules. System shall be complete with transformers, rectifiers, relays, switches, master selector switches, pilot lights, motor master controls, wall switch plates and wiring. All remote-control wiring components shall be of same manufacture and installed in accordance with the recommendations of the manufacturer. Remote control equipment shall be as manufactured by _____ or of equal quality, as approved by the engineer.

(B) Except where otherwise indicated, all such remote control wiring shall be in accordance with Article 725, Class 2, of the *NEC*.

(C) Relays: Remote control relays shall be heavy-duty, specification grade with clamp type screw terminals on high voltage side, rated .5 hp 125 V ac, 20 ampere 277 V ac. Relays shall be two-coil solenoid type with one coil for closing and the other for opening, designed for three-wire momentary contact control. The relay shall be manufactured by _____ catalog number _____ or approved equal.

(D) Switches: Remote control switches shall be heavy-duty, specification grade. They shall be single pole, double throw, momentary contact type. Both standard and pilot light type switches must be available.

(E) Master Selector Switches: Master selector switches shall be eight-position type with individual On-Off switches, built-in master locking switch, directory, and pilot lights.

(F) Motor Master Controls: Motor-driven controls shall have twenty-five circuits and use separate controls for On and Off.

(G) Transformers: Transformers shall be energy-limiting type approved for Class 2 systems, rated 115/24 V or 277/24 V as required.

(H) Rectifier Assembly: Rectifier assembly shall consist of a heavy-duty silicon rectifier, rated 20 amps intermittent duty, 7.5 amps continuous duty, 100V PR, mounted on a terminal strip and shall be manufactured by _____ catalog number _____, or approved equal.

(I) Component Cabinets: Component cabinets shall have a partition for mounting up to 24 relays, one transformer, two motor master controls and one rectifier assembly. They shall be equipped with suitable neutral high voltage terminal strips and low voltage terminal strips for all relays, motor masters and low voltage switch leads. Shall be UL listed for remote control wiring. In addition to relays required for installation, provide space for 25 percent additional relays for future use. The cabinet shall be manufactured by _____ catalog number _____ or approved equal.

(J) Wiring: Low-voltage wiring shall be of #18 AWG gauge thermoplastic insulated type with two, three or four conductors as required and shall be rated for 30 volt service with at least .0156-inch insulation as required by local codes and ordinances. All wiring shall be concealed wherever practical, run without conduit in ceiling spaces and between walls or partitions. Where wiring is installed in masonry partitions, or in exposed locations, conduit shall be used. Complete wiring systems shall be color coded according to the manufacturer's or engineer's recommendations and conductors must be tagged or identified at terminals.

(K) Where remote control switches are on metal partitions, wiring shall enter the top of the partition and be run concealed within the partitions. The proper size cutout shall be fabricated by the partition manufacturer for the electrical contractor to install switches.

16950 Testing

(A) As soon as electric power is available and connected to serve the equipment in the building, and everything is ready for final testing and placing in service, a complete operational test shall be made. The Contractor shall furnish all necessary instruments and equipment and make all tests, adjustments, and trial operations required to place the system in balanced and satisfactory operating condition; furnish all necessary assistance and instructions to properly instruct the Owner's authorized personnel in the operation and care of the system.

(B) Prior to testing the system, the feeders and branch circuits shall be continuous from main feeders to main panels, to subpanels, to outlets, with all breakers and fuses in place. The system shall be tested free from shorts and grounds. Such tests shall be made in the presence of the Engineer's representative.

(C) No circuits shall be energized without the Owner's approval.

(D) The right is reserved to inspect and test any portion of the equipment and/or materials during the progress of its erection. The Contractor shall further test all wiring and connections for continuity and grounds before connecting any fixtures or equipment.

(E) The Contractor shall test the entire system in the presence of the Architect or his engineer when the system is finally completed to insure that all portions are free from short circuits or ground faults.

16980 Demonstration of Electrical Equipment

(A) The Electrical Contractor shall provide the Architect with certification of the inspection and approval of an active member of the International Association of Electrical Inspectors of all work completed and included in the section, if required.

The Contractor shall be responsible for notifying the Inspector when work reaches inspection stage.

(B) The Electrical Contractor shall be responsible for notifying the local authority having jurisdiction in order that local inspection may be carried out at the proper stage.

(C) The Electrical Contractor shall pay for all permits, inspection fees, and installation fees as required to complete the work under this Section of the Contract.

(D) This Contractor shall guarantee the materials and workmanship for a period of twelve (12) months from the time the installation is accepted by the Owner. If, during this time, any defects should show up due to any defective materials, workmanship, negligence or want of proper care on the part of this Contractor, he shall furnish any new materials as necessary, repair said defects, and put the system in order at his own expense on receipt of notice of such defects from the Architect. This specification is not intended to imply that the Electrical Contractor shall be responsible for negligence of the Owner.

Index